Dynamic Analysis of Skeletal Structures

Force and Displacement Methods and Iterative Techniques

Dynamic Analysis of Skeletal Structures

Force and Displacement Methods and Iterative Techniques

Seetharamulu Kaveti
PhD, Fellow I.A. Struct E
Former Professor
Department of Civil Engg.
IIT, Delhi

Mc Graw Hill Education

New York Chicago San Francisco
Athens London Madrid
Mexico City Milan New Delhi
Singapore Sydney Toronto

1 2 3 4 5 6 7 8 9 0 QVS/QVS 1 9 8 7 6 5 4

ISBN 978-0-07-183585-5
MHID 0-07-183585-7

The sponsoring editor for this book was Bridget Thoreson and the production supervisor was Lynn M. Messina. The art director for the cover was Jeff Weeks.

Printed and bound by Quad Versailles Digital.

This book was previously published as *Analysis of Skeletal Structures: Force and Displacement Methods with an Introduction to Applied Dynamics* by Tata McGraw-Hill Education Private Limited, New Delhi, India, copyright © 2012.

This book is printed on acid-free paper.

To

My wife Kanchanamala

Contents

Preface

The proposed book, even though is essentially Analysis of Civil Engineering Skeletal Structures, the Design and Construction aspects are indicated for a proper appreciation, relevance and motivation for the analytical coverage. The Skeletal Structures constitute an idealization of Structures. The conceptual background of which simplifies the analysis without sacrificing accuracy required for the Design purposes. The skeletal modeling of the structures assumes the cross-sectional dimensions are 'small' compared to the length, and it covers practically all types of structures encountered in the design and construction practices of structures. The numerical examples are selectively chosen as illustrations to complement the theory. In the computer era, too many numerical examples obscure the basic concepts.

It was way back in 1864, J.C. Maxwell suggested 'Force Method', and in 1862, A. Clebsch presented Displacement Method. In the later years, because of over enthusiasm with various simplifications for design solutions, the emphasis shifted to different procedures of analysis suitable for the commonly used structures. In the case of structures with constraints more than the minimum required for stability (indeterminate structures), a justifiable attempt was to suitably locate points of inflection (zero curvature), points of zero moments, which facilitated the evaluation of internal forces from simple statics. The slope-deflection method is a simplified version of the displacement method where the skeletal members are assumed inextensible; the axial deformations have negligible effect on the flexural forces-bending moments and shears. The difficulty of evaluating the large number of unknown displacements was simplified by the iterative procedure like "The Moment Distribution" based on the assumption of inextensibility.The numerical algorithm of the moment distribution procedure is important for the structural understanding it gives from the structural behavior introduced by way of clamping of the joints, releasing and balancing of the joints in a sequence, and carrying over the effect of balancing.

The book is structured to highlight the unification of structural analysis based on, (1) Force Method, and (2) Displacement Method. The few examples chosen are complementary to the understanding of concepts introduced. The shear walls in the tall buildings are treated as a particular type of skeletal member. The member flexibility and the member stiffness are illustrated so as to prepare the reader for *finite element analysis*. There are a number of good books available on the finite element analysis and to limit the scope of the book, the finite element idealization has not been included. The modern trend of the earthquake design is to base the earthquake analysis on *capacity* considerations. The uniqueness theorem used for enabling plastic analysis is yet another great concept which is adequately explained by the Force Method. The plastic analysis has been included with an additional objective of preparing the readers for a proper appreciation of capacity of structures for the earthquake design of tall structures. The Force method is ideally suited for the comprehension of all these concepts. An important aspect of stiffening of structures for their increased capacities is included. The stiffening of the framed structures cannot be divorced from an important aspect of ductility. It is the ductility that ensures the structure reaching its full capacity, the collapse load. A good design limits the development of the plastic hinges in the girders. The shear effect in mobilizing the full plastic moment of resistance is not crucial. An Introduction to applied dynamics has also been included for preparing the students to earthquake design of tall buildings. A section has been devoted to the maintenance of large structural systems. The Tall Buildings constitute such a system.

Seetharamulu Kaveti

Introduction

The contents are broadly categorized under Force Method, Displacement Method, and Iterative Techniques; with a brief introduction to Structural Dynamics and its application to a dedicated structure, Tall Buildings. The force method is also known as compatibility method because compatibility conditions are used while forming the equations for the unknown forces. The number of (redundant) unknown forces is equal to the number of excess constraints over and above the minimum constraints required for achieving a stable structure. Whereas the displacement method, also known as equilibrium method, is used for solving the unknown displacements equal to the degrees of freedom of a chosen system. The system of equations obtained from equilibrium conditions of the free bodies, which are generally the joints or appropriately chosen parts of the total structure. Therefore, the displacement method is also known as the 'Equilibrium Method'. The force method is more conceptual as it helps to determine whether the structure to be analyzed is stable or unstable. It is only when the structure is stable; we proceed to find the unknown redundant forces. The number of the unknowns defines the degree of indeterminacy of any given structure. The degree of indeterminacy is *unique*. But the type of releases, their number equal to the degree of indeterminacy, depends on the choice by the engineer analyzing the structure. A proper choice of the releases results in the compatibility equations that are well conditioned (dominance of the leading diagonals of the equations). The method, therefore, demands a conceptual preparation from the engineers while choosing the system of releases. It is this aspect of analysis that prepares a thinking designer leading to good conceptualization.

In the case of displacement method, generally, the equilibrium equations do not result in any kind of ill conditioning of the stable structures. The method, however presupposes the structures are stable. The indicators whether the structure is stable or not, are provided by the unpredictable results; deflections, internal forces, and unusual values of frequencies. The method does not put any kind of demand on the engineers while making the choice of the nodes and the corresponding unknown displacements and their number. The degrees of freedom equal to the number of unknown displacements are essentially caused by the applied static loads or the externally applied forces. In the case of time varying loads, the analysis is referred as the dynamic analysis for determining the corresponding time dependent displacements. An important concept of geometric loading is introduced as one of the loading cases. It is sometimes expedient to use a mixed method which combines the two basic methods, i.e., force and displacement methods. These methods are developed. The examples chosen complement the presentation of the basic theories. The examples are selectively chosen which provide leads to the design of dedicated structures like tall buildings or bridges or large span structures, etc.

Appropriate techniques for evaluating the unknowns are developed if the solution of the equations is to be obtained from the manual calculations. These are essentially based on numerical techniques; iteration (Kani's Method) and incremental iteration (Hardy Cross Moment-Distribution Method). The conceptualization of structural analysis is incomplete without discussing the energy principles. The contents include: 1) Virtual work principles and 2) Complementary virtual work principles.

The entire book is based on linear analysis, i.e., Small Deformations. The other aspect of linearity pertains to the material linearity (elasticity). Except for the plastic analysis of structures where the plasticity is assumed to be localized at the plastic hinges, the skeletal members are assumed elastic. The basic concepts developed are important for the design of structures which are not only safe but result in structurally efficient and by and large economic structures. The topics pertaining to the structures which undergo large deformations, like cable structures (a typical case of skeletal configuration) and tension structures (a typical example is the textile tension membranes and the cutting patterns) being geometrically nonlinear, do not fall under the scope of the book.

The book comprises two parts. The part-A pertains to Force Method and it consists of five chapters.

Part A

Chapter 1–General Concepts and Energy Principles

After introducing general concepts pertaining to broad classification of structures and basic assumptions, the energy principles are introduced which are later used as virtual work principles. The principles are used in the development of analysis, particularly Chapter 4, based on the energy principles for presenting a unified force method.

Chapter 2–Force Method

The concepts pertaining to kinematical constraint conditions are fundamental to arriving at stable structures, trusses, beams, plane frames, grids, and space frames. A good understanding of stable and unstable structures is achieved while determining, Determinacy, and Indeterminacy. The contents of the chapter are oriented to lead the reader to plastic analysis in Chapter 3 and matrix approach in Chapter 4. A proper appreciation and understanding of the Chapter 2 are the essential requisites in the making of a Structural Engineer.

Chapter 3–Plastic Analysis

The scope of the plastic analysis is limited to skeletal structures and it is based on small deformations. The number of basic mechanisms is related to the degree of determinacy of skeletal structures. The lower bound approach is explained comprehensively with the help of technique based on virtual stiffening. The concept of the virtual stiffening is used for enhancing the collapse load by appropriately stiffening at the zones of maximum moments. This concept of stiffening, however, reduces the overall ductility which is an important parameter for the pushover design of earthquake resistant structures. The Uniqueness theorem and the Upper bound and Lower bound are the important conceptual contents.

Chapter 4–Matrix Approach for Force Method

It is an alternative to the scalar approach developed in the Chapters 2 and 3.

For the computer based analysis of structures, it is a requirement to present the steps of analysis in the matrix form. This has been achieved by presenting the matrix formulation analogous to scalar approach presented in Chapter 2. An alternative formulation has been presented as a unified force method based on energy principles. The typical examples of the indeterminate structures considered to illustrate the method are hinged trusses, pontoons treated as beams resting on interior supports as flexible and end supports hinged, rigid frames like plane frames, grids, space frames, rigid cum hinged frames likes trussed beams or trussed frames.

Chapter 5–Approximate Methods of Analysis

It is an illustration of various approximations introduced for the analysis of a dedicated structure. Here in, the dedicated structure chosen is a tall skeletal building frame; the concepts are introduced by way of appropriately assuming the inflexion points (zero bending moment). The location of inflexion points are enabled by sketching deflected shapes (elastic lines).The elastic line of tall skeletal structures is used not only for the bending moment diagrams but also for obtaining the influence lines. The concept is used for the loading patterns for the design purposes. To enable critical moments, shears and thrusts at chosen points, this important aspect remains totally obscured when a designer is totally dependent on the commercially available computer software. The principle of virtual work is employed to compute floor deflections.

Part B

It comprises Displacement Method, Iterative Techniques, and Introduction of Applied Dynamics and Earthquake Design of Tall Structures.

Chapter 6–Displacement Method and Chapter 7–Displacement Method-Illustrations

In as much as choice of release system is important for the force method, the modeling of a given structure before the analysis in respect of nodes and joints need to be done with care while using the Displacement Method. The degrees of freedom equal to the unknown displacements depends on the number of joints and locations of the nodes at locations other than the joints. The illustrations considered are mostly straight prismatic skeletal members, and these include trusses, beams, plane frames, grids, and space frames. The degrees of freedom are governed by approximation of inextensibility, that is negligible effect of axial deformations on bending moments and shears. For manual calculations, particularly the assumption of inextensibility reduces the computational effort quite significantly. Wherever inextensibility is used, the load matrix and the corresponding degrees of freedom need to be understood properly. All these aspects have been discussed while formulation of the equations for the unknown displacements. The slope deflection method is illustrated as a particular case of displacement method applied to small frames with the assumed inextensibility. In the case of beams, however, axial deformations get delinked from flexural deformations. While discussing trusses with hinged joints, the reader is cautioned against likely mechanisms. Sketching of such mechanisms of unstable trusses is also included. Sometimes a hybrid approach, combined force and displacement methods are computationally very efficient. The types of examples suitable for the mixed methods are illustrated.

Chapter 8–Iterative Techniques

The iterative techniques used for analysis are: Moment Distribution Method and Kani's Method .These methods are essentially numerical techniques for solving the equations of unknown displacements. The Kani's Method is an iterative technique and similarly, moment distribution method is an incremental iterative technique. These being the numerical techniques, typical numerical examples are presented for illustrating the techniques. It is particularly pointed out that the Hardy Cross Method from the considerations of incremental clamping and balancing of the moments at the ends of the members meeting the joint being released, makes it highly conceptual.

Chapter 9–Introduction to Applied Dynamics and Earthquake Design of Tall Frames

Tall buildings constitute an example of a dedicated structure. The various appropriate approximations are explained. The degrees of freedom being large in number, numericals are limited. The formulations are explained with the help of flow charts frequently used in the computer programming. The Shear walls are extensively used in the tall buildings and in the formulation of analysis for tall building frames using shear walls are explained as wide-column skeletal member. The floor slabs are assumed rigid in their own plain and therefore undergo in-plane rigid body displacements. Such a formulation is considered for writing down the dynamic equations of equilibrium and the mode shapes corresponding to horizontal degrees of freedom which are the dominant displacements for earthquake design. The numerical modeling for the analysis of tall buildings by, (i) Clough Wilson and King, and (ii) Weaver and Nelson are explained. Basics of Structural Dynamics are included with an objective of introducing earthquake design of tall buildings. Typical skeletal frames have been used as illustrations for computing mode shapes and arriving at the earthquake loads. The scope is limited to elastic analysis using Houser's Spectra. The push over analysis based on capacity of the structure and on the inelasticity/collapse is not included. However, a reader with the background of dynamics and plastic collapse, and the computations pertaining to angle discontinuities at the location of the plastic hinges should be able to comprehend capacities of the structures.

Appendix–Mathematical Preliminaries

Acknowledgments

To all my students who directly and indirectly contributed to the development of the book. The students of the IIT Roorkee contributed on the basics of Mechanics, IIT Bombay on the elasticity and structural systems, and IIT Delhi on the core of the book pertaining to the Skeletal Structures, Plastic Analysis, Tall Buildings and Dynamics. My wife, companion, Kanchanamala (1958 onwards), has been my inspiration and motivation. It is her desire that I am fulfilling by presenting the book.

The assistance rendered by K.N. Mehrotra, Samir Rawat and Yudhister Pal Sharma in the preparation of the book is gratefully acknowledged.

I gratefully acknowledge the contribution and assistance from Subhash C. Mehrotra, Principal Director, Mehro Consultants, Delhi.

Brief Profile of Mr. Subhash C. Mehrotra

Mr. Subhash Mehrotra Graduated in Civil Engineering in 1972 and subsequently did M. Tech. in Structural Engineering from IIT Delhi. His professional experience of 38 years includes structural designs for commercial, residential, multistoried buildings, hospitals, malls, industrial structures, and public infrastructures, including 600 tall buildings in the range of 10–40 storeys and 30 Airport terminal buildings in India. He has published many technical papers relating to civil engineering and consultancy profession and his designs have won several prestigious awards.

He is the Principal Consultant of M/s Mehro Consultants, which is one of the leading Structural Engineering Consulting Firms of India. He was the President of Consulting Engineers Association of India during the period December 2003–December 2005. He is presently Vice President of Indian Association of Structural Engineers (IAStructE) for the period of 2011–2013.

He is first Indian and first South Asian who got elected as Executive Committee member of FIDIC, for the period September 2005 to September 2009. He is currently Chairperson of Membership Committee of FIDIC for the period 2009–2012.

Seetharamulu Kaveti

PART A

1. General Concepts and Energy Principles
2. Force Method
3. Plastic Analysis
4. Matrix Approach for Force Method
   Unified Matrix Approach
5. Approximate Methods of Analysis

Chapter 1: General Concepts and Energy Principles

Chapter

1

General Concepts and Energy Principles

This is a brief chapter, the contents of which consist of the terminologies used, broad classification of structures presented as types of structures, and more importantly the concepts of Virtual Work Principle. Being a *foundation* material used practically for all the chapters; it is removed from the contents of the Chapter 2, where it belongs to. The Chapter 1 presents proper understanding of the following:

(a) Skeletal structures, and associated assumptions and approximations,

(b) Funicular shapes and directly stress members with minimum of flexure, and

(c) Virtual Work Principles. It is a necessary preparation for the understanding of the Structural Analysis presented under the Force Method, and Displacement Method.

(d) Inextensibility is a great assumption. It helps us to understand master and slave nodes, particularly in Tall Buildings. Even though theory of shells is not within the scope of the book, while discussing classification, stress-strain relations are presented as logical extensions of skeletal members. More as a further logical extension, it may be mentioned here, *Inextensional* deformation of the neutral surface is an important assumption used in bending deformations of Shells. It simplifies the bending theory of shells.

GENERAL CONCEPTS AND ENERGY PRINCIPLES

After introducing general concepts pertaining to broad classification of structures and basic assumptions, the energy principles are introduced, which are later used as virtual work principles in Chapters 2 and 3. The Chapter 4 uses the energy principles for presenting a unified force method.

A structure signifies its capability to perform efficiently and economically the functions for which it is designed without any failure or distress under the action of loads. It ceases to be a structure if it fails due to large stresses or large deformations. The different types of structures are: (a) Skeletal structures generally represented as line elements, (b) Surface structures like plates and shells, and (c) Solid bodies with finite lateral dimensions comparable to the longitudinal dimension. The Generalized Stress-strain relations or Force Deformation relations for the different elements are also given.

1. TYPES OF STRUCTURES

1.1 SKELETAL STRUCTURES

The cross sectional dimensions are assumed small as compared to the lengths of members.

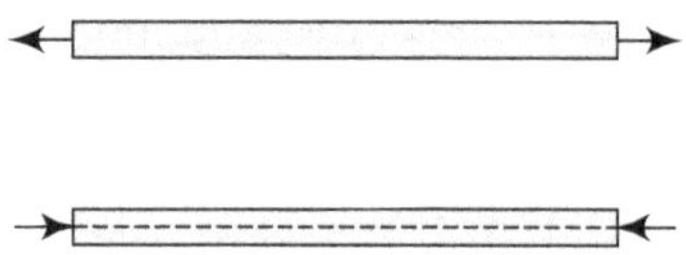

(a) Direct tension member and compression member

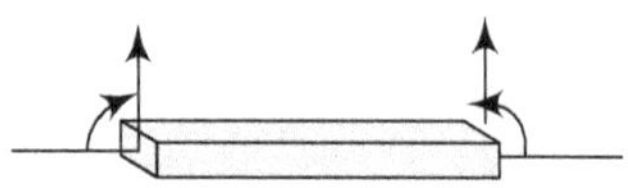

(b) Bending moment and shear (minimal axial force)

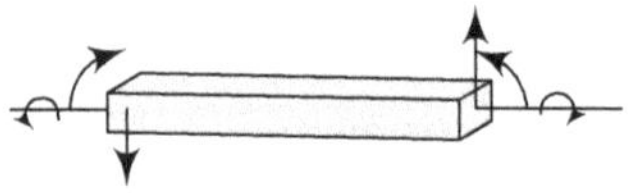

(c) Torsional member with moment and shear

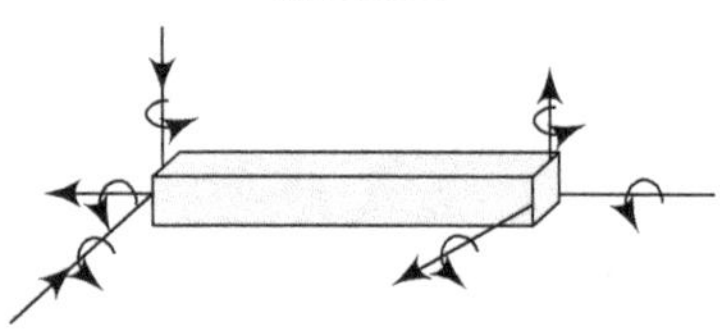

(d) Three dimensional system of stress resultants

Fig. 1.1 *Members*

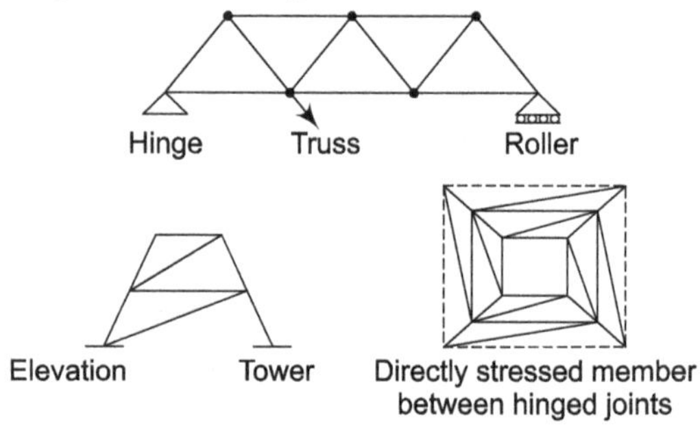

(a) Structure–Truss

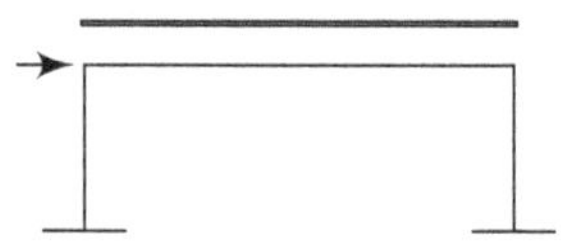

(b) Structure–Portal Frame

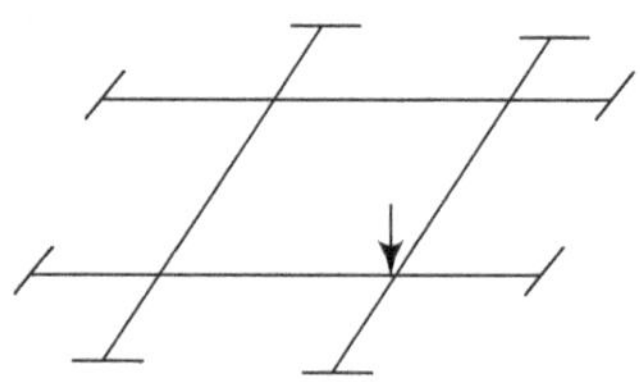

(c) Structure–Plane Grid

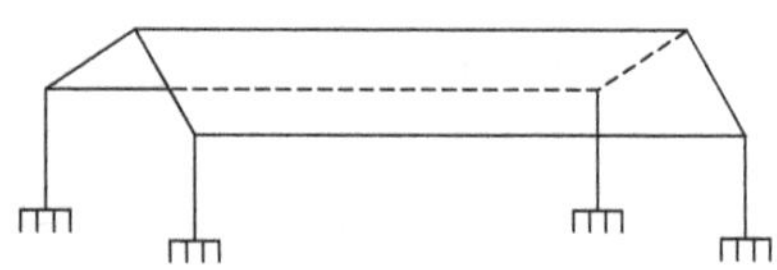

(d) Structure–Space Frame

Fig. 1.2 *Structure*

(a) Members under direct forces

Simple compression or tension members are under the action of axial forces only, Fig. 1.1(a). The plane trusses or space hinged towers, Fig. 1.2(a), with loads at joints only, develop direct compression or tension. Generalized stiffness or Force deformation relation, $P = (EA/L)\,\delta$

(b) Straight beams or beam segments

These are "flexure" members and a member requires three stress resultants to be specified at each section. The stress resultants are two forces and one moment. The forces and the axis of the member *(x-axis)* are in the same plane. Fig. 1.1 (b). The structure of Fig. 1.2 (b), a plane rigid jointed frame, consists of such structural members or elements.

Based on the assumption, "Plane section remains plane", Force-Deformation relation is given by, $M = EI\,(1/R)$ where M is Bending Moment; EI is Flexural Rigidity; $(1/R)$ is Curvature, rate of change of slope. And the flexural formula is $\frac{M}{I} = \frac{f}{y} = \frac{E}{R}$, where '$f$' is flexural stress at a distance 'y' from neutral axis. If flexure only, the neutral axis passes through the centroid of the cross section.

(c) Torsion members

A free body of a segment of a torsion member, Fig. 1.1(c), is a straight member which develops twisting moment m_x, bending moment m_z, and shear p_y under the action of transverse loading. A plane grid, Fig. 1.2(c), consists of such torsion members.

The torsional formula hollow cylinder is given by, $\frac{T}{J} = \frac{\tau}{r} = \frac{G}{L}\theta$

(d) Space member

A free body of such a skeletal member is shown in Fig. 1.1(d). It requires all the stress resultants, that is, three force components and three moment components, to be specified at any section taken. A space rigid jointed structure. Fig. 1.2(d), consists of such members.

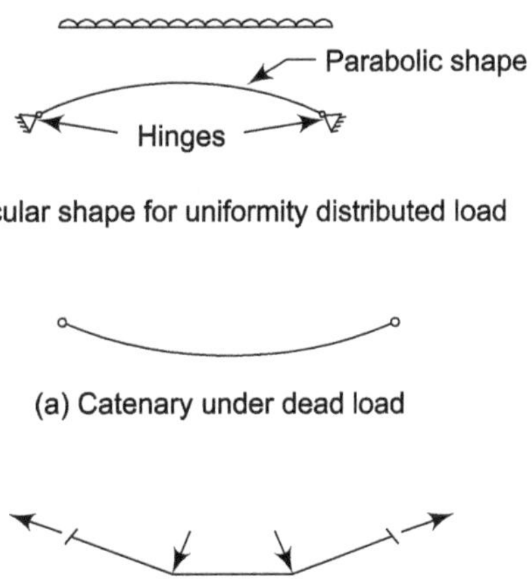

Funicular shape for uniformity distributed load

(a) Catenary under dead load

(b) Cable under concentrated loads

Fig. 1.3 *Funicular Shapes for Skeletal Members*

Structures which develop predominantly direct or uniform stresses under the action of loads are more efficient than the structures which carry loads essentially by flexure. The line elements are curved to coincide with the line of thrust, that is, the axis of a structure is given a "funicular shape". A structure with funicular shape develops direct stress only. The examples of funicular shapes are parabolic arches under uniformly distributed loads and cables, Fig. 1.3.

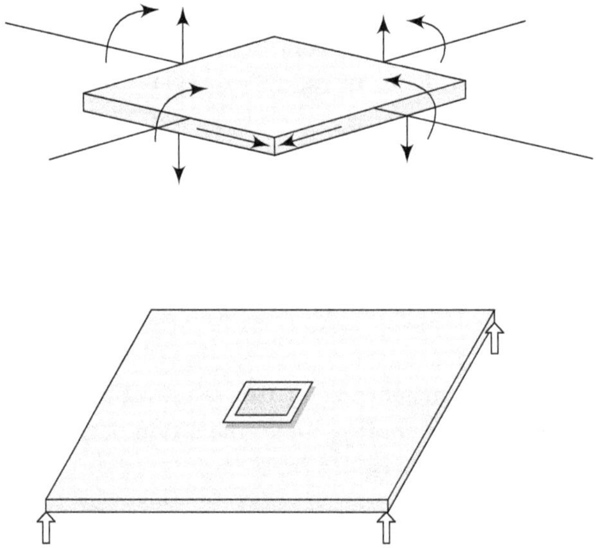

Fig. 1.4 *Plate Element with Flexural Stress Resultants and Corresponding Structure, a Plate Loaded Transversely*

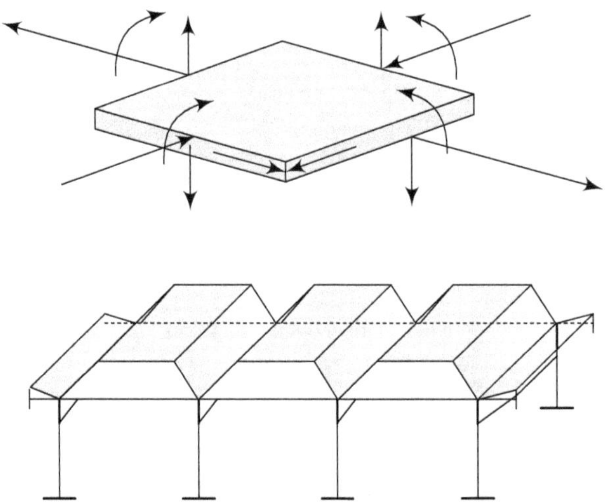

Fig. 1.5 *Plate Element with Flexural Stress Resultants and Membrane Stress Resultants (Shell Element) and Corresponding Structure, a Folded Plate Structure*

An important assumption made in the stress analysis of skeletal structures is that "plane sections" during deformation remain plane and "yield failure" is generally explained by "maximum stress theory". These assumptions are quite valid in the case of skeletal structures, which have small cross-sectional dimensions as compared to the length of constituent members. In this first course of structural analysis, skeletal structures only are considered. But it is not a serious limitation.

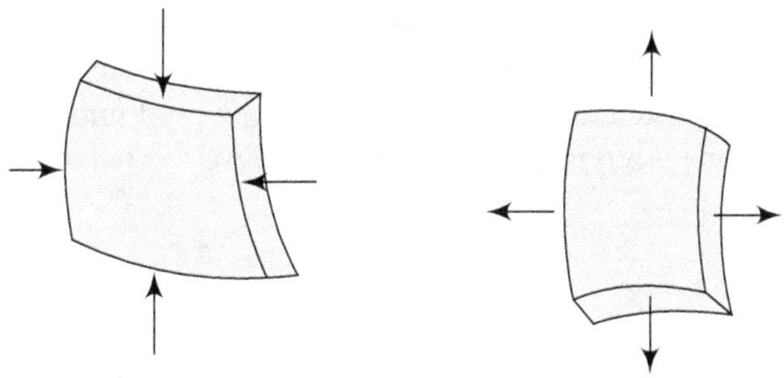

(a) Compressive member or funicular shape for compressive stress

(b) Tension member or funicular shape for tensile stress

Elements taken (a) from compression membrane, and (b) Textile membrance

Fig. 1.6 *Funicular Shapes for Surface Members–Principal Planes*

1.2 SURFACE STRUCTURES

Even though theory of plates and shells is not within the scope of the book, while discussing classification, stress-strain relations are presented as logical extensions of skeletal members. More as a further logical extension, it may be mentioned here, *Inextensional* deformation of the neutral surface is an important assumption used in bending deformations of Shells. It simplifies the bending theory of shells.

In the case of surface structures thickness is generally considered to be very small as compared to the surface dimensions. The analysis designates them as 'thin plates and thin shells'. The surface structures develop stress resultants on all the sides of any element considered and are, therefore, materially more efficiently utilized than the one-dimensional skeletal structures. In the case of plates, say when all the sides are simply supported and deformations are small, only *flexural stresses* are developed similar to Fig. 1.1(b), but in this case, it is stressed on all the sides of the element, Fig. 1.4. The Force Deformation–deformation relation is given by:

$M_x = D\left(\frac{1}{r_x} + \nu\frac{1}{r_y}\right)$ and $M_y = D\left(\frac{1}{r_y} + \nu\frac{1}{r_x}\right)$, where D is the Flexural rigidity of plate equal to

$\frac{E}{12(1-\nu^2)}h^3$, $\frac{1}{r_x}$ and $\frac{1}{r_y}$ are the curvatures.

Corresponding to the neutral axis of a beam, the neutral surface of a plate in flexure is the *centroidal* surface. The "*Inextensional* deformation of the neutral surface" is an important assumption in bending theory of shells.

Shells and Folded Plates The generalized cases of surface structures, mobilize bending as well as membrane stress resultants. The commonly used surface structures are: (a) thin shells, and thin prismatic folded plates. Figure 1.5 shows such an element and corresponding surface structure. A folded plate is a structure chosen as an illustration. The Plate element with flexural stress resultants and membrane stress resultants, is known as 'Shell Element'. And Shells are the corresponding Structures. A Folded Plate Structure is known as prismatic shell.

Funicular Surfaces By giving curvatures, based on funicular shapes which are *uniquely described* for a given set of loading, all the faces of any element develop only direct stresses. It is generally not possible to adopt funicular shapes from constructional considerations and the shapes are generally chosen to approximate funicular shapes such that the surface develops predominantly direct stresses, Fig. 1.6. However the textile membrane is a typical structure which develops tensile stresses only.

1.3 SOLID BODIES

In the case of solid bodies like large concrete dams, the assumption of plane section remaining plane is not valid and the methods of analysis are generally involved. These topics are studied under a separate subject of theory of elasticity and plasticity depending upon whether the material is elastic or inelastic. Gravity concrete dams are typical solid bodies. A 3-D element is used while idealizing for such structural systems. The six stresses fully describe stresses at appoint of a 3-D element. It is also known as a tensor. The maximum stress at a point is the major principle stress, and the other two are the minimum and intermediate principle stress. The principle planes do not have shear stresses. The direction cosines of the principle stresses at appoint are also known as the Eigen values, and the corresponding normal stresses are the principle stresses. The principle stress trajectory, the path of the major principle stress, is a good design aid for reinforcing the solid concrete bodies.

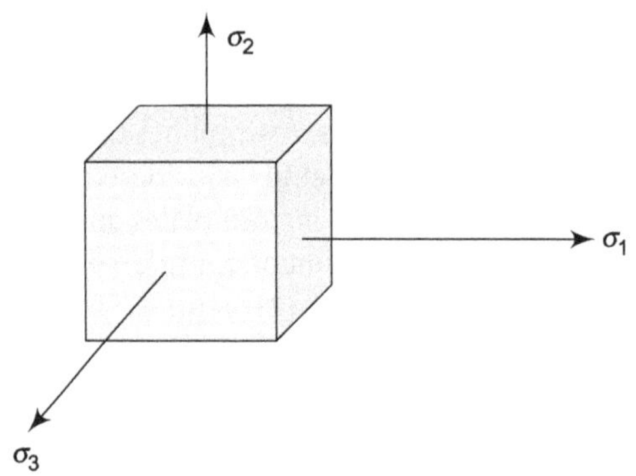

Fig. 1.7 *Principle Planes and Principle Stresses–Solid Element*

1.4 STRESS RESULTANTS AND DEFORMATIONS AT JOINTS OF SKELETAL STRUCTURES

From the foregoing definition of structures, it is obvious that structural analysis is required to find (i) stress resultants (forces and moments at a section so as to limit the stresses within yield stresses), and (ii) deformations to keep the maximum deformations within limits dictated by functional requirements. It may appear to be a difficult task to find stress resultants and deformations at every section of a structure.

Actually, to begin with, it is adequate to find stress resultants and deformations at joints only. Thereafter we consider the members separately to find, the stresses and deformations at any point desired, and then the position which it occupies in relation to other members in a structure is not important. The method, to first analyze stress resultants and then evaluate deformations, is discussed separately under the topic, "Force Method". Various methods of analysis can be grouped under the two basic methods; "Force Method" and "Displacement Method".

1.5 SUPERPOSITION AND NON-LINEARITY

Superposition is required to be used while analyzing structures for forces (see Chapter 3) which are indeterminate. In the case of displacement method also, the analysis is done by superposition in two stages (i) evaluation of joint forces due to given loading when the nodes or joints are clamped and (ii) analysis for joint loads, and allowing for joint deformations. Each stage of analysis uses superposition. In this first course of structural analysis, coverage is on linear analysis, which allows superposition. The non-linear analysis is discussed sparingly. However, in general, it is required to resort to non-linear analysis (which does not allow superposition) in the following cases:

1. Non-linear stress-strain relations of materials.
2. Non-linearity due to large deformations, which result into change in initial geometry.
3. Non-linearity due to large axial forces, which give rise to moments due to lateral deflection (henceforth referred as axial flexural interaction). This is synonym with the case 2.

In normal design practice, we account for non-linearity of stress-strain relations in a limited way. We assume, in the case of flexure members, idealized moment curvature relations like rigid plastic and "elastic and plastic" curves, Fig. 1.8. The analysis as it is generally done for complete collapse is known as limit or plastic analysis. Only in some simple cases of structures, it may not be difficult to consider bilinear moment curvature relations for non-linear analysis. This aspect of nonlinearity idealized by the bilinear relations is not discussed in this book.

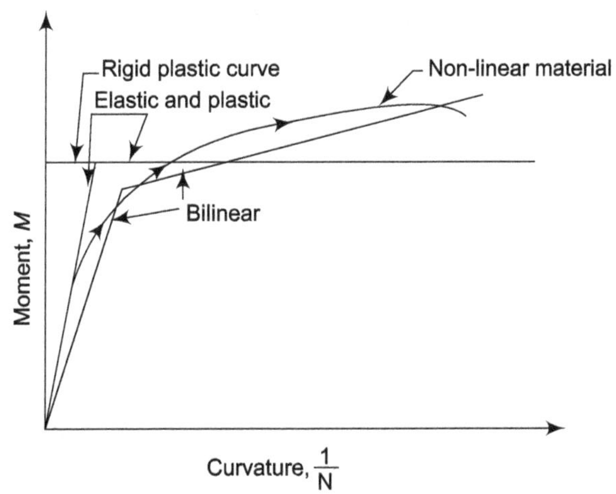

Fig. 1.8 *Idealized Moment Curvature Relations*

Large deformations are rarely permissible from utility and functional requirements of structures under working loads. It is, therefore, reasonable to exclude non-linearity of "Force-deformation" relations due to large deformations, excepting in the case of cable structures and compression instability referred in the following as P-δ effect.

The non-linearity, which is caused by large axial forces and their moments due to lateral deflections (P-δ effect), is covered to some details as an extension to linear stiffness analysis. The axial force—"lateral deflection" non-linearity (axial-flexural interaction), is not to be confused with the effect of "axial deformations" on moments and shears. The latter is strictly a linear analysis.

1.6 ENERGY PRINCIPLES*

Analysis of structures by using energy principles can be an alternative to the analysis by direct use of compatibility and static equilibrium. The principles of virtual work and complementary virtual work are used for evaluating unknown deformations and unknown forces respectively. The real merit of the method, however, lies in its being able to get approximate solutions to complex structural problems, where exact analysis is too tedious or not possible. Some obvious applications are in the stability and vibration problems where approximate deflected shapes are assumed to give good engineering results, or some guessed specification of the internal forces and their approximate evaluations.

In the case of minimum total potential method, it requires computation of energy integrals over the whole or part of the structure, followed by differentiation. But in the case of all engineering problems, it is possible to differentiate the energy expressions before the integration is carried out. The procedure of differentiation within the integral sign results in a formulation which is the same as the virtual work method. The evaluations of the integral even after differentiating under the integral sign is tedious and for surface structures, say finite element stiffness matrix evaluation, numerical integrations are commonly used. In the case of skeletal structures, however, the evaluation is direct and simple.

Energy methods give adequate engineering accuracy even for complex structures because the method averages out the errors made while assuming approximate deflected shape or approximate distribution of forces. In essence the local errors are smoothened.

The above aspects may not be obvious in the first reading. In the following the concepts are explained and derived which should make the above observations meaningful. The principle of virtual work is stated without any attempt to prove. The mathematical proof of the principle of virtual work is avoided deliberately not to confuse the reader at this stage of introduction to the topic. In contrast to its proof, the application of the principle, however, is simple and direct.

1.6.1 Principle of Virtual Work

Consider a structure in equilibrium under any given loads P_i. If the structure is given arbitrary displacements δu which need not be caused due to the applied load P_i, but the displacements do not violate the geometrical conditions, then the work done by P_i during the displacement δu_i (that is δu at the points corresponding to P_i and in the same direction) is equal to the internal work done by the internal stresses σ (due to P_i) during the strains $\delta\varepsilon$ caused by the arbitrarily imposed displacement δu.

* The section may be skipped by the readers who have not been introduced to the virtual work principle as substitute to writing equations of statical equilibrium.

The above principle is written in the form of an equation, Eq. (1.1)

$$\sum P_i\, \delta u_i = \iiint (\sigma_x\, \delta\varepsilon_x + \sigma_y\, \delta\varepsilon_y + \sigma_z\, \delta\varepsilon_z + \tau_{xy}\, \delta Y_{xy}) + (\tau_{yz}\, \delta Y_{yz} + \tau_{zx}\, \delta Y_{zx})\, dx\, dy\, dz \qquad \ldots(1.1)$$

Where σ_x, σ_y, σ_z, τ_{xy}, τ_{yz}, τ_{zx} and ε_x, ε_y, ε_z, Y_{xy}, Y_{yz}, Y_{zx} are the stress and strain components respectively in a three dimensional stress system. The loads are assumed to be point loads and the body forces are not considered. The formulation may be generalized but herein the numbers of parameters are kept to a minimum so as to emphasize the principle. The Eq. (1.1) is an alternative form of writing the static equations of equilibrium. Therefore, the method based on the principle of virtual work is also called "Equilibrium method".

The principle is restated in the context of flexural deformations, see Article 3.5 and Chapter 5.

1.6.2 Principle of Complementary Virtual Work

In the expression for Virtual work, $(=P_i\ \delta u_i)$, the arbitrary deformation δu_i may be considered as an incremental deformation, Fig. 1.7. Thus if P_i and u_i are the forces, and deformations of a structure and δu_i, is the incremental deformation due to the incremental load δ_p, the virtual, work equal to $P_i\ \delta u_i$ is the area below the cum OS and the area above the curve OS is equal to $u\delta P_i$ is called "complementary-virtual work". The total area above the curve is complementary to the total area below and together they add up to Pu. In general, δ_P represents a system of arbitrary forces, similar to the arbitrary displacements we had considered, while stating the principle of virtual work. In the case of complementary Virtual work, we consider deformations u_i and corresponding strains as prescribed (given). The loads δP_i and the corresponding stresses are arbitrary.

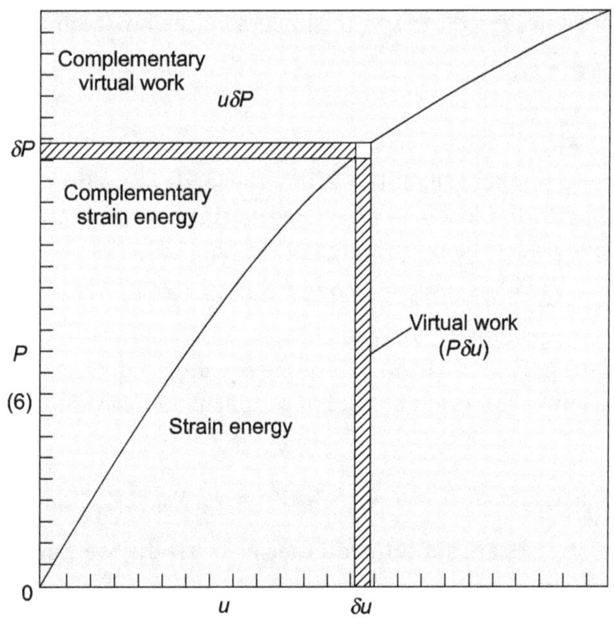

Fig. 1.9 *Virtual Work and Complementary Virtual Work*

The principle The complementary virtual work done by the arbitrary forces δP_i during the prescribed displacement *ui* is equal to the internal work done by the stresses δ (due to δP_i) during the prescribed strains ε.

When written in the form of an equation,

$$\sum u_i \,\delta P_i = \iiint (\varepsilon_x \,\delta\sigma_x + \varepsilon_y \,\delta\sigma_y + \varepsilon_z \,\delta\sigma_z + \gamma_{xy}\,\delta\tau_{xy} + \gamma_{yz}\,\delta\tau_{yz} + \gamma_{zx}\,\delta\tau_{zx}) \quad \ldots(1.2)$$

Also see Article 3.6 (Compatibility).

1.6.3 Minimum Potential Energy Principle

The minimum potential energy principle is derived from the virtual work Eq. (1.1).

$$\sum P_i \,\delta u_i = \iiint (\sigma_x \,\delta\varepsilon_x + \sigma_y \,\delta\varepsilon_y + \sigma_z \,\delta\varepsilon_z + \tau_{xy}\,\delta Y_{xy}) + (\tau_{yz}\,\delta Y_{yz} + \tau_{zx}\,\delta Y_{zx})_{dx\,dy\,dz}$$

In the above equation u_i and ε are the deformations and strains due to P_i and consider δu_i and $\delta\varepsilon$ as corresponding infinitesimal increments. The incremental values δu_i and $\delta\varepsilon$ are treated as arbitrary deformations.

Let us represent the left-hand side of the equation by – δW

$$\left.\begin{aligned} W_i &= -\sum P_i \,\delta u \\ \text{or} \quad \delta W_i &= -P_i\, u_i \end{aligned}\right\} \quad \ldots(a)$$

It should be particularly noted that in the virtual work formulation, the loads P_i are the prescribed values and therefore, they are not considered to vary. The integral of the Eq. (a) gives

$$\left.\begin{aligned} \delta W_i &= -\delta(P_i\, u_i) \\ \text{or} \quad W &= -\sum P_i\, u_i \end{aligned}\right\} \quad \ldots(b)$$

While solving the structural problems by this method, the loads P_i are prescribed and thus known. But the stresses, corresponding to the loads P_i are related to displacements u_i or strains ε. Therefore, we cannot write the expression under the integral sign of the Eq. (1.1),

$$\sigma_x\,\delta\varepsilon_x + \sigma_y\,\delta\varepsilon_y + \sigma_z\,\delta\varepsilon_z + \tau_{xy}\,\delta\gamma_{xy}\,\tau_{yz}\,\delta\gamma_{yz} + \tau_{zx}\,\delta\gamma_{zx} = \delta(\sigma_x\,\varepsilon_x + \sigma_y\,\varepsilon_y + \sigma_z\,\varepsilon_z + \tau_{xy}\,\gamma_{xy} + \tau_{yz}\,\gamma_{yz} + \tau_{zx}\,\gamma_{zx})$$

as we did while writing the equation (b) for loads.

Before further considering the right-hand side of the equation, Eq. (1.1), let us look at the expression for the strain energy per unit volume, *V* Eq. (c). For simplicity of derivation, the material is considered to be linearly elastic.

$$V = \frac{1}{2}(\sigma_x\varepsilon_x + \sigma_y\varepsilon_y + \sigma_z\varepsilon_z + \tau_{xy}\,\gamma_{xy} + \tau_{yz}\,\gamma_{yz} + \tau_{zx}\,\gamma_{zx}) \quad \ldots(c)$$

From Hookes law, when stresses are substituted in terms of **strain,** we can express *V* as a function of strain-components only. Then,

$$V = \frac{1}{2}\frac{Ev}{(1+v)(1-2v)}(\varepsilon_x + \varepsilon_y + \varepsilon_z)_2 + G(\varepsilon_x^2 + \varepsilon_y^2 + \varepsilon_z^2) + \frac{1}{2}G(\gamma_{xy}^2 + \gamma_{yz}^2 + \gamma_{zx}^2) \quad \ldots(d)$$

It can be shown that the derivative of V as given by Eq. (d) with respect to any strain component gives the corresponding stress component, i.e.,

$$\frac{\partial V}{\partial \varepsilon_x} = \sigma_x \qquad \text{...(e)}$$

Similar equations can be written for the other stress components

$$\sigma_y, \sigma_z, \tau_{xy}, \tau_{yz}, \tau_{zx}$$

Now consider the expression under the integral sign of the right-hand side of the Eq. (1.1), designated by Q.

$$Q = \sigma_x\,\delta\varepsilon_x + \sigma_y\,\delta\varepsilon_y + \sigma_z\,\delta\varepsilon_z + \tau_{xy}\,\delta\gamma_{xy}\,\tau_{yz}\,\delta\gamma_{yz} + \tau_{zx}\,\delta\gamma_{zx} \qquad \text{...(f)}$$

From Eq. (e) substitute for the stresses $\sigma_x, \sigma_y, \sigma_z$ ...
We get

$$Q = \frac{\partial V}{\partial \varepsilon_x}\delta\varepsilon_x + \frac{\partial V}{\partial \varepsilon_y}\delta\varepsilon_y + \frac{\partial V}{\partial \varepsilon_z}\delta\varepsilon_z + \frac{\partial V}{\partial \gamma_{xy}}\delta\gamma_{xy} + \frac{\partial V}{\partial \gamma_{yz}}\delta\gamma_{yz} + \frac{\partial V}{\partial \gamma_{xz}}\delta\gamma_{xz} \qquad \text{...(g)}$$

The right-hand side of the Eq. (g), is an expansion of dV

$\therefore$ $$Q = dV$$

and $$\iiint dV = dU$$

Symbolically also, referring to Fig. 1.7, the shaded area below the curve OS represents dU, where dU is the differential of the total strain energy, that is, considering the entire volume.

The Eq. (1.1) can be rewritten as

$$\sum P_i\,\delta u_i = dU$$

and from substituting Eq. (b)

$$-\,\delta W = U$$

or $$\delta(U + W) = 0 \qquad \text{...(1.3)}$$

The quantity $(U + W)$ is known as "Total Potential Energy", and U and W are the potentials due to strain energy and loads respectively. The equation, Eq. (1.3), implies a zero rate of change of total potential energy with variations of deformations, from the equilibrium position. For simplicity only, the derivation of the result, Eq. (1.3), assumed linear elasticity; otherwise the result can be shown to be valid for non-linear materials also. The result is equally valid for "large-deformation theory", provided that the stresses and strains are defined in such a way that Eq. (1.1) gives the correct work done due to the small change in strains. In the case of stability problems, even though a case of geometrical non-linearity, the strains are small. The case of large strains, even elastically, which may occur for material like rubber, is not within the scope of the above discussion.

1.6.4 Rayleigh-Ritz Method

The method is based on the minimum potential energy principle. It can be used to obtain deflected shapes, which may not be exact but give adequate accuracy in most of the engineering problems. Let the deflected shape be approximated by a series consisting of n finite number of terms.

$$u_n = \sum_{i=1}^{n} a_i \psi_i \quad \text{...(a)}$$

The shape functions ψ_i are such that they individually satisfy the given boundary conditions. The unknown coefficients a_i are obtained by minimizing the total potential Eq. (2.3) with respect to the unknown coefficients at a_1. That is

$$\frac{\partial}{\partial a_i}(U_n + W_n) = 0 \quad \text{...(1.4)}$$

Where U_n and W_n are the potentials due to the strain energy and loads based on the deflected shape approximated by the n terms in the series (a). The Eq. (1.4) results into n simultaneous equations for the n unknown coefficients ($a_1, a_2, \ldots, a_n$). The method is known as the "Rayleigh-Ritz method." The stresses are evaluated from the approximate solution for deflection, u.

The stiffness matrix of the skeletal members and the surface elements can be obtained by the Rayleigh-Ritz method. In the latter case, i.e., the "finite element method" for surface elements, it uses the shape functions and the Rayleigh-Ritz method. For obtaining the finite element stiffness, matrix a_i represents the end deformations at the nodes and ψ_i describes the corresponding variation of the deformations within the element.

1.6.5 Castigliano's Theorem (Part I)

In the equation $\delta(U + W) = 0$, if the total potential $(U + W)$ is minimized with respect to u_i, which is a deformation at any point of the structure, we get,

$$\frac{\partial U}{\partial u_i} + \frac{\partial W}{\partial u_i} = 0 \quad \text{...(a)}$$

But it was stated in Article 2.6.3, that the potential due to the loads is given by

$$\left.\begin{aligned} W_i &= -\,P_i u v_i \\ \text{and} \qquad W_i &= -\sum P_i u_i \end{aligned}\right\} \quad \text{...(b)}$$

As the loads P_i are specified and therefore, not varying, whereas u_i deformations are independent.

$$\frac{\partial W}{\partial u_i} = P_i \quad \text{...(c)}$$

From the Eqs (a) and (b)

$$\frac{\partial U}{\partial u_i} = P_i \quad \text{...(1.5)}$$

It is the Castigliano's Theorem (I). The Eq. (1.5) is used to obtain a set of simultaneous equations relating P_i to their corresponding displacements u_i. The coefficients of the displacements u_i are the stiffness coefficients. The Eq. (1.5) is valid for linear as well as non-linear systems. It may be obvious that instead of the theorem, Eq. (1.1), the principle of virtual work can be used directly. The minimum potential energy principle is the generalization of the Rayleigh-Ritz method and the Castigliano's theorem (I). All these methods which are based on the principle of virtual work are the same as the displacement method or the stiffness method.

1.6.6 Principle of Minimum Complementary Energy

The minimum complementary potential energy principle is derived from the complementary virtual work Eq. (1.2).

$$\delta\sum u_i P_i = \iiint(\varepsilon_x\,\delta\sigma_x + \varepsilon_y\,\delta\sigma_y + \varepsilon_z\,\delta\sigma_z) + (\gamma_{xy}\,\delta\tau_{xy} + \gamma_{yz}\,\delta\tau_{yz} + \gamma_{zx}\,\delta\tau_{zx})\,dx\,dy\,dz$$

The deformations u_i are treated as prescribed values and P_i values are treated as arbitrary (variable)

$$\sum u_i \delta P_i = \delta\sum u_i P_i$$

Or $$\sum u_i \delta P_i = -W$$

Where $$W = -\sum u_i P_i \qquad \ldots(a)$$

The right-hand side of the Eq. (1.2) is the complementary virtual work due to the internal stresses, treated as arbitrary, ride over the strains treated as constant. Following the similar reasoning as stated in the Article 1.6.3, the righthand side of the Eq. (1.2) is equal to the differential of the complementary strain energy, Fig. 2.7. That is

$$\iiint(\varepsilon_x\,\delta\sigma_x + \varepsilon_y\,\delta\sigma_y + \varepsilon_z\,\delta\sigma_z) + (\gamma_{xy}\,\delta\tau_{xy} + \gamma_{yz}\,\delta\tau_{yz} + \gamma_{zx}\,\delta\tau_{zx})\,dx\,dy\,dz = dU^* \qquad \ldots(b)$$

Symbolically, referring to Fig. 1.7, the shaded area above the curve **OS** represents dU^*.

From the Eqs (a) and (b) and the Eq. (1.2), we get

$$\delta(U^* + W) = 0 \qquad \ldots(1.6)$$

The Eq. (1.6) is valid for linear as well as non-linear structures. In the case of linearly elastic structures, total strain energy, U is equal to the total complementary strain energy U*. Therefore, for the linearly elastic structures, we may write

$$\delta(U + W) = 0 \qquad \ldots(1.6a)$$

Where $\delta W = -\sum u_i\,\delta P_i$, which is different from the δW of Article 1.6.3, Eqs (a) and (b).

The Eq.(1.6) is used for evaluating,

(i) The unknown redundant forces of statically indeterminate structures and
(ii) Deflection at any point of a structure.

While evaluating the unknown redundant forces, the given set of loads P_i are the prescribed loads and the redundant forces P_j are treated as "arbitrary forces." Considering W only from the Eq. (1.6)

$$-W = \underbrace{\sum P_i u_i}_{\text{Constant}} + \underbrace{\sum P_i u_i}_{\text{only } u_i \text{ are treated as constant}} \qquad \ldots(a)$$

$$-\delta W = 0 + \sum u_i\,\delta P_i$$

But as the deformations corresponding to the redundant P_j are equal to zero

$$-\frac{\delta W}{\delta P_j} = u_i = 0$$

Substituting $\dfrac{\delta W}{\delta P_j} = 0$ in the Eq. (2.6)

$$-\frac{\partial U^*}{\delta P_i} = 0 \qquad \dots(1.7)$$

The Eq. (1.7) in the case of linear structures

$$\left(-\frac{\partial U}{\partial P_i} = 0\right)$$

is used to generate a set of simultaneous equations for the unknown redundant forces P_i. It is also known as the *"method of least work."* Having evaluated *Pi*, the structure can be analyzed for forces. The deformations u_i at any point 1 of a structure can be found by treating P_t as an arbitrary force at the point 1. We write

$$-\delta W = \underbrace{\delta \sum P_i d_i}_{\text{Only } P_i \text{ is arbitrary}} + \underbrace{\delta \sum P_i u_i}_{\text{Constant}}$$

$$-\delta W = \delta P_i \delta_i + 0$$

The Eq. (1.6) is simplified to

$$\delta U^* = \delta P_i u_i$$

$$u_i = \frac{\partial U^*}{\partial u_i} \qquad \dots(1.8)$$

The Eq. (1.8) can be used for finding deflections at the points other than the load points by substituting $P_i = 0$, in the final form, i.e., after the partial differentiation. The procedure is the same as the "Dummy load method."

Illustrative examples based on the above principles are given under Article 1.5.3 and also see Article.

1.7 TOTAL STRAIN ENERGY FOR SKELETAL MEMBERS

While stating the virtual work and minimum potential principles, the strain energy or the internal work was described in terms of stresses and strains. In the case of skeletal members because of the assumption, that plane section remains plane, the internal stresses and strains over any cross-section are completely described by the resultant stresses, i.e., thrust, moment, shear and torsion, and the corresponding deformations. In the Eqs 1.3 to 1.6, the total strain energy *U* for the members of a skeletal frame, is found by using the following results,

$$\left.\begin{aligned}
&\text{Strain energy due to direct load} &&= \int_0^1 \frac{P_x^2}{2EA}\,dx \\
&\text{Strain energy due to flexural moment} &&= \int_0^1 \frac{M_x^2}{2EA}\,dx \\
&\text{Strain energy due to shear} &&= \int_0^1 \frac{P_y^2}{2G\bar{A}}\,dx \\
&\text{Strain energy due to torsional moment} &&= \int_0^1 \frac{M_x^2}{2GJ}\,dx
\end{aligned}\right\} \qquad \dots(1.9)$$

Where P_x, M_z, P_y and M_x are the defined in Fig. 1.1, *El* and *GJ* are flexural and torsional stiffnesses, and for *I* sections $\bar{A}$ is approximately equal to web area. The length *l* is measured along the axes.

In the majority of problems the contribution of shear and axial thrust to the total strain energy is neglected without any appreciable loss of accuracy.

The concepts introduced as the basics of Energy Principles are used extensively in Chapter 2, Unified Force Method. The use of the principle of virtual work is used for computing the unknown deformations as flexibilities while satisfying compatibility at the chosen releases, introduced in Chapter 2.

CHAPTER 2: FORCE METHOD

Chapter

2

Force Method

2.1 GENERALIZED FORCE METHOD

The evaluation of the internal forces developed under the action of applied loads on stable structures is the designer's primary requirement. Having designed, it is then followed up to ensure that the structure does not undergo excessive deformations. The Force method achieves the three objectives in the above sequence; (1) ensures the structure is stable, (2) evaluates the internal forces, and finally (3) ensures that the structure designed does not undergo excessive deformations during its stipulated functioning.

The concepts pertaining to kinematical constraint conditions are fundamental to arriving at stable structures; trusses, beams, plane frames, grids, and space frames. A good understanding of stable and unstable structures is achieved while determining

- Determinacy, and
- Indeterminacy

The Force and Displacement methods are considered to be the basic methods in structural analysis. The determinacy and indeterminacy are discussed based on (i) statical equations of equilibrium and also (ii) kinematic constraint conditions. The evaluation of the indeterminacy is shown for different structures, trusses, beams, plane frames, grids and space frames. The complete analysis, however, has been limited to some typical structures. Suitability of the method has been shown for grids also. The Force method of analysis is suitable if the number of unknowns obtained from the compatibility conditions are not many. The method has also been found to be suitable for "Elastic Plastic" analysis, developed in Chapter 3. The force method is also known as the Flexibility method.

The contents of the chapter are oriented to lead the reader to plastic analysis in Chapter 3 and matrix approach in Chapter 4.

2.1.1 Stable Structures and Indeterminacy

If a structure takes an equilibrium position under the action of static loads (gradually increased and not suddenly applied) without distorting its initial geometry, it is a "Stable" structure. Stability of a structure under dynamic loads or from buckling considerations is not included in this chapter. If a structure is stable and it is possible to find internal forces in all the members constituting the structure and support reactions from the statical equations of equilibrium only, the structure is "determinate". A determinate structure, therefore, (i) is a stable structure, and (ii) has the minimum number of constraints. In the case of an "indeterminate" structure, the internal or external constraints or both together are more than the minimum required. The number of excess constraints defines the degree of indeterminacy. For a complete solution, it is required to use deformation or compatibility conditions, their number being equal to the number of excess constraints.

2.1.1a Free Bodies and Statical Equations of Equilibrium

Therefore, to decide whether a structure is determinate or not, we draw appropriate free bodies and find, v, the total number of internal forces and support reactions and compare it with the number, s, equal to the total number of independent statical equations of equilibrium. If

$v = s$ the structure is determinate
$v > s$ the structure is indeterminate
$v < s$ the structure is unstable

In the following, the determinacy and the degree of indeterminacy, n, defined equal to $v - s$, are discussed. A systematic study to find whether a structure is determinate, indeterminate or unstable, based on the statical equations of equilibrium, is given later in Article 2.4. The above conditions are not *sufficient* for defining an unstable structure. The rank technique or kinematic considerations, discussed later, are necessary.

Before we find whether a structure is determinate or not, it is necessary to know the nature of internal forces and support reactions which are dependent on the type of a Structure. The classification of different types of structures based on the nature of internal forces is given in Fig. 2.1. Practically all the types are shown. Each type of structure, based on the nature of internal forces is described in the following.

2.1.2 Hinge-jointed Trusses

Consider Fig. 2.1(a). The number of internal forces per member are designated by i and the number of support reactions by r.

$i = 1$ per section of member, and it is an axial force
$r = 2$ for hinge support, and
$r = 1$ for roller (or link) support

When we are indicating the number of the support reactions, r, it is implied that the structure is freed from the supports.

With the sections shown on the members, the number of free bodies is equal to the number of joints. The forces acting on the members of a joint being concurrent, the equilibrium equation, $\sum M = 0$ being trivial, we get two equations of equilibrium namely $\sum p_x = 0$, and $\sum p_y = 0$ per free body or joint. If λ is the number of independent statical equations per free body, for a truss joint, $\lambda = 2$.

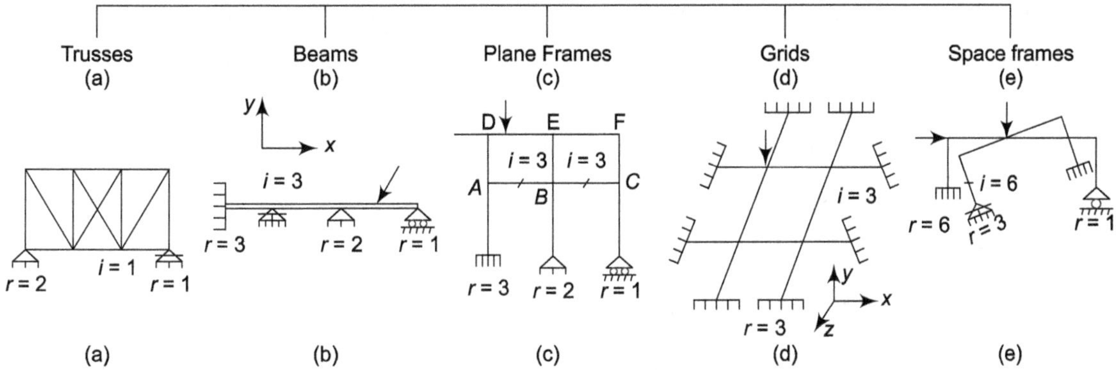

Fig. 2.1 *Values of i and r for Different Types of Structures*

Fig. 2.2 *Airport Roof Truss Under Fabrication*

2.1.3 Continuous Beams

Figure 2.1(b). The internal forces at a section consist of thrust, shear and moment.

$i = 3$ per section of a member

$r = 3$ for clamped support

$r = 2$ for hinge support

$r = 1$ for roller support

But the thrust values are evaluated independent of shear and moment. That is, the thrusts are 'delinked' from flexural forces, which are moments and shears, and the equilibrium equation $\sum p_x = 0$ is directly used for the evaluation of axial forces only, where x-axis is along the axis of the straight beam. Thus, we may consider $i = 2$ and consider two equations of statical equilibrium per free body, that is $\sum p_y = 0$ and $\sum M = 0$ for the evaluation of the two flexural internal forces at a section of a member. Thus in the case of straight continuous beams, we consider $\lambda = 2$.

2.1.4 Rigid-jointed Plane Frames

Figure 2.1(c). The numbers of support reaction per support are the same as given for continuous beam. In this case, however, axial forces are not delinked from shears and moments. The number of internal forces, $i = 3$ per section; thrust, shear and moment. The number of sections should be at least as many as to remove the closed ring or boxes. In Fig. 2.1(c), if the structure is freed from the supports, at least two internal sections should be taken so that the internal forces at any other section can be evaluated from statics alone in the terms of the internal forces at two sections and the support reactions. For enabling the computations of internal forces, the loops are to be eliminated. The number of equations per free body, $\lambda = 3$, that is, $\sum p_x = 0$, $\sum p_y = 0$ and $\sum M = 0$. Alternatively, *degree of Indeterminacy* is given by three times the number of closed loops or rings in the plane frames.

2.1.5 Grids

If the members of the grid are torsionally stiff, Fig. 2.1(d).

$r = 3$ for clamped support.

$i = 3$ and the internal forces consist of shear, bending moment and twisting moment.

If the torsional stiffness of member is neglected,

$r = 2$ for clamped support.

$i = 2$ and the internal forces consist of shear and bending moment only.

The number of independent equations per free body, $\lambda = 3$ in either case, that is,

$\sum p'_y = 0$, $\sum m'_x = 0$, $\sum m'_z = 0$ Fig. 2.1(d)

2.1.6 Space Frame

Figure 2.1(e)

$i = 6$

$r = 6$

The six member forces at a section are

$$\left[p_x \; p_y \; p_z \; m_x \; m_y \; m_z \right]^t$$

We write six equations of equilibrium per free body, i.e., $\lambda = 6$.

Degree of Indeterminacy Sufficient numbers of cuts are introduced within a structure so as to make sure that there are no closed rings. Corresponding to each cut, depending on the type of a structure, we know the value of i per cut. Let i' denote the sum of i values for all the cuts. Similarly let r' denote the sum of the support reactions, r. If f is the number of free bodies obtained by the cuts, and if the structure is statically stable, the degree of indeterminacy, n, is given by,

$$n = v - s$$

where $v = r' + i'$ and

$$S = \lambda f$$

The value of λ for each type of structure has been indicated separately.

As illustrations, some typical examples of trusses are considered for finding indeterminacy, Fig. 2.3. We will use the information regarding the values of r and i as stated earlier for the hinged trusses. Consider Fig. 2.3(a). Fifteen cuts are taken for the fifteen numbers. The cuts constitute the release system.

$$i = 1 \text{ and } i' = 15 \times i = 15.$$

$$r = 2 \text{ at hinge support, } r = 1 \text{ at roller support}$$

$$r' = 2 + 1 = 3,$$

$$v = r' + i' = 3 + 15 = 18. \quad ...(1)$$

Number of free bodies, f due to the cuts = number of joints = 9 equations of equilibrium per free body, l = 2

$$s = \lambda f = 2 \times 9 = 18 \quad ...(2)$$

Comparing Eqs (1) and (2) $v = s$

The structure is determinate.

Consider Fig. 2.3(b). There are no joints at the crossings of the diagonals.

$$i = 1,\ i' = 19$$

$$r' = 3$$

$$v = 22$$

$$f = 10 \quad ...(1)$$

$$s = 2 \times 10 = 20 \quad ...(2)$$

$$n = v - s = 2$$

The structure is indeterminate. Degree of indeterminacy = 2.

Consider Fig. 2.3(c)

$$i = 1,\ i' = 56$$

$$r' = 3$$

$$v = 59 \quad ...(1)$$

$$f = 21$$

$$s = 2 \times 21 = 42 \quad ...(2)$$

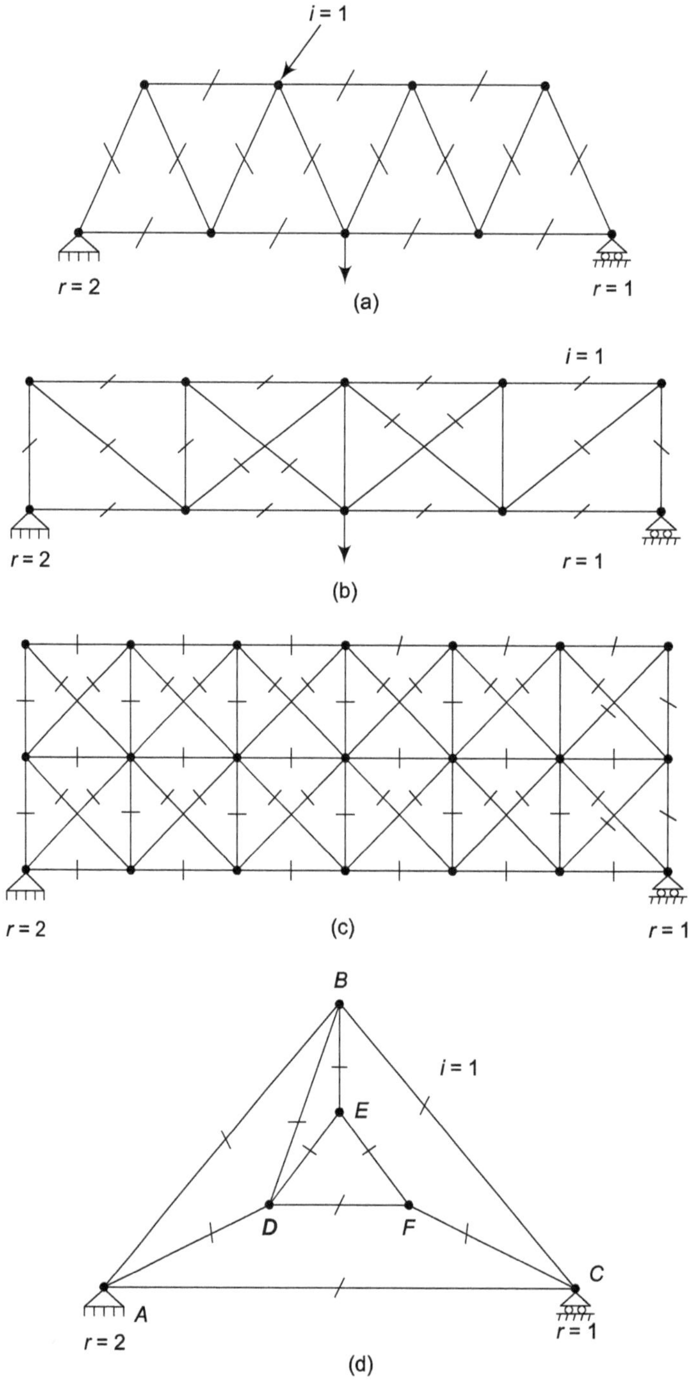

Fig. 2.3 *Hinge Jointed Trusses: i and r. Values*

The structure is indeterminate. Degree of indeterminacy = 59 – 42 = 17.

Consider Fig. 2.3(d)

$$i = 1, i' = 10$$
$$r' = 3$$
$$v = 13 \quad ...(1)$$
$$f = 6$$
$$s = 2 \times 6 = 12 \quad ...(2)$$

The truss is indeterminate, $n = 1$. To the reader it may not be obvious whether the truss is stable or not. The statical considerations sometimes leave this aspect of stability obscured, unless we use the rank technique, Article. 2.3. However, the kinematic considerations can be used more directly to find whether a truss for frame is stable or not and then to find the degree of indeterminacy for stable structures.

In the following, the kinematic considerations are used for finding the degree of indeterminacy for stable structures. It will be seen that sometimes it may be desirable that we use kinematic as well as statical considerations in order to not taking any chance while identifying stable structures.

2.2 KINEMATIC APPROACH FOR DETERMINING INDETERMINACY

A structure should have sufficient "Constraints" so that under the action of static loads (applied gradually with small increments), it takes a static position without distorting its initial geometry–small deformations are assumed. In the case of "Determinate" structures, the constraints or we may say "Kinematic constraints" are the minimum. The number of excess kinematic constraints gives the degree of indeterminacy. In the case of unstable structures, "motion" takes place and it is a problem of kinematics as static equilibrium is not possible.

2.2.1 Trusses and Kinematic Considerations

As illustrations, consider the examples in Fig. 2.4. We first check the adequacy of constraints provided by the support reactions. Under any applied loads, all the trusses of the Fig. 2.4 are adequately constrained at the supports. In the case of trusses with hinged joints, triangles are the stable units constituting the truss. Thus if we can cover all the given nodes with the basic triangular units, the structure is stable, that is, it has adequate internal constraints. The number of excess members when added to the number of excess support constraints, gives the degree of indeterminacy.

Consider Fig. 2.4(a). The truss has minimum external support constraints. It is determinate externally. It has ten nodes or joints, A, B, ..., J. We cover all the nodes by constructing the basic triangles such that there is at least one common side between the existing triangles and a new triangle added. For example, let us start constructing the triangles commencing from support A, and they are ΔABF, and with BF as a common side added ΔBFG, which introduces a new node G. The nodes H, D, I, J and E are covered by constructing the triangles GCH, HCD, HDI, IDJ and JDE. While constructing the above triangles, we have left out the sides BH and HD, which are the excess internal constraints. The degree of indeterminacy = 2.

Consider Fig. 2.4(b). Externally it has one excess support constraint. Regarding internal constraints, we can cover all the nodes with the triangles obtained by continuous units with one common side for every new node added. The members shown by the dotted lines are the excess constraints and their number is seventeen. The truss is stable and the degree of indeterminacy = 1 + 17 = 18. If support G is a roller, $N = 17$.

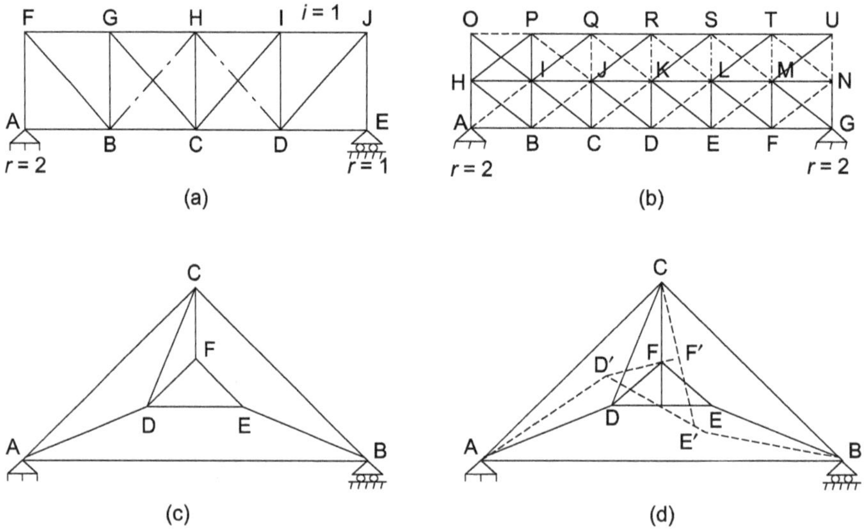

Fig. 2.4 *Degree of Indeterminacy from Kinematic Consideration*

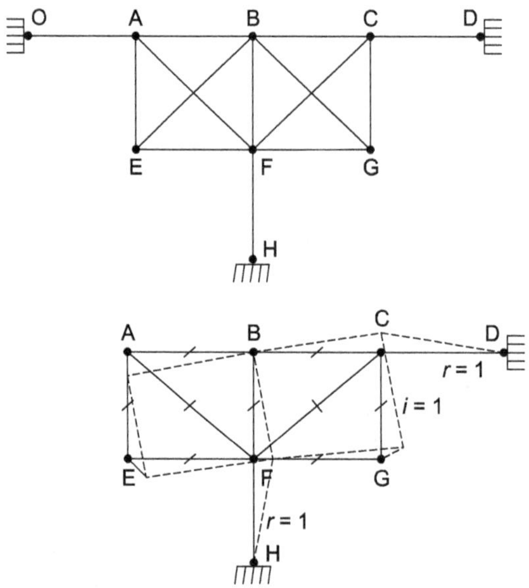

Fig. 2.5 *Unstable Truss*

Consider Fig. 2.4(c). Externally the truss is determinate. Construct triangles *ABC*, *ACD*, *CDF* and *FDE*. Thus we have included all the nodes by basic triangular units. The member EB is in excess, and therefore,

$$n = 1$$

The reader may verify that if we drop the member CD, Fig. 2.4, the truss is unstable. All the nodes *A*, *B*, *C*, *D*, *E*, *F*, cannot be covered by basic triangular units retaining one side common for every new triangle added. The kinematic rotational movement (instability) is also shown, Fig. 2.4(*d*).

Kinematic Approach Extended: Consider the truss, Fig. 2.4(b). There are several alternatives of building the truss using the triangles. A triangle is a stable configuration. Let us start with *HAB* as a basic triangle. With the 'common side and a new node' approach obtain bottom nodes *C* to *G*, and top nodes *H* to *N*. Obtain similarly nodes *O*, *P*, and *Q* using common sides *HI*, *HO*, and *IP* respectively. The nodes *J* and *Q* having been achieved from assembling the triangles, their positions get fixed. Therefore use the nodes *J* and *Q* to obtain a new node *R* by joining *JR* and *QR*. Similarly obtain the nodes, *S*, *T*, *U*. We have covered all the nodes. The number of the left over members, shown dotted, gives internal redundancy equal to 17. Adding the support redundancy, in this example one, the degree of indeterminacy, $n = 18$.

Consider Fig. 2.4. Where we combine kinematic as well as statical considerations for 'indeterminacy test', if only it is stable. The reader may verity that the truss is a mechanism. Node *B* is a center of rotation.

The member *OA* provides horizontal constraints, which is also being provided by the member *CD*. The member *OA* can be dropped without affecting stability in any way. Similarly, from the consideration of the triangles as basic units, the internal constraints *EB* and *BG* can also be dropped without affecting stability. The truss of Fig. 2.5 (*b*) is without the constraints provided by the members *OA*, *EB* and *BG*, and with the remaining members the truss can be shown to be unstable because from statical considerations,

$$v = 11, s = 12 \text{ and } s > v, \text{ see mechanism, Fig. 2.4}$$

In the following, different types of structures will be considered for the indeterminacy tests. Appropriate values of *i*, *r* and *s* will be used depending upon the type of structure. The examples of continuous beam being simple are excluded. Let us consider some examples of plane frames.

Mechanisms in Trusses The aspect is discussed under the Displacement method, Chapter 7.

2.2.2 Continuous Beams and Mechanisms

The support conditions and internal hinges are the important considerations. Further details are discussed after introducing concepts of hinges and mechanisms.

2.2.3 Plane Frames and Kinematics

The topic is deferred till we define plastic hinges and mechanisms. A mathematical presentation is given in Article 2.5.

Examples of plane frame Consider Fig. 2.6. The frame of Fig. 2.6(*a*) is externally determinate and internal constraints are in excess by three. If we provide a complete cut at *K*, the structure continues to be stable and the degree of indeterminacy, *n* is the number of excess constraints corresponding to the cut at *K*.

Therefore, $n = 3$

Following the same argument, the degree of indeterminacy for the frame, Fig. 2.6(*b*), is given by the five excess constraints consisting of two excess support constraints and three excess internal constraints.

$$n = 5$$

The same result can be obtained from statical consideration. It is enough to provide one cut at *K* to destroy the ring *CDEF*, so as to get one free body only. However, we may take more number of cuts, say, at *K* and *G*, Fig. 2.6(c)

$$i = 3$$

$$i' = 2 \times 3 = 6$$

$$r = 3 \text{ at clamped end } A$$

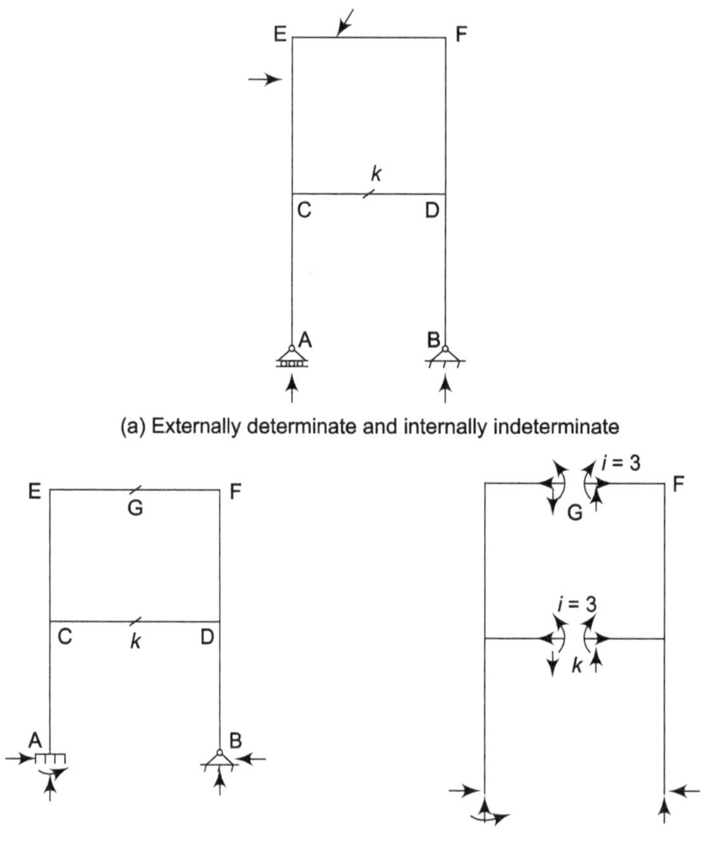

(a) Externally determinate and internally indeterminate

(b) Indeterminate internally and externally

(c) Free bodies with rings

Fig. 2.6 *Plane Frames*

$$r = 2 \text{ at hinged support } B$$

$$r' = 3 + 2 = 5$$

$$v = r' + i' = 5 + 6 = 11 \quad ...(1)$$

Equations of equilibrium per free body, $\lambda = 3$
Number of free bodies, $f = 2$

$$s = \lambda f = 3 \times 2 = 6 \quad ...(2)$$

The frame is indeterminate. Degree of indeterminacy,

$$n = v - s$$

$$= 11 - 6 = 5$$

If there are internal releases in the form of hinges in a given structure, a cut may be taken at the release itself. A hinge at a joint is equivalent to the number of hinges equal to the numbers of members meeting at the joint J less one, Fig. 2.7(b). If we insert a pin joint at the end of the member "3" also, the joint J, Fig. 2.7(c), becomes a joint mechanism, that is, it is free to rotate independent of the members meeting at the joint. This is prevented by providing one release less, Fig. 2.7(b). Therefore, if there are m number of members meeting at a hinged joint, it is equivalent of $m - 1$ hinges at the end because the value of i per hinge is 2, the number o releases i by

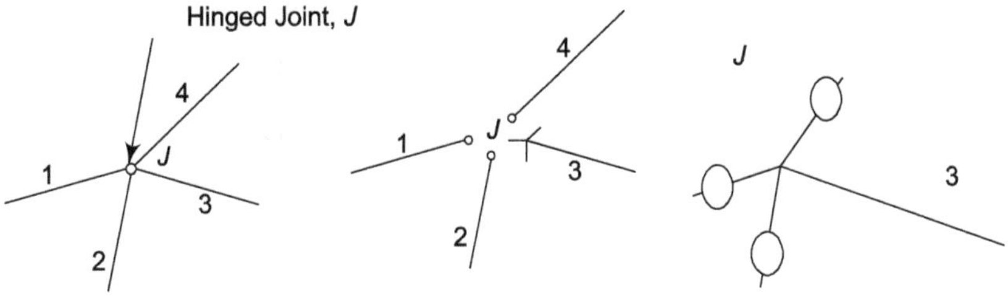

Fig. 2.7 *Member – end – releases of joint, J*

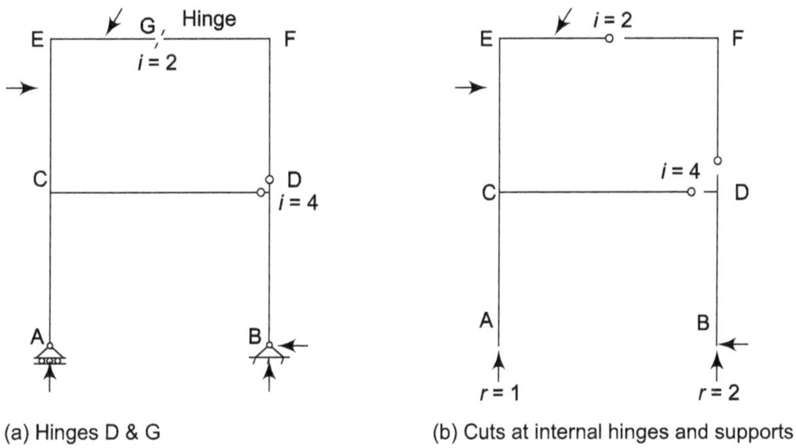

Fig. 2.8 *Cuts at Releases G, D and Supports*

Cutting a hinged joint is equal to $2\ (m - 1)$. It is reiterated in Fig. 2.8(a)
Consider the frame, Fig. 2.8(a). Provide cuts at the hinges D and G, Fig. 2.8(b).

At hinge $G, m = 2$ and $i = 2\ (m - 1) = 2$

At hinge D, $m = 3$ and $i = 2\ (m - 1) = 4$

$$i' = 6, r' = 3$$

$$v = r' + l' = 3 + 6 = 9 \quad ...(1)$$

Number of free bodies, Fig. 2.8(b), $f = 3$

$$s = \lambda f = 3 \times 3 = 9 \quad ...(2)$$

$$v = s$$

The frame is determinate.

The figure examples 2.1 to 2.11 consist of problems of mixed types, that is, hinged trusses, frames and space frames. The general approach of the indeterminacy test does not necessitate the segregation of the different type of structures, as long as we attribute proper values to the number of reaction forces, r and internal forces, i.

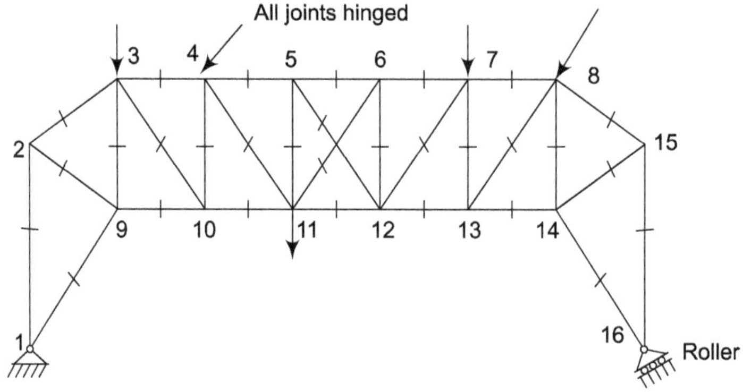

With all members cut, no. of free bodies = no. of joints = 16

$v = r' + i' = 3 + 30 = 33$

$s = h\,f = 2 \times 16 = 32, \quad n = 1$

Fig. Ex. 2.1

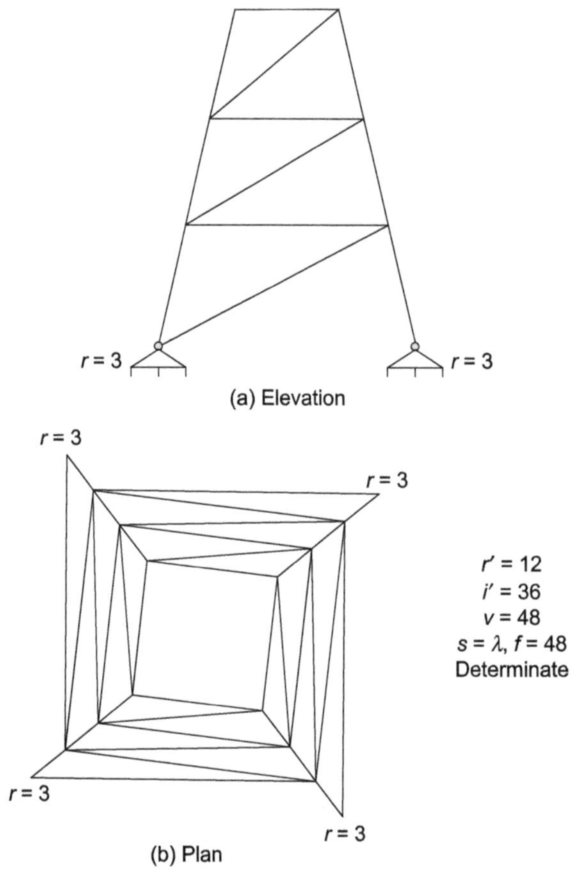

Fig. Ex. 2.2 *Hinged Transmission Tower*

Example 2.3 Find the degree of indeterminacy, *n*

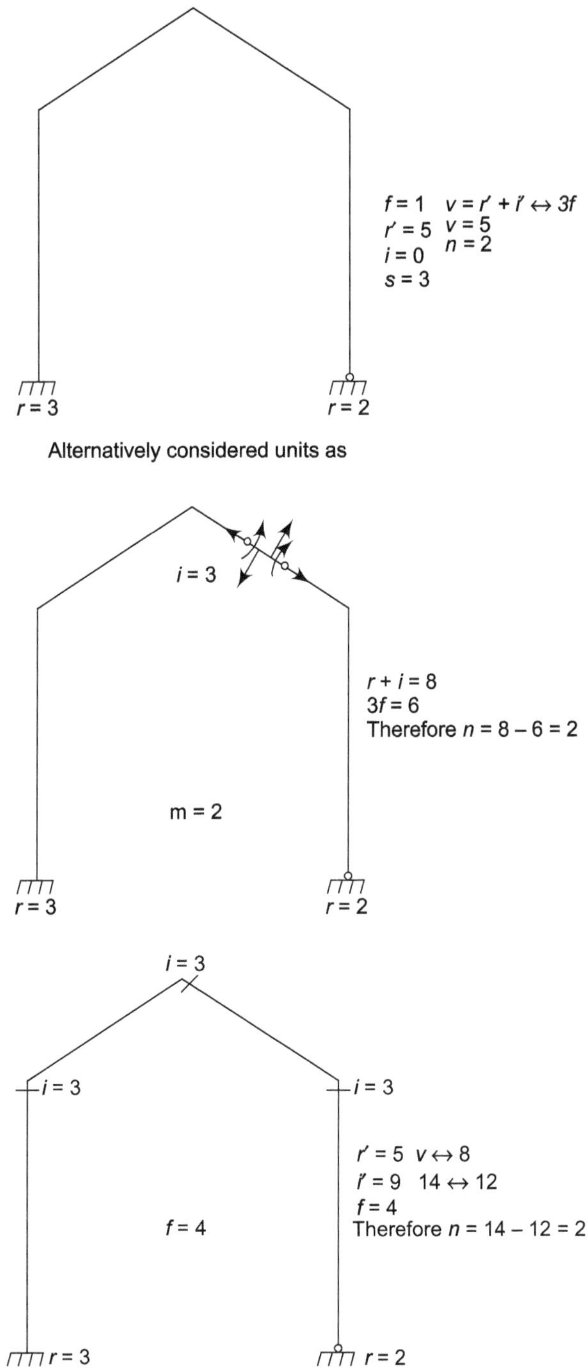

Fig. Ex. 2.3

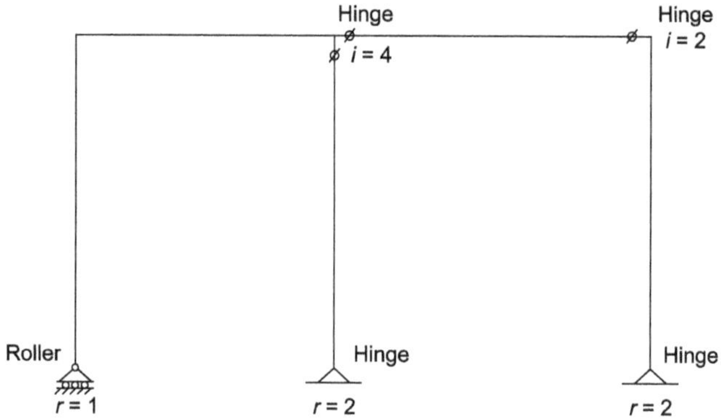

$r' + i' = 11$ $\quad 11 \leftrightarrow 12$

$f = 4$ $\quad (r' + i') < 3\,m$

$3f = 12$ $\quad$ Therefore Unstable

Fig. Ex. 2.4

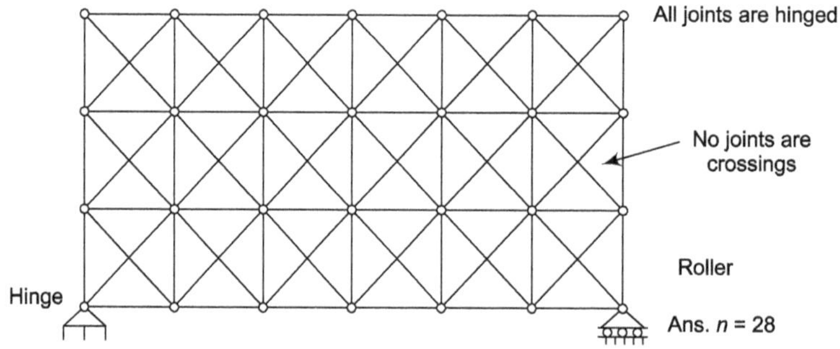

Fig. Ex. 2.5

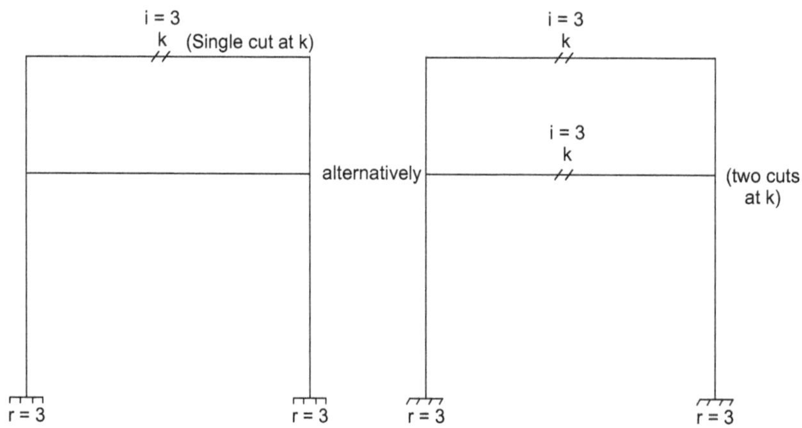

$r' + i' = 9$

$f = 1$

$v = r' + i' \leftrightarrow 3f$

$9 \leftrightarrow 3$ Therefore $n = 6$

$v = i' + r' = 12$

$f = 2$

$3f = 6$

Therefore $n = 12 - 6 = 6$

Fig. Ex. 2.6

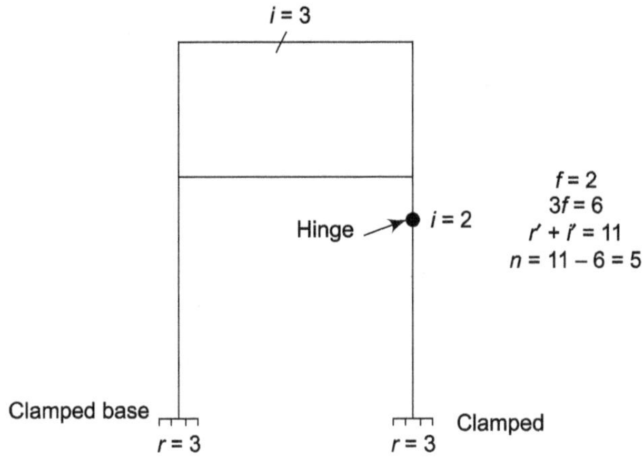

Fig. Ex. 2.7

Example 2.8 Find degree of indetermination (a) Plane Frame and (b) Space frame

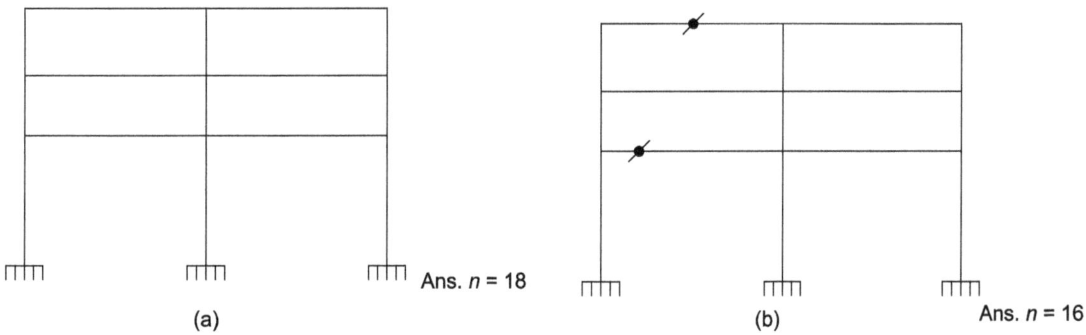

Fig. Ex. 2.8(a) *Plane Frame*

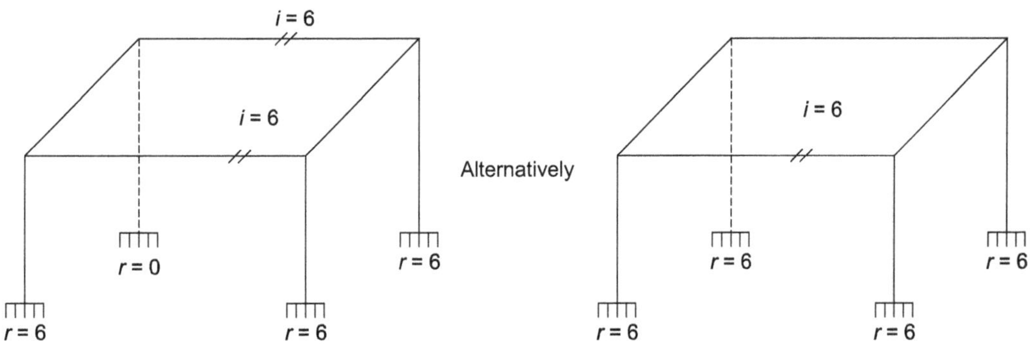

Solution. $r' = 24$ $i' = 12$ $f = 2$

$\therefore$ $r' + i' \leftrightarrow 6f$ $n = 36 - 12 = 24$

$r = 6$ $r = 6$

$r' = 24$ $i' = 6$ $f = 1$ $n = 30 - 6 = 24$

Fig. Ex. 2.8(b) *Space Frame*

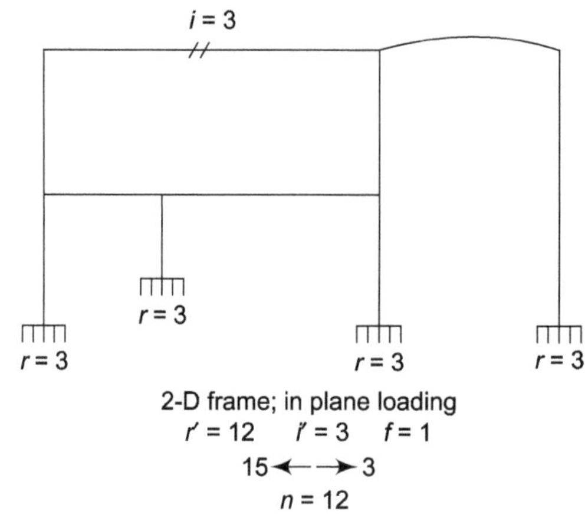

Fig. Ex. 2.9 *Including a Curved Member*

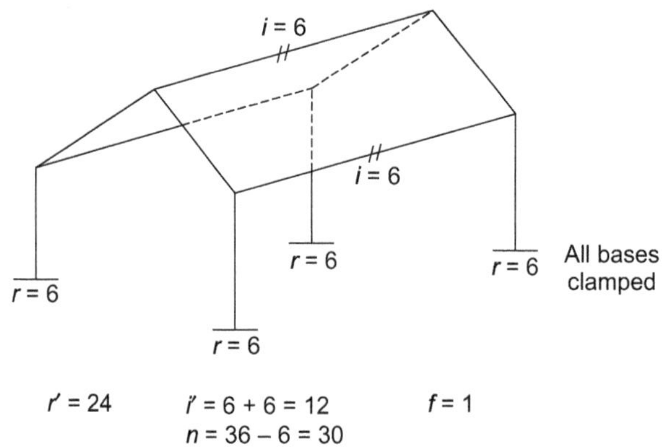

Fig. Ex. 2.10 *Space Frame*

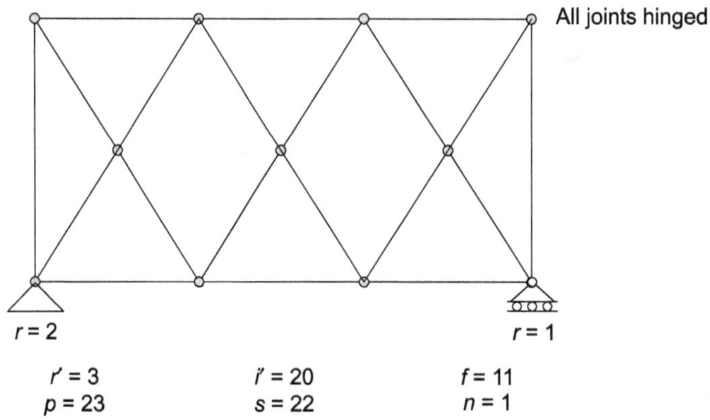

Fig. Ex. 2.11

2.3 PLANE GRIDS-TRANSVERSELY LOADED, DEGREE OF INDETERMINACY

Both the considerations, (i) statical as well as (ii) kinematic approaches are presented.

2.3.1 Indeterminacy of Grids from Statical Considerations

In the case of grids, the three internal stress resultants at a section consist of bending moment M_x, shear P_y and twisting moment M_z, where x, y and z are the member coordinates, Figs 2.1(c) and 2.1(d). The loads are in the direction perpendicular to the plane of the grid. The grid of Fig. 2.9 is supported at the corners only.

Referring to Fig. 2.9, total number of internal forces, $i' = 3 \times$ no. of grid members

$$= 3[m(n+1) + n(m+1)] \quad ...(2.1)$$

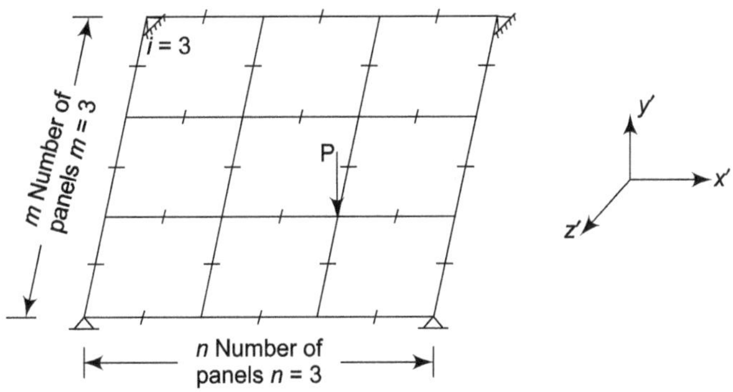

Fig. 2.9

In the case of four corner supports, Fig. 2.9,

$$i' + r' = 3[m(n+1) + n(m+1)] + 4 \quad ...(2.2)$$

According to the releases introduced, i.e., one cut per member, the number of free bodies is equal to the number of joints.

There are three equations of equilibrium per joint.

Total no. of equilibrium equations = 3 × no. of joints

$$= 3(m+1)(n+1) \quad ...(2.3)$$

From the above equations,

(a) Degree of indeterminacy,

$$N = 3[m(n+1) + n(m+1)] + r' - 3(m+1)(n+1)$$

$$N = 3(mn-1) + r' \quad ...(2.4)$$

For the grid of Fig. 2.9, $m = 3$, $n = 3$ and $r' = 4$

∴ Degree of indeterminacy = 28

If torsional stiffness of the members is neglected, the number of internal forces at a section gets reduced from 3 to 2.

$$\therefore \quad N = 2\,[m\,(n+1) + n\,(m+1) + r' - 3\,(m+1)\,(n+1)\,]$$

$$N = mn - m - n - 3 + r'$$

$$N = (m-1)\,(n-1) - 4 + r' \qquad \text{...(2.5)}$$

In the case of four corner supports, $r' = 4$. And therefore

$$N = (m-1)\,(n-1) \qquad \text{...(2.5}a\text{)}$$

For the grid of Fig. 2.9, substituting the values of m and n
$N = 4$, when torsional stiffness is ignored.
If the loads applied are inclined, the axial forces so developed get delinked from shear and moments.

Ex. 2.12. Find degree of indeterminacy of a roof grid supported on all sides.

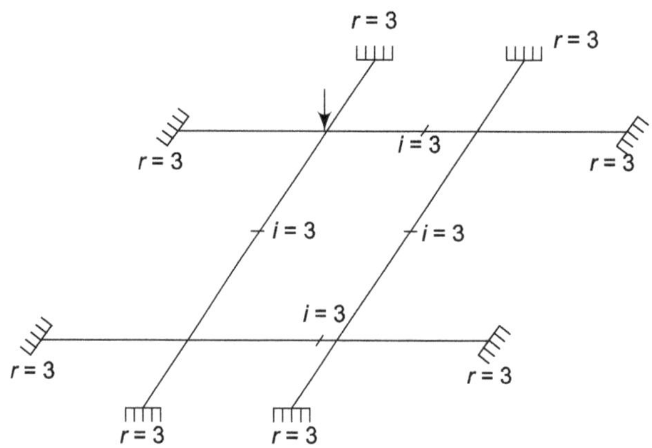

Fig. Ex. 2.12

$$i' = 4 \times 3 = 12; \qquad r' = 8 \times 3 = 24$$

$$v = i' + r' = 36 \qquad \text{...(1)}$$

No. of free bodies, $f = 4$, $\lambda = 3$
No. of equations of equilibrium,

$$s = 3 \times 4 = 12 \qquad \text{...(2)}$$

$\therefore$ Degree of indeterminacy = 36 – 12 = 24
If torsional stiffness is ignored,

$$i' = 4 \times 2 = 8$$

$$r' = 8 \times 2 = 16$$

$$v = i' + r' = 24$$

No. of equations of equilibrium, $s = 3 \times 4 = 12$
Degree of indeterminacy, $n = 24 - 12 = 12$

2.3.2 Kinematic Considerations, Degree of Indeterminacy of Grids

As an illustration, let us consider the grid, Fig. 2.10, with zero torsional stiffness for all the members. The grid of Fig. 2.10 consists of the minimum number of internal as well as external constraints. The constraints are kinematically sufficient, i.e., the structure takes a stationary position after application of loads. The beams *AB, CD, AC* and *BD,* Fig. 2.10, are resting on the corner supports *A, B, C* and *D*. The beams *JK* and *LM* are resting on the beams *AB* and *CD*. The beams *EF* and *GH* are such that they rest on the beams *AC* and *BD* but not on the beams *LM* and *JK*. Such an assembly of beams for the grid can be analyzed from equations of equilibrium only, i.e., the constraints are kinematically "just" sufficient. Now, if we allow the beams *EF* and *GH* to rest on the interior transverse beams *LM* and *JK* at the points *R, S,* and *P, Q,* the additional constraints are i.e., one per internal support. The degree of indeterminacy is equal to the number of "extra" constraints in addition to the minimum required for kinematic stability. The number of extra constraints in this example is four which agrees with the result obtained in the Article 2.2.1.

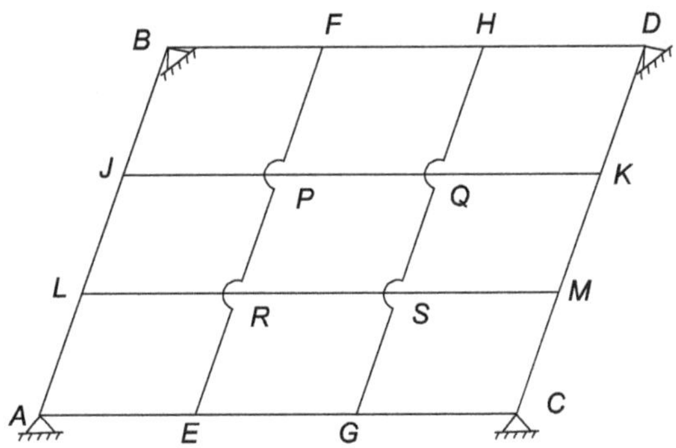

All members of zero torsional stiffness

Fig. 2.10 *Plane Grid with the Minimum Constraints*

2.4 UNSTABLE FRAMES, RANK OF EQUATIONS OF EQUILIBRIUM

Consider equations of equilibrium for the free bodies of Fig. 2.11(*b*). For the free body of the member "*a*":

Horizontal force equilibrium,

$$\Sigma H = 0, p_1 + p_4 = 0 \qquad ...(a)$$

Vertical force equilibrium,

$$\Sigma V = 0, p_2 + p_5 = 0 \qquad ...(b)$$

Moment equilibrium,

$$\Sigma M = 0, 4p_1 + p_3 + p_6 = 0 \qquad ...(c)$$

Similarly for the free body of the member "*c*" :

$$\Sigma H = 0 \text{ gives } p_4 + p_7 - 2p = 0 \qquad ...(d)$$

$$\Sigma V = 0 \text{ gives } p_5 + p_8 - 5p = 0 \qquad ...(e)$$

$$\Sigma M = 0 \text{ gives } p_6 + 6p_8 + p_9 + 5p = 0 \qquad ...(f)$$

And for the free body of the member "b" we get

$$p_7 + p_{10} = 0 \qquad ...(g)$$

$$p_8 + p_{11} = 0 \qquad ...(h)$$

$$p_9 + 4p_{10} + p_{12} = 0 \qquad ...(i)$$

In the previous 9 equations, there are 12 unknown from p_1 to p_{12} . We cannot obviously solve the equations. It may prompt one to take a free body for the entire structure and thus get 3 more equations given in the following without adding to the number of unknowns.

$$p_1 + p_{10} + 2p = 0 \qquad ..(j)$$

$$p_2 + p_{11} - 5p = 0 \qquad ...(k)$$

$$p_3 + 6p_{11} + p_{12} - 13p = 0 \qquad ...(l)$$

Even though we have 12 equations for the 12 unknowns, we may show by rank technique that all of them are not independent. Let us write the 12 equations in the matrix form, Eq. 2.6, Fig. 2.11.

In the equilibrium equation, we have a coefficient matrix (12 × 12), and the column matrix (for which P is a scalar multiplier) may be considered as an augmenting column matrix. Now we can apply the fundamental theorem (stated in Chapter 1).

(1) If $r = r' = 12$, then there is unique solution for the unknowns p_1 to p_{12}.
(2) If $r = r' < 12$ then the independent equations are not adequate to solve for the unknowns. The difference of the number of unknowns and the number of independent equations gives the degree of indeterminacy.
(3) If $r^1 \neq r'$ (r can be only less than r) then no solution is possible, which should suggest that the structure is unstable, i.e., the constraints are not sufficient kinematically.

In the following, a systematic procedure is given for evaluating the ranks r and r' for a system of equations. The problem of a portal is taken as an example to illustrate the procedure. Changing signs of all the elements of augmenting column or any other column does not change the rank of a matrix. If a square coefficient matrix can be reduced to a lower or upper triangular matrix (by linear combinations of the rows in the above augmented matrix) with the non-zero diagonal elements, then all the equations (rows) should be independent, that is, $r = r'$ = number of unknowns and there is unique solution for the unknowns. In the case of an upper triangular matrix (Chapter 1), the last row has only one unknown in the equation form and as we go upward each equation has an additional unknown associated with it, thus all the equations are independent. In the process of linear combinations of rows, before we get a complete triangular form of coefficient matrix, if the last rows are reduced to all elements equal to zero (including the elements of augmented column), then the number of rows, say p rows, with zero elements give the number of "dependent equations". This suggests that the rank $r = r'$ = total number of equations less 'p'. If in the process, the last rows of coefficient matrix have zeros but not the last element (of augmenting column) then $r \neq r'$ and there is no solution possible, and the system is unstable.

The above procedure is further explained by considering the matrix (2.6), which is the equilibrium matrix of the portal, Fig. 2.11. While finding the rank, it is not essential that it should be reduced to a triangular form because sometimes the leading diagonal element may not be non-zero. For "independence", all that is required is to associate a new unknown for every equation. However, for the matrix considered, all the diagonal elements are non-zeros and we can attempt to reduce it to a triangular form.

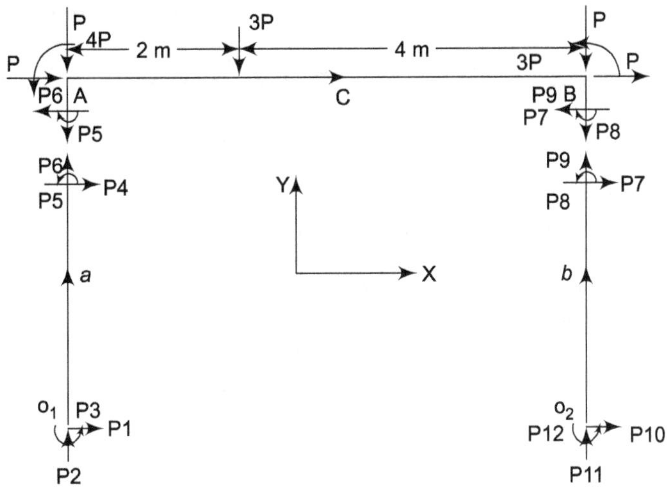

(a) Portal frame under given loads

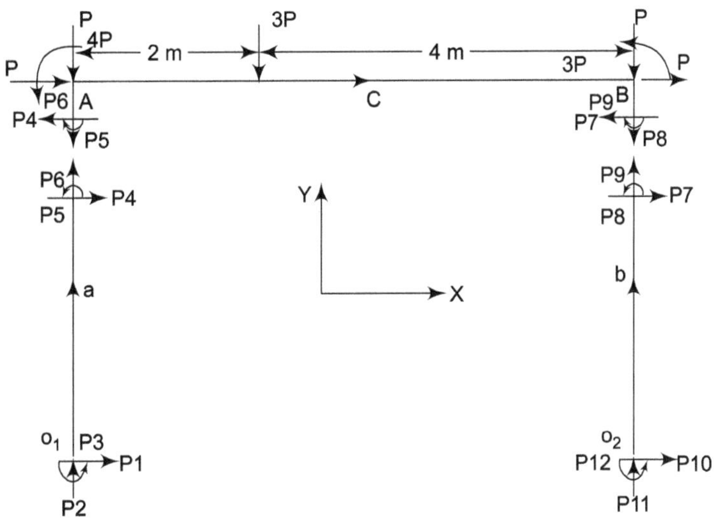

(b) Free bodies of the members a,b and c

Fig. 2.11 *Free Bodies for Equilibrium Equations*

Scanning the first column, we have three non-zeros belonging to first, third and 10th rows. We can take any one of these rows to be associated with p unknown. A more logical step would be to take the first row itself. For the sake of generality, let us take the third row. As the first element of the third row is four, it is convenient to normalize the row, i.e., divide the elements of the row by four so that the first element is unity.

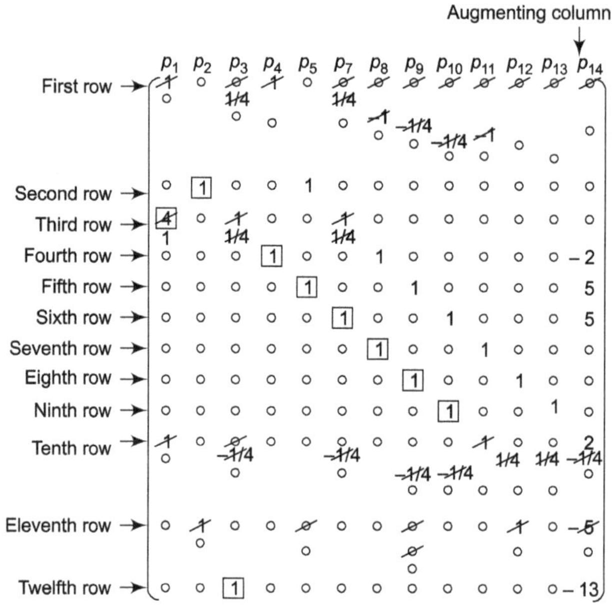

Determining rank of equilibrium equations ....(2.6)

3rd row	(4 0 1 0 0 1 0 0 0 0 0 0 0)
Normalized 3rd row	(1 0 ¼ 0 0 ¼ 0 0 0 0 0 0 0)

By linear combination (i.e., by addition or subtraction) with the 1st and 10th rows, the other non-zeros in the 1st column are reduced to zero. The pivotal element in the 3rd row is enclosed in a square.

1st row	(1 0 0 1 0 0 0 0 0 0 0 0 0)
Subtract the 3rd row	(– 1 0 – ¼ 0 0 – ¼ 0 0 0 0 0 0 0)
Modified 1st row	(0 0 – ¼ 1 0 – ¼ 0 0 0 0 0 0 0)

The same procedure is followed for the 10th row. These are marked directly on the matrix (2.6). It is now the turn for other rows to have a pivotal element in the second column. For the unknown p_2, let us associate the 2nd row, and the pivotal element of row is indicated by the enclosing square. The pivotal element being unity, it is in a normalized form. Scan the second column. The 11th row has a non-zero.

11th row	(0 1 0 0 0 0 0 0 0 0 1 0 –5)
Subtract 2nd row	(0 –1 0 0 –1 0 0 0 0 0 0 0 0)
Modified 11th row	(0 0 0 0 –1 0 0 0 0 0 1 0 –5)

Scan the 3rd column for non-zeros. Excluding 3rd and 2nd rows (as they have already been associated with p_1 and p_2) the other rows having non-zeros are the 1st, 10th and 12th. Let us choose the 12th row to be associated with p. Enclose the unity of the 12th row as a pivotal element. In the third column the other rows containing non-zero are the 1st, 3rd and 10th rows. As the 3rd row has already been associated with pivotal element, we combine the 1st row with the multiplies of the 12th row so that –¼ of the first row is reduced to zero.

1st Row	(0 0 –¼ 1 0 –¼ 0 0 0 0 0 0 0)

Add ¼ of 12th row (0 0 ¼ 0 0 0 0 0 0 0 6/4 ¼ –13/4)
Modified 1st row (0 0 0 1 0 –¼ 0 0 0 0 6/4 ¼ –13/4)

These new values are marked directly on the first row. Follow the same procedure for the 10th row.

Scan the 4th column. We may choose the 4th row and mark the pivotal element associated with the unknown p_4. Only the 4th column element of the 1st row is to be reduced to zero. Subtract the 4th row the 1st. The modified 1st row is marked directly on the matrix.

Associate 5th row with the unknown p_5 and combine with the 11th row. Modified 11th row is marked directly on the matrix. Associate 6th row with the unknown p_6 and combine with the 1st and 10th. Modified 1st and 10th rows are marked on the matrix.

Associate 7th row with the unknown p_7 and combine with the 1st row only.

Associate 8th row with unknown p_8 and combine with the 1st, 10th and 11th rows.

Associate 9th row with the unknown p_9 and combine with the 1st and 10th rows.

In this process before we come to an end, it may be observed that the "first row", "10th row" and "11th row" elements are all zeros (including the elements of the augmenting column). $p = 3$, is the number of dependent equations.

Therefore, the rank of the number of independent equations is equal to

$$12 - 3 = 9$$

The numbers of unknowns are excess of three. Therefore, the degree of indeterminacy = 3.

Effect of Releases Reduced indeterminacy: Consider Fig. 2.12(a). The portal has two hinges at two ends 2 of the members "*a*" and "*b*", due to the two releases, $p_6 = p_9 = 0$. In the equilibrium matrix (2.6) for the portal, Fig. 2.10, we account for this by cancelling the column 6 and 9. It can be verified that this cancellation enables us to associate 6th and 9th rows with other unknowns, say p_{11} and p_{10} respectively with the result that we have 9 independent equations (unchanged) and the number of unknowns are 10 (p_6 and p_9 excluded). The degree of indeterminacy is thus one.

Let us now consider the scheme of releases, Fig. 2.12(b). In this case we have $p_6 = 0$ and a new set of additional unknowns p_{13} and p_{14}. Considering the free body, Fig. 2.12(b), we get three more independent equations. It can be verified that we have 15 equations out of which three will be zero rows (i.e., three dependent equations) and the number of the unknowns, $n = 13$. The number of independent equations, $r = 12$. The degree of indeterminacy is, $n - r = 13 - 12 = 1$.

2.4a Unstable Structures Consider Fig. 2.12(*c*). Moment load, *m*, is shown acting on the joint free-body. We get the following three additional equations which include the new unknowns p_{15} and p_{16}.

$$P_4 - p_{15} = 0$$

$$P_5 - p_{16} = 0$$

$$0 = m_A$$

From the third equation above, it is obvious that moment load m_A cannot be balanced. Mathematically, it shows that $r < r'$ and thus no solution is possible. The structure is therefore unstable, even though the member of releases is only two.

Similarly consider Fig. 2.12(*d*). In this case, we cancel the 6th and 9th rows due to $p_6 = p_9 = 0$ and we consider one of the two free bodies shown in the same figure. For the free body containing joint B, we write the following three additional equations involving new unknowns p_{17} and p_{18}.

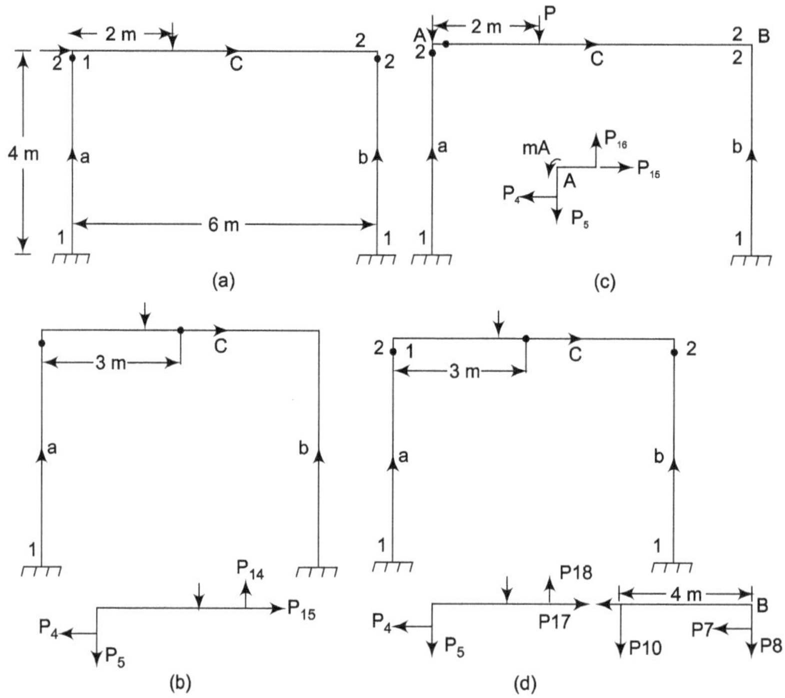

Fig. 2.12 *Portal with Releases*

$$P_7 - p_{17} = 0$$

$$P_8 - p_{18} = 0$$

$$4p_8 = 0$$

We can combine these three equations with the other equations in the equilibrium matrix for the portal, Fig. 2.11, and noting that $p_6 = p_9 = 0$, it can be shown that $r < r'$ and therefore, the structure is unstable.

The above generalized considerations are very helpful for deciding whether a structure is stable or not. The degree of indeterminacy, however, is found by the methods described in the Articles 2.1 and 2.2.

In the case of hinged trusses, it is more expedient to decide its stability by considering the constituent "triangular cells". A triangle is a stable configuration.

In the case of rigid frames, the proposed method is quite useful. While considering complex rigid frames, local instability can be checked by taking a limited number of suitable free bodies as was shown in the case of joint instability of Fig. 2.12(*c*).

2.5 BASIC STEPS OF ANALYSIS BY FORCE METHOD

In the case of determinate structures, material and cross-sectional properties are not required to find the stress resultants. Only in the case of indeterminate structures, it is required to know the material and cross-sectional properties to evaluate deformations, essential for satisfying compatibility conditions.

The method of analysis is presented considering the examples of plane frames. To a limited extent, the analysis of continuous beams and grids also is included. However, a detailed study for all the types of structures based on the matrix approach for the Force method is given in Chapter 4. In the case of continuous beams, plane frame and grids, it is generally adequate to ignore shear and axial deformation of the members as compared to their deformations due to bending moment (flexure).

The structure, Fig. 2.13(a), is determinate and the support forces H_A, V_A and V_D can be found from the three equations of equilibrium of the 2D system.

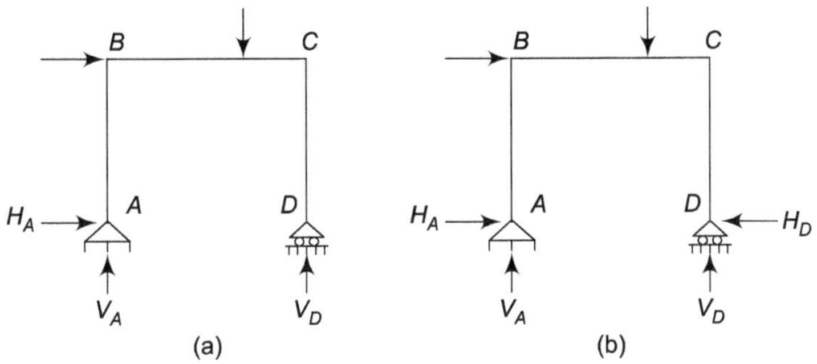

Fig. 2.13

If the support constraints are more, say hinged at *D*, Fig. 2.13(*b*), then we have four unknown forces H_A, V_A, H_D, V_D which cannot be evaluated from the three equations of equilibrium alone that are available for the 2*D* system considered. If we chose to consider the determinate system to be Fig. 2.13(*a*), then the excess constraint is the horizontal defection of end *D* to be equal to zero. Therefore, for the structure, Fig. 2.13(*b*), which is indeterminate to the order 1, the property of the excess constraint, i.e., horizontal deflection of end *D*, (d), equal to zero is utilized as a compatibility condition of joint *D* to get an additional equation. The number of excess constraints or the compatibility conditions is equal to the degree of indeterminacy, *n*. In the following, the sequence of steps of analysis is considered for a structure, Fig. 2.14. The degree of indeterminacy is equal to three. It should be possible to extend the application to all the types of structures. There are only three distinct parts or steps in the analysis, which presupposes that superposition is possible and the steps of analysis are.

2.5.1 Step 1—Equilibrium

Equilibrium of determinate structures under different cases of loading: Choose an appropriate determinate system, and introduce constraining forces corresponding to the additional constraints in excess of the minimum constraints required for the determinate system. Consider each case separately. For the structure, Fig. 2.14(*a*), consider a determinate structure 2.14(*b*), and the excess constraints removed are, horizontal and vertical deflections of *D* equal to zero and also rotation of *D* equal to zero; and the corresponding stress resultants H_D, V_D, M_D are the unknowns equal in number to the degree of indeterminacy, *n* (for this example, $n = 8$). There may be alternative choice of determinate systems, for example, Fig. 2.13(*a*) and Fig. 2.14(*b*) are two such choices. Corresponding to Fig. 2.14(*a*), taking advantage of superposition satisfy equilibrium for each case separately as shown in Fig. 2.14 (b), (c), (d), (e), i.e., H_D, V_D, M_D can be found for each case as they are separately determinate and are evaluated from equilibrium conditions. The first step of analysis is to satisfy equilibrium for each case separately. As their actual values are not known, superposition becomes necessary and is considered separately assuming the constraining forces at releases to be equal to unity.

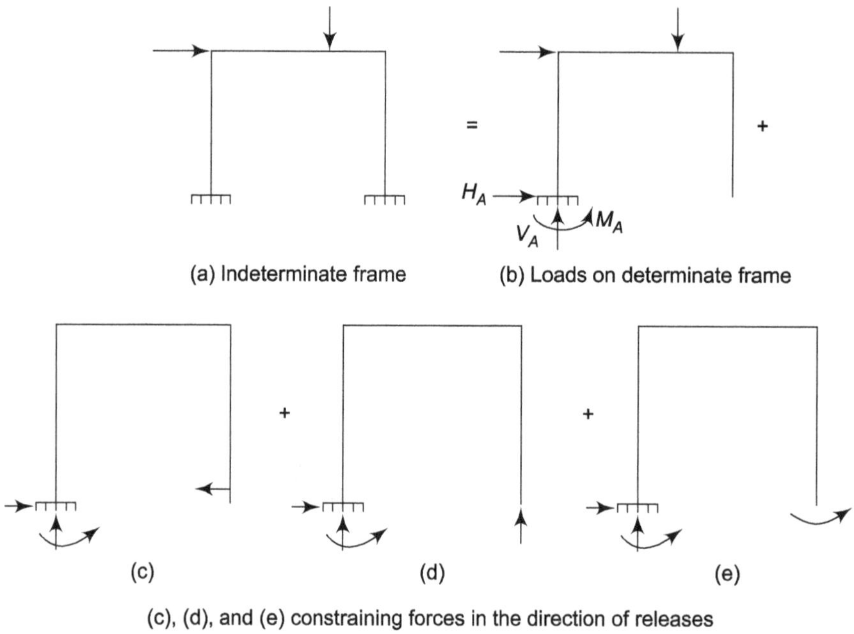

Fig. 2.14 *Indeterminate Frame and Superposition*

2.5.2 Step 2—Deformation–Force Relations: Computation of Deformations

For the example, Fig. 2.14, for each case we are required to compute horizontal and vertical deflections of *D* and rotation of *D*. It is while computing the deformations that material and cross-sectional properties are made use of. The deformations are of three kinds,

(a) Axial force deformation relations,

$$\delta = \frac{L}{AE} P$$

where *P* is an axial force.

(b) Flexural deformations, Fig. 2.15,

$$\frac{1}{R} = \frac{M}{EI}$$

Where curvature, $\frac{1}{R}$, rate of change of slope, is the measure of deformation for flexural members. In the above relation, an assumption of "plane sections remain plane" is made. The neutral axis and corresponding deformations based on the assumption are shown for arriving at $\frac{1}{R} = \frac{M}{EI} = \frac{\sigma}{y}$. It is a significant assumption that helped in the simplifications for computing flexural deformations. It is known as 'flexural' formula in mechanics. The assumption is used in Chapter 4, pertaining to plastic analysis.

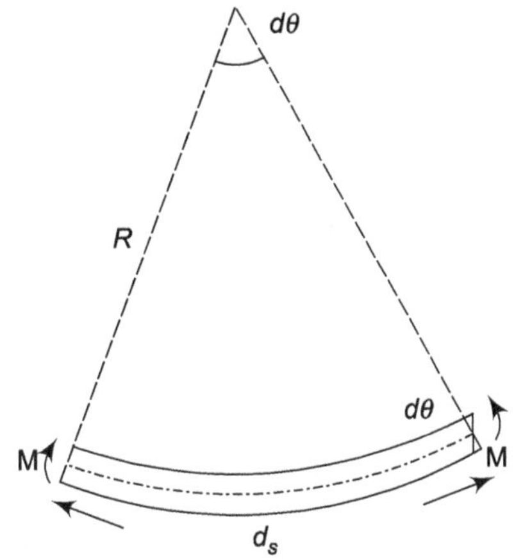

Fig. 2.15 *Flexural Deformation, Curvature 1/R*

(c) Torsional deformation,

$$\frac{\theta}{L} = \left(\frac{1}{GJ}\right)T$$

where angle of twist per unit length $= \frac{\theta}{L}$, is measure of deformation or torsional members.

The computation of deformations under axial force and torsional are explained in any elementary course of strength of materials. In majority of the structural problems, flexural deformations are important.

2.5.3 Step 3—Compatibility

The final part of the analysis consists of formulating simultaneous equations based on restoring the constraint conditions corresponding to the unknown chosen redundant. For example, Fig. 2.13(*b*), *A* and *D* hinged, the chosen redundant was H_D and the compatibility condition was $(\delta h)\ D = 0$.

Similarly for the example, Fig. 2.13, the chosen redundant were H_D, V_D and M_D and the corresponding compatibility conditions to formulate three simultaneous equations were

$$(\delta h)D = 0,\ (\delta v)D = 0,\ \theta D = 0,$$

The deformations, i.e., $(\delta h)D$, $(\delta v)D$ and θD due to the given loads and due to the unit constraining forces for each case having been evaluated in the preceding step 2 (and Article 2.5), constitute the constant elements and the coefficients for the unknowns H_D, V_D and M_D appearing in the three simultaneous equations based on compatibility. At the point *D*, when written in the form of equations,

$$\left[\delta_h\right]_{\text{due to given load}} + \left[\delta_h\right]_{\text{due to } H_D = 1} \quad H_D + \left[\delta_h\right]_{\text{due to } V_D = 1} \quad V_D + \left[\delta_h\right]_{\text{due to } M_D = 1} \quad M_D = 0$$

$$\left[\delta_v\right]_{\text{due to given load}} + \left[\delta_v\right]_{\text{due to } H_D = 1} \quad H_D + \left[\delta_h\right]_{\text{due to } V_D = 1} \quad V_D + \left[\delta_h\right]_{\text{due to } M_D = 1} \quad M_D = 0$$

$$\left[\theta\right]_{\text{due to given load}} + \left[\theta\right]_{\text{due to } H_D = 1} \quad H_D + \left[\theta\right]_{\text{due to } V_D = 1} \quad V_D + \left[\theta\right]_{\text{due to } M_D = 1} \quad M_D = 0 \qquad \text{...(2.7)}$$

2.6 FLEXURAL DEFORMATIONS

The evaluation of flexural deformations is presented from the first principles, moments area theorems, Article 2.5.1. A more generalized method for finding deformations is based on the principle of virtual work Article 2.5.3. Evaluation of deformations is necessary for satisfying compatibility, Step-3 of Force method.

2.6.1 'Moment Area' Method

Based on the flexural formula, Step 2 of Article 2.4, change of slope of the angle θ, between tangents at ends 1 and 2 of a member, Fig. 2.16(a), can be shown to be equal to the area enclosed by M/EI diagram.

$\Delta\theta$ = change in slope between the ends of element dx

$$\therefore \qquad \theta = \underbrace{\int_1^2 \frac{M}{EI}\,.\,dx}_{\text{Rate of change of slope}} \qquad \text{...(2.8)}$$

The Eq. (2.8) gives the first theorem of the moment area method. In the Fig. 2.16, x is taken positive to the right and deflection δy positive downwards. The bending moment, M, is positive when it is acting clockwise on the portion of the beam to the left of the section and counter-clockwise to the right of the section. Similarly, deflection of end 1 with respect to the tangent at end 2 is given by

$$\delta y_1 = \underbrace{\int_2^1 \frac{\bar{x}M}{EI}\,.\,d\bar{x}}_{\text{deflection of end 1 due to } \Delta\theta \text{ at } x} \qquad \text{...(2.9a)}$$

And the value deflection of end 2 is given by

$$\delta y_2 = \int_2^1 x\frac{M}{EI}\,.\,dx \qquad \text{...(2.9b)}$$

The Eq. (2.9) gives the second theorem of the moment area method for computing tangential deflections, δy_1 and δy_2. It may be found that if rate of change of slope or curvature $1/R$ is considered as intensity of (angle) loading then for such an analogous beam with the angle-load intensity, 1/R = M/EI, shear gives the slope and moment at a point gives the deflection. Such a beam is called a conjugate beam.

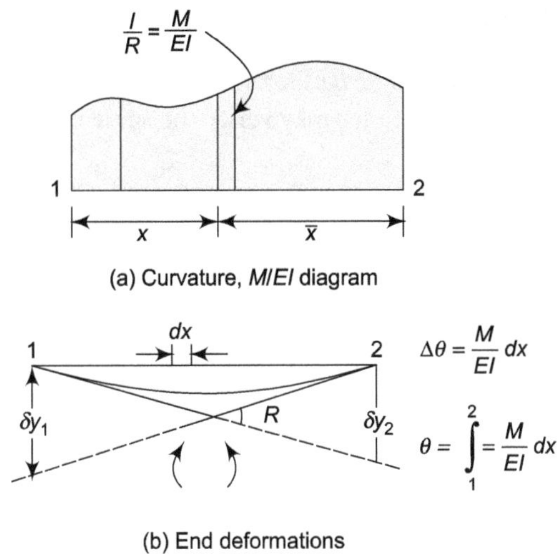

Fig. 2.16

The Fig. 2.17 suggests a procedure for obtaining conjugate beams. The similarities of analogous expression are to be noted for the points *A*, *A*, *C* and *D*, Fig. 2.17(*a*) and (*b*). The conjugate beam method is suitable for computing deflections and rotations by graphical methods as used for drawing bending moment and shear force diagrams. In general, complementary virtual work method of Chapter 2, sometimes referred simply as virtual work, is more suitable for computing rotations and deflections. To substantiate the statement, for the beam of Fig. 2.17, the deflection at *C*, δC, is found by the two methods.

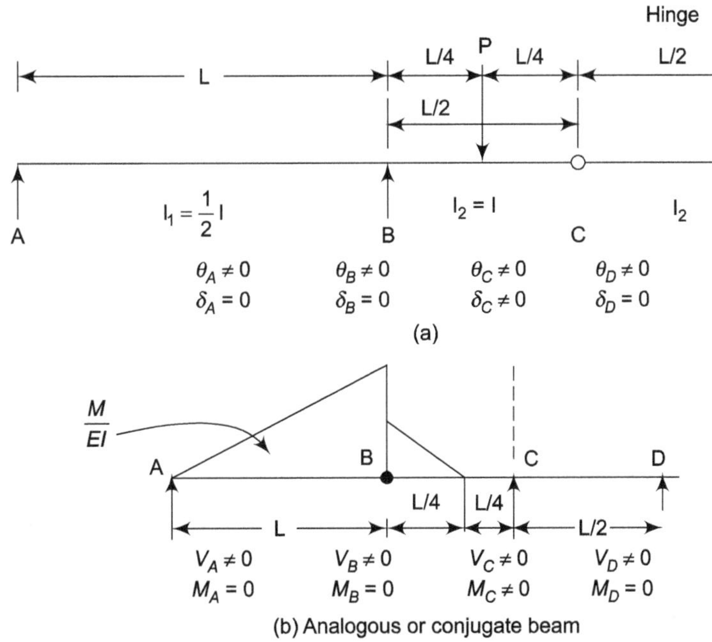

Fig. 2.17 *Real and Conjugate*

2.6.2 Conjugate Beam Method

For the conjugate beam, Fig. 2.18(*a*), under "*M*/*EI* load", L_1 and L_2 are the angle loads due to the loading (*M*/*EI*) on the segments *AB* and *BE*. The reader may verify the value of moment and *C* for the conjugate beam to be equal to[1]

$$\frac{31PL^3}{192EI}$$

Therefore, for the real beam

$$\delta_c = \frac{31PL^3}{192EI}$$

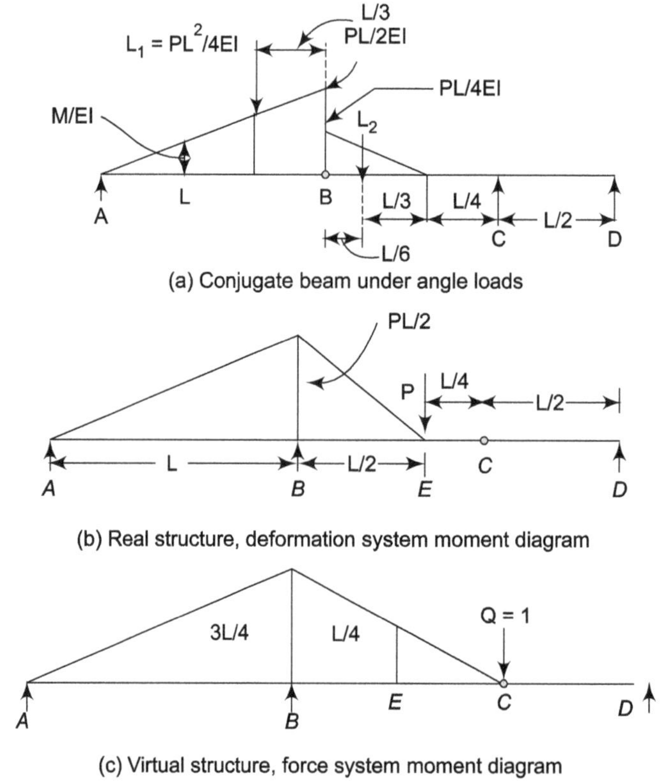

Fig. 2.18 *Conjugate Beam Method and Virtual Work Method*

2.6.3 Virtual Work Method

The deflection and deformations (curvature - *M*/*EI*) defined by bending moment, *M*, due to given loading are referred to as the real beam, Fig. 2.18(*b*). If the deflection is desired at *C*, a virtual beam

[1]Hint: Compute simple supported reaction for free body *AB*.
The reaction at *B* when reversedis the shear on the free body *BCD*. The moment at *C* is computed considering the free body *BEC* under the loads loads $PL^2/6EI$ at *B* and the angle load $PL^2/16EI$ at a distance $7L/12$ from *C* equal to $31PL^3/192EI$.

Fig. 2.18(c) is considered under the load $Q = 1$ at C where the deflection is required to be computed. Using the principle of complementary virtual work,

$$\delta c = \int n_2 \frac{M}{EI} ds$$

Where m is the moment due to $Q = 1$, and the integration is to be performed over the entire length. By direct integration or by using the integration table, Table-1, the reader may verify $\delta c = 31PL^3/192\, EI$, the same value as obtained by the conjugate beam method. The virtual work method is more general and the advantage of the virtual work method will be more obvious when required to compute deformations or more complex structure like frames. The virtual work method is equally useful for trusses, Chapter 4.

2.6.4 Conjugate Structure for Frames, Modeling an Analogous Structure

Consider a typical frame, Fig. 2.18. Change in slope for the element ds due to angle-load-intensity M/EI, is equal to $M\, ds/EI$. If the end A is fixed, change in the slope of D is the actual angle of rotation and it is given by,

$$\theta_D = \int_A^D \frac{M}{EI} ds \qquad \text{...(2.10)}$$

Similarly, deflection of D due to $\Delta\theta = \left(M \dfrac{ds}{EI} \text{ at } X \right)$, in the direction normal to the straight line joining X and D is given by,

$$\Delta_n = (X\,D)\,\Delta\theta \text{ and}$$

$$\Delta_x = \Delta_n \cos\phi = \underbrace{(XD)\cos\phi}_{\substack{Y \text{ coordinates measured}\\ \text{in the direction perpen-}\\ \text{dicular to } \Delta x}} \quad \Delta\theta \qquad \text{...(2.11)}$$

$$\Delta_y = \Delta_n \sin\phi = \underbrace{(XD)\sin\phi}_{\substack{x \text{ coordinates of length measured}\\ \text{in the direction perpendicular to } \Delta y\\ \text{direction, i.e., the direction of}\\ \text{deflection desired}}} \quad \Delta\theta \qquad \text{...(2.11a)}$$

The same expression or Δx and Δy can be obtained by referring to Fig. 2.19(c), where end D is considered fixed for the skeletal frame in plan with $1/EI$ as width. For such a conjugate structure, the M diagram can be considered as vertical loading per unit area and M/EI diagram as the vertical loading per unit length along s and or deflection at D take moments about D.

Therefore, $\Delta x = M\, ds/EI\, y$, and

$\Delta y = M\, ds/ME\, x$ which are same as given by Eqs (2.11) and (2.11a). Rotation and net deflection of D are,

$$\theta_D = \int_A^D \frac{M}{EI} ds$$

$$(\delta x)_D = \int_A^D y \frac{M}{EI} ds \quad \text{and} \quad (\delta x)_D = \int_A^D x \frac{M}{EI} ds$$

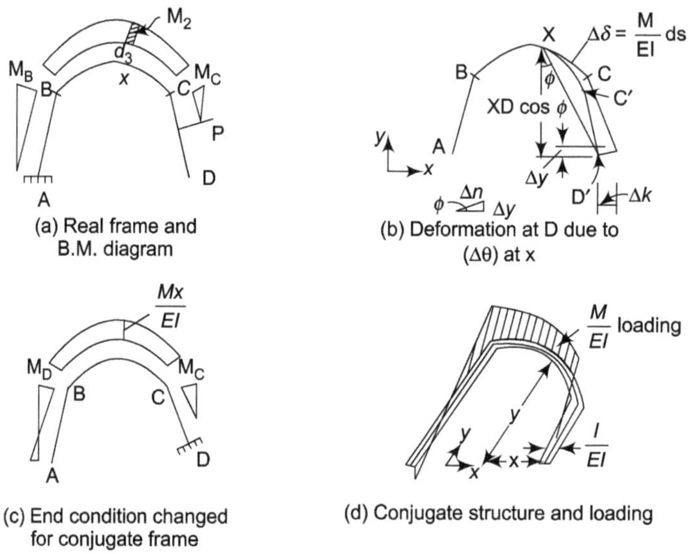

Fig. 2.19 *Real and Conjugate Plane Skeletal Structure*

The following example illustrates the procedure outlined.

Ex. 2.13 Find deformations at D, Fig. Ex. 2.13.

Angle Load,

$$L_1 = \frac{\text{Area of } (3-4-5)}{EI} = \frac{(1/2 \times 15 \times 7.5)}{EI} = \frac{56.20}{EI}$$

Angle Load,

$$L_2 = \frac{\text{Area of } (2-3-4)}{0.6\,EI} = \frac{(1/2 \times 15 \times 10)}{0.6\,EI} = \frac{125.0}{EI}$$

Angle Load,

$$L_3 = \frac{\text{Area of } (1-2-3)}{0.6\,EI} = \frac{(1/2 \times 33.66 \times 10)}{0.6\,EI} = \frac{280.5}{EI}$$

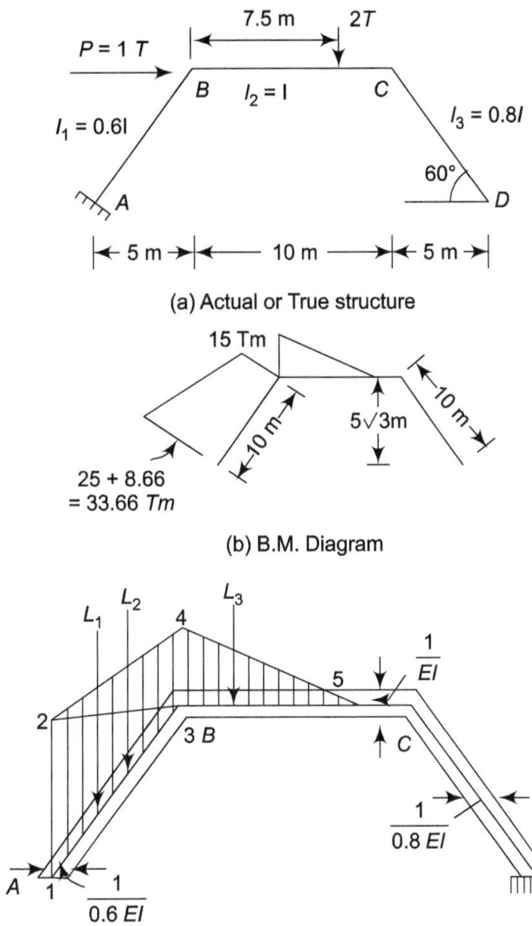

Fig. Ex. 2.13

θ_D = Vertical shear for the conjugate structure shown in horizontal plane and loaded in vertical plane as shown in Fig. 2.13(c)

$$= L_1 + L_2 + L_3$$

$$= \frac{(56.2 + 125.0 + 280.5)}{EI}$$

$$= \frac{461.7}{EI} \text{ radians}$$

Direction of deformations can be decided by inspection.

$(\delta_h)_D$ = Moment at D, Fig. 2.13(c), of angle loads L_1, L_2, L_3 measuring distance indirection perpendicular to the direction of the displacement desired. In this case y coordinates are required to be considered for moments.

$$(\delta_h)_D = L_1\,(5\sqrt{3}\,) + L_2\,(2/3 \times 5\sqrt{3}\,) + L_3\,(1/3 \times 5\sqrt{3}\,)$$

$$= \left(\frac{56.2}{EI}\right)(8.66) + \left(\frac{125.0}{EI}\right)(5.77) + \left(\frac{280.5}{EI}\right)(2.89)$$

$$= \left(\frac{487}{EI}\right) + \left(\frac{721}{EI}\right) + \left(\frac{810}{EI}\right)$$

$$= \left(\frac{2028}{EI}\right)$$

where EI is in tm^2 units.

$$(\delta_h)_D = \text{Moment at } D, \text{ Fig. (c), of angle loads measuring } x \text{ distances}$$

$$= L_1\,(12.5) + L_2\,(15 + 1/3 \times 5) + L_3\,(15 + 2/3 \times 5)$$

$$= \left(\frac{702}{EI}\right) + \left(\frac{2085}{EI}\right) + \left(\frac{5140}{EI}\right)$$

$$= \left(\frac{7927}{EI}\right)$$

where EI is in tm^2 units.

The concept of conjugate structure can be extended to the structures with the two ends not necessarily free and fixed. Such a case is considered in the following example, Ex. 2.14.

Ex. 2.14:

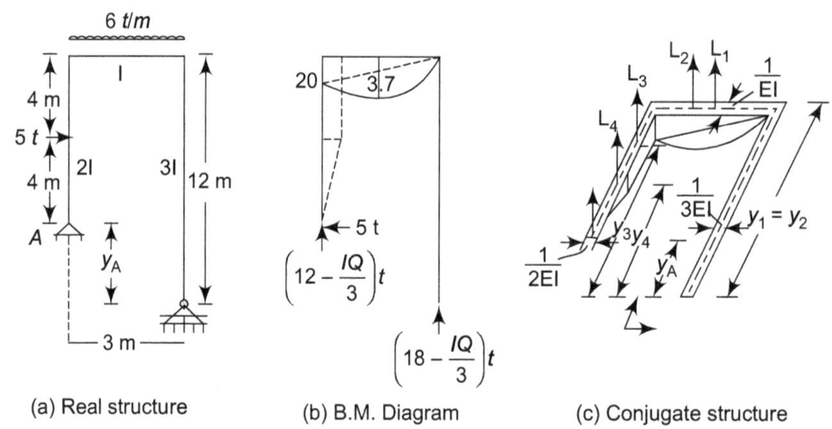

Fig. Ex. 2.14

Consider the structure, Fig. Ex. 2.14(a)

Angle loads,

$$L_1 = 2/3 \times 27 \times 6/EI = 108/EI$$

$$L_2 = 1/2 \times 20 \times 6/EI = 60/EI$$

$$L_3 = 20 \times 4/2EI = 40/EI$$

$$L_4 = 1/2 \times 20 \times 4/2EI = 20/EI$$

The Fig. Ex. 2.14(c) is a representative conjugate structure, if (i) free end has an angle load equal to slope at A and (ii) moment of all the angle loads ($L_1 + L_2 + L_3 + L_4 + \theta_A$) with multiplying distance (x_1, x_2, x_3, x_4, x_A) equal to x coordinates measured from D, gives δ_v at D, which should be equal to zero. The condition (ii) gives the following equation, Eq. (a), from which the value of θ is evaluated.

$$\left(\frac{108}{EI}\right)3+\left(\frac{60}{EI}\right)4+\left(\frac{40}{EI}+\frac{20}{EI}\right)6-6\,\theta_A=0 \qquad \text{...}(a)$$

$$\therefore \qquad \theta_A = \frac{154}{EI}$$

And δx of D can be evaluated by taking moment's angle loads about D, but this time the distance will be y coordinates measured from D.

$$(\delta x)D = \left(\underset{L_1}{\frac{108}{EI}}+\underset{L_2}{\frac{60}{EI}}\right)\underset{y_1,\,y_2}{\times 12}+\left(\underset{L_3}{\frac{40}{EI}}\right)\underset{y_3}{\times 10}+\left(\underset{L_4}{\frac{20}{EI}}+\underset{y_4}{\frac{20}{3}}\right)-\underset{\theta_A}{\frac{154}{EI}}\underset{y_1}{\times 4}$$

$$(\delta x)D = \frac{1}{EI}(2016+400+133.30\,e)=1933.\frac{3}{EI}$$

Ex. 2.15: Consider the loading system consisting of a single load at D in horizontal direction, Fig. Ex. 2.15.

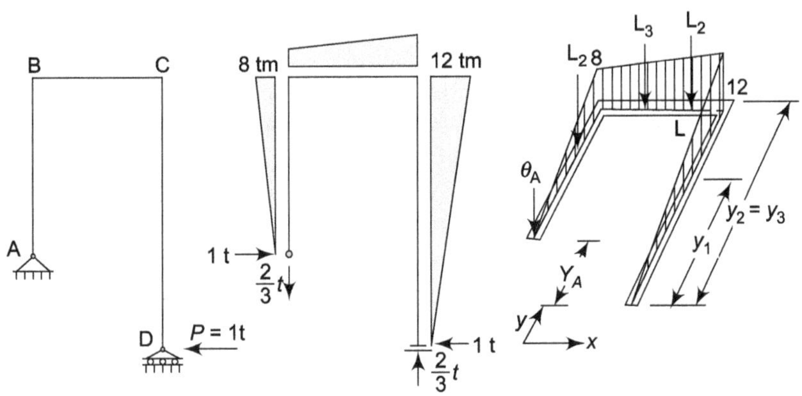

(a) Real structure (b) B.M. Diagram (c) Conjugate structure

Fig. Ex. 2.15

Angle loads,

$$L_1 = \frac{1}{2}\times\frac{12\times 12}{3EI} = \frac{24}{EI}$$

$$L_2 = \frac{1}{2}\times\frac{12\times 6}{EI} = \frac{36}{EI}$$

$$L_3 = \frac{1}{2}\times\frac{8\times 6}{EI} = \frac{24}{EI}$$

$$L_4 = \frac{1}{2} \times \frac{8 \times 8}{2EI} = \frac{16}{EI}$$

θ_A is evaluated from the following equation based on vertical deformation constraint at end D.

$$\left(\frac{16}{EI}\right)\left(\frac{1}{3} \times 6\right) + \left(\frac{24}{EI}\right)\left(\frac{2}{3} \times 6\right) + \left(\frac{16}{EI}\right)(6) = \theta_A \times 6$$

$\therefore \quad (1/EI)\,(72 + 96 + 96) = 6\,\theta_A$

$$\therefore \qquad \theta_A = \frac{44}{EI}$$

Following the reasoning of Ex. 2.14

$$(\delta x)D = \left(\frac{24}{EI}\right)\left(\frac{2}{3} \times 12\right) + \left(\frac{36}{EI}\right)(2) + \left(\frac{24}{EI}\right)\frac{1}{2} + \left(\frac{16}{EI}\right)\left(\frac{4+2}{3} \times 8\right) + \left(\frac{2}{3}EI\right)(4)$$

$$= \left(\frac{192}{EI}\right) + \left(\frac{432}{EI}\right) + \left(\frac{288}{EI}\right) + \left(\frac{149}{EI}\right) - \left(\frac{176}{EI}\right)$$

$$(\delta x)D = \frac{885}{EI}$$

where EI is in tm^2

2.6.5 Elastic Centre

Consider a structure with the loading the releases, Fig. 2.20(a) and (b). Also consider one redundant and let that be P_x. The expression for deformations can be found from the forgoing results and Fig. 2.20.

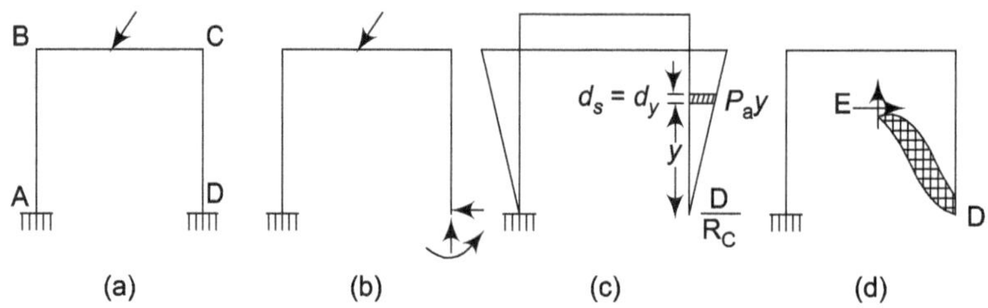

Fig. 2.20 *Symmetrical Portal and Elastic Centre*

$$\theta_D = \int_D^D \frac{(P_x Y)}{EI} ds = p_x \int_A^D \frac{Y ds}{EI} \qquad \ldots(a)$$

$$(\delta x)D = \int_D^D \frac{y(P_x Y)}{EI} ds = p_x \int_A^D \frac{Y^2 ds}{EI} \qquad \ldots(b)$$

$$(\delta x)D = \int_D^D \frac{x(P_x Y)}{EI} ds = p_x \int_A^D \frac{x\, y\, ds}{EI} \qquad \ldots(c)$$

From the equations (a), (b) and (c), it is obvious that if the end "D" happens to be a point such that it coincides with the centroid of the geometry of the frame with $1/EI$ as width, Fig. 2.20(c), and P_x is applied along a principal axis of such a configuration then

$$\int y\frac{ds}{EI}=0 \text{ and } \int x\,y\frac{ds}{EI}=0$$

and therefore, $\theta_D = (\delta_Y)_D = 0$ and deflection takes place only along the direction of P, Eq. *(b)*. Such a structure is shown in Fig. 2.20(*d*). The point E located at the "centroid" is called, "Elastic Centre". As the elastic centre does not necessarily lie on the structure it is then connected by a fictitious rigid link such that for the links $1/EI = 0$ and no integration is required along the length of the rigid link. Similarly it can be shown that due to moment, M, applied at elastic centre, $\theta_E \neq 0$ and $\delta_y = 0$; and for P alone applied at elastic centre in the direction of principal axis y,

$$[\delta x]_E = 0,\ [\theta]_E = 0,$$

and deflection takes place only in the direction of P_y. For symmetrical 2-*D* frames, axis of symmetry is one of the principle axes and a perpendicular line drawn from the centroid will be the other principle axis. The method of "elastic centre" is used for single span frames or hingeless arches which can be made determinate by cutting one of the ends and makes it free, while the other end remains fixed. The conjugate structure methods, however, can be used for multiple span frames made determinate by introducing releases, but the method is seldom used for such structures.

2.7 INDICIAL NOTATION

Let the stress resultant P_x, P_y and M be designated by P_1, P_2 and P_3 (generalized coordinates 1, 2 and 3) respectively and deflection in the direction of the stress resultant be designated as δ_1, δ_2 and δ_3. If P_1, P_2 and P_3 are not necessarily applied at elastic centre, then the deformations for each case of loading may be designated by,

f_{11}, f_{21}, f_{31} deformations along the directions 1, 2 and 3 (i.e., directions of P_1, P_2 and P_3) and caused by the load $P = 1$.

f_{12}, f_{22}, f_{23} the second index is indicative of P alone acting, and its magnitude is considered unity.

f_{13}, f_{23}, f_{33} deformations (due to $P_3 = M = 1$) along the directions P_1, P_2 and P_3.

The first subscript or index is the direction of deformation and the second subscript is the direction along which the unit force is applied.

Following the indicial notation, for the case of end loading at elastic centre and in the directions of principal axes,

$$f_{ij} = 0 \quad \text{if} \quad i \neq j \text{ and}$$
$$f_{ij} \neq 0 \quad \text{if} \quad i = j$$

In the theory of structural analysis, the deformations f_{ij} caused by unit loads are called flexibility coefficients. The flexibility coefficient of a structure is the deformation at a point and along a specified direction governed by compatibility conditions, caused by a unit load acting on the structure at the point and direction governed by the redundant release chosen.

Ex. 2.16. Analyze the frame, Fig. Ex. 2.16, by considering the elastic centre.
Computation for locating the elastic centre, *EC*

$$2\left(6\times\frac{1}{EI}\right)3+\left(8\times\frac{1}{2EI}\right)\times 6=\left(\frac{16}{EI}\right)\bar{y}$$

$$\bar{y} = 3.75 \text{ m}$$

For the releases, the structure is deliberately cut at the point of symmetry so that the number of unknowns will be only two (horizontal thrust and moment).

(*i*) **Equilibrium:** Equilibrium is satisfied for each case, Fig. (b), (d) and (e).

(*ii*) **Deformations:** From Fig. (*c*), deformations due to given loading are designated by U_1^0, U_2^0 and U_3^0 which are caused due to curvature, M^0/EI.

$$U_1^0 = (\delta_h)^0 = \left(\frac{8}{EI}\right) 2.25\, a\, \frac{48}{EI} \times 0.75 = \frac{-18}{EI}$$

$$U_2^0 = (\delta_h)^0 = \left(\frac{8}{EI} \times \frac{2}{3} \times 4\right) + \frac{48}{EI} \times 4 = \frac{213\frac{1}{3}}{EI}$$

(not required for solution)

$$U_3^0 = (\theta_0) = \frac{8}{EI} + \frac{48}{EI} = \frac{56}{EI}$$

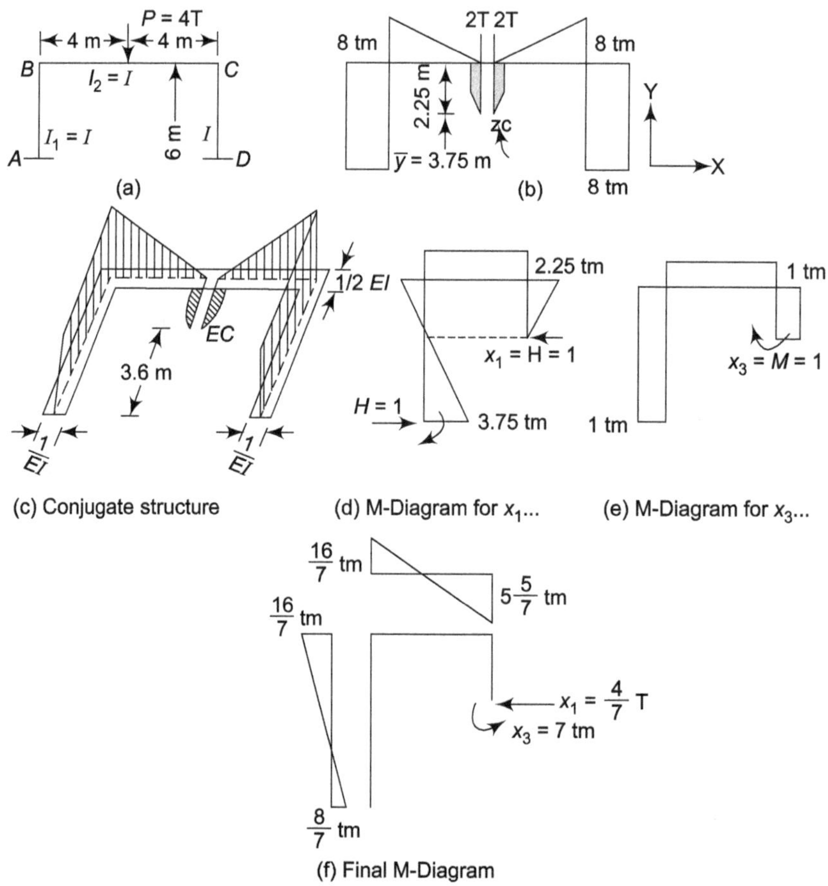

Fig. Ex. 2.16 *Elastic Center*

$$f_{11} = (\delta_\lambda)_{H-1} = \left(\frac{2.25 \times 4}{2EI}\right) 2.25 + \left(\frac{2.25 \times 2.25}{2EI}\right) \frac{2}{3} \times 2.25 - \left(\frac{3.75 - 3.75}{2EI}\right) \times \frac{2}{3} \times 3.75$$

$$f_{11} = \frac{31.5}{EI}$$

$$f_{33} = \frac{2}{EI} + \frac{6}{EI} = \frac{8}{EI}$$

(*iii*) Compatibility

$$X_1 f_{11} + U_1^0 = 0$$

$$X_3 f_{33} + U_3^0 = 0$$

$$\therefore \qquad X_1 = -\frac{U_1^0}{f_{11}} = \frac{18}{31.5} = \frac{4}{7}$$

$$\text{and} \qquad X_3 = -\frac{U_3^0}{f_{11}} = \frac{56}{8} = 6$$

Final moment diagram is shown in Fig. Ex. 2.16(*f*) for half the structure.

2.8 Virtual Work Method

The moment area and conjugate structure methods are used for finding deformations of 2*D* – skeletal frames with inplane loading and are not very useful for 3*D*-skeletal frames. And also that the method is not direct even for 2D frames, consisting of multiple bays with ends not hinged. The principle of virtual work proves to be a very useful technique for such cases for finding the deformations. Herein, only the virtual work principle is restated without proof. A generalized description of the principle is given in Article 2.6. A and B are two structures, Fig. 2.20, such that the skeletal geometry of one is either the same as the other or consists of only a part of the entire structure.

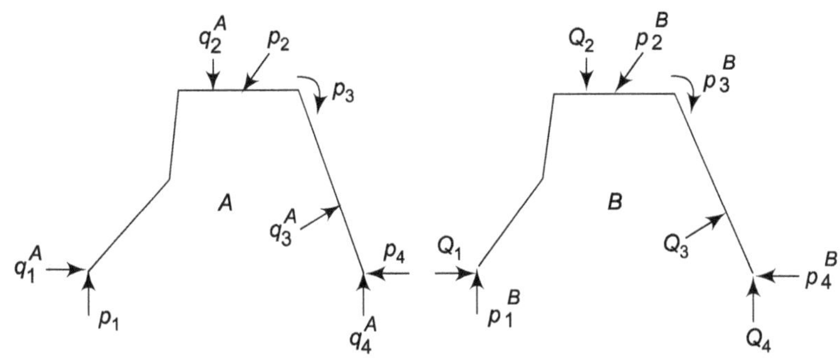

Fig. 2.21

There is no restriction that the support conditions of *A* and *B* need be the same. Let the structures *A* and *B* under the load system *P* and *Q* respectively be in equilibrium; and the deformations of the structure *A* under *P* loading be $q_1^A\ q_2^A\ \ldots,\ q_5^A$ (in the directions of Q loading) and the deformation of the structure *B* under *Q* loading be $P_1^B, P_2^B, P_3^B, \ldots, P_6^B$ (in the directions of P loading).

The statement of principle will then be:

For the structure, '*A*' under *P* loading, "the work done by *P* forces, acting on '*A*' structure in moving through the p_B displacements of '*B*' structure under Q loading" is equal to internal work due to internal stress resultants of structure '*A*' in moving through the corresponding deformations of the structure, '*B*'. The statement may be expressed by the following equation, Eq. (2.12), when flexure (moment) is predominant.

$$P_1(p_1^B) + P_2(p_2^B) + P_3(p_3^B) + P_4(p_4^B) = \int_{S_1}^{S_2} M_A \frac{M_B}{EI} ds \quad ...(2.12)$$

Where M_A is flexural stress resultant of the structure *A* and curvature M_B/EI is the flexural deformation of the structure *B* at the corresponding points.

If the structure *B* is a segment of the structure *A*, then the forces to be considered on the left hand side of the Eq. (2.12) and the limits of integration on the right hand side will be only for the common length, i.e., length of the structure which is a segment of the other. The internal stress resultant at the section of '*A*' corresponding to the end of the segmental structure '*B*', requires to be considered as an external force.

Alternatively it is possible to consider '*B*' structure for loads and internal stress resultant system, and '*A*' structure for displacements and curvatures. In this case, the following virtual work equation, Eq. (2.13), is obtained considering only flexural deformation.

$$Q_1(q_1^A) + Q_2(q_2^A) + Q_3(q_3^A) + Q_4(q_4^A) + Q_5(q_5^A)$$
$$= \int_{S_1}^{S_2} M_B \frac{M_A}{EI} ds \quad ...(2.13)$$

In a general case where axial and torsional deformations are required to be considered then we need to add two more integrals or internal direct stress resultant and twisting moment – on the right hand side the equations, Eqs (2.12) and (2.13). Similarly shears can also be considered.

The generalized definitions of virtual work Eqs (2.12) and (2.13) together establish the Maxwell Betti reciprocal theorem, Eq. (2.14), the right hand side integrals being the same.

$$P_1(p_1^B) + P_2(p_2^B) + \ldots + P_i(p_i^B)$$
$$= Q_1(q_1^A) + Q_2(q_2^A) + \ldots + Q_i(Q_i^A) \quad ...(2.14)$$

Application of the virtual work theorem for computing deformations is illustrated by solving the examples which have been solved elsewhere by using moment area method. A detailed description of the energy methods is given in the Article 2.6.

Ex. 2.17(a). For the given loading find horizontal deflection at *E* by the virtual work method.

Based on Eq. (2.13), we consider Fig. 2.17(*b*) for Q load system and resulting internal stress system; and Fig. 2.17(a) for deformation system due to given loading, which is a *P* loading. d_q is the desired deflection.

Therefore, Eq. (2.13) when written for the Figs 2.17(*a*) and 2.17(*b*) of this example

$$(Q)(q)^A = \int_A^E \frac{mq\,Mp}{EI} ds = Q\int_A^E \frac{mq\,Mp}{EI} ds = Q\left[\int_A^B \frac{mq\,Mp}{EI} ds + \int_B^C \frac{mq\,Mp}{EI} ds + \int_C^D \frac{mq\,Mp}{EI} ds\right]$$

Where $m_q = \dfrac{Mq}{Q}$ is the moment diagram for $Q = 1$, and *q* is the displacement in the real structure (under given loading) in the direction corresponding to *Q*.

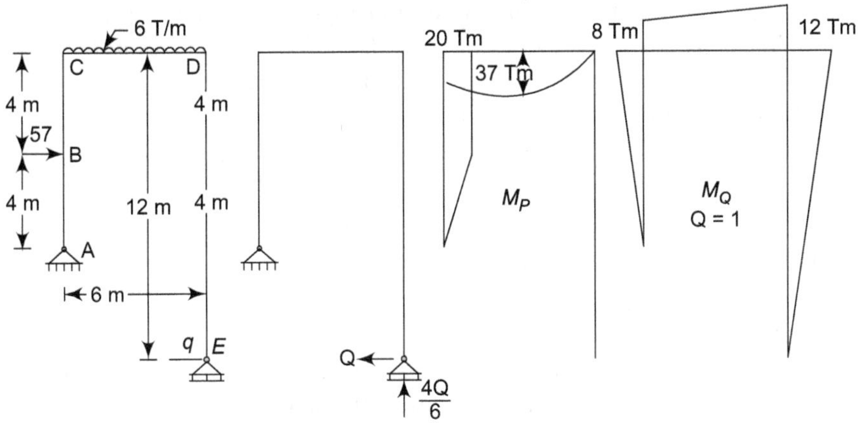

(a) Actual structure deformation system

(b) Force 'Q' applied in the direction of deformation desired force system

(c) B.M. diagram deformation system

(d) B.M. diagram force system

Fig. Ex. 2.17

Making use of the integration Table 2.1, for the above integrations

$$q = -\left(\underbrace{\frac{4}{3}\times\frac{20}{2EI}\times\frac{8}{2}}_{A\to B}+\underbrace{\frac{1}{2}\frac{1}{2EI}(20)(4+8)\,4}_{B\to C}+\underbrace{\frac{1}{EI}\left[\frac{6}{6}(20)(16+12)+\frac{1}{3}(27)(20)\right]}_{C\to D}\right)$$

$$= -53.33 - 240 - (560 + 1080)$$

$$\therefore \qquad q = \frac{-1933.33}{EI}$$

Ex. 2.17 (b). Find the deflection of end E in the horizontal direction due to $Q = 1$ at E, Fig. Ex. 2.17 (b).

In this case the moment diagram of Q loading is used for deformation system as well as force system. Based on the Eq. (2.12) or (2.13)

$$Q_q = Q\left[\int_A^C (Mq)2\frac{ds}{EI}+\int_C^D (mq)^2\frac{ds}{EI}+\int_D^E (mq)^2\frac{ds}{EI}\right] \qquad \text{....(a)}$$

Making use of the integration table and Fig. (b) for B.M. diagram

$$q = \frac{1}{3}\times(8)^2\times\frac{\delta}{2EI}+\frac{6}{6}[2(8)^2+(8\times 12)2+2(12)^2]+\frac{1}{EI}+\frac{1}{3}+(12)^2\times\frac{12}{3EI}$$

$$q = \frac{1}{EI}[85.33+608+192]+\frac{1}{EI}(885.33)$$

It may be verified that the method based on the principle of conservation of energy,

Work done = Strain energy stored,

gives the equation identical to the Eq. (a), with the only change that a factor ½ is not appearing on either side of the Eq. (a). The conservation of energy principle is used when there is one load and deflection is required under it. The virtual work principle gives a general approach.

The above results may be used to analyze an indeterminate structure, Fig. (a), with the end E hinged. The compatibility condition for the horizontal thrust release is given by

$$\delta_h = \frac{-1933.33}{EI} + Q\frac{1}{EI} + (885.33) = 0$$

∴ The redundant reaction, Q at E is equal to

$$\frac{1933.33}{885.33} = 2.18\ t$$

Before proceeding to illustrate solution of indeterminate structures based on the three basic steps of the force method of analysis, let us consider a typical example suggesting a direct use of deformations to obtain influence lines.

2.9 Deformations and Influence Lines

Consider parabolic arch, Fig. 2.22 (a), hinge at end A and roller supported at B, let us find horizontal deflection at B in the direction of the applied force, R.

We use the principle of virtual work. The real and arbitrary (or virtual) systems are shown in Fig. 2.22 (a). Moment due to given load R is given by

$$m_0 = R\,Y \qquad \text{...(i)}$$

where

$$Y = h\left(1 - 4\frac{x^2}{L^2}\right)$$

As the real structure is deformation structure, the deformation are $(\delta_h)_B$ in the direction of R and curvatures given by

$$\frac{m_0}{EI} = \frac{RY}{EI}$$

The arbitrary force system or statical system belonging to the virtual structure, consists of unit load applied at the place and in the direction given by the position and direction of $(\delta_h)_B$. Therefore, the chosen statical system of virtual structure consists of unit load at B and internal moments given by

$$m_1 = y, \qquad \text{...(ii)}$$

Corresponding to $(\delta_h)_B$ and $m\,/EI$ of given structure (real structure), Fig. 2.22(a)

From the principle of complementary virtual work,

$$1 \times (\delta_h)_B = \int m_1 \frac{m_0}{EI}\,ds$$

Substituting for m_1 and m_1, Eqs (i) and (ii),

$$(\delta_h)_B = \int \frac{Ry^0}{EI}\,ds = R\int \frac{y^2}{EI}\,ds \qquad \text{...(a)}$$

Or the value of force R casing unit displacement along its direction is given by

$$\frac{R}{(\delta_h)_B} = \frac{1}{\int \frac{y^2}{EI}\,ds}$$

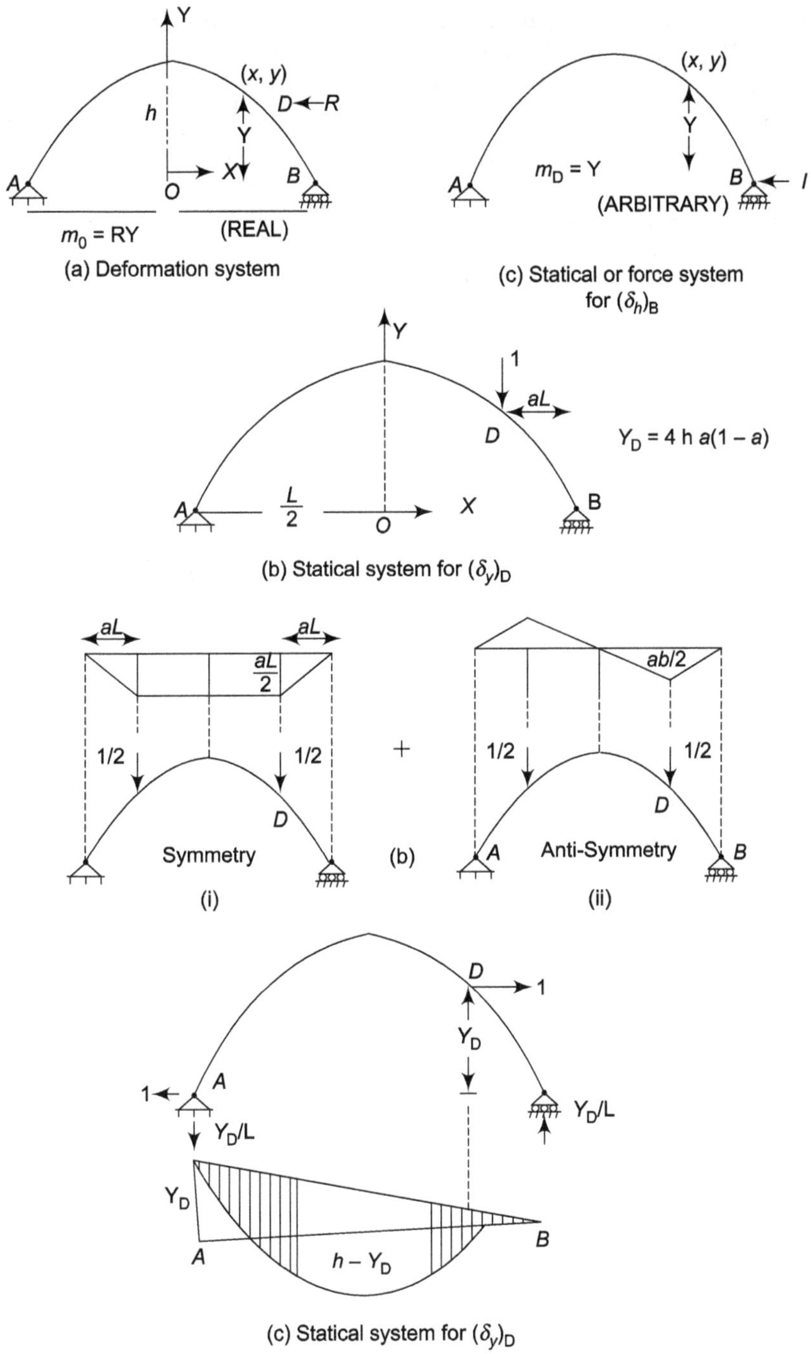

Fig. 2.22 *Principle of Virtual Work for Computing Deformations*

Let us find vertical deflection, Fig. 2.22(*b*), at a point *D*, $(\delta_h)_D$, distance aL from the roller support *B*, due to force *R* as in Fig. 2.22(*a*). For this purpose the chosen statical system, Fig. 2.22(*b*), consists of a unit load applied in the direction of the desired deflection, $(\delta_h)_D$. The statical *m* diagram is obtained from the superposition of symmetrical and anti-symmetrical loading, Fig. 2.22(*b*) (*i*) and (*ii*). The anti-symmetrical part of the statical system does not contribute to the external or internal work due to the vertical deflections of the real system caused by *R*. We consider symmetrical part only and use the principle of complementary virtual work.

Deformation of real system: $(\delta_h)_D = m_0/EI$

Force or statical system: unit load at *D*, and

$$m = \frac{1}{2}\left(\frac{L}{2} - x\right) \text{for} \left(\frac{L}{2} - aL\right) < x < \frac{L}{2}$$

and
$$m = \frac{aL}{2} \text{ for } 0 < x < \left(\frac{L}{2} - aL\right)$$

Taking advantage of symmetry:

$$\frac{1}{2} \times (\delta_v)_D = \iint_{\frac{L}{2}-aL}^{x=\frac{1}{2}} \frac{1}{2}\left(\frac{L}{2} - x\right) \frac{Ry}{EI}\, ds + \int_0^{x=\frac{L}{2}-aL} \frac{aL\, Ry}{2\, EI}\, ds$$

$$(\delta_v)_D = R\left[\int_{\left(\frac{L}{2}-aL\right)}^{x=\frac{1}{2}} \left(\frac{L}{2} - x\right) \frac{Y}{EI}\, ds\right] + aL \int_0^{x=\frac{L}{2}-aL} y\, ds \qquad \text{...(a)}$$

Re-arranging the limits of integral, we may modify the above Eq. (b), to give

$$(\delta_v)_D = R\left[\int_{\left(\frac{L}{2}-aL\right)}^{x=\frac{1}{2}} \left(\frac{L}{2} - x - aL\right) \frac{Y}{EI}\, ds\right] + aL \int_0^{\frac{L}{2}} \frac{Y}{EI}\, ds \qquad \text{...(b)}$$

From the Muller-Breslau's Principle in the Eq. (b) if *R* is a force per unit deflection, i.e., $(\delta_h)_B = 1$, the value of $(\delta_y)_D$, gives the horizontal reaction due to the unit load applied at *D*. If we substitute the values of *R* from right hand side of Eq. (*a*), in place of *R* appearing in Eq. (*b*), the ordinates of influence line for horizontal reaction of 2-hinged arch due to vertical loads, are given by $(\delta_v)_D$ for different values of (*a*). Thus the Eqs (*a*) and (*b*) can be used to prepare a table of influence values or horizontal reaction *H*, due to vertical loading.

Similar procedure is used to obtain influence line or a table of influence values at suitable intervals for load positions, when the loads are applied horizontally. Consider Fig. 2.22(*c*), where chosen statical system consists of unit horizontal load at *D*. From deformation system of Fig. 2.22(*a*) and statical force system of Fig. 2.22(*c*),

$$1\, X\, (\delta_h)_D = \int m \frac{R\, y}{EI}\, ds$$

Where
$$m = \frac{y_D}{L}\left(\frac{L}{2} - x\right) \text{for} \left(\frac{L}{2} - aL\right) < x < \frac{L}{2}$$

and
$$m = \frac{y_D}{L}\left(\frac{L}{2} - x\right) + (y - y_D) \text{ for } -\frac{L}{2} < x < \left(\frac{L}{2} - aL\right)$$

Substituting the values of m,

$$(\delta_h)_D = \int_{-\frac{L}{2}}^{x=L} \frac{y_D}{L}\left(\frac{L}{2}-x\right)\frac{Ry}{EI}\,ds + \int_{-\frac{L}{2}}^{x=\frac{L}{2}-aL} \frac{R}{EI}(y-y_D)\,ds \qquad \ldots(c)$$

In the Eq. (c), if R is the force per unit horizontal deflection at B, Eq. (c) gives influence line ordinates for horizontal reaction, R, due to horizontal loads. The value of R per unit horizontal deflection is obtained from Eq. (a).

2.10 CONTINUOUS BEAM ON FLEXIBLE SUPPORTS

So far, the structures analyzed were essentially flexural members. For such members, the effect of axial deformations is ignored except in the case of shallow arch analysis, Article 2.12. However, in the case of continuous beam, Fig. 2.23, or the beam members AB and BC, we need to consider moments only and for the springs d, e and f we consider axial forces only. The analysis is explained with the help of the numerical example, Fig. 2.23.

The degree of indeterminacy is one*. A hinge release is introduced at support B, Fig. 2.23(b). The m_0 and m_i diagrams are drawn, Figs 2.23(c) and (d).

From the principle of complementary virtual work, Article 2.6.2, in the equation

$$f_{11}\,q_1 + u_1{}^0 = 0 \qquad \ldots(1)$$

$u_1{}^0$ is evaluated by treating Fig. 2.23(c) as deformation structure and Fig. 2.23(d) as statical or force structure. Allowing $q = 1$ and the internal forces consisting of moments given by m_1 diagram and axial forces, $(p_1)_d$,$(p_1)_b$, $(p_1)_f$, Fig. 2.23(c), to ride over the corresponding deformations $u_1{}^0$ and the internal generalized strains consisting of curvature m/EI and axial deformations, $(p_0/K)_d$, $(p_0/K)_b$, $(p_0/K)_f$

$$u_1^0 = \int \frac{m_0}{EI} m_1\,dx + (p_1)_d\left(\frac{p_0}{K}\right)_d + (p_1)_e\left(\frac{p_0}{K}\right)_f$$

Using integration chart, Table 1 or Simpson's rule (Figs 2.23(c) and (d) for m_0 and m_1, diagrams) and substituting the values of p_0, p_1,

$$u_1^0 = \frac{ph^2}{16\,EI} - \frac{P}{2Kh} \qquad \ldots(2)$$

Similarly f_{11} is computed from the virtual work principle

$$f_{11} = \int \frac{m_1}{EI} m_1\,dx + \left(\frac{p_1}{K}\right)_d (p_1)_d + \left(\frac{p_1}{K}\right)_b (p_1)_e + \left(\frac{p_1}{K}\right)_f (p_1)_f$$

Substituting the values given in Fig. 2.23(a), and integrating the first term on the right-hand side of the before-mentioned equation

$$f_{11} = \frac{2h}{3\,EI} + \frac{6}{Kh^2} \qquad \ldots(3)$$

Substituting the results of Eqs (2) and (3) in Eq. (1)

$$q_1 = \frac{P\left(\dfrac{h^2}{16\,EI} - \dfrac{1}{2Kh}\right)}{\dfrac{2h}{3\,EI} + \dfrac{6}{Kh^2}}$$

* If analyzed by the Displacement Method, the number unknown displacement is 6, that is, (three translations and three rotations)

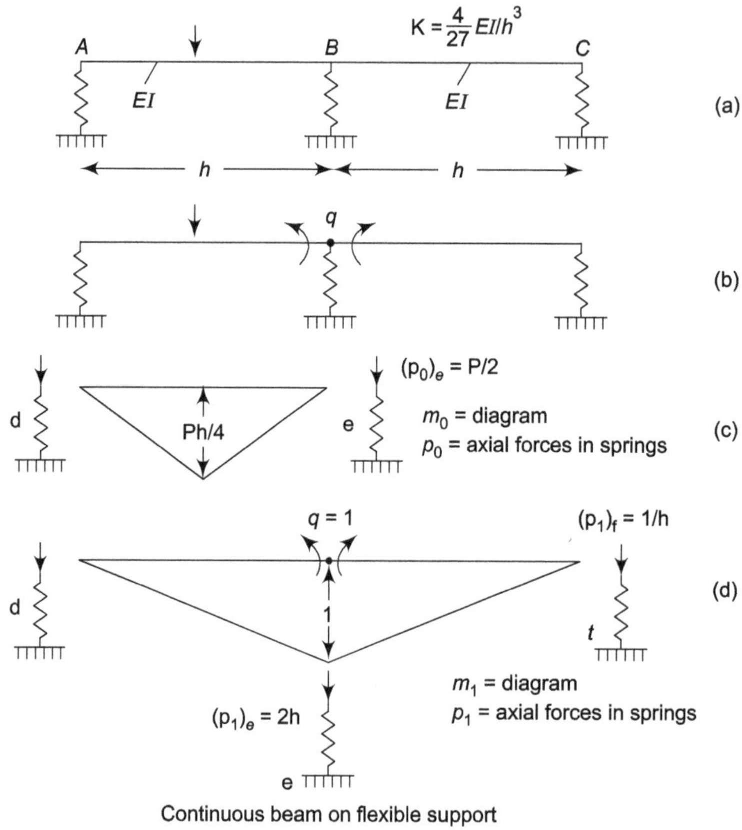

Fig. 2.23 *Continuous Beam on Flexible Support*

If $$K = 4/27\ EI/h^3$$

$$q_1 = .079\,ph$$

and the reactions of the springs are

$$p_d = P/2 + .079\ P = 0.579\ P \text{ compression}$$

$$p_e = P/2 - .158\ P = 0.342\ P \text{ compression}$$

$$p_f = 0 + .079\ P = 0.079\ P \text{ compression}$$

The reader may like to complete the solution by drawing the final bending moment diagram.

More generalized problems of this type are included in Chapter 4, Article 4.4.3, by using matrix approach.

The force method for the beams on spring supports has been used for the harmonic analysis of grids.

2.11 FRAMES

The following examples illustrate analysis of indeterminate frame structures based on the sequence of the three essential parts of the force method of analysis.

2.11a Superposition of Symmetrical and Anti-symmetrical Loadings on a Symmetrical Frame

It is explained through the following illustration of a symmetrical frame. The same is illustrated while applying to the Analysis of Grids, 2.15. This super position is valid for the symmetrical structures.

Ex. 2.18. Analyze a symmetrical frame, Fig. Ex. 2.18, under the given loading. Introduce the releases at G.

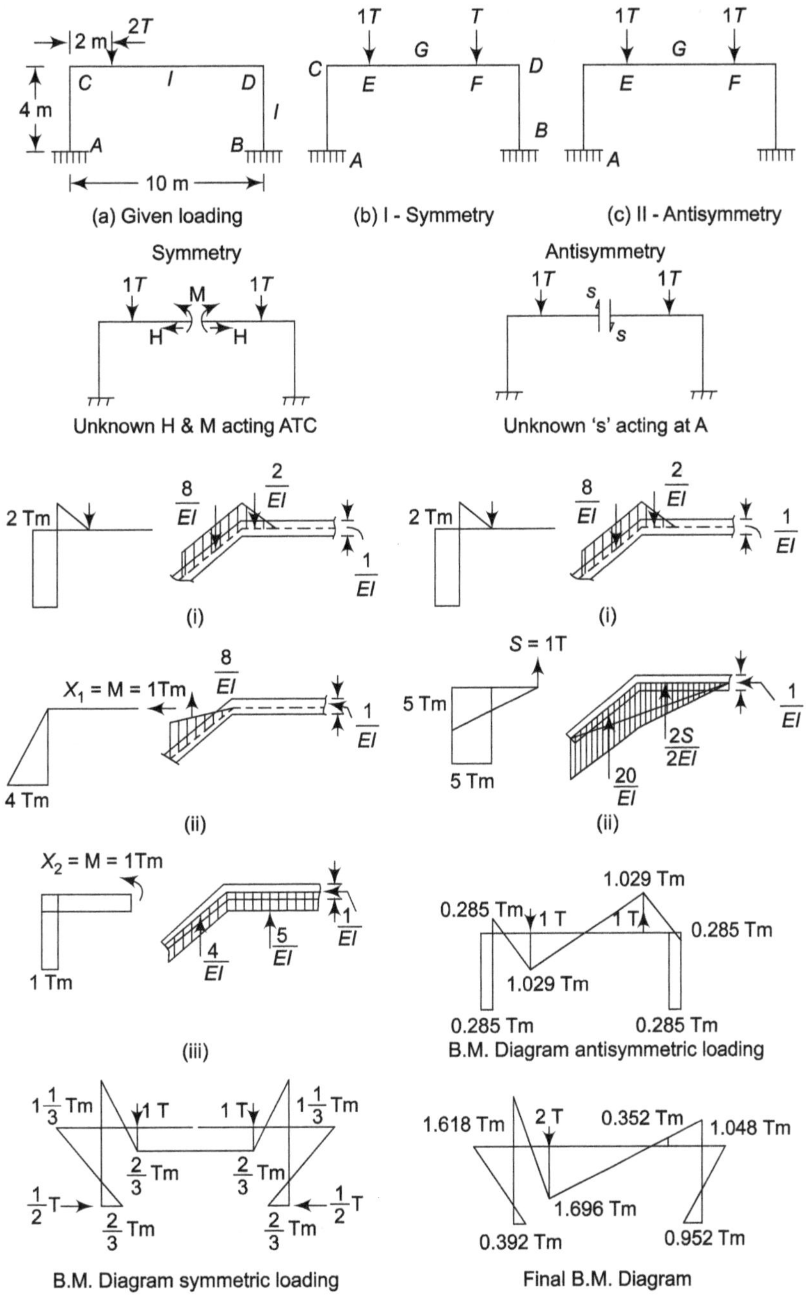

Fig. Ex. 2.18

I. Symmetrical structure under symmetrical loading: In the case of symmetry, the unknown forces redundant are the thrust *H* and moment *M* only.

Step 1. For the first part of analysis, equilibrium is satisfied for each case separately, Fig. Ex. 2.18, bending moment diagrams (*i*), (*ii*) and (*iii*).

Step 2. Evaluation of u_i^0 and f_{ij}, the deformations due to given loading and unit redundants respectively is shown by the two methods, (*i*) moment area method and (*ii*) virtual work method.

Moment area method

u_i^0 : Take moments about G, treating $\frac{M^0}{EI}$ diagrams as the loading. The angle loads shown in the Figs (i), (ii) and (iii)

$$u_1^0 = \frac{8}{EI} \times 2 = \frac{16}{EI}$$

$$u_2^0 = \frac{2}{EI} + \frac{8}{EI} = \frac{10}{EI}$$

f_{11} : Consider Fig. (ii)

$$f_{11} = \left(\frac{8}{EI}\right)\left(\frac{2}{3} \times 4\right) = \frac{64}{3\,EI}$$

$f_{21} = f_{12}$: Consider Fig. (ii)

$$= \frac{-8}{EI}$$

f_{22} : Consider Fig. (iii)

$$= \int_0^5 1 \times \frac{1}{EI} dx + \int_0^4 1 \times \frac{1}{EI} dy = \frac{9}{EI}$$

Virtual work method

Consider Figs. (i) and (ii) for

$$u_i^0 = +\int_0^4 \frac{2y}{EI} dy$$

$$u_i^0 = +\frac{16}{EI}$$

for u_2^0 = consider Figs (i) and (iii)

$$\text{for } u_2^0 = \frac{-2}{EI} - \frac{8}{EI} = \frac{-10}{EI}$$

f_{11} : Fig. (ii)

$$f_{11} = \frac{64}{3\,EI}$$

f_{12} : Consider Figs (ii) and (iii)

$$f_{12} = -\int_0^4 y\, dy = \frac{-8}{EI}$$

f_{22} = Consider Fig. (iii)

$$= \int_0^5 1 \times \frac{1}{EI} dx + \int_0^4 1 \times \frac{1}{EI} dy = \frac{9}{EI}$$

Step 3. Substituting the value of f_{11}, f_{12}, f_{13} in the compatibility equations,

$$f_{11} x_1 + f_{12} x_2 = -u_1^0 \quad x_1 = H = -1/2$$

$$f_{21} x_1 + f_{22} x_2 = -u_2^0 \quad x_2 = M = 2/3 \text{ tm}$$

II. Anti-symmetrical loading: In this case, the only redundant is the shear s.

$$fS = -u^0$$

Moment Area

Fig. (ii)

$$u_2^0 = \frac{2}{EI}\left(3+\frac{2}{3}\times 2\right) + \frac{8}{EI}(5)$$

$$= \frac{146}{3EI}$$

Flexibility

Consider Figs (i) and (ii)

$$u^0 = \frac{26}{3EI} + \frac{40}{EI}$$

$$= \frac{146}{3EI}$$

From Fig. (ii),

$$f = \frac{25}{2EI}\left(\frac{2}{3}\times 5\right) + \frac{20}{EI}(5)$$

$$f = \frac{125}{3EI} + \frac{100}{EI} = \frac{425}{3EI}$$

Substituting the values in the equation,

$$fS = -u^0$$

$$S = 0.343$$

The bending moment diagrams for the symmetrical and Anti-symmetrical cases, and combined bending diagram obtained by superposition of the two cases are given in the Fig. Ex. 2.18. Reader may solve the same example by the elastic centre method.

Gable Portal ***Ex. 2.19.*** Analyze the frame for the following cases:

(a) Ends *A* and *E* fixed,

(b) Ends *A* and *E* hinged,

(c) A fixed and *E* hinged.

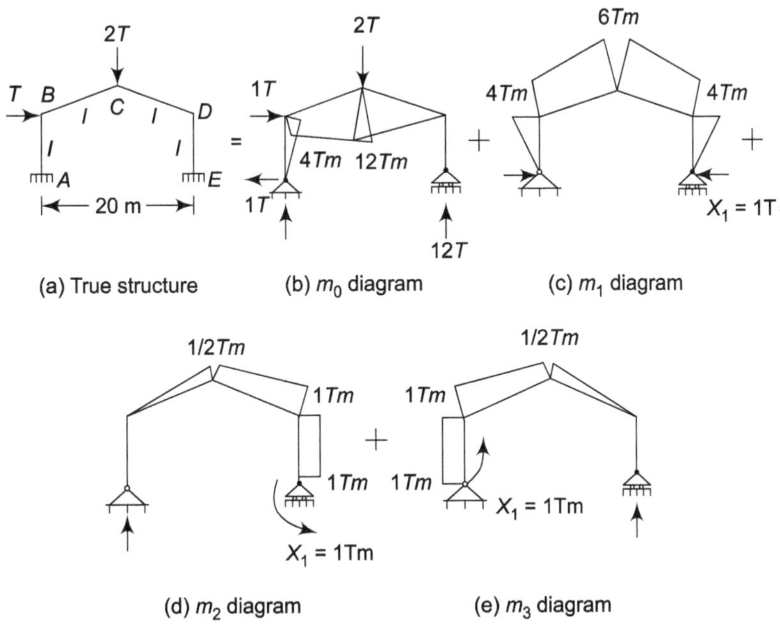

Fig. Ex. 2.19

Choose a determinate structure such that it is valid for each of the three cases: Equilibrium and superposition:

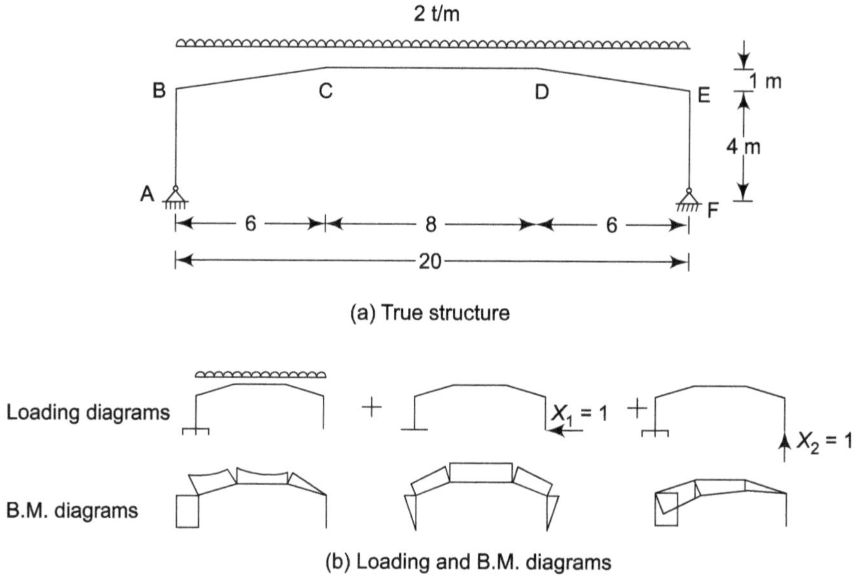

Fig. Ex. 2.20

For the case (a), all the four superposition are required to be considered.

For the case (b) m_0, m_1 diagrams are required to be considered.

For the case (c) m_0, m_1 and m_2 diagrams are required to be considered

Based on m_0 and m_1 diagrams,

$$u_1^0 = \int_A^B m_1 \frac{m_0}{EI}\, ds + \int_B^C m_1\, ds + \int_C^D m_1 \frac{m_0}{EI}\, ds$$

The integrals when evaluated,

$$u_1^0 = -\frac{64}{3EI} - \frac{1264.8}{3EI} - \frac{326.6}{EI} = \frac{-769.33}{EI}$$

Similarly U_2^0 is computed with the help of m_0 and m_2 diagrams.

$$u_2^0 = -\frac{71.4}{3EI} - \frac{40.8}{EI} = \frac{-64.60}{EI}$$

Considering m_0 and m_3 diagrams,

$$u_3^0 = -\frac{1}{EI}\left[+8 + \frac{173.4}{3} + 20.4\right] = \frac{-86.20}{EI}$$

f_{11} is obtained from m_1 diagram only and is equal to

$$f_{11} = \int_A^B \frac{m_1^2\,ds}{EI} + \int_B^C \frac{m_1^2\,ds}{EI} + \int_C^D \frac{m_1^2\,ds}{EI} + \int_D^E \frac{m_1^2\,ds}{EI}$$

$$= \frac{64}{3EI} + \frac{775.2}{3EI} + \frac{775.20}{3EI} + \frac{64}{3EI}$$

$$= \frac{559.47}{EI}$$

Similarly f_{22} from m_2 diagram is equal to

$$\frac{1}{EI}\left(\frac{10.2}{12} + 5.95 + 4\right) = \frac{10.80}{EI}$$

In this case, m_3/EI diagram being same as m_2/EI diagram,

$$f_{33} = f_{22} = \frac{10.80}{EI}$$

f_{12} is evaluated from the m_1 and m_2 diagrams and is given by,

$$f_{12} = (13.60 + 37.40 + 8)\ \frac{1}{EI} = \frac{59}{EI}$$

and from m_2 and m_3 diagrams,

$$f_{23} = \frac{1.70}{EI} + \frac{1.70}{EI} = \frac{3.40}{EI}$$

From the study of m_2/EI and m_3/EI diagrams and symmetry of m_1 diagram,

$$f_{13} = f_{12} = \frac{59}{EI}$$

and from reciprocal theorem,

$$f_{12} = f_{21}, f_{23} = f_{32} \text{ and } f_{13} = f_{31}$$

The simultaneous equations are:

$$f_{ij}\,x_i = U_1^{\,0}$$

$$559.47\ x_1 + 59\ x_2 + 59\ x_3 = 769.33$$

$$59\ x_1 + 10.80\ x_2 + 3.40\ x_3 = 64.60$$

$$59\ x_1 + 3.40\ x_2 + 10.80\ x_3 = 86.20$$

Solving the above three equations:

$$x_1 = 2.054t,\ x_2 = -4.750 \text{ tm and } x_3 = -1.730 \text{ tm}$$

for both ends hinged $x_1 = \dfrac{-U_1^0}{f_{11}} = \dfrac{769.33}{559.47} = 1.375\,t$

x_1 is the only redundant for both column ends hinged.

For A fixed and E hinged:

$$f_{11}x_1 + f_{13}x_3 = 769.33$$
$$f_{31}x_1 + f_{33}x_3 = 86.20$$

f_{11}, f_{13}, f_{31} and f_{33} remain unaltered and the solution gives the redundants.

Ex. 2.20. Analyze the portal frame by the flexibility method, Fig. Ex. 2.20.

(i) *Equilibrium.* Based on static equilibrium only, the bending moment diagrams for the given loading and the chosen redundants are drawn, Fig. Ex. 2.20 (*b*).

(ii) *Deformation.* The flexibility coefficients f_{ij} and u_{10} are the deformations due to the unit redundant and applied loading. The following gives the table for evaluating f_{ij} and u_{10}.

Member	u_{10}	u_{20}	f_{11}	f_{22}	$f_{12} = f_{21}$
AB	3200	-32000	64/3	1600	-160
BC	7118	-27360	370.88/3	1775	-462
CD	10720/3	-8000	200	843	-400
DE	346	-328	370.88/3	72.96	-85.12
EF	0	0	64/3	0	0
Total	14237.3	-67688	489.92	4290.96	-1107.12

(iii) Compatibility conditions:

$$f_{11}x_1 + f_{12}x_2 + u_{10} = 0$$
$$f_{11}x_1 + f_{12}x_2 + u_{10} = 0$$

Substituting the value from the table,

$$x_1 = 14.6\,t \text{ and } x_2 = 19.37t$$

Ex. 2.21. Given the shape and loads on a box frame as shown in Fig. Ex. 2.21 (a). Draw B.M. diagram.

The structure is externally determinate. It has one ring. Therefore, it is 3 times indeterminate, being a plane structure. Introduce the cut along the centre line on top deck. Further, we divide the problem into symmetrical and antisymmetrical halves and analyze separately.

$$f_{11} = \frac{1}{EI}\left[\frac{1}{3}\times(2.0)^2\times2\times2+4\times3\right] = \frac{17.33}{EI}$$

$$f_{12} = -\frac{1}{EI}\left[2\times\frac{1}{3}\times2\times2+3\times2\right] = \frac{-10}{EI}$$

$$f_{22} = 10\,EI$$

$$f_{33} = \frac{1}{EI}\left[\frac{4}{3}\times1.5\times1.5^2+2\times1.5^2\times2\right] = \frac{13.5}{EI} \quad \text{Anti-symmetry}$$

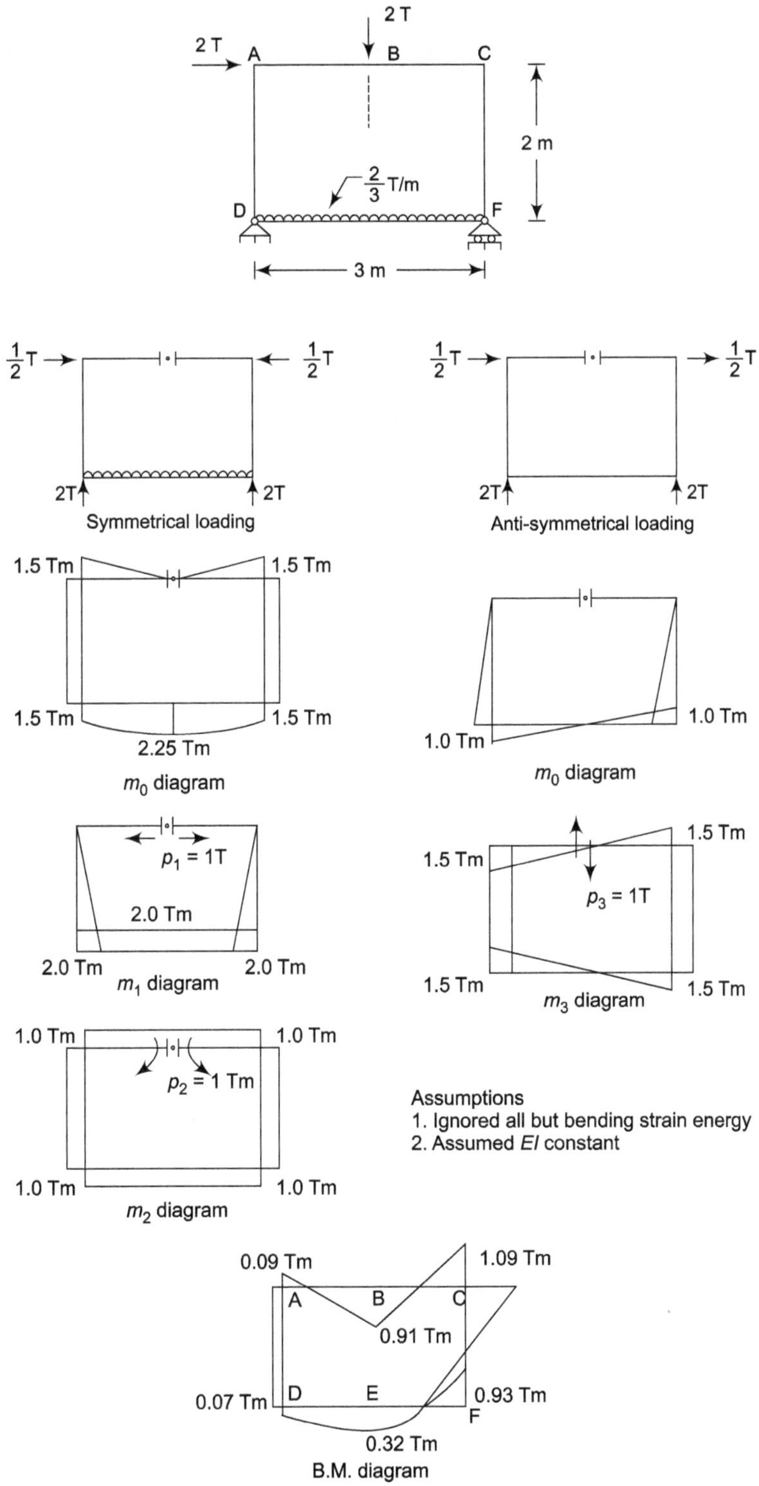
2 T
2 T
A
B
C
2 m
$\frac{2}{3}$ T/m
D
F
3 m
$\frac{1}{2}$T
$\frac{1}{2}$T
2T
2T
Symmetrical loading
$\frac{1}{2}$T
$\frac{1}{2}$T
2T
2T
Anti-symmetrical loading
1.5 Tm
1.5 Tm
1.5 Tm
1.5 Tm
2.25 Tm
m_0 diagram
1.0 Tm
1.0 Tm
m_0 diagram
p_1 = 1T
2.0 Tm
2.0 Tm
2.0 Tm
m_1 diagram
1.5 Tm
1.5 Tm
p_3 = 1T
1.5 Tm
1.5 Tm
m_3 diagram
1.0 Tm
1.0 Tm
p_2 = 1 Tm
1.0 Tm
1.0 Tm
m_2 diagram
Assumptions
1. Ignored all but bending strain energy
2. Assumed EI constant
0.09 Tm
1.09 Tm
A
B
C
0.91 Tm
0.07 Tm
D
E
0.93 Tm
F
0.32 Tm
B.M. diagram

Fig. Ex. 2.21

$$u_1^0 = -\left(1.5\times\frac{2\times2\times2}{2}+1.5\times2.0\times3+\frac{2}{3}\times3\times.75\times2\right)\frac{1}{EI}=\frac{-18}{EI}$$

$$u_2^0 = 2\left(\frac{1.5}{2}\times1.5+1.5\times2+1.5\times1.5+\left(\frac{2}{3}\right)\times.75\times1.5\right)\frac{1}{EI}=+\frac{14.25}{EI}$$

$$u_3^0 = 2\left(1.5\times2\times\frac{1}{2}+\frac{1.5}{2}\times1.5\right)\frac{2}{EI}=\frac{-4.5}{EI}\quad \text{Anti-symmetry}$$

For the case of symmetry,

$$F=\begin{bmatrix}17.33 & -10\\ -10 & 10\end{bmatrix}\frac{1}{EI}\quad u^0=\begin{bmatrix}-18\\ 14.25\end{bmatrix}\frac{1}{EI}$$

$$F^{-1}=\frac{EI}{(17.33\times10-10\times10)}\begin{bmatrix}10 & 10\\ 10 & 17.33\end{bmatrix}$$

$$=\frac{EI}{73.33}\begin{Bmatrix}10 & 10\\ 10 & 17.33\end{Bmatrix}$$

Symmetrical Set

The symbol p is being used for the redundants.

Since $\quad Fp + u^0 = 0$

$$P = -F^{-1}u^0$$

$$P=\frac{1}{73.33}\begin{Bmatrix}10 & 10\\ 10 & 17.33\end{Bmatrix}\begin{Bmatrix}18\\ -14.25\end{Bmatrix}$$

$$=\begin{Bmatrix}0.51\\ -0.91\end{Bmatrix}$$

Anti-symmetrical Set

$$P_3=\frac{+4.5}{13.5}$$

$$P_3=+\frac{1}{3}$$

	A	B	C	D	E	F
M^0	+ 1.5	0	+1.5	2.5	2.25	.05
p_1m_1	0	0	0	−1.02	−1.02	−1.02
p_2m_2	-.91	-.91	-.91	-.91	-.91	-.91
p_3m_3	-.5	0	.5	-.5	0	+.5
Total B.M.	+.09	-.91	1.09	+.07	+.32	-.93

The following are two more examples, for which, again from the considerations of the number of unknowns, force method is more suitable.

Fig. Ex. 2.21(A-1). The data for the example is given in Fig. Ex. 2.21(A-1) (*a*). A single release required to make the frame determinate is chosen to be a hinge at *D*. The statical m_0 and m_1 diagram are given in Figs. Ex. 2.21(A-1) (*b*) and (*c*). The reader may verify the following results. Use Simpsons' rule or integration Table 2.1.

$$EI\ u_1^0 = \frac{L^3}{6}(1.438),\ EI f_{11} = \frac{L}{6}(13.961),\ \text{and}\ q_1 = -.103\ L^2\ tm$$

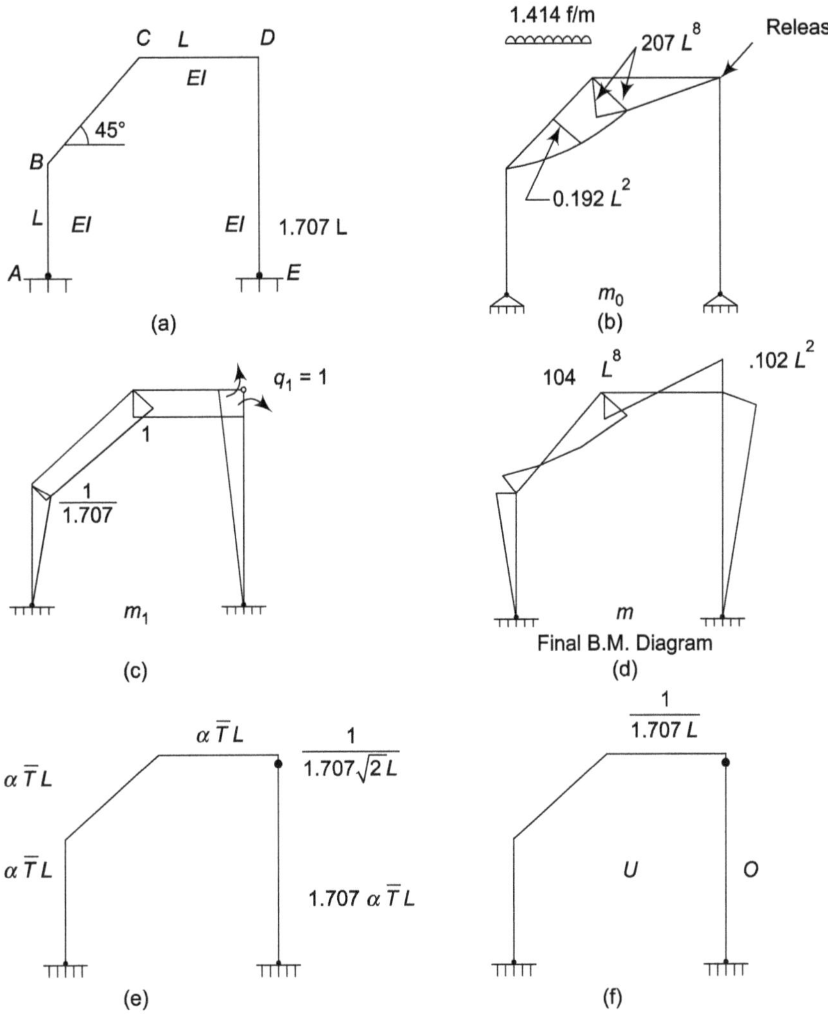

Fig. Ex. 2.21(A-1)

The final bending moment diagram is given in Fig. (*A*-1) (*d*). The effect of axial deformations has been neglected.

Temperature effect: Analyze the same frame, for uniform rise in temperature, $\overline{T} = 100$ C and thermal coefficient of expansion, $a = 1 \times 10^{-5}$ per degree centigrade.

The value of the flexibility $EIf_{11} = L/6\,(13.961)$ computed above remains unaltered. For the evaluation of u_1^0, refer Figs A-1(e) and (*f*). The axial deformations (equivalent to) $\frac{m_0 L}{EA}$ due to rise in temperature as shown against the members, Fig. A-1(e). The virtual statical diagram giving axial forces $q = 1$ is given in Fig. A-1(f). Combining the results given in Figs. A-1(e) and (f).

$$u_1^0 = \alpha \bar{T}_L \left(\frac{1}{1.707} + \frac{1}{1.707\sqrt{2L}} \right)$$

Substituting $\alpha = 1 \times 10^{-5}$ and $T = 10°C$

$$u_1^0 = 1 \times 10^{-4}$$

$$\therefore \quad q_1 = -\,u_0^1/f_{11} = 0.43 \times 10^{-4}\ EI/L$$

Multiply the ordinates of m diagram, Fig. A-1(c), with the q value computed to get the bending moment diagram due to rise in temperature.

Ex. 2.21 (A-2). Refer Fig. Ex. 2.21 A-2 (a) for data. The single release chooses is the introduction of a link or roller at support D, which removes the constraint against horizontal movement. The reader may verify the following:

$$u_0^1 = -\,81 \times 68$$

$$f_{11} = 81 \times 9, \text{ and } q = 7.555\ t$$

the final moment diagram is given in Fig. Ex. 2.21 A-2 (b).

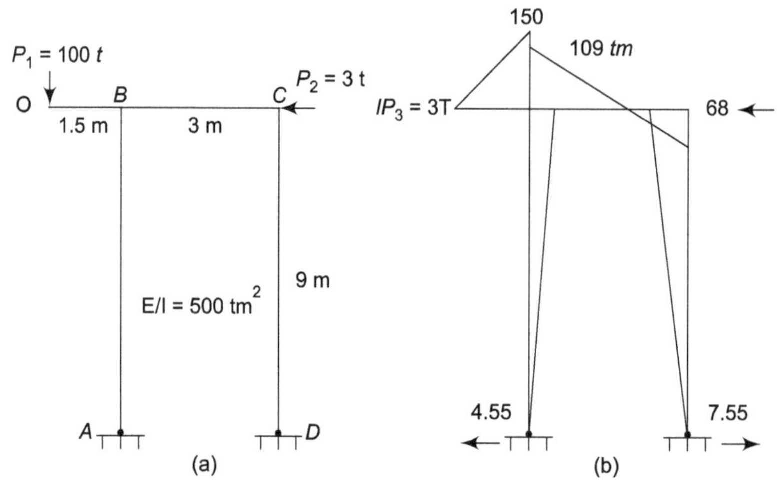

Fig. Ex. 2.21(A-2)

Ex. 2.22. Analysis of frames based on minimum energy principles: In the Article 1.6.6, it was stated that the Eq. (1.6) is used for evaluating redundant forces as well as deflections at any point of a given structure. Consider the example of Fig. Ex. 2.16. Let the three releases be introduced at E, Fig. Ex. 2.22.

Neglecting the strain energy contribution due to normal stress and shear, the strain energy U due to flexure only is given. The structure being symmetrical, there is no shear at the releases. The two redundant are X_1 and X_3.

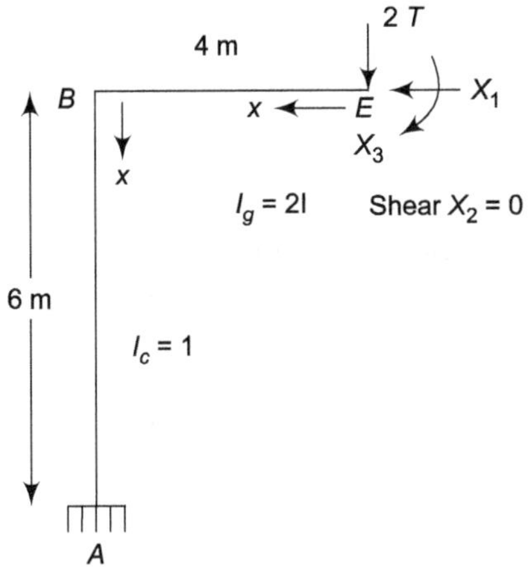

Fig. 2.22 *(Also see Fig. Ex. 2.16)*

$$U = \int \frac{m^2}{2EI} dx = \underbrace{\int_0^4 \frac{(2x + X_2)^2}{2EI_g} dx}_{\text{Due to beam EB}} + \underbrace{\int_0^6 \frac{(X_3 - x\,X_1 + 8)^2}{2EI} dx}_{\text{Due to column BA}}$$

Considering the redundant only as the arbitrary forces, Eq. (2.7) gives

$$\frac{\partial U}{\partial X_1} = 0 \text{ and } \frac{\partial U}{\partial X_3} = 0$$

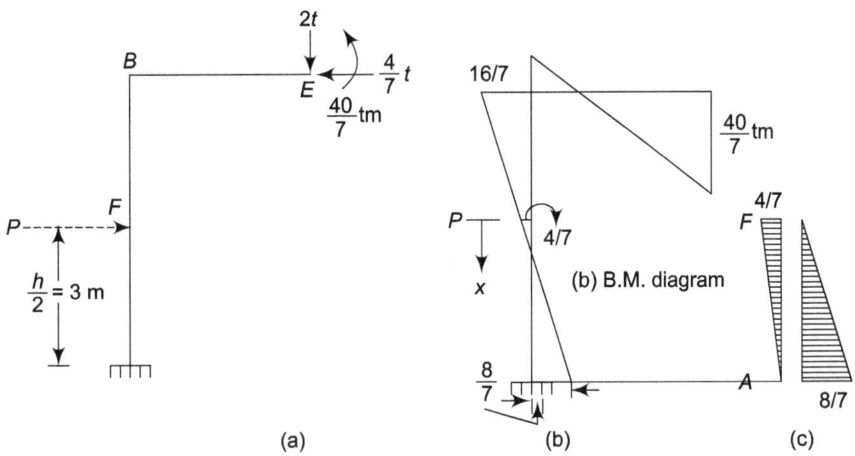

Fig. Ex. 2.23

Substituting for *u*, we get

$-(9/4)X_1 + X_3 + 7 = 0$ and $4\,X_1 - X_3 - 8 = 0$

The unknowns are:

$$X_1 = 4/7\ t$$

$$X_3 = 40/7\ \text{tm}$$

Ex. 2.23. From Figs Ex. 2.16 and Ex. 2.22, find deflection at the point *F*, i.e., at mid height of the column AB, Fig. Ex. 2.23. As stated in the Article 2.6.6 introduce an arbitrary force.

Based on the Eq. (2.8) and Eq. (2.9), the required deflection, from the Fig. (c)

The relative deflection at *P* is the horizontal deflection. Take moment of *M*/*EI* loading at *P*.

Consider length *FA* only. The above moment area method gives $\delta_F = \dfrac{18}{7EI}$ deflects left.

Evaluation of deflection using Virtual Work Principle, Art. 2.8: We will continue with the same example, Ex. 2.23 for illustrating the use of virtual work principle for computing the deflection at a chosen point, dF as an alternative to the 'moment area method'. The problem is redrawn in the following for ready reference.

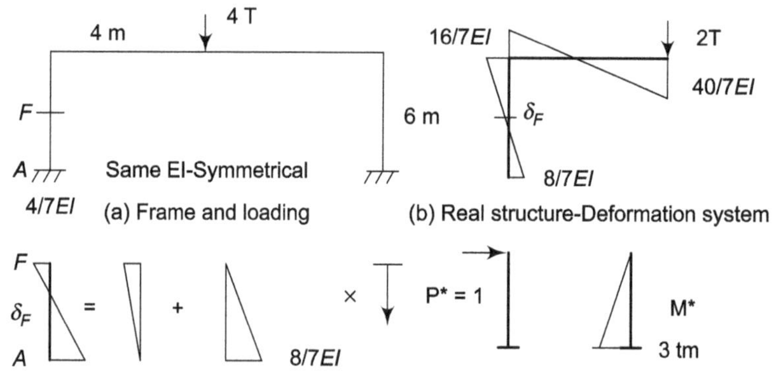

Fig. Ex. 2.23 *Extended*

'Riding' the Force System on the Deformation System, we get

$$P^* \times \delta_F = \int_0^3 M \frac{M^*}{EI} dx$$

Substituting, $$M = \frac{4}{7EI}\frac{(3-x)}{3} + \frac{8}{7EI}\frac{(x)}{3}$$

And $$M^* = x,$$

The reader may verify the result:

$$\delta_F = \frac{-18}{7EI}, \text{ that is, deflection is to the left.}$$

Table 2.1 Integration Table

<table>
<tr><td>m_q / m_p</td><td>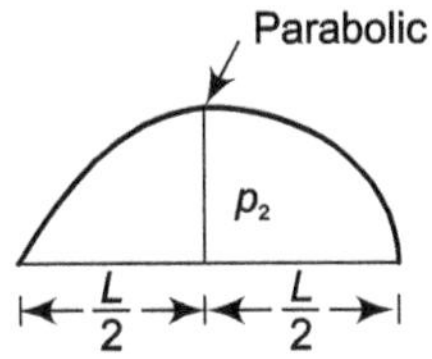</td><td>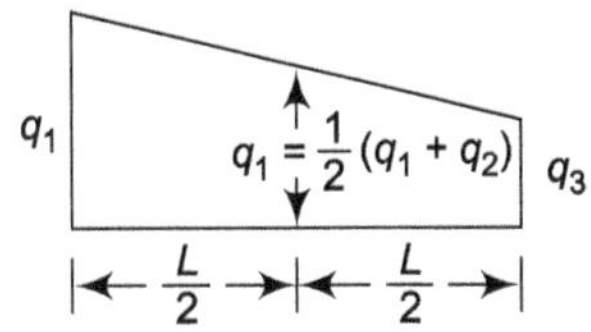</td></tr>
<tr><td>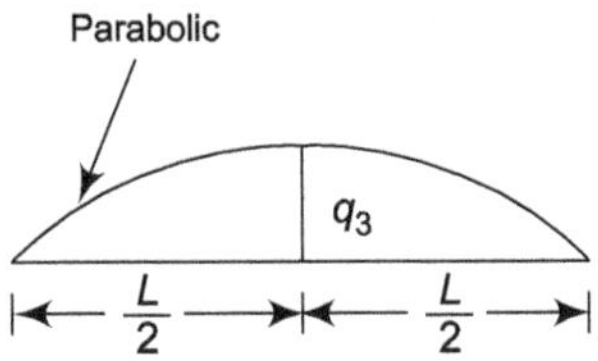</td><td>$\frac{8L}{15}p_1q_1$</td><td>$\frac{L}{3}p_1(q_1+q_2)$</td></tr>
<tr><td>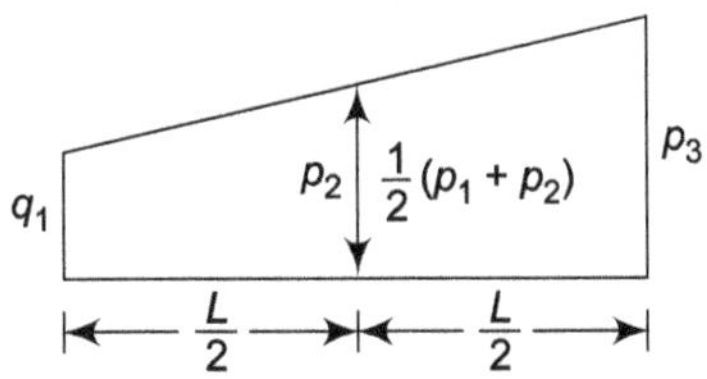</td><td>$\frac{L}{3}q_3(p_1+p_3)$</td><td>$\frac{L}{6}(2q_1p_1+p_1q_3+p_3q_1+2p_3q_3)$</td></tr>
<tr><td>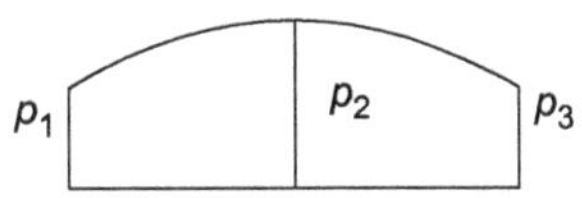</td><td>Use combinations</td><td>$\frac{L}{6}[p_1q_1+2p_3(p_1+q_3)+p_3q_3]$</td></tr>
</table>

Numerical integration of Product terms, $m_p \times m_q$:

Other cases of triangular and simple parabolic distributions being special cases or combinations of the above integrations are not included in the integration Table 2.1. In a general cause, Simpson's rule of integration can be used directly. An application of the rule is explained while analyzing a non-prismatic arch with built-in ends. The Simpson's integration is accurate, if the integrands ($m_p\, m_q$) are upto or less than third degree.

2.12 PARABOLIC ARCH – NON-PRISMATIC

Arch is on efficient structure for long span bridges.

Consider a parabolic with built-in ends, Fig. 2.24(a). The arch is divided into 12 segments for the purpose of numerical integration, Fig. 2.24(b). A cut at the crown results in three releases. However, because of symmetry, there is no relative vertical displacement at the crown, C, and therefore the redundants get reduced to only two, X_1 and X_2, corresponding to the constraints against relative horizontal displacement and relative rotation at the crown. The final equation is of the form

$$\begin{pmatrix} f_{11} & f_{12} \\ f_{21} & f_{22} \end{pmatrix}\begin{pmatrix} X_1 \\ X_2 \end{pmatrix}+\begin{pmatrix} u_1^0 \\ u_2^0 \end{pmatrix}=0 \qquad \text{...(A)}$$

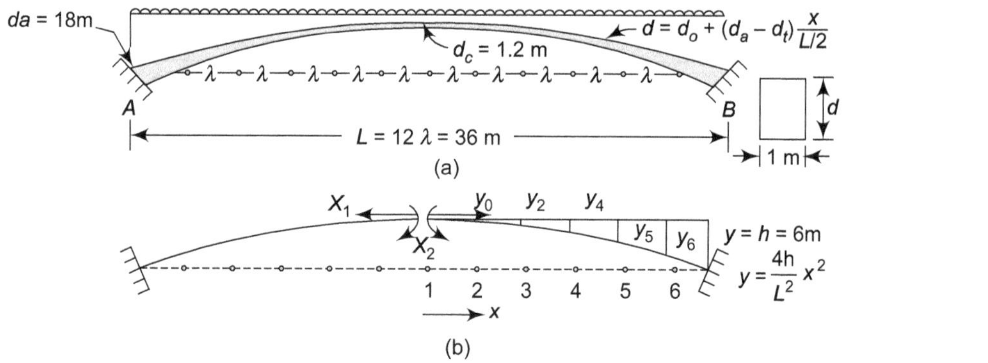

Fig. 2.24 *Non-prismatic Parabolic Arch*

Irrespective of the nature of loading or geometric symmetry, the flexibility matrix is always symmetric, i.e., $f_{ij} = f_{ij}$. Thus we are required to evaluate f_{11}, f_{12}, f_{22}, u_1^0 and u_2^0. The arch is non-prismatic and its variation of depth is given by the relation,

$$d = d_c + (d_a - d_c)\,\frac{x}{L/2}$$

Where d_c and d_a are the depths at crown and abutment respectively.

Rib shortening: In the case of shallow arches with rise to span ratio less than 1/5, they require to be analyzed or for flexural as well as axial deformations. The latter consideration (i.e., axial deformation) is known as "rib shortening" effect. And the flexibility coefficients are,

$$f_{ij} = \int \frac{m_i\, m_j}{EI}\, ds + \int \frac{n_i\, n_j}{EA}\, ds$$

In the Fig. 2.24 the equal segments are taken for horizontal projection of the arch. To account for this the dummy variable is x in place of s. The effect of shear contributing to f_{ij} is neglected, Refer Eq 1.9 also.

$$f_{ij} = \int \frac{m_i\, m_j}{EI}\, \frac{ds}{dx}\, dx + \int \frac{n_i\, n_j}{EA}\, \frac{ds}{dx}\, dx$$

or

$$f_{11} = \frac{1}{E}\int (m_1)_2 \left(\frac{\frac{ds}{dx}}{I}\right) dx + \frac{1}{E}\int (n_1)_2 \left(\frac{\frac{ds}{dx}}{A}\right) dx \qquad \text{...(a)}$$

$$f_{12} = \frac{1}{E}\int m_1\, m_2 \left(\frac{\frac{ds}{dx}}{I}\right) dx + \frac{1}{E}\int n_1\, n_2 \left(\frac{\frac{ds}{dx}}{A}\right) dx \qquad \text{...(b)}$$

Table 2.2

Point i	x_i	d_i	A_i	I_i	Y_i	$\left(\frac{ds}{dx}\right)$	m_1	n_1	m_2	n_2	m_1^2	n_1^2	m_2^2	n_2^2	$m_1 m_2$	$n_1 n_2$	n_0
1	0	1.20	1.2	.1440	0	1	0	1.00000	1.0	0	0	1.00000	1.0	0.0	0.0	0.0	0.00
2	3	1.30	1.3	.1831	1/6	1.006154	.166667	0.99388	1.0	0	0.27778	.98780	1.0	0.0	.16667	0.0	9.00
3	6	1.40	1.4	.2287	2/3	1.024394	.666667	0.97618	1.0	0	.444444	.95292	1.0	0.0	.66667	0.0	36.00
4	9	1.50	1.5	.2813	3/2	1.054093	1.500000	0.94868	1.0	0	2.2500	.89999	1.0	0.0	1.5000	0.0	81.00
5	12	1.60	1.6	.3413	8/3	1.094940	2.666667	0.91370	1.0	0	7.11111	.83485	1.0	0.0	2.66667	0.0	144.00
6	15	1.70	1.7	.4094	25/6	1.143959	4.166667	0.87415	1.0	0	17.36111	.76414	1.0	0.0	4.16667	0.0	225.00
7	18	1.80	1.8	.4860	6	1.201850	6.00000	0.83205	1.0	0	36.000	.69231	1.0	0.0	6.00000	0.0	324.00

$$f_{22} = \frac{1}{E}\int (m_2)_2 \left(\frac{\frac{ds}{dx}}{I}\right) dx + \frac{1}{E}\int (n_2)_2 \left(\frac{\frac{ds}{dx}}{I}\right) dx \qquad \ldots(c)$$

As the redundant moment X_2 does not axial forces, $n = 0$, which eliminates the quantities under the second integral sign of f_{12} and f_{22}. For the released arch the normal stresses due to vertical loading are not large and can be ignored while evaluating $u_1{}^2$ and $u_2{}^0$. Therefore,

$$u_1{}^0 = \frac{1}{E}\int m_0\, m_1 \left(\frac{\frac{ds}{dx}}{I}\right) dx \qquad \ldots(d)$$

$$u_2{}^0 = \frac{1}{E}\int m_0\, m_2 \left(\frac{\frac{ds}{dx}}{I}\right) dx \qquad \ldots(e)$$

Where, $$\frac{ds}{dx} = \left(1 + \frac{64h^2}{L^2}x^2\right)^{1/2}$$

Table 2.3 *Evaluation of* $m_1{}^2\left(\frac{ds}{dx}\right)/I$ *and* $n_1{}^2\left(\frac{ds}{dx}\right)/I$

Point, i	I_i	$\left(\frac{ds}{dx}\right)$	$\left(\frac{ds}{dx}\right)/I$	$m_1{}^2$	$m_1{}^2\left(\frac{ds}{dx}\right)/I$	A_1	$\left(\frac{ds}{dx}\right)/A$	$n_1{}^2$	$n_1{}^2\left(\frac{ds}{dx}\right)/A$
1	2	3	4	5	6	7	8	9	10
1	.1440	1.0	6.94444	0.000000	0	1.2	.83333	1.00000	.8333
2	.1830	1.00615	5.49808	0.027778	.15273	1.3	.77396	.98750	.76445
3	.2287	1.02439	4.47918	0.444444	1.98974	1.4	.73170	.95292	.69725
4	.2813	1.05409	3.74720	2.25	8.4312	1.5	.70272	.89999	.63244
5	.3413	1.09444	3.20668	7.111111	22.27394	1.6	.68402	.83485	.57105
6	.4094	1.4396	2.79423	17.341111	48.50729	1.7	67291	.76414.	.51419
7	.4860	1.20185	2.47294	36.00000	89.02584	1.8	.66749	.69231	.46224

All the integrations of the Eqs (a) to (e) of the form

$$\int F\, dx,$$

Require numerical evaluation of F_i at difference stations, I, obtained by dividing the arch into 12 equal segments, that is $\Delta x = \lambda = 3$ m. The F values for the different cases are plotted, Fig. 2.24. The expressions for F are given by the Eqs (a) to (e). the numerical values of depth d, area A, $\frac{ds}{dx}$, m_1, n_1, m_2, n_2, m_0 at the stations I = 1, 2, ..., 7 are given in Table 2.2

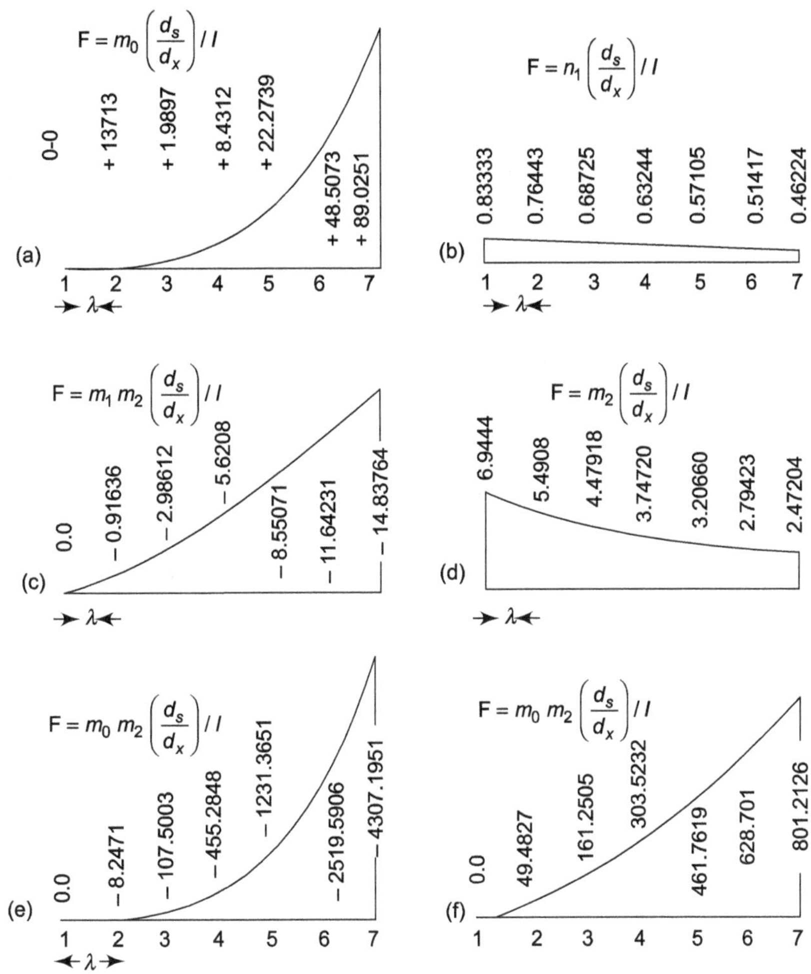

Fig. 2.25 *Ordinates, F, for numerical integration* $\int F\,dx$

The values of F for the integrals of the Eq. (a) are given in the column (6) and (10) of Table 2.3 and also see Fig. 2.25 (a) and (b). the numerical integration by Simpson's rule for the seven station points is given by the relations, Eq. (*f*),

$$\therefore \quad \int F\,dx = \frac{\lambda}{3}(F_1 + 4F_2 + 2F_3 + 4F_4 + 2F_5 + 4F_6 + F_7) \quad \text{...(B)}$$

Substituting, the values of F_i, Fig. 3.25(a) and (b)

$$f_{11} = \frac{\lambda}{3E}(365.9181 + 11.4765)$$

The second term on the right hand side is the contribution due to rib shortening.

$$f_{11} = \frac{\lambda}{3E}(377.3946)$$

The F values, Eq. (b), for the evaluating f_{12} are given in the column (6) of Table 2.4 and Fig. 2.25(c). The values when substituting in the Eq. (b) give

$$f_{12} = \frac{\lambda}{3E}(-109.6292)$$

Similarly, the F values, Fig. 2.25(d), (e), (f), i.e., column (7), (3) and (4) of Table 2.3 are used to evaluate f_{22}, $u_1{}^0$ and $u_2{}^0$ respectively.

$$f_{22} = \frac{\lambda}{3E}(72.9471)$$

$$u_1^0 = \frac{\lambda}{3E}(-19817.6162)$$

and

$$u_2^0 = \frac{\lambda}{3E}(5974.0878)$$

substituting the above values of f_{ij} and u_1^0, the unknown redundant are evaluated, X_1 = 51.10 t, X_2 = – 4.4 tm.

final moments m_i at the stations i = 1, …, 7 are obtained by superposition, Eq.(C)

$$m_i = (m_0 + m_1X_1 + m_2X_2)i \qquad \text{...(C)}$$

Table 2.4 *Values of F for evaluating u_1^0, u_2^0, f_{12} and f_{22}*

S. No.	$m_0 m_1$	$m_0 m_2$	$m_0 m_1\left(\frac{ds}{dx}\right)/I$	$m_0 m_2\left(\frac{ds}{dx}\right)/I$	$m_1 m_2$	$m_1 m_2\left(\frac{ds}{dx}\right)/I$	$m_2{}^2\left(\frac{ds}{dx}\right)/I$
	1	2	3	4	5	6	7
1	-0.00	+0.00	-0.00	0.00	-0.00	-0.00	6.94444
2	-1.50	+9.00	-8.24712	49.48273	-0.16667	-0.91636	5.49808
3	-24.00	36.00	-107.50032	161.25048	-0.66667	-2.98612	4.47918
4	-121.50	81.00	-455.28480	303.52320	-1.50000	-5.62080	3.74720
5	-384.00	144.00	-1231.36512	461.76192	-2.66667	-8.55071	3.20668
6	-937.50	225.00	-2609.59062	628.70175	-4.16667	-11.64231	2.79423
7	-1944.00	324.00	-4807.39536	801.23756	-6.00000	-14.83764	2.47294

Analytical Check. The Eq. (2.7) based on minimum complementary potential energy, can be used as an analytical check.

$$\frac{\partial U}{\partial P_j} = 0 \qquad \text{...(2.7)}$$

In the present example of an arch $P_j = X_j$ (j = 1,2), and

$$U = \int \frac{m^2}{EI} ds$$

$$\frac{\partial U}{\partial X_j} = \int \frac{m m_j}{EI} ds = 0 \qquad \text{...(f)}$$

Substituting for m, Eq. (C)

$$\int m_j \frac{(m_0 + m_1 X_1 + m_2 X_2)}{EI} ds = 0 \qquad \text{...(g)}$$

Or

$$\int m_j \frac{(m_0 + m_1 X_1 + m_2 X_2)}{EI} \frac{ds}{dx} dx = 0 \qquad \text{...(g)}$$

If axial deformations and shear deformation are considered to contribute to the total strain energy, U, the Eq. (f) is modified to give (Refer Eq. 1.9 also)

$$\int \frac{m_i m_j}{EI} ds + \int \frac{p_y (p_y)_j}{k\,GA} ds + \int \frac{p_x (p_x)_j}{EA} ds = 0 \qquad \text{...(h)}$$

After the complete analysis, as all the stress resultants are known, when they are substituted, the Eq. (h) should be satisfied. The stress resultants are the same as used in Eq. (2.9). While doing the arch analysis symbol n was used instead of p_x or normal stress. The Eq. (h) can be used for checking the analytical results of any indeterminate structure. The reader may verify the arch results by this method.

2.13 GEOMETRIC LOADING–TEMPERATURE STRESSES IN INDETERMINATE STRUCTURES

Indeterminate structure develop stresses caused by

(a) Lack of Fit, (b) Temperature Stresses, and (c) Settlement of Supports.

2.13.1 Temperature Stresses

In the case of determinate structures, temperature variations do not give rise to stresses. Indeterminate structures develop stresses due to temperature, which may be significant from design considerations. This is a particular case of Lack of Fit. Consider a cross section of member of a planar frame, Fig. 2.26.

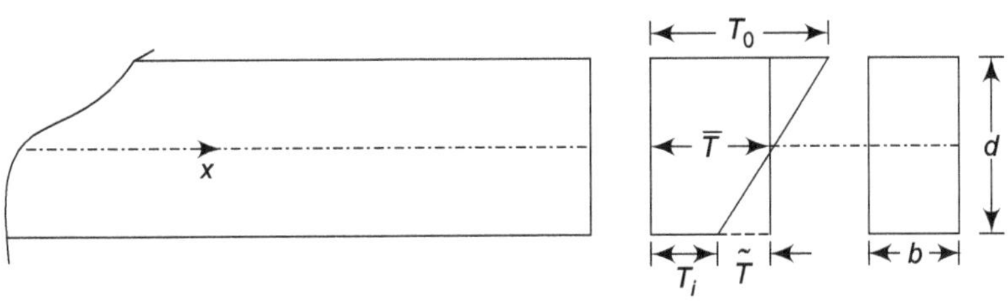

Fig 2.26 *A Cross Section of Member of a Planar Frame*

The inside and outside variation in temperatures be designated by T_i and T_0 respectively. It is equivalent to (*i*) uniform change in temperature equal to $\overline{T}$ and (*ii*) a varying temperature with zero change at mid depth and maximum value T at extreme fibres, Fig. 2.26. The effects on stresses will be studied separately.

(*i*) Uniform temperature rise = $\overline{T}$: If α is the coefficient of thermal expansion, the expansion of the member per unit-length is equal to a T which may be correlated with the direct strain ε_x

$$\varepsilon_x = \alpha \overline{T} \qquad \text{...(a)}$$

With this equivalence established, the deformations u_1^0, at releases due to the uniform temperature rise can be obtained from the virtual work consideration.

$$u_i^{\overline{t}} = \int_x n_1 \alpha \overline{T}\, ds \qquad \text{...(2.15)}$$

The evaluation of $u_i^{\overline{t}}$, is similar to the determination of u_i^0, except that n_i^0 is to be substituted by a $\alpha\overline{T}$. If a structure is to be analyzed for uniform rise in temperature only, then

$$FX + U^{\bar{t}} = 0$$

Where $\qquad U^{\bar{t}} = \bar{t} = [(u_1)]^{\bar{t}}, u_2^{\bar{t}}, u_3^{\bar{t}} \ldots\}$

and f_{ij} do not get altered due to the temperature rise.

(*ii*) Effect of temperature gradient defined by $\tilde{T}$: The effect of temperature changes at different fibres due to $\overline{T}$ is to cause an equivalent maximum strain e_x

$$(e_x)_{\max} = \alpha\tilde{T}$$

and corresponding curvature $= \dfrac{(\varepsilon_x)_{\max}}{d/2} = \dfrac{2\,\alpha\tilde{T}}{\alpha}$

From the virtual work consideration the corresponding deformations at releases are obtained from the Eq. (2.16)

$$u_i^{\tilde{t}} = 2\int_s m_i \frac{2\alpha\tilde{T}}{d}\, ds$$

or

$$u_i^{\tilde{t}} = 2\int_s m_i \frac{\alpha\tilde{T}}{d}\, ds \qquad \text{...(2.16)}$$

The redundant forces due to T can be determined from the equation,

$$FX + U^{\bar{t}} = 0$$

f_{ij} do not get altered due to any temperature change.

If the stresses are to be evaluated due to the temperature rise only,

$$FX + U^{\bar{t}} + U^{\bar{t}} = 0$$

and the stress resultant m, in a statically indeterminate structure of degree, n, is given by

$$m = m_1 X_1 + m_2 X_2 + \ldots + m_n X_n$$

It is particularly to be noted that m_0 term does not appear. In a general case n_0 and $(P_x)_0$ also do not appear

Ex. 2.24 Analyze the arch, Fig. 2.24, due to temperature rise only.

$$T_0 = 30_0 \text{ and } T_i = 10^0$$
$$\alpha = 1 \times 10^{-5} \text{ per degree centigrade}$$
$$E_c = 3 \times 10^5 \text{ kg/sq cm}$$

Table 2.5 *Values of F for Evaluating u_1^0, u_2^0, f_{12} and f_{22}*

X	$\frac{ds}{dx}$	m_1	m_2	$\alpha \bar{T} n_1 \frac{ds}{dx}$	$\frac{\alpha \tilde{T}}{d} \frac{ds}{dx}$	$m_1 \frac{\alpha \tilde{T}}{d} \frac{ds}{dx}$	$m_2 \frac{\alpha \tilde{T}}{d} \frac{ds}{dx}$
1	*2*	*3*	*4*	*5*	*6*	*7*	*8*
0	1	0	1.0	-20×10^{-5}	0.8333×10^{-4}	-0.0	$+0.8333 \times 10^{-4}$
3	1.0061	1/6	"	"	0.7697×10^{-4}	-0.1283×10^{-4}	7697×10^{-4}
6	1.0244	2/3	"	"	0.7317×10^{-4}	-0.4878×10^{-4}	0.7317×10^{-4}
9	1.0541	3/2	"	"	0.7027×10^{-4}	-1.0540×10^{-4}	0.7027×10^{-4}
12	1.0949	8/3	"	"	0.6843×10^{-4}	-1.8248×10^{-4}	0.6843×10^{-4}
15	1.1440	25/6	"	"	0.6730×10^{-4}	-2.8042×10^{-4}	0.6730×10^{-4}
18	1.2018	6	"	"	0.6677×10^{-4}	-4.0062×10^{-4}	0.6677×10^{-4}

(*i*) Uniform rise in temperature. The values of the column 5 of Table 2.5 are used to evaluate u_1^0 due to uniform rise in temperature, $\bar{T} = 20°C$.

u_1^0 due to uniform rise in temperature $= 18 \times (-20 \times 10^{-5})$

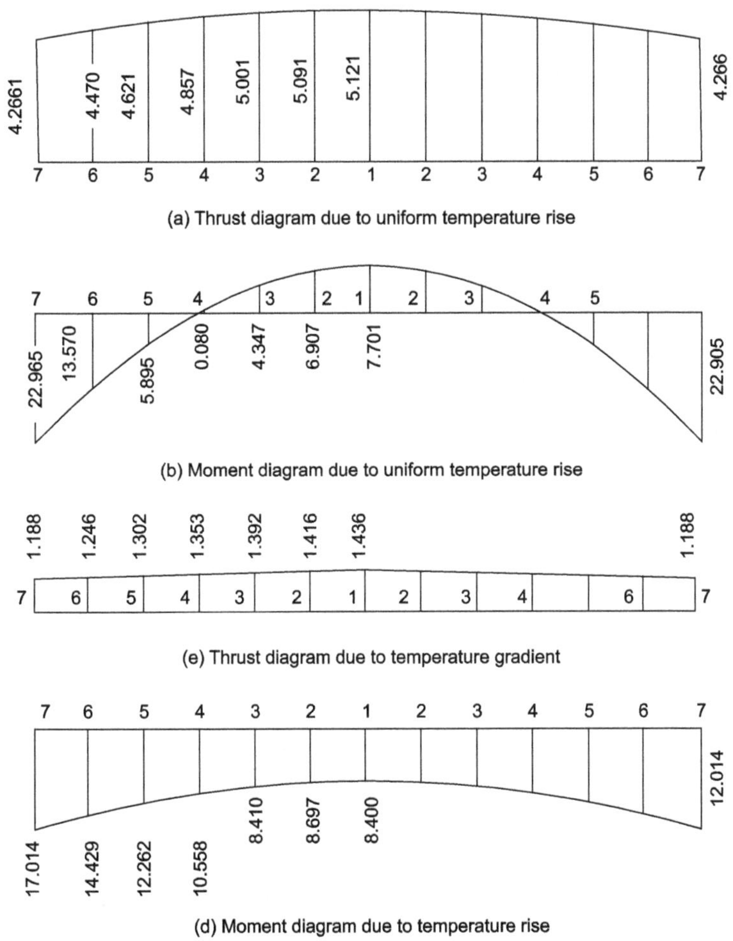

Fig. Ex. 2.24

u_1^0 due to uniform rise in temperature = -3.6×10^{-3}
u_2^0 due to $\overline{T}$ = 20 is equal to zero.

The coefficients, f_{ij} have already been computed. Substituting the value of $E_c = 3 \times 10^5$ kg/sq cm and solving the equations for the unknown X_1 and X_2 due to rise in temperature

$$X_1 = 5.121t,\ X_2 = 7.761\ tm.$$

The net moment and thrust values at the points 1 to 7 due to the combined effect of X_1 and X_2 are computed to draw the moment and thrust diagrams, Fig. Ex. 2.24.

(*ii*) *Temperature gradient.* The values of columns 7 and 8 of Table 2.5 are used to evaluate u_1^0 and u_2^0 respectively due to the temperature gradient $\overline{T}$ = 10°C.

$$u_1^0 = \text{due to temperature gradient} = 2\int m_1 \frac{\alpha \overline{T}}{d}\frac{ds}{dx}$$

$$u_1^0 = -2 \times 10^{-4}\ [0.0 + 4.0062 + 4\ (.1283 + 1.0540 + 2.8042) + 2\ (0.4878 + 1.8248)]$$

$$u_1^0 = -49.1548 \times 10^{-4} \text{ similarly,}$$

$$u_2^0 = +\ 2 \times 10^{-4}\ [0.8333 + 0.6677 + 4\ (.7697 + .7027 + .6730) + 2(.7317 + .6843)]$$

$$u_2^0 = 25.8212 \times 10^{-4}$$

The redundant due to the temperature gradient are,

$$X_1 = 1.436\ t,\ X = -8.46\ \text{tm}$$

The net moment and thrust diagrams for the case of temperature gradient are also drawn separately, Fig. Ex. 2.24.

It may be noted that the values of the redundants due to the temperature effects are comparable to the values due to the given loading, Fig. 2.24.

In the case of design of an arch, the critical design moments and shears and thrusts are obtained by suitable combinations of the results of the analyses of the arch due to dead load, live load, uniform temperature rise and temperature gradient.

2.14 CHOICE OF RELEASE SYSTEM AND FLEXIBILITY ANALYSIS

While analyzing by the flexibility method, the important aspects of analysis are choice of release system and static analysis for drawing m_0 and m_i diagrams. Thereafter it is simple integrations to evaluate the deformations, f_{ij} and u_0 to obtain the compatibility equations, and solution of simultaneous equations for the unknown redundant x_i corresponding to the releases chosen. The integrations and solution of simultaneous equations are simple mathematical processes.

In the majority of problems where releases consist of complete "cuts" m_0 and m_i diagrams are obtained from equilibrium equations of free bodies and in this process, we evaluate shears and thrusts also even though these values are not required for the approximate analysis, where flexural deformations only are considered. When the releases consist of hinges only, it is generally convenient to draw m_0 and m_i diagrams using the principle of virtual work instead of writing the statical equations of equilibrium for different free bodies. These aspects will be discussed in the following.

2.14.1 Choice of Proper Release System

While analyzing by the force method, we choose a number of releases equal to the degree of determinacy, n, and the basic steps of the force method results in a set of simultaneous equations for the redundants, $x_1, x_2, \ldots, x_n$ Eq. (a)

$$\begin{pmatrix} f_{11} & f_{12} & \cdots & \cdots & f_{1n} \\ f_{21} & f_{22} & \cdots & \cdots & f_{2n} \\ - & - & \cdots & \cdots & - \\ - & - & \cdots & \cdots & - \\ f_{n1} & f_{n2} & \cdots & \cdots & f_{nn} \end{pmatrix} \begin{pmatrix} x_1 \\ x_2 \\ - \\ - \\ x_n \end{pmatrix} + \begin{pmatrix} u_1^0 \\ u_2^0 \\ - \\ - \\ u_n^0 \end{pmatrix} = \begin{pmatrix} 0 \\ 0 \\ - \\ - \\ 0 \end{pmatrix} \qquad \ldots(a)$$

Or in the matrix form

$$F X + U^0 = 0 \qquad \ldots(a)$$

After having solved for the redundant, X we find moment m_i at any point i from the following Eq. (b)

$$m_1 = m_0 + m_1 x_1 + m_2 x_2 + \ldots + m_n x_n \qquad \ldots(b)$$

In the above analysis, for each set of releases choose we get corresponding flexibility coefficients f_{ij} and u_1^0, and the solution which gives the redundants. Even though moment and m_0, m_1, m_2 ..., m_n and x_1, x_2 ..., x_n appearing in Eq. (b) are different for different release systems, but the final moments m_i at any point i, Eq. (b), are independent of the release system chosen. However, we have to observe the following while choosing a proper release system such that the matrix is not ill conditioned and the numerical computations are minimum, i.e. the matrix F should have

(*i*) diagonal elements which are dominant, and

(*ii*) minimum of the non-zero elements, close to the diagonal elements.

Such a flexibility matrix is not sensitive to the round off errors. In the case of single ring structures, i.e., frames, portals, arches and boxes, etc., if we choose releases' elastic centre, it results in a diagonal flexibility matrix F of size 3×3. If the number of redundants is not large, choice of release system is not important. (In the case of arch and portal frames, we did not care to introduce releases at elastic centre.) Proper choice of redundants is an important consideration if the degree of indeterminacy is large. In the following, the necessity of proper choice of releases and corresponding redundants is explained with the help of two typical structures,

1. Continuous beam, and
2. Multistorey frame.

1. Continuous Beams: Consider a continuous beam, Fig. 2.27. The first set of redundants is the interior reactions and the corresponding releases consist of removal of the interior supports. The deflected shape and bending moment diagrams for the structure with the releases are drawn for given loads, Fig. 2.27(*b*) and due to a redundant X Fig. 2.27(*b*). It is obvious from the deflected shape and bending moment diagrams that all the elements of the elements of F and U matrices are non-zeros and integrations are to be carried out for the entire length. And from the Fig. 2.27(*b*) the f_{22} value need not be the largest. The following matrix equation for the release system b is fully populated (non-zero elements are shown by the symbol x). The second row elements and due to symmetry the elements of the second Columns are the deformations due to $X_2 = 1$.

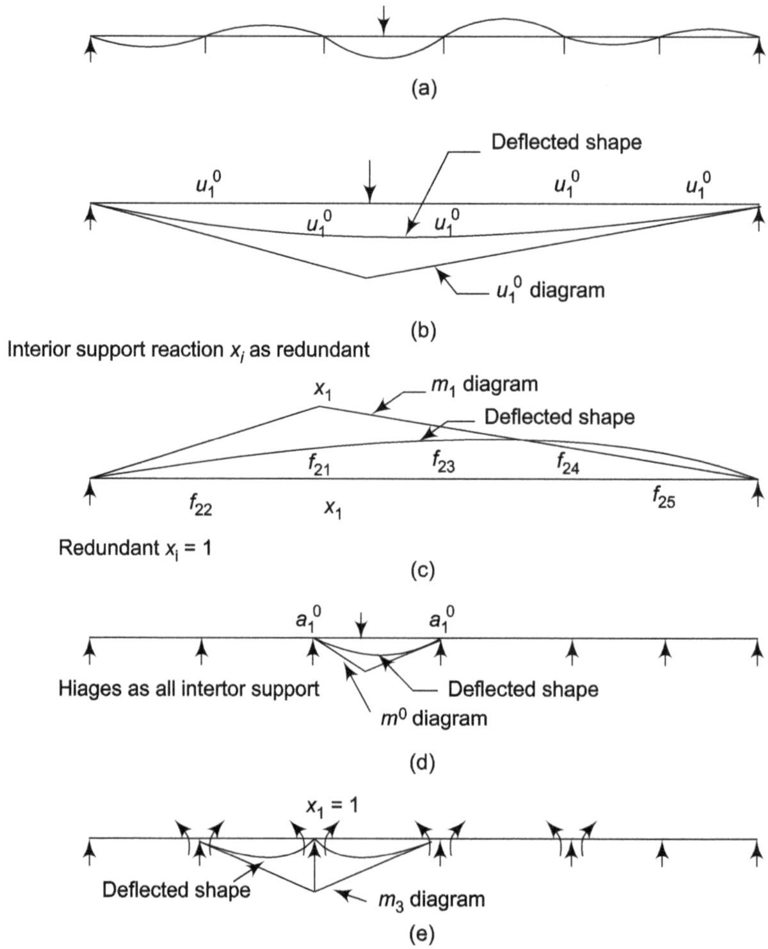

Fig. 2.27 *Continuous Band Two Typical Release Systems*

$$\begin{pmatrix} x & x & x & x & x \\ x & x & x & x & x \\ x & x & x & x & x \\ x & x & x & x & x \\ x & x & x & x & x \end{pmatrix}\begin{pmatrix} x_1 \\ x_2 \\ x_3 \\ x_4 \\ x_5 \end{pmatrix} + \begin{pmatrix} x \\ x \\ x \\ x \\ x \end{pmatrix} = \begin{pmatrix} 0 \\ 0 \\ 0 \\ 0 \\ 0 \end{pmatrix} \quad \ldots\text{(i)}$$

From the Fig. 2.27(c), drawn for the release system consisting of hinges at interior supports. It is obvious that integrations are to be carried out over limited number of spans. The diagonal elements of the F matrix are the largest, and the matrix F is a tridiagonal matrix. The non-zero elements for the release system, Fig. 3.26(c), are marked x in the following equation

$$\begin{pmatrix} x & x & 0 & 0 & 0 \\ x & x & x & 0 & 0 \\ 0 & x & x & x & 0 \\ 0 & 0 & x & x & x \\ 0 & 0 & 0 & x & x \end{pmatrix} \begin{pmatrix} x_1 \\ x_2 \\ x_3 \\ x_4 \\ x_5 \end{pmatrix} + \begin{pmatrix} 0 \\ x \\ x \\ 0 \\ 0 \end{pmatrix} = \begin{pmatrix} 0 \\ 0 \\ 0 \\ 0 \\ 0 \end{pmatrix} \quad \text{...(ii)}$$

From the Eqs (i) and (ii) it is obvious that the release system, Fig. (*c*) is superior to the release system (b). The following may be used as guidelines for choosing release systems.

From deformation considerations: Under given loads, the deflected shape with releases should be as close as possible to the deflected shape without releases. Compare the Figs (*b*) and (*c*) with (*a*). In other words, in the Eq. (*b*), the maximum contribution to m_i is mainly due to m_0.

From considerations of moment diagrams: The moment diagrams m_1, m_2, ... should be limited to restricted lengths, i.e., limited number of spans. Limiting the spreading of m^0 diagrams reduces the computation work for obtaining u^0 values.

The latter consideration will be used for choosing releases for multistory frames, discussed in the following.

2. Multistorey Frames: Consider the frame, Fig. 2.28(A). The degree of indeterminacy is 45. Out of several possible release systems, only four typical release systems are considered for comparison, Fig. 2.28(B). The release system (a) consists of (when generalized for any number of storeys and bays):

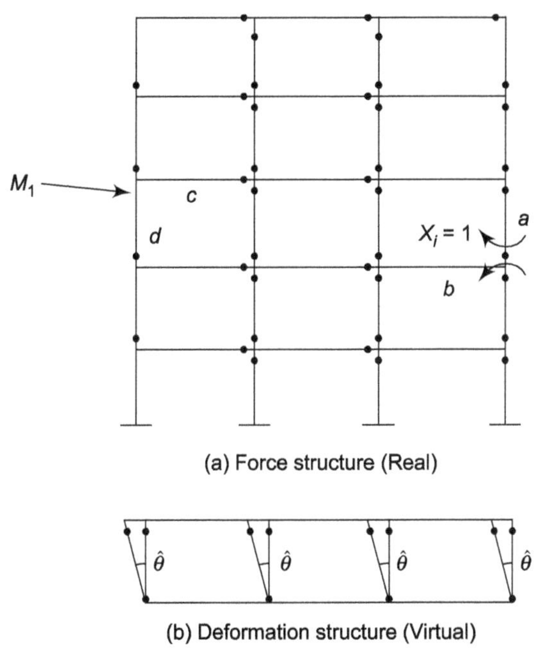

Fig. 2.28(A)

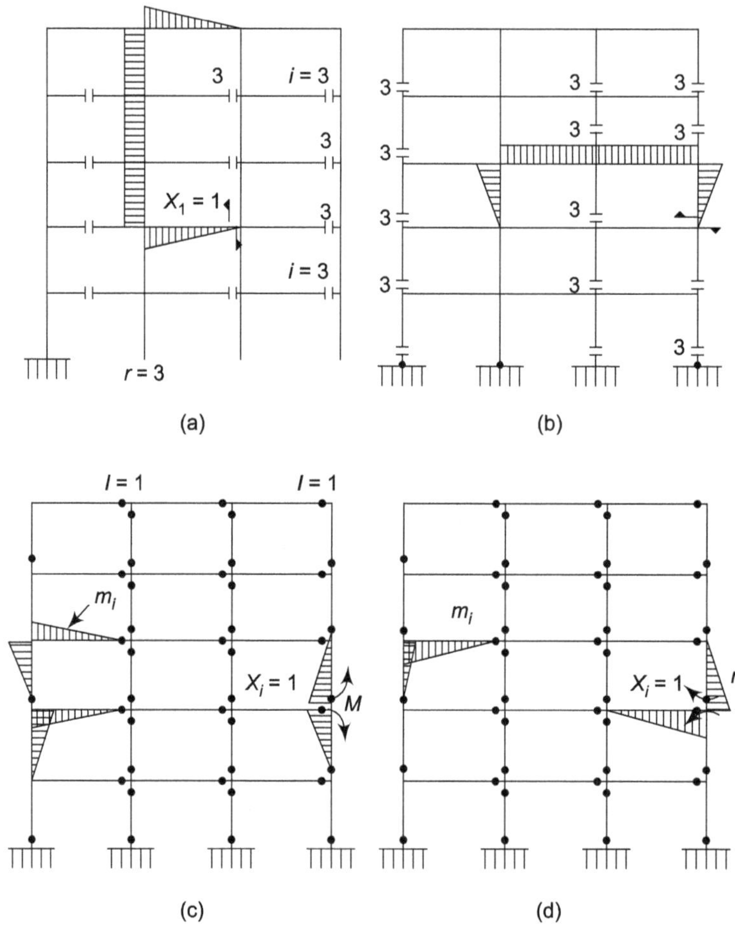

(a) Multistorey frame

(b) Release systems and m_i diagrams

Fig. 2.28(B) *Release System for Multistorey Frames*

1. Complete cuts for all girder except the top.
2. Complete cuts for all the bottom columns except one column line.

"The moment diagram, m_i due to $X_i = 1$, extends to all the storeys above the girder."

The release system (b) consists of cuts at the bottoms of all columns except one column line. In the example considered column line 2 is without cuts. For the chosen X_i of the release system (b),

"The moment diagram, m_i is limited to the storey where the redundant X_i is considered act."

Obviously from the moment diagram considerations, system (b) is to be preferred over system (a).

Consider the release system (c). In this system the m diagram extends to one storey below. The system is particularly suitable for elastic-plastic analysis, as the positions of the hinge release are the possible points of maximum moments where plastic hinges are likely to develop. The elastic-plastic considerations are explained with the help of an example, Article 2.6.4.

The release system (d) is an improvement to the extent of limiting the m_1 diagram to the particular storey, where the redundant X_i is acting.

The above considerations should help to arrive at a good release system, even though it may not be the best.

Ex. 2.25 illustration of application of virtual work principle. Draw moment diagram for $X_i = 1$ for the frame, Fig. 2.28A(a). The Fig. 2.28B(d) gives moment diagram for the frame Fig. 2.28A(a). The frame of Fig. 2.28A(a) is same as Fig. 2.28B(d). Due to the unit "bimoment" redundant $X_i = 1$, the frame Fig. 2.28A(a), the members *a*, *b*, *c*, and *d* develop bending moments, m_i. The mi diagram can be drawm if the moment m_r at the top end of member '*d*' is evaluated. Assume the member develops tension at the left. For applying the principle of virtual work, the 'real' frame Fig. 2.28B(d) is treated as 'force' structure. Assume a virtual structure such that it has an angle discontinuity at the point. The sway mechanism of Fig. 2.28A(b) achieves the objective and therefore chosen as a deformation (virtual) structure. As there is no external loads except the bimoment X_i, the virtual work equation is written in the following:

$X_i\hat{\theta} + m_r\hat{\theta} = 0$, $\hat{\theta}$ being arbitrary we obtain: $m_r = -X_i = -1$,

The moment causes tension on the right side of the member (not on the left as assumed).

The moment diagram can be drawn as shown in Fig. 2.28 B(d)

In fact the m_0 and m_i diagrams can be obtained either by writing equilibrium equations for different free bodies or by the virtual work method. The above example shows the superiority of the virtual work method of satisfying equilibrium and draws the "*m*" diagrams, for the frames with hinges. With the m_0 and m_i values known, the flexibilities f_{ij} and u_i^0 are evaluated to get simultaneous equations for the unknown redundants (in this case the number of the unknown redundants is equal to 45). While analyzing examples of this kind, if elastic analysis is desired, stiffness or displacement method is more suitable. As started earlier for elastic-plastic analysis the hinge releases introduced at the points of maximum moments render a viable flexibility method which is generally to be preferred over the displacement method.

2.14.2 Generalized Analysis: Single, Double and Triple Release System

Before the generalized analysis is developed, consider a simple portal, Fig. 2.29. Introduce three releases by a complete cut to the left of the load point *C*. The m_0 and m_i diagrams for such a release system can be easily drawn, Fig. 2.29(a), (b), (c) and (d). The compatibility equations for such a release system are:

$$42.66\,X_1 + 48.0\,X_2 + 16\,X_3 - 128 = 0$$
$$48.0\,X_1 + 445.3\,X_2 + 54\,X_3 - 854 = 0$$
$$46\,X_1 + 54X_2 + 18\,X_3 - 128 = 0$$

The solution of the above equation gives

$$X_1 = 0.502,\ X_2 = 1.657 \text{ and } X_3 = 1.692$$

and
$$m = m_0 + m_1x_1 + m_2x_2 + m_3x_3$$

The redundants are the thrust, shear and moment at the section where the releases are introduced.

An alternative release system is considered in Fig. 2.30, which consists of releases introduced at section *C*, *E* and *B*, which are the three out of the total five points of the maximum moments. The other two points of the maximum moments are *A* and *D*. The m_0 and m_i diagrams for this system are also drawn, Fig. 2.30(*a*), (*b*), (*c*) and (*d*). The angle discontinuities at the releases introduced give u_1^0 and f_{ij}. The compatibility equations for this release system can be formed to solve for the redundants which are the moments at *C*, *E* and *F*. The final *m* values will be the same as obtained for the release system, Fig. 2.29.

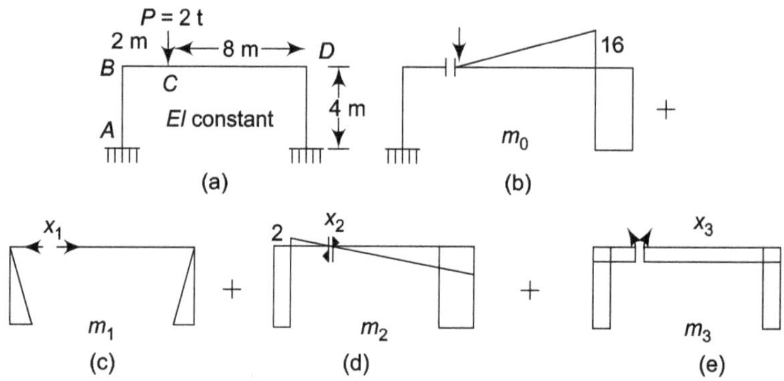

Fig. 2.29

For drawing m_0 and m_i diagrams, it is more expedient to use virtual work principle. As an illustration, for drawing m diagram, Fig. 2.30(a), we need to know the moment M_D and M_A. For evaluating M_D consider the deformation structure (virtual), Fig 2.30(a). The virtual work equation is

$$P\,(8\theta) = M_D(\theta)$$

$$M_D = 16 \text{ tm}$$

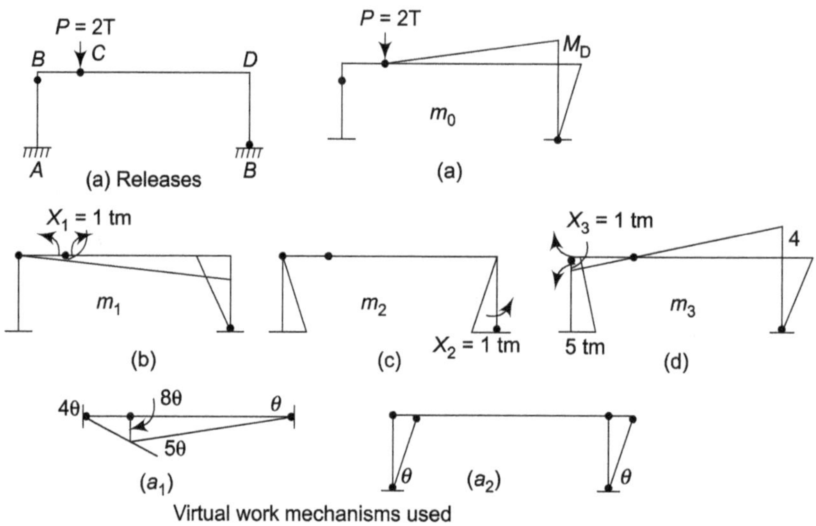

Fig. 2.30 *A Single Release System*

Having evaluated M_D for the evaluation of M_A considers the deformation structure (virtual), Fig. 2.30 (a_2). The virtual work equation is

$$M_A\,\theta - M_D\,\theta = 0$$

$$M_A = M_D = 16 \text{ tm}$$

Similarly, m_1, m_2 and m_3 diagrams are drawn. It is left to the reader to obtain u_i^0 and f_{ij} and thus evaluate the redundant.

$$X_1 = 1.6695 \text{ tm}, X_2 = 0.952 \text{ tm, and } X_3 = -1.618 \text{ tm}$$

The final moments $m = m_0 + m_1 x_1 + m_2 x_2 + m_3 x_3$ will be the same as obtained for the release system, Fig. 2.29. The above two release systems are called "Single Release Systems", because the release systems for m_0 and m_i diagrams are the same, and to obtain the deformations f_{ij} and $u_1^{\,0}$ of the real structure, the statical force diagrams (virtual) correspond to the same release system as used for m_i diagrams.

It is possible to use two different release systems for m_0 and m_i diagrams. And if the release system for statical force diagram (virtual) is the same as used for m_i diagram but different from the release system used for m_0 diagram, such a system is called "Double Release System".

The further generalization of the above is the "Triple Release System" where the releases for the statical force diagrams (virtual) are different from the releases chosen for m_i diagrams and the two release systems are separately different from m_0 diagram. As the Double Release System is a particular case of the triple release system, the theoretical formulation for the latter only is given in the following.

2.14.3 Theoretical Formulation for Triple Release System

The formulation in the following is developed through an example. Suppose we consider the releases consisting of hinges at *B* and *D* for m_0 diagram, and also that we consider separate releases for m_i diagram consisting of hinges at *C*, *E* and *B*. The two release systems and the corresponding m_0 and m_i diagrams are shown in Fig. 2.31.

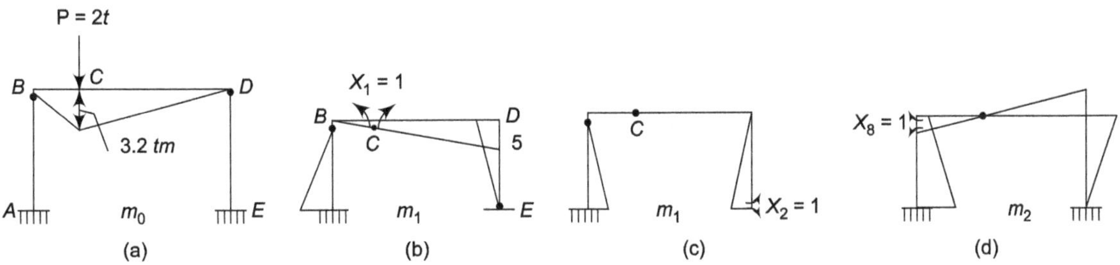

Fig. 2.31 *Deformation System (Real Structure)*

The m_i diagrams are linearly independent, *i.e.*, one cannot be obtained from any linear combinations of the other m_i diagrams. It is possible to express m equal to the sum of (*i*) particular solution m_0 due to given loading and for the assumed releases, which means individually equilibrium is satisfied but, due to the hinge releases at *B* and *D*, compatibility is not satisfied, and (*ii*) a linear combination of the independent homogeneous solutions i.e., m_i diagrams, which also satisfy equilibrium but individually violate compatibility. Let the following be the solution for m, Eq. (*i*).

$$m = m_0 + X_1 m_1 + X_2 m_2 + X_3 m_3 \qquad \ldots\text{(i)}$$

We treat the unknown X_1, X_2 and X_3 as unknown coefficients (their number equal to the degree of indeterminacy) to be chosen such that compatibility is satisfied for the final m diagram.

The compatibility condition is proposed to be satisfied from virtual work considerations. Consider any independent set of statical $\bar{m}_i$ diagrams (virtual force system), i.e., the $\bar{m}_i$, diagrams satisfy equilibrium independently and their number is also equal to the degree of indeterminacy. If the $\bar{m}_i$, diagrams are the same as m_i diagrams then we have Double release System. If we wish to consider a general case of Triple Release System, let the statical diagrams, Fig. 2.29 (*b*), (*c*) and (*d*), be treated as

$\bar{m}_i$, diagrams. Now we allow the virtual force system ($\bar{m}_i$) to ride on the real deformations due to m (equal to $m_0 + \Sigma m_i X_i$), and if the real deformations (m/EI), if by (flexural deformations only considered) are compatible then the complementary virtual work principle gives the following equations.

$$\int \bar{m}_1 \frac{(m_0 + \Sigma m_i X_i)}{EI} ds = 0 \qquad \ldots\text{(iia)}$$

$$\int \bar{m}_2 \frac{(m_0 + \Sigma m_i X_i)}{EI} ds = 0 \qquad \ldots\text{(iib)}$$

$$\int \bar{m}_3 \frac{(m_0 + \Sigma m_i X_i)}{EI} ds = 0 \qquad \ldots\text{(iic)}$$

The Eqs (ii) involve simple integrations and result into simultaneous equations for the evaluation of the unknown coefficients X_i. The solution for X_i together with the equation (i) complete the analysis.

Alternatively, we may minimize the total complementary potential with respect to the arbitrary coefficients X_i to obtain simultaneous Eqs (iii) for the evaluation of the unknown coefficients X_i, which when substituted in the Eq. (i), result in m diagram which satisfied compatibility also (equilibrium has already been satisfied individually).

$$\left.\begin{aligned} \frac{\partial}{\partial X_1}\int \frac{(m_0 + \Sigma m_i X_i)^2}{2EI} ds = 0 \\ \frac{\partial}{\partial X_2}\int \frac{(m_0 + \Sigma m_i X_i)^2}{2EI} ds = 0 \\ \frac{\partial}{\partial X_3}\int \frac{(m_0 + \Sigma m_i X_i)^2}{2EI} ds = 0 \end{aligned}\right\} \qquad \ldots\text{(iii)}$$

Ex. 2.26. Analyze the frame, Fig. Ex. 2.26, by triple release system, that is,

(i) Hinge releases at *B* and *D* for m_0 diagram, Fig. 2.31(*a*),

(ii) Hinge releases at *C*, *E* and *B* for m_i diagrams, Fig. 2.31(*b*), (*c*), (*d*) and

(iii) Releases caused by a complete cut to the left of *C*, for statical system, i.e., force structure (virtual), for $\bar{m}_i$ diagrams, Fig. 2.29 (*b*), (*c*), (*d*)

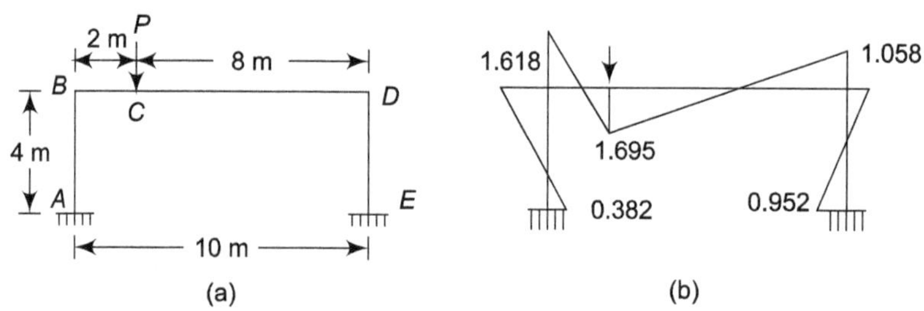

Fig. Ex. 2.26

From the Eq. (iia),

$$\int \frac{\bar{m}_1 (m_0 + \Sigma m_i X_i)}{EI} ds = 0$$

$$\int \bar{m}_1 \frac{m_0}{EI} ds = u_1^0 = 0$$

$$\int \bar{m}_1 \frac{m_1}{EI} ds = f_{11} = -\frac{40}{3}$$

$$\int \bar{m}_1 \frac{m_2}{EI} ds = f_{12} = \frac{32}{3}$$

$$\int \bar{m}_1 \frac{m_3}{EI} ds = f_{13} = \frac{56}{3}$$

and

and thus we get the first equation

$$-\left(\frac{40}{3}\right)X_1 + \left(\frac{32}{3}\right)X_2 + \left(\frac{56}{3}\right)X_3 + 0 = 0 \qquad \text{...(a)}$$

Similarly from the Eq. (iib),

$$\int m_2 \frac{(m_0 + \Sigma m_i x_i)}{EI} ds = 0$$

we get $(650/3)\, X_1 + 12\, X_2 - (524/3)\, X_3 + 32 = 0$ ...(b)

and finally from the Eq. (iic),

$$25\, X_1 + 4\, X_2 - 11\, X_3 + 16 = 0 \qquad \text{...(c)}$$

When written in the matrix form,

$$\begin{bmatrix} -40/3 & 32/3 & 56/3 \\ 650/3 & 12 & -524/3 \\ 6 & 4 & -11 \end{bmatrix} \begin{pmatrix} X_1 \\ X_2 \\ X_3 \end{pmatrix} + \begin{pmatrix} 0 \\ 32 \\ 16 \end{pmatrix} = \begin{bmatrix} 0 \\ 0 \\ 0 \end{bmatrix} \qquad \text{...(iv)}$$

It may be noticed particularly that the coefficient matrix in the above matrix equation is not symmetrical. While solving by using single release system, Article 2.62, the coefficient matrix is always symmetrical.

The solution of the simultaneous Eqs. (a), (b) and (c) gives

$$X_1 = 1.505,\ X_2 = 0.952 \text{ and } X_3 = -1.618$$

and the final moment diagram, Fig. Ex. 2.26 (b), is obtained by linearly combining the m_i diagrams of Fig. 2.31, i.e.,

$$m = m_0 + m_1 X_1 + m_2 X_2 + m_3 X_3$$

2.14.4 Generalization of Triple Release System for Frames

While developing the theoretical formulation or triple release system, we express the final moment, m, equal to (i) a particular solution m and (ii) linear combination of independent homogeneous solutions, m_i, their number being equal to the degree of indeterminacy, n. In a general case we have

$$m = m_0 + m_1 x_1 + m_2 x_2 + \ldots\ m_i x_i \ldots + m_n x_n \qquad \text{...(i)}$$

We are required to choose $x_1, x_2 \ldots, x_n$ such that the final moment satisfies compatibility also. Consider an indeterminate frame, Fig. 2.32(a). The degree of indeterminacy is equal to 29. A simple particular solution m_0 may be obtained by taking complete cuts in the girder members. For the m_i diagrams, it is not necessary that we take the same release system only. In fact for the m_i diagrams we may consider any other release system. What all that is required is to choose an independent set of self-equilibrating m_i diagrams. As each ring introduces three degrees of indeterminacy [the effect of hinges like the one, Q, Fig. 2.32(a), will be considered separately], we may draw three self-equilibrating m_i diagrams per ring. For example, the three typical m_i diagrams m_{10}, m_{11}, m_{12} for the ring 4, are shown in Fig. 3.31(d). The frame under consideration consists of ten rings, Fig. 2.32(c). Thirty m_i diagrams can be drawn, that is, starting from the open ring 1, which gives m_1, m_2, m_3 to the open ring 10, which gives m_{28}, m_{29} and m_{30}. The functional values m to m are one such set which may be used in the Eq. (i).

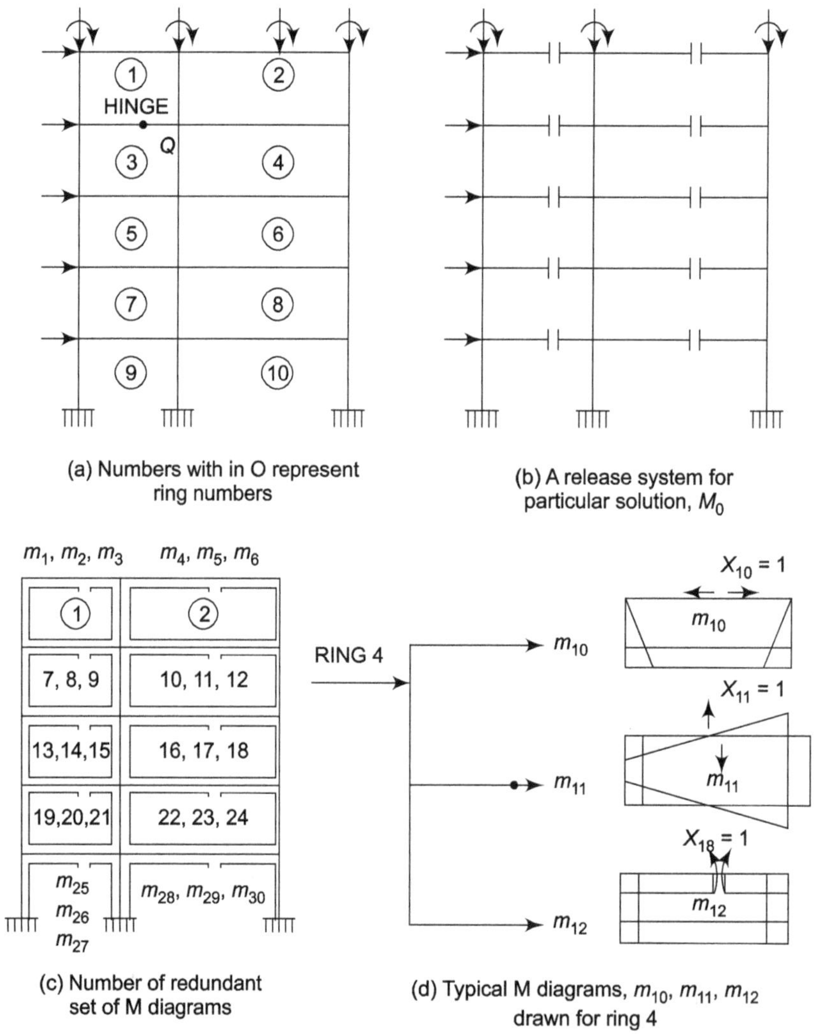

(a) Numbers with in O represent ring numbers

(b) A release system for particular solution, M_0

(c) Number of redundant set of M diagrams

(d) Typical M diagrams, m_{10}, m_{11}, m_{12} drawn for ring 4

Fig. 2.32 *Choice of m_0, m_i and Statical Diagrams*

Due to the hinge, Q, the degree of indeterminacy is reduced from 30 to 29 and thus one of the x_i of Eq. (*i*) is in excess and that requires to be dropped. To achieve this, we use an additional conditional that the moment at $Q = 0$ which gives the following equation, see also Fig. 2.32(*c*).

$$(m_0 + m_1x_1 + m_2x_2 + m_3x_3 + m_7x_7 + m_8x_8 + m_8x_9)\,Q = 0 \qquad \ldots\text{(ii)}$$

It is to be noted that only m_1, m_2, m_3, m_7, m_8, m_9 diagrams cause moments at Q. If the cut of the ring 3 is taken corresponding to the hinge position then only m_1, m_2, m_3 and m_9 diagrams cause moment at Q. We may express say x in terms of other x appearing in Eq. (ii) and when substituted for x in Eq. (i) the unknowns x appearing in the equation (i) are reduced from thirty to twenty nine.

Compatibility. We may satisfy compatibility by using the principle of complementary virtual work or simply from the principle of minimum complementary potential. Based on the principle of the minimum complementary potential

$$\frac{\partial}{\partial X_i}\int \frac{m^2}{2EI}\,dx = 0 \qquad \ldots\text{(iii)}$$

We get as many equations as the number of the unknown co-efficient X_i. The procedure is the same as the one based on the complementary virtual work, where the statical system (virtual force system) is the same as the m_i diagrams (or functions) used in Eq. (i).

If the rings are numbered storey-wise as shown in Fig. 2.32(*a*), we get a well conditioned set of Eq. (iii) for the unknown X_i. The coefficient matrix consists of sub matrices which are positioned in a tridiagonal form as indicated below.

$$FX + U_0 = 0$$

Coefficient matrix,
$$F = \begin{pmatrix} x & x & 0 & 0 & 0 \\ x & x & x & 0 & 0 \\ 0 & x & x & x & 0 \\ 0 & 0 & x & x & x \\ 0 & 0 & 0 & x & x \end{pmatrix}$$

The symbol x represents a non-zero sub matrix of size 6×6 if there are no internal releases in the frame, Fig. 2.32.

Alternatively, from other considerations like elastic-plastic analysis, the homogeneous solution, m_i may be obtained based on appropriate release system, and Statical diagrams $(\overline{m}_1)$ for writing complementary virtual work equations may be any other independent set of self-equilibrating virtual forces. The $(\overline{m}_1)$ diagrams as obtained from the rings, Fig. 2.32(*d*), may be recommended to be used, being the simplest, in conjunction with the deformation system (m/EI of the real structure). The m values of the real structure defined by Eq. (i), may be obtained from m_i corresponding to Baker's or Edward's release system and particular solution m_0 for a given loading from any other release system like Fig. 2.32(c). If the m_i diagrams are based on the release system, with this suggested procedure we once again get a tridiagonal flexibility matrix. It is illustrated in the following.

Ex. 2.27. Analyze the frame, Fig. Ex. 2.27(*a*). Consider m_0 and m_i diagrams based on the release, Fig. Ex. 2.27(*b*). For the statical or virtual system, $(\overline{m}_1)$ diagrams, consider self-equilibrating statical system corresponding to the release of Fig. (*c*). It may be noted that the statical diagrams from Fig. Ex. 2.27(*c*) are similar to the ring system, Fig. 2.32 (*c*), with the only change that the cuts of the rings are introduced at the right end column bases instead of the cuts in girders.

Solution. Let $EI = 4$ for columns and $EI = 10$ for girders.

Given $P = 2t$

The m_0 and m_i diagrams for the release system, Fig. Ex. 2.27(b), are drawn, Figs (*d*) and (*e*). These diagrams constitute the kinematic or deformation system. As stated elsewhere, for hinge release systems and where it is required to find moments only, it is more convenient to use virtual work equations than writing equations of equilibrium for different free bodies. In the Example 2.27, the procedure was applied to an example, when a unit moment was applied at a hinge. In the Fig. Ex. 2.27(*d*), moment M_r at the top of the column *AC* is evaluated based on virtual work equation. The moment M_r is assumed to cause tension to the right o the column *AC*, and the moment has been shown to be *r* equal to 24 tm. Similarly other moments are evaluated to complete m_0 diagram.

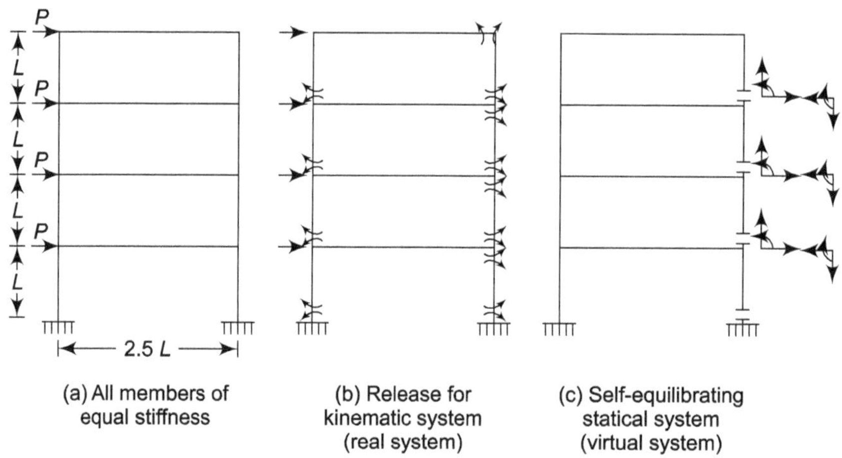

Fig. Ex. 2.27 *(a), (b), (c)*

The Statical $\bar{m}_i$ diagram (virtual force system for satisfying compatibility) for the release system, Fig. Ex. 2.27(c), are much simpler to draw, Fig. Ex. 2.27(f).

The u_i^0 are computed from m_0 and $\bar{m}_i$ diagram, Figs (d) and (f)

$$U_i^0 = \int \bar{m}_1 \frac{m_0}{EI} ds$$

$$U_i^0 = \left\{\frac{80}{3}, 0, \frac{40}{3}, \frac{160}{3}, 4, \frac{160}{3}, 80, 8, \frac{280}{3}, \frac{320}{3}, 32, \frac{800}{3}\right\}$$

The flexibility coefficients f_{ij} are computed from the m_i *and* $\bar{m}_1$ diagram, Figs (e) and (f).

$$f_{ij} = f_{ji} = \int \bar{m}_1 \frac{m_j}{EI} ds$$

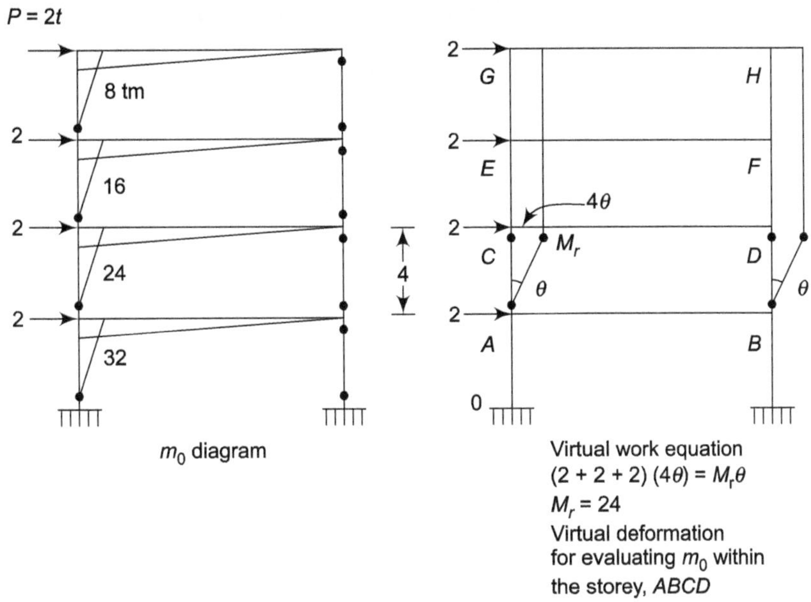

Fig. Ex. 2.27 *(d) m_0 Diagram*

The flexibility matrix *F*, so obtained is given in the following, Eq. (1). It may be observed again that kinematical and statical system being different. The *F* matrix is not symmetrical. The diagonal elements are not dominant. The solution of the equation $FX + U_i^0$, gives

$$X = \{1.43, -2.57, 1.43, 3.32, -4.82, 3.32, 5.89, -6.14, 5.89, 10.80, -5.13, 10.80\}$$

From the above result, the anti-symmetrical nature of the moments may be noticed.

$$F = \begin{bmatrix} \begin{pmatrix} -4 & 20/2 & -8/3 \\ -2 & +2 & 0 \\ -50/3 & 10 & -20/3 \end{pmatrix} & \begin{pmatrix} 0 & 0 & 0 \\ 1/2 & -1 & 1/2 \\ 10/3 & -5 & 10/3 \end{pmatrix} & (C)_{3\times 3} & (0)_{3\times 3} \\ \begin{pmatrix} 2 & -1/2 & -10/3 \\ 1/2 & 0 & -1/2 \\ 10/3 & 0 & -5/3 \end{pmatrix} & \begin{pmatrix} -4 & 20/2 & -8/3 \\ -2 & 2 & 0 \\ -50/3 & 10 & -20/3 \end{pmatrix} & \begin{pmatrix} 0 & 0 & 0 \\ 1/2 & -1 & 1/2 \\ 10/3 & -5 & 10/3 \end{pmatrix} & (0)_{3\times 3} \\ (0)_{3\times 3} & \begin{pmatrix} 2 & -1/2 & -10/3 \\ 1/2 & 0 & -1/2 \\ 10/3 & 0 & -5/3 \end{pmatrix} & \begin{pmatrix} -4 & 20/3 & -8/3 \\ -2 & 2 & 0 \\ -50/3 & 10 & -20/3 \end{pmatrix} & \begin{pmatrix} 0 & 0 & 0 \\ 1/2 & -1/2 & 1/2 \\ 10/3 & -5 & 10/3 \end{pmatrix} \\ (0)_{3\times 3} & (0)_{3\times 3} & \begin{pmatrix} 2 & -1/2 & -10/3 \\ 1/2 & 0 & -1/2 \\ 10/3 & 0 & -5/3 \end{pmatrix} & \begin{pmatrix} -4 & 20/3 & -8/3 \\ -3/2 & 2 & -1/2 \\ -40/3 & 10 & -25/3 \end{pmatrix} \end{bmatrix} \quad \ldots(1)$$

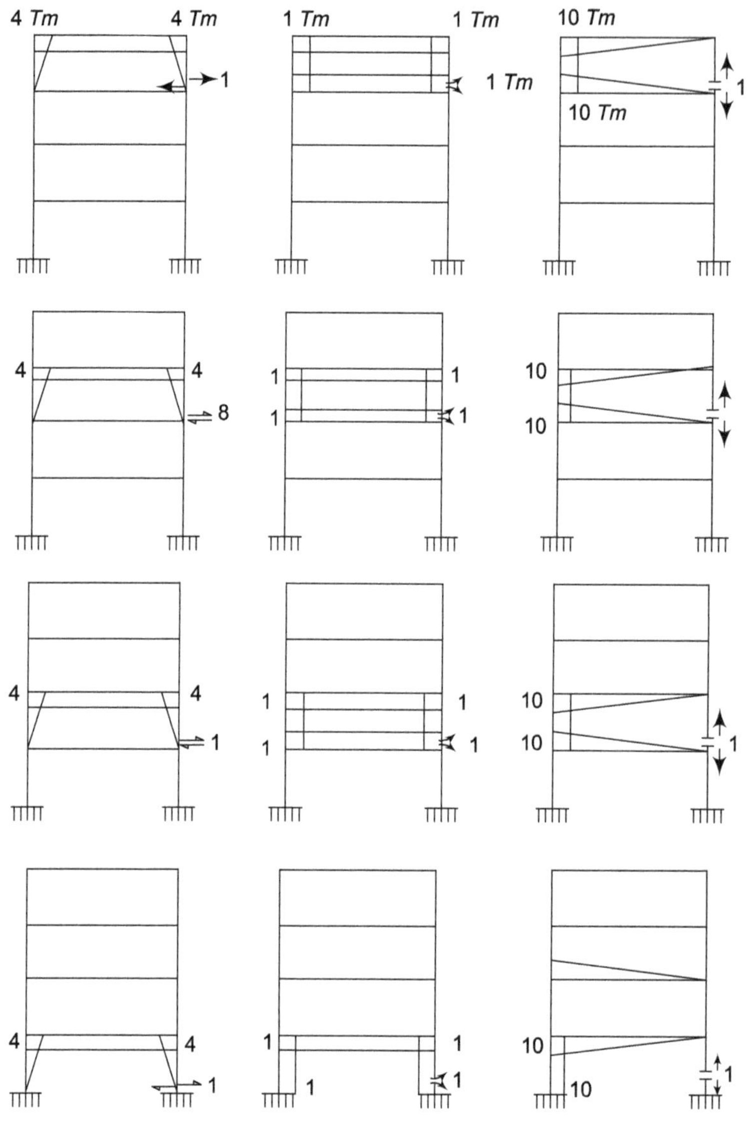

Fig. Ex. 2.27 *(e) m_i Diagrams for the Deformation System*

Alternative solution of Ex. 2.27

It was observed that the F matrix for the different release systems used for m_i and $\bar{m}_i$ diagrams, even though may result in simple computations of U_i^0, and $f_{ij,}$ the flexibility matrix is not symmetrical and the diagonal elements may not be dominant. These shortcomings are not to be found if we use single release system, say, Fig. Ex. 2.27(b) for m_0, and m_i diagrams as well. The resulting U_i^0 and the flexibility matrix are given in the following equation, Eq. (2). The equation may be solved taking advantage of anti-symmetry, i.e.,

$$X_1 = X_3, X_4 = X_6, X_7 = X_9, X_{10} = X_{12}$$

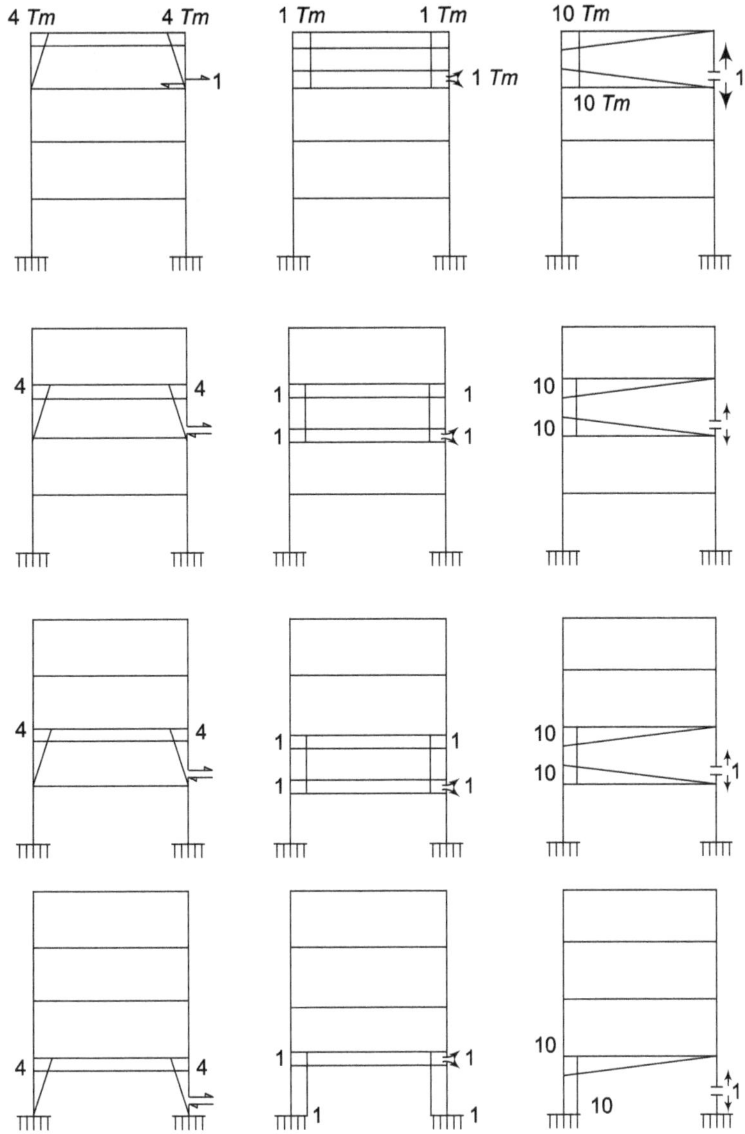

Fig. Ex. 2.27 *(f) Statical, $\bar{m}_i$ Diagrams, Force System*

The numbers of unknowns to be computed are thus reduced from 12 to 7 only. The values of the unknowns X_i as given earlier were computed for the symmetric F matrix, Eq. (2). The present flexibility matrix has dominant diagonal elements. The unknowns were evaluated by iteration (two cycles) with side rule accuracy only. The first trial values were obtained from an approximate method of analysis (see Chapter V), instead of assigning zero initial values. The readers may use other methods of solution as given in Chapter I, for the solution of unknowns. The computer results are:

$$X = \{1.424, -2.577, 1.424, 3.386, -4.613, 3.386, 5.667, -6.333, 5.667, 9.953, -6.048, 9.953\}$$

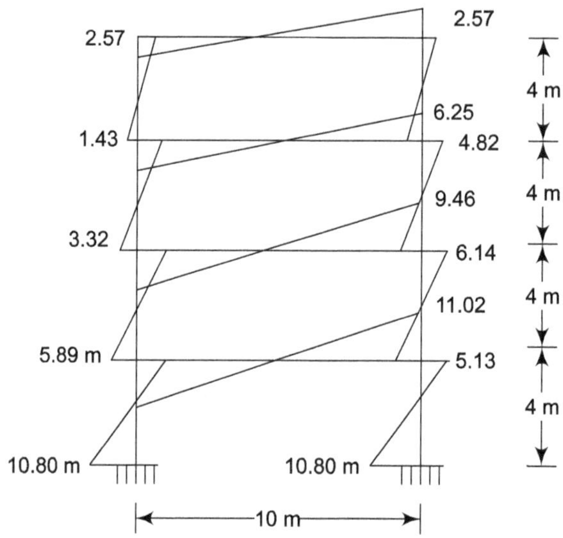

Fig. Ex. 2.27 *(g) Final M-diagram*

$$\begin{bmatrix} \begin{pmatrix} 5/3 & -1 & 2/3 \\ -1.0 & 5/3 & -2/3 \\ -2/3 & -2/3 & 4/3 \end{pmatrix} & \begin{pmatrix} -1/3 & 1/2 & -1/3 \\ 0 & 0 & 0 \\ 1/6 & -1/2 & 1/6 \end{pmatrix} & (0)_{3\times 3} & (0)_{3\times 3} \\ \begin{pmatrix} -1/3 & 0 & 1/6 \\ 1/2 & 0 & -1/2 \\ -1/3 & 0 & 1/6 \end{pmatrix} & \begin{pmatrix} 10/6 & -1 & 2/3 \\ -1 & 5/3 & -2/3 \\ -2/3 & -2/3 & 4/3 \end{pmatrix} & \begin{pmatrix} -1/3 & 1/2 & -1/3 \\ 0 & 0 & 0 \\ 1/6 & -1/2 & 1/3 \end{pmatrix} & (0)_{3\times 3} \\ (0)_{3\times 3} & \begin{pmatrix} -1/3 & 0 & 1/6 \\ 1/2 & 0 & -1/2 \\ -1/3 & 0 & 1/6 \end{pmatrix} & \begin{pmatrix} 5/3 & -1 & 2/3 \\ -11 & 5/3 & -2/3 \\ 2/3 & -2/3 & 4/3 \end{pmatrix} & \begin{pmatrix} -1/3 & 1/2 & 1/3 \\ 0 & 0 & 0 \\ 1/6 & -1/2 & 1/6 \end{pmatrix} \\ (0)_{3\times 3} & (0)_{3\times 3} & \begin{pmatrix} -1/3 & 0 & 1/6 \\ +1/2 & 0 & -1/2 \\ -1/3 & 0 & 1/6 \end{pmatrix} & \begin{pmatrix} 4/3 & -1 & 2/3 \\ -1 & 5/3 & -2/3 \\ 0.86 & -2/3 & 1 \end{pmatrix} \end{bmatrix} X = \begin{pmatrix} -4/3 \\ -20/3 \\ +8 \\ 16/3 \\ -40/3 \\ +44/3 \\ +28/3 \\ -20 \\ +64/3 \\ +80/3 \\ -80/3 \\ +64/3 \end{pmatrix} \quad \ldots(2)$$

The final bending moment ($m = m_0 + m_i X_i$) diagram is given in Fig. Ex. 2.27(g). Hand computed values (slide rule accuracy) are used for drawing the final bending moment diagram.

2.15 GRID ANALYSIS

The two applications are,

(a) Building Floor Decks, and (b) Bridge decks. The building floor decks are generally supported on all the four sides. The bridge decks, however, are supported on two parallel abutments, and the skew decks are not unusual. In the following, the example chosen is a typical building grid with symmetry about both the axes.

Flexibility method may be preferred over other methods for the analysis of grids consisting of members with negligible torsional stiffness. Consider a simple grid, Fig. 2.33(a). For zero torsional stiffness, the

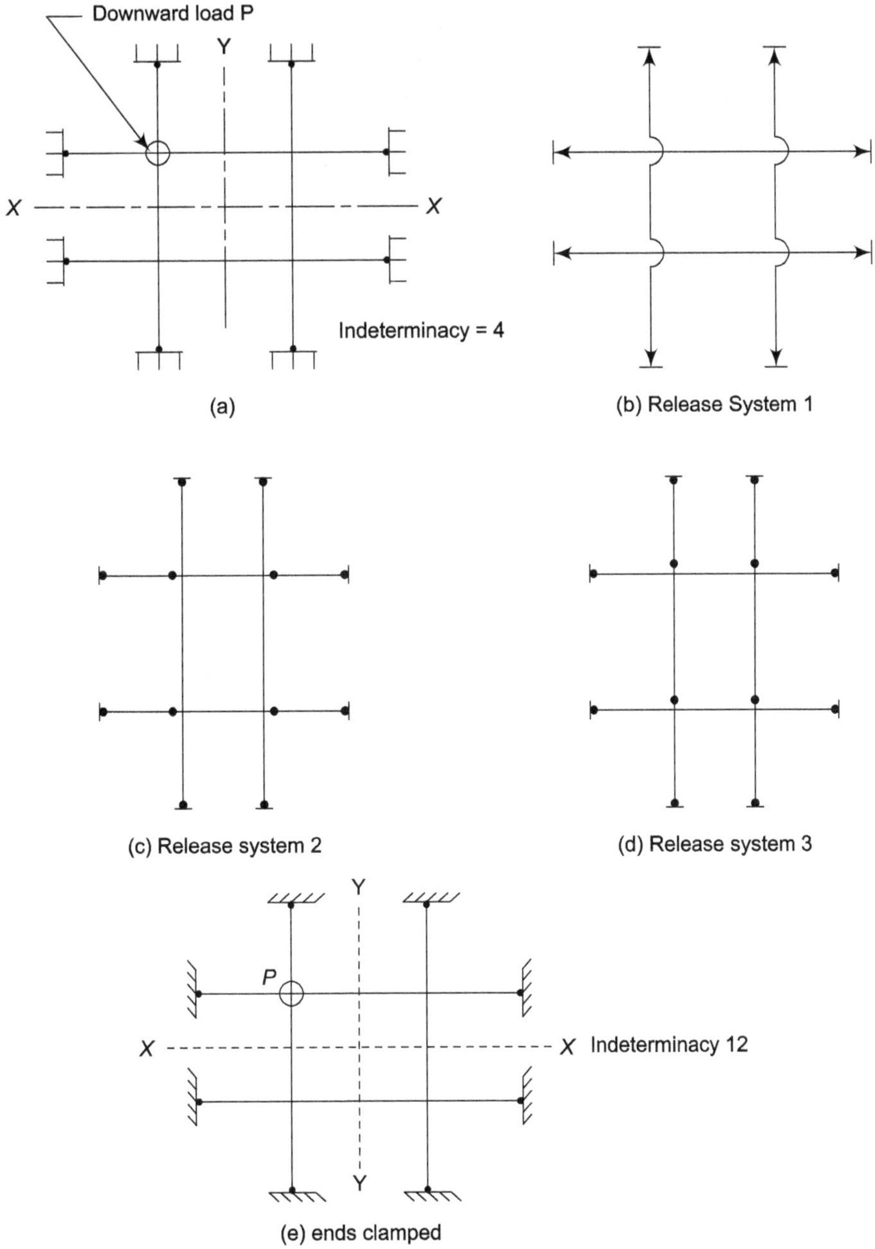

Fig. 2.33 *Grids with Zero Torsional Stiffness*

degree of indeterminacy is equal to 4. In the case of stiffness or displacement method of analysis, the number of unknown displacements (considering modified stiffness of end members) is twelve and does not depend upon the torsional stiffness of the members. Generally, the assembly of flexibility matrix is more tedious than the assembly of stiffness matrix. However, with proper choice of release system the number of non-zero elements is reduced and by proper numbering of the chosen redundants, we can get a well banded flexibility matrix. Some of the possible release systems for the grid, Fig. 2.33(*a*), are shown, Figs. 2.33 (b), (c) and (d). The release system (b) can be shown to be a better system, and if the grid is symmetrical about *x*, *y* axes and diagonals, then the release system retains diagonal symmetry also.

Consider also a grid with ends built in, under a concentrated load, Fig. 2.33(*e*). The number of unknowns by both the methods, flexibility and stiffness, is twelve. As an illustration, the problem is solved by the flexibility method, Ex. 2.28. If the frame is symmetrical about *x* and *y* axes and about diagonal, the number of unknowns for the diagonal symmetrical loading can be reduced from 12 to 5 not more than three unknowns for the case(iii).

If advantage is taken of expressing the given loading as a superposition of four cases,

(i) Symmetry about *x*-axis and *y*-axis,

(ii) Symmetry about *x*-axis and anti-symmetry about *y*-axis,

(iii) Anti-symmetry about *x*-axis and symmetry about *y*-axis, and

(iv) Anti-symmetry about *x*-axis and anti-symmetry about *y*-axis,

The number of unknowns for the four cases and based on diagonal symmetry are one, three, three and one. The results of the case (*ii*) are directly usable for the case (*iii*) and, therefore, no separate analysis is required for the case (*iii*).

Ex. 2.28. Analyze the grid, Fig. 2.33(*e*), by superposition of the four cases of symmetry and anti-symmetry. All members are of the same length, *L* and have the same flexural rigidity. The loading and *m* diagrams are given in the first row of the figures of the Fig. Ex. 2.28 for the four cases based on symmetry and anti-symmetry consideration**. The analysis or each case is given in the following. The grid is clamped at the supports.

Case (i). Symmetry-symmetry. Refer Fig. 2.34 taking advantage of diagonal symmetry, the unknown redundants are only two interacting reactions and end moments. Further in this case, the interacting reactions X_1 can be anticipated to be equal to zero, that is, due to $f_{12} = f_{21} = U_2^{\,0} = 0$. From the m_0, m_1 and m_2 diagrams.

$$f_{12} = f_{21} \int \frac{m_1\, m_2}{EI}\, ds = 0; \quad u_1^{\,0} = \int \frac{m_0\, m_1}{EI}\, ds = 0$$

$$f_{11} = \int \frac{(m_1)2}{EI}\, ds = \frac{5L^3}{3EI}; \quad f_{22} = \int \frac{(m_2)2}{EI}\, ds = \frac{3L}{EI}; \text{ and } u_0 = \frac{PL^2}{4}$$

And the solution of the equations for the unknown redundant gives:

$X_1 = 0$ and end moment $X_2 = -PL/12$

The deflected shape and bending moment diagrams are shown in Fig. 2.35(i).

Case (ii). Symmetry-antisymmetry. Fig. 2.34 In this case, as there is no diagonal symmetry, three

** In such an analysis there is considerable saving in the computations but it requires effort to visualize the behavior. Saving in the computational effort is not any claim in the 'computer era' but more important is the comprehension of the subject of analysis-it helps to build confidence in the subject.

unknown redundants are to be evaluated. The m_0, m_1, m_2 and m_3 diagrams are drawn for this case. From the m_0, m_1, m_2 and m_3 diagrams, we get

$$u_1^0 = \frac{7\,PL^3}{72\,EI} \qquad u_2^0 = \frac{PL^2}{8\,EI} \qquad u_3^0 = \frac{PL^2}{72\,EI}$$

$$f_{11} = \frac{8\,L^3}{9\,EI} \qquad f_{12} = \frac{L^2}{EI} \qquad f_{13} = \frac{-\,L^2}{9\,EI}$$

$$f_{22} = \frac{3\,L}{2\,EI} \qquad f_{23} = 0\,; \qquad f_{33} = \frac{L}{2\,EI}$$

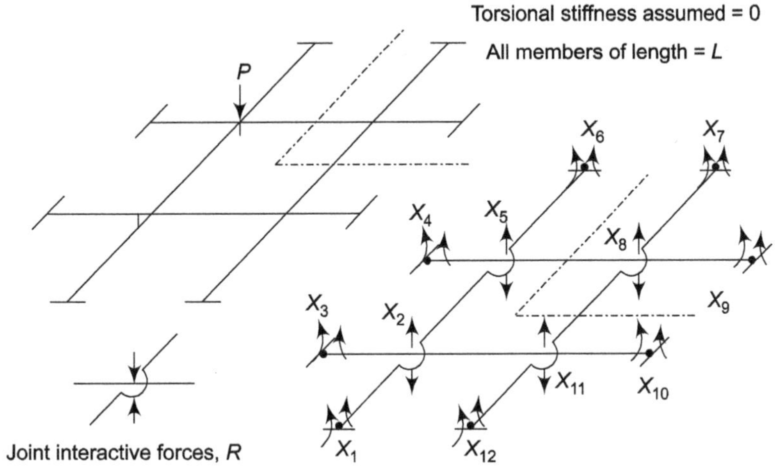

Fig. Ex. 2.28 *Problem Statement and Redundants*

Case (i) Loading and m_0 Diagram

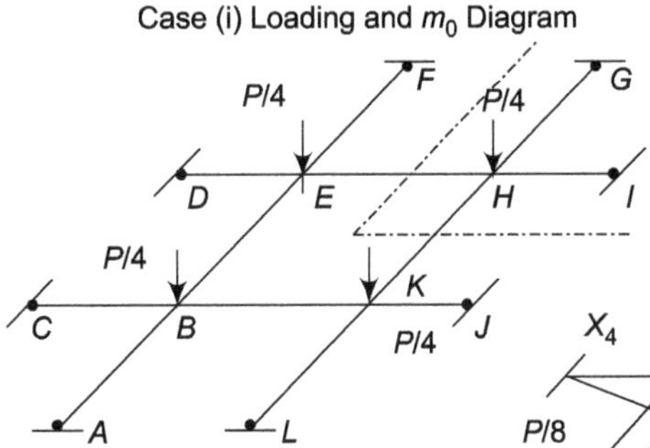

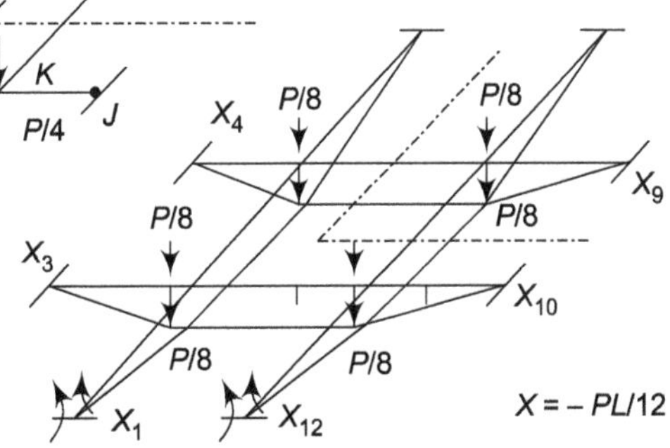

$X_1 = X_3 = X_4 = X_6 = X_7 =$
$X_9 = X_{10} = X_{12} = X_1$
$X_2 = X_5 = X_8 = X_{11} = 0.$

Case (i) Loading and m_0 Diagram with Releases

Case (ii) Loading and m_0 Diagram

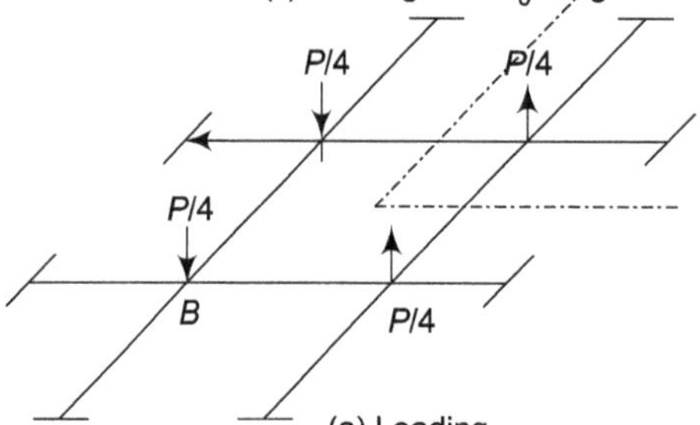

(a) Loading

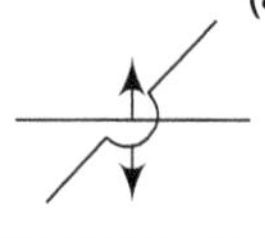

(b) R direction assumed

All interacting forces at the joints are same = R_1

$X_1 = -X_6 = -X_7 = X_4 = R_2$

$X_3 = X_4 = -X_9 = -X_{10} = R_3$

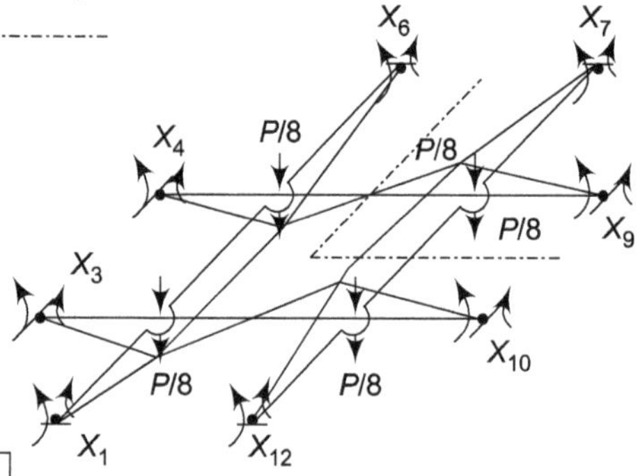

(c) Moment direction assumed

Case (ii) Loading and m_0 Diagram with Releases

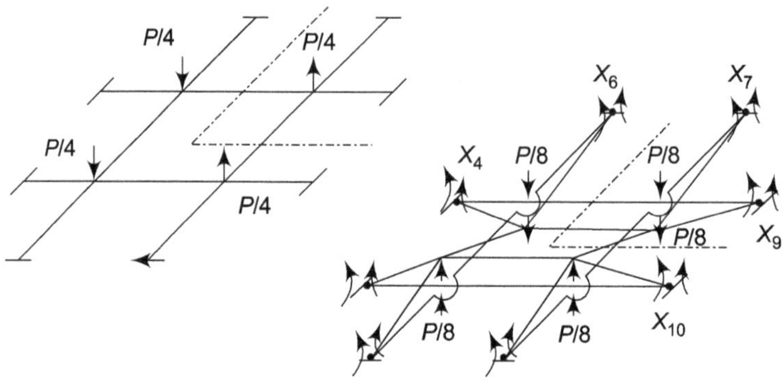

Case (iii) Loading and m_0 Diagram

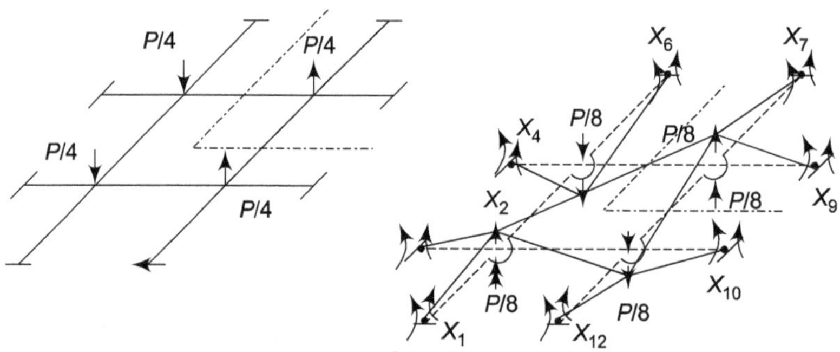

Case (iv) M_0 Diagram

Fig. 2.34

When written in the matrix form, and cancelling common terms, we get

$$\begin{pmatrix} 8/9 & 1 & -1/9 \\ 1 & 3/2 & 0 \\ -1/9 & 0 & 1/2 \end{pmatrix}\begin{pmatrix} X_1 L \\ X_2 \\ X_3 \end{pmatrix}+\begin{pmatrix} 7\,PL/72 \\ PL/8 \\ PL/72 \end{pmatrix}=\begin{pmatrix} 0 \\ 0 \\ 0 \end{pmatrix}$$

The solution gives the unknown redundant,

$$X_1 L = \frac{-11}{128}PL, \qquad X_2 = \frac{-5}{192}PL \qquad X_3 = -\frac{3}{64}PL$$

The deflected shape and bending moment diagrams are shown in Fig. 2.35(ii).

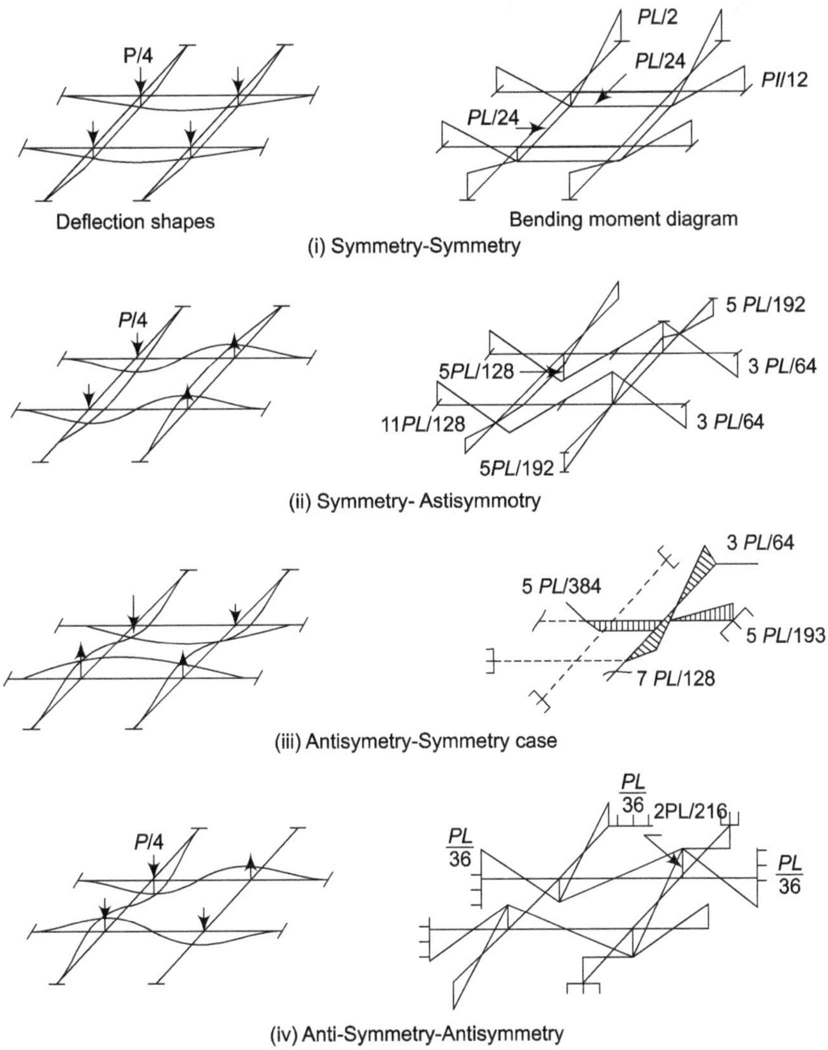

Fig. 2.35 *Results of four Cases given separately*

Case (iii) Antisymmetry-symmetry. As this case is similar to the case 2, only loading-cum-m_0 diagram is shown. The deflected shape and bending moment diagrams are shown for this case also, Fig. 2.35(ii).
Case (iv) Antisymmetry-antisymmetry. Refer Fig. Ex. 2.28(*d*) and Fig. Ex. 2.28(*b*). As in the case 1, because of diagonal symmetry there are only two unknown redundants to be evaluated but in this case also the interacting redundant $X = 0$. From the m_0, m_1, m_2 diagrams, Fig. 2.35(iv),

$$u_1^{\,0} = 0,\ u_2^{\,0} = \frac{1}{36}\frac{PL^2}{EI}$$

$$f_{11} = \frac{1}{9}\frac{L^3}{EI}\ f_{12} = 0,\ f_{22} = \frac{L}{EI}$$

and the solution for the unknown redundant gives

$$X_1 = 0,\ X_2 = -\frac{1}{36},\frac{PL}{EI}$$

Refer Fig. 2.24(iv) for the deflected shape and bending moment diagram.

The superposition of the four cases gives the final moment diagram which is the same as given in Fig. Ex. 2.29 (*c*).

Ex. 2.29. Formulate the flexibility matrix for the grid of Ex. 2.28, without taking advantage of symmetry and anti-symmetry.

Sequence the numbering of the redundants as shown in Fig. 2.36 which gives a good banded matrix. For the release system chosen, (Fig. 2.36), the following flexibility matrix equation is obtained.

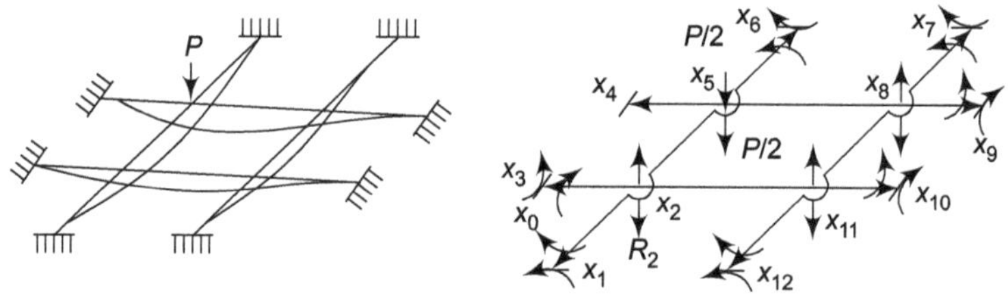

Release system and numbering of redundant

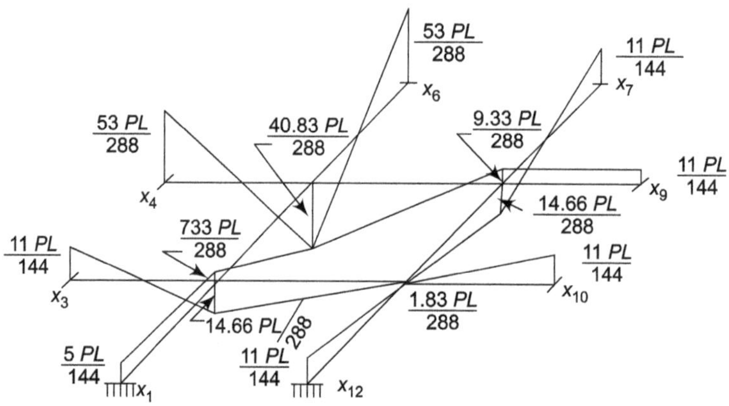

Final moment diagram

Fig. 2.36 (Ex. 2.28) Contd.

$$
\begin{array}{r}
1\\2\\3\\4\\5\\6\\7\\8\\9\\10\\11\\12
\end{array}
\begin{bmatrix}
1 & 5/9 & 0 & 0 & 0 & 0 & 0 & 0 & 0 & 0 & 4/9 & 1/2 \\
5/9 & 8/9 & -5/9 & -4/9 & 7/18 & 0 & 0 & 0 & 0 & 0 & 7/18 & +4/9 \\
0 & -5/9 & 1 & 1/2 & -4/9 & 0 & 0 & 0 & 0 & 0 & 0 & 0 \\
0 & -4/9 & 1/2 & 1 & 5/9 & 0 & 0 & 0 & 0 & 0 & 0 & 0 \\
0 & 7/18 & -4/9 & -5/9 & 8/9 & 5/9 & 5/9 & 7/18 & 0 & 0 & 0 & 0 \\
0 & 0 & 0 & 0 & +5/9 & 1 & 1 & 4/9 & 0 & 0 & 0 & 0 \\
0 & 0 & 0 & 0 & 4/9 & 1/2 & 1/2 & 5/9 & 0 & 0 & 0 & 0 \\
0 & 0 & 0 & 0 & 7/18 & 4/9 & 4/9 & 8/9 & -5/9 & -4/9 & 7/18 & 0 \\
0 & 0 & 0 & 0 & 0 & 0 & 0 & -5/9 & 1 & 1/2 & -4/9 & 0 \\
0 & 0 & 0 & 0 & 0 & 0 & 0 & -4/9 & 1/2 & 1 & 5/9 & 0 \\
4/9 & 7/18 & 0 & 0 & 0 & 0 & 0 & 7/18 & -4/9 & 5/9 & 8/9 & 5/9 \\
+1/2 & 4/9 & 0 & 0 & 0 & 0 & 0 & 0 & 0 & 0 & 5/9 & 1
\end{bmatrix}
\begin{bmatrix}
X_1L^2 \\ X_2L^3 \\ X_3L^2 \\ X_4L^2 \\ X_5L^3 \\ X_6L^2 \\ X_7L^2 \\ X_8L^3 \\ X_9L^2 \\ X_{10}L^2 \\ X_{11}L^3 \\ X_{12}L^2
\end{bmatrix}
=
\begin{bmatrix}
0 \\ \frac{7}{36}PL^3 \\ -\frac{2}{9}PL^2 \\ -\frac{5}{18}PL^2 \\ 0 \\ -\frac{5}{18}PL^2 \\ -\frac{2}{9}PL^2 \\ -\frac{7}{36}PL^3 \\ 0 \\ 0 \\ 0 \\ 0
\end{bmatrix}
$$

Fig. 2.36 (Ex. 2.28)

The solution of which gives:

$$X_1 = X_9 = -\ 11\ PL/144 \qquad X_2 = 11\ P/64$$

$$X_6 = X_4 = -\ 53\ PL/288 \qquad X_5 = 0$$

$$X_3 = X_7 = -\ 5\ PL/144 \qquad X_8 = -\ 11\ P/64$$

$$X_{10} = X_{12} = -\ 11\ PL/288 \qquad X_{11} = 0$$

The final bending moment, $m = m_0 + \Sigma m_i X_i$

The final bending moment diagram is shown in Fig. 2.36(c)

Revision Example on Conceptualization

Ex.1 Propped cantilever under the action of a moment (couple) loading: Degree of indeterminacy or number of excess constraints is equal to one.

Draw qualitatively BM diagram and corresponding elastic line.

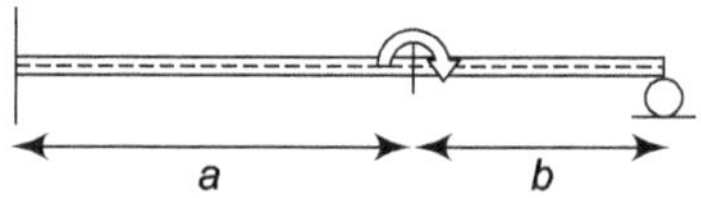

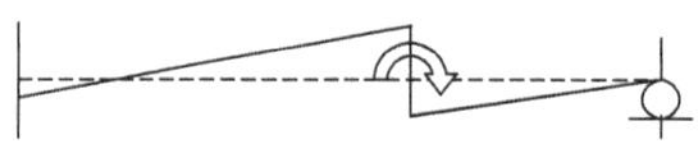

Likely BM diagram

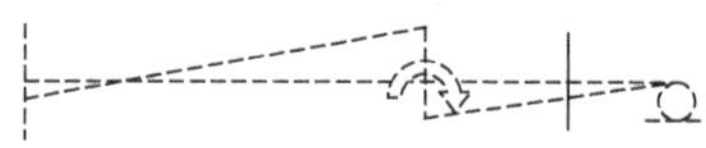

Sketch Likely Elastic line? While Sketching notice inflection points

Ex.2 Propped cantilever system under the action of a Single load, P.
Represent by a single equivalent spring
(Number of excess constraints is equal to one.)

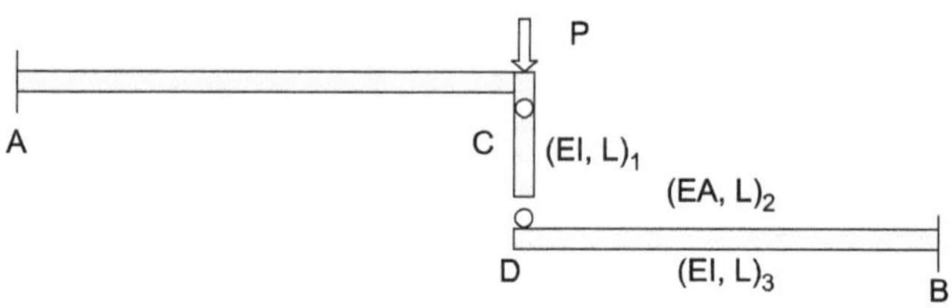

Joints A and B are fixed ends. C and D are pin connection
Show the above system as a Load P resting on a single spring. If $L_1 = L_3 = L$, and $EI_1 = EI_2 = EI$
Assume $I/A = rL$, $L_2 = L/2$ P

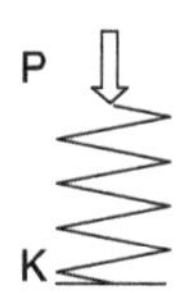

CHAPTER 3: PLASTIC ANALYSIS

Chapter

3

Plastic Analysis

3.1 PSEUDO NON-LINEAR ANALYSIS, UNIQUENESS THEOREM, AND STIFFENING

The Force Method developed already provides an adequate background and preparation for introducing Plastic Analysis. The scope of the plastic analysis is limited to skeletal structures and it is based on small deformations. The number of basic mechanisms is related to the degree of determinacy of skeletal structures. The inelasticity in skeletal structures is localized as angle discontinuities at the points of maximum moments, $(M_{\max})_i \le (M_P)_i$. The skeletal members between such maximum moment points are assumed to remain elastic. Therefore, it is a **pseudo non-linear analysis.** In the case of indeterminate structures the plastic hinges that are formed, first undergo angle discontinuities while the other impending plastic hinges are developing or yet to form. The pseudo non-linear analysis for a given load, P,

$P_1 < P < P_c$, (where P_1 and P_c are the loads corresponding to first plastic hinge formation and collapse load respectively) is achieved by

1. Increments of ΔP, till we achieve the given load P, or
2. Find P_c and track backwards by locating 'the last plastic hinges' till the given load is reached.

The approaches are explained with the help of illustrations.

The uniqueness theorem Upper bound and Lower bound are the important conceptual contents. The lower bound and upper bound are directly related to the uniqueness theorem. The concepts are better explained by the Force method. The virtual work principle is used for satisfying equilibrium and also for the computation of angle discontinuities of the true structure (in contrast to virtual structure). The lower bound approach is explained with the help of technique based on artificial stiffening.

The Concept of the Artificial Stiffening is used for enhancing the partial collapse load by appropriately stiffening at the zones of maximum moments. The concept can be used for actual stiffening of the partially collapsing frames. The stiffening however reduces the overall ductility which is an important parameter for the pushover design of earthquake resistant structures. An important aspect of stiffening of structures for their increased capacities is suggested. The stiffening of the framed structures cannot be divorced from the important aspect of ductility. It is the ductility that ensures the structure its energy absorbing capacity, and it provides prior warning before it collapses.

The recent (2010) practice of the earthquake design is to base the earthquake analysis on *capacity* considerations. The plastic analysis has been included with an additional objective of preparing the readers for a proper appreciation of pushover analysis for the earthquake design of tall structures. As far as possible, the examples chosen develop plastic hinges in the girders. It is a good design practice. The shear effect in mobilizing the full plastic moment of resistance being not crucial is not included herein.

3.2 ELASTIC-PLASTIC ANALYSIS AND "HINGE RELEASE" SYSTEM

In Chapter 2, while discussing about superposition and non-linearity, different idealization for material properties of flexural members were suggested in Fig. 1.8. One such idealization or the moment curvature relationship was elastic and plastic moment curvature relationship. As an illustration, the frame of Ex. 2.26, is analyzed at high loads above elastic range. Let the idealized elastic plastic moment curvature relationship be given by Fig. 3.1, where $M_p = 2$ tm. From the moment diagram, Fig. Ex. 2.27(*b*), it may be observed that the section *C*, under the load, is the first to reach the value of M_p as the load *P* is increased.

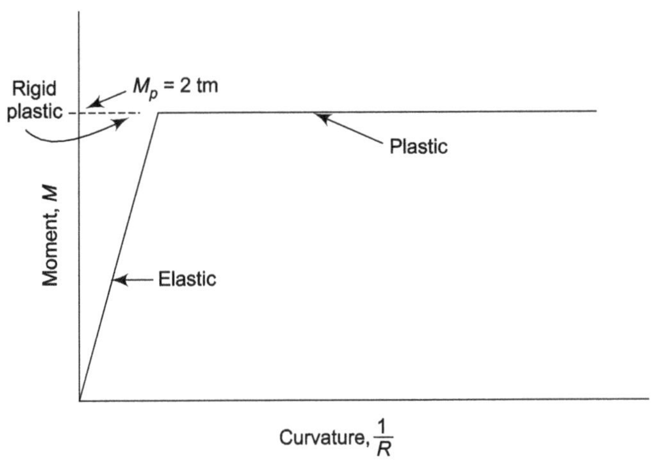

Fig. 3.1 *Elastic-Plastic Moment Curvature*

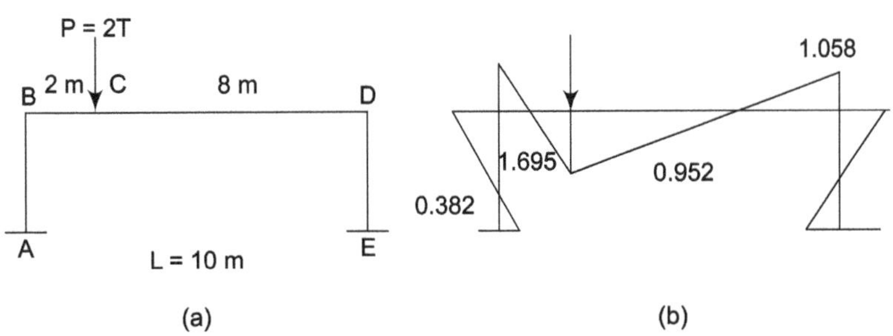

Fig. 3.2 *The Problem and Results of Ex. 2.26*

The Problem and moment diagram are reproduced in the following as Fig. 3.2.

The moments are linearly proportional to the applied load, $P < P_1$. The load P_1 corresponding to the moment $M_c = M_p$ is $P_1 = (2/1.695)\ P = 2.36$ tons.

Stage 1. The moments of the Fig. 3.2 (*b*) are also to be multiplied by this factor = 2/1.695 = 1.18 to get the moments corresponding to the load P_1. The moments corresponding to P_1 are shown in Fig. 3.3 (*a*). The elastic analysis for $P \leq P_1$ is the first stage of analysis. Beyond this load moment, M_c does not get altered with the increase in P and the point C behaves like a hinge for load $P > P_1$. Such a hinge with the fixed value of moment = M_p, is called "plastic hinge". Thus for the incremental loads, $\Delta D\ P = P - P_1$, the degree of indeterminacy is reduced from three to two. It is, therefore, desirable to assume hinge releases at the points of maximum moments, so that at higher loads ($P > P_1$) the assumed hinge releases become the actual hinges (plastic hinges). It is possible to continue the elastic-plastic analysis till the frame collapses. For the example considered, the second stage of analysis is for ΔP only, the degree of indeterminacy is two and the releases are introduced at B and E only, Fig. 3.3.

In order to enable an appreciation of the generalization of "triple release system", we continue with the same.*

The reader may recall the generalized triple system comprising of:

1. Release system for m_0 diagram,
2. Release system for unit redundant moments, m_i, and finally
3. Release system for $\bar{m}_1$ diagrams for satisfying compatibility, using virtual work principle.

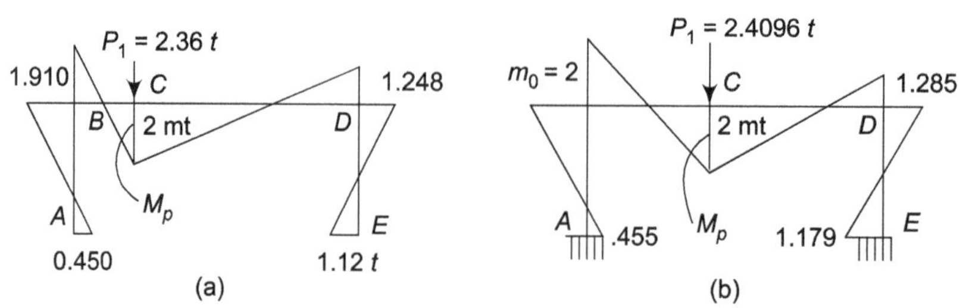

Fig. 3.3 *Moment at the Formation of (1) One Plastic Hinge (2) Two Plastic Hinges*

Stage 2. Let us assume $P = 2t$, so that we may make use of the earlier results, Eq. (*iv*). The following are the two important observations, for the second stage of analysis which allows elastic analysis or ΔP only.

(1) While formulating the equations (*iv*), the m_0 diagram gives 3.2 tm under load which is not admissible at the second stage because we now have a hinge at C. To get zero moment at C due to ΔP assign $X_1 = -3.2$ tm. For the second stage, degree of indeterminacy being two only, we need to evaluate the corresponding two redundants, that is, X_2 and X_3.

(2) Now we have to decide which one of the three equations, Eq. (iv), should be dropped. The final moment $m = m_0 + m_1 X_1 + m_2 X_2 + m_3 X_3$ (where $X = -3.2$) is so distributed that it allows angle discontinuity at C, while satisfying compatibility at other releases. Thus the third equation, which is obtained by statical diagram which has unit Biomet at C, is no more valid directly, because

*If the reader wishes to avoid flipping the pages back and forth (back to Ex. 2.26), he is advised to analyze the problem using the conventional 'single release system', and do the computations for all the stages based on the 'single release system'. Introduce the hinge releases at C, B, and E and analyze for the unknown redundant (moments) at the chosen releases.

the solution allows angle discontinuity at C equal to θC in the corresponding real deformation structure.

We, therefore, consider the first two equations of Eq. (iv) for compatibility, that is,

$$(-40/3)\,X_1 + (32/3)\,X_2 + (56/3)\,X_3 + 0 = 0$$

$$(650/3)\,X_1 + 12\,X_2 = (524/3)\,X_3 + 32 = 0$$

Substituting for $X_1 = -3.2$, we get

$$32\,X_2 + 56\,X_3 = -128$$

$$36\,X_2 + 524\,X_3 = 1984$$

The solution of the above equations gives $X_2 = 2.350$ tm and $X_3 = -3.628$ tm. The third equation can be used for evaluating the angle discontinuity, θ_c.

The above analysis (the second stage) is valid only till the formation of next or a second plastic hinge. From the Fig. 3.3(a), the second plastic hinge is likely to form at B. The moment at B, due to $P = 2t$, is given by $X_3 = -3.628$.* The limiting value of P, up to which is second stage of analysis holds good, is given by,

$$(\Delta P)_{\text{limit}} = \frac{2}{(-3.628)}(-2+1.910)$$

$$(\Delta P)_{\text{limit}} = .0496\text{t}$$

The X_i values corresponding to $P = .0496$t, are

$$X_2 = \frac{.0496}{2}\times 2.35 = .0584$$

$$X_3 = \frac{.0496}{2}\times(-3.628) = -.090$$

$$X_1 = \frac{.0496}{2}\times(-3.2) = .0794$$

And $P_2 = 2.360 + .0496 = 2.4096\ t$

Where P_2 is the load corresponding to the formation of the second plastic hinge. The moments corresponding to the load P_2 are computed in Table 3.6. The moments under the column (8) of the Table 3.1 are plotted, Fig. 3.3(b).

Table 3.1 *Moments due to P_1 and the redundants combined*

Points	*Moments due to P_1*	*Moments due to P = .0496t*					*Moments due to P = P_2 Col. (2) + Col. (7)*
		m_0	X_1m_1	X_2m_2	X_3m_3	$m_0+\Sigma X_im_i$	
1	2	3	4	5	6	7	8
A	+.450	–	+0.397	.058	–.450	+.005	+.455
B	–1.910	–	–	–	–.090	–.090	–2
C	+2	.0784	–0.0784	–	–	–	+2
D	–1.248	–	–.397	–	+.360	–.037	1.285
E	+1.121	–	–	.058	–	+.058	1.179

* If X_3 does not give moment at C which is of the same sign as that of the moment at C due to P_1, then the reversal suggests that the plastic hinge is to be expected at the other point of maximum moment.

Stage 3. Similarly the analysis is extended to the third stage, that is for $P > P_2$, and $\Delta P = P - P_2$. At this stage, the degree of indeterminacy is one. This analysis is valid till a third plastic hinge is formed. From the Fig. 3.3 (b), the third plastic hinge is likely to form at D, where under the load P_2 the moment, $M_D = -1.285$ tm. For the third stage of analysis, we once again use the same Eq. (iv) for $P = 2t$. We assign $X_1 = -3.2$ tm, so that $M_C = 0$ and further we have to assign values either X_2 or X_3 such that we get the M_B also equal to zero (the second elastic hinge). In this case, we choose $X_3 = 0$, which ensure $X_B = 0$. We have to choose one of the first two equations of Eq. (iv) for the evaluation of X_2 under $\Delta P = 2t$ for the third stage of analysis. The second equation of Eq. (iv) is not valid because the corresponding statical diagram, Fig. 3.23, has a non-zero moment at B. We therefore, choose the first equation only for the unknown X_2.

$$-(40/3)\,X_1 + (32/3)\,X_2 + (56/3)\,X_3 + 0 = 0$$

where

$$X_1 = -3.2 \text{ and } X_3 = 0. \text{ Thus we get}$$

$$X_2 = -4 \text{ and the moment at } D, \text{ Fig. 3.24,}$$

$$m_D = m_0 + m_1X_1 + m_2X_2 + m_2X_2 + m_3X_3 = m_1X_1 = 5 \times (-3.2)$$

$$= -\ 16 \text{ tm}$$

In the second stage of the analysis, maximum limiting value ΔP for the third stage is obtained from the moment considerations at D, that is, $\Delta M = (M_P - 1.285)$, see Fig. 3.34(b).

$$(\Delta P)_{\text{limit}} = \frac{2}{M_D}(\Delta M)$$

Here $M_D = -\ 16$ tm as completed and $\Delta M = (-2 + 1.285)$

$$\Delta M = -0.715 \text{ tm}$$

$$(\Delta P)_{\text{limit}} = -\frac{2}{16}(-0.715) = .0894 \text{ t}$$

M_D The corresponding moments due to $\Delta P = .0894$ t are given

$$M = m_0 + m_1x_1 + m_2x_2 + m_3x_3$$

$$M_A = [0 - 5(-3.2) - 4 + 0]\ \frac{.0894}{2} = +.536 \text{ tm}$$

$$M_B = M_C = 0$$

$$M_D = [0 + 5\,(-3.2) + 0 + 0]\ \frac{.0894}{2} = -.715 \text{ tm and}$$

$$M_E = [0{+}0\ {-4}\ {+0}]\ \frac{.0894}{2} = -.179 \text{ tm}$$

The moments for the third stage $\Delta P = 0.0894$, and the final moments at $P_3 = P_2 + (\Delta P = 0.0894) = 2.499$t, are shown in the Fig. 3.4(a) and (b). Figure 3.4(b) is obtained by combining the results of the Fig. 3.3(b)

and Fig. 3.4(a). By now it must be obvious to the reader, the justification for assuming hinge releases at the points of maximum moments. The Eq. (iv) written for elastic analysis could be used till the end of elastic-plastic analysis that is up to the collapse.

3.3 MECHANISM AT COLLAPSE

It may be observed that with the development of the third plastic hinge of *D*, Fig. 3.4(b), the beam *BCD* becomes unstable and the members *BC* and *BD* continue to deform and the deflection δ under the load does not take a stationary value. If we plot the load deflection curve, Fig. 3.4(*c*), the stiffness of the structure (slope of *P-d* curve) is different at different stages. At the end of stage 3, the structure collapses, that is the structure does not come to rest or a stationary position.

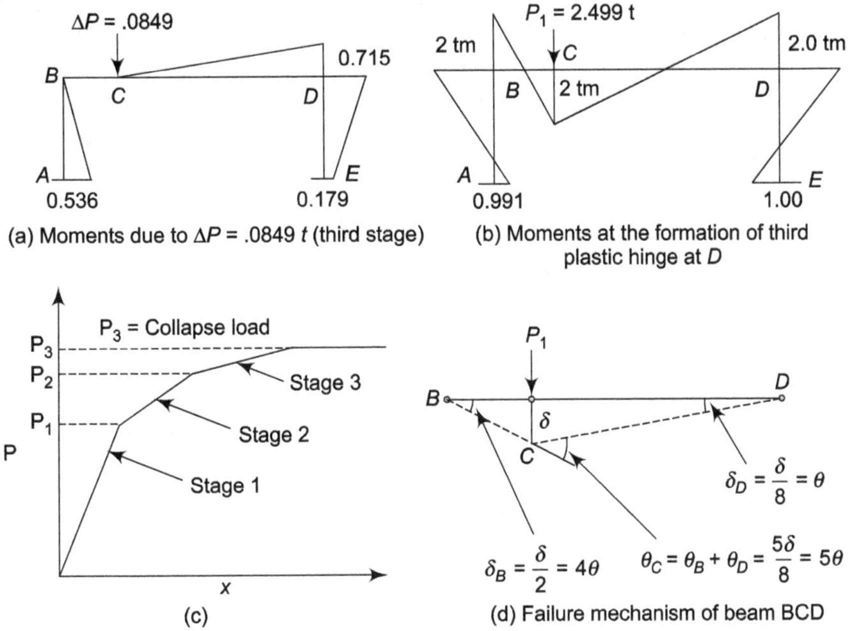

(a) Moments due to ΔP = .0849 *t* (third stage) (b) Moments at the formation of third plastic hinge at *D*

(c) (d) Failure mechanism of beam BCD

Fig. 3.4 *Mechanism of Collapse*

The continuous deformation without an application of loads is called Mechanism. In this example, the mechanism is restricted to the girder only which is called *Beam Mechanism*. For the beam mechanism, Fig. 3.4(*d*), the collapse load *Pc* can be obtained from equilibrium of the girder. If we write equilibrium equation from virtual work consideration, we get

$$P_C(\delta) = M_B\,\theta_B + M_C\,\theta_C + M_D\,\theta_D$$

Substituting the values of θ_B, θ_C and θ_D from the Fig. 3.35(d),

and $$|M_B| = |M_B| = |M_D| = M_P,$$

$$P_C = M_P = \left(\frac{1}{2}+\frac{5}{8}+\frac{1}{8}\right) = \frac{10\,M_P}{8}$$

$$M_P = 2 \text{ tm},$$

$$P_C = \frac{10 \times 2}{8} = 2.5 \text{ tons},$$

Which is the same as we obtained at the end of the elastic-plastic analysis. In the above analysis for the collapse load, because we had known the failure mechanism (beam mechanism), the failure load is 'exact'. In the general case and in the absence of any elastic analysis, it is not so direct to decide the failure mechanism. In the mechanism approach, the plastic hinges are assumed to be rigid plastic, Fig. 3.6.

3.4 PARTIAL COLLAPSE

In the example considered, the girder is undergoing mechanism and column members remain stationary. Such a collapse is called *Partial Collapse*. For the complete collapase to take place, the number of plastic hinges are equal to $n + 1$, where n is the degree of indeterminacy. For the frame considered, $n = 3$, the number of plastic hinges is four for a complete collapse. In the case of partial collapse by the mechanism method, we do not know individually the moment values at all those points that are not plastic. Before the development of the last plastic hinge, the structure being indeterminate, the moment diagram cannot be drawn from statics.

3.5 ELASTIC AND RIGID PLASTIC ANALYSIS, COLLAPSE LOAD

3.5.1 General

For materials like steel, it is justified to idealize the stress strain curve by elastic and plastic segments as shown in Fig. 3.5(a). The stress developed at a section at different stages of load for the idealized stress strain relationship is shown in Fig. 3.5(b).

Discussion on Fig. 3.5(b) : Flexural Stresses and Strains

The assumption of plane section remaining plane is considered to be applicable in the inelastic range also. The stress distribution Fig. (ii) is valid in the elastic range only and the stress distribution given in Fig. (iii) is corresponding to the limit of elastic range, i.e., maximum stress = yield stress. And the moment corresponding to this distribution is called yield moment, M_y. The stress distributions, Fig. (iv) and (v), are or the case of yielding penetrating towards the zero stress axis. In the Fig. (v) both the extreme fibers have already yielded and for this case only the corresponding strain distribution is also shown based on the assumption, "plane section remaining plane". With the increase in bending moment, the value of $2y_0$, defining the depth of elastic portion of the cross-section, gets reduced and in the limits it is neglected. The limiting or plastic moment or resistance, M_P is computed for the stress distribution, Fig. (vi).

Based on the typical stress distribution Fig. 3.5(b) and the corresponding strains, moment curvature relationships can be drawn as shown in Fig. 3.5(c). The curvature is computed from the relation

$$x = \frac{\varepsilon_y}{y_0}$$

and x_y is corresponding to the extreme fiber reaching the yield strain, Fig. 3.5(b) (iii). Based on the Fig. 3.5(c), we make the following observations.

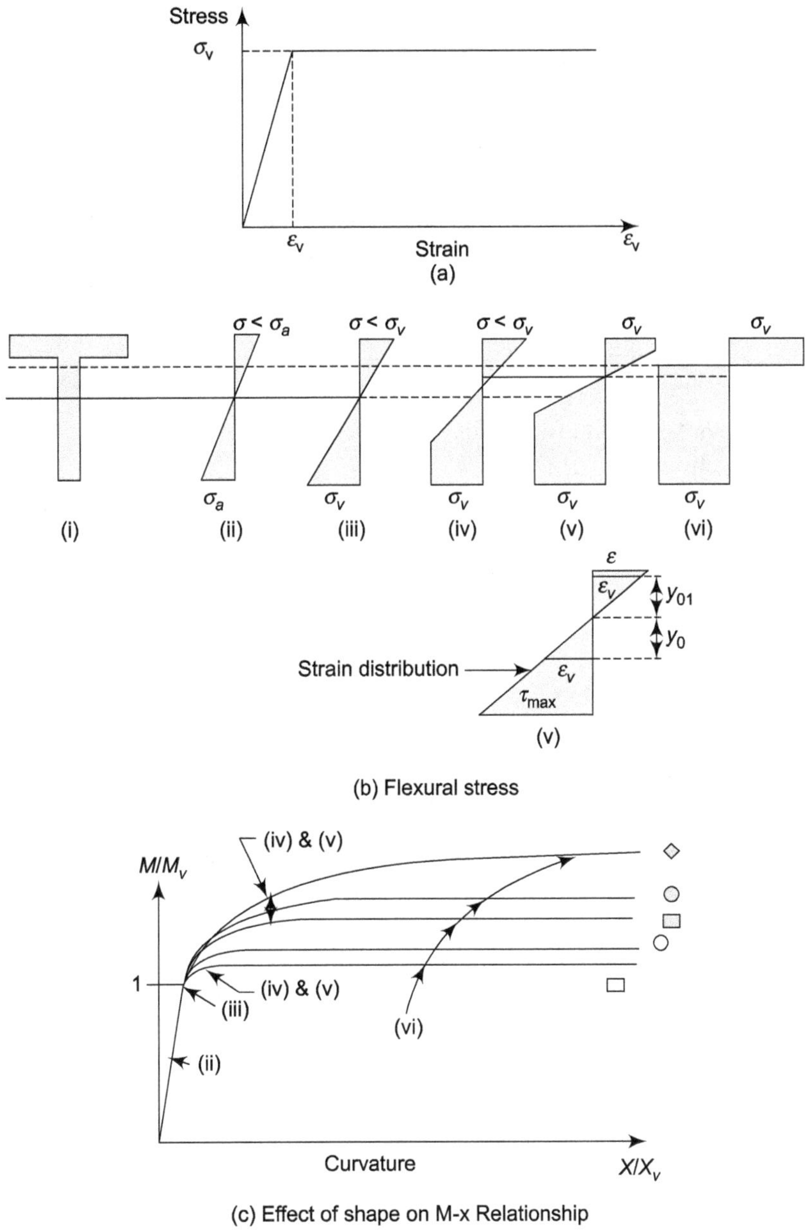

Fig. 3.5 *Stress distribution based on approximated moment-curvature relationship and linear strain distribution across the beam cross section*

The steel structural section generally used for flexure are I-Joints, channels, hollow sections and for such sections, Fig. (c), "Shape Factor" defined by the ratio, $f = M_p / M_v$ is close to unity. In the case of I-Sections the ratio is about 1.15. the value of shape factor for some typical sections may be verified to be the following:

Hollow the section

$$f = \frac{3}{2}\frac{1-\dfrac{b_1 h_1^2}{bh^2}}{\dfrac{b_1 h_1^3}{bh^3}}$$

where b and b_1 are outside and inside widths, and h and h_1 are outside and inside heights respectively. An I-section with fillet area ignored can be shown as an equivalent hollow box section.

Hollow circular section

$$f = \frac{16r_1}{3\pi}\cdot\frac{r_1^3 - r_2^3}{r_1^4 - r_2^4}$$

where r_1 and r_2 are the outside and inside radii.

Solid rectangular section,

$$f = 1.5$$

Solid circular section,

$$f = \frac{16}{3\pi}$$

Solid rhombus section,

$$f = 2$$

In the case of determinate structures, there is no significant advantage by considering elastic plastic behavior as the advantage gained thereby is marginal (the shape factors in the case of usual structural sections or flexure are not much different from unity). However, as was shown in the case of portals in the case of more complex structures (high redundancy), the loads higher than P_y were possible due to 'redistribution' even after the formation of the first plastic hinge.

From the moment curvature relationships, Fig. 3.5(*c*) and referring to the sections used as flexural members, it is justified to ignore the transition from elasticity to plasticity and idealize it by elastic-plastic moment curvature relationship, Fig. 3.6. Such an 'elastic-plastic' moment curvature relationship was used for the incremental load analysis in Article 3.2. The effects of shear and axial forces were ignored in the analysis.

In the following, Article 3.5.2, elastic and rigid plastic analysis is shown to be adequate for materials like steel. It is a full load analysis, in contrast to the incremental load analysis for different stages based on sequence of formation of plastic hinges. The collapse or plastic analysis, when sufficient hinges have formed to give rise to mechanism (Article 3.5.3), is based on the upper bound method. It is generally preferred over the lower bound method for complex structures. However, in the case of partial collapse of highly redundant structures, the upper bound method does not easily lend itself to obtain statically admissible moment diagram. This difficulty can be overcome by the method of artificial stiffening as suggested by Heyman.*

* Heyman, J., "Bending Moment Distribution in Collapsing Frames," Engineering Plasticity, Cambridge University Press, 1968.

3.5.2 Elastic and Rigid Plastic Analysis

In the Article 3.1, the inelastic analysis was done for increments of loads because in the inelastic range, the material properties of the structure change with the load increments. That is, the previous loading history has to be tracked before analyzing the structure for the latest or running load increment. Up to the load P_1, Article 3.1, the entire structure was elastic, and stage-2 was valid for the increment ΔP above the load P_1 (when the section *C* becomes plastic). In the elastic-plastic analysis, Article 3.1, under the load $P = P + \Delta P$, we considered 'elastic-plastic' *M-x* relationship, Fig. 3.1, for the section C of the portal, i.e., elastic upto P_1 and plastic for the load increment, ΔP. Similarly for the stage-3, the 'elastic-plastic' *M-x* relationship was used for the section *B* also. Such an analysis is tedious. It is possible to do the 'full load' analysis, without having to consider increments to add upto the given (full) load, if we attribute 'rigid-plastic' *M-x* relationship, Fig. 3.6, to the sections that have reached plastic moment, M_p. In fact we used this property while computing collapse load at mechanism condition, Article 3.1. This idealization consists of 'elastic-plastic' *M-x* relationship to the elastic *M-x* relationship for all sections, where $M < M_p$ and rigid plastic *M-x* relationship for the discrete sections which have developed moments equal to M_p and ignore the previous history of elastic behaviour of these discrete sections. The assumption of rigid plasticity at the plastic hinges does not make any significant difference from the results obtained by the 'incremental load analysis'. As an illustration consider the same frame of Fig. 3.2, for the 'full load' analysis.

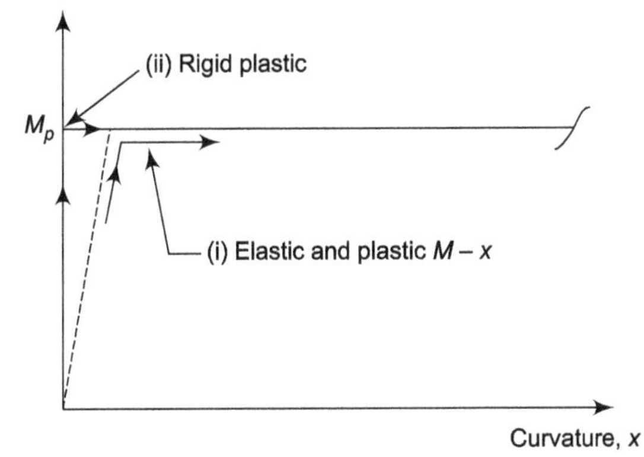

Fig. 3.6 *Two Idealizations of M-x*

Ex. 3.1. Analyze the frame, Fig. Ex. 3.1 for the given full load, $P = 2.5$ tons. Assume elastic and rigid plastic moment curvature relationship. The figure is reproduced in the following for ready reference. We had then used the triple release system.

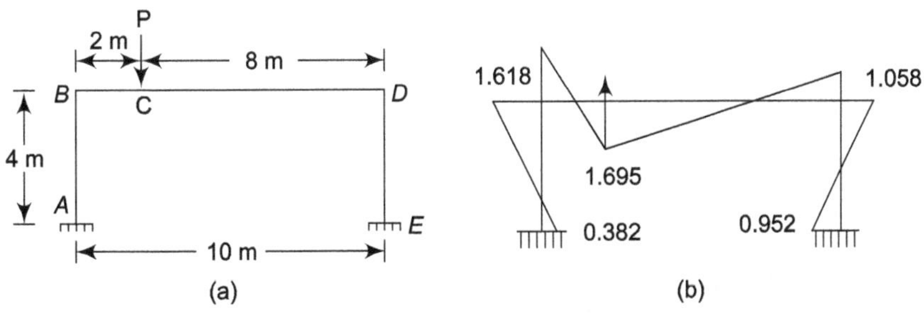

Fig. Ex. 3.1

Let us, this time, use double release system. The m_0 and m_i diagrams of Fig. 3.2(b), are used for deformation (real) system, and the releases of Fig. 3.2(c) for statical (virtual) system to obtain $\bar{m}_i$ diagrams. These systems are redrawn, Fig. Ex. 3.2.

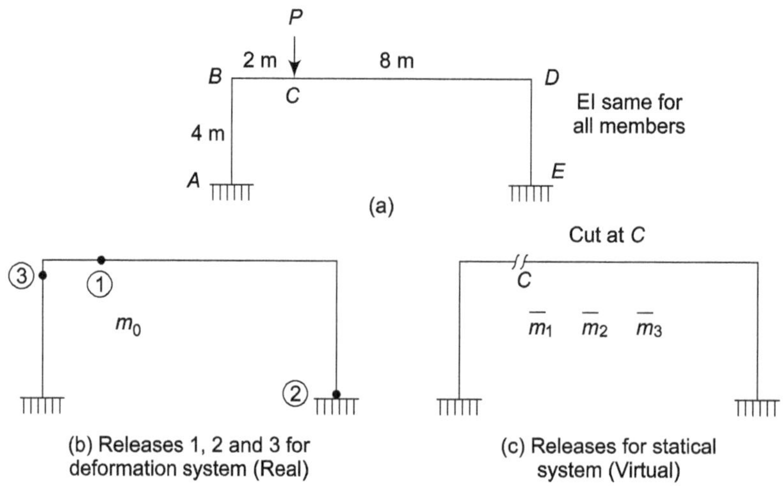

Fig. Ex. 3.2 *Double Release System*

Assume M_p at section D, $(M_p)_D = 2.0 + \varepsilon$, where ε is a small value to be introduced to inhibit formation mechanism. Thus, for the example considered at the load, $P = 2.5$ tons, only two plastic hinges develop at the sections B and C. The plastic analysis for the chosen double release system gives the following equations. The flexibilities f_{ij} are the same as computed in Chapter 2. The deformations $u_1^{\,0}$ are different and are based on $\bar{m}_i$, and m_0 diagram.These are redrawn in Fig. Ex. 3.1.

$$\begin{pmatrix} -40/3 & 32/3 & 56/3 \\ 650/3 & 12 & -524/3 \\ 25 & 4 & -11 \end{pmatrix}\begin{pmatrix} X_1 \\ X_2 \\ X_3 \end{pmatrix}\begin{pmatrix} 21.333 \\ -330.667 \\ -32 \end{pmatrix} P = \begin{pmatrix} 0 \\ 0 \\ 0 \end{pmatrix}$$

For the load, $P = 2.5$ tons, plastic hinges are developed at the section B and C, and the equations based on the statical diagrams $\bar{m}_2$ and $\bar{m}_3$ are not valid. The dropping of the equations given by the second and third rows are due to $\bar{m}_2$ and $\bar{m}_3$ diagrams given non-zero moments at the plastic hinge B in the case of $\bar{m}_2$ diagram at the plastic hinge B and C in the case of $\bar{m}_3$ diagram. The two equations, however, can be used to compute angle discontinuities due to the plastic deformations at the hinges B and C. We thus use the equation given by the first row to evaluate the only unknown redundant, X_2. Substituting $X_1 = X_3 = 2$ tm, in the first equation.

$$-\frac{40}{3}X_1 + \frac{32}{2}X_2 + \frac{56}{3}X_3 + 21.33P = 0$$

We evaluate

$$X_2 = 1 \text{ tm}$$

The above is a "full load analysis" for the given load, $P = 2.5$ t. The result may be verified to be the same as obtained by the "incremental load analysis" assuming elastic-plastic moment curvature relationship.

The result of the above example justifies the plastic analysis assuming rigid M-x at plastic hinges. The analysis can be used to compute angle discontinuities at the plastic hinges and deflections. To facilitate the deflection study, we should know which was the last plastic hinge to develop before collapse. That was the reason, in the above analysis or $P = 2.5$ t, we excluded the last plastic hinge formation by letting $(M_p)_D > 2$, or alternatively we could have consider $P = 2.5 - \varepsilon$, which justifies to consider $M_D \leq M_P$.

We thus have the choice of doing 'full load' analysis in place of 'incremental load' analysis without any significant loss of accuracy.

3.5.3 Plastic Analysis and Proportional Loading

We consider "proportional loads" only, that is, the loads acting simultaneously on a structure and described as multiples of a typical load P. Thus while increasing the load of magnitude P, the other loads are also increased proportionally. Such a proportional loading system may be referred as a generalized load, P. The elastic analysis for a given proportional loading system, as we are familiar, consists of

(i) **Equilibrium** — statical moment diagrams, m_0 and m_1.
(ii) **Deformations** — based on material properties (M-x relation) and assuming moment $m < m_p$, the deformations are computed at the chosen releases and compatibility at all other sections is not violated, and
(iii) **Compatibility** — the redundants x_i are so chosen that the final moment $m = m_0 + x_i m_i$ restores compatibility at the chosen releases.

If the given generalized load, P, is increased to the load P_1 such that the largest of all (local) maximum moments is equal to M_P, we know from the elastic plastic analysis that

$P_1 <$ collapse load, P_c

If $\lambda_1 P = P_1$ and $\lambda_c P = P_c$

$\lambda_1 < \lambda_c$

Further if due to a generalized load λP, we find a statical m-diagram such that we satisfy

(i) EQUILIBRIUM
(ii) YIELD CONDITION, $m \leq m_p$

Without paying any attention to compatibility, then,

$$\lambda \leq \lambda_c$$

The above is the statement of the "Lower Bound" theorem. The plastic moment distribution* which is based on the above steps (i) and (ii), is a lower bound method. It is obvious that a lower bound method gives a safe load.

However, if a statical m-diagram due to λ_P results in sufficient number of plastic hinges so as to cause failure mechanism, the λP is the actual collapse load P_c, that is,

$$\lambda = \lambda_c$$

the lower bound method of finding collapse load $\lambda_c P$ based on the sequence, (i) equilibrium, (ii) yield condition, $m \leq m_p$ and (iii) collapse mechanism, is tedious. Generally the following upper bound method is preferred.

Upper Bound Method In the case of the "Upper Bound" method, we assume a possible failure mechanism, and we find the corresponding failure load λ_P which satisfies equilibrium. The failure load λ_P so found is such that

$$\lambda \geq \lambda_c$$

Thus the computed collapse load ensured

(i) Collapse MECHANISM and

(ii) EQUILIBRIUM

However, it a bending moment field, m-diagram (which satisfies equilibrium), can be obtained such that it satisfies one more condition,

(iii) YIELD CONDITION, $m \leq m_P$, then

the λP is the actual collapse load P_c, that is, $\lambda = \lambda_c$

In the following, upper bound method is used for finding collapse load, $\lambda_c P$. As stated already, Article 3.5, the checking of the third step, YIELD CONDITION $m \leq m_p$, is tedious in the case of partial collapse. In the case of complete collapse the moment field, m-diagram, is obtained uniquely from static equilibrium. We, therefore, start with the examples of complete collapse to be followed by the examples of partial collapse.

3.6 BASIC MECHANISMS

We are familiar with (i) Beam mechanisms and (ii) Sway mechanisms for evaluating moments directly while satisfying equilibrium. Following the approach, **joint mechanisms** are used for satisfying joint equilibrium, and **sway** mechanism is used in the case of gable frames for obtaining equilibrium equation for gable frames. We will use these mechanisms while analyzing frames and gables. The above four types of mechanisms that are 'basic' because they are used for writing 'independent' equilibrium equations, which directly evaluate moments without having to find shears or thrusts. These basic mechanisms that is beam mechanism for every load point, sway mechanism for every storey, joint mechanisms for every joint and gable mechanisms, are independent because any one of these mechanism cannot be obtained by combining the remaining. The equations of equilibrium obtained by considering the basic mechanisms as virtual deformations are independent equations of equilibrium.

The maximum moments in a frame occur at the ends of the members meeting at the joints and under the loads. In the case of joints with only two members meeting, we consider only one maximum moment which implies that joint equilibrium is automatically satisfied. If m is number of maximum moments and if the degree of indeterminacy is n, the number of independent equations of equilibrium involving the unknown maximum moments are $m - n$. While writing an equation of equilibrium using any of these basic mechanisms as virtual deformation systems, the maximum moments in real structure corresponding to the angle discontinuities in virtual structure appear in the equation. *If the number of maximum moments are equal to the number of basic mechanisms, the structure is determinate*. However, if corresponding to a mechanism, all the maximum moments corresponding to the hinges causing mechanism are specified equal to m_p, it is possible only for the fixed value of λP and this gives the collapse load $P_c = \lambda_c P$. In this case, the 'virtual' mechanism is also a collapse mechanism of 'real' structure.

3.7 COMBINED MECHANISMS

The independent mechanisms are useful as these mechanisms together with their combinations, are tried to explore true collapse load, which is the minimum of all possible collapse loads computed. The combinations of the basic mechanisms are the "Combined Mechanisms". Quite often, it may be necessary to combine the basic mechanisms with a combined mechanism so as to find yet another combined mechanism. It is a process of exploration till we get the minimum collapse load. Because of a collapse load may not be 'the true collapse load', we draw corresponding moment diagram for checking $M_{max} \leq M_P$. Artificial stiffening suggested by Heyman is discussed elsewhere.

In the examples considered, effects of thrust and shear are neglected.

The concept of Instantaneous Centre of Rotation (ICR) is explained while solving the examples. The understanding of ICR in the present context is useful for finding the deformations of assumed mechanisms. However, the ICR is useful in the displacement method also, when the members are assumed to be inextensible.

Examples of Complete Collapse

Ex. 3.2. Ref. Fig. Ex. 3.2(a) for data. M_p is same for all members.

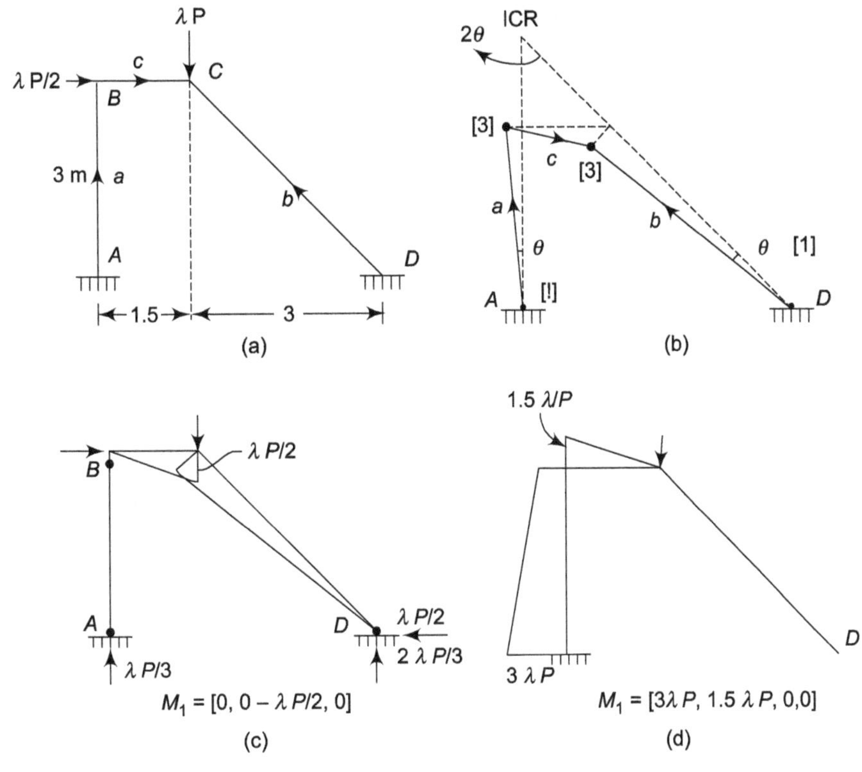

Fig. Ex. 3.2

The number of local maximum moment points = 4, that is, sections A, B, C and D.

The degree of indeterminacy = 3.

The number of basic mechanisms = 4 – 3 = 1.

(i) Collapse MECHANISM. The sway mechanism, Fig. Ex. 3.2(b), is the only mechanism. Wherever possible, it is a good practice to show hinges at the end of the members meeting the joint.

(ii) EQUILIBRIUM. The collapse mechanism is also considered to be the virtual deformation system for writing the equation of equilibrium.

The members a, b and c, Fig. (b), undergo rigid body rotations and displacements. For the members a and b the support A and D are the centers of rotation. The center of rotation of the member c changes its position. As we are considering small deformations, the initial geometry is considered to locate the

Instantaneous Center of Rotation (ICR) or the member *c*. Making use of the centers of rotations, *A*, *D* and ICR, the angles of rotation at the corresponding points are $[\theta_i, \theta, 2\theta]$ and the angle discontinuities at the points *A*, *B*, *C* and *D* are

$$\theta_i = [-\theta, 3\theta, -3\theta, \theta] \quad \text{...(a)}$$

or simply $\theta_i = [-1, 3, -3, 1]$. θ being arbitrary is omitted.

Sign Convention The angle discontinuities corresponding to tension on outside fibres are positive. The values of angle discontinuities are shown within parentheses [ ] and on tension side. The internal work is due to the above angle discontinuities only, as elsewhere it is rigid body deformation. All the terms of the internal work are positive at the plastic hinge at internal moment = M_p.

$$\text{Internal work} = \Sigma(M_p)\ \theta_i = 8\ M_p\ \theta \quad \text{...(b)}$$

Similarly, deformations in the direction of applied loads are -3θ at *B* and 3θ at *C*, and

$$\text{External work} = \frac{\lambda P}{2}(-3\theta) + P(3\theta)$$

$$= (3\lambda P/2)\theta \quad \text{...(c)}$$

We may obtain the same result (*c*) in a different way, which is based on virtual work. Introduce any release system to make the given structure determinate, Fig. Ex. 3.2(c). Draw the corresponding moment determinate moment diagram. In the *M* values from the statical diagram, at the points *A*, *B*, *C* and *D* are

$$M_i = [0, 0, -\lambda P/2, 0]$$

We get the external work from the Eqs (a) and (d),

$$\text{External work} = \sum M_i\ \theta_i$$

$$= 0 + 0 + -\lambda P/2\ (-3\theta) + 0 = (3\lambda P/2)\theta,$$

which is the same result as given by Eq. (*c*). The reader may verify the same result of external work from other statical diagrams, say, Fig. Ex. 3.2(d). The equilibrium is satisfied from virtual work principles. From the result of the Eqs (*b*) and (*c*)

$$\frac{3\lambda P}{2}\theta = 8\ M_p\theta$$

$$\therefore \quad \lambda P = 16/3\ M_p$$

As all the local maximum moments are equal to M_p, the yield condition $M_i \le M_p$ is satisfied.

$$\therefore \quad \lambda = \lambda_c = 16/3\ M_p/P$$

Ex. 3.3. Determine the ratios of *W* to *H* for all possible modes of plastic collapse for the frame, Fig. Ex. 3.3.

The possible modes of collapse are: one beam mechanism, Fig. (*b*) (*i*), one sway mechanism, Fig. (*b*) (*ii*) and one combined mechanism, Fig. (*b*) (*iii*). The equilibrium equations (based on principle of virtual work) for the three mechanisms are

$$M_p = \frac{W}{4} \text{ (beam mechanism)} \quad \text{...(i)}$$

$$H - \frac{W}{2} = \frac{7}{3}M_p \text{ (sway mechanism)} \quad \text{...(ii)}$$

$$3H + W = 12\ M_p \text{ (combined mechanism)} \quad \text{...(iii)}$$

Case I: If the failure is simultaneously taking place due to beam mechanism as well as combined mechanism, both the equations (*i*) and (*iii*) are valid, and eliminating M_p we get

$$3W = 3H + W \text{ or } H = \frac{2}{3}W$$

∴ For the range $0 < H \leq 2/3\ W$ for Beam mechanism.

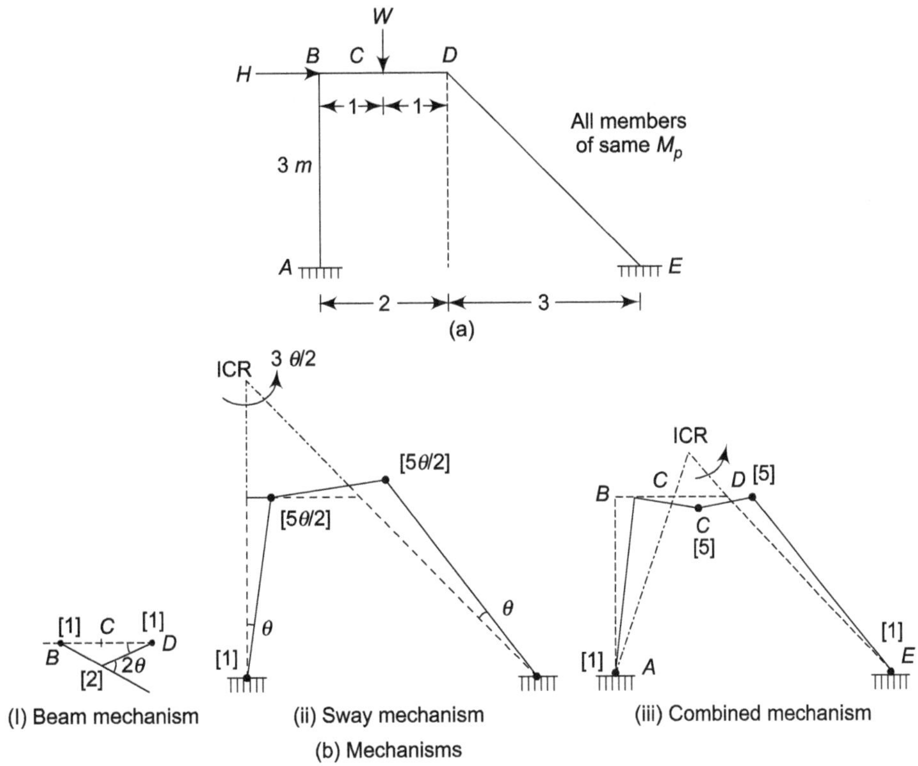

Fig. Ex. 3.3

Case II: Similarly, considering simultaneous failure due to combined and sway mechanisms, Eqs (*iii*) and (*ii*)

$$\frac{1}{12}(3H + W) = \frac{3}{7}\left(H - \frac{W}{2}\right)$$

$$H = \frac{5}{3}W \text{ for the simultaneous failure}$$

$$? \leq H \leq \frac{5}{3}W \text{ Combined mechanism}$$

$$\frac{5}{3}W \leq H \text{ Sway mechanism}$$

Summarizing the results

$$0 < H \le \frac{2}{3}W \text{ beam mechanism}$$

$$\frac{2}{3}W \le H \le \frac{5}{3}W \text{ combined mechanism}$$

$$\frac{5}{3}W \le H \text{ sway mechanism}$$

Ex. 3.4. Find collapse load, λ_c, Fig. Ex. 3.4(a). All members are of same M_p.
The number of local maximum points = 7
Degree of indeterminacy, $n = 3$
Number of basic mechanisms = 7 − 3 = 4

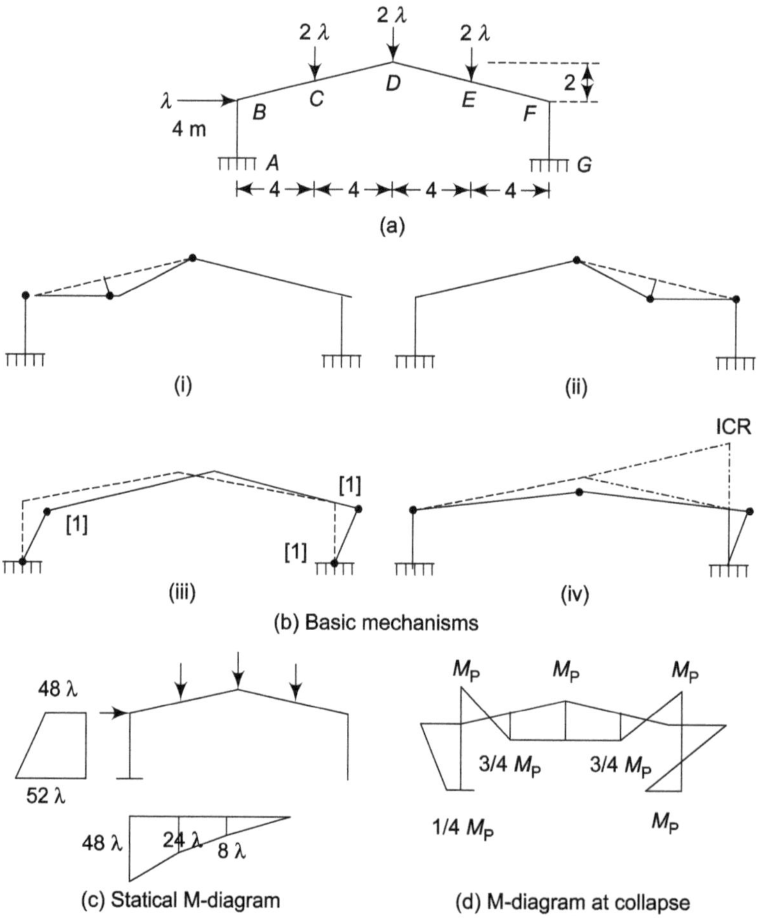

Fig. Ex. 3.4

The basic mechanisms are shown in Fig. (*b*). The Fig. (*b*) (*iv*), is the gable mechanism. The deformation sets for the basic mechanism are

$\theta_i = [0, 1, -2, 1, 0, 0, 0]$ beam mechanism, Fig. (b) (i)

$\theta_i = [0, 0, 0, 1, -2, 1, 0]$ beam mechanism, Fig. (b) (ii)

$\theta_i = [1, -1, 0, 0, 0, 1, -1]$ sway mechanism, Fig. (b) (iii)

$\theta_i = [0, 1, 0, -2, 0, 2, -1]$ gable mechanism, Fig. (b) (iv)

For computing external work, a statical m-diagram is suggested, Fig. (c).

Statical set $M_i = \lambda\,(52, 48, 24, 8, 0, 0, 0]$ ...(a)

Collapse loads:

(i) Beam mechanism, $48\lambda - 48\lambda + 8\lambda = 4\,M_p$

$$\lambda = \frac{1}{2}M_p$$

(ii) Beam mechanism, $\lambda = (1/2)\,M_p$

(iii) Sway mechanism, $52\lambda - 48\lambda = 4M_p$

$$\lambda = M_p$$

(iv) Gable mechanism, $48\lambda - 16\lambda = 6M_p$

$$\lambda = \frac{3}{16}M_p$$

The mechanism (iv) appears to be the collapse load. To make sure, check the yield condition

$$(M_A, M_C, M_E) \le M_P$$

Moment set based on mechanism (iv),

$$M_i\,[M_A, M_P, M_C, -M_P, M_E, M_P, -M_P] \quad \text{...(b)}$$

In the above, moment causing tension on outside fibers is considered positive.

Use virtual work method to find M_A, M_C, M_E

Consider mechanism *(iii)* for virtual deformations to evaluate M_A

Virtual deformation set $\theta_i = [1, -1, 0, 0, 0, 1, -1]$ ...(c)

From Eqs (b) and (c)

Internal work $= M_A - M_P + M_P = M_A + M_P$ ...(d)

Similarly from Eq. (a) and virtual deformation set (c) of mechanism (iii), we write

External work $= 52\lambda - 48\lambda = -4\lambda$

$= 4 \times 3/16\,M_p = 3\,M_p/4$ ...(e)

From virtual work principle, we equate the results from Eqs. *(d)* and *(e)*

$$M_A + M_p = 3\,M_p/4$$

$$M_A = -\,M_p/4$$

Similarly, consider virtual deformation set corresponding to the mechanism *(i)*, $\theta_i = [0, 1, -2, 1, 0, 0, 0]$. Ride the statical set Eq. (a) and the internal forces set (b), on the chosen virtual deformation set, θ_I, we get

$$M_C = -\frac{3}{4}M_p$$

The reader may verify, $M_E = \left(-\frac{3}{4}\right)M_P$ by considering the deformation set corresponding to the mechanism (ii).

The set M_p, Eq. (b), is drawn as moment diagram, Fig. (*d*). As the yield condition

$$M \le M_p \text{ is satisfied}$$

$$\lambda_c = \frac{3}{16} M_p$$

Ex. 3.5. Find the collapse load λ_c for the frame, Fig. Ex. 3.5(a)

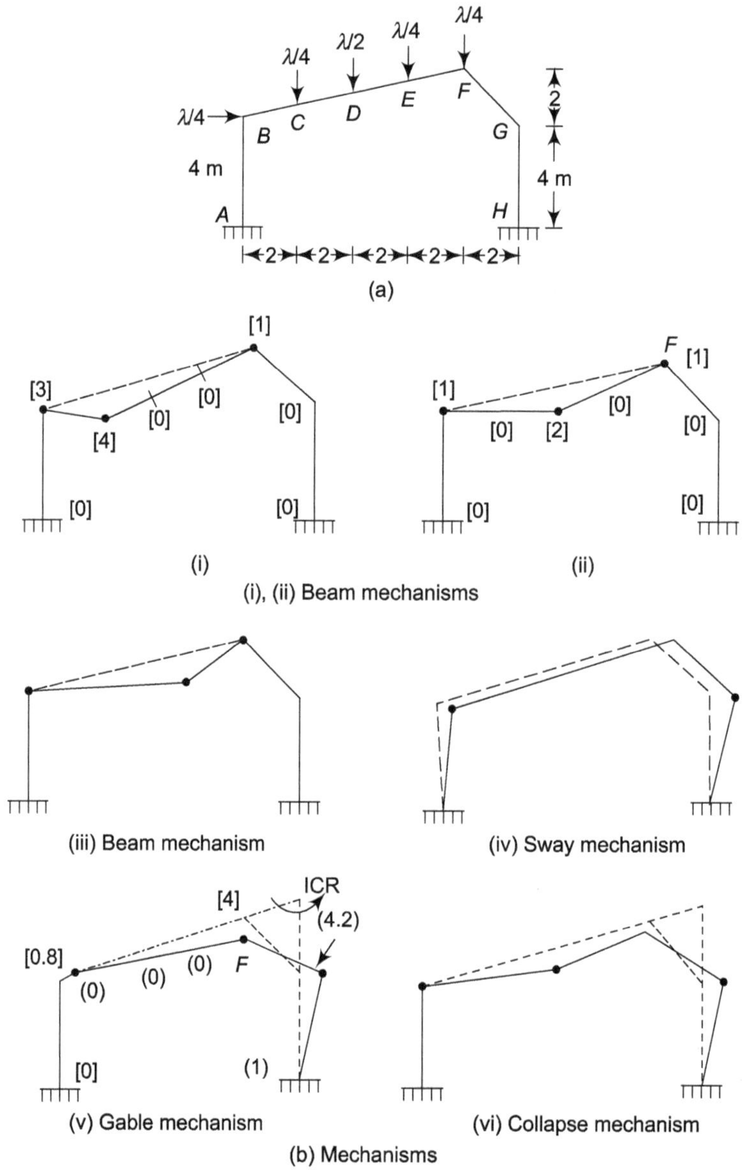

Fig. Ex. 3.5 *Mechanisms*

Number of maximum moment points, $m = 8$
Degree of indeterminacy, $n = 3$
Number of basic mechanisms = 8 – 3 = 5
The deformation sets for the basic mechanisms, Fig. (b), are given below:-

(i) $\theta_i = [0, 3, -4, 0, 0, 1, 0, 0]$ Beam mechanism,
(ii) $\theta_i = [0, 1, 0, -2, 0, 1, 0, 0]$ Beam mechanism,
(iii) $\theta_i = [0, 1, 0, 0, -4, 3, 0, 0]$ Beam mechanism,
(iv) $\theta_i = [1, -1, 0, 0, 0, 0, 1, -1]$ Sway mechanism
(v) $\theta_i = [0, 0.8, 0, 0, 0, -4, 4.2, -1]$ Gable mechanism

For the statical set, M_i, introduce the three releases by cutting at the base, H

$$M_i = \left[7\lambda, 6\lambda, \frac{7}{2}\lambda, \frac{3}{2}\lambda, \frac{1}{2}\lambda, 0, 0, 0\right]$$

The collapse loads for the five basic mechanisms are

$$\lambda_1 = 2M$$
$$\lambda_2 = 1.333\ M_p$$
$$\lambda_3 = \lambda_1$$
$$\lambda_4 = 4\ M_p$$
$$\lambda_5 = 2.08\ M_p$$

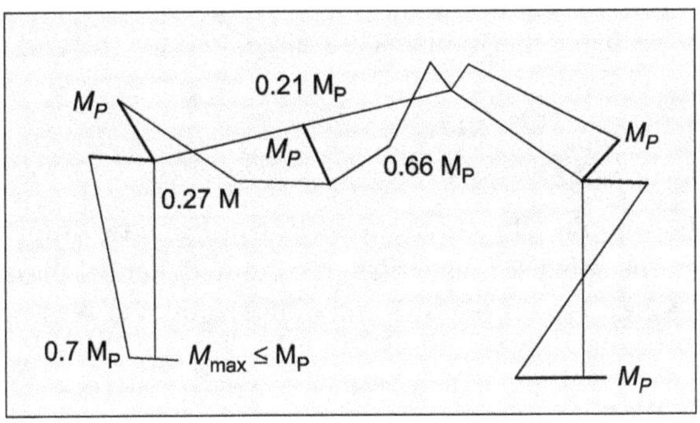

Fig. Ex 3.6 *Bending Moment Diagram at Collapse*

For mechanism *(vi)* combine mechanisms *(ii)* and *(v)*. The plastic hinges are at *B*, *D*, *G* and *H*. the hinge at F is cancelled, because of opposing deformations at F.

(ii) $4 \times (3\ \lambda = 4\ M_p)$

(v) $\dfrac{4.8\lambda}{16.8\lambda} = \dfrac{10M_p}{26M_p}$

Less 8 M_p due to hinge cancellation at *F*. The reader may verify that the equation for the mechanism (ii) is multiplied by 4 to cancel the hinge at *F*.

$$16.8\ \lambda = 18\ M_p$$
$$\lambda_6 = 1.071\ M_p$$

λ_6 appears to be the lowest. It is now required to check the yield condition $M_i \leq M_p$. The M_i set correspondining to the mechanism (vi) is

$$M_i = [M_A, M_P, M_C, -M_P, M_E, M_F, M_P, -M_P]$$

In the above, moment causing tension outside is considered positive.

The unknown moments M_A, M_C, M_E and M_F can be obtained by choosing suitable mechanisms to satisfy equilibrium and thus evaluate the unknown moments as it was done in the example, Ex. 3. 4.

The final m-diagram is shown in Fig. Ex. 3. 5(c).

From the moment field,

$$M_i = [.07\ M_p, M_p, -.27\ M_p, -.66\ M_p, .21\ M_p, M_p - M_p]$$

The individual values of M_i satisfy $M_i \leq M_p$

$$\therefore \qquad \lambda_6 = \lambda_c = 1.071\ M_p$$

3.8 VIRTUAL STIFFENING

The virtual stiffening may be used as 'real stiffening. The virtual 'K' factor is then used as real the implication of using virtual as real stiffening. The implication of converting 'virtual of real' is the reduction in ductility an essential feature for the design against earthquake. A 'k' factor inihibits the impending collapse by appropriately stiffening the cross section. It is further explained with the help of an illustration.

Consider continuous beam, Fig. Illustration (a). the values of M_p at different sections are shown in the figure. To draw a statical m-diagram by the virtual stiffening.

Solution: The number of probable mechanisms are three, two for the end spans and one for the mid span.

Collapse load for end span, $\lambda_1 = 32/3$, and

For mid span, $\lambda_2 = 20/3$

Inhibition of Collapse It is required to stiffen the section at F and two support sections, to the left of B and to the right of C, so that the collapse mechanism is inhibited. Let the plastic moment of resistance of these sections be k_1 at F and $2\ k_1/3$ to the left of B and to the right of C. writing the equilibrium equation for the stiffened segment,

$$\frac{\lambda}{2} = \frac{10}{3}k_1$$

$$\therefore \text{ stiffening factor,} \quad k_1 = \frac{3\lambda}{20}$$

Collapse loads for the stiffened continuous beam

Consider central span with plastic hinges entirely within the beam BFC, that is at F (stiffened) and to the right of B and to the left of C (no stiffened).

$$\frac{\lambda}{2} = \frac{3\lambda}{20}(2) + 2k_1$$

$$\therefore \qquad \lambda_3 = 10$$

The stiffening factor at this stage is: $K_1 = \dfrac{3\lambda}{20} = 1.5$, if it is a true collapse load.

Considering end spans, we have two cases,

(i) Plastic hinges at A, E and left of B, which gives, $\lambda_4 = 40/3$ and

(ii) Plastic hinges at A, E and right of B; $\dfrac{\lambda}{2} = (3)\dfrac{2}{3} + 1 = 3\,;\ \lambda_5 = 12$

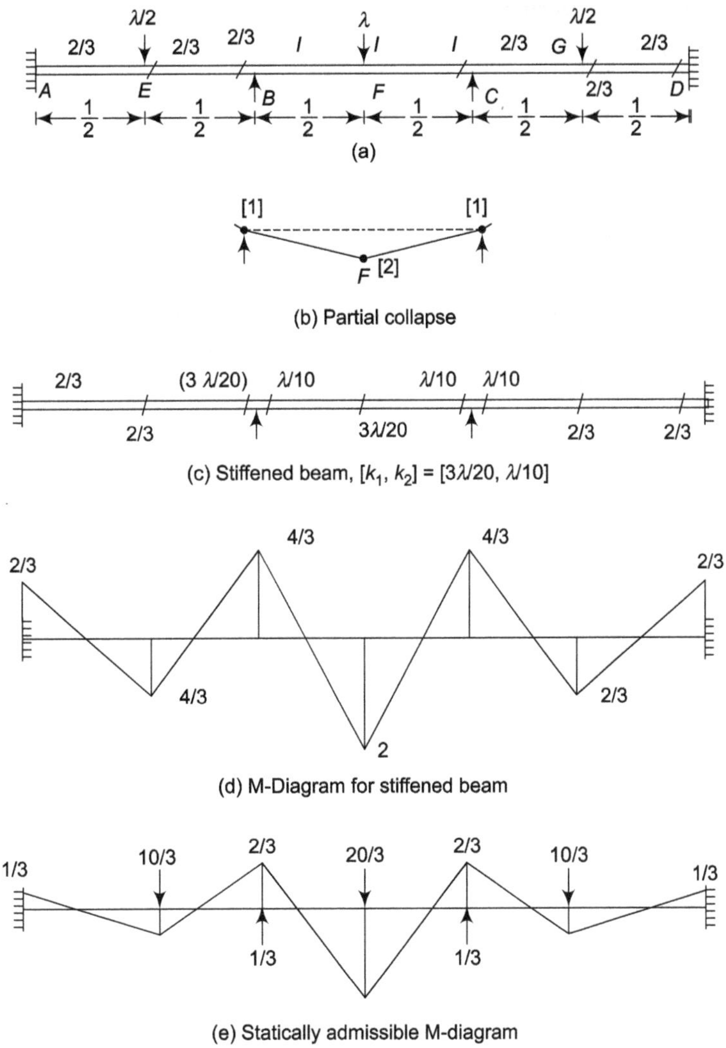

Partial Collapse

For the stiffened beam λ_3 gives the lowest collapse load. Stiffening against this collapse,

$$\frac{\lambda}{2} = \frac{3\lambda}{20} + 2k_2$$

$$k_2 = \frac{\lambda}{10}$$

Collapse loads after the second stiffening defined by the factor k_2

We consider end spans only.

(i) Plastic hinges at A, E and left of B (same as λ_4)

$$\lambda_4 = 40/3$$

(ii) Plastic hinges at A, E and right of B

$$\frac{\lambda}{4} = (3)\frac{2}{3} + k_2$$

Substituting for k_2

$$\lambda_6 = 40/3$$

As it is to be expected

$$\lambda_6 = \lambda_4$$

gives the collapse load for the stiffened beam and the corresponding stiffened beam and M-diagram are shown in Figs. (*c*) and (*d*) respectively. The Fig. (*d*) is the unique – diagram for the stiffened beam.

Scaling down the M_i diagram of Fig. (*d*), that is, dividing the moment ordinates by 2, so that $\lambda = 20/3$, we get a statically admissible M-diagram for unstiffened beam, Fig. 3.36(*e*). This is a virtual statical diagram which is used as a statically admissible diagram.

Ex. 3.7. Find collapse load for the frame Fig. 3.7(*a*) the values of plastic moment of resistance of columns = 8 *d* for the beams $m_p = 12$.

Solution: Basic mechanisms

Number of maximum moment points, $m = 11$.

Degree of determinacy, $n = 6$.

∴ Number of basic mechanism = 11 – 6 = 5

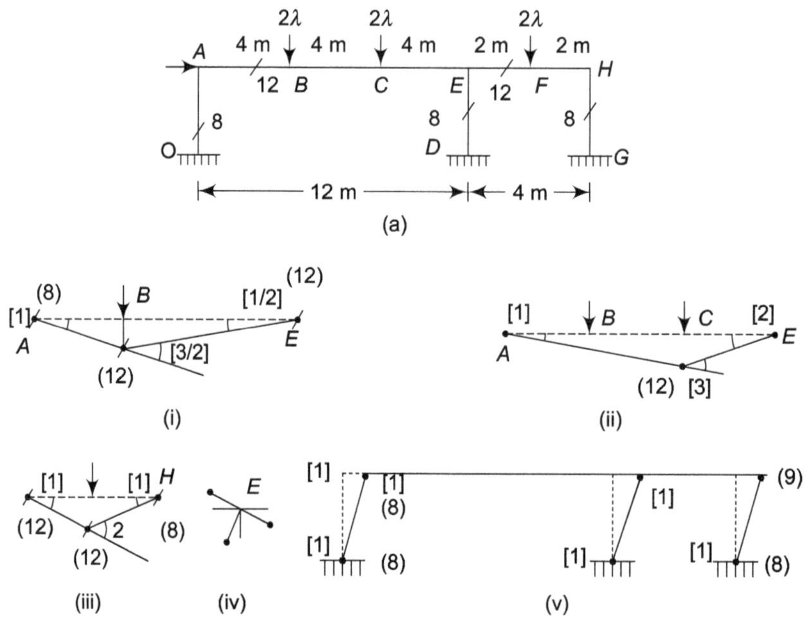

Fig. Ex. 3.7 *Contd.*

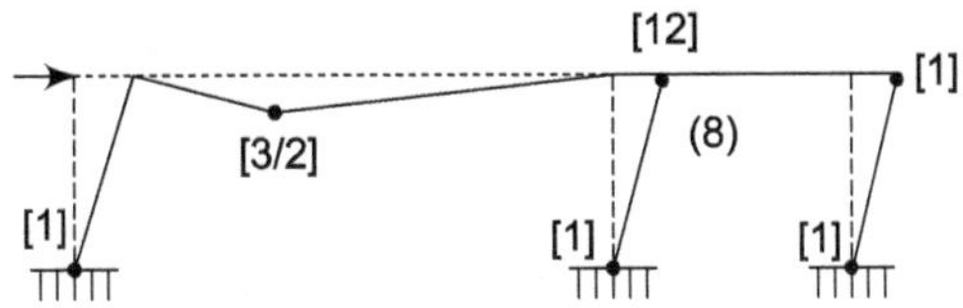

(vi)

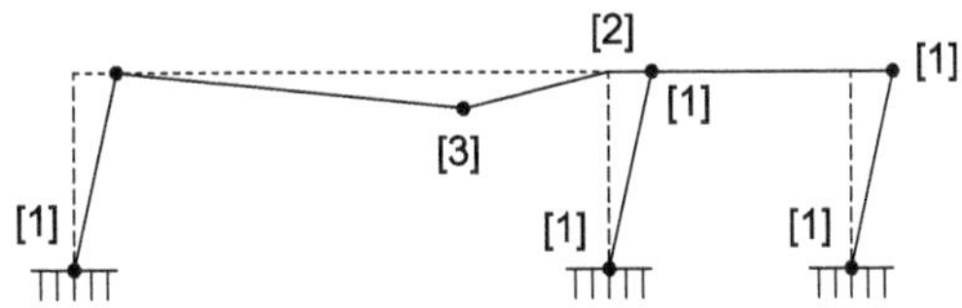

(vii)

(b) Mechanisms of unstiffened frame (i) to (vii)

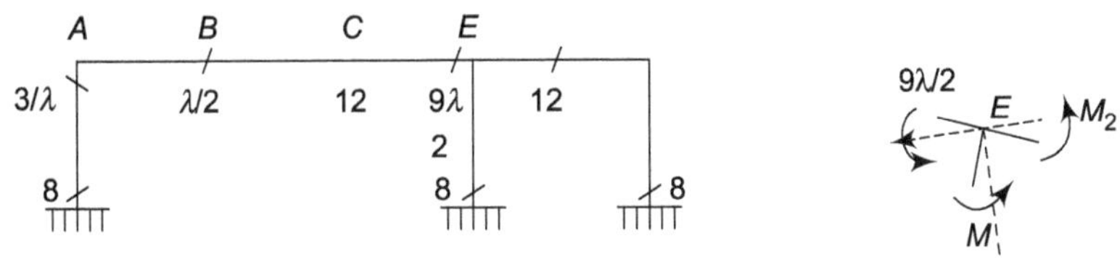

(c) Stiffening – 1

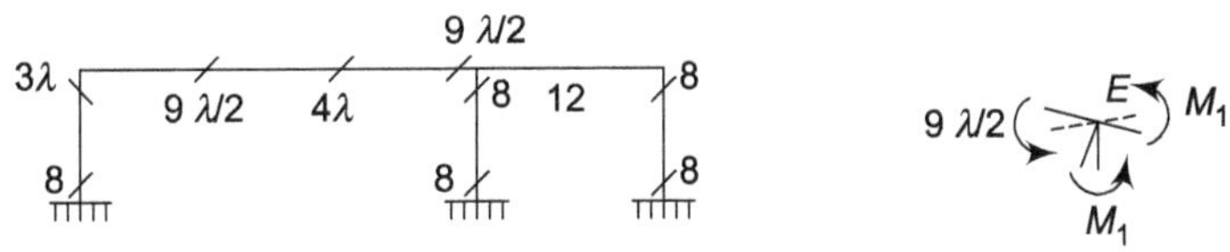

(d) Stiffening – 2

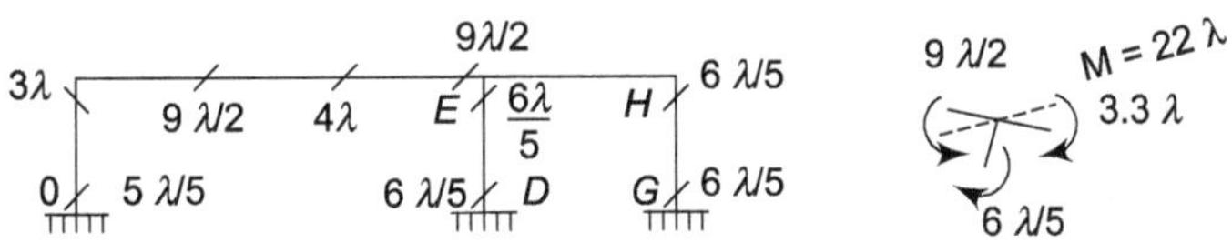

(e) Stiffening – 3

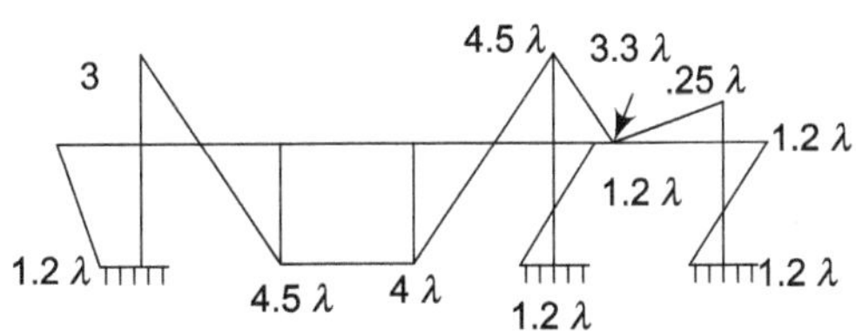

(f) M-Diagram for Stiffening – 3

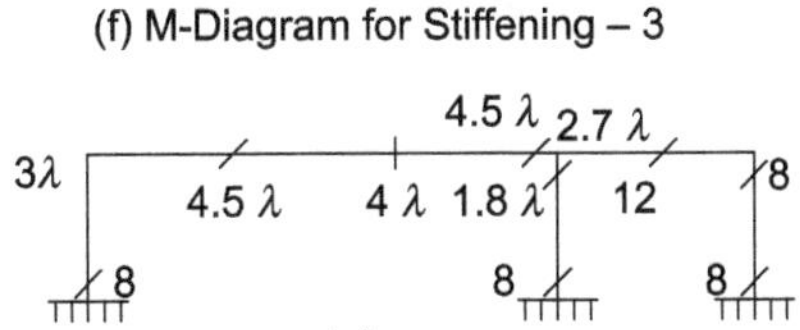

(g)

Fig. Ex. 3.7 *Contd.*

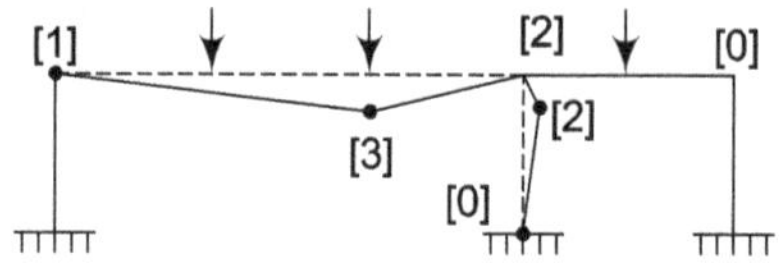

(h) Corrected stiffening – 3

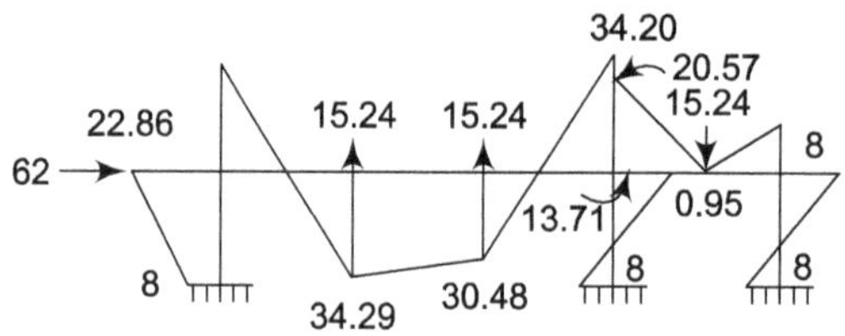

(i) M-Diagram for new stiffening – 3

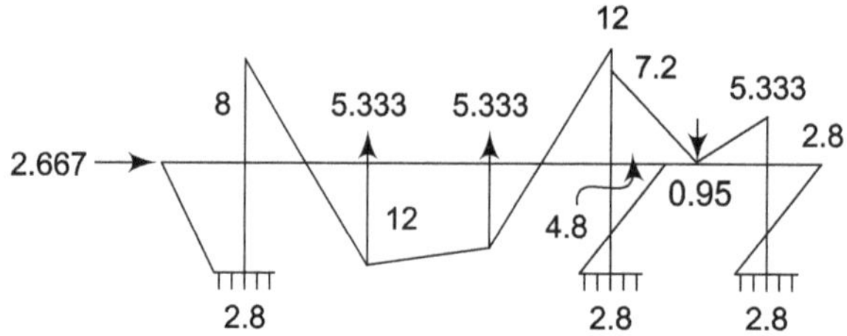

(j) Statical M-diagram, check M ≤ M_P

Fig. Ex. 3.7

That is, three beam mechanisms, Figs (*b*) (i) to (iii), one joint mechanism, Fig. (*b*) (iv), and one sway mechanism, Fig. (*b*) (v). Corresponding failure loads are

$$\left.\begin{aligned} \lambda_1 &= 2.667 \\ \lambda_2 &= 2.883 \\ \lambda_3 &= 11 \end{aligned}\right\} \qquad \text{Beam mechanism, Figs } (b) \text{ (i), (ii) and (iii)}$$

$$\lambda_4 = 16 \qquad \text{Sway mechanism, Fig. } (b), \text{ (v).}$$

The joint mechanism is used for equilibrium only and not for collapse as there is no moment load at the joint.

Combined mechanism

(v) = (i) + (iv) gives $\lambda_6 = 4.267$, Fig. (*b*) (vi)

(vii) = (ii) + (iv) gives $\lambda_7 = 3.703$, Fig. (*b*) (vii).

Out of the mechanisms tried, the mechanism (i) giving $\lambda_1 = 2.667$ appears to be the collapse mechanism. In the following, we stiffen the frame with a hope to get statically admissible M-diagram ($M_i \le M_p$).

Virtual stiffening is given in the following

Stiffening–I

To inhibit the mechanism (i) stiffen the section A of column and the section B and E of the left girder. Let the corresponding M_p values after the artificial stiffening be $8k_1$, $12k_1$ and $12k_1$, which gives the equilibrium equation.

$$12\lambda = 8k_1 + 12k_1\,(2)$$

$\therefore$ $k_1 = 3\lambda/8$, that is,

$(M_p)_A$ at the column end $= 3\lambda$, $(M_p)_B = 9\lambda/2$ and

$(M_p)_E$ at the and of girder $AE = 9\lambda/2$

The frame as stiffened above is shown in Fig. (c). Collapse loads corresponding to the first stiffening:
Mechanism (i), inhibited, $12\lambda = 12\,\lambda$

(ii) $\lambda_2 = 3$

Mechanism (iii) remains unaltered, $\lambda_3 = 11$

(iii) $3\lambda = 5(8) + 3\lambda$, not possible.

(v) Joint equilibrium, $9\lambda/2 + M_1 + M_2 = 0$, Fig. (c).

(vi) $\lambda_6 = 6.667$

(vii) $\lambda_7 = 4.22$

Other mechanisms are not considered hopefully. The mechanism (ii) appears to be the new collapse mechanism for the case of first stiffening.

Stiffening–2

Set the M_P value at C be $12k_2$, and corresponding to the mechanism (ii), we get the equilibrium equation,

$$24\lambda = 3\lambda + 12k_2\,(3) + 9\lambda$$

$$K_2 = \lambda/3$$

To the second stiffening, $(M_P)_C = (12)\,\lambda/3 = 4\lambda$
A stiffened frame is shown in Fig. (d).
Collapse loads corresponding to the second stiffening:

(i) Mechanism inhibited, $12\lambda = 12\lambda$

(ii) Mechanism inhibited, $24\lambda = 24\lambda$

(iii) Remains unaltered, $\lambda = 11$

(iv) Remains unaltered, $3\lambda = 40 + 3\lambda$, not possible

(v) Joint equilibrium, $9\lambda/2 + M_1 + M_2 = 0$, Fig. (*d*)

(vi) Remains unaltered, $\lambda = 6.667$

(vii) $27\lambda = 9\lambda + 76 + (4\lambda \times 3 - 36)$

$27\lambda = 21\lambda + 40$

$\lambda = 6.667$, as to be expected, it is same as (vi)

Out of the mechanisms tried, the mechanism (vi) being the lowest among the collapses computed, let us inhibit it in the following as stiffening–3.

Stiffening–3

We stiffen the five sections of column meeting at *O, D, E, G* and *H*, Fig. (*e*).

Collapse loads corresponding to the third stiffening:

(vii) $27\lambda = 21\lambda + 40k_3$

$K_3 = 3\lambda/20$

$\therefore$ $(M_P)_{O,\,D,\,E,\,G,\,H}$ of the column $= (3\lambda/20)\,8 = 6\lambda/5$, Fig. (*e*)

Collapse loads corresponding to the third stiffening:

(i) $12\lambda = 12\lambda$

(ii) $24\lambda = 24\lambda$
(iii) $4\lambda = 12 \times 3 + 6\lambda/5$
$\lambda = 12.86$
(iv) Not possible
(v) Joint equilibrium, Fig. (*e*), gives $M = 3.3\lambda$
(vi) $15\lambda = 15\lambda$
(vii) $27\lambda = 27\lambda$

Figure (*f*) is the corresponding M-diagram
Corresponding of $\lambda = 12.86$

$M = 3.3\lambda$ (joint equilibrium) exceeds M_P value, or alternatively if we limit M to M_P

$M = M_P = 12 = 3.3\lambda$

$\therefore$ $\lambda = 4/1.1 = 3.654$ which is less than 6.667.

Corrollary: If the value of λ subsequent to stiffening is less than the previously computed λ, then the assumed mechanism for the stiffening is not the true mechanism to be inhibited.

Based on the above corollary, we try some more mechanisms before we do the third stiffening. Let us try the mechanism (viii), Fig. (g), which is suggested to us from the result, $\lambda = 3.654$ computed above.

(viii) $24\lambda = 3\lambda\,(1) + 4\,\lambda(3) + 8\,(2) + 12(2)$
$\lambda = 4.444 < 6.667$

New Stiffening-3, based on the mechanism (viii) which is to be inhibited:

$$24\lambda = 15\lambda + 16k_3 + 24k_3$$

$$k_3 = 9\lambda/40$$

We stiffen the end E of girder EH and the end E of the column ED and the frame so stiffened is shown in Fig. (h)

Collapse loads corresponding to the 'new' third stiffening:

(i) $12\lambda = 12\lambda$
(ii) $24\lambda = 24\lambda$
(iii) $\lambda = 11$
(iv) Not possible
(v) Joint equilibrium is satisfied
(vi) $15\,\lambda = 4 \times 8 + 1.8\,\lambda + (3/2) \times 4.5 + (1/2) \times 4.5$
$\therefore \quad \lambda = 7.62$

The unique moment diagram for the stiffened frame, Fig. (h), can be drawn easily, Fig. (i) for the $\lambda_c = 7.62$. The yield condition $M \le M_P$ is satisfied for the stiffened frame.

The moments are scaled down corresponding to the $\lambda_c = 2.667$ of the given frame, without stiffening. The reduction factor = 2.667/7.62. Figure 3.37(*j*) is the statically admissible M-diagram, and the yield conditions is automatically satisfied.

$\therefore$ $\lambda_c = 2.667$ is the true collapse load

Ex. 3.8. Find collapse load, λ_c, for the frame, Fig. Ex. 3.8(a). The plastic moment of resistance M_P, is same for all members. The number of basic mechanism = 8.

Excluding the two joint mechanisms, for the remaining six basic mechanisms, the collapse loads are shown in Figs. (b) (i) to (vi). Some typical combinations are also shown, Figs. (b) (vii) to (x). Out of the mechanisms tried, the joint mechanisms (iii) and (iv) occurring simultaneously give the collapse load, that is,

$$\lambda_3 = \lambda_4 = 0.889\ M_P$$

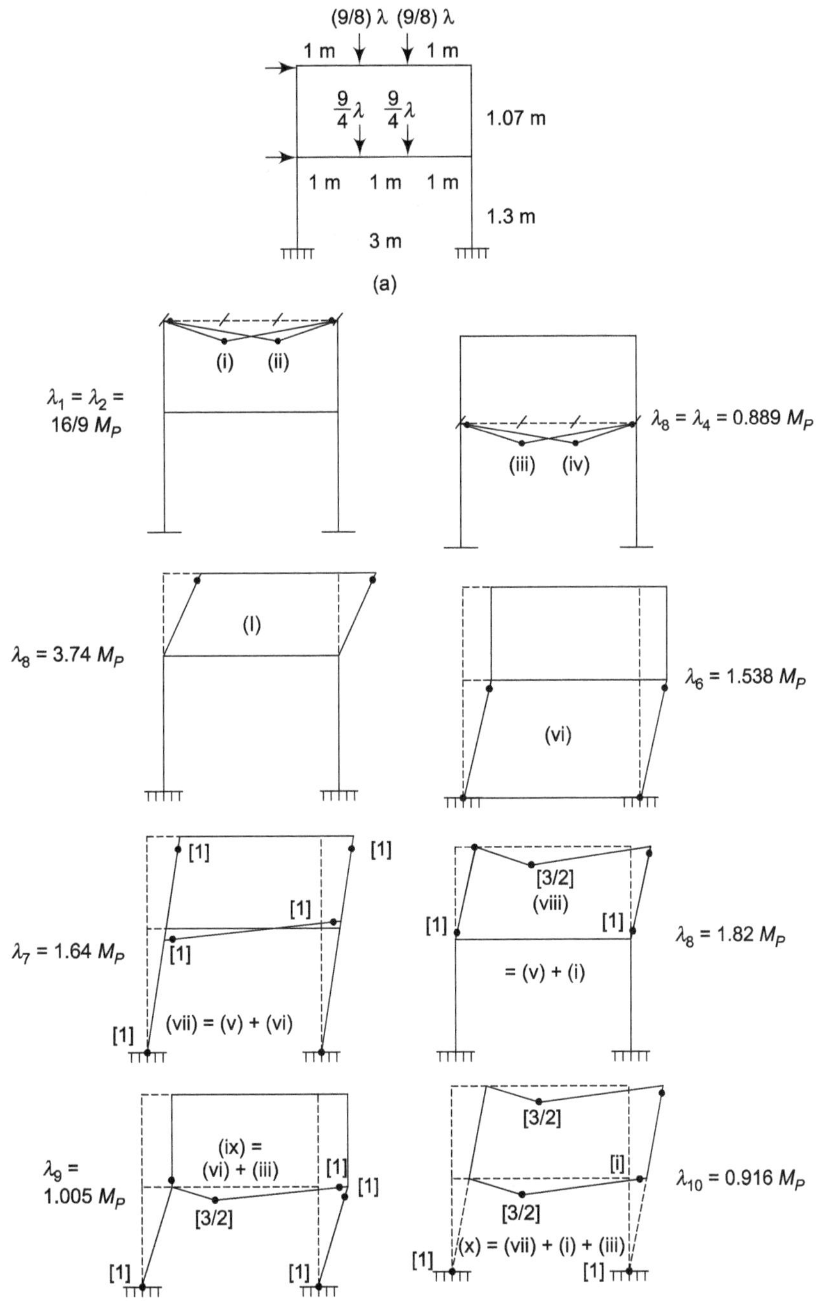

Fig Ex. 3.8 *Contd.*

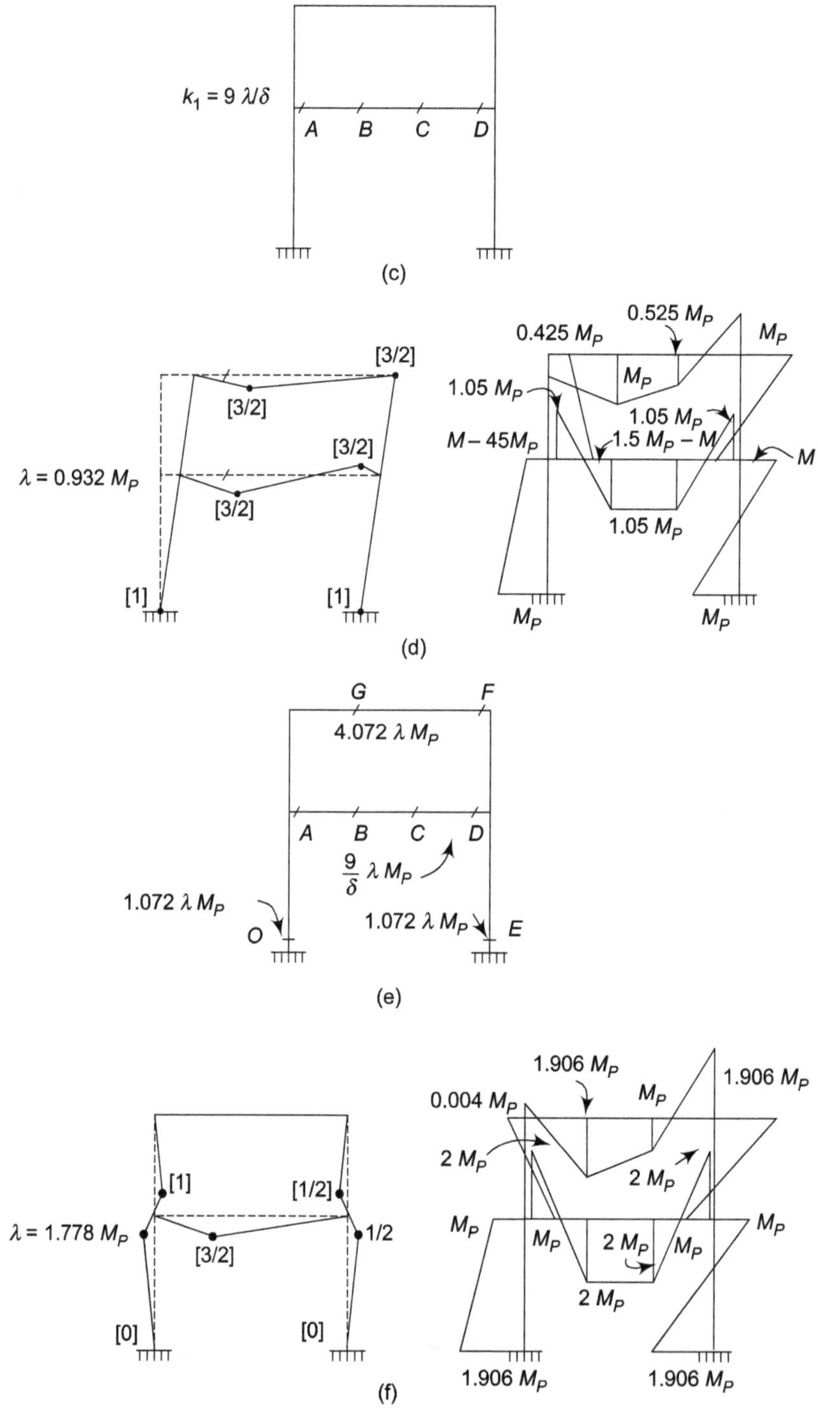

$k_1 = 9\,\lambda/\delta$
A B C D
(c)
0.425 M_P
0.525 M_P
M_P
[3/2]
[3/2]
[3/2]
[3/2]
$\lambda = 0.932\,M_P$
[1]
[1]
1.05 M_P
1.05 M_P
$M - 45M_P$
1.5 $M_P - M$
M
1.05 M_P
M_P
M_P
(d)
G F
4.072 $\lambda\,M_P$
A B C D
$\frac{9}{\delta}\,\lambda\,M_P$
1.072 $\lambda\,M_P$
O
1.072 $\lambda\,M_P$
E
(e)
1.906 M_P
1.906 M_P
0.004 M_P
M_P
2 M_P
2 M_P
[1]
[1/2]
$\lambda = 1.778\,M_P$
1/2
[3/2]
M_P
M_P
2 M_P
M_P
M_P
2 M_P
[0]
[0]
1.906 M_P
1.906 M_P
(f)

Fig Ex. 3.8

Being the case of partial collapse we obtain statical M-diagram by artificial stiffening.

Stiffening–1. To inhibit the mechanism (*iii*) and (*iv*) the sections A, B, C, D are stiffened.

$$(M_P)_{A,\,B,\,C,\,D} = k_1 M_P$$

Where $\quad k_1 = 9\lambda/8$, Refer Fig. (c)

The reader may find collapse loads for the stiffened frame. One such mechanism which is hopefully considered to be the collapse load is shown in Fig. (*d*). For the mechanism, Fig. (*d*), the collapse load is shown in Fig. (d). For the mechanism, Fig. (d) the collapse load

$$\lambda = 0.932\ M_P$$

With the six plastic hinges developed of the mechanism, bending moment diagram can be drawn with only one arbitrary moment, M, Fig. (*d*). It may be verified that we can obtain a statically admission moment diagram by choosing arbitrary M, such that

$$0.5\ M_P < M < M_P$$

The collapse load $\lambda_c = 0.889\ M_P$

However, the reader may verify the stiffening to inhibit the mechanism, Fig. (*d*), results in $k_2 = 1.072\lambda$. Such a stiffened frame is shown in Fig. (*e*), for which the collapse mechanism gives $\lambda = 1.778\ M_P$, Fig. (*f*).

The moment diagram, Fig. (*f*), is the unique diagram for the stiffened frame, which undergoes complete collapse.

For the given frame, Fig. (*a*), a statically admissible m-diagram is obtained by scaling down the m-diagram (*f*) by a factor = 1/2, which gives $M = M_P$ at the point of maximum moments in the girder $ABCD$, and elsewhere $M < M_P$

Ex. 3.9. Find collapse load for the frame, Fig. 3.9(a), for the following two cases.

(i) $H = 0.6\lambda\ P$ and (ii) $H = 0.5\ \lambda\ P$.

Draw statically admissible m-diagram for each case, using artificial stiffening.

(*i*) Beam mechanism is the lowest.

$$\lambda_c = 2M_P / P$$

To inhibit the mechanism, artificially stiffen the sections B, C and D

$$k\ M = P/2$$

For the stiffened frame, sway mechanism is not possible. And combined mechanism is the collapse load for the stiffened frame.

$$\therefore \quad \lambda_P = 5M_p/P$$

The m-diagram for the stiffened frame is drawn, Fig. (b). When the moments, Fig. (b), are reduced by the scaling factor = 0.4, such that the absolute maximum moment = M_P, we get a statically admissible m-diagram at collapse load of the given frame.

(*ii*) In this case $H = 0.5\ \lambda\ P$, the collapse load is given by the beam mechanism only. (The reader may verify that for $0 < H < 0.75\ \lambda_P$, it fails by beam mechanism only.)

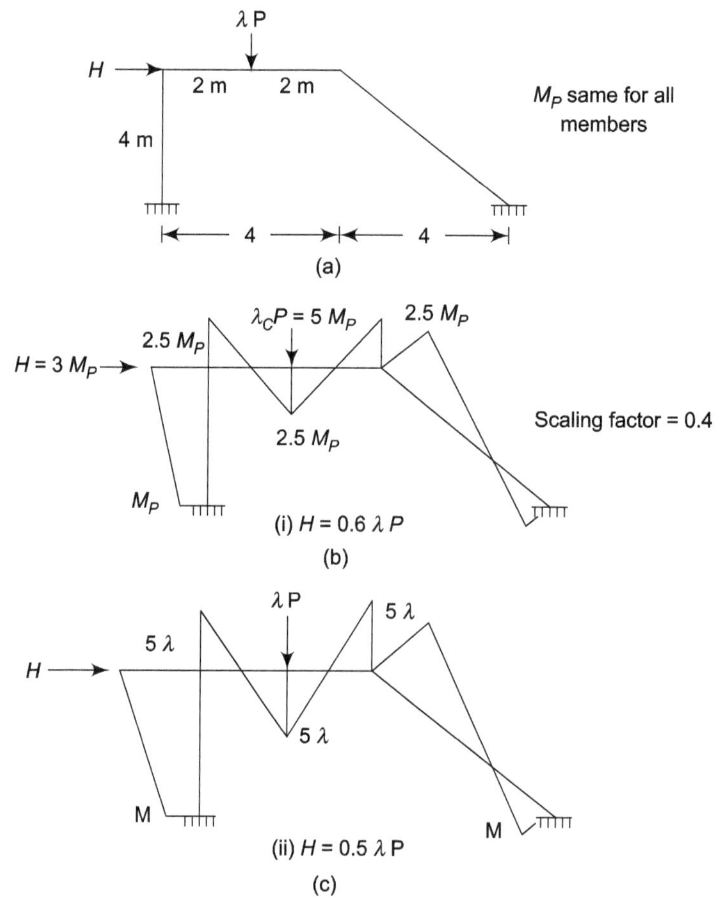

Fig. Ex. 3.9

For the stiffened frame with

$$(M_P)_{B,C,D} = k\,M_P = 0.5\,\lambda\,P$$

The sway and combined mechanisms are not possible (because λ is negative for the sway mechanism and it is infinity for the combined mechanism) suggesting that with the artificial stiffening the frame does not fail at all. The statically admissible moment diagram is given in Fig. (*c*). The value of moment M at the bases A and E may be any value between $-M_P$ and $+M_P$.

Ex. 3.40. Obtain statically admissible *m*-diagram or the frame of Ex. 3.27.
The basic mechanisms and one combined mechanism are shown, Fig. Ex. 3.40. It is suggested that the reader computes external work while evaluating collapse load, by using a statical diagram, Fig. (*b*).
Beam mechanism gives the collapse load, $2\,\lambda_c = 2.5$.

With artificial stiffening, the frame does not fail at all as in the case of Ex. 3.37 (*ii*). The statically admissible *M*-diagram for the given frame is shown in Fig. Ex. 3.40 (*e*), where $-M_P < M < +M_P$.

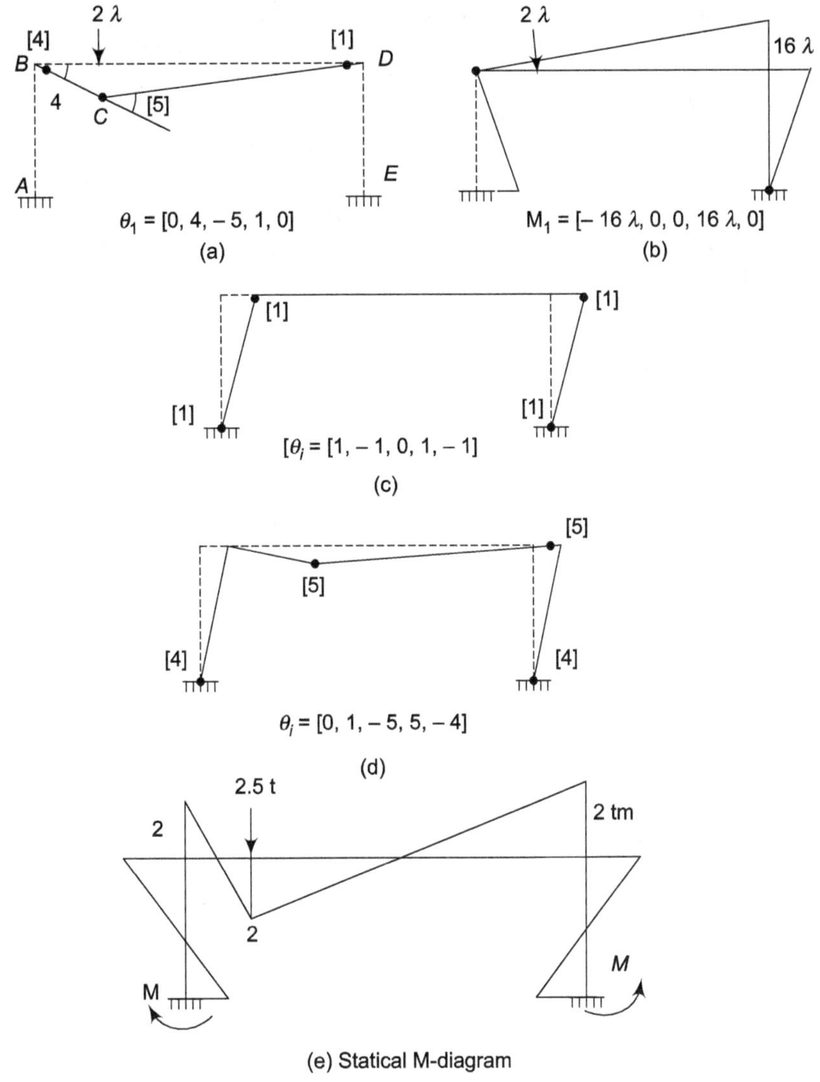

Fig. Ex. 3.10

3.9 FULL LOAD ANALYSIS AND 'LAST PLASTIC HINGES' BEFORE COLLAPSE

In the case of 'incremental load' analysis of frames due to a generalized load P, such that $P_i < P < P_{i+1}$ where P_i is the corresponding to the i-th plastic hinge, we start with the elastic analysis—the first stage. We keep tracking the development of i plastic hinges. The analysis for the i-th stage is done for the load, $\Delta P = P - P_i$. Obviously this method is recommended if only i is much smaller than the number of plastic hinges r, causing complete collapse. If, however, the load P_i is very close to the collapse load P_c, it is more expedient to use 'full-load' analysis, that is, 'elastic and rigid plastic' analysis, Article 3.5.2. But in the case of full load analysis, it is necessary to know which of the plastic hinges have already formed corresponding to the given load, which suggests that we should know the value of i. Alternatively, we may say which are the $r–i$ plastic hinges required for the collapse mechanism. $r–i$, is the number of last plastic hinge to form. In the case of plastic analysis, it is also necessary to know the angle discontinuities at the plastic hinges so as to check that the magnitudes of the angle discontinuities do not exceed the permissible values. In the example, Ex. 3.1, we implied the hinge at D to be the last to form. The following 'extra example' is chosen to illustrate to procedure for finding out the sequence of last plastic hinges to form. Some aspects have been repeated only to reiterate the basic concepts of inelastic analysis. The example Extra Ex. (b) (1) develops the theory for computing angle discontinuities $\hat{\theta}$ at the plastic hinges.

Extra Examples (*a*). Analyze the frame. Fig. (*a*), by 'incremental load' analysis for $P = 1.17$ ton, $K = 0.4$. Given: $M_p = 1.2$ tm, EI = constant, $L = 4$ m.

Stage 1: Elastic analysis for $K = 0.4$ and under generalized load P, gives the following maximum moments sequenced based on the magnitudes.

$$[M_C, M_D, M_E, M_B, M_A] = PL\ [0.3, 0.275, 0.225, 0.125, 0.025], \text{ Ref. Fig. } (b).$$

Maintaining proportional loading, with increasing P, one may anticipate development of plastic hinges in the sequence C, D, E, B, A, Fig. (*a*). The first plastic hinge develops at C at the load $P_1 = 1$ ton for $L = 4$ m and $M_p = 1.2$ tm. The values in the parentheses, Fig. (*b*), are the moments for $P = P_1$. As $P_1 < 1.17$ we proceed for the second stage of analysis.

Stage 2, analysis for $P > P_1$, is performed for $\Delta P = P–P_1$. The M-diagram for the reduced indeterminacy, $n = 2$, is given in Fig. (*c*) for $\Delta P = 1$. Anticipating the second plastic hinge at D,

$$1.10 + 2.3\ \Delta P_2 = 1.2 \text{ from which } \Delta P_2 = 0.0435 \text{ t}$$

The linear moment load relationship within the stage is given by

$$m = m_1 + \Delta P_2\, m_2$$

where m_1 and m_2 are the moments due to $P = P_1$ and $\Delta P = 1$ respectively, and the corresponding moment diagrams are as per Fig. (*b*) and Fig. (*c*). The incremental rate being the largest at D, $(= 2.3 > 1.7 > 1.5 > 0.5)$, other maximum moment points need not be tested for possible second plastic hinge. $P_2 = P_1 + \Delta P_2 =$ 1.0435 t. Figure (*d*) gives the corresponding moments. As P_2 is also less than the given load $1.17t$, we proceed for the third stage of analysis.

Stage 3: The results for the stage-3, $\Delta P = P – P_2$ are given in Fig. (*e*). If the moment at B is assume to reach $M_p = 1.2$ tm

And if the moment of E is assumed to reach $M_p = 1.2$ tm,

$$0.965 + 3.8\ \Delta P_3 = 1.2, \text{ from which } \Delta P_3 = 0.0618 \qquad \text{...(ii)}$$

From the values of the computed ΔP, (i) and (ii), the third plastic hinge forms at E and $P_3 = P_2 + \Delta P_3$ = 1.1053. Refer Fig. (*f*) for M-diagram corresponding. which continues to be less than the given load = $1.17t$. As three hinges have formed (at C, D and E), the structure is determinate for the fourth stage.

Stage 4: Refer Fig. (*g*), for the fourth stage. The linear equations for this stage, giving maximum moments are:

$$m_A = 0.191 + 5.6\ \Delta P \text{ and}$$

$$m_B = 0.822 + 4\ \Delta P$$

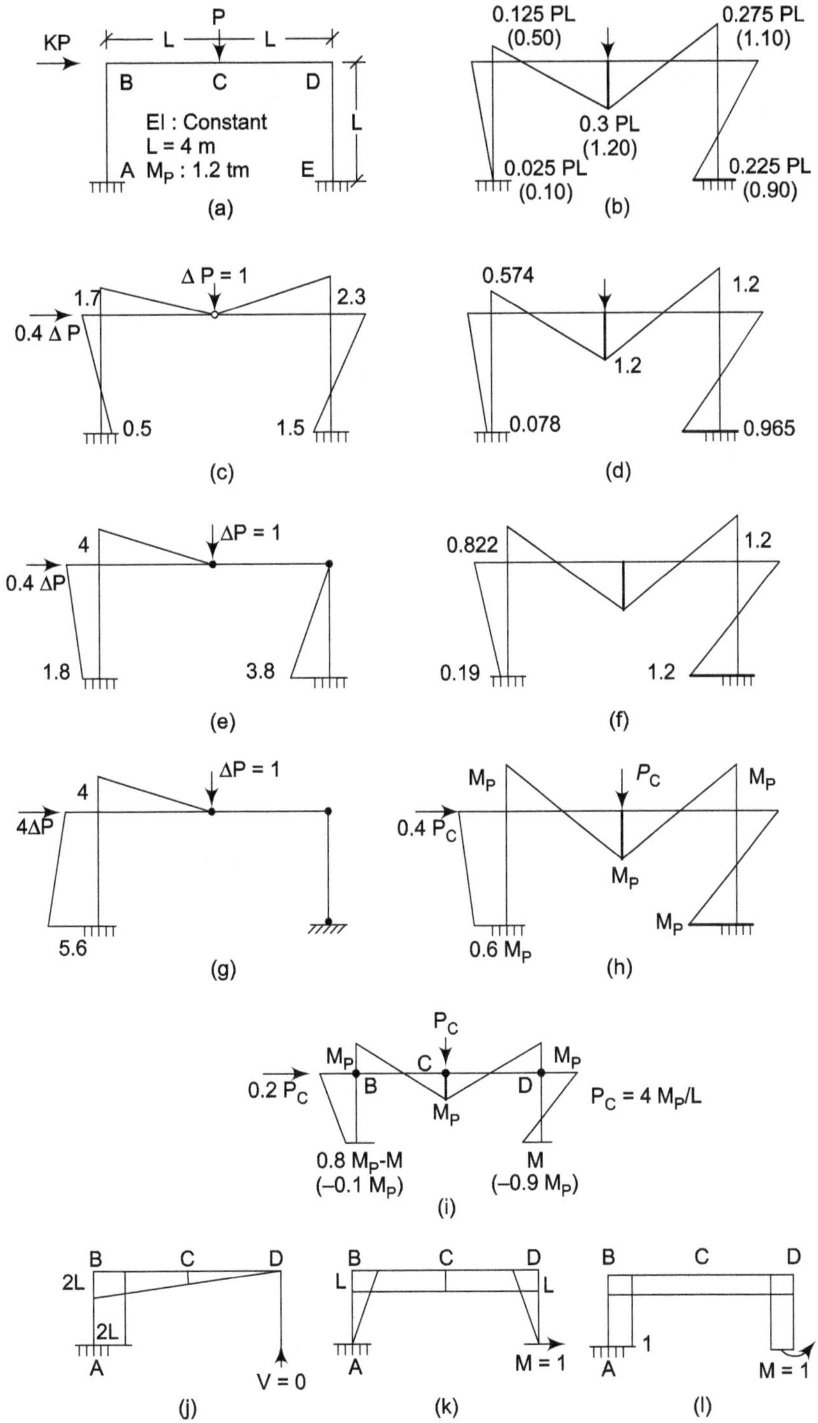
KP
P
L
L
B
C
D
EI : Constant
L = 4 m
A M_P : 1.2 tm
E
(a)
0.125 PL
(0.50)
0.275 PL
(1.10)
0.3 PL
(1.20)
0.025 PL
(0.10)
0.225 PL
(0.90)
(b)
Δ P = 1
1.7
2.3
0.4 Δ P
0.5
1.5
(c)
0.574
1.2
1.2
0.078
0.965
(d)
ΔP = 1
4
0.4 ΔP
1.8
3.8
(e)
0.822
1.2
0.19
1.2
(f)
ΔP = 1
4
4ΔP
5.6
(g)
M_P
P_C
M_P
0.4 P_C
M_P
M_P
0.6 M_P
(h)
P_C
M_P
C
M_P
0.2 P_C
B
D
P_C = 4 M_P/L
M_P
0.8 M_P-M
(–0.1 M_P)
M
(–0.9 M_P)
(i)
B
C
D
2L
2L
A
V = 0
(j)
B
C
D
L
L
A
M = 1
(k)
B
C
D
1
A
M = 1
(l)

(Contd.)

(Contd.)

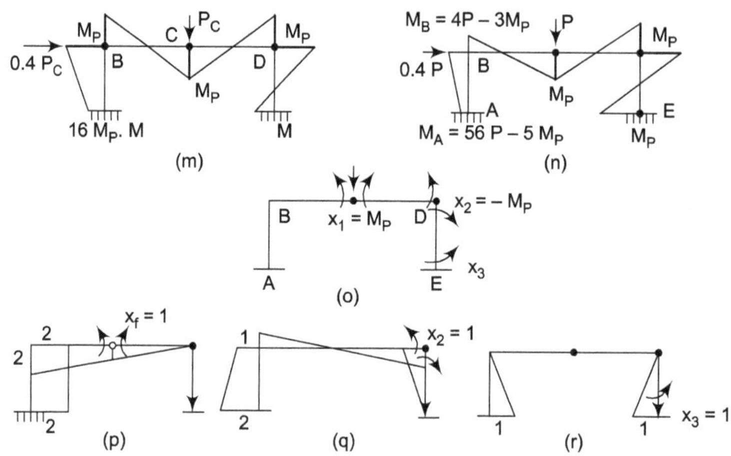

Illustration for full load analysis

and for ΔP = given load (= 1.17) – P_3 = 1.17 – 1.1053

$$\Delta P = 0.0647 \text{ t.}$$

Substituting ΔP = 0.0647 t in the above equations we get m_A = 0.553 and m_B = 1.08, we obtain the remaining moment and the analysis is complete.

It may be verified that ΔP_4 corresponding to the last plastic hinge at B is equal to 0.0945 which is larger than the increment ΔP = 0.0647 over the load P_3 i.e.,

$$1.17 < P_4, \quad P_4 = P_C = 1.105 + .095 = 1.2\ t$$

and it fails by beam mechanism. The m-diagram at collapse load is given, Fig. (h).

Angle Discontinuities at Plastic Hinges and Last Plastic Hinge*

The theory is developed with the help of following example

Illustration (b) (1). Full load analysis at collapse for K = 0.2, Fig. (a). Find collapse load and also the last plastic hinge to form. Also find angle discontinuities at collapse ($P = Pc - \varepsilon$). ε is a positive infinitesimal value. Collapse is due to beam mechanism. Pc = 1.2 t. The failure being a partial collapse, from statics alone, m-diagram is not unique, and an admissible m-diagram at collapse is shown in Fig. (i). The evaluation of the unique value of m in the Fig. (i) and the value of angle discontinuities is explained in the following.

Evaluation of angle discontinuities: From the example considered for $P < Pc$ the structure is stable. Just before collapse and $P > P_2$, the degree of indeterminacy is 'one' and thus there is correspondingly only one undetermined value of m. Consider a set of three statical diagrams for the evaluation of two angle discontinuities and the value of m corresponding to the single indeterminacy in this case. Treat Fig. (i) as deformation structure and Figs (j), (k) and (l), as the arbitrary statical systems for writing complementary virtual work equations, Eqs (i), (ii) and (iii). Just before collapse, one of the angle discontinuities from among, $\hat{\theta}_B, \hat{\theta}_C$ and $\hat{\theta}_D$ should be equal to zero as one of the three sections (B, C, D), is yet to be converted into a plastic hinge. It is the task to know which one of them is equal to zero, and that determines the last hinge to form. An angle discontinuity is considered positive if tension due

* The development is different and unique from the other derivations. The topic is developed based on derivations with logic. The reader may like this approach of the subject development

to flexure which causes the angle discontinuity is amount inside ring to the along convention for positive angle discontinuity).

From Figs (*i*) and (*j*)

$$2\hat{\theta}_B + \hat{\theta}_C - \frac{L}{6EI}(10.8\,M_P - 6m) = 0 \qquad \text{...(i)}$$

From Figs (*i*) and (*k*)

$$\hat{\theta}_B + \hat{\theta}_C + \hat{\theta}_D - \frac{L}{6EI}(4.8\,M_p - 2m) = 0 \qquad \text{...(ii)}$$

Lastly from Figs (*i*) and (*l*)

$$\hat{\theta}_B + \hat{\theta}_C + \hat{\theta}_D - \frac{L}{6EI}(8.4\,M_p - 6m) = 0 \qquad \text{...(iii)}$$

Always in the case of equations formed as above, Eqs. (i) to (iii), the angle discontinuities $\hat{\theta}$ can be delinked from the unknown moment *m*. Thereby the unknown moment *m* (their number equal to degree of indeterminacy just before collapse) can be evaluated without having to find the numerical values of the angle discontinuities, $\hat{\theta}$.

From Eqs (ii) and (iii)

$$3.6\,Mp, - 4m = 0$$

$\therefore$ $m = 0.9\,M_p$ and thus the base moments shown within parentheses, Fig. (i), are determined.

Assuming last plastic hinge to form at *B*, that is $\hat{\theta}_B = 0$ we get from Eqs (i) and (ii) $\hat{\theta}_C = 0.9\,M_p\,L/EI$ (it agrees with the sign of plastic hinge moment, that is, tension inside) and ($\hat{\theta}_D = 0.4\,M_p\,L/EI$ (it agrees with the corresponding plastic hinge moment causing tension outside – as per sign convention for angle discontinuities stated earlier).

As the plastic moment and corresponding angle discontinuities are consistent, the assumption of $\hat{\theta}_B$ equal to zero is correct and thus the last hinge forms at *B*. If it was wrongly assumed, the plastic moments and the corresponding angle discontinuities will not be consistent.

Illustration (b) (2) As an additional Illustration, let us analyse for $K = 0.4$, other data remaining the same.The collapse load $P_c = 1.2$ t remains unaltered.

Refer Fig. (*m*) for deformation structure and Figs. (*j*), (*k*), (*l*), for statical diagrams to obtain the following equations based on complementary virtual work.

$$2\hat{\theta}_B + \hat{\theta}_C - \frac{L}{6EI}(15.6\,M_P - 6m) = 0 \qquad \text{...(iv)}$$

$$\hat{\theta}_B + \hat{\theta}_C + \hat{\theta}_D - \frac{L}{6EI}(5.6\,M_P - 2m) = 0 \qquad \text{...(v)}$$

$$\hat{\theta}_B + \hat{\theta}_C + \theta_D - \frac{L}{6EI}(10.8\,M_P - 6m) = 0 \qquad \text{... (vi)}$$

From Eqs (v) and (vi), $m = 1.3\,M_p$, i.e.,

$M_E = 1.3\,M_P$ being larger M_P, suggests that before the last plastic hinge of the beam mechanism (hinges at *B*, *C* and *D*) develops, a plastic hinge at *E* is formed and that we should treat $M_E = M_P$. This observation corroborates with the solution by the incremental load analysis. Corresponding to the plastic hinge at E, angle discontinuity $\hat{\theta}_E$ is also required to be evaluated. The bending moment diagram is defined completely from statics only *($m = M_P$)*. We are, thus, required to find the angle discontinuities, i.e., $\hat{\theta}_B$, $\hat{\theta}_C$, $\hat{\theta}_D$ and $\hat{\theta}_E$ and out of which one angle discontinuity is equal to zero. The equations (iv), (v) and (vi) are modified to yield the following for $P = P_c - \varepsilon$.

$$2\hat{\theta}_B + \hat{\theta}_C - \frac{L}{6EI}(9.6\, M_P) = 0 \qquad \text{...(vii)}$$

$$\hat{\theta}_B + \hat{\theta}_C + \hat{\theta}_D - \frac{L}{6EI}(3.6\, M_p) = 0 \qquad \text{...(viii)}$$

$$\hat{\theta}_B + \hat{\theta}_C + \hat{\theta}_D + \hat{\theta}_E - \frac{L}{6EI}(4.8\, M_p) = 0 \qquad \text{... (ix)}$$

Assuming $\hat{\theta}_B = 0$

$\hat{\theta}_C = 1.6\, M_p\, L/EI$	Consistent	O.K.
$\hat{\theta}_D = -\, M_p\, L/EI$	Consistent	O.K.
$\hat{\theta}_E = 0.2\, M_p\, L/EI$	Consistent	O.K.

The signs of the angle discontinuities agree with the corresponding bending moments. The assumption of $\hat{\theta}_B = 0$ is justified and the above solution does not require to be revised.

Illustration (c) (1). Full load analysis of the frame for $K = 0.4$ and $P = 1.17$ ton. Compute angle discontinuities.

From (b) (2), the last plastic hinge is formed at B for $P = 1.2$, and for $P = 1.17$ the structure is determinate with plastic hinges at C, D and E. Being a determinate system the results obtained by 'incremental load' analysis based on elastic plastic behaviour and full load analysis on 'elastic and rigid plastic' idealization (Article 3.5.2) will exactly be the same. Besides the plastic hinges, the other "maximum moment-points" are A and B, and these moments are

$$M_A = 5.6\, P - 5\, M_p = 5.6 \times 1.17 - 5 \times 1.2 = 0.552 \text{ tm}$$

$$M_B = 4\, P - 3\, M_p = 1.08 \text{ tm}$$

(The results are the same as computed for stage-3 of incremental load analysis)

Hopefully, we assumed in the above that $P > P_3$ which we will verify in the following, while computing the angle discontinuities.

The moment diagram for $P_3 < P < P_c$ is shown in Fig. (n). Treating Fig. (n) as deformation structure and Figs (j), (k) and (l) as the statical system, the complementary virtual work equations are

$$\hat{\theta}_C + \frac{L}{6EI}(68\, M_p - 77.6\, P) \qquad \text{...(x)}$$

$$\hat{\theta}_C + \hat{\theta}_D + \frac{L}{6EI}(22\,M_p - 25.6\,P) \qquad ...(xi)$$

$$\hat{\theta}_C + \hat{\theta}_D + \hat{\theta}_E + \frac{L}{6EI}(36\,M_p - 40.8\,P) \qquad ...(xii)$$

From the Eqs (x), (xi) and (xii),

$$\hat{\theta}_C = -\frac{L}{6EI}(68\,M_p - 77.6\,P),$$

$$\hat{\theta}_D = -\frac{L}{6EI}(-46\,M_p + 52\,P) \text{ and}$$

$$\hat{\theta}_E = -\frac{L}{6EI}(14\,M_p - 15.2\,P) \qquad ...(xiii)$$

If among the hinges at C, D and E, the hinge at E is the last to form, we equate $\hat{\theta}_E = 0$ to compute the load, P_3.

Thus $14\,M_p - 15.2\,P = 0$ from which

$$P_3 = 14\,M_p / 15.2 = 1.1053 \text{ t}$$

To substantiate that the computed $P = 1.1053$ is the load P_3, the reader may verify that the values of P computed for $\hat{\theta}_C = 0$ or $\hat{\theta}_D = 0$ are smaller than 1.1053.

Substitution of $P = 1.17$ t and $M_p = 1.2$ tm in the equation (xiii), gives the values of the angle discontinuities, $\hat{\theta}_C$, $\hat{\theta}_D$ and $\hat{\theta}_E$.

Illustration (c) (2). Indeterminacy or more correctly reduced indeterminacy and full-load analysis for $K = 0.4$ and $P = 1.1$ t. In this case $P_2 < P < P_3$ and therefore the plastic hinges are at C and D only and the degree of indeterminacy is equal to one. Let us use this solution for some more generalizations. The 'elastic' analysis of the frame, (Fig. a), gives the following equations for the unknown redundants x_1, x_2 and x_3

$$\underbrace{\frac{L}{EI}\begin{pmatrix} \frac{20}{3} & -\frac{11}{3} & 1 \\ -\frac{11}{3} & \frac{10}{3} & -\frac{2}{3} \\ 1 & -\frac{2}{3} & \frac{2}{3} \end{pmatrix}}_{F} \underbrace{\begin{pmatrix} x_1 \\ x_2 \\ x_3 \end{pmatrix}}_{X} + \underbrace{\frac{PL^2}{6EI}\begin{pmatrix} -19.4 \\ +13 \\ -3.8 \end{pmatrix}}_{U^0} = \begin{pmatrix} 0 \\ 0 \\ 0 \end{pmatrix} \qquad ...(xiv)$$

For the given case, $P = 1.1$ t, when plastic hinges are formed at the sections C and D, $x_1 = M_p$ and $x_2 = -M_p$, and there are corresponding unknown angle discontinuities $\hat{\theta}_C$ and $\hat{\theta}_D$.

We get the following equation, Eq. (xv)

$$[F] = \begin{bmatrix} M_p \\ -M_p \\ \hdashline X_3 \end{bmatrix} + U^0 = \begin{bmatrix} -\hat{\theta}_C \\ -\hat{\theta}_D \\ \hdashline 0 \end{bmatrix} \quad \text{...(xv)}$$

Where F and U^0 are defined in Eq. (xiv)

From the third scalar equation of (xv), and substituting $M_p = 1.2$, from statics we find other moments.

$$X_3 = M_E = 3.8\,P - 3$$
$$M_A = 1.8\,P - 1.8$$
$$M_B = 4\,P - 3.6 \quad \text{...(xvi)}$$

The results, Eq. (xvi), are valid for $P_2 < P < P_3$ and substituting $P = 1.1\ t$

$$M_E = 1.18 \text{ tm},\ M_A = 0.18 \text{ tm},\ M_B = 0.8 \text{ tm}$$

The first two scalar equations of Eq. (xv) give

$$-\hat{\theta}_C = \left(\frac{31}{3}M_P - 9.132P - 3\right)\frac{L}{EI} \quad \text{...(xvii a)}$$

$$-\hat{\theta}_D = \left(-7M_P + 6.133P + 2\right)\frac{L}{EI} \quad \text{...(xvii b)}$$

Substituting for $P = 1.1$ and $M_p = 1.2$ tm

$$\hat{\theta}_C = 0.645\ L/EI \quad \text{and} \quad \hat{\theta}_D = -\ 0.34\ L/EI$$

To find load P_2, equate $\hat{\theta}_D$ equal to zero, Eq. (xvii b)

$$P_2 = (7\,M_p - 2)/6.133 = 1.0435 \text{ t}$$

The value of P_2 from Eq. (xvii a) any be verified to be smaller and therefore computed $P_2 = 1.0435$ needs no revision and we have verified that the given load $P = 1.1$ t is larger than P_2. Following a similar argument, we can find P_1 for the reduced indeterminacy equal to two and only one plastic hinge. From the above example, the following important observations are made for the idealized inelastic analysis.

1. For a given frame find whether the given load P is closer to the load P_c or P_1. If it is closer to P_c use 'full load analysis' as suggested in *(b)* and *(c)*, otherwise use incremental load analysis as in *(a)*.
2. For the case of 'full load analysis' which assumes rigid or perfect plasticity at plastic hinges only and elasticity at all other sections (elastic and rigid plastic analysis), the following is the generalization of Eq. (xv).

$$[F]\left[\frac{\bar{X}_1}{X_2}\right] + \left[\frac{\bar{U}_1^0}{U_2^0}\right] = \left[\frac{-\hat{\theta}}{0}\right] \quad \text{...(xviii)}$$

Where

$$\bar{X}_1 = [(M_p)_1, (M_p)_2, (M_p)_i]^t$$

i is the number of plastic hinges that have formed for

$$P_i < P < P_{i+1}$$
$$X_2 = [R_1, R_2, R_3 \ldots\ldots, R_{n-1}]^t$$

n is the initial degree of indeterminacy, and the total number of possible plastic hinges is less than or equal to $n+1$

$$\theta = \left[\hat{\theta}_1, \hat{\theta}_2, \ldots, \hat{\theta}_i\right]^t$$

are the angle discontinuities at the plastic hinges that have formed.

We either anticipate the i plastic hinges, or the collapse load P_c is evaluated and then proceed in the sequence (b) and (c) till the load P_{i+1} is evaluated to establish the i plastic hinges.

3. For structural safety, besides limiting maximum stress resultants and permissible deflections, it is necessary to limit the maximum angle discontinuity which is like to be the $\hat{\theta}_1$ (the one that has formed first and the maximum angle discontinuity should be less than the permissible rotational capacity).

   It is hoped that the reader has enjoyed the topic of 'Last Plastic Hinge and Angle discontinuations'.

3.10 REVISION EXAMPLE OF CHAPTER 2 AND 3

Analyse the following examples, Ex. R.1 to Ex. R.16, by force method and draw the bending moment and elastic line diagrams. For the example Ex. R.2 to Ex. R.7, consider the members to be inextensible. The reader may like to solve the Ex. R.5 and R.6 after going through the Ex. R.12. The examples Ex. R.13 to R.16 illustrate settlement analysis.

Ex.R.1: (a) Refer Fig. Ex. R.1 for data

(b) Same as Ex. R.1 (a), but prismatic over the entire length $ABCD$

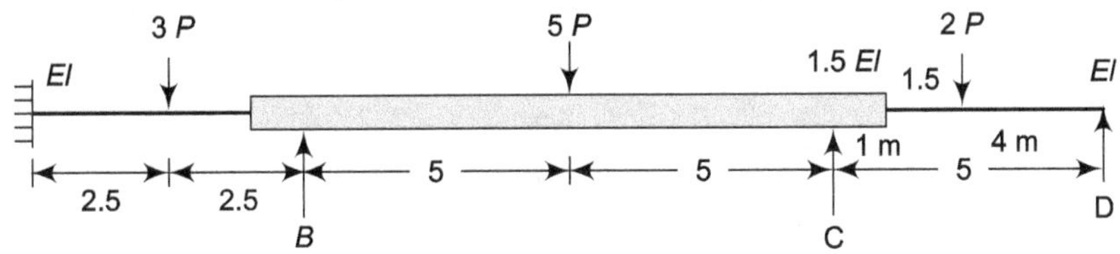

Fig. Ex. R.1

Ex.R.2: (a) Refer Fig. Ex. R.2

(b) Analyse the same frame for uniform rise in temperature, $\overline{T} = 10°C$. $\alpha = .00012$

(c) Analyse if the support at E settles by 5 cm.

$$EI = 8 \times 10^7 \text{ t cm}^2$$

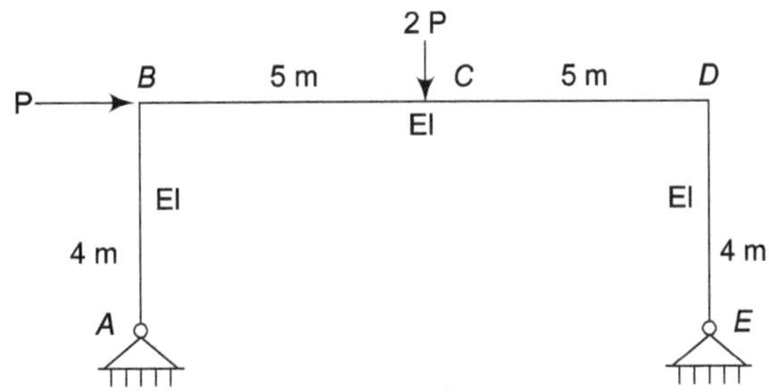

Fig. Ex. R.2

Ex.R.3: Analyse the frame, Fig. Ex. R.3, for the following releases

(a) Cut at *E*,

(b) Hinges at *A* and *D*

(c) Do the elastic-plastic analysis for different stages and compare the collapse load by the upper bound method. Choose a suitable release system

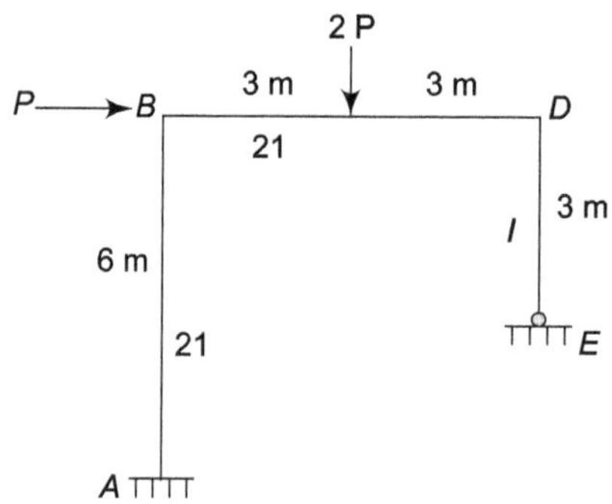

Fig. Ex. R.3

Ex.R.4: (a) Refer Fig. Ex. R.4

(b) Consider same *EI* for all members

Based on the results of (a) and (b), comment on the modifications in the moment diagram due to stiffening at the joints and bases.

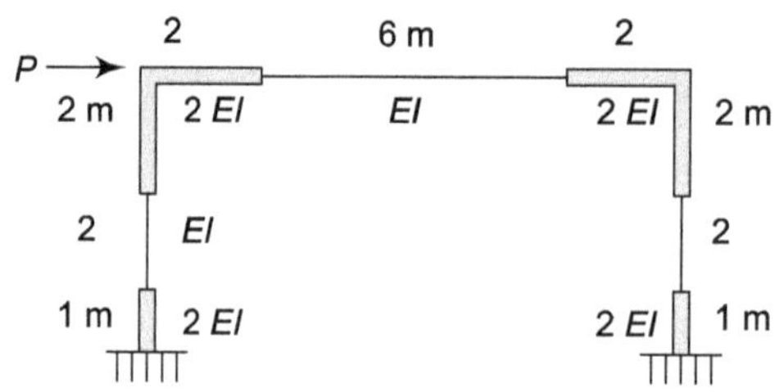

Fig. Ex. R.4

Ex.R.5: Refer Fig. Ex. R.5

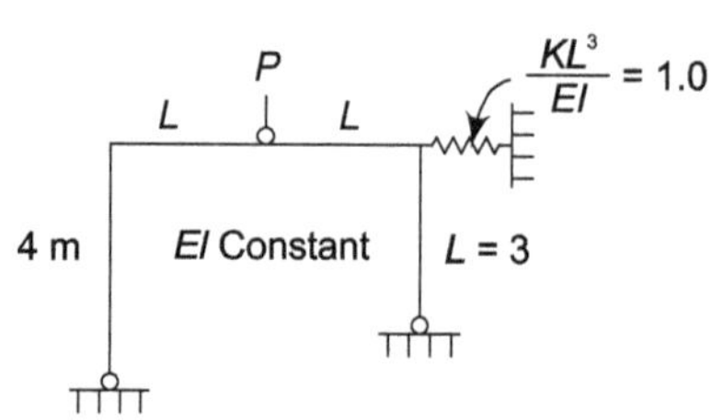

Fig. Ex. R.5

Ex.R.6: Refer Fig. Ex. R.6. Introduce hinges at *O*, *D* and *E* as the releases.

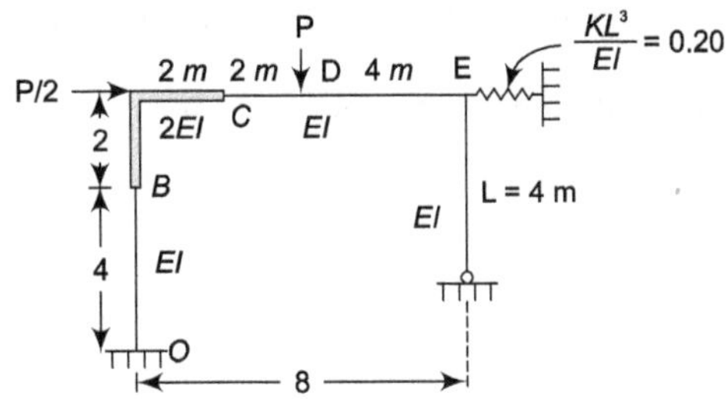

Fig. Ex. R.6

Ex.R.7: Refer Fig. Ex. R.7. Introduce hinge at *C* at the release

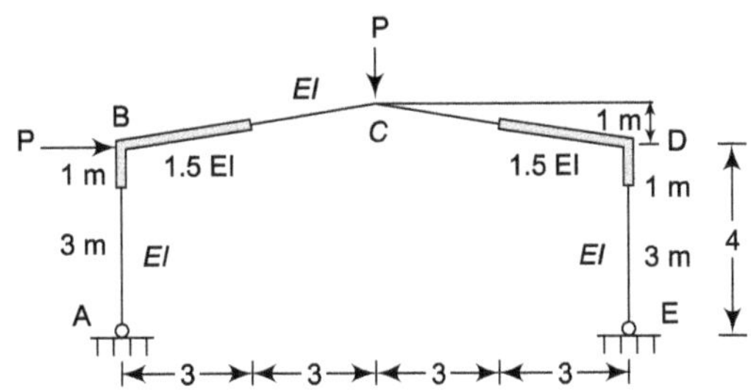

Fig. Ex. R.7

Ex.R.8: Refer Fig. Ex. R.8. Consider the members to be extensible.

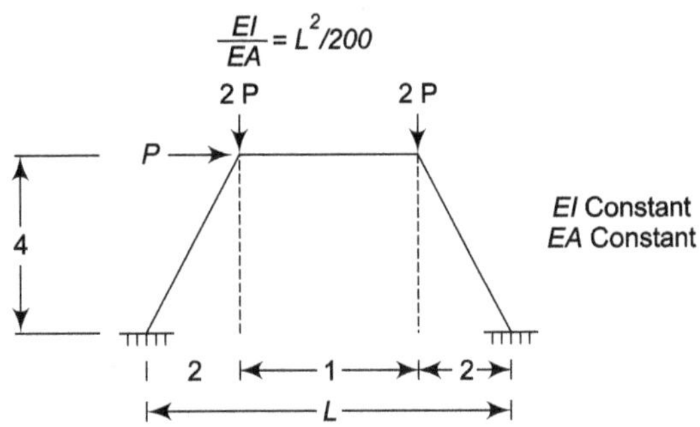

Fig. Ex. R.8

Ex.R.9: Divide the arch. Fig. Ex. R.9, into 12 equal projected segments and consider each segment prismatic. It consists of rectangular cross-section of constant width. Consider concentrated load, *P* at the crown.

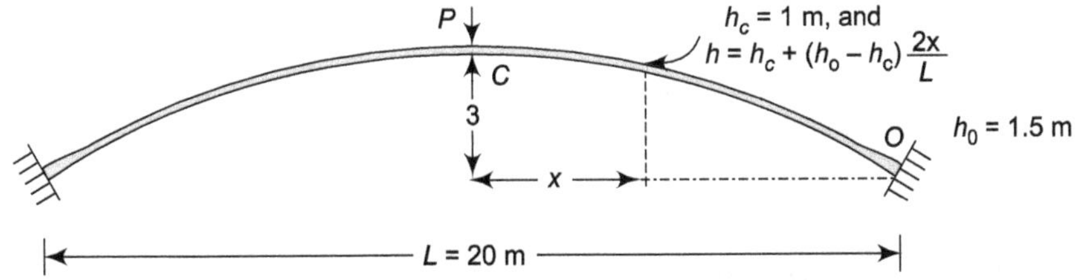

Fig. Ex. R.9

Ex.R.10:(a) Ref. Fig. Ex. R. 10

(b) Replace the end supports ***A*** and ***D*** by springs of; stiffness ***K***. The spring stiffness given by $KL^3/EI = 0.2$

Do the elastic plastic analysis for different stages and compare the collapse load by the upper bound method. Plastic moment of resistance, $M_p = 4$ tm.

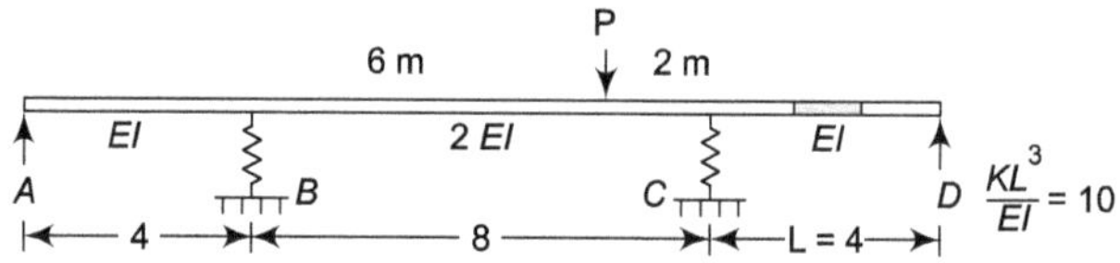

Fig. Ex. R.10

Ex.R.11: Refer Fig. Ex. R.11. Consider members inextensible choose 'double release system', Figs. (b) and (c).

Solution: The releases at *B* and *E* for the deformation system are the impending plastic hinges. The flexibility equation for the first stage (elastic) is given below.

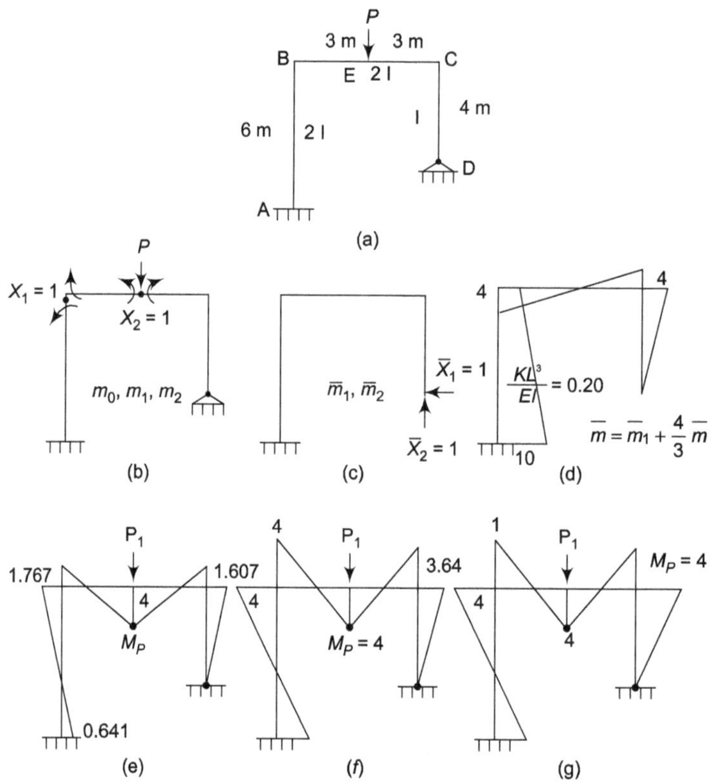

Fig. Ex. R.11

$$2.333\,x_1 - 21.666\,x_2 + 25.00\,P = 0 \qquad \text{...(i)}$$

$$34.5000\,x_1 - 21.000\,x_2 + 38.25\,P = 0 \qquad \text{...(ii)}$$

$$x_1 = 0.467\,P,$$

$$x_2 = 1.0549\,P.$$

The load corresponding to the first plastic hinge at *E*,

$$P_1 = 3.791\,t$$

For the second stage, $\Delta P_1 = P - P_1$, the release at E is a natural hinge and the hinge at B is the only release. The statical diagrams $\overline{m}_1$ and $\overline{m}_2$ are not suitable individually because they give rise to moments at E. This difficulty can, however, be overcome by suitably combining $\overline{m}_1$ and $\overline{m}_2$ diagrams, so that the superposition of $\overline{m}_1$ and $\overline{m}_2$ gives zero moment at E. Therefore, we choose the statical $\overline{m}$ diagram.

$\overline{m} = \overline{m}_1 + \frac{4}{3}\overline{m}_2$, Fig. Ex. R.11 (d), which should suggest:

we add the Eq. *(i)* to $\frac{4}{3}$ times the equation

(ii) Because for this stage $x_2 = 0$, we get

$$48.333\, x_1 + 76\, \Delta P_1 = 0, \; x_1 = 1.57\, \Delta P_1$$

The load corresponding to the second plastic hinge at B,

$$P_2 = P_1 + 1.422$$

For the third stage, $\Delta P_2 = P - P_2$, the frame is determinate and the collapse load, $P_C = 5.333$ t.

The moment diagrams corresponding to P_1, P_2 and P_3 are given, Figs (e), (f), (g). The reader may verify the collapse load by the upper bound method (beam mechanism).

Ex. R.12. Analyse the frame, Fig. Ex. R.12 (a). All members are assumed inextensible excepting the spring at D. Given

$$\frac{KL^3}{EI} = 1$$

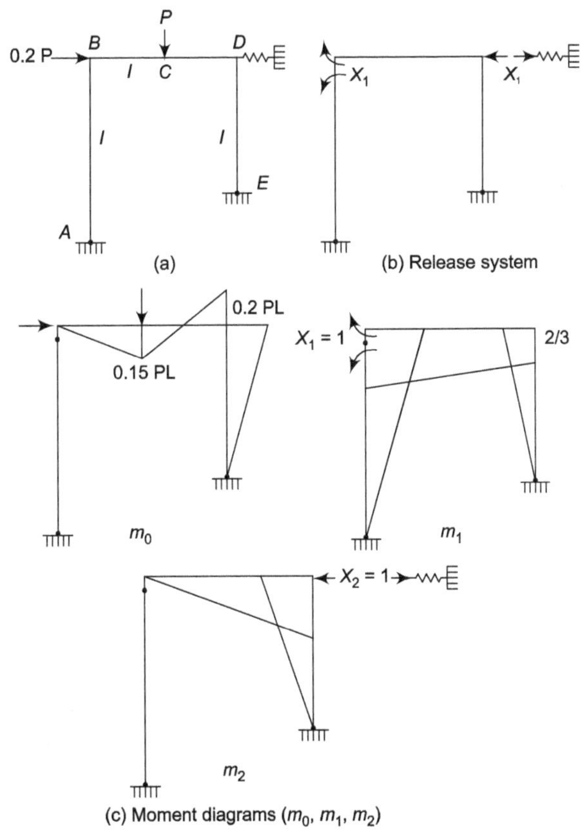

Fig. Ex. R.12

For releases:

(i) Introduce a hinge at B and a cut at the spring, Fig. (b)

(ii) Introduce hinges at the joints B and D

Solution for Ex. R.12 (i). For the chosen release system, m_0, m_1 and m_2 diagrams are drawn, Fig. (c). The reader may verify the following results.

$$U_1^0 = -0.018\ PL^2/EI, \quad U_2^0 = -0.071\ PL^3/EI,$$

$$f_{11} = 1.35\ L/EI, f_{12} = f_{21} = 0.607\ L^2/EI$$

$$f_{22} = 2\ L^3/3EI + 1/K$$

Substituting the values of $K = EI/L^3$

$$f_{22} = 5\ L^3/3EI$$

solving for the unknown redundant,

$$x_1 = -0.007\ PL \text{ and } x_2 = 0.045\ P$$

The reader may analyse the frame by choosing the release system (ii).

Settlement analysis for indeterminate structures.
Refer Ex. R. 13 in the following.

Ex. R.13. Analyse the frame, Fig. Ex. R.13(a). Assume members inextensible.

It was stated elsewhere that temperature effects, lack of fit and settlements give rise to internal stresses in the case of indeterminate structures only. The settlement effect is almost identical to lack of fit of members, if the settlement is in the direction of the members. For the example considered, Fig. Ex. R.13(a), the settlement of support E by δ, gives the same effect as the lack of fit of member DE (i.e. length of DE equal to $4 - \delta$ instead of 4 specified.

Refer Fig. (b). Using the internal release system and signs specified by the assumed directions of x_1 and x_2 and corresponding positive deformations, Fig. (c), one can obtain the following directly,

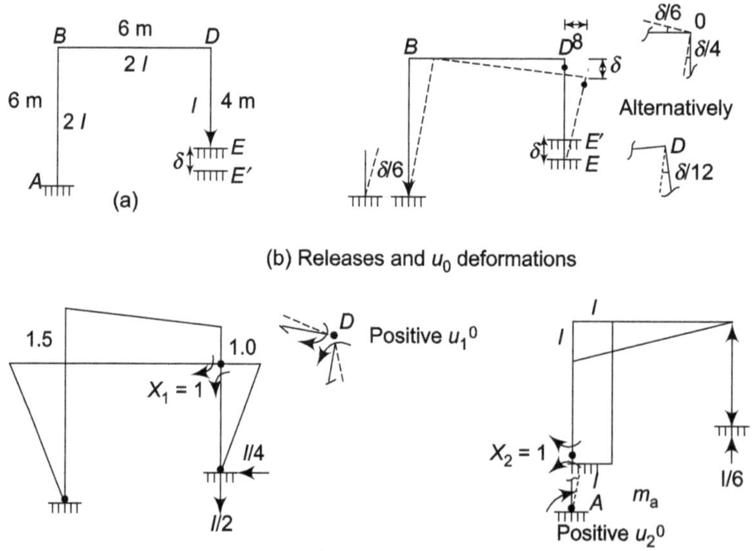

(b) Releases and u_0 deformations

(c) Moment diagrams

Fig. Ex. R.13

$$U_1^0 = -\ \delta/12 \text{ angle discontinuity at hinge } D\text{, and}$$

$$U_2^0 = +\ \delta/6 \text{ angle discontinuity at hinge } A.$$

And from the m_1 and m_2 diagrams, we obtain

$$f_{11} = 3.333/EI, f_{12} = f_{21} = -4.25/EI$$
$$f_{22} = 4/EI$$

solving for the unknown redundants,

$$x_1 = -EI\ \delta/40.7 \text{ and } x_2 = -\ EI\ \delta/14.8$$

with the statical evaluation of moment at B equal to $1.5\ x_1 - x_2$, complete bending moment diagram can be drawn.

The reader may analyze the same problem by choosing a different release system, say by cutting at the hinge E. Let the redundant to be horizontal reaction X_1 from right to left and vertical reaction X_2 acting upward. The following may be verified to be compatibility equations.

$$81.33\ X_1 - 54.0\ X_2 = 0$$
$$-54.0\ X_1 + 144\ X_2 = -EI\ \delta \text{ and}$$
$$X_1 = -EI\ \delta/162.5, X_2 = -EI\ \delta/108$$

The reader may complete the analysis and compare with the bending moment diagram obtained by the hinge release system. It is reiterated that temperature effects, settlement, and lock of file are grouped under geometric loading. The geometric loading as in any other case of loading is relevant for computing u^0.

Ex. R.14. Find u_1^0 and u_2^0 at the hinge releases of Ex. R.13, by using complementary virtual work principle.

Treat m_1 diagram, Fig. (c), as virtual force system. Ride the virtual force system (in this case the two forces are external *) on the real deformations, Fig. (b).

$$\frac{1}{12}(\delta) + 1 \times (\theta_D)^* = 0$$

$$\theta_D = u_1^0 = -\frac{\delta}{12} \qquad \text{....(1)}$$

Similarly riding m_2 force system, Fig. (c), on the real deformations, Fig. (b),

$$-1/6\ (\delta) + 1 \times (\theta_A)^* = 0 \qquad \text{.....(2)}$$

$$\theta_A = u_2^0 = \frac{\delta}{6}$$

Ex. R.15. For the frame $ACDB$, the ends A and B are hinged, Fig. Ex. R.15(a). If the end B settles horizontally by δ, evaluate u^0 corresponding to the redundatns X acting at hinge release D. Find u_1^0.

Refer Fig. Ex. R.15 (b). Locate instantaneous centre of rotation $(ICR)_b$ for the member b, by extending AD and intersecting it by the line drawn perpendicular to the displacement δ.

*It may be noted that the forces, X_1 and X_2, are treated as external moments. Form the free body of the hinge, D, the hinge is moment free, as it should be, and the moment free, as it should be, and the moments, $X_1 = 1$, are acting at the ends of two members meeting at the joint D. The moments, say $X_1 = 1$, do not constitute bimoment as in the case of internal stress resultants. In the Fig. (c), corresponding to the external moments X_1 and X_2 as redundanats, positive deformations at D and A respectively are also shown. The reader may pause here to verify that in the case of internal forces (bimoment at plastic hinges) the positive deformations corresponding to the assumed directions of the moments are opposite to those shown in Fig. (c).

From geometrical considerations:

$$\theta_b = \delta / \left(\frac{L}{2} + \frac{\sqrt{3}}{2}L\right) = \frac{\delta}{1.366\,L}$$

Horizontal displacement of D, $(\delta_h)_D = \dfrac{\delta}{1.366L} \times \dfrac{L}{2}$

$$(\theta)_{ACD} = (\delta_h)_D/L$$

$$\therefore \qquad |u^0| + \theta_b + (\theta)_{ACD} = \delta/(0.91L)$$

Comparing with the positive u^0, Fig. Ex. R.15 (b),

$$u^0 = -\delta/(0.91L)$$

Alternatively:

From virtual work considerations :

From statics the reaction at B are,

$$H = 1/(0.91L) \text{ and } V = 0.0984/L.$$

Writing the virtual work equations,

External work = internal work

$$H \times \delta + 1 \times u^0 = 0$$

$$u^0 = -\delta/(0.91\,L)$$

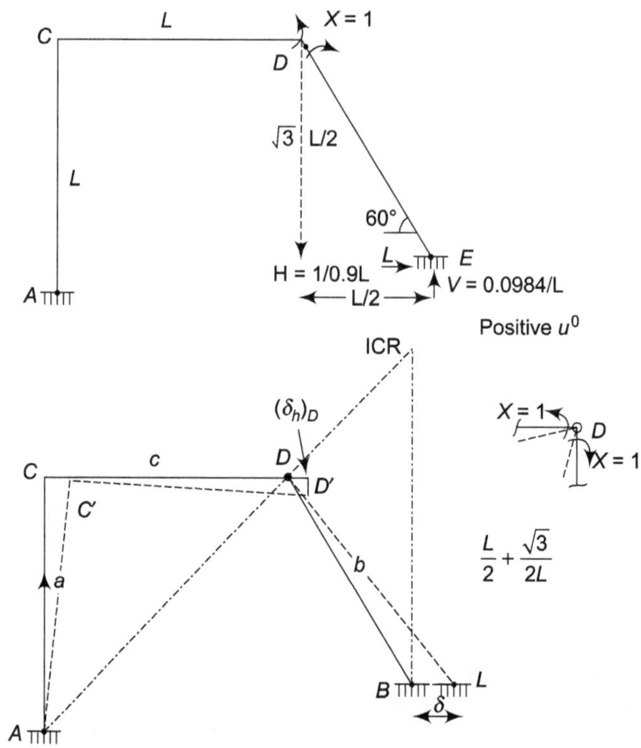

Fig. Ex. R.15

Ex.R.16: Find u^0 if the frame of Ex. R.15, settles vertically by δ_v.

From the virtual work equation

$$-V \times \delta_v + 1 + u^0 = 0, \text{ and } V = .0984/L.\ u^0 \text{ can be evaluated.}$$

■ *"A human being is part of a whole, called by us the 'Universe,' a part limited in time and space.*
He experiences himself, his thoughts and feelings, as something separated from the rest _ a kind of optical delusion of his consciousness.
This delusion is a kind of prison for us, restricting us to our personal desires and to affection for a few persons nearest us.
Our task must be to free ourselves from this prison by widening our circles of compassion to embrace all living creatures and the whole nature in its beauty."

Albert Einstein

■

Chapter 4: Matrix Approach for Force Method

Chapter

4

Matrix Approach for Force Method

4.1 GENERAL

The three basic steps, (i) equilibrium, (ii) deformation – force relations, and (iii) compatibility constituting the force method described in Chapter 3, are repeated in this chapter. This time, matrix operations are used for the basic steps. The temperature effects and the stresses due to lack of fit of members of indeterminate structures are important. In Chapter 3, temperature stresses developed in a fixed arch were evaluated. In this chapter, matrix force method is used to find the stresses developed in frames due to temperature and lack of fit. As far as possible, the attempt in the presentation has been to choose examples for which the force method is more suitable than displacement methods. No particular preference of the matrix approach over the conventional integration approach evaluating f_{ij} and u_i^0 (See Chapter 3) can be claimed. However, the force method when systematized by the matrix operations for the three basic steps, makes computer programming easier and simpler to check. We will be using the energy principles in a generalized form of 'virtual work' principles which are stated as law of 'conragradience'. By making use of the contragradience equilibrium equations lead us to obtain compatibility conditions. The typical examples of the indeterminate structures considered to illustrate the method are:

Hinged trusses, pontoons treated as beams resting on interior supports as flexible and end supports hinged, rigid frames like plane frames, grids, space frames, etc., rigid cum hinged frames like trussed beams or trussed frames.

The bridge decks idealized by a slab monolithic with longitudinal girders are treated as beams on flexible supports. A harmonic analysis for any given loading combined with the force methods for such bridge decks is presented.

4.2 JOINTS AND STRUCTURAL NODES

We define the joints as the points where idealized straight skeletal members meet. Wherever the members are not straight, like arches and beams curved in plan, such members can be geometrically approximated by straight segments, as we did in the case of an arch problem, Fig. 2.4 and Fig. Ex.2.24. the loads need not necessarily be applied at the joints. If the loads are applied not only at the joints but also at some discrete points, we may introduce the "discrete load points" as nodes. The joints and the "discrete load points" will be referred as structural nodes or simply nodes. The number of structural nodes may be in excess of the number of joints in the case of rigid frames only. In the case of hinged trusses, structural nodes consist of joints only. If a load is applied between the two joints, it is converted to joint loads as explained in Fig. 4.1

4.2.1 Hinged Trusses

The loads are generally transferred directly to the joints of hinged trusses. However, when the loads act in between the joints or nodes equivalent loads acting on the adjoining nodes are computed considering the loaded member as a beam with hinged supports at the nodes. If a load P is applied between the joints B and C, Fig. 4.1 (a), the member BC is considered as a beam with hinged ends to compute end reactions, Fig. 4.1(b), and these reactions are transferred back to the truss, Fig. 4.1(c), as equivalent nodal loads at the joint B and C. While designing the member BC, however, we have to consider simple support moments of the member together with the axial forces caused by the equivalent nodal loads.

Similar procedure is adopted in the case of roof trusses if the purlins do not rest directly on the joints.

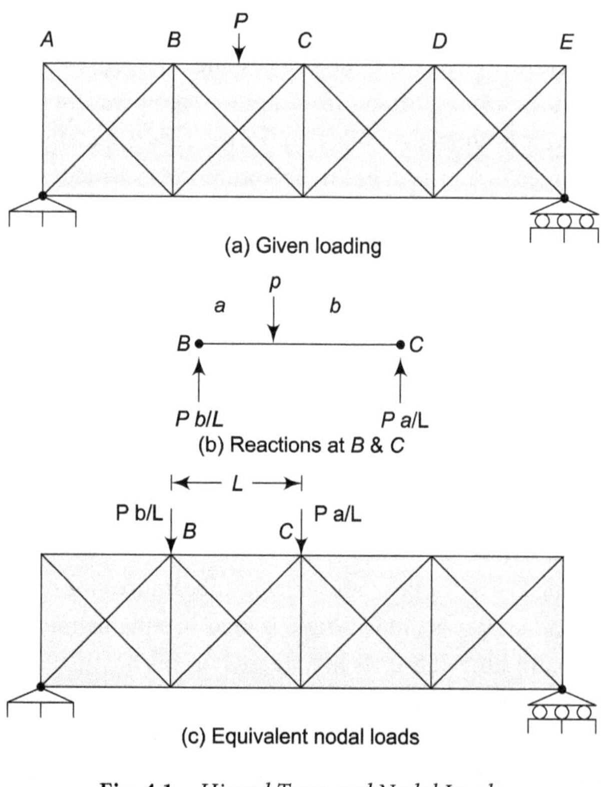

Fig. 4.1 *Hinged Truss and Nodal Loads*

4.2.2 Rigid Frames and Nodes

In the case of rigid frames it is sometimes more convenient to introduce additional nodes at the load point. Consider a frame, Fig. 4.2 (a). Out of the two loads (P_1 and P_2), the load P_2 is applied between the joints A and B. The load point D may be considered as an additional node, and thus resulting into three structural nodes, A, D and B and four members OA, AD, DB and CB. Alternatively, the number of structural nodes can be limited to the number of joints by doing the analysis in two stages, in a way similar to the truss example, Fig. 5.1. In the stage – I, the joints are clamped and member end forces are found for no "joint-deformations".

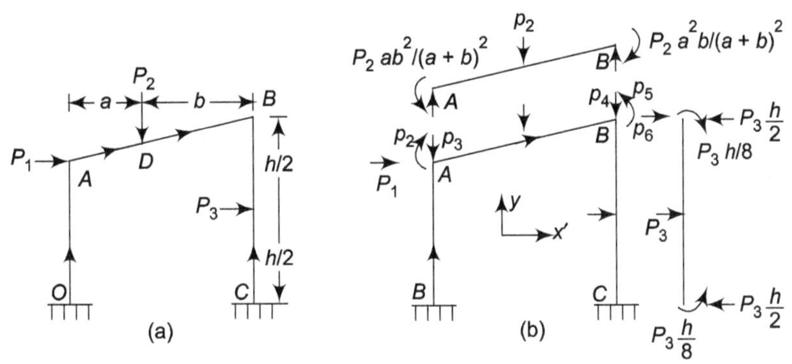

Fig. 4.2 *Rigid Frame and Structural Nodes*

In the Fig. 4.2, the joints A and B are constrained against any deformation and thus we find fixed-end-forces for the members OA, AC and CD. For the example, Fig. 4.2, there are no fixed end forces developed for the member OA, as it does not carry load within its length. The fixed-end-forces for the concentrated load P_2 and P_3 are found for the members AB and CB respectively. In a similar way fixed-end-forces are computed when the load is uniformly distributed the fixed end forces so computed are then transferred as equivalent loads to the joints A and B, Fig. 4.2 (b), e.g., generalized load $P_5 = P_2\, a^2b/(a + b)^2 + p_3\, h/8$. The equivalent loads also called as generalized loads consist of forces and moments. The directions of the generalized loads are shown referring to the system coordinates, x', y', Fig. 4.2(b). The member coordinates refer to the appropriate members, whereas the system coordinates are common for all the nodes of the entire structure. In the stage-2, the structure under equivalent loads $P_1 \ldots P_6$, Fig. 4.2(b), is analysed allowing the joints to deform. The superposition of the results of the stages 1 and 2, gives the final solution. The first stage is simple to analyse, and the analysis presented is mainly for the second stage, for which matrix approach is used.

Consider also a frame, Fig. 4.3(a), which consists of two joints A, B and the member AB is non-prismatic, because of the stiffened segment EF. In such cases, finding fixed end forces for a non-prismatic member is tedious. Instead, it is definitely more expeditious to introduce additional nodes at E and F, resulting in four structural nodes and five members. The node under the load P is avoided by finding equivalent or generalized loads at the nodes E and F for the member EF, Fig. 4.3(b).

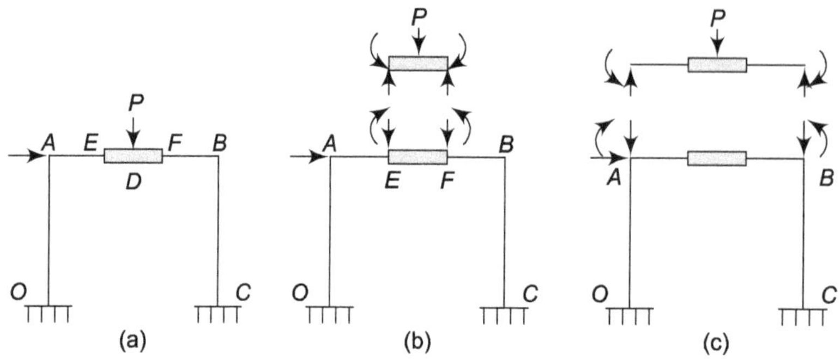

Fig. 4.3 *Rigid Frame with Non-prismatic Member*

If we have readily available formulae for finding fixed end moments and shears for the non-prismatic members like AB, we may directly compute the equivalent loads for the joint A and B and thus limit the number of structural nodes to the number of joints, Fig. 4.3(c).

4.3 THE BASIC STEPS

The three basic steps are repeated in the following, but this time with matrix approach. As we had not analysed any truss problem in Chapter 3, let us compensate it by illustrating the basic steps of the force method of analysis, with the help of an example of a plane hinged truss which will be followed by other structural problems. The basic theory is explained in the Article 4.3.1 using the illustration of a hinged truss. In the subsequent articles, the application of the theory is extended to other structural problems.

4.3.1 Basic Theory

We first find degree of indeterminacy of a given structure. Suitable releases are introduced so that the structure is not only determinate and stable, but also the resulting equations are well conditioned. The redundants equal in number to the degree of indeterminacy are applied at the chosen releases and the values of the redundants are to be determined such that they restore compatibility at the release points.

1. Equilibrium

The determinate structure is analysed to give member forces for (i) the given nodal loads, and (ii) for unit redundant forces applied at the releases. The results are put in the matrix form,

$$\begin{aligned} P_{m^0} &= B_o P_j \quad \text{(i)} \\ P_{m^q} &= B\,Q \quad \text{(ii)} \end{aligned} \qquad \ldots(4.1)$$

Where $P_j = (P_1, P_2, \ldots, P_r)$ are the r number of joint loads or nodal loads,

P_{m^q} = member force due to the nodal loads

$Q = (q_1, q_2, \ldots, q_n)$, are the number of redundatns.

The matrices B_0 and B are the forces due to unit nodal loads (each load considered separately) and unit redundatns (each redundant considered separately) respectively. In the case of nodal loads, their values being known, the member forces P_{m^0} can be found for all the loads together and not separately, in which case B_0 will be a column matrix and P_j takes the value of scalar unity. This aspect is further elaborated foregoing.

Combining the Eqs (4.1) (i) and (ii), we write

$$P_m = (B_0\ B)\begin{bmatrix} P_j \\ Q \end{bmatrix} \qquad \text{...(4.2)}$$

Where $\qquad P_m = (P_{m^0} + P_{m^c})$

The evaluation for the elements B_0 and B matrices of the Eq. 4.2, completes the equilibrium analysis– the first step of the force method.

Let us apply the above description of equilibrium, which is general and applicable to any structure, to a simple truss problem as an illustration. Consider a truss, Fig. 4.4. The degree of indeterminacy is three. The diagonal member J, m and r are cut near the joints B, C and D respectively for making the structure statically determinate. The member axis, x and the ends 1 and 2 marked along the direction of the axis, are shown separately, Fig. 4.4. The forces and deformations of member are positive if they occur along the chosen x-axes. The member force acting along the axis of a member is described by the force at the end 2 only. The force at the other end, that is end 1 can be obtained by the equilibrium relation $(P_x)_1 + (P_x)_2 = 0$. The corresponding axial deformation for a member is $(\delta x_2 - \delta x_1)$, which eliminates rigid body displacement and elastic deformation alone is considered. With the sign convention of positive forces and relative deformations, tension and corresponding deformation (that is elongation) are positive. The relevance of the member axes becomes more important in the case of the members of rigid frames, where a Cartesian system x, y, is used to define the positive directions of axial forces, shears and moments at the end 2 and the corresponding relative deformations.

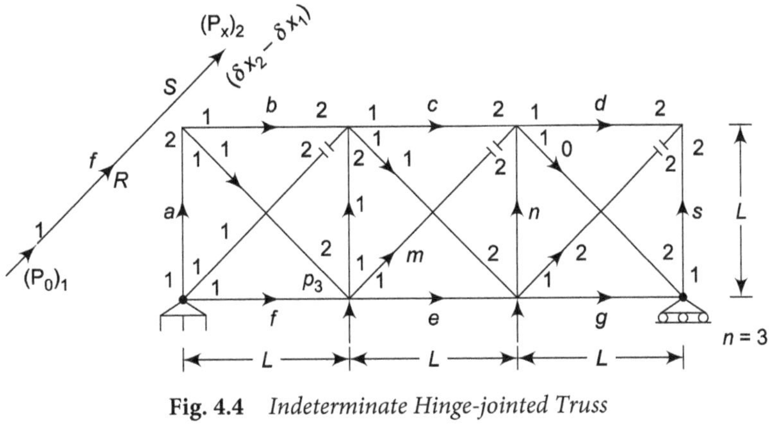

Fig. 4.4 *Indeterminate Hinge-jointed Truss*

The member forces for the truss with the releases are obtained by satisfying equilibrium of the joints as free bodies or by the "method of sections". For the truss under the action of loads, $P = \{P_1, P_2\}$ and the chosen redundants, $Q = \{q_1, q_2, q_3\}$, the equilibrium equation (4.2), is given in expanded form, Fig. 4.5. The elements of the first and second columns of the B_0 matrix are the member forces due to $P_1 = 1$ and $P_2 = 1$, respectively. Similarly, the elements of three columns, corresponding to the three redundants, are the member forces due to the three unit redundants. The elastic properties of the members are not used in the statical solution given in the form of B_0, B matrices, Fig. 4.5. The elastic properties are used in the second step, which establishes deformations-force relation of members.

2. Deformation-Force relation

The 'deformation-force' relation is also known as generalized strain-stress relation or flexibility of a member. In the case of the members of the hinged trusses, say considering the member j, Fig. 4.4, the member flexibility or generalized strain-stress relation is given by a simple relation,

$$(d_m)_j = (f_m)_j\,(p_m)_j$$

Where $(d_m)_j = (\delta x_2 - \delta x_1)$; $(p_m)_j = (p_x)_{2,j}$ and

$$(f_m)_j = (L/AE)_j$$

Which is elongation of the member due to unit axial force.

The flexibilities of all the member, when assembled

$$D_m = F_m p_m \qquad \ldots(4.3)$$

Where $d_m = [(d_m)_1, (d_m)_2, \ldots, (d_m)_3]^{i}$,

$p_m = [(p_m)_1, (p_m)_2, \ldots, (p_m)_3]^{i}$, and

$F_m = [(f_m)_1, (f_m)_2, \ldots, (d_m)_3]$

In the case of the members of hinged trusses, the diagonal flexibilities given by (f_m), are scalars. It will be seen later that in the case of the members of the rigid frames, member flexibility $(f_m)_j$ is a submatrix positioned diagonally in the assembled F_m matrix. The equation (4.3) is the result of the step (2). This leads us to the final step, the compatibility relations.

3. Compatibility

The matrix operations given in the following are general without any bias to any particular type of structure.

When the constraints are eliminated by introducing releases to make a given structure determinate, the given loading and the redundants individually give rise to relative deformations at the releases. We then choose the values of the redundants such that the original compatibility conditions, violated by the chosen releases, are satisfied. If the net relative deformations due to the loads and the redundants at the releases, are designated by U, then the original restraint conditions are restored by satisfying the matrix equation,

$$U = 0$$

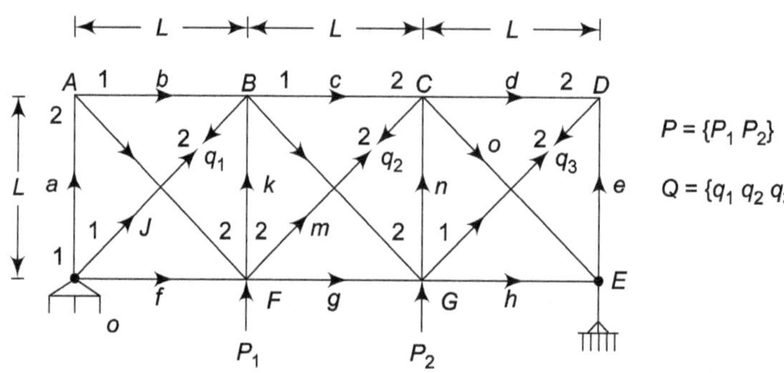

Fig. 4.5 *Releases and Equilibrium Solution of Hinged Truss*

$$
\begin{array}{c}
\\
\begin{bmatrix}
(P_x)_a \\ (P_x)_b \\ (P_x)_c \\ (P_x)_d \\ (P_x)_f \\ (P_x)_g \\ (P_x)_h \\ (P_x)_i \\ (P_x)_j \\ (P_x)_k \\ (P_x)_l \\ (P_x)_m \\ (P_x)_n \\ (P_x)_o \\ (P_x)_r \\ (P_x)_s
\end{bmatrix}
\end{array}
=
\begin{array}{c}
\boxed{\begin{array}{cccccc} P_1 & P_2 & & q_1 & q_2 & q_3 \end{array}} \\
\begin{bmatrix}
\frac{2}{3} & \frac{1}{3} & . & -\sqrt{\frac{2}{2}} & 0 & 0 \\
\frac{2}{3} & \frac{1}{3} & . & -\sqrt{\frac{2}{2}} & 0 & 0 \\
\frac{1}{3} & \frac{2}{3} & . & 0 & -\sqrt{\frac{2}{2}} & 0 \\
0 & 0 & . & 0 & 0 & -\sqrt{\frac{2}{2}} \\
0 & 0 & . & -\sqrt{\frac{2}{2}} & 0 & 0 \\
-\frac{2}{3} & -\frac{1}{3} & . & 0 & -\sqrt{\frac{2}{2}} & 0 \\
-\frac{1}{3} & -\frac{2}{3} & . & 0 & 0 & -\sqrt{\frac{2}{2}} \\
-2\sqrt{\frac{2}{3}} & -\sqrt{\frac{2}{3}} & . & 1 & 0 & 0 \\
0 & 0 & . & 1 & 0 & 0 \\
-\frac{1}{3} & \frac{1}{3} & . & -\sqrt{\frac{2}{2}} & -\sqrt{\frac{2}{2}} & 0 \\
\sqrt{\frac{2}{3}} & -\sqrt{\frac{2}{3}} & . & 0 & 1 & 0 \\
0 & 0 & . & 0 & 1 & 0 \\
-\frac{1}{3} & -\frac{2}{3} & . & 0 & -\sqrt{\frac{2}{2}} & -\sqrt{\frac{2}{2}} \\
\sqrt{\frac{2}{3}} & 2\sqrt{\frac{2}{3}} & . & 0 & 0 & 1 \\
0 & 0 & . & 0 & 0 & 1 \\
0 & 0 & . & 0 & 0 & -\sqrt{\frac{2}{2}}
\end{bmatrix}
\end{array}
\begin{bmatrix}
\begin{bmatrix} P_1 \\ P_2 \end{bmatrix} = P_j \\
\begin{bmatrix} q_1 \\ q_2 \\ q_3 \end{bmatrix} = Q
\end{bmatrix}
$$

For imposing the above compatibility condition, we find an expression giving deformations U at the releases. This is done in the following by using the principle of complementary virtual work, Article. 2.6.2.

Let d_j be the nodal deformations and when these are put together with U, we have $\{d_j\ U\}$, which are corresponding to the member deformations d_m. Let us choose any arbitrary load system (δP_j) and let (δP_m) be the corresponding equilibrating member forces. Allow the arbitrary force system $(\delta P_j, \delta P_m)$ to ride on to the corresponding real deformation system, that is, $\{d_j, U\}$ and d_m, the complementary virtual work equation gives (Refer Article. 2.6.2)

$$(\delta P_m)^i\, d_m = \begin{bmatrix} P_j \\ Q \end{bmatrix}_i \begin{bmatrix} d_j \\ U \end{bmatrix}$$

For the given structure to be analysed, if the deformations are small, we may substitute p_m and $[P_j, Q]$ in place of (δP_m) and (δP_j). Thus we have

$$(P_m)^i\, d_m = \begin{bmatrix} P_j \\ Q \end{bmatrix}^i \begin{bmatrix} d_j \\ U \end{bmatrix}$$

But we know from step 1, equilibrium condition,

$$p_m = (B_0\, B) \begin{bmatrix} P_j \\ Q \end{bmatrix} \qquad \text{...(a)}$$

$$\therefore \qquad p_{m^i} = \begin{bmatrix} P_j \\ Q \end{bmatrix}^t (B_0\, B)^i$$

Substituting from p_{m^i} in the virtual work equation, we get

$$= \begin{bmatrix} P_j \\ Q \end{bmatrix}^t (B_0\, B)^i\, d_m = \begin{bmatrix} P_j \\ Q \end{bmatrix}^t \begin{bmatrix} d_j \\ U \end{bmatrix}$$

Since δP_m and δP_j arbitrary and we had substituted them by one of the many possible sets of values, i.e., by p_m and $\{P_j\, Q\}$ of the actual structure), we may drop $\{P_j\, Q\}$ from either side of the equation to give

$$(B_0\, B)^i\, d_m = \begin{bmatrix} P_j \\ U \end{bmatrix}$$

Or

$$\begin{bmatrix} d_j \\ U \end{bmatrix} = \begin{bmatrix} B_0^t \\ B^t \end{bmatrix}_{d_m} \qquad \text{...(b)}$$

The Eqs (a) and (9B) establish the law of "Contragradience" substituting the values of d_m and also noting that

$$P_m = (B_0\, B) \begin{bmatrix} P_j \\ Q \end{bmatrix} \text{ from equilibrium, we get}$$

$$\begin{bmatrix} d_j \\ U \end{bmatrix} = \begin{bmatrix} B_0^t \\ B^t \end{bmatrix} F_m\, P_m$$

or

$$\begin{bmatrix} d_j \\ U \end{bmatrix} = \begin{bmatrix} B_0^t \\ B^t \end{bmatrix} F_m\, (B_0\, B) \begin{bmatrix} P_j \\ Q \end{bmatrix}$$

From the above matrix equation, we get the following equations for the deformations d_j and U.

$$d_j = B_o{}^t F_m P_m$$

or $$d_j = B_o{}^t F_m (B_0 P_j + B Q) \qquad \text{...(c)}$$

and $$U = B^t F_m (B_0 P_j + B Q)$$

or $$U = B^t F_m B_0 P_j + B^t F_m B Q \qquad \text{...(d)}$$

The Eq. (d) is the desired expression for the relative deformations U.

The compatibility condition gives*

$$U = 0 = B^t F_m B_0 P_j + B^t F_m BQ$$

or $$Q = -(B^t F_m B)^{-1} (B^t F_m B_0) P_j \qquad \text{...(4.4)}$$

This is the required solution for redundants Q.

After the evaluation of the redundants Q, the evaluation of the other important results, member forces and joint deformations, is simple and direct.

$$P_m = (B_0 B) \begin{bmatrix} P_j \\ Q \end{bmatrix}$$

All the quantities on the right side are known.

From the Eq. (c) we get

$$d_j = B_0{}^t F_m P_m$$

The above results of the Force Method are applied to the hinged truss. In the case of the truss, Fig. 4.4, U consists of the relative axial displacement of each of the members, j, m and r at the end of each member. The nature of the U displacements in the case of rigid frames will be described separately.

For the hinged truss, Fig. 5.4 m the redundants

$$Q = -(B^t F_m B)^{-1} (B^t F_m B_0) P_j$$

The intermediate steps are

$$B^t F_m = \begin{bmatrix} -\frac{\sqrt{2}}{2}Z_a & -\frac{\sqrt{2}}{2}Z_b & 0 & 0 & -\frac{\sqrt{2}}{2}Z_i & o & o & Z_i & Z_i & -\frac{\sqrt{2}}{2}Z_k & o & o & o & o & o & o \\ 0 & 0 & -\frac{\sqrt{2}}{2}Z_c & 0 & 0 & -\frac{\sqrt{2}}{2}Z_g & o & o & o & -\frac{\sqrt{2}}{2}Z_k & Z_i & Z_m & -\frac{\sqrt{2}}{2}Z_n & o & o & o \\ 0 & 0 & 0 & -\frac{\sqrt{2}}{2}Z_d & 0 & 0 & -\frac{\sqrt{2}}{2}Z_h & 0 & 0 & 0 & 0 & 0 & -\frac{\sqrt{2}}{2}Z_h & Z_0 & Z_r & -\frac{\sqrt{2}}{2}Z_g \end{bmatrix}$$

Where $$(Z)_i = \left(\frac{L}{LA}\right)_i$$

If EA is taken same for all the members

$$(Z)_i = \left(\frac{L_i}{EA}\right)_i$$

And $B^t F_m B_0$

$$= 1/EA$$

* The deformations due to temperature and 'lock of fit' are considered in Art. 4.6.

$$\left[\begin{array}{c|c} -\frac{2}{3}L_a \;\; -\frac{2}{3}L_b \;\; -\frac{2\sqrt{2}}{3}L_i \;\; +\frac{\sqrt{2}}{6}L_k & -\frac{\sqrt{2}}{6}L_a \;\; -\frac{\sqrt{2}}{6}L_b \;\; -\frac{\sqrt{2}}{3}L_i \;\; -\frac{\sqrt{2}}{6}L_k \\ -\frac{\sqrt{2}}{6}L_c \;\; \frac{\sqrt{2}}{3}L_g \;\; +\frac{\sqrt{2}}{6}L_k \;\; +\frac{\sqrt{2}}{3}L_l \;\; +\frac{\sqrt{2}}{6}L_n & -\frac{\sqrt{2}}{3}L_c \;\; +\frac{\sqrt{2}}{6}L_g \;\; -\frac{\sqrt{2}}{6}L_k \;\; -\frac{\sqrt{2}}{3}L_l \;\; +\frac{\sqrt{2}}{3}L_n \\ +\frac{\sqrt{2}}{6}L_h \;\; +\frac{\sqrt{2}}{6}L_n \;\; +\frac{\sqrt{2}}{3}L_o & -\frac{\sqrt{2}}{3}L_h \;\; +\frac{\sqrt{2}}{3}L_n \;\; +\frac{2\sqrt{2}}{3}L_o \end{array}\right]$$

$$B^t F_m B_0 = \frac{\sqrt{2}}{3}\frac{L}{EA}\begin{bmatrix} \left(-1\frac{1}{2}-2\sqrt{2}\right) & \left(-\frac{3}{2}-\sqrt{2}\right) \\ \left(+\frac{3}{2}+\sqrt{2}\right) & -\sqrt{2} \\ \left(+1+\sqrt{2}\right) & \left(2+2\sqrt{2}\right) \end{bmatrix} = \frac{L}{EL}\begin{bmatrix} -2.041 & -1.372 \\ +1.372 & -0.667 \\ +1.137 & +2.277 \end{bmatrix}$$

$$B^t F_m B = \frac{1}{EA}\left[\begin{array}{c|c|c} +\frac{1}{2}L_a+\frac{1}{2}L_b+\frac{1}{2}L_f+L_i+L_j+\frac{1}{2}L_k & +\frac{1}{2}L_k & 0 \\ +\frac{1}{2}L_k & +\frac{1}{2}L_a+\frac{1}{2}L_g+\frac{1}{2}L_k+L_l+\frac{1}{2}L_n & \frac{1}{2}L_n \\ 0 & \frac{1}{2}L_n & \frac{1}{2}L_d+\frac{1}{2}L_h+\frac{1}{2}L_n+L_o+L_r+\frac{1}{2}L_g \end{array}\right]$$

$$= \frac{L}{EA}\begin{bmatrix} 2+2\sqrt{2} & \frac{1}{2} & 0 \\ \frac{1}{2} & 2+2\sqrt{2} & \frac{1}{2} \\ 0 & \frac{1}{2} & 2+2\sqrt{2} \end{bmatrix}$$

$$= \frac{L}{EA}\begin{bmatrix} 4.828 & 0.5 & 0 \\ 0.5 & 4.828 & 0.5 \\ 0 & 0.5 & 4.828 \end{bmatrix}$$

Let us use permutation-augmentation approach for inversion, Chapter 1. In the above matrix no permutation is required because the diagonals as obtained are dominant. Augmenting 4th column.

$$\begin{bmatrix} \boxed{4.828} & 0.5 & 0 & \boxed{1} \\ 0.5 & 4.828 & 0.5 & 0 \\ 0 & 0.5 & 4.828 & 0 \end{bmatrix}$$

Divide the first row by 4.828. Then 0.5 times the modified first row when subtracted from the second row gives

$$\begin{bmatrix} 1 & 0.1035 & 0 & 0.207 \\ 0 & 4.776 & 0.5 & -0.104 \\ 0 & 0.500 & 4.828 & 0 \end{bmatrix}$$

Delete the first column and augment {0 1 0}

$$\begin{bmatrix} \boxed{0.1035} & 0 & 0.207 & \boxed{0} \\ 4.776 & 0.5 & -0.104 & 1 \\ 0.500 & 4.828 & 0 & 0 \end{bmatrix}$$

Divide the second row by 4.776; multiples of the modified second row are combined with first and third rows so that the first column is reduced to {0 1 0}. Deleting the first column so obtained and augmenting by {0 1 0}

$$\begin{bmatrix} -0.011 & 0.209 & -0.022 & 0 \\ 0.105 & -0.022 & +0.209 & 0 \\ 4.775 & +0.011 & -0.105 & 1 \end{bmatrix}$$

Finally divide the last row by 4.775; multiples of the modified last row are combined with the first and second rows to reduce the first column to {0 0 1}. After the deletion of the first column, so obtained gives the inverse of the matrix considered.

$$\begin{bmatrix} 4.828 & 0.5 & 0 \\ 0.5 & 4.828 & 0.5 \\ 0.0 & 0.5 & 4.828 \end{bmatrix}^{-1} = \begin{bmatrix} 0.209 & -0.022 & +0.0023 \\ -0.022 & -0.211 & -0.022 \\ +0.0023 & -0.022 & +0.209 \end{bmatrix}$$

$$A_{qj} = (B^t F_m B)^{-1} (B^t F_m B)$$

$$= \begin{bmatrix} 0.209 & -0.022 & 0.0023 \\ -0.022 & +0.211 & -0.022 \\ +0.0023 & -0.022 & +0.209 \end{bmatrix} \begin{bmatrix} -2.041 & -1.372 \\ +1.372 & -0.667 \\ +1.137 & +2.277 \end{bmatrix}$$

$$A_{qj} = \begin{bmatrix} -0.4541 & -2.668 \\ +0.3094 & -0.1590 \\ +0.2028 & +0.4874 \end{bmatrix}$$

∴ The redundants $(q_1 \quad q_2 \quad q_3) = Q = -A_q j\, P_j = -A_{qj} \begin{bmatrix} P_1 \\ P_2 \end{bmatrix}$

These can be evaluated for the numerical values of

$$P_j = [P_1 P_2]^t$$

and $\quad p_m = (B_0 - B\,A_{qj})\,P_j = A_{mj}\,P_j \quad \text{where } A_{mj} = B_0 - B\,A_{qj}$

and $\quad d_j = B_0{}^t\,F_m\,(B_0 - B\,A_{qj})\,P_j = B_0{}^t\,F_m\,A_{mj}\,P_j$

or $\quad d_j = F\,P_j \quad \text{where } F = B_0{}^t\,F_m\,A_{mj}$

4.3.2 Comments on Matrix Method

1. Ill - conditioning of $B^T F_m B$: In chapter 3, the redundants Q, were evaluated from the flexibility equation,

$$F\,Q + U^0 = 0 \qquad \text{...(i)}$$

The elements f_{ij} and U^0 were obtained by integrations involving m_0 and m_i diagrams. Based on the matrix method, Eq. (4.4) may be written as

$$(B^T F_m B) Q + (B^T F_m B_0) P_j = 0 \quad \ldots \text{(ii)}$$

Comparing the Eqs (i) and (ii)

$$\left.\begin{aligned} F &= B^T F_m B, \\ \text{and} \quad U^0 &= B^T F_m B_0 P_j, \end{aligned}\right\} \quad \ldots (4.5)$$

The elements of the matrices B_0 and B are not unique. They depend on the release system chosen. Therefore, a proper choice of release system is important to avoid ill-conditioning of the flexibility matrix, $B^T F_m B$. These aspects have already been discussed, Article 2.14.1. In the case of the example of the hinged truss, the matrix $B^T F_m B$, was well-conditioned with the dominant diagonal elements. Introduction of releases in the top or bottom members may result in an ill-conditioned flexibility matrix $(B^T F_m B)$.

2. Reduced size of B_0 matrix: We observe that the size of matrix $B^T F_m B$ is given by the degree of indeterminacy, and the size of $B^T F_m B_0$ depends on the number of load elements of the load matrix P_j. It is possible to reduce the size of the B_0 matrix to a single column. In the example considered, Fig. 4.5, the B_0 matrix consisted of two columns corresponding to the two load elements P_1 and P_2. Consider another example of tower, Fig. 4.6. The load matrix, P_j, consist of eight load elements $(P_1, P_2, \ldots, P_8)$ and degree of indeterminacy, is equal to six. The size of the B_0 matrix is 46×8, which is given by the number of rows equal to the number of members (=46) and the number of columns equal to the number of loads (=8). The intersections of diagonals are considered as node. If the member forces are computed for all the loads acting together, then these member forces constitute the elements of B_0 matrix, which is now a column matrix of 46 elements, and the corresponding equilibrium equation is given by

$$P_m = (B_0 \; B) \begin{bmatrix} 1 \\ Q \end{bmatrix}$$

In the above equation, we notice that the place of P_j is taken by unity. The one disadvantage of merging all the loads together is that we are not in a position to interpret d_j, which is a single element corresponding to P_j = unity. The results for this case are given by

$$Q = -(B^T F_m B)^{-1} (B^T F_m B_0),$$

$$P_m = B_0 + B\,Q \text{ and}$$

The scalar $\quad d_j = B_0^{\,T} p_m$

It is evident that at the expense of the reduced B_0 matrix, we are not able to find real deflections. Instead, we have a generalized scalar deflection d_j which is not amenable to any physical interpretation.

3. Evaluation of critical deflection: At a second look, it may be observed that this shortcoming of not being able to find deflections can be overcome. In a design to check against deflections, we are required to evaluate critical or maximum deflections only. Consider the example of the antenna tower, Fig. 4.6, which is a plane truss idealized from a space truss. In this case, a designer would like to limit the deflection at the top where antenna is installed.

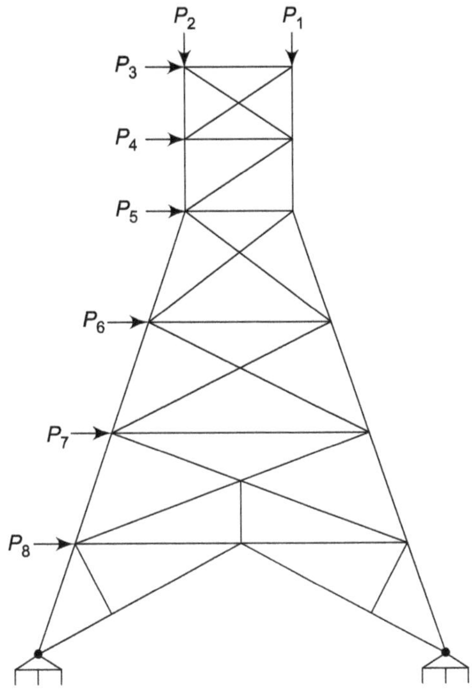

Fig. 4.6 *Antenna Tower*

Let us assume that we are required to know the deflection in the direction of the load P_3. To achieve this, the matrix B_0 can be written in two columns only, where the elements of the first column are the member forces due to all the loads excepting P_3, and the second column is obtained by evaluating member forces due to $P_3 = 1$. The equilibrium equation with this compromise is given by

$$P_m = (B_{01}\ B_{02}\ B)\begin{bmatrix} 1 \\ P_3 \\ Q \end{bmatrix}$$

and the deflections d_j, evaluated from the relation, $d_j = (B_{01}\ B_{02})^T P_m$ consist of two elements only and the second element is the required deflection in the direction of P_3. In this process, we have not only limited the size of B_0 matrix but also computed the critical deflection.

4. Hinged-Space truss: The space-truss can also be analysed in a similar manner. The (B_0, B) matrices are obtained by static analysis of the trusses with the suitable releases, by using method of joints or tension coefficient. The truss, Fig. Ex. 3.2, was an example of a determinate space truss.

   When additional diagonals, 3 on each face are introduced, we have twelve redundatns. Once the (B_0, B) matrices are written for the truss with releases, under the action of any given loading and chosen redundants, the evaluation of Q, P_m, and d_j consists of simply a set of steps of matrix operations. As an illustration, consider a 20 meter high space antenna tower, Fig. 4.7.

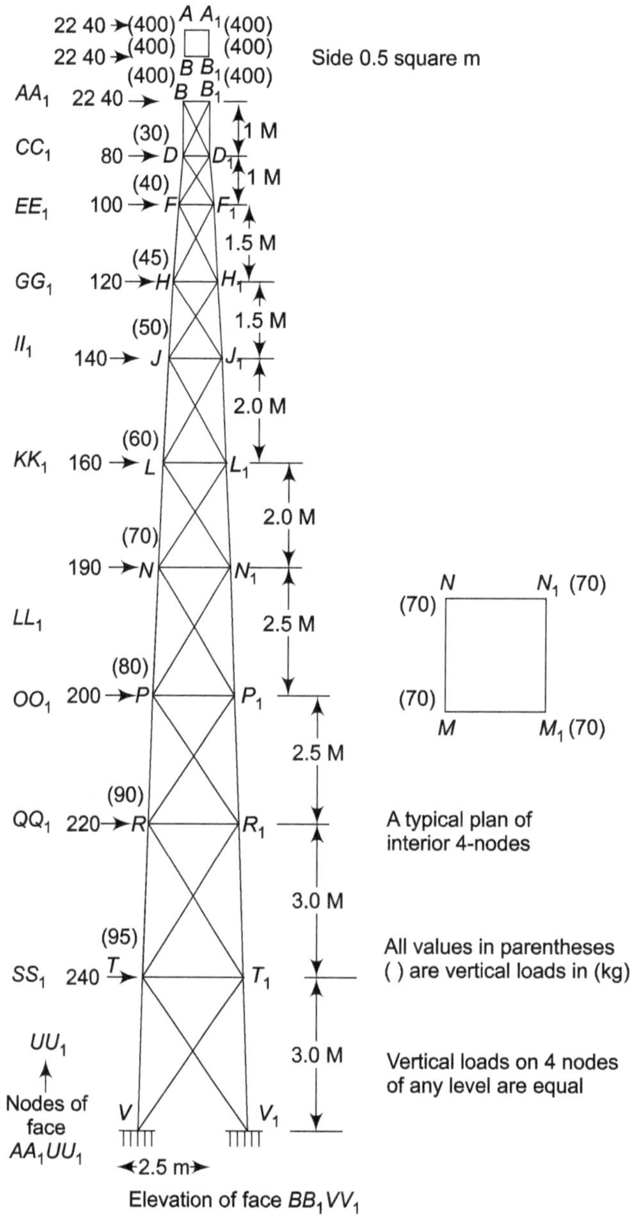

Fig. 4.7 *Symmetrical Space truss—An Antenna*

An isometric view of the top panel is shown in Fig. 4.8. To avoid congestion, only one diagonal per face is shown. The degree of indeterminacy is forty. If we take advantage of symmetry and introduce releases by cutting the diagonals not shown in Fig. 4.8, we are left with twenty unknown redundants. Further, for the frame, Fig. 4.7, under the loads as shown, the diagonals and horizontal members on the two faces defined by corner nodes A, B, D, C and A_1, B_1, D_1, C_1 develop negligible forces. It may, therefore, be justified to idealize the truss as a plane truss, considering the face BB_1DD_1 only, under the loads as shown. The vertical loads, however, may be resolved vectorially to obtain the components in the plane of the truss. The problem is thus reduced to a plane truss involving ten redundants only. Similar idealization for wind in y-direction the critical deflection is at the top where antenna is installed. The large value of the horizontal forces at level B (=2240 kg.) is due to the wind pressure on the antenna. The B_0 matrix consists of two columns, where the second column is due to the unit horizontal columns, where the second column is due to the unit horizontal load at the level B.

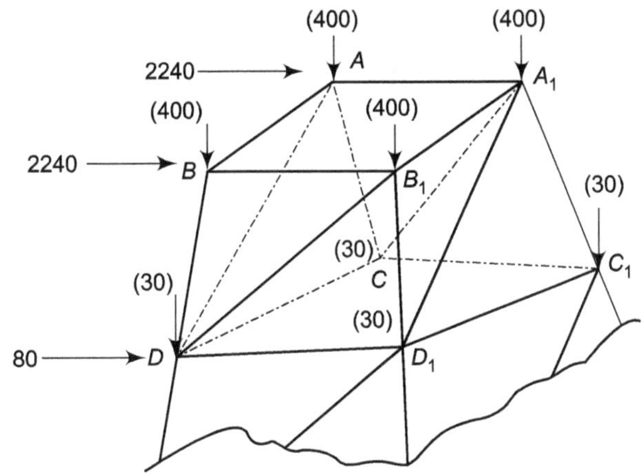

Fig. 4.8 *Top Panel of Space Truss*

If one wishes to analyse it as a space frame, the determinate truss under the action of loads and unit redundants is analysed starting from the top, Fig. 4.8, commencing from the node B. The rest of the procedure is too simple to explain further. In fact a determinate statical analysis for given loads may be treated as a first order solution for design.

ADDITIONAL EXAMPLES

The readers may like to attempt some simple examples of plane trusses, Figs Ex. 4.1, Ex. 4.2, Ex. 4.3. For the example, Fig. Ex. 4.2, the results of the important steps are given below.

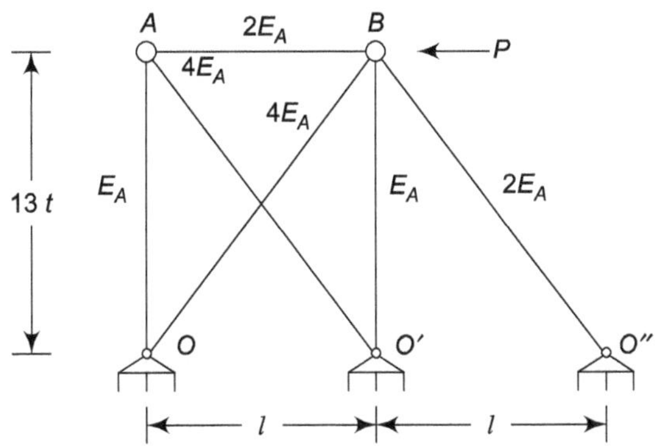

Fig. Ex. 4.1

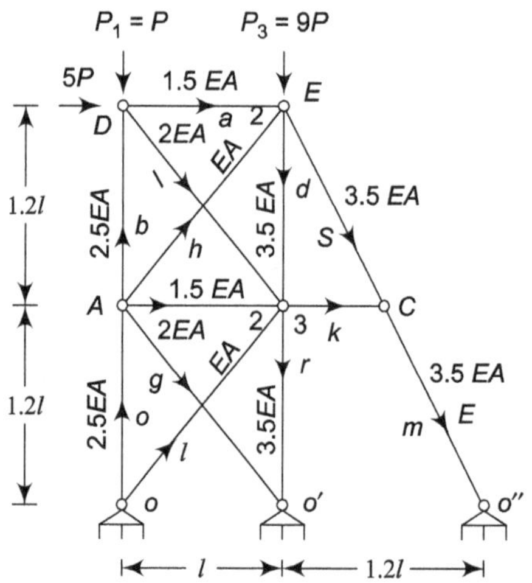

Fig. Ex. 4.2

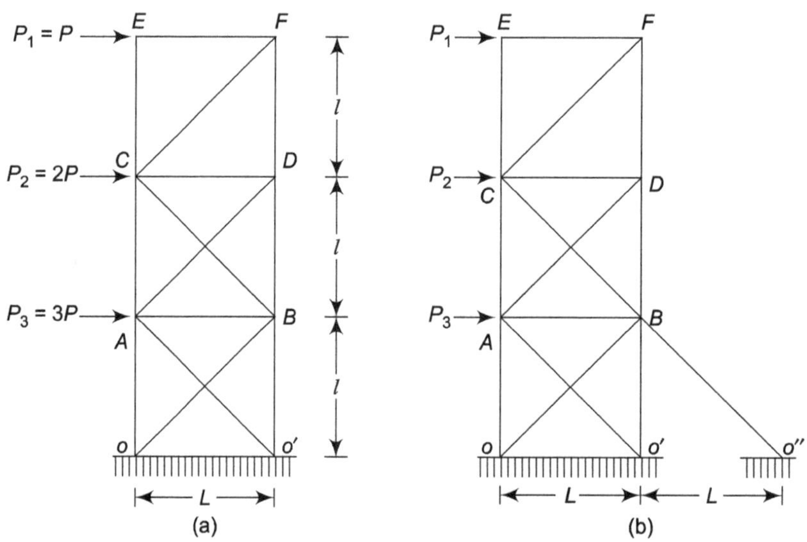

Fig. Ex. 4.3

Ex. 4.2. The degree of indeterminacy is equal to three. The releases are introduced by cutting the members *I*, *h* and *m*. For the member sequence, *a* to *n* and loads sequence P_1, P_2 and P_3 the B_0 and B matrices are given.

And the diagonal member flexibility matrix is also given.

$$F_m = [0.48 \quad 0.48 \quad 0.67 \quad 0.343 \quad 0.343 \quad 0.78 \quad 0.78 \quad 1.56 \quad 1.56 \quad 0.383 \quad 0.6 \quad 0.383 \quad 0.67]$$

The B_0 matrix can be reduced to a column matrix by considering the effects of all loads together as explained before. In the equilibrium equation, the only changes will then be in B_0 matrix and the load column matrix. The B_0 column matrix is computed for $P_1 = P$, $P_2 = 5P$ and $P_3 = 9P$.

$$B_0 = P[11 \quad 50 \quad -9 \quad -15 \quad -7.8 \quad 0 \quad 0 \quad 0 \quad 0 \quad 0 \quad 5]^t$$

And the modified right hand side column matrix is

$[1\ q_1\ q_2\ q_3]^t$

$$\underset{p_m}{\begin{bmatrix} (p_x)_a \\ (p_x)_a \\ (p_x)_a \\ (p_x)_a \\ (p_x)_a \\ (p_x)_a \\ (p_x)_a \\ (p_x)_a \\ (p_x)_a \\ (p_x)_a \\ (p_x)_a \\ (p_x)_a \\ (p_x)_a \end{bmatrix}} = \left[\begin{array}{ccc|ccc} P_1 & P_2 & P_3 & q_1 & q_2 & q_3 \\ -1 & +2.4 & 0 & 01.078 & 0 & -0.769 \\ -1 & +1.2 & 0 & +0.539 & -0.769 & 0 \\ 0 & 0 & 0 & +0.448 & -0.641 & 0 \\ 0 & 0 & -1 & -0.896 & -0.769 & 0 \\ 0 & -1.2 & -1 & -1.435 & 0 & -0.769 \\ 0 & -1.56 & 0 & -0.700 & +1 & 0 \\ 0 & 1.56 & 0 & -0.700 & 0 & +1 \\ 0 & 0 & 0 & 0 & +1 & 0 \\ 0 & 0 & 0 & 0 & 0 & +1 \\ 0 & 0 & 0 & +1 & 0 & 0 \\ 0 & 0 & 0 & 0 & 0 & 0 \\ 0 & 0 & 0 & +1 & 0 & 0 \\ 0 & +1 & 0 & +0.448 & -0.641 & -0.641 \\ & B_0 & & & B & \end{array}\right] \begin{bmatrix} P_1 \\ P_2 \\ P_3 \\ \cdots \\ q_1 \\ q_2 \\ q_3 \end{bmatrix}$$

4.4 CONTINUOUS BEAMS ON RIGID AND FLEXIBLE SUPPORTS

Consider the continuous beams of 5 spans each, Fig. 4.9. The degree of indeterminacy (equal to four) is the same for the three beams. However, if one decides to analyse by displacement methods, the number of unknown deformations (or independent degrees of freedom) is different for each beam. Considering modified stiffnesses for end spans and supports as nodes, the unknown deformations are:

Fig. 4.9 (a) 4 interior rotations

Fig. 4.9 (b) 4 interior rotations and 4 displacements, total = 8

Fig. 4.9 (c) 4 interior rotations and 5 relative displacements, total = 9

It is obvious that for this type of structures, force method is preferred for static analysis. The continuous beams on rigid supports, Fig. 4.9(a), are very familiar. A typical example of the beam, Fig. 4.9(b), is a pontoon bridge where interior supports are the floats. The beams of Fig. 4.9 (c) occur: (i) as beams on elastic bed, where the continuous bed has been idealized by discrete spring supports, (ii) Hendry-Jaeger idealization of bridge decks with zero torsional stiffness, where longitudinal girders are idealized by spring supports. In this section the beams of the types (b) and (c) are taken as illustrations for the analysis. The basic steps of the force method remaining unaltered, we are required to obtain member flexibility which is the generalized strain-stress relation.

Member flexibility: Consider a typical member, a beam element, Fig. 5.10(a). In the case of beam, internal forces are shear and moment. The undeformed position is shown in the figure. We define the member end forces at the end 2 as $(p_m)_2 = (p_y m)_2$. The member forces at the end 1 can be obtained from static equilibrium of the member. The corresponding member-end-deformations are

$$(d_m)_1 = \{\delta_{y1}\ \theta_1\},\ (d_m)_2 = \{\delta_{y2}\ \theta_2\},$$

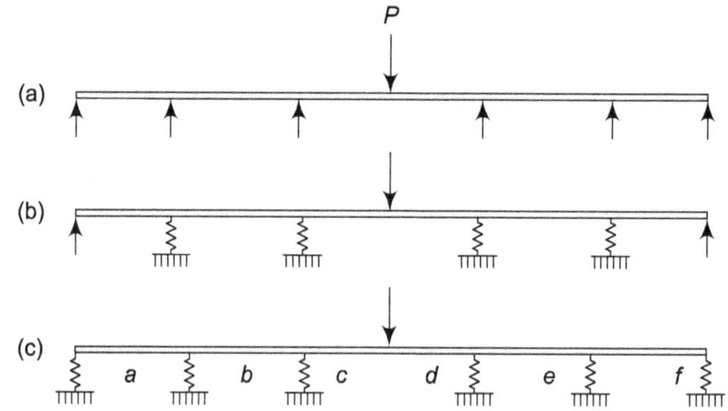

Fig. 4.9 *Continuous Beam*

But if we are considering $(p_m)_2$ only as independent member forces, the corresponding two independent deformations are the relative deformations, which in this case are the deformations at the end 2 with respect to the tangent at the end 1.

The tangent at end 1 in the deformed position is the member axis, $\bar{x}$. The deformations measured referring to $\bar{x}\,\bar{y}$ axis, exclude rigid body displacements which in this case are the deformations at the end 1, i.e. rigid body displacement δ_{y1} and rigid body rotation θ_1. Rigid body displacements do not result into any internal forces. The deformations and are equivalent to simple cantilever deformations of the member at the free end 2, due to the member-end-forces P_{y2} and m_2. Henceforth, the bar above the deformation symbols will be dropped even though they are the relative deformations. The member flexibility matrix for such a member is obtained from Fig. 4.10(b) and the resulting deformation force relation is given by Eq. (4.7). The first

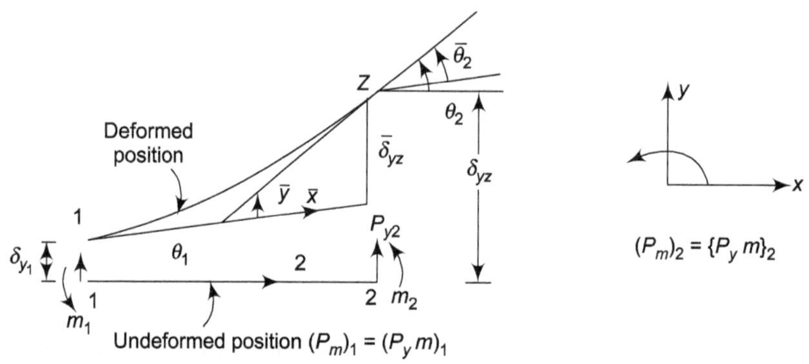

(a) Reference line, tangent at end 1 in deformed position

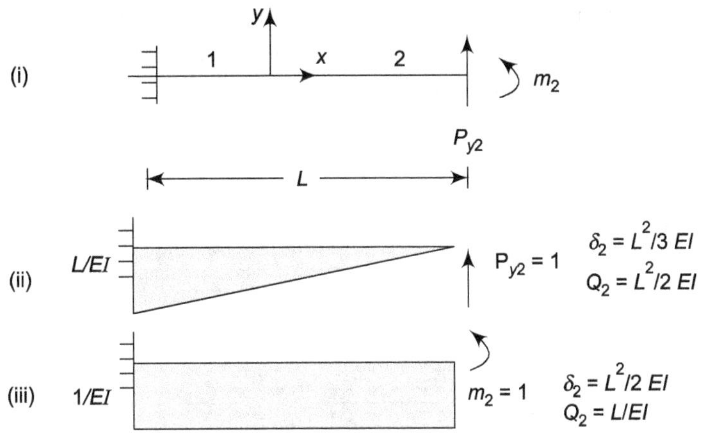

(b) Member flexibilities

Fig. 4.10 *Relative Deformations and Corresponding Member Flexibilities*

$$\begin{bmatrix} \delta y \\ \theta \end{bmatrix}_2 = \begin{bmatrix} \dfrac{L^3}{3EI} & \dfrac{L^2}{2EI} \\ \dfrac{L^2}{2EI} & \dfrac{L}{EI} \end{bmatrix} \begin{bmatrix} Py \\ m \end{bmatrix}_2 \qquad \text{(columns: } P_{y2},\ m_2\text{)} \qquad \ldots\ldots(4.7)$$

$$\delta_{2i} = (f_m)_I\, p_{2i}$$

Where δ_{2i} consists of relative deformations of member i.

With the member flexibility given by Eq. (4.7), the basic steps of analysis are repeated to solve the continuous beam problems.

Alternatively the relative deformations may be referred with respect to the line joining the end points in the deformed position instead of the tangent at the end –1, for which equation (4.7) is obtained. Member flexibility for such a reference line is obtained while solving the example, Ex. 4.7.

Solved Example

Ex. 4.4. Consider a continuous beam, Fig. Ex. 4.4(a), under a load P = 45.4 T and moment = 11.5 tm. The beam is resting on six flexible supports at a spacing of 2/3 m and constants $K_f = K_n$ = 354 t/m and $K_g = K + K$ = 708 t/m.

Flexural rigidity of the beam, EI = 324 tm^2. The flexibility of the springs, say, spring I is given by, $(f_m)i = 1/K_i$

The continuous beam consists of 11 members, out of which five are beam members (a, b, c, d, e) and six are spring members. Like the hinged truss members, the springs developed axial forces only and the member forces and deformation are defined by p_{x2} and δ_{x2}.

As the degree of indeterminacy is four, hinge releases are introduced at the four interior supports (B, C, D and E), Fig. 4.4(b). The evaluation of member forces for the beams (P_{y2}, m_2) and for springs (P_{x2}) caused by the given loads P and M is shown diagrammatically in the same figure. Based on the member forces, the equilibrium equation is given below, where the elements of B_0 are computed for the load.

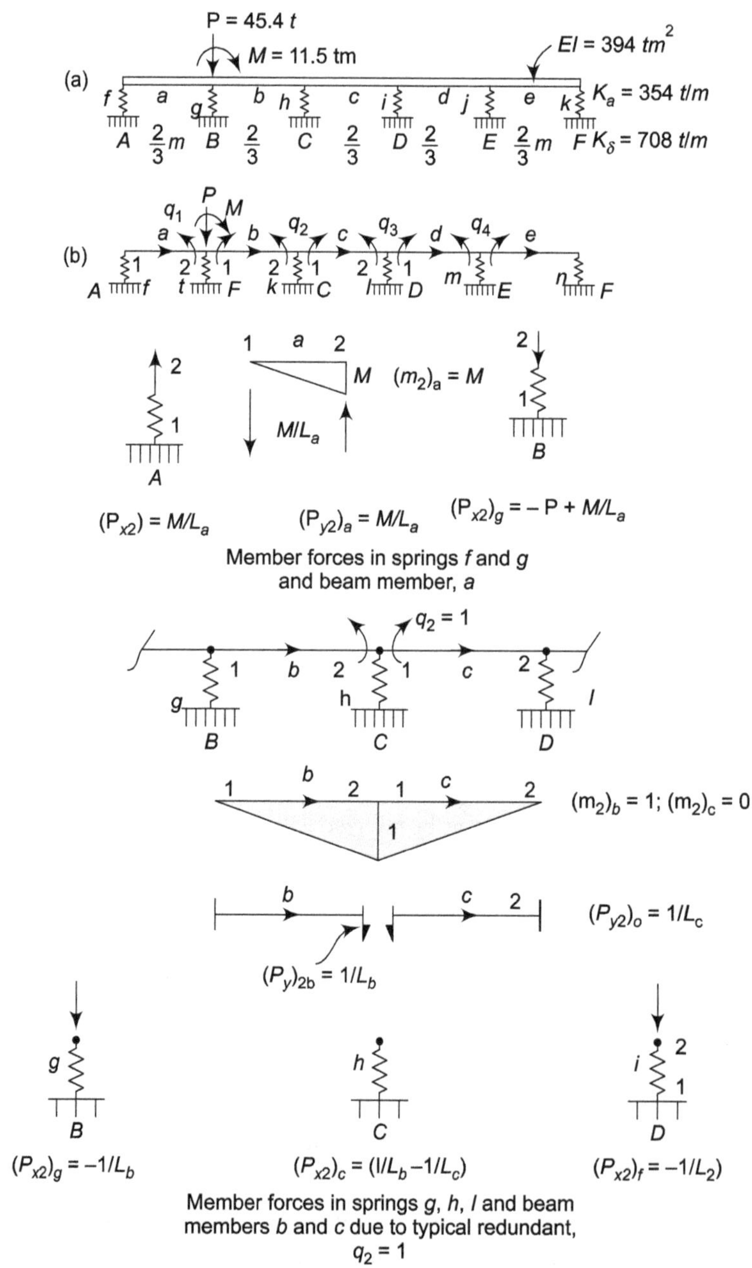

Fig. Ex. 4.4

$$P = 45.4 \text{ t and } m = 11.5 \text{ tm}$$

The elements of B matrix are computed for unit redundants. The computations of member forces for a typical redundant are shown diagrammatically, Fig. Ex. 4.4 (c). The chosen redundant gives rise to forces in the beam members b and c, and spring members g, h and i.

$$P_m = (B_0\, B)\begin{bmatrix} 1 \\ Q \end{bmatrix} \quad \text{Article 5.32, Eq. (5.6)}$$

$$\begin{bmatrix} \begin{bmatrix} P_{y2} \\ m_2 \end{bmatrix}_a \\ \begin{bmatrix} P_{y2} \\ m_2 \end{bmatrix}_b \\ \begin{bmatrix} P_{y2} \\ m_2 \end{bmatrix}_c \\ \begin{bmatrix} P_{y2} \\ m_2 \end{bmatrix}_d \\ \begin{bmatrix} P_{y2} \\ m_2 \end{bmatrix}_e \\ (P_{x2})_f \\ (P_{x2})_g \\ (P_{x2})_h \\ (P_{x2})_i \\ (P_{x2})_j \\ (P_{x2})_n \end{bmatrix} = \begin{array}{c} \begin{array}{ccccc} 1 & q_1 & q_2 & q_3 & q_4 \end{array} \\ \begin{bmatrix} 17.15 & -1.5 & 0 & 0 & 0 \\ -11.50 & 1 & 0 & 0 & 0 \\ 0 & 1.5 & -1.5 & 0 & 0 \\ 0 & 0 & 1 & 0 & 0 \\ 0 & 0 & 1.5 & -1.5 & 0 \\ 0 & 0 & 0 & 1 & 0 \\ 0 & 0 & 0 & 1.5 & -1.5 \\ 0 & 0 & 0 & 0 & 1 \\ 0 & 0 & 0 & 0 & 1.5 \\ 0 & 0 & 0 & 0 & 0 \\ 17.15 & -1.5 & 0 & 0 & 0 \\ -62.65 & 3 & -1.5 & 0 & 0 \\ 0 & -1.5 & 3 & -1.5 & 0 \\ 0 & 0 & -1.5 & 3 & -1.5 \\ 0 & 0 & 0 & -1.5 & 3 \\ 0 & 0 & 0 & 0 & -1.5 \end{bmatrix} \end{array} \begin{bmatrix} 1 \\ q_1 \\ q_2 \\ q_3 \\ q_4 \end{bmatrix}$$

The generalized strain-stress relation is given by

$$d_m = F_m\, p_m$$

where $F_m = [(f_m)_a, (f_m)_b, (f_m)_c, (f_m)_d, (f_m)_e, (f_m)_f, (f_m)_g, (f_m)_h, (f_m)_i, (f_m)_n]$

From the member properties given,

$$(f_m)_a = (f_m)_b = (f_m)_c = (f_m)_d = (f_m)_e = \begin{bmatrix} 2.425 & 5.60 \\ 5.60 & 16.75 \end{bmatrix} 10^{-4}$$

$$(f_m)_f = (f_m)_n = 2.84 \times 10^{-3} \text{ and}$$

$$(f_m)_g = (f_m)_h = (f_m)_i = (f_m)_f = 1.42 \times 10^{-3}$$

The subsequent steps are simple matrix operations to evaluate the redundatns $Q = [q_1\ q_2\ q_3\ q_4]^t$.

The salient intermediate results are

$$B^T F_m B = 10^{-3}\begin{bmatrix} 235 & -124.86 & 32.45 & 0 \\ -124.86 & 203 & -124.86 & 32.45 \\ 32.45 & -124.86 & 235 & -124.84 \\ 0 & 32.45 & -124.86 & 235 \end{bmatrix} \quad \ldots(1)$$

$$B^T F_m B_0 = 10^{-3}\begin{bmatrix} -3466.4 \\ 1337.5 \\ 0 \\ 0 \end{bmatrix} \quad \ldots(2)$$

Redundant Q are obtained from the equation

$$(B^T F_m B)\, Q + B^T F_m B_0 = 0 \quad \ldots(3)$$

Solution of equation for unknown, Q:

As the releases have been chosen properly, the diagonal elements are quite dominant. Instead of inverting the flexibility matrix, $B^T F_m B$, we may solve the equations by iteration. However, the unknowns are evaluated by block iterations combined with elimination, as explained in the following. It will be found that reasonably accurate results are obtained in the first cycle itself. For better explanation of the method let us write the Eq. (3), with the numerical values from Eqs (1) and (2), in the expanded form.

Block 1
$$\left|\begin{array}{l} 2.35q_1 - 1.25q_2 + 0.325q_3 + 0 = 34.66 \\ -1.25q_1 + 2.03q_2 - 1.25q_3 + 0.325q_4 = -13.375 \end{array}\right.$$

Block 2
$$\left|\begin{array}{l} 0.325q_1 + 1.25q_2 + 2.350q_3 - 1.25q_4 = 0 \\ 0 + 0.325q_2 - 1.25q_3 + 2.350q_4 = 0 \end{array}\right.$$

Cycle–1

Assume $q_3 = q_4 = 0$ and by using the first block, that is, the first two equations only, solve for q_1 and q_2 by elimination.

$$2.35q_1 - 1.25q_2 = 34.66$$

$$-1.25q_1 + 2.35q_2 = -13.375$$

By elimination procedure $q_1 = 16.75$ and $q_2 = 3.70$

With these values of q_1 and q_2 substituted in the second block of equations, that is, the third and fourth equations, we modify the right-hand side constant (in this case, zeros). The modified equations of the second block are:

$$2.35q_3 - 1.25q_4 = -0.8$$

$$-1.25q_3 + 2.35q_4 = -1.2$$

By elimination procedure, $q_3 = -0.87$, $q_4 = -1$.

Cycle–2

The values of q_3 and q_4 of the first cycle are substituted in two equations of the first block. The resulting equations with the modified right-hand side constants are :

$$2.35q_1 - 1.25q_2 = 34.94$$

$$-1.25q_1 + 2.03q_2 = -14.14$$

Solving the elimination

$$q_1 = 16.45,\ q_2 = 3.03$$

similarly with the q_1 and q_2 values substituted in the second block, we obtain:

$$q_3 = -1.24,\ q_4 = -1.08$$

The above values of q_1, q_2, q_3, q_4 are not much different from the values of cycle–1, and we may treat the values of the cycle–2 as the final values.

The redundant moments being known, final bending moment diagram can be drawn directly, Fig. Ex. 4.5 (c).

Ex. 4.5. Solve the same example (Ex. 4.4), but this time take advantage of symmetrical structure and analyse for symmetrical and antisymmetrical loadings.

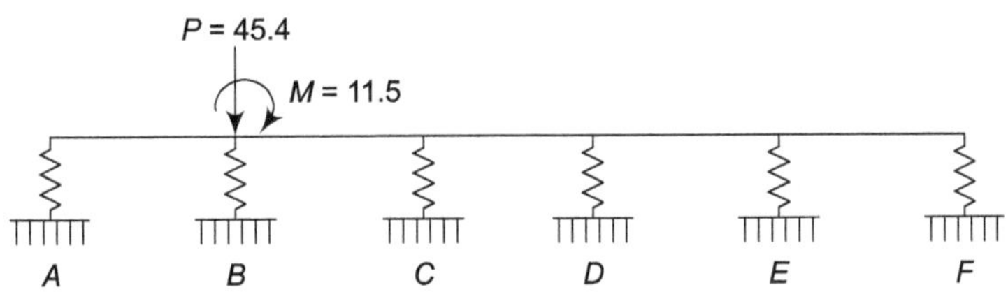

Fig. Ex. 4.5 *Given Loading*

The given loading, Fig. Ex. 4.5 is expressed as symmetrical and antisymmetrical loadings, Fig. Ex. 4.5. With the result, for each case we solve for two unknown redundants. Superposition of the results gives the final values of redundants for the given loading. Figure Ex. 4.5(a) and (b) are the symmetrical and antisymmetrical loadings respectively for the given loads, *P* and *M* at support *B*.

(a) Symmetrical loading: Fig. Ex. 4.5(a)

Taking advantage of symmetry, the four redundants are defined by q_1 and q_2 as shown on the supports *B*, *C* and *B′*, *C′*. Further we consider only three members *a*, *b* and *c*, lying within the left half *AD* of the beam. The moments due to the symmetrical loads and redundant q_1 at support *B* are contained within the left half only. The moments due to the redundants q_2 applied on the supports C and *C′* are given. BM-diagram is shown in the Fig. Ex. 4.5 (a). We use the three basic steps to solve the symmetrical case.

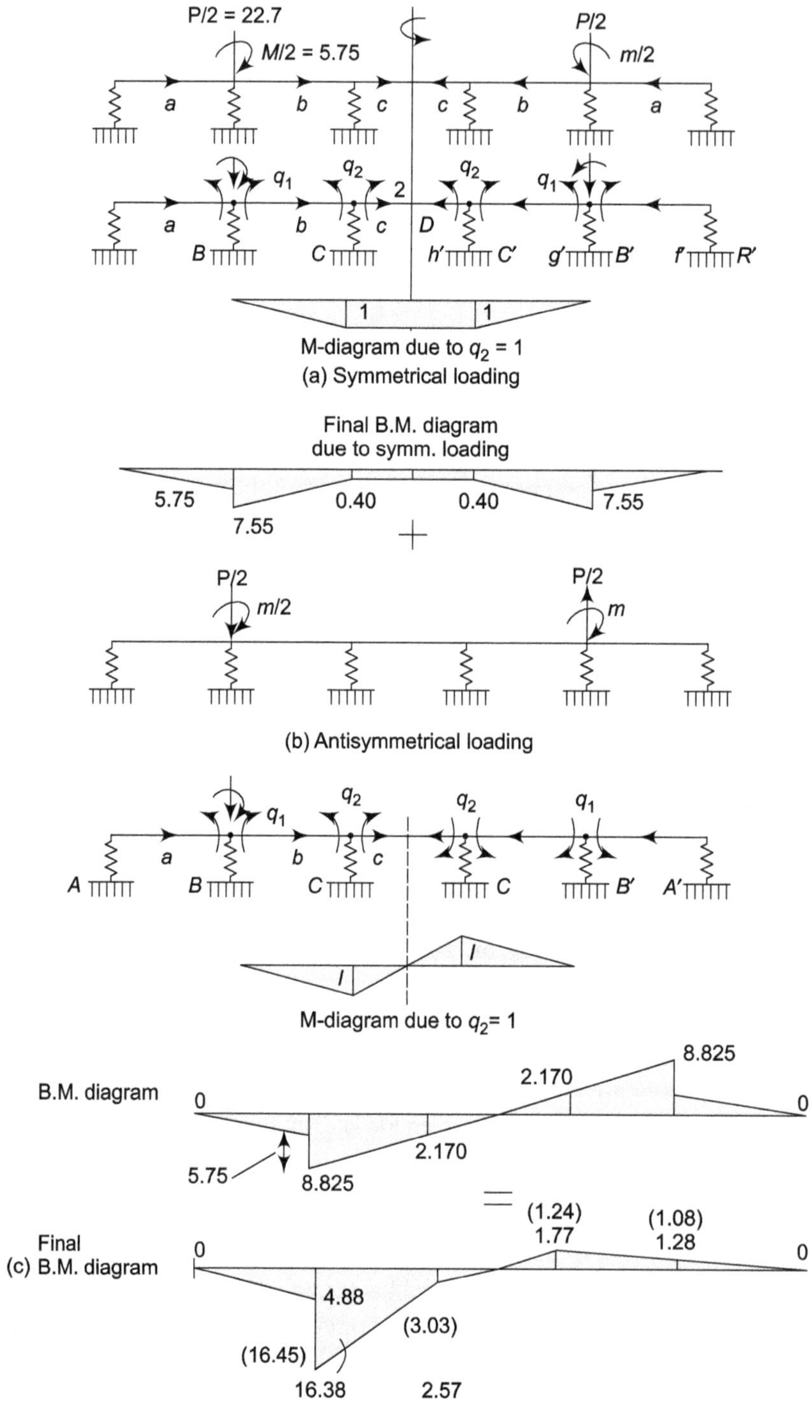

Fig. Ex. 4.5 *Superposition of Symmetry and Antisymmetry*

1. Equilibrium:

$$\begin{bmatrix} \begin{bmatrix} P_{y2} \\ m_2 \end{bmatrix}_a \\ \begin{bmatrix} P_{y2} \\ m_2 \end{bmatrix}_b \\ \begin{bmatrix} P_{y2} \\ m_2 \end{bmatrix}_c \\ (P_{x2})_s \\ (P_{x2})_q \\ (P_{x2})_h \end{bmatrix} = \left[\begin{array}{c|cc} 8.575 & -1.5 & 0 \\ -5.75 & 1 & 0 \\ 0 & 1.5 & -1.5 \\ 0 & 0 & 1 \\ 0 & 0 & 0 \\ 0 & 0 & 1 \\ 8.575 & -1.5 & 0 \\ -31.325 & 3 & -1.5 \\ 0 & 1.5 & 1.5 \end{array}\right] \begin{bmatrix} 1 \\ q_1 \\ q_2 \end{bmatrix} \quad \ldots\text{(i)}$$

2. Generalized strain-stress relation is given by assembling the member flexibilities.

$$F_m = [(f_m)_a \, (f_m)_b \, (f_m)_c \, (f_m)_f \, (f_m)_g \, (f_m)_h] \quad \ldots\text{(ii)}$$

Where $(f_m)_a = (f_m)_b = \begin{bmatrix} 2.425 & 5.6 \\ 5.6 & 16.75 \end{bmatrix}$

$$(f_m)_c = \begin{bmatrix} .303 & 1.40 \\ 1.40 & 8.375 \end{bmatrix}$$

$(f_m)_f = 28.4$ and $(f_m)_g = (f_m)_h = 14.2$

3. $$F = B^T F_m B = 10^{-4} \begin{bmatrix} 234.53 & -92.91 \\ -92.91 & 77.68 \end{bmatrix}$$

$$U^0 = B^T F_m B = 10^{-4} [-1729 \quad 667]^t$$

$$Q = -F^{-1} U^0 = -\begin{bmatrix} 8.10 & 9.67 \\ 9.67 & 24.40 \end{bmatrix} \begin{bmatrix} -1.729 \\ 0.667 \end{bmatrix}$$

$Q = [7.55 \quad 0.40]^t$ Symmetry.

The bending moment diagram based on the redundants Q evaluated and symmetrical loading is also drawn, Fig. Ex. 4.5 (a)

(b) Antisymmetrical loading: Fig. Ex. 4.5 (b)

As argued in the case of symmetrical loading, because the moments due to loads at B and redundants q_1 are contained within the left half only, the B_0 and the first column of B matrix appearing in Eq. (i) remain unaltered. The M-diagram due to q_2 acting in the direction as shown at supports C and C' is drawn in the Fig. Ex. 4.5(b). The equilibrium equation for the antisymmetrical case based on moments due to antisymmetrical loads and redundants $(q_1 q_2)$ is given below:

$$P_m = \begin{bmatrix} 8.575 & -1.5 & 0 \\ -5.75 & 1 & 0 \\ 0 & 1.5 & -1.5 \\ 0 & 0 & 1 \\ 0 & 0 & 3 \\ 0 & 0 & 0 \\ 8.575 & -1.5 & 0 \\ -31.325 & 3 & -1.5 \\ 0 & -1.5 & 4.5 \end{bmatrix} \begin{bmatrix} 1 \\ q_1 \\ q_2 \end{bmatrix}$$

The F_m matrix as given for the symmetrical case remains unaltered. Thus we get

$$F = B^T F_m B = 10^{-4} \begin{bmatrix} 234.53 & -156.80 \\ -156.80 & 330.58 \end{bmatrix}$$

$$U^0 = B^T F_m B_0 = 10^{-4} \begin{bmatrix} -1729 & 667 \end{bmatrix}^t$$

$$Q = -F^{-1} U^0 = -\begin{bmatrix} 6.24 & 2.96 \\ 2.96 & 4.42 \end{bmatrix} \begin{bmatrix} -1.729 \\ 0.667 \end{bmatrix}$$

$$Q = [\ 8.825 \quad 2.170]^t \text{ antisymmetry}$$

The bending moment diagram based on the above redundants and loading for the antisymmetrical case is also drawn, Fig. Ex. 4.5 (b).

The final values of the redundants due to given loads, $P = 45.4$ t. $M = 11.5$ tm, are obtained by combining the results of the symmetrical and antisymmetrical cases.

$$Q = \begin{bmatrix} 7.55 \\ 0.40 \\ \hline 0.40 \\ 7.55 \end{bmatrix} + \begin{bmatrix} 8.83 \\ 2.17 \\ \hline -2.17 \\ -8.83 \end{bmatrix} = \begin{bmatrix} 16.38 \\ 2.57 \\ \hline -1.77 \\ -1.28 \end{bmatrix}$$

The final bending moment diagram obtained by combining the results of symmetrical and antisymmetrical cases is drawn, Fig. Ex. 5.5 (c). The numerical values are shown underlined. The reader may like to compare the results obtained by block iteration cum elimination given within the parentheses in the same figure.

Comment: The example, Fig. 5.4, was actually an idealization of the continuous foundation bed by discrete springs. The foundation modulus of the bed equal to 1062 t/m^2 when multiplied by the spacing of discredited springs gives the spring constant 708 t/m. The spring constant of the end support is equal to 708/2 = 354 t/m. The reader may like to know the error in the value of maximum moment due to discretization of continuous foundation bed.

Maximum moment to the right of support:

Discrete spring supports, $M_B = 16.38$ tm

Continuous foundation bed, $M = 16.52$ tm (analysis not presented here)

Error is less than one per cent.

Ex. 4.6. Analysis a continuous beam on three flexible supports, Fig. Ex. 4.6

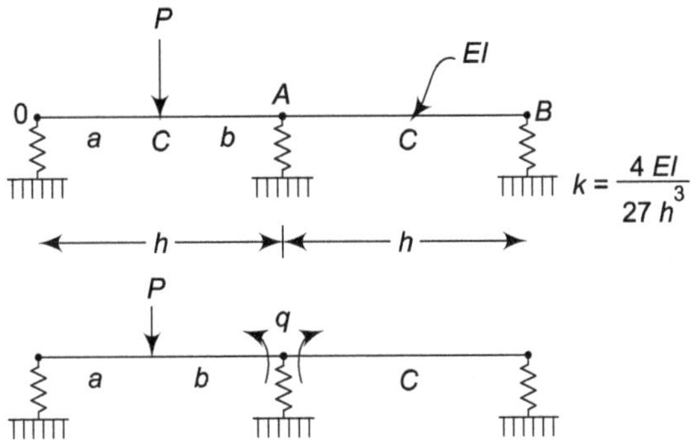

Fig. Ex. 4.6

Solution. Degree of indeterminacy = 1. Introduce hinge releases at support B. Equilibrium equation

$$p_m = [B_0\, B]\begin{bmatrix} 1 \\ Q \end{bmatrix}$$

$$\begin{bmatrix} \begin{bmatrix} p_{y2} \\ m_2 \end{bmatrix}_a \\ \begin{bmatrix} p_{y2} \\ m_2 \end{bmatrix}_b \\ \begin{bmatrix} p_{y2} \\ m_2 \end{bmatrix}_c \\ (p_{x2})_d \\ (p_{x2})_e \\ (p_{x2})_f \end{bmatrix} = \overset{\begin{matrix} B_0 & \quad B \end{matrix}}{\begin{bmatrix} -\frac{P}{2} & -\frac{1}{h} \\ \frac{ph}{4} & \frac{1}{2} \\ \frac{P}{2} & -\frac{1}{h} \\ 0 & 1 \\ 0 & \frac{1}{h} \\ 0 & 0 \\ -\frac{P}{2} & -\frac{1}{h} \\ -\frac{P}{2} & \frac{2}{h} \\ 0 & -\frac{1}{h} \end{bmatrix}}$$

$$F_m = [(f_m)_a (f_m)_b (f_m)_c (f_m)_d (f_m)_e (f_m)_f]$$

Where $$(f_m)_a = (f_m)_b = \begin{bmatrix} \frac{\left(\frac{h}{2}\right)^2}{3EI} & \frac{\left(\frac{h}{2}\right)^2}{2EI} \\ * & \frac{\frac{h}{2}}{EI} \end{bmatrix}$$

$$(f_m)_c = \begin{bmatrix} \frac{h^3}{3EI} & \frac{h^2}{2EI} \\ * & \frac{h}{EI} \end{bmatrix}$$

$$(f_m)_d = (f_m)_e = (f_m)_f = \frac{27}{4}\frac{(h)^3}{EI}$$

The reader may like to complete the remaining steps to evaluate the single unknown redundants.

Comments: The same example has been solved in Chapter 3, Article 3.5.4. In the case of the single unknown redundant, the virtual method discussed in Chapter 3, without using matrix approach, may be preferred.

This example is actually a structural idealization of a bridge deck, where the slab and transverse girders are replaced by the beam *OB* and the longitudinal girders are replaced by the springs at *O*, *A* and *B* with appropriate spring constants related to the flexural rigidities of the corresponding three longitudinal girders. A detailed description of such a structural idealization and analysis are given Art, 4.11.4.

4.5 PONTOON BRIDGES

A pontoon bridge girder, Fig. 4.7(a), rests on abutments *O* and *E* and interior supports are provided by the floats *A*, *B*, *C* and *D*. The spring action is provided by the buoyancy of the floats, Fig. (a). It is required to draw bending moment diagram for the following data: $E = 10^6$ t/m^2, $I = 2 \times 10^{-3}$ m^4. Horizontal area of each float at water level = ab = 20 m^2. The pontoons provide the same effect as springs, spring constant, k = 20 t/m.

The degree of indeterminacy is four. The releases consist of four hinges at the interior support points, Fig. (b). Taking advantage of symmetry, we consider the members *a*, *b* and *c* only and solve for the redundants, q_1 and q_2, Fig. (b).

As stated earlier, we will develop a member-flexibility matrix for a different reference line. In the beginning of the Article 4.4, we established a generalized strain-stress relation for the relative deformations and forces at the end 2, with tangent at end 1 as reference line. In essence what we did as:

While recognizing deformations d_1 and d_2 at the ends 1 and 2, we give a rigid body displacement, $-d_1$ (equal and opposite to the deformation components at 1). The resulting end deformations are:

$$\overline{d_1} = 0 \text{ and } \overline{d_2} = d_2 - d_1$$

The rigid body displacement components given by $-d_1$ do not result into any internal forces. The consideration of d_2 equal to $(d_2 - d_1)$ implies a simple cantilever as a basic member with deformations only at the end 2 of the cantilever, Fig. 4.10. We can use the same flexibility relations for the analysis of the pontoon bridge. However, for a change, we may consider a possibility of a different reference line.

Let us consider a reference line obtained by joining the ends 1 and 2 in the deformed position of the member. Fig. 4.11 (a).

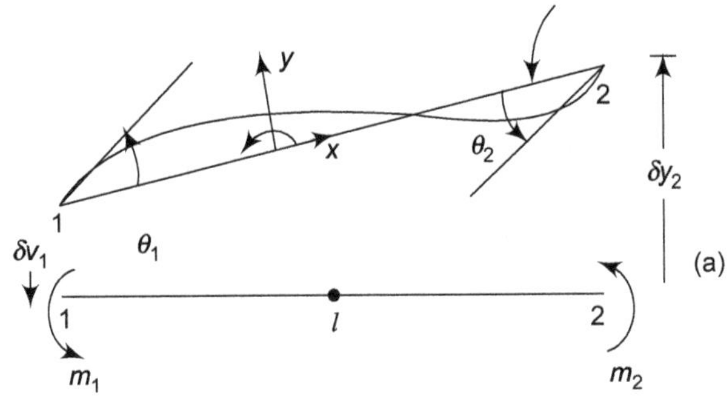

(a) Reference line, line joining end points in deformed position

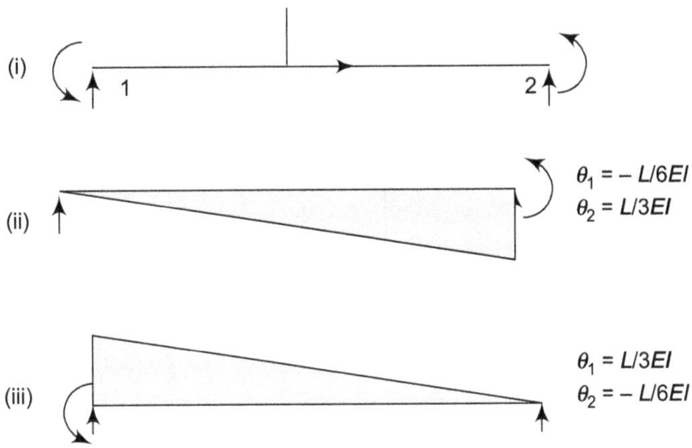

(b) Member flexibilities

Fig. 4.11 *Relative Deformations and Member Flexibilities*

By this, we imply that from the total deformations at the ends 1 and 2, we eliminate a different set of rigid body displacements. From the Fig. 4.11(a), the rigid body displacements chosen are, (i) rigid body displacement δ_{v1} and (ii) rigid body rotation $(\delta_{v2} - \delta_{v1})/L_i$ about the end 1. With these rigid body displacements eliminated, we are measuring relative deformations with respect to the line joining the ends 1 and 2 in the deformed position, and the relative deformations are $\overline{\theta_1}$ and $\overline{\theta_2}$, we consider the member forces m_1 and m_2.

Henceforth, we will drop bars and refer θ_1 and θ_2 as relative deformations. The consideration of relative deformations of such a beam member implies end rotation only without any displacement components, i.e., it is a beam with simple supports at the ends under the action of end moments, Fig. 4.11 (b). The member deformations due to $m_1 = 1$, are obtained by conjugate beam method, Fig. 4.11 (b) (ii)

$$\begin{bmatrix} \theta_1 \\ \theta_2 \end{bmatrix}_{\text{due to } m_1 = 1} = \begin{bmatrix} \frac{L}{3} & EI \\ -\frac{L}{6} & EI \end{bmatrix}$$

Similarly from Fig. 4.11 (b) (iii)

$$\begin{bmatrix} \theta_1 \\ \theta_2 \end{bmatrix}_{\text{due to } m_2} = 1 = \begin{bmatrix} -\frac{L}{6} & EI \\ \frac{L}{3} & EI \end{bmatrix}$$

Combining the above results, we get Eq. (4.8)

$$\begin{bmatrix} \theta_1 \\ \theta_2 \end{bmatrix}_i = \begin{bmatrix} \frac{L}{3EI} & -\frac{L}{6EI} \\ -\frac{L}{6EI} & \frac{L}{3EI} \end{bmatrix} \begin{bmatrix} m_1 \\ m_2 \end{bmatrix} \qquad \ldots(4.8)$$

If we are using the above generalized strain-stress relation while writing equilibrium equation $p_m = (B_0\ B) \begin{bmatrix} p_j \\ Q \end{bmatrix}$, the member forces $(p_m)_i$ consist of end moments m_1 and m_2 of the member, i.

Solution Ex. 4.7.

1. Equilibrium equation:

$$\begin{bmatrix} \begin{bmatrix} m_1 \\ m_2 \end{bmatrix}_a \\ \begin{bmatrix} m_1 \\ m_2 \end{bmatrix}_b \\ \begin{bmatrix} m_1 \\ m_2 \end{bmatrix}_c \\ (p_{x_2})_d \\ (p_{x_2})_e \end{bmatrix} = \begin{bmatrix} 0 & 0 & 0 \\ 0 & 1 & 0 \\ 0 & -1 & 0 \\ 0 & 0 & 1 \\ 0 & 0 & -1 \\ 15 & 0 & 1 \\ 0 & \frac{1}{3} & -\frac{1}{6} \\ -5 & -\frac{1}{6} & \frac{1}{6} \end{bmatrix}$$

2. Generalized strain stress relation:

$$F_m = [(f_m)_a (f_m)_b (f_m)_c (f_m)_d (f_m)_e]$$

$$(f_m)_a = (f_m)_b = \begin{bmatrix} \frac{6}{3EI} & -\frac{6}{6EI} \\ -\frac{6}{6EI} & \frac{6}{3EI} \end{bmatrix} = 1/2000 \begin{bmatrix} -2 & -1 \\ -1 & 2 \end{bmatrix}$$

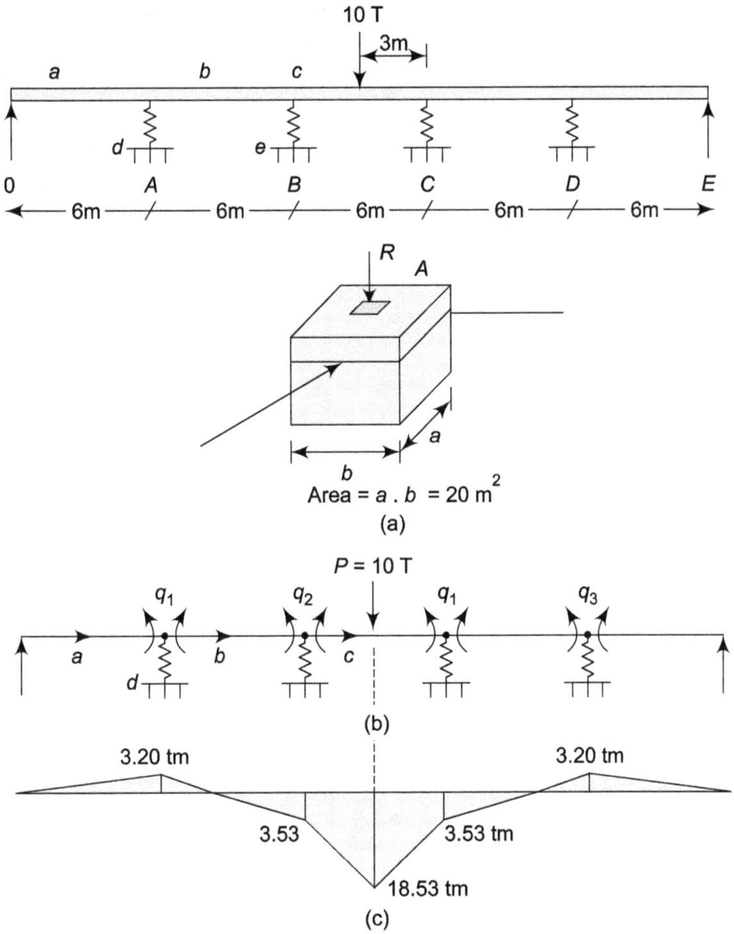

Fig. Ex. 4.7 *Pontoon Bridge (Resting on Floats)*

$$(f_m)_c = 1/2000 \begin{bmatrix} 1 & -\frac{1}{2} \\ -\frac{1}{2} & 1 \end{bmatrix}$$

$$(f_m)_d = (f_m)_e = 1/k = 1/20$$

3. Compatibility $$F = B^T F_m B = 1/720 \begin{bmatrix} 6.44 & -2.64 \\ -2.64 & 3.80 \end{bmatrix}$$

$$U^0 = (B^T F_m B_0) = 1/120 \; [5 \quad 3.65]^t$$

$$Q = F^1 U^0 = -6/17.5 \begin{bmatrix} 6.80 & 2.64 \\ 2.64 & 6.44 \end{bmatrix} \begin{bmatrix} 5 \\ 3.65 \end{bmatrix}$$

$$Q = [-3.20 \quad 3.53]^t$$

Fig. Ex. 4.7 (c) gives the final bending moment diagram.

Comment: If the problem is solved by a displacement method, say moment distribution, the moments have to be distributed for given loading and two cases of spring settlements. If the simultaneous equations for four unknowns, i.e., for two rotations and two springs settlements in this particular example, are solved by iteration, it hardly converges. The force method for such problems results not only in lesser number of unknowns but also gives rise to a flexibility matrix $F = B^T F_m B$ which has diagonally dominant elements.

Ex. 4.8. Analysis of pontoon bridge, Fig Ex. 4.8. It has two floats with horizontal area of each float at water level = 40 m^2. Other properties are same as given for the Ex. 4.7.

$$\begin{bmatrix} \begin{bmatrix} m_1 \\ m_2 \end{bmatrix}_a \\ \begin{bmatrix} m_1 \\ m_2 \end{bmatrix}_b \\ \left(p_{x_2}\right)_c \end{bmatrix} = \begin{bmatrix} 0 & 0 \\ 0 & 1 \\ 0 & -1 \\ 30 & 1 \\ -5 & \frac{1}{9} \end{bmatrix} \begin{bmatrix} 1 \\ q \end{bmatrix}$$

$$P_m \qquad\qquad B_0 \qquad B$$

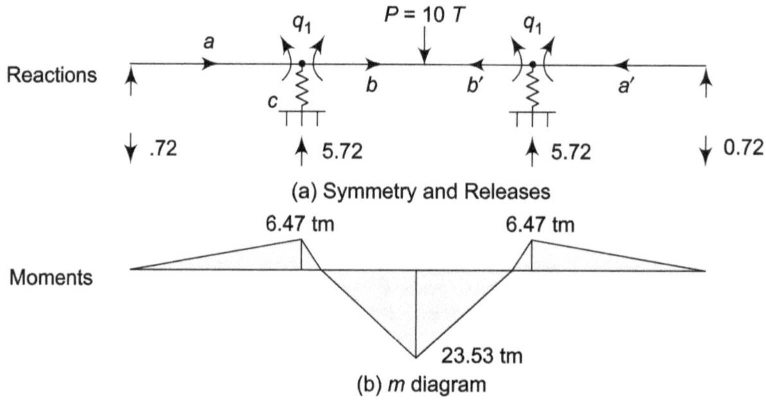

Fig. Ex. 4.8 *Pontoon Bridge Resting on Two Floats*

1. Member flexibilities:

$$(f_m)_a = 1/EI \begin{bmatrix} 3 & -1.5 \\ -1.5 & 3 \end{bmatrix}$$

$$(f_m)_b = 1/EI \begin{bmatrix} 2 & -1 \\ -1 & 2 \end{bmatrix}$$

$$(f_m)_c = 1/40 = 50/EI$$

2. The matrix operations give, $q = -6.47$ tm.

Fig. Ex. 4.8(b) gives the bending moment diagram.

4.5.1 Plane Frame

In the case of hinged trusses, member forces are defined by axial force P_{x2}. In the case of continuous beams discussed in the preceding Article, the member forces consisted of shear, P_y and moment, m. If there are any axial forces, their effects are to be considered independently without being linked with shears and moments.

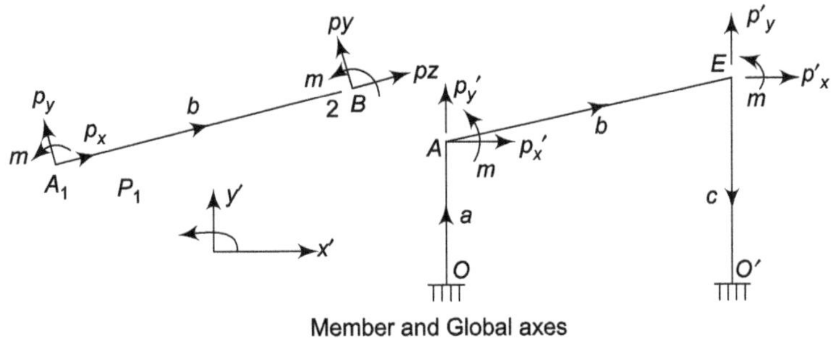

Fig. 4.12 *Member Free Body of the Structure*

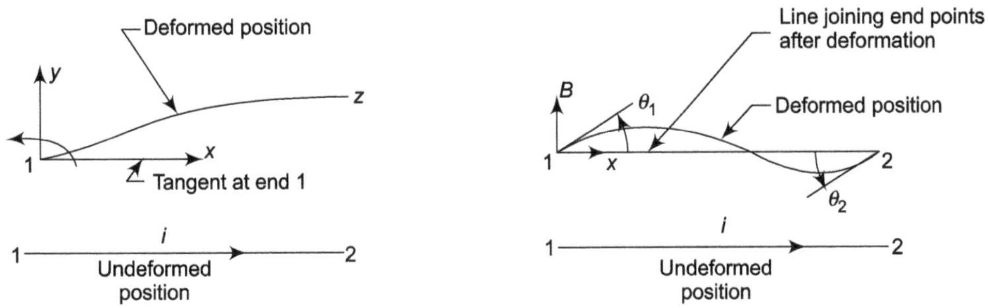

(a)

$(p_m)_i = \{p_x, p_y, m\}_{2i}$

$(\bar{d}_m)_i$ = Relative deformations or deformations w.r.t. tangent at 1

$(\bar{d}_2)_i = (\delta_x, \delta_y\, \theta)_{2i}$

Deformation-force relations:

$$(\bar{d}_2)_i = \begin{pmatrix} \delta_x \\ \delta_y \\ \theta \end{pmatrix}_{2i} = \begin{pmatrix} \frac{L}{EA} & 0 & 0 \\ 0 & \frac{L^3}{3EI} & \frac{L^2}{2EI} \\ 0 & \frac{L^2}{2EI} & \frac{L}{EI} \end{pmatrix} \begin{pmatrix} p_x \\ p_y \\ m \end{pmatrix}_{2t}$$

$(d_m)_i = (f_m)_i (p_m)_{2i}$

(b)

$(p_m)_i = \{p_x, m_1, m_2\}$

$(\bar{d}_m)_i$ = Deformations with respect to the x-axis as shown

$= (\delta_{x2}, \theta_1\, \theta_2)$

Deformation-force relations:

$$\begin{pmatrix} \delta_{x2} \\ \theta_1 \\ \theta_2 \end{pmatrix} = \begin{pmatrix} \frac{L}{EA} & 0 & 0 \\ 0 & \frac{L}{3EI} & -\frac{L}{6EI} \\ 0 & -\frac{L}{6EI} & \frac{L}{3EI} \end{pmatrix} \begin{pmatrix} p_{x_2} \\ m_1 \\ m_2 \end{pmatrix}$$

$(d_m)_i = (f_m)_i (p_m)_i$

Fig. 4.13 *Member Forces and Deformations*

In the case of plane frames, however, as the members constituting a frame (Fig. 4.12) have different inclinations, the axial forces are linked with the flexural forces, P_y and m. But while writing flexibility matrix of a straight single skeletal member, the axial actions, forces and deformations continue to be delinked with the flexural actions. The member flexibility relations, so obtained, are given in Fig. 4.13 (a) and (b). The member flexibility relations, so obtained, are given in Fig. 4.13 (a) and (b). The member flexibilities or generalized strain stress relations for a member, of a plane frames are rewritten, Eqs (4.9) and (4.10). Taking the tangent at the end 1 as a reference line,

$$\begin{bmatrix} \delta_x \\ \delta_y \\ \theta \end{bmatrix}_{2i} = \left[\begin{array}{c|cc} \frac{L}{EA} & 0 & 0 \\ 0 & \frac{L^3}{3EI} & \frac{L^2}{2EI} \\ 0 & \frac{L^2}{2EI} & \frac{L}{2EI} \end{array}\right]_i \begin{bmatrix} p_x \\ p_y \\ m \end{bmatrix}_{2i} \quad \ldots(4.9)$$

Also refer Fig. 4.13(a). The first diagonal element L/EA is the flexibility of axial actions as used for truss members, and the sub matrix of size 2×2 for the flexural actions is the same as used in the case of a member of a continuous beam, Eq. (4.7). Similarly if we consider the line joining the end points in the deformed position as a reference line, Fig. 4.13(b).

$$\begin{bmatrix} \delta_x \\ \theta_1 \\ \theta_2 \end{bmatrix}_{2i} = \left[\begin{array}{c|cc} \frac{L}{EA} & 0 & 0 \\ 0 & \frac{L}{3EI} & -\frac{L^2}{6EI} \\ 0 & -\frac{L}{6EI} & \frac{L}{3EI} \end{array}\right]_i \begin{bmatrix} p_{x2} \\ m_1 \\ m_2 \end{bmatrix}_i \quad \ldots(4.10)$$

The submatrix $\begin{bmatrix} \frac{L}{3EI} & -\frac{L}{6EI} \\ \frac{L}{6EI} & \frac{L}{3EI} \end{bmatrix}$ in the Eq. (4.10)

may be identified to be the same as the member flexibility appearing in the Eq. (4.8). In either case the generalized strain stress relation for the members of a frame is given by

$$F_m = [(f_m)_a (f_m)_b \cdots (f_m)_i \cdots (f_m)_n]$$

Where n is the number of members.

The other basic steps based on equilibrium and compatibility are obvious and are illustrated by the typical examples given in the following. The linking of axial actions and flexural actions is achieved while obtaining the flexibility matrix, F,

$$F = B^T F_m B$$

And the relative deformations U^0 at the releases chosen,

$$U^0 = B^T F_m B_0$$

Ex. 4.9. Analyse the frame, Fig. Ex. 4.9(a). The geometrical nodes are A and B. We have two alternatives, (i) treating the load point as an additional node or (ii) converting the given load to equivalent loads at the nodes A and B. Based on the second alternative and for $P_1 = 2P_2 = 2t$, generalized forces are shown in Fig. 4.9 (b) and (c). Member x-axes are directed in the ascending order of nodes, i.e., $0ABC$ and the members are designated by a, b and c. The chosen release forces $q = [q_1, q_2, q_3]^t$ are shown at the end 2 of c, Fig. 4.9 (c).

In the case of the example considered, $\beta = 20\,\alpha$ suggests a very small value of EA as compared to EI value. This has been done deliberately to make the axial deformations significantly large so as to effect moments and shears. In the "Force method" the axial deformation effects, when included, do not change the number of unknowns. In the case of the "Displacement method" or stiffness approach, the effect of axial deformations on moments and shears, increases the number of unknown independent displacements.

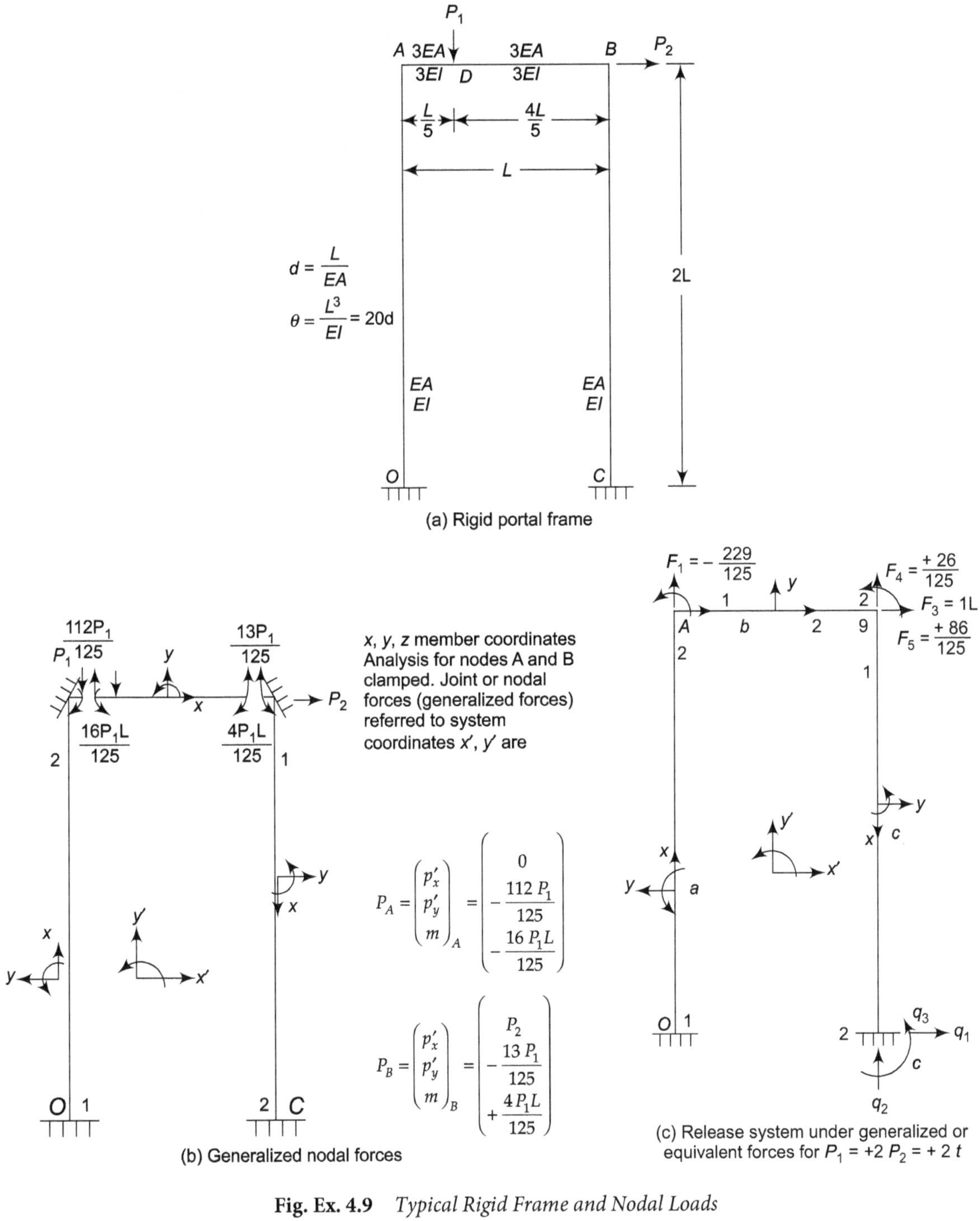

Fig. Ex. 4.9 *Typical Rigid Frame and Nodal Loads*

Tangent at the end 1 is chosen as a reference line for the member flexibilities. Therefore, the member forces in the equilibrium equation given below are the forces at the end 2.

Step 1. Equilibrium: Based on equilibrium equations for the structure made determinate by the three releases, the evaluation of member forces may be done by inspection, and it results into following equation:

$$\begin{bmatrix} \begin{pmatrix} p_x \\ p_y \\ m \end{pmatrix}_{2a} \\ \begin{pmatrix} p_x \\ p_y \\ m \end{pmatrix}_{2b} \\ \begin{pmatrix} p_x \\ p_y \\ m \end{pmatrix}_{2c} \end{bmatrix} = \left[\begin{array}{ccccc|ccc} 1 & 0 & 0 & 1 & 0 & 1 & 0 & 0 \\ 0 & 0 & -1 & 0 & 0 & -1 & 0 & 0 \\ 0 & L & 0 & (L)_b & L & (L)_c & (L)_b & L \\ 0 & 0 & 1 & 0 & 0 & 1 & 0 & 1 \\ 0 & 0 & 0 & 0 & 0 & 0 & 1 & 0 \\ 0 & 0 & 0 & 0 & L & (L)_c & 0 & L \\ 0 & 0 & 0 & 0 & 0 & 0 & -1 & 0 \\ 0 & 0 & 0 & 0 & 0 & +1 & 0 & 0 \\ 0 & 0 & 0 & 0 & 0 & 0 & 0 & L \end{array}\right] \begin{bmatrix} P_j = \begin{bmatrix} (F_1) \\ (\overline{F_2}) \\ (F_3) \\ (F_4) \\ (\overline{F_5}) \end{bmatrix} \\ q = \begin{bmatrix} (q_1) \\ (q_2) \\ (\overline{q_3}) \end{bmatrix} \end{bmatrix}$$

$$P_m \quad = \quad B_0 \qquad\qquad B \qquad \begin{bmatrix} p \\ q \end{bmatrix}$$

Where $[p_x \quad p_y \quad m]^t_{2i}$ are the forces in the respective directions acting at the end 2 of member i. Ends 1,2 for each member are marked, Fig. Ex. 4.9 (c). Because numerical value of L is not given, let $\overline{F_2} = F_2/L$, $\overline{F_5} = F_5/L$ and $\overline{q_3} = (q_3/L)$. The units of moments defined by $\overline{F_2}, \overline{F_5}, \overline{q_3}$ are the same as for forces.

Step 2. Deformation-force relations of members:

$$\begin{bmatrix} \begin{pmatrix} \delta_x \\ \delta_y \\ \theta \end{pmatrix}_{2a} \\ \begin{pmatrix} \delta_x \\ \delta_y \\ \theta \end{pmatrix}_{2b} \\ \begin{pmatrix} \delta_x \\ \delta_y \\ \theta \end{pmatrix}_{2c} \end{bmatrix} = \begin{bmatrix} \begin{pmatrix} \frac{L}{EA} & 0 & 0 \\ 0 & \frac{L^3}{3EI} & \frac{L^2}{2EI} \\ 0 & \frac{L^2}{2EI} & \frac{L}{EI} \end{pmatrix}_a & (0)_{3\times 3} & (0)_{3\times 3} \\ (0)_{3\times 3} & \begin{pmatrix} \frac{L}{EA} & 0 & 0 \\ 0 & \frac{L^3}{3EI} & \frac{L^2}{2EI} \\ 0 & \frac{L^2}{2EI} & \frac{L}{EI} \end{pmatrix}_b & (0)_{3\times 3} \\ (0)_{3\times 3} & (0)_{3\times 3} & \begin{pmatrix} \frac{L}{EA} & 0 & 0 \\ 0 & \frac{L^3}{3EI} & \frac{L^2}{2EI} \\ 0 & \frac{L^2}{2EI} & \frac{L}{EI} \end{pmatrix}_c \end{bmatrix} \begin{bmatrix} \begin{pmatrix} p_x \\ p_y \\ m \end{pmatrix}_{2a} \\ \begin{pmatrix} p_x \\ p_y \\ m \end{pmatrix}_{2b} \\ \begin{pmatrix} p_x \\ p_y \\ m \end{pmatrix}_{2c} \end{bmatrix}$$

Step 3. Compatibility: $q = (B^t F_m B)^{-1} (B^t F_m B_0)$

$$B^t F_m = \begin{bmatrix} 0, & -\left(\frac{L^3}{3EI}\right)_a + L_c\left(\frac{L^2}{2EI}\right)_a, & -\left(\frac{L^2}{2EI}\right)_a + L_c\left(\frac{L}{EI}\right)_a, & \left(\frac{L}{EA}\right)_b & L_c\left(\frac{L^2}{2EI}\right)_b, L_c\left(\frac{L}{EI}\right)_b, & 0 & \left(\frac{L^3}{3EI}\right)_c, & \left(\frac{L^2}{2EI}\right)_c \\ \left(\frac{L}{EA}\right)_a, & L_b\left(\frac{L^2}{2EI}\right)_a, & L_b\left(\frac{L}{EI}\right)_a, & 0, & \left(\frac{L^3}{3EI}\right)_b, \left(\frac{L^2}{2EI}\right)_b, & -\left(\frac{L}{EA}\right)_c, & 0 & 0 \\ 0, & L\left(\frac{L^2}{2EI}\right)_a, & L\left(\frac{L}{EI}\right)_a, & 0, & L\left(\frac{L^2}{2EI}\right)_b, L\left(\frac{L}{EI}\right)_b, & 0 & L\left(\frac{L^2}{2EI}\right)_c, & L\left(\frac{L}{EI}\right)_c \end{bmatrix}$$

$$B^t F_m B_0 = \begin{bmatrix} 0, & L\left(-\left(\frac{L^2}{2EI}\right)_a + L_c\left(\frac{L}{EI}\right)_a\right), & \left(\frac{L^3}{3EI}\right)_a - L_c\left(\frac{L^2}{2EI}\right)_a + \left(\frac{L}{EA}\right)_b, & 0 + L_b\left(\frac{L^2}{2EI}\right)_a + L_c\left(\frac{L}{EI}\right)_a + L_c\left(\frac{L^2}{2EI}\right)_b, & L\left(-\left(\frac{L^2}{2EI}\right)_a + L_c\left(\frac{L}{EI}\right)_a\right) + L_c\left(\frac{L^2}{EI}\right)_b \\ \left(\frac{L}{EA}\right)_a, & L\left(L_b\left(\frac{L}{EI}\right)_a\right), & -L_b\left(\frac{L^2}{2EI}\right)_a, & \left(\frac{L}{EA}\right)_a + L_b^2\left(\frac{L}{EI}\right)_a + \left(\frac{L^3}{3EI}\right)_b, & \left(L_b\left(\frac{L}{EI}\right)_a + \left(\frac{L^2}{2EI}\right)_b\right)L \\ 0, & L^2\left(\frac{L}{EI}\right)_a, & -L\left(\frac{L^2}{2EI}\right)_a, & L\left(L_b\left(\frac{L}{EI}\right)_a + \left(\frac{L^2}{2EI}\right)_b\right), & \left(\left(\frac{L}{EI}\right)_a + \left(\frac{L}{EI}\right)_b\right)L^2 \end{bmatrix}$$

$B^t F_m B =$

$+\left(\frac{L^3}{3EI}\right)_a - L_c\left(\frac{L^2}{2EI}\right)_a + L_c\left(\frac{L^2}{EI}\right)_a + L_c\left(\frac{L}{EI}\right)_a + \left(\frac{L}{EA}\right)_b + L_c^2\left(\frac{L}{EI}\right)_b + \left(\frac{L^3}{3EI}\right)_c$	$L_b\left(-\left(\frac{L}{2EI}\right)_a + L_c\left(\frac{L}{EI}\right)_a + L_c\left(\frac{L}{2EI}\right)_b\right)$	$\left(-\left(\frac{L^3}{2EI}\right)_a + L_c\left(\frac{L}{EI}\right)_a\right) + L_c\left(\frac{L}{EI}\right)_b + \left(\frac{L^2}{2EI}\right)_c\right)$
$-L_b\left(\frac{L^2}{2EI}\right)_a + L_c\left(L_b\left(\frac{L}{EI}\right)_a\right) + L_c\left(\frac{L^2}{2EI}\right)_b$	$\left(\frac{L}{EA}\right)_a + L_b^2\left(\frac{L}{EI}\right) + \left(\frac{L^3}{3EI}\right)_b + \left(\frac{L}{EA}\right)_a$	$L_b\left(\frac{L}{EI}\right)_a + \left(\frac{L^2}{2EI}\right)_b L$
$L\left(-\left(\frac{L^2}{2EI}\right)_a\right) + L_c\left(\frac{L}{EI}\right)_b + L_c\left(\frac{L}{EI}\right)_b + L_c\left(\frac{L^2}{2EI}\right)_c$	$\left(L_b\left(\frac{L}{EI}\right)_a + \left(\frac{L^2}{2EI}\right)_b\right)L$	$\left(\left(\frac{L}{EI}\right)_a + \left(\frac{L}{EI}\right)_b + \left(\frac{L}{EI}\right)_c\right)L^2$

For the frame, Fig. Ex. 4.9 (a), if we substitute the values of *EI*, *EA* of different members

$$U^0 = (B^t F_m B_0) = \alpha \begin{pmatrix} 0 & 40 & -\frac{79}{3} & \frac{140}{3} & \frac{160}{3} \\ 2 & 40 & -40 & \frac{398}{9} & \frac{130}{3} \\ 0 & 40 & -40 & \frac{130}{3} & \frac{140}{3} \end{pmatrix}$$

$$F = B^t F_m B = \alpha \begin{pmatrix} \frac{401}{3} & \frac{140}{3} & \frac{280}{3} \\ \frac{140}{3} & \frac{416}{9} & \frac{130}{3} \\ \frac{280}{3} & \frac{130}{3} & \frac{260}{3} \end{pmatrix} = \frac{\alpha}{9} \begin{pmatrix} 1203 & 420 & 840 \\ 420 & 416 & 390 \\ 840 & 390 & 780 \end{pmatrix}$$

The unknown redundants $Q = [q_1, q_2, q_3]^t$ are obtained by solving the equation

$$F\,Q = -U^0$$

In the following the unknowns, Q, are evaluated by inverting the *F* matrix, i.e.,

$$Q = -F^{-1}\,U^0$$

$$F^{-1} = (B^t F_m B)^{-1} = \frac{9}{\alpha} \begin{pmatrix} 1203 & 420 & 840 \\ 420 & 416 & 390 \\ 840 & 390 & 780 \end{pmatrix}^{-1}$$

The matrix when inverted gives

$$F^{-1} = \frac{0.09}{\alpha} \begin{pmatrix} 0.33513 & 0 & -0.36091 \\ 0 & 0.45249 & -0.22624 \\ -0.36091 & -0.22624 & 0.62999 \end{pmatrix}$$

Thus $$A_{gj} = (B^t F_m B)^{-1}\,(B^t F_m B_0)$$

$$= 0.09 \begin{pmatrix} 0.33513 & 0 & -0.36091 \\ 0 & 0.45249 & -0.22624 \\ -0.36091 & -0.22623 & +0.62999 \end{pmatrix} \times \begin{pmatrix} 0 & 40 & -\frac{79}{3} & \frac{140}{3} & \frac{160}{3} \\ 2 & 40 & -40 & \frac{398}{9} & \frac{130}{3} \\ 0 & 40 & -40 & \frac{130}{3} & \frac{140}{3} \end{pmatrix}$$

$$A_{qj} = 0.09 \begin{bmatrix} 0 & -1.02596 & +5.61131 & -0.31462 & 2.23990 \\ 0.90497 & +9.04980 & -9.0498 & +10.20609 & 9.04954 \\ -0.45246 & 1.71388 & -6.64652 & 0.45228 & 0.34793 \end{bmatrix}$$

$$Q = A_{qj}\,P_j$$

Where $$p_j = (1/125)\,[-224 \quad -32 \quad +125 \quad -26 + 8]^t$$

$$q = (-.09/125)\ [+760.344 \quad -1816.4939 \quad -793.29]^t$$
$$q = [-.547 \quad +1.305 \quad +.571]^t$$

Computer results give, $q = [-.5347 \quad +1.3079 \quad .5712]^t$

Member Forces:

Forces at end 2 of members a, b and c are given by the relation

$$p_m = A_{mj}\, p_j$$

Where $\quad A_{mj} = B_0 - B\, A_{qj}$

$$p_m = [-.692 -.465\ .409, .465 -.692\ .548, .465\ 1.308 -.498, -1.308 -5.5\ .571]^t$$

To the p_m so computed, fixed end forces found from the first part of the analysis are required to be algebraically added. In the case of the example solved only for the member, b, the fixed and forces are required to be added. The final member forces are given at the end of comment-I of this solution.

Joint deformations:

$$d_j = B_0^{\,t}\, F_m\, A_{mj} P_j$$

or $\quad d_j = F\, P_j$

where $\quad F = B_0^{\,t}\, F_m\, A_{mj}$

It is left to the reader to complete the problem by evaluating numerically the values of p_m and d_j

Comment – 1

In the example solved, B_0 matrix consists of five columns corresponding to the five loads, $p_j = [F_1\ F_2\ F_3\ F_4\ F_5]^t$. However, we may find the member forces of the determinate structure for all the loads considered together, so as to reduce the previous B_0 matrix to a column matrix. Further, if we treat the load point as a node, the number of members gets increased to four, i.e., OA, AD, DB and BC, Fig. Ex. 4.9(a), and let these members be designated as a, b, c and d respectively. Let $L = 2$ m. With these alternations in the procedure, the following is equilibrium matrix

$$\begin{bmatrix} \begin{pmatrix} p_{x2} \\ p_{y2} \\ m_2 \end{pmatrix}_{2a} \\ \begin{pmatrix} p_{x2} \\ p_{y2} \\ m_2 \end{pmatrix}_{2b} \\ \begin{pmatrix} p_{x2} \\ p_{y2} \\ m_2 \end{pmatrix}_{2c} \\ \begin{pmatrix} p_{x2} \\ p_{y2} \\ m_2 \end{pmatrix}_{d} \end{bmatrix} = \left[\begin{array}{r|rrr} -2 & 0 & 1 & 0 \\ -1 & -1 & 0 & 0 \\ -.4 & 2 & 1 & 1 \\ 1 & 1 & 0 & 0 \\ -2 & 0 & 1 & 0 \\ 0 & 2 & 0.8 & 1 \\ 1 & 1 & 0 & 0 \\ 0 & 0 & 1 & 0 \\ 0 & 2 & 0 & 1 \\ 0 & 0 & -1 & 0 \\ 0 & 1 & 0 & 0 \\ 0 & 0 & 0 & 1 \end{array}\right] \begin{bmatrix} 1 \\ q_1 \\ q_2 \\ q_3 \end{bmatrix}$$

$$P_m = [B_0 \quad B] \begin{bmatrix} 1 \\ Q \end{bmatrix}$$

The reader may like to complete the problem and check the results with those obtained previously. With the present form of B_0 matrix (column matrix) we cannot find the nodal deformations. As the load point has been considered as a node, no superposition is required for finding the final member forces as in the previous solution. The final p_m may be verified to be

$$P_m = [-.690\ -.465\ .410,\ .462\ -.692\ .548,\ .465\ .131\ -.498,\ -.131\ -.535\ .571]^t$$

Comment – 2

It may be observed that the F matrix was not diagonally dominant. The F matrix could have been improved for diagonal dominance by choosing a better release system. Another example is solved in the following by introducing hinges as releases.

Ex. 4.10. Analyse the frame, Fig. Ex. 4.10(a). Introduce 3 hinges at the points A, C and D as releases. The values of EI and EA for the members are shown in the figure. The moment of inertial I and area A are related as $AL^2 = 20\ I$. The mid span load equal to $2t$ is transformed to equivalent nodal loads F_1, F_2, F_3 and F_4. The nodal load F_5 is equal to the given horizontal load = 1 t, Fig. Ex. 4.10(b). Let us consider deformations at the end 2 of a member as the relative deformations with respect to the tangent at end 1. With this reference line assumed (same as Ex. 5.9) the member force are defined by $[p_{x2}, p_{y2}, m_2]^t$, and the corresponding flexibility matrix f_m is as given in Eq. (4.9).

For the chosen redundants acting at the hinge releases, Fig. (c), the m_i diagrams are drawn for $q_1 = 1$, $q_2 = 1$ and $q_3 = 1$, Fig. (d) (i), (ii) and (iii). The moment values for each case being known, thrusts and shears at the end 2 of the member can be evaluated. Thus

No moments and shears are caused due to the nodal loads F_3 and F_4. The moment diagrams for F_1, F_2 and F_3 are drawn, Fig. (e) (i), (ii) and (iii), from which thrusts and shears also can be evaluated to write the B_0 matrix.

And $\qquad p_j = [F_1\ F_2\ F_3\ F_4\ F_5]^t = [2\ 2\ 1\ 1\ 1]^t$

The B_0 and B matrix constitute the equilibrium equation,

$$\begin{array}{ccc} q_1 = 1 & q_2 = 1 & q_3 = 1 \end{array}$$

$$\begin{bmatrix} -\frac{1}{8} & 0 & -\frac{1}{8} \\ 0 & -\frac{1}{3} & \frac{1}{3} \\ -1 & \frac{4}{3} & -\frac{4}{3} \\ \frac{1}{8}\sin\theta & \frac{1}{3}\cos\theta & \left(\frac{1}{3}\sin\theta - \frac{1}{3}\cos\theta\right) \\ -\frac{1}{8}\cos\theta & \frac{1}{3}\sin\theta & \left(\frac{1}{8}\cos\theta - \frac{1}{3}\sin\theta\right) \\ 0 & 1 & 0 \\ \frac{1}{8} & 0 & \frac{1}{8} \\ 0 & \frac{1}{3} & -\frac{1}{3} \\ 0 & 0 & 1 \end{bmatrix}$$

$$\begin{array}{ccccc} F_1 & F_2 & F_3 & F_4 & F_5 \end{array}$$

$$\begin{bmatrix} 1/8 & 0 & -1 & 0 & 1/2 \\ 0 & -1/3 & 0 & 0 & -1 \\ 0 & 4/3 & 0 & 0 & -4 \\ 1/8\sin\theta & 1/3\sin\theta & 0 & 0 & -1/2\sin\theta \\ -1/8\cos\theta & 1/3\cos\theta & 0 & 0 & 1/2\cos\theta \\ 0 & 1 & 0 & 0 & 0 \\ 1/8 & 0 & 0 & -1 & -1/2 \\ 0 & 1/3 & 0 & 0 & 0 \\ 0 & 0 & 0 & 0 & 0 \end{bmatrix}$$

$$P_m = (B_0\ B)\begin{bmatrix} P_j \\ Q \end{bmatrix}, \text{ where } p_m = \begin{bmatrix} (p_m)_a \\ (p_m)_b \\ (p_m)_c \end{bmatrix}$$

$(p_m)_i = [p_{x2}\, p_{y2}\, m_2]_i^t$,
$P_j = [F_1\, F_2\, F_3\, F_4\, F_5]^t$, and
$Q = [q_1\, q_2\, q_3]^t$.

The generalized strain-stress relation is given by

$$F_m = \left[(f_m)_a \quad (f_m)_b \quad (f_m)_c\right]$$

$$(f_m)_a = \begin{bmatrix} \frac{2}{EA} & 0 & 0 \\ 0 & \frac{10.65}{EI} & \frac{4}{EI} \\ 0 & \frac{4}{EI} & \frac{2}{EI} \end{bmatrix}$$

$$(f_m)_b = \begin{bmatrix} \frac{5.37}{EA} & 0 & 0 \\ 0 & \frac{87.50}{EI} & \frac{16.25}{EI} \\ 0 & \frac{16.25}{EI} & \frac{4.03}{EI} \end{bmatrix}, \text{ and}$$

$$(f_m)_c = \begin{bmatrix} \frac{3}{EA} & 0 & 0 \\ 0 & \frac{9}{EI} & \frac{4.5}{EI} \\ 0 & \frac{4.5}{EI} & \frac{3}{EI} \end{bmatrix}$$

Substituting $A\ L^2 = 20\ I$ or $A = 1.25\ I$, the flexibility matrix F and relative deformation U^0 are obtained.

$$F = B^t F_m B = \frac{1}{EI}\begin{bmatrix} 3.4055 & -3.770 & 3.1615 \\ -3.770 & 8.163 & -4.395 \\ 3.1615 & -4.395 & 5.0455 \end{bmatrix}$$

$$U^0 = (B^t F_m B_0)\ P_j$$

where $$B^t F_m B_0 = \frac{1}{EI}\begin{bmatrix} -13.7666 & 22.691 & -16.313 \end{bmatrix}_t$$

$$F^{-1} = (B^t F_m B)^{-1} = EI\begin{bmatrix} 0.8490 & 0.1985 & -0.3610 \\ 0.1985 & 0.2800 & 0.1191 \\ -0.3610 & 0.1191 & 0.5290 \end{bmatrix}$$

$\therefore$ $$Q = -(B^t F_m B)^{-1}(B^t F_m B_0)\ P_j$$

$$Q = [1.30 - 1.66 + 96]^t$$

$$P_m = [B_0 B]\begin{bmatrix} P_j \\ Q \end{bmatrix}$$

$$P_m = [-.783\ -.127\ -.80,\ -.083\ .107\ 0.34,\ -1.217 - 0.873\ 0.96]^t$$

The final member forces are obtained by superposing the P_m values computed above and the member forces P_m evaluated for the clamped or fixed nodes. The final bending moment diagram is also drawn, Fig. Ex. 4.10 (Cont.).

$$P_m = [0\ 0\ 0\ \sin\theta \cos\theta - 2\ 0\ 0\ 0]^t \qquad \ldots\text{(ii)}$$

Adding the results (i) and (ii) the final member forces,

$$P_m = [-.783 - .127 - .80 - .954\ 1.10 - .1\ .217 - .873\ .96]^t$$

Comments: The first diagonal element (3.4055) even though not the largest in the row, is not much different from the neighboring element (3.770). The other diagonal elements are the largest in their respective rows or columns.

If the three releases are introduced by cutting the end 2 of the member c, the F matrix for such a release system is equal to

$$F = \frac{1}{EI}\begin{bmatrix} 73.6193 & 90.5847 & 22.6090 \\ 90.5847 & 217.8288 & 32.1245 \\ 22.6090 & 32.1245 & 9.0311 \end{bmatrix}$$

It may be seen that the last row is worst place in respect of diagonal dominance.

Ex. 4.11. Solve the frame, Fig. Ex. 4.10 (a), by

(i) Introducing the three releases at the base, D, Fig. Ex. 4.11 and

(ii) Considering the line joining end points in the deformed position as reference line, Eq. (4.10).

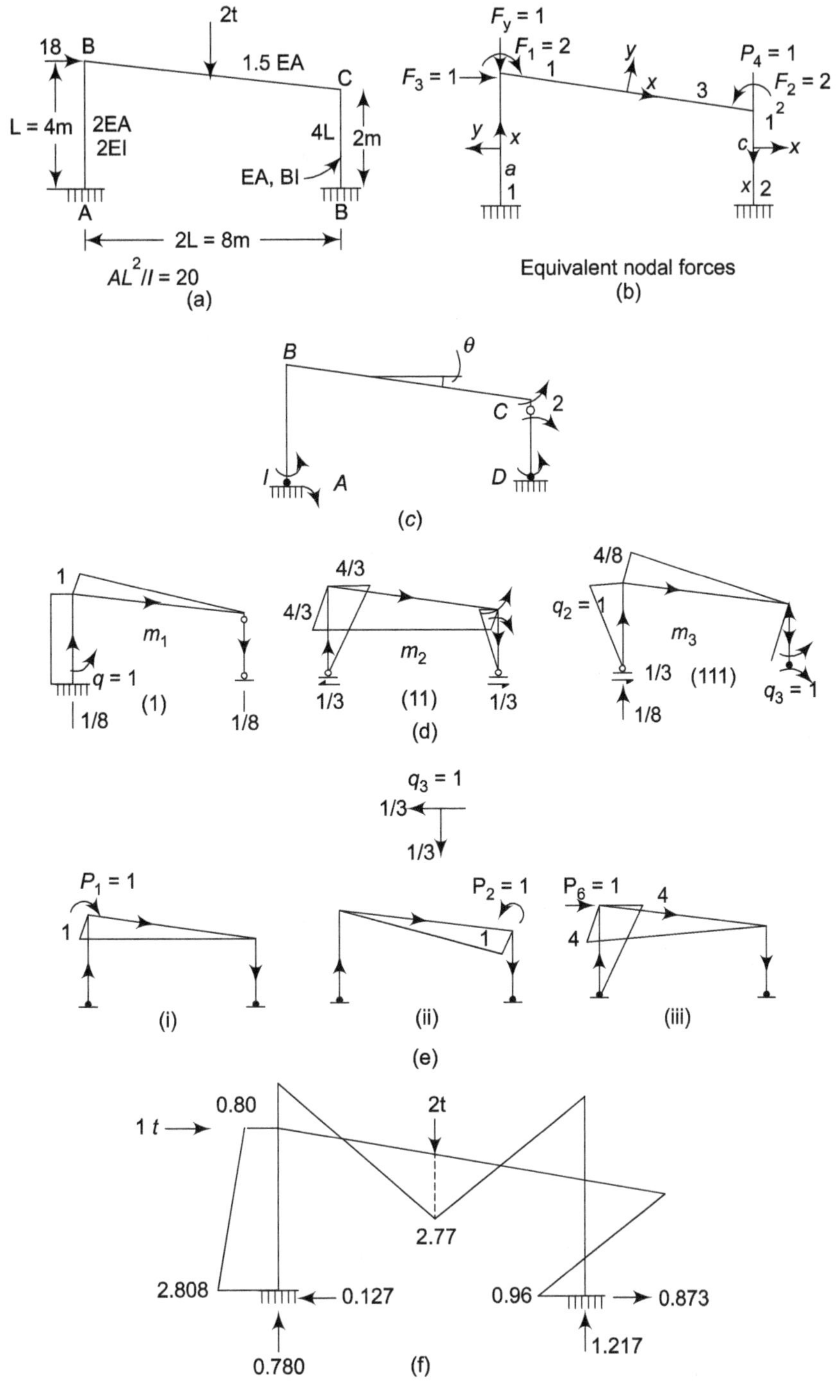

Fig. Ex. 4.10 *BM Diagram*

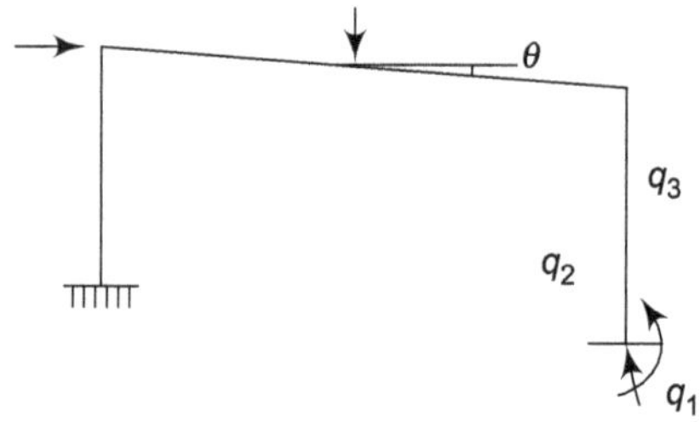

Fig. Ex. 4.11 *Release System*

For such a release system and reference line, the reader may verify the following results.

$$\begin{bmatrix} \begin{pmatrix} p_{x2} \\ p_{y2} \\ m_2 \end{pmatrix}_{2a} \\ \begin{pmatrix} p_{x2} \\ p_{y2} \\ m_2 \end{pmatrix}_{2b} \\ \begin{pmatrix} p_{x2} \\ p_{y2} \\ m_2 \end{pmatrix}_{2c} \end{bmatrix} = \begin{bmatrix} 0 & 0 & -1 & -1 & 0 & 0 & 1 & 0 \\ 1 & -1 & 0 & 8 & 4 & 0 & -8 & -1 \\ 1 & 1 & 0 & -8 & 0 & 4 & 8 & 1 \\ 0 & 0 & 0 & \sin\theta & 0 & \cos\theta & -\sin\theta & 0 \\ 0 & -1 & 0 & 8 & 0 & -4 & -8 & -1 \\ 0 & 1 & 0 & 0 & 0 & 3 & 0 & 1 \\ 0 & 0 & 0 & 0 & 0 & 0 & -1 & 0 \\ 0 & 0 & 0 & 0 & 0 & -3 & 0 & -1 \\ 0 & 0 & 0 & 0 & 0 & 0 & 0 & 1 \end{bmatrix} \begin{bmatrix} 2 \\ 2 \\ 1 \\ 1 \\ 1 \\ q_1 \\ q_2 \\ q_3 \end{bmatrix}$$

$$P_m = \underbrace{\qquad B_0 \qquad}\ \underbrace{\qquad B \qquad} \begin{bmatrix} P_j \\ Q \end{bmatrix}$$

The F matrix for such a release system has already been given (see comment Ex. 4.10). The results given below are obtained with numerical accuracy better than "slide rule" accuracy relied upon for solution of Ex. 4.10.

$$U^0 = (B^T F_m B_0)\, P_j = \frac{1}{EI}\,[-\,67.710 - 217.026 - 28.062]^t$$

$$Q = -\,F^{-1}U^0 = [-\,0.875\ 1.215\ 0.977]^t$$

$$p_m = [-\,0.7850\ 1.3030\ 0.803, -\,0.8950 - 1.1970\ 0.3520, -\,1.2150\ 1.6480\ 0.9770]^t$$

$$d_j = B_o^{\,t}\, F_m\, p_m$$

$$d_j = \frac{1}{EI}\,[4.545\ 1.256\ 2.968\ 2.106\ 1.000]^t$$

The directions of the deformations are the same as that of F_1, F_2, F_3, F_4 and F_5.

4.5.2 Non-Prismatic Arches

The proposed analysis for the arches replaces the curved profile by straight prismatic segments, and thus it is reduced to a plane frame. A non-prismatic arch was analysed by the conventional flexibility method,

Article 3.5.6. The arch was divided into twelve segments and each segment was considered prismatic. Let us analyse the same arch by the matrix flexibility method. For the symmetrical arch the uniformly distributed load 2 t/m is replaced by approximate concentrated loads at the discretised nodes *O*, *A*, ..., *E*, *F*, Fig. Ex. 4.12 (a), over half its length. The discredited members *a*, *b*, ..., *f* and the member axes are marked on the axis of the arch, Fig. (b). The releases are introduced by cutting the arch at the crown. Due to symmetry, the redundant consist of moment q_1 and thrust q_2 acting in the direction as shown in Fig. (b). The idealized member properties, area, moment of inertia and length for the member are given in the table of member properties, Fig. Ex. 4.12. The member properties are obtained from the data given in Article 3.5.6. The basic steps and procedure given for the plane frames, Article 4.5.1, remain unaltered when applied to the idealized arch. For the generalized strain-stress relations, tangent at the end of 1 of a member is taken as reference line, and, therefore, the member forces and relative member deformations are for the end 2.

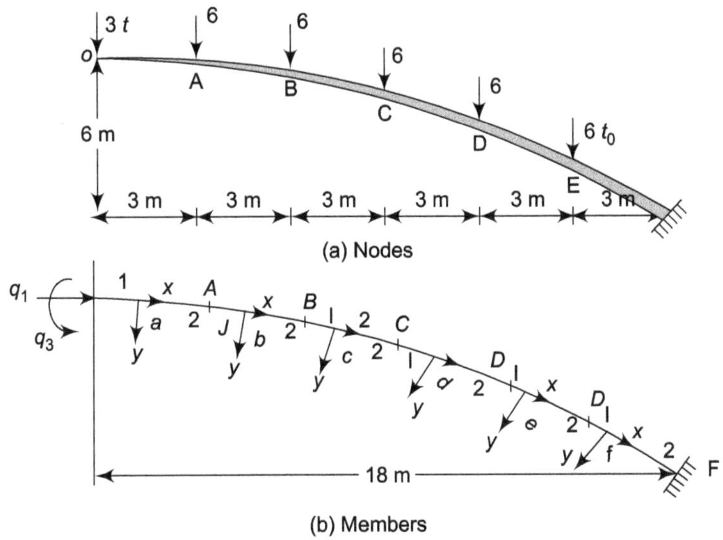

Member properties

Member	*a*	*b*	*C*	*d*	*E*	*F*
$A_i/10$	0.125	0.135	0.145	0.155	0.165	0.175
$I_i/10$	.0162	.0205	.0255	.0310	.0375	.0445
L_i	3.00	3.05	3.12	3.22	3.35	3.52

Area $= A_i\, m^2$

Moment of inertia $= I_i\, m^4$

Length $= L_i\, m$

Fig. Ex. 4.12 *Non-prismatic Arch Idealized by Prismatic Segments*

1. Equilibrium Equations:

$$\begin{bmatrix} \begin{pmatrix} p_{x2} \\ p_{y2} \\ m_2 \end{pmatrix}_a \\ \begin{pmatrix} p_{x2} \\ p_{y2} \\ m_2 \end{pmatrix}_b \\ \begin{pmatrix} p_{x2} \\ p_{y2} \\ m_2 \end{pmatrix}_c \\ \begin{pmatrix} p_{x2} \\ p_{y2} \\ m_2 \end{pmatrix}_d \\ \begin{pmatrix} p_{x2} \\ p_{y2} \\ m_2 \end{pmatrix}_e \\ \begin{pmatrix} p_{x2} \\ p_{y2} \\ m_2 \end{pmatrix}_f \end{bmatrix} = \begin{bmatrix} -0.165 & 0 & -0.998 \\ -2.994 & 0 & 0.055 \\ 9.000 & 1 & -0.167 \\ -1.480 & 0 & -0.985 \\ -8.850 & 0 & 0.164 \\ 36.000 & 1 & 0.667 \\ -4.000 & 0 & -0.960 \\ -14.400 & 0 & 0.266 \\ 81.000 & 1 & -1.500 \\ -7.600 & 0 & -0.930 \\ -19.500 & 0 & 0.362 \\ 144.000 & 1 & -2.660 \\ -12.100 & 0 & -0.895 \\ -24.200 & 0 & 0.448 \\ 225.000 & 1 & -4.170 \\ -17.200 & 0 & -0.852 \\ -28.000 & 0 & 0.520 \\ 324.000 & 1 & -6.000 \end{bmatrix} \begin{bmatrix} 1 \\ Q \end{bmatrix}$$

$$P_m = [B_0 \quad B] \begin{bmatrix} 1 \\ \hline Q \end{bmatrix}$$

2. Generalized strain-stress relation:

$$F_m = [(f_m)_a \, (f_m)_b \cdots (f_m)_f]$$

The $(f_m)_i$ for a member is according to Eq. (4.10), and the member properties $(L, EA, EI)_i$ are given in Fig. Ex. 4.12.

3. $$F = B^t F_m B = \frac{1}{E}\begin{bmatrix} 729 & -1130 \\ -1130 & 3880 \end{bmatrix}$$

$$U^0 = B^T F_m B_0$$

$$Q = -F^{-1} U^0 = [-5.1055 + 50.738]^t$$

The above values agree with the results given in Article 3.5.6

$$P_m = [B_0 \quad B] \begin{bmatrix} 1 \\ Q \end{bmatrix}$$

$$P_m = [-50.801 - 0.203 - 4.579, -51.457 - 0.529 - 2.948, -52.708 - 0.904 - 0.212, -54.786 - 1.133\ 3.932, -57.510 - 1.469\ 8.318, -60.429 - 1.616\ 14.467]^t$$

Ex. 4.13, The symmetrical non-prismatic parabolic arch of Article 3.5.6 is idealized into a six-segment prismatic rigid frame, Fig. Ex. 4.13. The member properties of such an idealized frame are given in the table, Fig Ex. 4.13 (c). Based on the member properties for a concrete arch, a first order approximate solution is attempted by assuming the dead loads to be acting as point loads at the nodes. The loads acting on the symmetrical half are shown in Fig. Ex. 4.13 (a). Members axes are designated differently from that in Ex. 4.12, that is, starting from the fixed end to the crown, Fix. Ex. 4.13 (b). Analyse such an idealized symmetrical frame by introducing releases at the crown only, and consider tangent at the end 1 of each member as a reference line.

Solution:

1. Equilibrium Equation:

$$\begin{bmatrix} \begin{pmatrix} p_{x2} \\ p_{y2} \\ m_2 \end{pmatrix}_a \\ \begin{pmatrix} p_{x2} \\ p_{y2} \\ m_2 \end{pmatrix}_b \\ \begin{pmatrix} p_{x2} \\ p_{y2} \\ m_2 \end{pmatrix}_c \end{bmatrix} = \begin{bmatrix} -27.00 & 0.875 & 0 \\ 48.60 & 0.485 & 0 \\ 238.00 & 2.670 & 1 \\ -9.60 & 0.947 & 0 \\ 28.70 & 0.317 & 0 \\ 56.52 & 0.670 & 1 \\ -1.045 & 0.992 & 0 \\ 9.35 & 0.111 & 0 \\ 0 & 0 & 1 \end{bmatrix} \begin{bmatrix} 1 \\ q_1 \\ q_2 \end{bmatrix}$$

$$P_m = [B_0 \qquad B] \begin{bmatrix} 1 \\ Q \end{bmatrix}$$

2.

$$F_m = [(f_m)_a\ (f_m)_b,\ \ldots,\ (f_m)_f]$$

The reader may like to verify the following results.

$$F = B^T F_m B = \frac{1}{E}\begin{bmatrix} 412 & 120 \\ 120 & 71.8 \end{bmatrix}$$

$$Q = -F^{-1}(B^T F_m B_0) = [-88.655 - 3.7316]^t$$

$$P_m = [-104.57\ 5.602 - 2.442, -93.557\ 0.596 - 6.611, -88.991 - .491 - 3.732]^t$$

Additional example. Analyse the four storey single bay frame, Ex. 2.27, by the matrix flexibility method. The degree of indeterminacy is 12 and the number of members = 12. For the hinge-release-system chosen, write B_0, B matrices.

A computer program for the rigid frame is given in the appendix based on the matrix method. The reader should be able to appreciate the matrix approach which makes the program so simple and direct. It is a sample program. The reader may like to generalize it so that B_0 and B matrices are generated of being given as data.

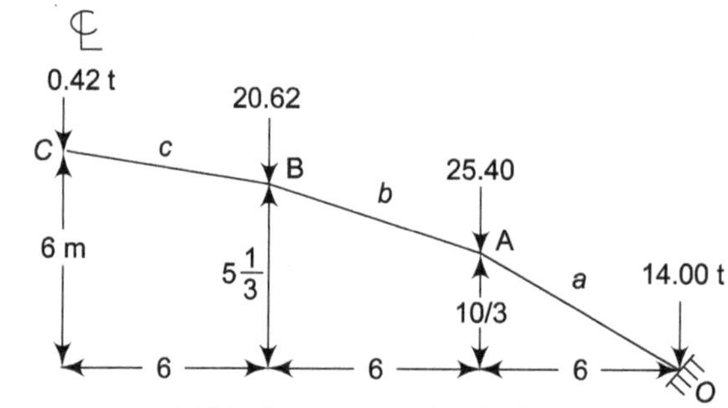

(a) Idealised arch and nodal loads

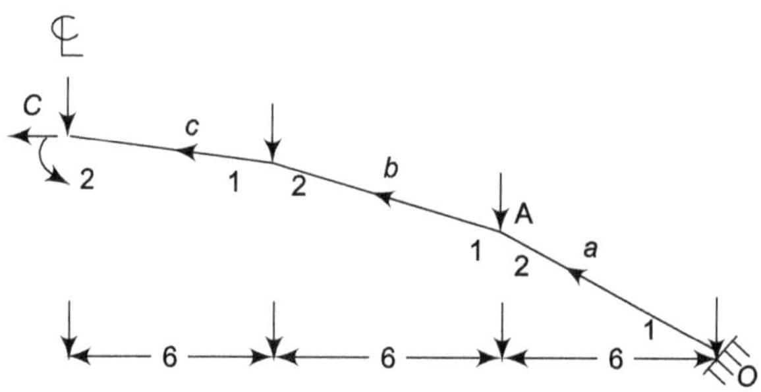

(b) Member axis for a, b and c

Member i	*Length L_i*	*Average Area A_i*	*Average I_i*	*sin θ_i*	*cos θ_i*
a	6.86	1.7	0.415	0.485	0.875
b	6.32	1.5	0.284	0.317	0.947
c	6.04	1.3	0.183	0.111	0.992

(c) Table of member properties. All units in metres

Fig. Ex. 4.13 *Symmetrical Arch Idealized by Six Prismatic Segments*

4.6 TEMPERATURE EFFECTS AND LACK OF FIT

4.6.1 General

Determinate structures do not develop internal stresses due to change in temperature. Similarly in the case of steel or prefabricated structure, if the structural members are over sized or undersized, which we call as "Lack of Fit", no internal stresses are developed when the structures are determinate. In the case of indeterminate structures, however, internal stresses are caused due to the temperature changes and lacks of fit are self-equilibrating. Settlements of supports may be treated as 'lock of fit' of the members, whose ends are support points. All these effects are grouped under 'Geometric loading'.

Recall the equation, Eq. (4.2)

$$P_m = (B_0 B)\begin{bmatrix} p_j \\ Q \end{bmatrix}$$

Using the principle of complementary virtual work, we had established the equation, Eq. (b),. The equation is rewritten below:

$$\begin{bmatrix} d_j \\ Q \end{bmatrix} = \begin{bmatrix} B_0^T \\ B^T \end{bmatrix}_{dm} \quad \text{...(a)}$$

In the preceding Articles, the member deformations, d_m, were considered to the caused by the external loads P_j and the redundatns Q, as described below by Eq. (b).

$$d_m = F_m p_m = F_m(B_0\ B) \begin{bmatrix} p_j \\ Q \end{bmatrix} \quad \text{...(b)}$$

To this we add the member deformations $(d_m)_T$ and $(d_m)_L$ caused by the temperature and lack of fit respectively. Therefore, the total member deformation is given by

$$d_m = F_m = F_m(B_0\ B) \begin{bmatrix} p_j \\ Q \end{bmatrix} + (d_m)_T + (d_m)_L \quad \text{...(4.11)}$$

or $$d_m = F_m B\, Q + F_m\, B_0\, P_j + (d_m)_T + (d_m)_L \quad \text{...(4.11)}$$

Let $$(d_m)_P = F_m\, B_0\, P_j \quad \text{...(c)}$$

Where $(d_m)_P$ is visualized as member deformations of the released structure, caused by the joint loads, P_j

From Eqs (4.11) and (c),

$$d_m = F_m B\, Q + (d_m)_P + (d_m)_T + (d_m)_L \quad \text{...(4.12)}$$

From Eqs (4.12) and (a),

$$\begin{bmatrix} d_j \\ Q \end{bmatrix} = \begin{bmatrix} B_0^T \\ B^T \end{bmatrix} (F_m B\, Q + (d_m)_P + (d_m)_T + (d_m)_L) \quad \text{...(4.13)}$$

The above equation may be split into the following equations:

$$d_j = B_0^{\ T} (F_m B\, Q + (d_m)_P + (d_m)_T + (d_m)_L) \quad \text{...(4.13a)}$$

and $$U = B^T (F_m B\, Q + (d_m)_P + (d_m)_T + (d_m)_L) \quad \text{...(4.13b)}$$

or $$U = FQ + U_P^{\ 0} + U_T^{\ 0} + U_L^{\ 0} \quad \text{...(4.13b)}$$

where

$$F = B^T F_m B$$

$$U_P^{\ 0} = B^T F_m B_0 P_j,$$

$$U_T^{\ 0} = B^T (d_m)_T \text{ and } U_L^{\ 0} = B^T (d_m)_L$$

From the compatibility condition,

$$U = 0 = FQ = U_P^{\ 0} + U_T^{\ 0} + U_L^{\ 0}$$

Or $$Q = -F^{-1} (U_P^{\ 0} + U_T^{\ 0} + U_L^{\ 0}) \quad \text{...(4.14)}$$

With the redundants Q evaluated, we find d_j.

$D_j = B_0^{\ T} F_m B\, Q$ due to the redundants

$+ B_0^{\ T} F_m\, B_0\, P_j$ due to the applied loads

$+ B_0^{\ T} (d_m)_T$ due to temperature change

$+ B_0^{\ T} (d_m)_L$ due to lack of fit.

By using Eq.(4.2), the member forces are evaluated.

Sometimes from design consideration, it may be desirable to evaluate the redundants separately for given loads, temperature and lack of fit. In which case

$$Q_P = - F^{-1} U_P^{\ 0} \qquad \ldots(4.14a)$$

The problems of this kind have been solved in the Articles 4.1 to 4.5

$$Q_T = - F^{-1} U_T^{\ 0} \qquad \ldots(4.14b)$$

An example of this kind was solved in Chapter 3, and some more examples will be solved in the article following.

$$Q = - F^{-1} U_L^{\ 0} \qquad \ldots(4.14c)$$

The above equation, Eq. (4.14c), gives an analysis for lack of fit where $U_L^{\ 0}$ was shown to be equal to B^T $(d_m)_L$. The values of (d_m) are known directly as initial geometrical imperfections. Such problems being simple and direct, it is not felt necessary to included example for illustration.

4.6.2 Member Deformations $(d_m)_T$ due to Temperature

The parabolic arch analysed for temperature stresses, Chapter 3, was of rectangular section and prismatic between nodes. Consider a more general cross-section but symmetrical about the y-axis, Fig. 4.14(a). Let the rise in temperature be taken as positive. The outside and underside temperatures are designated by T_0 and T_i, and the variation across the depth D is assumed to be liner. Let $T_0 > T_i$. Let the neutral or centroidal axis* be located by y_t measured from the top fibre, Fig. 4.14(a). From the figure, the temperature $\bar{T}$ at the neutral axis level gives rise to simple elongation, and

$$\bar{T} = T_i + \frac{Y_b}{D}(T_0 - T_i)$$

Where y_b is the distance of the bottom fibre from the neutral axis. In the case of uniform rise in temperature $T_0 = T_i = \bar{T}$. The member deformation d_m for uniform rise in temperature, $\bar{T}$, is given by Fig. 4.14 (b) (i)

$$\bar{d}_m = [\alpha L\, \bar{T}, 0, 0]^t \qquad \ldots(a)$$

Where a is the thermal coefficient of expansion. The effect of temperature gradient $(T_0 - T_i)$ gives rise to curvature,

$$\frac{1}{R} = \alpha\frac{T_0 - T_i}{D}$$

And from Fig. 4.14 (b) (ii), the corresponding member deformations,

$$\bar{d}_m = [0, \delta_{y2}, \theta_2]^t$$

Where $$\theta_2 = \frac{-L}{R} = -\frac{T_0 - T_i}{D} \qquad \ldots(b)$$

and $$\delta_{y2} = -\frac{L^2}{2R} = -\alpha L^2\frac{T_0 - T_i}{D}$$

---

* In the case of members with large initial curvatures strain variation across the depth may be assumed linear but stresses dc not vary linearly, and the neutral axis and censorial axis are not coincident. Such cases are excluded. The arches in practice do not have large initial curvatures to justify non linear stress variation across the depth.

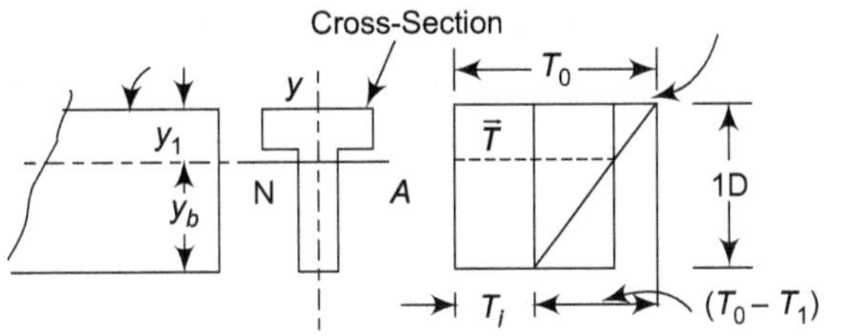

$$\bar{T} = T_i + \frac{y_b}{D}(T_0 - T_i)$$

Temperature-change and its variation along depth. D

$T > T_i$

(a) Prismatic member

(i) Uniform rise in temperature, $\bar{T}$

$$\bar{d}_m = \left[\alpha L \bar{T}, 0, 0\right]^t$$

(ii) Effect of temperature gradient curvature $= \dfrac{1}{R} = \alpha(T_0 - T_i)/D$

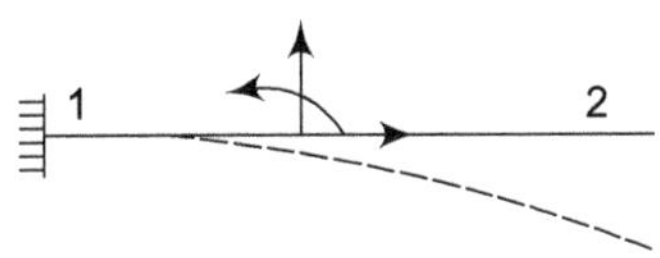

$$\theta_2 = -\frac{L}{R} = -\alpha L(T_0 - T_i)/D$$

$$\tilde{d}_m = \left[0, \quad \delta_{y2}, \quad \theta_2\right]^t \qquad \delta_{y2} = -\frac{L^2}{2R} = -\alpha L^2(T_0 - T_i)/2D$$

$$d_m = \bar{d}_m + \tilde{d}_m = [\alpha L \bar{T}, -\alpha L^2(T_0 - T_i)/D; -\alpha L(T_0 - T_i)/D]^t$$

(b) Reference axis, tangent at end 1

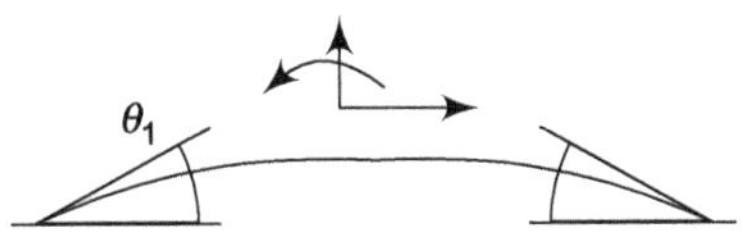

$$\theta_2 = -\theta_1 = -\frac{L}{2R}$$

The Fig. 4.14 is continued in the next page

(i) Uniform rise $\bar{T}$,

$\overline{d_m}$ = [a L $\bar{T}$, 0, 0]^t same as (b) (i)

(ii) $\tilde{d}_m$ = [0, a $L[T_0 - T_i]/2D$, – a $L[T_0 - T_i]/2D]^t$

(c) Reference axis, line joining end points

Fig. 4.14. *Temperature changes and member deformations*

It must be obvious to the reader that the result (b) is for the case of the tangent at end 1, in the deformed position, as the reference axis. Combining the results (a) and (b) for such a reference axis.

$$(d_m)_T = \bar{d}_m + \bar{d}_m$$

$$(d_m)_T = [\alpha L\, \bar{T}, -\alpha L^2 (T_0 - T_i)/2D, -\alpha L (T_0 - T_i)/D]^t \qquad \ldots(4.15)$$

And from Eq. (4.13b)*

$$U_T^0 = B^T (d_m)_T$$

and from Eq. (4.14) the redundant Q_T and member forces due to temperature $(P_m)_T = BQ_T$, are evaluated.

In the case of the reference axis obtained by the line joining the end points in the deformed position, Fig. 4.14 (c).

$$\bar{d}_m = [\alpha L\, \bar{T}, 0, 0]^t$$

$$\bar{d}_m = [0, \alpha L (T_0 - T_i)/2D, -\alpha L (T_0 - T_i)/2D]^t,$$

and

$$(d_m)_T = [\alpha L\, \bar{T}, \alpha L (T_0 - T_i)/2D, -\alpha L (T_0 - T_i)/2D]^t \qquad \ldots(4.16)$$

Ex. 4.14. Analyse the frame, Ex. 4.10, if all the members experience $T_0 = 25°C$ and $T_i = 5°C$ and thermal coefficient of expansion $= 1 \times 10^{-5}/°C$. Assume depth of cross-section of all the members, $D = 0.5$ m. the cross section is symmetrical about its axes.

Solution. Retaining the same release system (3 hinges) of the Ex. 4.10, the flexibility matrix $F = B^T F_m B$ remains unaltered.

$$(F)^{-1} = EI \begin{bmatrix} 0.849 & 0.1985 & -0.361 \\ 0.1985 & 0.280 & 0.1191 \\ -0.361 & 0.1191 & 0.529 \end{bmatrix}$$

The cross-section being symmetrical about the centroidal axis also, $y_b = y_t = 0.5D$

$$\bar{T} = 15°C \text{ and } T_0 - T_i = 20°C$$

Using the tangent at end 1 as reference axis, we obtain $(d_m)_T$ from the Eq. (4.15). By substituting values of L, D, $\bar{T}$ and $(T_0 - T_i)$.

$$(d_m)_T = 10^{-5} [(60, -320, -160)_a, (121, -1300, -324)_b, (45, -180, -120)_c]^t$$

From Eq. (4.13b),

$$U_T^0 = B^T (d_m)_T$$

The B matrix was obtained while solving Ex. 4.10 and with $(d_m)_T$ obtained above.

$$U_T^0 = 10^{-3} [3.20 - 5.04\ 2.212]^t,$$

* In the case of temperature stresses,
$(p_m) = BQ$, and following contragradience,
$(U_c)_T = B^T (d_m)$

And if the analysis is for temperature effects only, from Eq. (4.14b)

$$Q_T = -F^{-1} U_T^0$$

And substituting the values of F^{-1} and U_T^0

$$Q_T = 10^{-3} EI [-0.917\ 0.513\ 0.583]^t.$$

The member forces $p_m = B^T Q_T$. However for drawing the bending moment diagram, we need to know the moment at B only, which can be found expeditiously by using the principle of virtual work. Let M_B be causing tension outside.

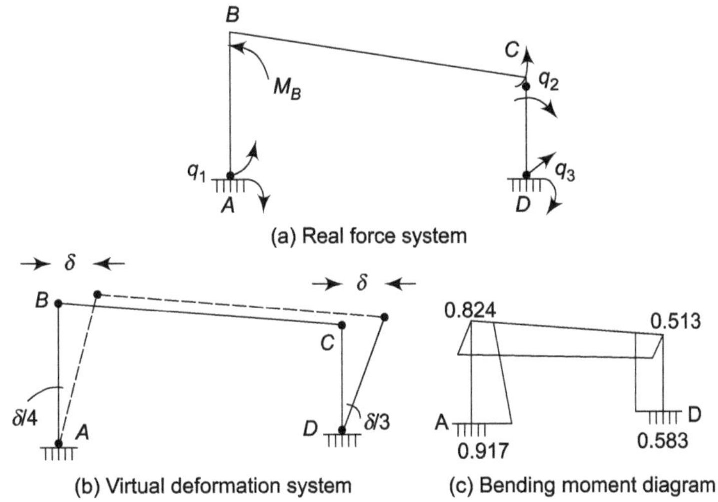

Fig. Ex. 4.14

From Fig. Ex. 4.14 (a) and (b)

$$M_B(-\delta/4) + (10)^{(-3)} EI [-.917(\delta/4) + .513(I\delta/3) + .568(\delta/3)] = 0$$

$$M_B = -0.824 \times 10^{-3} E \text{ tm}$$

Negative sign suggests tension outside. The final bending moment diagram is given in Fig. Ex. 4.14 (c).

Ex. 4.15. The reader may like to solve the Ex. 4.15, but this time with the release system and reference axes as considered in the Ex. 4.11. Release system: complete cut at the base of the right hand side column. Reference axes: Line joining end points as reference axis for each member.

Based on the release system, we know $F = B^t F_m B$ as computed while solving Ex. 4.11. We also know B matrix for the release system and chosen reference axis from the same example, Ex. 4.11. The reader may verify the following values. From Eq. (4.16),

$$(d_m)_T = 10^{-5} [(60, -80, -80)_a, (121, 161.12, -161.12)_b, (45, 60, -60)_c]^t$$

$$B^T (d_m)_T = -10^{-3} [15.1, 25.7, 6.025]^t$$

And $$Q_T = 10^{-3} EI [-0.025, 0.042, 0.575]^t$$

The redundant for the chosen release system are shear thrust and moment at the end of member c, and bending moment diagram can be drawn from simple statics. Alternatively complete set of internal forces (thrust, shear and moment force each member) can be determined by the relation, $(p_m) = B^t Q_T$

The results obtained may be checked with those already obtained in the previous example, Ex. 4.14.

4.7 RIGID CUM HINGED STRUCTURES

From design considerations, it may be economical to have a structure consisting of flexurally stiff members as well as members with hinged ends which are directly stressed. In such structures the flexural members are the main members and strength of the structures in considerably enhanced by introducing pin ended may decide to connect the columns of a bay by lacing members with the ends hinged at floor levels, Fig. 4.15. The heavy columns OG and OG_1, which are designed to carry no-sway gravity loads, limit the say or lateral deflections when stiffened by lacing.

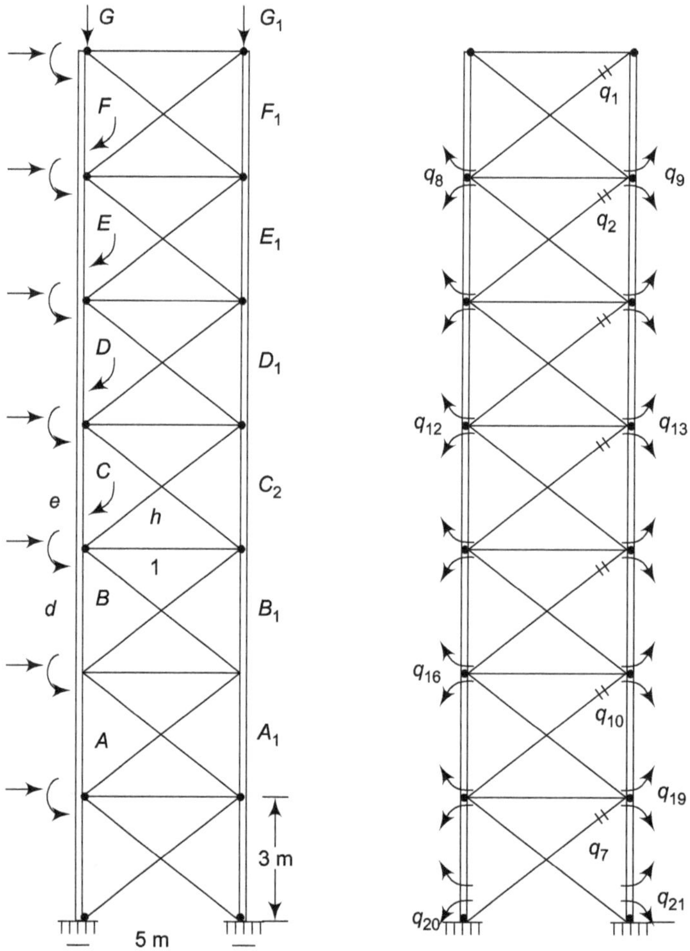

(a) Trussed shear wall (b) A release system

Fig. 4.15 *Hinge Release System at One of the Ends of Each Beam Member*

The pin ended members connect the columns so that they act as unit. Such paneled walls are called shear walls. Consider a joint, C, Fig. 4.15. The directly stressed members f, h and i are hinged at C. The flexural members d and e are rigidly connected at C, and also rigidly connected to the other members at the far joints B and D. The flexibility method may be preferred for reasons of reduced number of unknowns as compared to the stiffness method. The degree of indeterminacy is 21, whereas if the structure is to be analysed by the stiffness method, the column members also should be considered extensible and thus the unknown deformations are 42. A possible release system is shown for the shear wall, Fig. 4.15 (b).

Similarly, in Fig. 4.16(a), a beam *OD* has been stiffened by the pin ended member in the form of a truss. The flexural members, *OA*, *AB*, *BC* and *CD* of the trussed beam are rigidly connected at the nodes *A*, *B*, and *C*, and the other members are hinged at the joints. Figure 4.16(b) gives a possible release system corresponding to the five redundant.

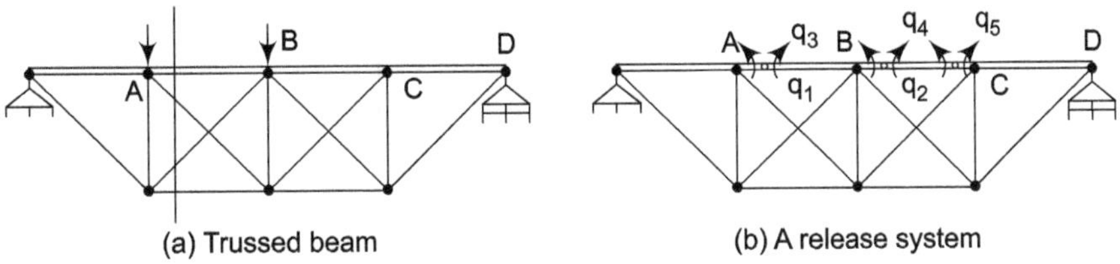

Fig. 4.16 *Truss-Girder with Releases in the Beam Members*

The member forces $(p_m)_i$ for a flexural member *i*, consist of $(p_m, p_y, m)_{2i}$ and for a pin ended member axial force p_m only. Similarly for a flexural member $(f_m)_i$ is given by Eq. (4.9) or (4.10), and for a pin ended member $f_m = L/EA$.

A numerical example of a rigid cum hinged frame is solved in the following.

Ex. 4.15. Analyse the frame, Fig. 4.15(a). The degree of indeterminacy is two. Choose the release system, Fig. 4.15(b). Consider tangent at end 1 as reference axis.

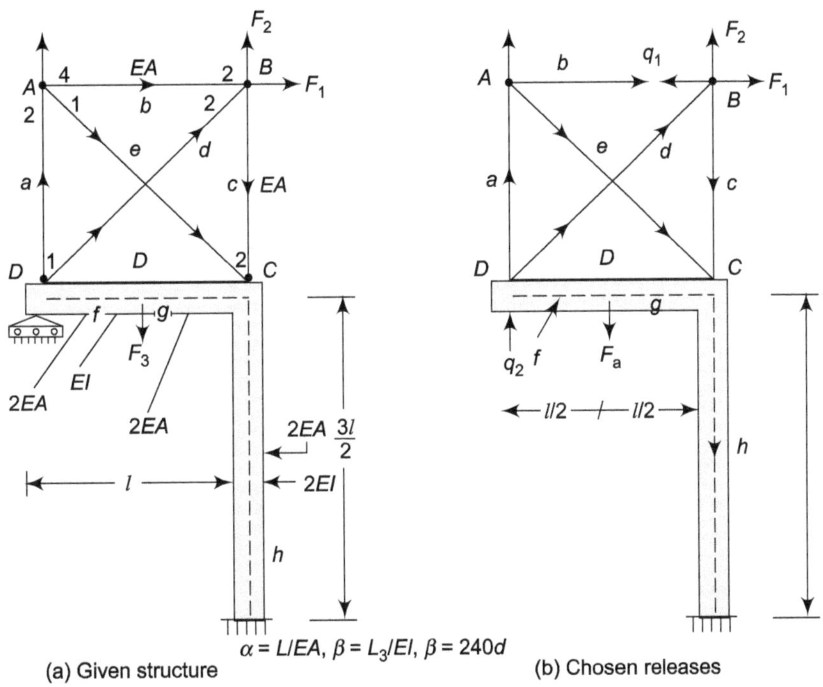

Fig. Ex. 4.15 *Release Introduced in a Truss Member and Roller Support*

Step 1: Equilibrium:

$$\begin{bmatrix} (p_x)_{2a} \\ (p_x)_{2b} \\ (p_x)_{2c} \\ (p_x)_{2d} \\ (p_x)_{2e} \\ \begin{pmatrix} p_x \\ p_y \\ m \end{pmatrix}_{2f} \\ \begin{pmatrix} p_x \\ p_y \\ m \end{pmatrix}_{2g} \\ \begin{pmatrix} p_x \\ p_y \\ m \end{pmatrix}_{2h} \end{bmatrix} = \left[\begin{array}{ccc|cc} 0 & 0 & 0 & 1 & 0 \\ 0 & 0 & 0 & 1 & 0 \\ -1 & 1 & 0 & 1 & 0 \\ \sqrt{2} & 0 & 0 & -\sqrt{2} & 0 \\ 0 & 0 & 0 & -\sqrt{2} & 0 \\ -1 & 0 & 0 & 1 & 0 \\ -1 & 0 & 0 & 0 & -1 \\ \frac{l}{2} & 0 & 0 & 0 & \frac{l}{2} \\ -1 & 0 & 0 & 1 & 0 \\ -1 & 0 & 1 & 0 & -1 \\ l & 0 & -\frac{l}{2} & 0 & l \\ 0 & 1 & -1 & 0 & 1 \\ -1 & 0 & 0 & 0 & 0 \\ \frac{5l}{2} & 0 & -\frac{l}{2} & 0 & l \end{array}\right] \begin{bmatrix} \begin{pmatrix} F_1 \\ F_2 \\ F_3 \end{pmatrix} = P_i \\ \begin{pmatrix} q_1 \\ q_2 \end{pmatrix} = Q \end{bmatrix}$$

$$P_m = \qquad B_0 \qquad B \qquad \begin{pmatrix} P_i \\ Q \end{pmatrix}$$

Step 2. Deformation force relation, F_m

$$F_m = \left[\left(\frac{L}{EA}\right)_a, \left(\frac{L}{EA}\right)_b, \left(\frac{L}{EA}\right)_c, \left(\frac{L}{EA}\right)_d, \left(\frac{L}{EA}\right)_e, \begin{pmatrix} \frac{L}{EA} & 0 & 0 \\ 0 & \frac{L^3}{2EI} & \frac{L^2}{2EI} \\ 0 & \frac{L^2}{2EI} & \frac{L}{EI} \end{pmatrix}_f, \right.$$

$$\left. \begin{pmatrix} \frac{L}{EA} & 0 & 0 \\ 0 & \frac{L^2}{2EI} & \frac{L^2}{2EI} \\ 0 & \frac{L^2}{2EI} & \frac{L}{EI} \end{pmatrix}_g \begin{pmatrix} \frac{L}{EA} & 0 & 0 \\ 0 & \frac{L^2}{2EI} & \frac{L^2}{2EI} \\ 0 & \frac{L^2}{2EI} & \frac{L}{EI} \end{pmatrix}_h \right]$$

Step 3: Compatibility

$$U = 0 = B^t F_m B_0 P_j + B^t F_m B Q$$

or

$$Q = -(B^t F_m B_0)^{-1} (B^t F_m B_0) P_J$$

$$B^t F_m B_0 = \begin{bmatrix} \begin{matrix}(-(L/EA)_c - 2(L/EA)_d) \\ -(L/EA)_f - (L/EA)_g \\ = -7/2\propto\end{matrix} & \begin{matrix}(L/EA)_c \\ = \propto\end{matrix} & 0 \\ \begin{matrix}\left((-L^3/6EI)_f + \frac{l}{2}(2L^2/2EI)_f\right) \\ -(2L^3/3EI)_g + l(3L^2/2EI)_g \\ +(L^3/3EI)_h + \frac{5l}{2}(2L^3/3EI)_i \\ = 395\propto\end{matrix} & (L/EA)_h = \propto/2 & \begin{matrix}\left((2L^3/3EI)_g - \frac{l}{2}(3L^2/2EI)_g\right) \\ -(L/EI)_h - \frac{l}{2}((2L^2/3EI)_h) \\ -115.5\propto\end{matrix} \end{bmatrix}$$

$B^t F_m B =$

$\left(\frac{L}{EA}\right)_a + \left(\frac{L}{EA}\right)_b + \left(\frac{L}{EA}\right)_c + \left(\frac{L}{EA}\right)_d + \left(\frac{L}{EA}\right)_e + \left(\frac{L}{EA}\right)_f + \left(\frac{L}{EA}\right)_g = 15\frac{\alpha}{2}$	0
0	$\left(\frac{L^2}{6EA}\right)_f + \frac{1}{2}\left(\frac{L^2}{2EA}\right)_f + \left(\frac{2L^3}{3EA}\right)_g + l\left(\frac{3L^2}{2EA}\right)_g + \left(\frac{L}{EA}\right)_h + l\left(\frac{2L}{3EA}\right)_h = 15\frac{\alpha}{2}$

$$(B^tF_mB)^{-1} = 1/\propto \begin{bmatrix} \frac{2}{15} & 0 \\ 0 & \frac{2}{521} \end{bmatrix}$$

$$A_g = (B^tF_mB)^{-1}(B^tF_mB_0)$$

$$A_{gf} = \begin{bmatrix} \frac{2}{15} & 0 \\ 0 & \frac{2}{521} \end{bmatrix} \begin{bmatrix} -\frac{7}{2} & 1 & 0 \\ 395 & \frac{1}{2} & -\frac{231}{2} \end{bmatrix}$$

$$A_{gj} = \begin{bmatrix} -\frac{7}{15} & \frac{2}{15} & 0 \\ \frac{790}{521} & \frac{1}{521} & -\frac{231}{521} \end{bmatrix}$$

The redundant q are obtained by the relation $q = -A_{qj}P$ where $P = P_j = [F_1\ F_2\ F_3]^t$
The member forces are obtained by the relations,

$$P_m = (B_0\ B)\begin{bmatrix} P_j \\ q \end{bmatrix}$$

Joint deformations, d_j

$$d_j = B_0^{\,t}\, F_m\, (B_0 P + B\, q)$$

The same example has been solved by Pestel and Leckie* but using a different release system.

Additional Illustration: The example has members under transversely loaded Beam member and truss members. Refer Fig. Additional Illustration. We do the first part of the analysis to convert the member load to nodal Loads by clamping the joints A, B, C, and D

The degree of Indeterminacy is equal to three. For the force method of analysis, three hinge releases are shown in Fig. (c).

(a) Write B_0 and B matrices. For the member flexibility matrix, use the line joining the end points in deformed position as member axis.
(b) Compute the redundant moments at the chosen releases, and
(c) Draw the final BM diagram.

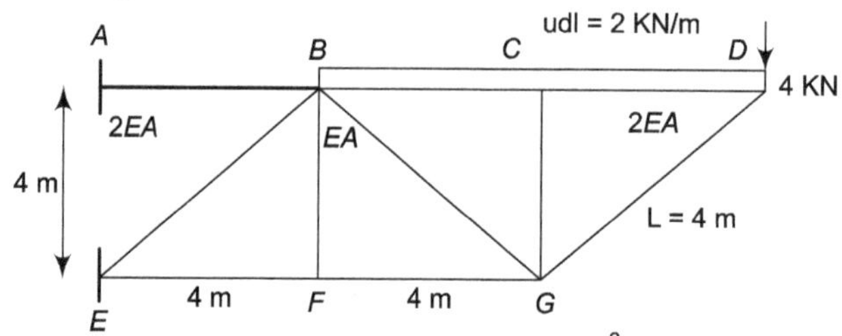

We need to define the ratio of EI to EA. $EA\,L^2/EI = 20$

Additional Illustration

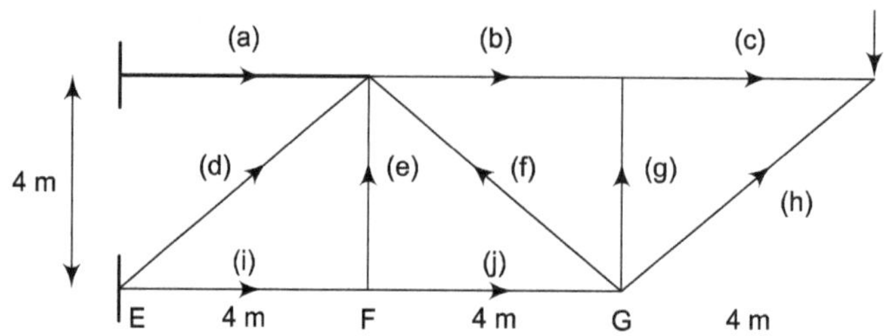

The fixed end support moments are frozen

Fig. (a) *Member Axes*

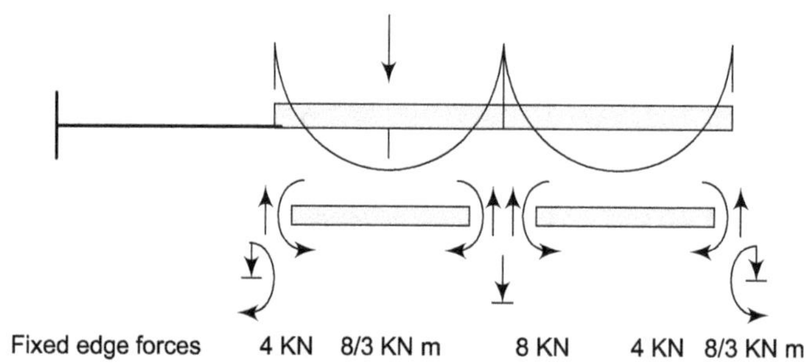

The fixed edge forces at nodes B and C are considered together with 4 KN load at node E

Fig. (b) *Nodal Loads from Fixed Edge Forces*

* "Matrix Methods is Elasto Mechanics", by Pestel E.C., and Leckie F.A. McGraw Hill Book Company, 1963, pp. 258.

The analysis is to be carried out for the following loading:

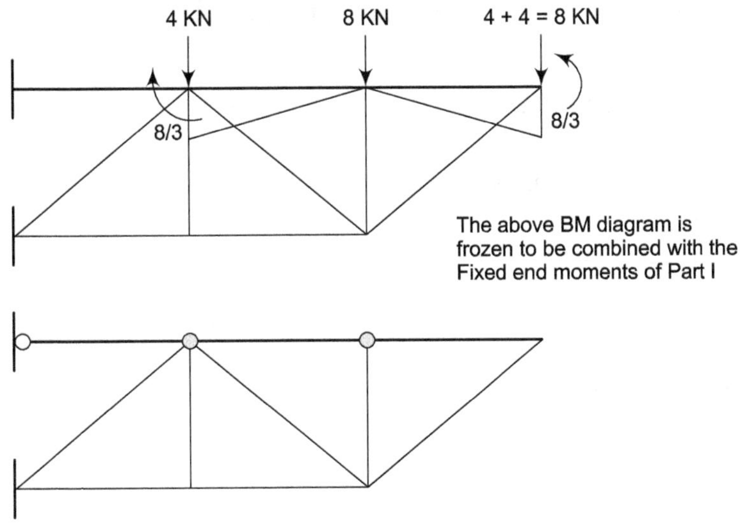

Fig. (c) *Loading and Release System*

Part II for the solution of the Illustration

Step 1. Equilibrium matrices B_0 and B matrices:

$$\begin{bmatrix} \begin{pmatrix} p_x \\ m_1 \\ m_2 \end{pmatrix}_a \\ \begin{pmatrix} p_x \\ m_1 \\ m_2 \end{pmatrix}_b \\ \begin{pmatrix} p_x \\ m_1 \\ m_2 \end{pmatrix}_c \\ (Px_2)_d \\ (Px_2)_e \\ (Px_2)_f \\ (Px_2)_g \\ (Px_2)_h \\ (Px_2)_i \\ (Px_2)_j \end{bmatrix} = \{B_0 \quad B\} \begin{pmatrix} 1 \\ q_1 \\ q_2 \\ q_3 \end{pmatrix}$$

Step 2. Deformation–Force Relations, Flexibilities

$$F_m = [f_{ma}, f_{mb}, f_{mc}, f_{md}, f_{me}, f_{mf}, f_{mg}, f_{mh}, f_{mi}, f_{mj}]$$

f_m for the first three (beam) members is given by $\begin{pmatrix} L/EA & 0 & 0 \\ 0 & L/3EI & -L/6EI \\ 0 & -L/6EI & L/3EI \end{pmatrix}$ member

And for the other members, the flexibilities are Scalars equal to L/EA

Step 3. Contragradience: Compatibility

$$\begin{pmatrix} d_0 \\ dq_1 \\ dq_2 \\ dq_3 \end{pmatrix} = \begin{Bmatrix} B^T \\ BT_0 \end{Bmatrix} dm$$

For $dq_i = 0$, and $Pj = 1$,

$$Q = [B^T F_m B]^{-1} [B^T F_m B_0]$$

Results: The final bending moment diagram for the members (a), (b), and (c) is drawn in the following:

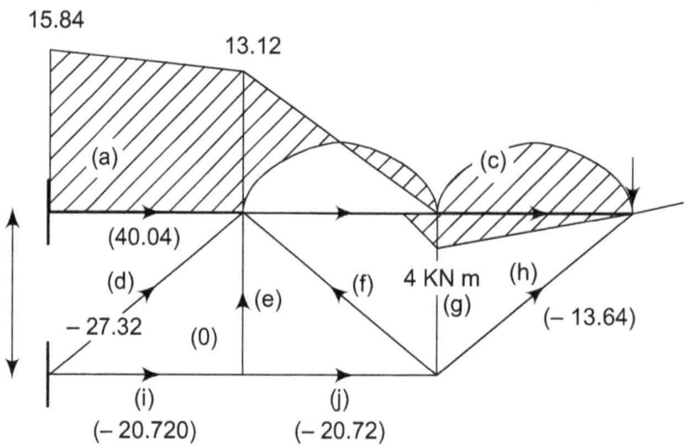

Fig. (d) *Final Bending Moment Diagram*

Comment: If one attempts to solve the problem by the Displacement method, the number of unknown degrees of freedom is thirteen. (Displacement method Chapters 7 and 8).

4.8 GRIDS AND BEAMS CURVED IN PLAN (BOW GIRDERS)

4.8.1 General

In Article 4.5.1, we considered the plane frames where the members of a frame and the applied loads were in the same vertical plane. Such plane frames member developed axial forces (P_x), transverse shears P_y and transverse moment m directed along the z-axis.

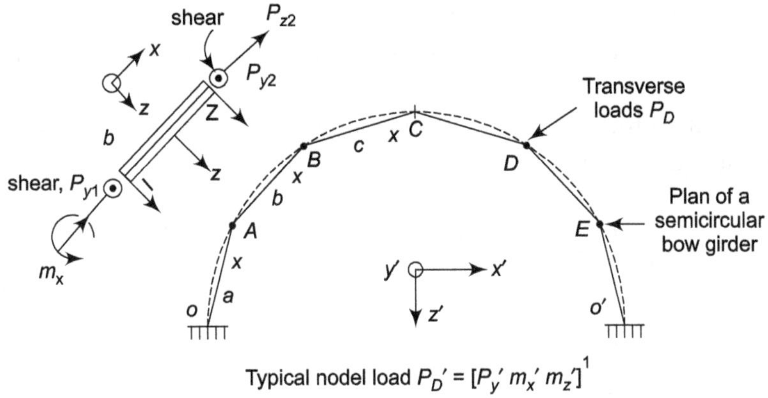

Fig. 4.17 *Bow Girder Approximated by Prismatic Members Connected by Rigid Joints*

In the case of grids, all the members of a grid are also in a plane but transverse to the applied loads. Such grid frames consisting of members with zero torsional stiffness were analysed in Article 2.15 and Examples 2.28 and 2.30. In the following, matrix approach will be used for the analysis. The grids consisting of members without torsional stiffness will be considered first, followed by grids consisting of torsionally stiff members. The member forces given by the P_2, reference line tangent at end 1, consist of (P_y, m_{x2}, m_{z2}) for torsionally stiff members, Fig. 4.17. The moments m_{x2} and m_{z2} are the moment vectors directed along x and z axes respectively at the end 2, Fig. 4.17. The moment m_x is the twisting moment or torsion and m_z is the bending moment accompanied by the flexural shear p_{y2}. The beams curved in plan, which are also known as how girders, are analysed similar to grids if we replace the curved length OO' by a DE, EO', Fig. 4.17, as we did in the case of arches. The analysis of bow girders may also be looked upon as a particular case of a generalized space frame analysis, like spiral staircase with end-supports only, Article 4.9.

4.8.2 Grids with Members of Zero Torsional Stiffness

In the case of members with zero torsional stiffness, the member flexibility matrices given by Eqs (4.7) and (4.8), remain unaltered. If we consider the tangent at the end 1 as a reference axis, Eq. (4.7), reproduced in the following, is applicable to a grid member I with a zero torsional stiffness. The grid is hinged at supports

$$\begin{bmatrix} \delta_{y2} \\ \theta_{x2} \end{bmatrix} = \begin{bmatrix} \dfrac{L^3}{3EI} & \dfrac{L^2}{2EI} \\ \dfrac{L^2}{2EI} & \dfrac{L}{EI} \end{bmatrix} \begin{bmatrix} P_{y2} \\ m_{z2} \end{bmatrix} \qquad \ldots(4.7)$$

A complete analysis is simple and direct, and it is better explained with the help of numerical examples.

Ex. 4.16. Analyse the grid, Fig. Ex. 4.16 (a), consisting of members with zero torsional stiffness. All members are of equal lengths L, and of same EI values. All end-supports are hinged.*

* A similar example but with ends-clamped was solved in chapter 2 Refer Fig. Ex. 2.28.

The degree of indeterminacy is four. The interacting forces at the nodes A, B, C and D are chosen as redundant. For such a released or determinate structure, the moments due to applied loads at A and unit redundant are given in Fig. Ex. 4.16 (b), (c), (d), (e) and (f). The applied load P on the node A is assumed to be acting as $P/2$ on the girder $OACO$ and $P/2$ on the girder $OABO$, Fig. Ex. 4.16 (b). Based on the M_0 and m_i diagrams, the equilibrium matrix is written.

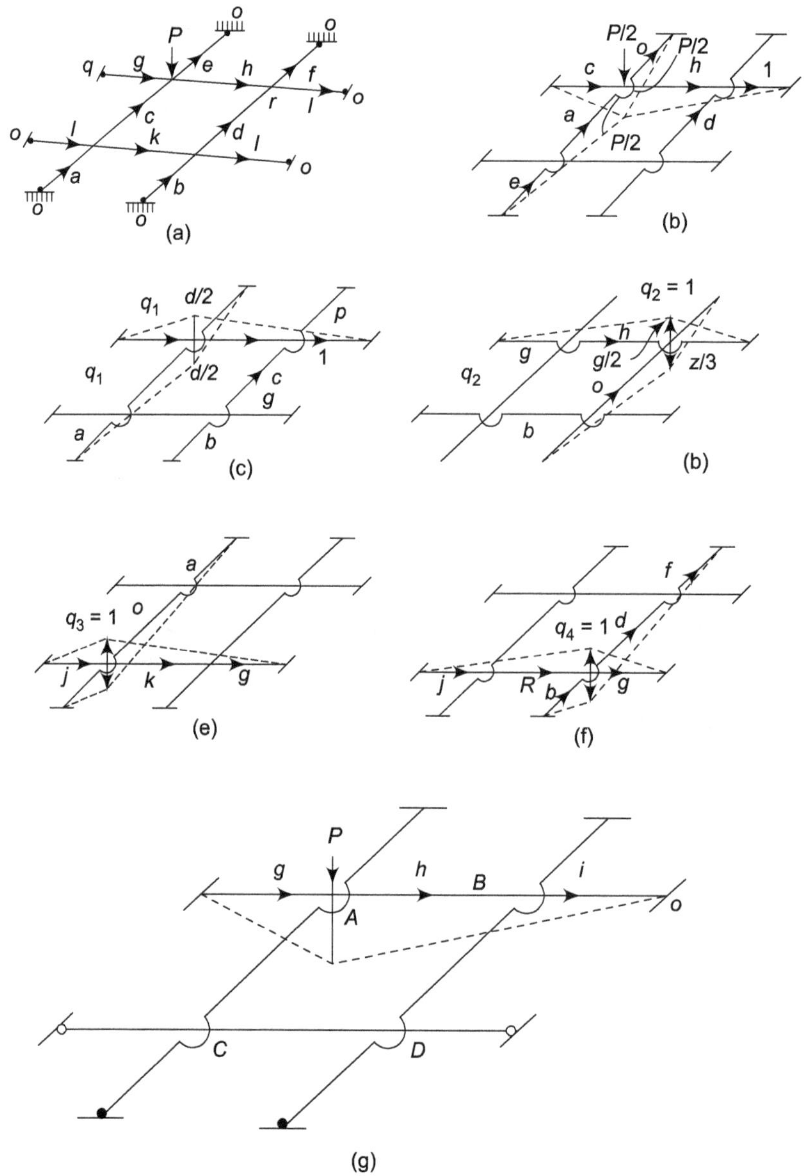

Fig. Ex. 4.16 *Grid Members with Zero torsional stiffness*

Step 1. Equilibrium equation

$$
\begin{bmatrix}
\begin{pmatrix} p_{y2} \\ m_{z2} \end{pmatrix}_a \\
\begin{pmatrix} p_{y2} \\ m_{z2} \end{pmatrix}_b \\
\begin{pmatrix} p_{y2} \\ m_{z2} \end{pmatrix}_c \\
\begin{pmatrix} p_{y2} \\ m_{z2} \end{pmatrix}_d \\
\begin{pmatrix} p_{y2} \\ m_{z2} \end{pmatrix}_e \\
\begin{pmatrix} p_{y2} \\ m_{z2} \end{pmatrix}_f \\
\begin{pmatrix} p_{y2} \\ m_{z2} \end{pmatrix}_g \\
\begin{pmatrix} p_{y2} \\ m_{z2} \end{pmatrix}_h \\
\begin{pmatrix} p_{y2} \\ m_{z2} \end{pmatrix}_i \\
\begin{pmatrix} p_{y2} \\ m_{z2} \end{pmatrix}_j \\
\begin{pmatrix} p_{y2} \\ m_{z2} \end{pmatrix}_k \\
\begin{pmatrix} p_{y2} \\ m_{z2} \end{pmatrix}_l
\end{bmatrix}
=
\underbrace{\begin{bmatrix}
-1/6 & -1/3 & 0 & -2/3 & 0 \\
1/6 & 1/3 & 0 & 2/3 & 0 \\
0 & 0 & -1/3 & 0 & -2/3 \\
0 & 0 & 1/3 & 0 & 2/3 \\
-1/6 & -1/3 & 0 & 1/3 & 0 \\
1/3 & 2/3 & 0 & 1/3 & 0 \\
0 & 0 & -1/3 & 0 & 1/3 \\
0 & 0 & 2/3 & 0 & 1/3 \\
1/3 & 2/3 & 0 & 1/3 & 0 \\
0 & 0 & 0 & 0 & 0 \\
0 & 0 & 2/3 & 0 & 1/3 \\
0 & 0 & 0 & 0 & 0 \\
-1/3 & 2/3 & +1/3 & 0 & 0 \\
1/3 & -2/3 & -1/3 & 0 & 0 \\
1/6 & -1/3 & 1/3 & 0 & 0 \\
1/6 & -1/3 & -2/3 & 0 & 0 \\
1/6 & -1/3 & -2/3 & 0 & 0 \\
0 & 0 & 0 & 0 & 0 \\
0 & 0 & 0 & 2/3 & 1/3 \\
0 & 0 & 0 & -2/3 & -1/3 \\
0 & 0 & 0 & -1/3 & 1/3 \\
0 & 0 & 0 & -1/3 & -2/3 \\
0 & 0 & 0 & -1/3 & 2/3 \\
0 & 0 & 0 & 0 & 0
\end{bmatrix}}_{B}
\begin{bmatrix} P \\ q_1 \\ q_2 \\ q_3 \\ q_4 \end{bmatrix}
$$

$$
P_m \quad = \quad B_0 \qquad\qquad B \qquad\qquad \begin{bmatrix} P \\ Q \end{bmatrix}
$$

Step 2. Generalized strain-stress relation

$$
F_m = [(f_m)_a, (f_m)_{b1} \cdots (f_m)_i \cdots (f_m)_l]
$$

Step 3. Compatibility. Evaluation of redundant and member forces:

$$F = B^T F_m B = \frac{L}{EI} = \begin{bmatrix} \frac{8}{9} & \frac{7}{18} & \frac{7}{18} & 0 \\ \frac{7}{18} & \frac{8}{9} & 0 & \frac{7}{18} \\ \frac{7}{18} & 0 & \frac{8}{9} & \frac{7}{18} \\ 0 & \frac{7}{18} & \frac{7}{18} & \frac{8}{9} \end{bmatrix}$$

$$U^0 = B^T F_m B_0 P = \frac{PL}{EI} [0 - 7/36\ 7/36\ 0]^t$$

$$Q = -F^{-1} U^0$$

$$Q = P[0 \quad 7/32 \quad -7/32 \quad 0]^t \qquad \text{...(a)}$$

Comment: The advantage of splitting P into $P/2$ loads acting separately on the two members, is to predict $q_1 = q_4$ and $q_2 = q_3$, and thus the Q is achieved by evaluating q_2 only.

Alternatively if we consider the entire load P to be acting on the member *OABO* only Fig. Ex. 5.16 (g), the B_0 matrix gets altered and B matrix remains unaltered.

$$B_0 = [0\ 0, 0\ 0, 0\ 0, 0\ 0, 0\ 0, 0\ 0, -2/3\ 2/3, 1/3\ 1/3, 1/3\ 0, 0\ 0, 0\ 0, 0\ 0]^t$$

F remains unaltered.

$$U^0 = B^T F_m B P = \frac{PL}{EI} [-4/9 - 7/18\ 0\ 0]^t$$

$$Q = P[1/2\ 7/32 - 7/32\ 0]^t$$

With the redundant evaluated, the final bending moment diagram can be drawn directly. It is also possible to obtain p_m from the equilibrium equation.

$$F_m = (B_0\ B)\begin{bmatrix} P \\ \hline Q \end{bmatrix}$$

$$\begin{aligned} P_m = P[&-0.0204 \quad 0.0204, \quad -0.0728 \quad 0.0728, \\ &-0.2391 \quad 0.2603, \quad -0.0728 \quad 0.1459, \\ &0.2603 \quad 0, \quad 0.1459 \quad 0, \\ &-0.2610 \quad 0.2610, \quad 0.2395 \quad 0.0208, \\ &0.0208 \quad 0, \quad -0.1459 \quad 0.1459, \\ &-0.0728 \quad 0.0728, \quad 0.0728 \quad 0]^t \end{aligned}$$

Ex. 4.17. Analyse the grid, Fig. Ex. 4.16(a), by introducing 4 hinge releases as shown in Fig. Ex. 4.17(b). The m_0 and m_i diagrams are drawn for such a release system. Fig. Ex. 4.17(b), (c), (d), (e) and (f).

Result. The first row of $(B^T F_m B)\,Q + U^0 = 0$ is

$52/18\, q_1 - 29/18\, q_2 + 35/18\, q_3 - 14/9\, q_4 = -8/9$

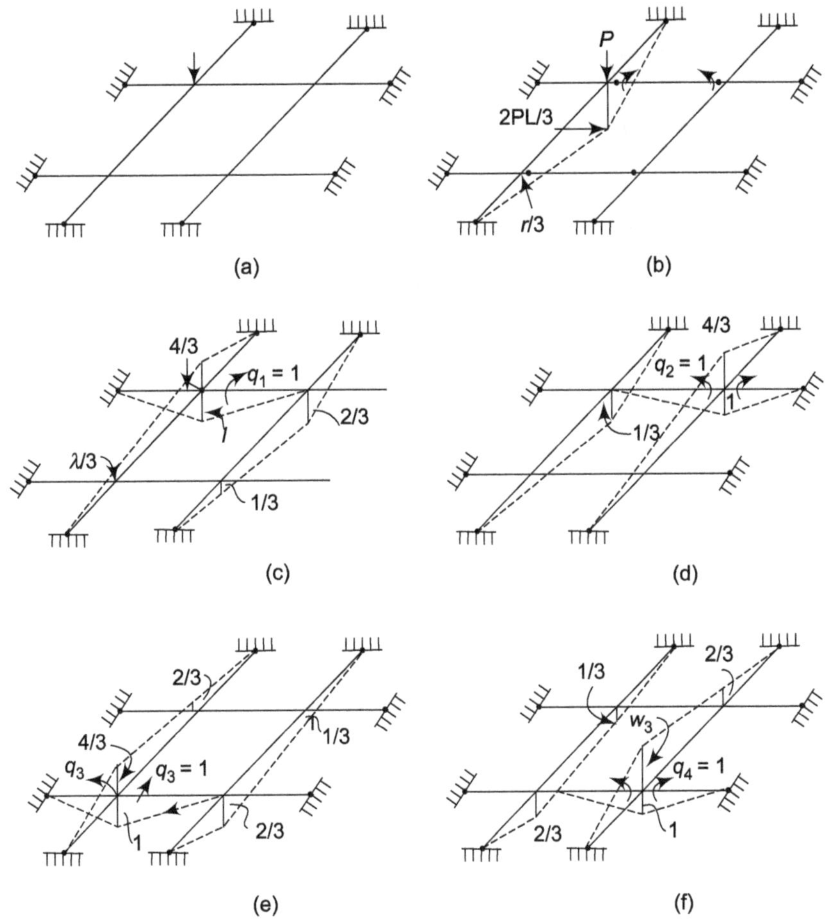

Fig. Ex. 4.17 *Hinge Releases in the Members*

The reader would like to complete the solution, and verify the final results with the Q and p_m already evaluated in the Ex. 4.16.

4.8.3 Grids of Torsionally Stiff Members

Let us continue to consider tangent at the end 1 of a member as reference axis. To arrive at the member flexibility for the chosen reference axis, the member is considered clamped at the end 1, Fig. Ex. 4.18(a). For a member, the torsional flexibility L/GJ does not get linked with flexural flexibilities, analogous to delinking of axial flexibility L/EA in the case of plane frames. Referring to Fig. 4.18, θ_{x2} is caused by m_{x2} only, and it does not influence the flexural deformations, δ_{y2} and θ_{x2}. Thus putting together the flexural flexibility given by Eqs (4.7) and (4.17) with delinked torsional flexibility, we get Eq. (4.18a)

$$\begin{pmatrix} \delta_{y2} \\ \theta_{x2} \\ \theta_{z2} \end{pmatrix}_i = \begin{pmatrix} \frac{L^3}{3EI} & 0 & \frac{L^2}{2EI} \\ 0 & \frac{L}{GJ} & 0 \\ \frac{L^2}{2EI} & 0 & \frac{L}{EI} \end{pmatrix}_i \begin{pmatrix} P_{y2} \\ m_{x2} \\ m_{z2} \end{pmatrix}_i \quad \ldots(4.18a)$$

If one chooses to consider the alternative reference axis obtained by joining the end points in deformed position, we obtain the member flexibility, Eq. (4.18b), by combining the torsional flexibility along and the flexural flexibility given by Eq. (4.8).

$$\begin{pmatrix} \theta_{z2} \\ \theta_{x2} \\ \theta_{z2} \end{pmatrix}_i = \begin{pmatrix} \frac{L}{3EI} & 0 & \frac{L}{6EI} \\ 0 & \frac{L}{GJ} & 0 \\ \frac{L}{6EI} & 0 & \frac{L}{3EI} \end{pmatrix}_i \begin{pmatrix} m_{z2} \\ m_{x2} \\ m_{z2} \end{pmatrix}_i \quad \ldots(4.18b)$$

Unlike axial stiffness *EA*, consideration of torsional stiffness *GJ* increases the number of unknown redundant considerably.

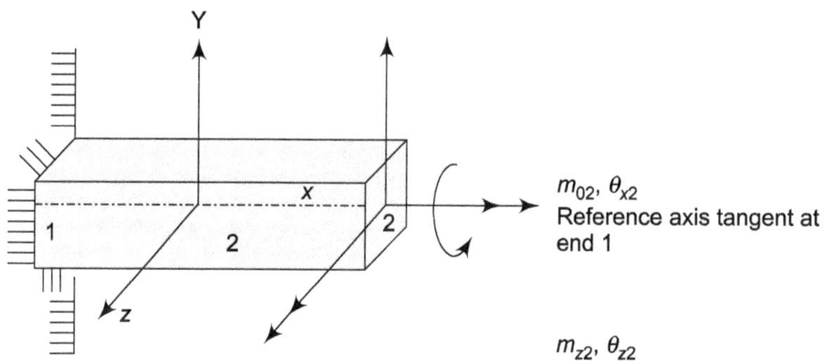

Fig. Ex. 4.18 *(contd.)*

The member properties and geometry are given in Fig. Ex. 4.18(a)

Solution for releases system 1 – refer Fig. Ex. 4.18 (b)

For a given problem with $(f_m)_I$ defined by Eq. (4.18), the statical analysis is done to define B_0 and B. Thereafter, evaluation of the redundant and member forces is direct matrix operation. In the following, and example is solved for two different release systems, Fig. Ex. 4.18. The reference axis chosen for both the solutions is the tangent at end 1.

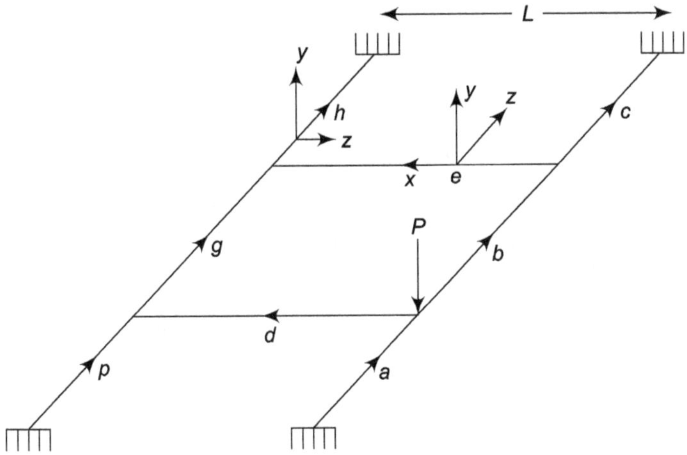

x, y, z member-axes
[All members of equal lengths]
EI = 10 GJ
EI and GJ same for all members

(a)

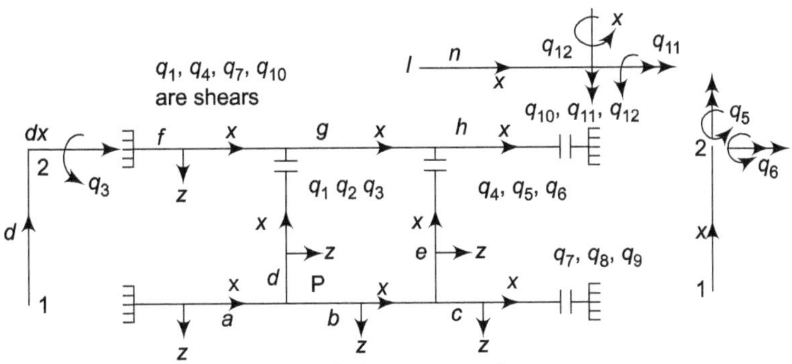

(b) Release system-1

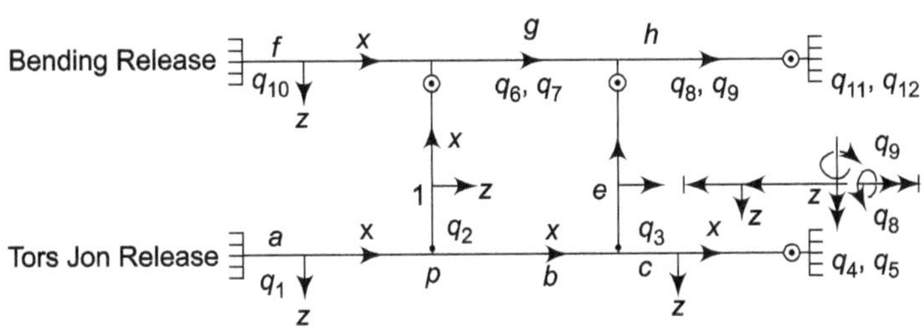

(c) Releases system – II, hinges for B matrix only

Fig. Ex. 4.18

Step 1. Equilibrium matrix

$$P_m = B_0 \begin{bmatrix} P \\ Q \end{bmatrix}$$

$$
\begin{bmatrix}
\begin{pmatrix} P_y \\ m_x \\ m_z \end{pmatrix}_a \\
\begin{pmatrix} P_y \\ m_x \\ m_z \end{pmatrix}_b \\
\begin{pmatrix} P_y \\ m_x \\ m_z \end{pmatrix}_c \\
\begin{pmatrix} P_y \\ m_x \\ m_z \end{pmatrix}_d \\
\begin{pmatrix} P_y \\ m_x \\ m_z \end{pmatrix}_e \\
\begin{pmatrix} P_y \\ m_x \\ m_z \end{pmatrix}_f \\
\begin{pmatrix} P_y \\ m_x \\ m_z \end{pmatrix}_g \\
\begin{pmatrix} P_y \\ m_x \\ m_z \end{pmatrix}_h
\end{bmatrix}
=
\overbrace{\begin{bmatrix}
-1 & 1 & 0 & 0 & 1 & 0 & 0 & 1 & 0 & 0 & 0 & 0 & 0 \\
0 & L & 0 & 1 & L & 0 & 1 & 0 & 1 & 0 & 0 & 0 & 0 \\
0 & 0 & -1 & 0 & 0 & -1 & 0 & 2L & 0 & 1 & 0 & 0 & 0 \\
0 & 0 & 0 & 0 & 0 & 0 & 0 & 1 & 0 & 0 & 0 & 0 & 0 \\
0 & 0 & 0 & 0 & 0 & 0 & 1 & 0 & 1 & 0 & 0 & 0 & 0 \\
0 & 0 & 0 & 0 & 0 & -1 & 0 & L & 0 & 1 & 0 & 0 & 0 \\
0 & 0 & 0 & 0 & 0 & 0 & 0 & 1 & 0 & 0 & 0 & 0 & 0 \\
0 & 0 & 0 & 0 & 0 & 0 & 0 & 0 & 1 & 0 & 0 & 0 & 0 \\
0 & 0 & 0 & 0 & 0 & 0 & 0 & 0 & 0 & 1 & 0 & 0 & 0 \\
0 & 1 & 0 & 0 & 0 & 0 & 0 & 0 & 0 & 0 & 0 & 0 & 0 \\
0 & 0 & 1 & 0 & 0 & 0 & 0 & 0 & 0 & 0 & 0 & 0 & 0 \\
0 & 0 & 0 & 1 & 0 & 0 & 0 & 0 & 0 & 0 & 0 & 0 & 0 \\
0 & 0 & 0 & 0 & 1 & 0 & 0 & 0 & 0 & 0 & 0 & 0 & 0 \\
0 & 0 & 0 & 0 & 0 & 1 & 0 & 0 & 0 & 0 & 0 & 0 & 0 \\
0 & 0 & 0 & 0 & 0 & 0 & 1 & 0 & 0 & 0 & 0 & 0 & 0 \\
0 & -1 & 0 & 0 & -1 & 0 & 0 & 0 & 0 & 0 & 1 & 0 & 0 \\
0 & 0 & 0 & -1 & 0 & 0 & 1 & 0 & 0 & 0 & 0 & 1 & 0 \\
0 & 0 & 1 & 0 & -L & 1 & 0 & 0 & 0 & 0 & 2L & 0 & 1 \\
0 & 0 & 0 & 0 & -1 & 0 & 0 & 0 & 0 & 0 & 1 & 0 & 0 \\
0 & 0 & 0 & 0 & 0 & 0 & -1 & 0 & 0 & 0 & 0 & 1 & 0 \\
0 & 0 & 0 & 0 & 0 & 0 & 0 & 0 & 0 & 0 & L & 0 & 1 \\
0 & 0 & 0 & 0 & 0 & 0 & 0 & 0 & 0 & 0 & 1 & 0 & 0 \\
0 & 0 & 0 & 0 & 0 & 0 & 0 & 0 & 0 & 0 & 0 & 1 & 0 \\
0 & 0 & 0 & 0 & 0 & 0 & 0 & 0 & 0 & 0 & 0 & 0 & 1
\end{bmatrix}}^{B}
\begin{bmatrix} P \\ q_1 \\ q_2 \\ q_3 \\ q_4 \\ q_5 \\ q_6 \\ q_7 \\ q_8 \\ q_9 \\ q_{10} \\ q_{11} \\ q_{12} \end{bmatrix}
$$

Step 2

$$F_m = [(f_m)_a, (f_m)_a, \ldots, (f_m)_h]$$

Where (f_m) of a member is given by Eq. (5.18a)

Step 3

Assume

$$F = B^T F_m B$$

$$F = \frac{L}{EI}$$

$$\begin{bmatrix} 11.0 & -1.0 & 10.5 & 11.2 & -1.0 & 10 & 1.33 & 10 & .50 & -1.33 & 0 & -.5 \\ -1.0 & 12 & 0 & -2 & 2 & 0 & -2.5 & 0 & -1 & 2.5 & 0 & 1 \\ 10.5 & 0 & 21 & 10 & 0 & 20 & 0 & 10 & 0 & 0 & -10 & 0 \\ 11.2 & -2 & 10 & 23.7 & -3 & 20.5 & 2.17 & 20 & 1 & -4.67 & 0 & -2 \\ -1.0 & 2 & 0 & -3 & 14.2 & 0 & 4 & 0 & -2 & 4 & 0 & 2 \\ 10 & 0 & 20 & 20.5 & 0 & 41 & 0 & 20 & 0 & 0 & -20 & 0 \\ 1.33 & -2.5 & 0 & 2.17 & -4 & 0 & - & 0 & 4.5 & 0 & 0 & 0 \\ 10 & 0 & 10 & 20 & 0 & 20 & 0 & 30 & 0 & 0 & 0 & 0 \\ .50 & -1 & 0 & 1 & -2 & 0 & 4.5 & 0 & 3 & 0 & 0 & 0 \\ 1.33 & 2.5 & 0 & -4.67 & 4 & 0 & 0 & 0 & 0 & 9 & 0 & 4.5 \\ 0 & 0 & -10 & 0 & 0 & -20 & 0 & 0 & 0 & 0 & 30 & 0 \\ -.5 & 1 & 0 & -2 & 2 & 0 & 0 & 0 & 0 & 4.5 & 0 & 3 \end{bmatrix}$$

It may be observed that the F matrix consist of the diagonal elements which are not dominant.

$$Q = -F B^T F_m B_0$$

$$Q = P[-.0355 \quad .0063 \quad .0177 \quad .0185 \quad .0082 \quad -0092 \quad -.2564 \quad -.0003$$
$$-.2147 \quad -.0053 \quad -.0003 \quad -.0018]^t$$

From the Q evaluated, we may directly draw the bending moment and torque diagrams. Alternatively we find,

$$P_m = [B_0 \, B] \begin{bmatrix} P \\ Q \end{bmatrix}$$

Comments. The F matrix could have been improved by choosing a release system consisting of hinges for obtaining elements of B matrix, Fig. Ex. 4.18(c). It is possible to retain the first release system, Fig. Ex. 4.18 (b), for the B_0 matrix. In the following such a "two-release" system (mixed release system) is used (Art. 3.6.3 & 3.6.4).

Alternative Solution Ex. 4.18. Release system-2, Fig. Ex. 4.18 (c), is used for m_i diagrams. For the m_0 diagram, the release system, Fig. Ex. 4.18 (b), is retained to define the B_0 matrix. The solution is a particular case of a triple release system discussed in Chapter 3, Article 3.63. Reference axis considered is the tangent at end 1. In the system of the redundant, q_5, q_7, q_9, and q_{12}, are torques and $q_1, q_2, q_3, q_4, q_6, q_8, q_{10}, q_{11}$ are the flexural moments.

$$
\begin{bmatrix}
\begin{pmatrix} p_y \\ m_x \\ m_z \end{pmatrix}_a \\
\begin{pmatrix} p_y \\ m_x \\ m_z \end{pmatrix}_b \\
\begin{pmatrix} p_y \\ m_x \\ m_z \end{pmatrix}_c \\
\begin{pmatrix} p_y \\ m_x \\ m_z \end{pmatrix}_d \\
\begin{pmatrix} p_y \\ m_x \\ m_z \end{pmatrix}_e \\
\begin{pmatrix} p_y \\ m_x \\ m_z \end{pmatrix}_f \\
\begin{pmatrix} p_y \\ m_x \\ m_z \end{pmatrix}_g \\
\begin{pmatrix} p_y \\ m_x \\ m_z \end{pmatrix}_h
\end{bmatrix}
=
\underbrace{\begin{bmatrix}
-1 & \frac{-1}{3L} & \frac{-2}{3L} & \frac{-1}{3L} & \frac{-1}{3L} & 0 & \frac{-2}{3L} & \frac{1}{3L} & \frac{-1}{3L} & \frac{1}{3L} & 0 & 0 & 0 \\
0 & 0 & -1 & -1 & 0 & 1 & 0 & 0 & 0 & 0 & 0 & 0 & 0 \\
0 & \frac{-2}{3} & \frac{2}{3} & \frac{1}{3} & \frac{1}{3} & 0 & \frac{2}{3} & \frac{-1}{3} & \frac{1}{3} & \frac{-1}{3} & 0 & 0 & 0 \\
0 & \frac{-1}{3L} & \frac{1}{3L} & \frac{-1}{3L} & \frac{-1}{3L} & 0 & \frac{1}{3L} & \frac{1}{3L} & \frac{-1}{3L} & \frac{1}{3L} & 0 & 0 & 0 \\
0 & 0 & 0 & -1 & 0 & 1 & 0 & 0 & 0 & 0 & 0 & 0 & 0 \\
0 & \frac{-1}{3} & \frac{1}{3} & \frac{2}{3} & \frac{2}{3} & 0 & \frac{1}{3} & \frac{1}{3} & \frac{2}{3} & \frac{-2}{3} & 0 & 0 & 0 \\
0 & \frac{-1}{3L} & \frac{1}{3L} & \frac{2}{3L} & \frac{2}{3L} & 0 & \frac{1}{3L} & \frac{1}{3L} & \frac{2}{3L} & \frac{-2}{3L} & 0 & 0 & 0 \\
0 & 0 & 0 & 0 & 0 & 1 & 0 & 0 & 0 & 0 & 0 & 0 & 0 \\
0 & 0 & 0 & 0 & 1 & 0 & 0 & 0 & 0 & 0 & 0 & 0 & 0 \\
0 & 0 & \frac{-1}{L} & 0 & 0 & 0 & \frac{-1}{L} & 0 & 0 & 0 & 0 & 0 & 0 \\
0 & 0 & 0 & 0 & 0 & 0 & 0 & 1 & 0 & 0 & 0 & 0 & 0 \\
0 & 0 & 0 & 0 & 0 & 0 & 1 & 0 & 0 & 0 & 0 & 0 & 0 \\
0 & 0 & 0 & \frac{-1}{L} & 0 & 0 & 0 & 0 & \frac{-1}{L} & 0 & 0 & 0 & 0 \\
0 & 0 & 0 & 0 & 0 & 0 & 0 & 0 & 0 & 1 & 0 & 0 & 0 \\
0 & 0 & 0 & 0 & 0 & 0 & 0 & 0 & 1 & 0 & 0 & 0 & 0 \\
0 & 0 & \frac{2}{3L} & \frac{1}{3L} & 0 & 0 & \frac{2}{3L} & \frac{-1}{3L} & \frac{1}{3L} & \frac{-1}{3L} & \frac{-1}{3L} & \frac{-1}{3L} & 0 \\
0 & 0 & 0 & 0 & 0 & 0 & -1 & 0 & -1 & 0 & 0 & 0 & 1 \\
0 & 0 & \frac{-2}{3} & \frac{-1}{3} & 0 & 0 & \frac{-2}{3} & \frac{1}{3} & \frac{-1}{3} & \frac{1}{3} & \frac{-2}{3} & \frac{1}{3} & 0 \\
0 & 0 & \frac{-1}{3L} & \frac{1}{3L} & 0 & 0 & \frac{-1}{3L} & \frac{-1}{3L} & \frac{1}{3L} & \frac{-1}{3L} & \frac{-1}{3L} & \frac{-1}{3L} & 0 \\
0 & 0 & 0 & 0 & 0 & 0 & 0 & 0 & -1 & 0 & 0 & 0 & 1 \\
0 & 0 & \frac{-1}{3} & \frac{-2}{3} & 0 & 0 & \frac{-1}{3} & \frac{-1}{3} & \frac{-2}{3} & \frac{2}{3} & \frac{-1}{3} & \frac{2}{3} & 0 \\
0 & 0 & \frac{-1}{3L} & \frac{-2}{3L} & 0 & 0 & \frac{-1}{3L} & \frac{-1}{3L} & \frac{-2}{3L} & \frac{-1}{3L} & \frac{-1}{3L} & \frac{-1}{3L} & 0 \\
0 & 0 & 0 & 0 & 0 & 0 & 0 & 0 & 0 & 0 & 0 & 0 & 1 \\
0 & 0 & 0 & 0 & 0 & 0 & 0 & 0 & 0 & 0 & 0 & 1 & 0
\end{bmatrix}}
\begin{pmatrix} P \\ q_1 \\ q_2 \\ q_3 \\ q_4 \\ q_5 \\ q_6 \\ q_7 \\ q_8 \\ q_9 \\ q_{10} \\ q_{11} \\ q_{12} \end{pmatrix}
$$

$$
p_m = [B_0 \quad B] \begin{bmatrix} P \\ Q \end{bmatrix}
$$

Step 2. The F_m matrix remains unaltered

Step 3.

$$F = B^T F_m B$$

$$F = \frac{1}{EI}\begin{bmatrix} 1.0 & -.555 & -.444 & -.5 & 0 & -.555 & .166 & -.444 & .334 & 0 & 0 & 0 \\ -.555 & 11.2 & 10.8 & .444 & -10 & .721 & .443 & .778 & -.556 & .555 & .555 & 0 \\ -.444 & 10.8 & 21.2 & .556 & -20 & .778 & .556 & .723 & -.446 & .444 & .444 & 0 \\ -.5 & .444 & .556 & 1.0 & 0 & .444 & .333 & .556 & -.168 & 0 & 0 & 0 \\ 0 & -10 & -20 & 0 & 30 & 0 & 0 & 0 & 0 & 0 & 0 & 0 \\ -.555 & .721 & .778 & .444 & 0 & 11.2 & .443 & 10.8 & -.556 & .555 & -.444 & -10 \\ .166 & .443 & .556 & .333 & 0 & .443 & 10.7 & .556 & -.334 & .166 & -.333 & 0 \\ -.444 & .778 & .723 & .556 & 0 & 10.8 & .556 & 21.2 & -.446 & .444 & -.556 & -20 \\ .334 & -.556 & -.446 & -.168 & 0 & -.556 & -.334 & -.446 & 10.7 & -.333 & .167 & 0 \\ 0 & .555 & .444 & 0 & 0 & .555 & .166 & .444 & -.333 & 10 & -.5 & 0 \\ 0 & -.44 & -.556 & 0 & 0 & -.444 & -.333 & -.556 & .167 & -.5 & 10 & 0 \\ 0 & 0 & 0 & 0 & 0 & -10 & 0 & -20 & 0 & 0 & 0 & 30 \end{bmatrix}$$

$$Q = -F B^T F_m B_0$$

$$Q = -[-.5703 \quad .0117 \quad .0032 \quad .2119 \quad .0060 \quad -.0017 \quad -.0064 \quad -.0082 \quad -.0146 \quad -.0102 \quad -.0060]^t$$

$$P_m = [-.7212 \quad .0089 \quad .2915, \quad .2554 \quad -.0028 \quad .0289, \quad .0489 \quad -.0060 \quad .2119, \quad .0234 \quad -.0064 \quad -.0117, \quad .0064 \quad .0082 \quad -.0032, \quad -.0198 \quad .0089 \quad .0052, \quad 0036 \quad -.0028 \quad .0081, \quad .0101 \quad -.0060 \quad -.0102]^t$$

It may be noticed that the F matrix is much improved in respect of diagonal dominance.

The redundant $q_2, q_3, q_5, q_6, q_8, q_9, q_{11}, q_{12}$ of the release system – 1 are the same as the redundant $q_7, q_6, q_9, q_8, q_5, q_4, q_{13}, q_{11}$ of the second system respectively. Comparing the numerical values of q_2, q_5 and q_9 of system-1 with the values of q_7, q_9 and q_4 of system-2, it may be observed that they are in good agreement. The variation in the results, in the case of the other redundant, may be attributed to the shortcoming of the first release system. The reader may like to evaluate the member forces for the first release system, Fig. Ex. 4.18(b) and compare the member forces given above for the second release system.

4.9 SPACE FRAMES

4.9.1 General, Single Line Frames

Generally we are required to analyse the types of structures that have been discussed in Articles 4.1 to 4.8. The complex space frames are idealized into simple plane frames, continuous beams, etc. In the following, simple space frames, which cannot possibly be idealized into simple plane frames, are considered, Fig. 4.19. The frame of Fig. 4.19(a) is a case of simple space frame consisting of two portals rigidly connected at B. The frames of Fig. 4.19(b), (c), and (d) are the examples of staircases supported at the ends only.

As we have seen, for any analysis we have to (i) introduce suitable releases system so at to write equilibrium matrix based on statical analysis, i.e., we define B_0 B and (ii) define generalized strain stress relation, i.e., the member flexibility matrix, $(f_m)_i$. Thereafter we simply apply the basic steps for a complete analysis.

Let us start with the description of the end-flexibility of a member. The frames of the type, Fig. 4.19 (b), (c), (d), are considered, which are "single line frames", continuous between the end supports, and without interior supports. The arches, Article 3.12, are the examples of single line frames. The bow girder, Fig. 4.17, is also an example of a single line frame idealized by prismatic members continuous between the end supports in the horizontal plane and without any interior supports.

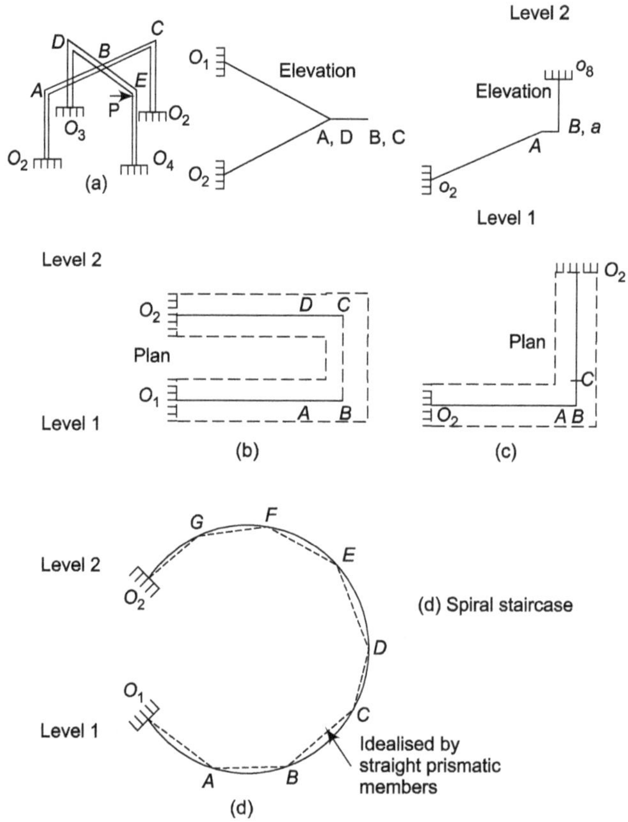

Fig. 4.19 *Simple Examples of Space Frames*

4.9.2 Member Flexibility

Consider tangent at the end 1 in deformed position as reference axis and accordingly obtain end flexibility of a member.

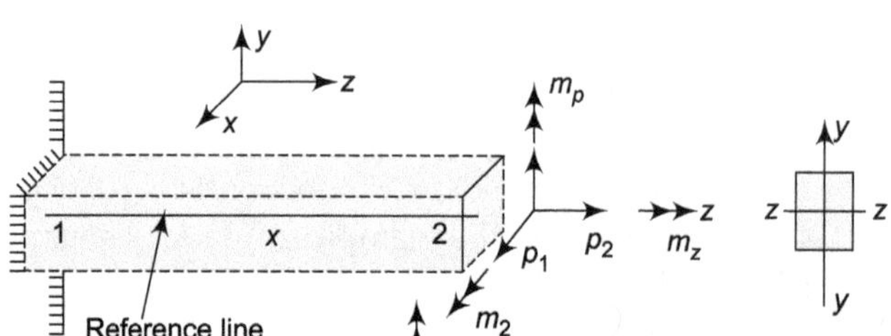

Fig. 4.20 *(Contd.)*

$$\begin{pmatrix} \delta_x \\ \delta_y \\ \delta_z \\ \hline \theta_x \\ \theta_y \\ \theta_z \end{pmatrix} = \left(\begin{array}{c|c|c|c|c|c} L/EA & 0 & 0 & 0 & 0 & 0 \\ 0 & L^3/3EI_x & 0 & 0 & 0 & L^2/2EI_z \\ 0 & 0 & L^3/3EI_y & 0 & -L^2/2EI_y & 0 \\ 0 & 0 & 0 & L/GJ & 0 & 0 \\ 0 & 0 & -L^2/2EI_y & 0 & L/EI_y & 0 \\ 0 & L^2/2EI_z & 0 & 0 & 0 & L/EI_z \end{array}\right) \begin{pmatrix} p_x \\ p_y \\ p_z \\ \hline m_x \\ m_y \\ m_z \end{pmatrix}$$

$$(\delta_m)_2 = \qquad f_m \qquad (p_m)_2$$

Fig. 4.20 *End Flexibility of a Member of a Space Frame*

For the assumed reference axis the member flexibility is obtained by evaluating δ_2 which are relative deformations or the deformations with respect to the tangent at end 1, Fig. 4.20. The reader may like to verify the elements of the f_m matrix as given in Fig. 4.20, where the elements of each column are the deformations at the end 2 due to the corresponding force element equal to unity. The complete end-flexibility matrix is given in Fig. 4.20. The end flexibility of a member of a plane frame. The Eqs (4.9) and (4.10), and grid, Eq. (4.18), can be obtained from the end-flexibility given in Fig. 4.20.

All along we ignored the shear-deformations. If the deformations due to shear are also considered, the elements f_{22} and f_{33}, Fig. 4.20, get modified and we get $f_{22} = \dfrac{L}{GA_y} + \dfrac{L^3}{3EI_x}$ and $f_{23} = \dfrac{L}{GA_y} + \dfrac{L^3}{3EI_y}$

4.9.3 Equilibrium Matrix, Static Analysis

Single line space frames, Article 4.9.1, are considered in the following. The six releases are introduced by a cut at one of the ends. Consider a space frame AN, Fig. 4.21, idealized by a set

$$P_H = [P_x', P_y', P_z', m_x', m_y', m_z']$$
$$[\Delta x', \Delta y', \Delta z']_{DE}$$

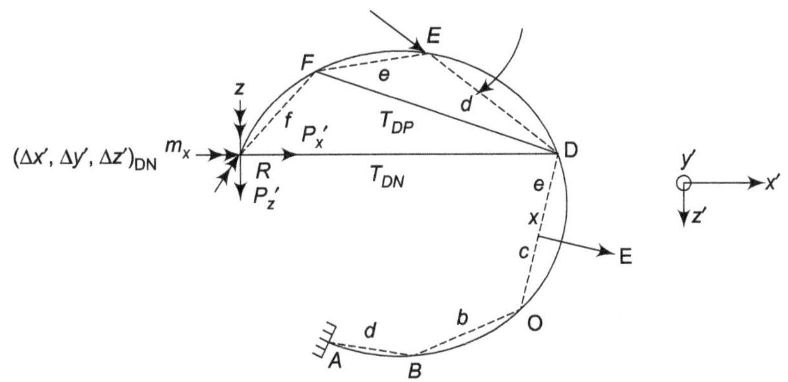

$$Q = [P_x', P_y', P_z', m_x', m_y', m_z']^t_N$$

Fig. 4.21 *Plan of a Single Line Frame*

of straight prismatic members *AB*, *BC*, *CE*, *DE*, *EF* and *FN*. The six releases are introduced at the end, and the corresponding redundants, Q, are shown. The evaluation of the member forces, say for a member *c*, caused by the redundants acting at the end N, defines the six rows (corresponding to the member c) of all the columns of B matrix. The evaluation of such member forces for all the members a, b, c, d, e and f, completely defines the B matrix.

Evaluation of B matrix: Consider a member c for the evaluation of member forces caused by the redundants acting at N, Fig. 4.21. The member forces at the end 2 will be evaluated in two transformations, (i) translational transformation T_r and (ii) rotational transformation $T\theta$. The pre-multiplication of Q with T_r gives the member end forces referring to the system coordinates x', y', z', that is,

$$(P_m') = (T_r)_{DN} Q \qquad \text{...(4.19)}$$

where

$$(T_r)_{DN} = \begin{bmatrix} \multicolumn{3}{c|}{I_{3\times3}} & 0_{3\times3} \\ \hline 0 & -\Delta z' & -\Delta y' & I_{3\times3} \\ \Delta z' & 0 & -\Delta x' & \\ -\Delta y' & \Delta x' & 0 & \end{bmatrix}_{DN}$$

I and o are unit and null matrices respectively. $\Delta x'$, $\Delta y'$, $\Delta z'$ are the distances measured form D to N for the translational transformation and therefore the suffix DN has been used.

The second transformation $T\theta$, transforms p_m' to the p_m referred to the member coordinates, that is,

$$(P_m)_{2c} \text{ due to } Q = (T\theta)_D\,(p_m')_{2c} \qquad \text{...(4.20)}$$

Where

$$(T\theta)_D = \left[\begin{array}{ccc|ccc} \cos xx' & \cos xy' & \cos xz' & & & \\ \cos yx' & \cos yy' & \cos yz' & & 0_{3\times3} & \\ \cos zx' & \cos zy' & \cos xz' & & & \\ \hline & 0_{3\times3} & & \cos xx' & \cos xy' & \cos xz' \\ & & & \cos yx' & \cos yy' & \cos yz' \\ & & & \cos zx' & \cos zy' & \cos xz' \end{array}\right]$$

The matrix $(T\theta)_D$ consists of directions cosines of the member c. From Eqs (4.19) and (4.20),

$$(B)_c = (T\theta)_D\,(T_r)_{DN}$$

The product $T\theta\, T_r$ gives $(B)_c$ a submatrix of B and the submatrix consists of all the six columns but the six rows are corresponding to the member c only.

Evaluation of B_0 matrix: We may continue to keep the B_0 matrix reduced to a single column. Referring to Fig. 4.21 and once again considering the end 2 of the element c, its member forces are due to the nodal loads acting at the nodes D, E and F only. The nodal loads P are described with reference to the system coordinates as we did for the redundants Q. Thus the member forces $(p_m')_{2c}$ due to the external loads only give a sub-column matrix $(B_0)_c$.

$$(P_m)_{2c} \text{ due to loads} = [T_{DF}\,P_F + T_{DE}\,P_E + P_D]$$

and

$$(B_0)_c = (p_m)_{2c} = (T\theta)_D\,[T_{DF}\,P_F + T_{DE}\,P_E + P_D] \qquad \text{...(4.21)}$$

'Complete B_0 matrix' is obtained by assembling (B_0), of different members.

With the B and B_0 matrices known and the member flexibility $(f_m)_i$ already specified, the remaining analysis is simple and direct.

Ex. 4.19. Analyse the staircase, Fig. Ex. 4.19(a). The transverse thickness of the slab = 0.15 m. The geometrical details are given in Fig. Ex. 4.19(a) and line elevation in Fig. Ex. 4.19(c). Live load = 845 Kg/sqm. The staircase, Fig. Ex. 4.19(a), has been idealized to consist of three members a, b and c as shown in Fig. Ex. 4.19 (b), (c) and (d). For the given loading as shown the frames is first analysed for fixed edge condition by clamping the nodes A and B, which gives the nodal loads with respect to system coordinates x', y', z'.

$$P_A = [0 - 2277\ 0 - 289\ 0 + 12.6]^t$$
$$P_B = [0 - 2277\ 0 - 286\ 0 - 12.6]^t$$

The system axes are chosen to coincide with the member axes of the member b, Fig. Ex. 4.19 (c).

Step 1 Equilibrium

$$(B_0)_a = (T\theta)_A\ [(T_r)_{AB}\, P_B + P_A] \quad \text{and} \quad (B_0)_b = (T\theta)_B\, P_B = P_B,$$

Since $(T\theta)_B$ is a unit matrix

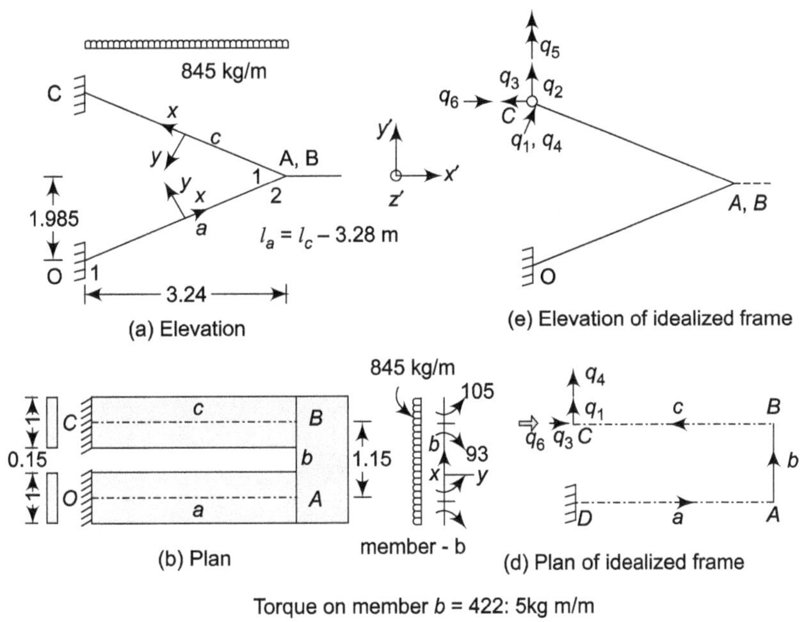

Fig. Ex. 4.19

The redundants are shown in Fig. Ex. 4.19(c) and (d)

$$(B)_a = (T\theta)_A\, (T_r)_{AN}$$
$$(B)_b = (T\theta)_B\, (T_r)_{BN} = (T_r)_{BN}$$
$$(B)_c = (T\theta)_N$$

$$P_m = (B_0\ B) \begin{bmatrix} 1 \\ Q \end{bmatrix}$$

Where $$B_0 = \begin{bmatrix} (B_0)_a \\ (B_0)_b \\ (B_0)_c \end{bmatrix}_{18\times 1} \quad \text{and } B = \begin{bmatrix} (B)_a \\ (B)_b \\ (B)_c \end{bmatrix}_{18\times 6}$$

The reader may verify the following transformation matrices in the sequence of their appearance above. Refer Fig. Ex. 4.19, and Eqs (4.19), (4.20) and (4.21)

$$(T\theta)_A = \left[\begin{array}{ccc|ccc} 0 & \sin\alpha & \cos\alpha & & & \\ 0 & \cos\alpha & -\sin\alpha & & & \\ -1 & 0 & 0 & & 0_{3\times 3} & \\ \hline & 0_{3\times 3} & & 0 & \sin\alpha & \cos\alpha \\ & & & 0 & \cos\alpha & -\sin\alpha \\ & & & -1 & 0 & 0 \end{array}\right]$$

$$(T\theta)_B = \left[\begin{array}{ccc|c} & I_{3\times 3} & & 0_{3\times 3} \\ \hline 0 & 0 & 0 & I_{3\times 3} \\ 0 & 0 & -1.15 & \\ 0 & 1.15 & 0 & \end{array}\right]$$

$$(T_r)_{AB}\, P_B + P_A = [0, -4554, 0, -572, 0, -2277 \times 1.15]^t$$

$$(B_0)_b = [(-4554 \sin\alpha), (-4554 \cos\alpha), 0,$$
$$(-2277 \times 1.15 \cos\alpha), (2277 \times 1.15 \sin\alpha)572]^t$$

$$(B_0)_b = P_B = [0 - 2277\ 0 - 286\ 0 - 12.5]^t$$

And $(B_0)_c = [0\ 0\ 0\ 0\ 0\ 0]^t$ together with $(B_0)_a$ and $(B_0)_b$ specifies completely the B_0 matrix.

$$(T\theta)_B = \text{Unit matrix}$$

$$(T_r)_{AN} = \left[\begin{array}{ccc|c} & I_{3\times 3} & & 0_{3\times 3} \\ \hline 0 & 3.24 & -1.985 & I_{3\times 3} \\ -3.24 & 0 & -1.15 & \\ -1.985 & 1.15 & 0 & \end{array}\right]$$

$$(B)_a = (T\theta)_A\,(T_r)_{AN}$$

$$= \left[\begin{array}{ccc|ccc} 0 & \sin\alpha & \cos\alpha & 0 & 0 & 0 \\ 0 & \cos\alpha & -\sin\alpha & 0 & 0 & 0 \\ -1 & 0 & 0 & 0 & 0 & 0 \\ \hline -3.384 & 1.15\cos\alpha & -1.15\sin\alpha & 0 & \sin\alpha & \cos\alpha \\ -1.722 & -1.15\sin\alpha & -1.15\cos\alpha & 0 & \cos\alpha & -\sin\alpha \\ 0 & -3.24 & -1.985 & -1 & 0 & 0 \end{array}\right]$$

$$(B)_b = (T_r)_{BN} \left[\begin{array}{ccc|c} & I_{3\times 3} & & 0_{3\times 3} \\ \hline 0 & 3.24 & -1.985 & I_{3\times 3} \\ -3.24 & 0 & 0 & \\ -1.985 & 0 & 0 & \end{array}\right]$$

Lastly,

$$(B)_c = (T\theta)_N = (T\theta)_A = \left[\begin{array}{ccc|ccc} 0 & \sin\alpha & -\cos\alpha & & & \\ 0 & -\cos\alpha & -\sin\alpha & & & \\ -1 & 0 & 0 & & 0_{3\times 3} & \\ \hline & 0_{3\times 3} & & 0 & \sin\alpha & -\cos\alpha \\ & & & 0 & -\cos\alpha & -\sin\alpha \\ & & & -1 & 0 & 0 \end{array}\right]$$

$$B = \begin{bmatrix} (B)_a \\ (B)_b \\ (B)_c \end{bmatrix}$$

And the equilibrium matrix is thus fully specified as given below.

$$\begin{bmatrix} (p_m)_a \\ (p_m)_b \\ (p_m)_c \end{bmatrix} = \begin{bmatrix} -2380.9 & 0 & 0.523 & 0.852 & 0 & 0 & 0 \\ -3882.8 & 0 & 0.852 & -0.523 & 0 & 0 & 0 \\ 0 & -1 & 0 & 0 & 0 & 0 & 0 \\ -2293.0 & -3.384 & 0.980 & -0.601 & 0 & 0.523 & 0.852 \\ 1408.0 & -1.722 & -.601 & -0.980 & 0 & 0.852 & -.523 \\ 572.0 & 0 & -3.24 & -1.985 & -1 & 0 & 0 \\ - & - & - & - & - & - & - \\ 1 & 1 & 0 & 0 & 0 & 0 & 0 \\ -227.0 & 0 & 1 & 0 & 0 & 0 & 0 \\ 0 & 0 & 0 & 1 & 0 & 0 & 0 \\ -286.0 & 0 & 3.24 & 1.985 & 1 & 0 & 0 \\ 0 & -3.24 & 0 & 0 & 0 & 1 & 0 \\ -12.5 & -1.985 & 0 & 0 & 0 & 0 & 1 \\ - & - & - & - & - & - & - \\ 0 & 0 & 0.823 & -.852 & 0 & 0 & 0 \\ 0 & 0 & -0.852 & -.523 & 0 & 0 & 0 \\ 0 & -1 & 0 & 0 & 0 & 0 & 0 \\ 0 & 0 & 0 & 0 & 0 & 0.823 & -.852 \\ 0 & 0 & 0 & 0 & 0 & -.852 & -.523 \\ 0 & 0 & 0 & 0 & -1 & 0 & 0 \end{bmatrix}$$

P_m $\qquad [B_0 \qquad\qquad B] \qquad\qquad \begin{bmatrix} 1 \\ Q \end{bmatrix}$

Step 2

$$F_m = \left[(f_m)_a \quad (f_m)_b \quad (f_m)_c\right]$$

Flexibility matrix of a member is given in Fig. 4.20.

From the properties given in Fig. Ex. 4.19

$$\left.\begin{aligned} EA &= 0.15\text{E} \\ GJ &= .0006\text{E} \\ EI_z &= .00028\text{E} \\ EI_y &= .012\text{E} \end{aligned}\right\} \text{same for all the members}$$

$$L_a = L_c = 3.8 \text{ m}, \qquad L_b = 1.15 \text{ m}$$

Step 3

$$F = B^T F_m B = \frac{10^5}{E} \times \begin{bmatrix} .916 & -.257 & .130 & 0 & -.120 & -.268 \\ -.257 & 1.23 & .961 & .502 & .308 & .0775 \\ .130 & .961 & 1.53 & .577 & -.0231 & -.0308 \\ 0 & .502 & .577 & .291 & 0 & 0 \\ -.120 & .0308 & -.0231 & 0 & .0402 & 0 \\ -.268 & .0775 & -.0308 & 0 & 0 & .135 \end{bmatrix}$$

$$Q = -F^{-1}(B^T F_m B_0)$$

$$Q = -10^4\,[.00268 - .22776\ .34379\ .02659\ .20085\ .08105]^t$$

Member forces

$$\begin{aligned} P_m = 10^4\,[&-.41211 \quad -.01440 \quad .00268, \\ &.03550 \quad .21659 \quad .02827, \\ &-.00268 \quad .00006 \quad -.34379, \\ &.00033 \quad -.19217 \quad -.07698, \\ &.41214 \quad -.01444 \quad .00268, \\ &-.03591 \quad .21359 \quad .02659]^t \end{aligned}$$

To the member forces computed above, the fixed edge forces due to the given loading are to be added to get the final member forces.

Ex. 4.20. Analyse the frame, Fig. Ex. 4.20. The members a and b have the same cross-sectional properties.

$EA = 1.6 \times 10^5$, $GJ = 1065$, $EI_x = 2130$, $EI_y = 2130$, $L_a = 5$ m, $L_b = 2$ m. The system coordinate are chosen to be the same as the member coordinates of b. The six releases are introduced at the end of the member b.

Results:

$$B_0 = [-2.1\ -2.8\ 0\ -.27\ .2\ 1.67\ 0\ 0\ 0\ 0\ 0\ 0]^t$$

$$B = \begin{bmatrix} 0 & 0.6 & 0.8 & 0 & 0 & 0 \\ 0 & 0.8 & -.6 & 0 & 0 & 0 \\ -1 & 0 & 0 & 0 & 0 & 0 \\ 0 & 1.6 & -1.2 & 0 & 0.6 & 0 \\ 0 & -1.2 & -1.6 & 0 & 0.8 & -.6 \\ 0 & 0 & 0 & -1 & 0 & 0 \\ 1 & 0 & 0 & 0 & 0 & 0 \\ 0 & 1 & 0 & 0 & 0 & 0 \\ 0 & 0 & 1 & 0 & 0 & 0 \\ 0 & 0 & 0 & 1 & 0 & 0 \\ 0 & 0 & 0 & 0 & 1 & 0 \\ 0 & 0 & 0 & 0 & 0 & 1 \end{bmatrix}$$

$$Q = -\,[.00084 \quad -1.8959 \quad 1.1777 \quad -.12689 \quad 2.1558 \quad 3.1971]^t$$

$$p_m = [-1.9046 \quad -.5766 \quad .0008, \quad .3256 \quad .0028 \quad 1.5431, \quad -.0008 \quad 1.8959 \quad -1.1777, \quad .1269 \quad -2.1558 \quad -3.1971]^t$$

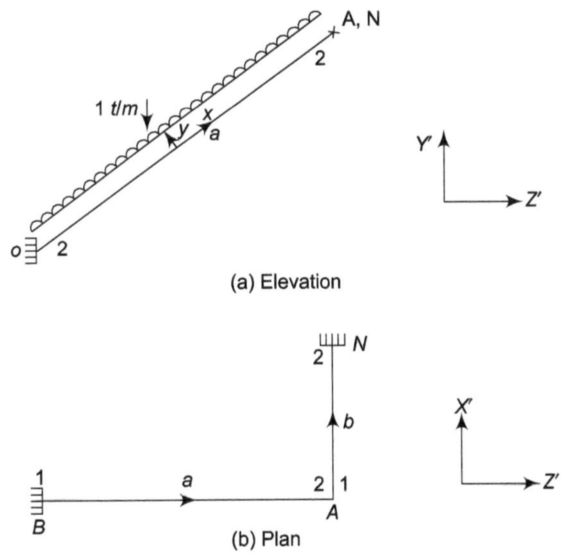

Dead load = 1 t/m over entire length

Fig. Ex. 4.20

To the member forces computed above, add the fixed edge forces and draw the moment, torque and thrust diagram.

4.10 UNIFIED FORCE METHOD–STATICS AND ENERGY PRINCIPLES

The method given in the following is essentially based on the 'Rank-Force method' given by Robinson*. The difference is in the more economical procedure, while obtaining the equations of equilibrium. For the plane frames, Robinson includes shears and thrusts along with moments in the equations of equilibrium for all the members and joints constituting the frame. He reduces these equations to an independent set. This technique has been discussed in Article 2.4. In the following, however, the independent equations of equilibrium are written directly, by making use of the basic mechanisms discusses in the preceding articles. The internal forces appearing in the equilibrium equations are only the moments $q_1, q_2, ..., q_m$ acting at the ends of different members. If a given structure is indeterminate the number of independent equations of equilibrium are less than the number of internal moments, q_i. The additional equations, their number equal to the degree of indeterminacy, are obtained from energy principle, 'minimum complementary'. The second half, that is, formulation of n equations based on the minimum complementary potential is essentially same as given by Robinson. These two halves of the unified force method are given in the following Articles, 4.10.1 and 4.10.2. The unknown's q_1 to q_m are evaluated by Jordan elimination procedure.

4.10.1 Independent Equations of Equilibrium

In the Article 3.6 and elsewhere we have observed that if the number of maximum moment points is m and degree of indeterminacy is n, the number of independent equations of equilibrium, $r = m - n$. And therefore the corresponding number of basic mechanisms for writing equations of equilibrium is $m - n$. However, this observation is valid when the members are inextensible and flexural. In the case of extensible members like springs used as support constraints in addition to all other inextensible members, we modify the resulting expression giving the number of independent equilibrium equations by the following.

Number of independent equations of equilibrium,

$$r = m - n + e$$

where e is the number of springs.

Consider the frame, Fig. 4.22. The number of maximum moment point is 16. Here, even or the joints, D, F, L and H each connecting two flexural members only, the end members are specified separately, and we connected by joint equilibrium.

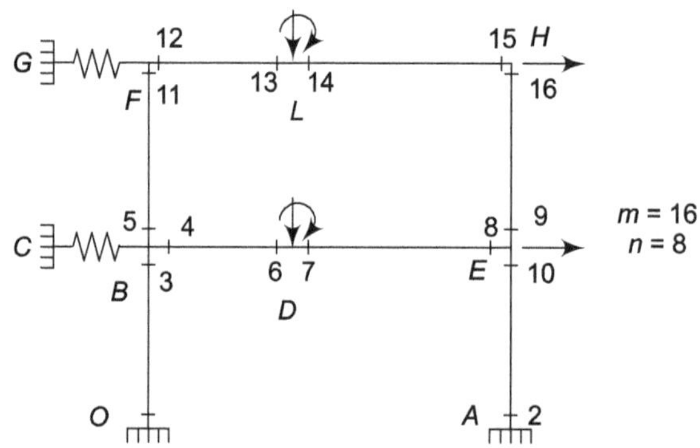

Fig. 4.22 *Rigid Jointed Frame with Lateral Spring Supports*

*"Structural Matrix Analysis for the Engineers", John Robinson, John Wiley & Sons, 1966.

Degree of indeterminacy, $n = 8$, and number of springs, $e = 2$ at G and C.

Number of independent equations of equilibrium, $r = 16 - 8 + 2 = 10$ and they are obtained by considering,

Six joint mechanisms corresponding to the end moments meeting at the joints B, D, E, F, L and H,

Two beam mechanisms corresponding to the moments of the beams BDE and FLH, and

Two sway mechanisms.

Total number of mechanisms = 6 + 2 + 2 = 10.

Also consider a frame, Fig. 4.23.

Number of independent equations of equilibrium,

$$r = m - n + e = 6.$$

where $m = 6, n = 2, e = 2$

Three joint mechanisms (treat E as joint) and one beam mechanism are obvious. The fifth, sway mechanism, is shown dotted which includes the constraining force due to the springs CB and EF. The sixth mechanism is sway of member DE, which is similar to the sway of member FF, Fig. 4.24.

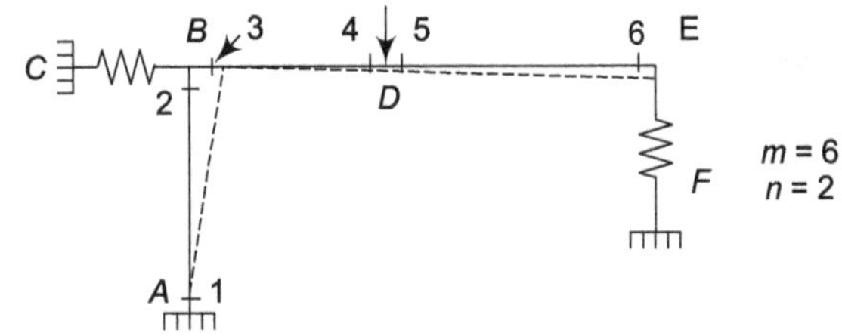

Fig. 4.23 *Lateral Spring Supports and Rigid Support*

As a further illustration, consider a frame, Fig. 4.24.

$$m = 17, n = 9, e = 3$$

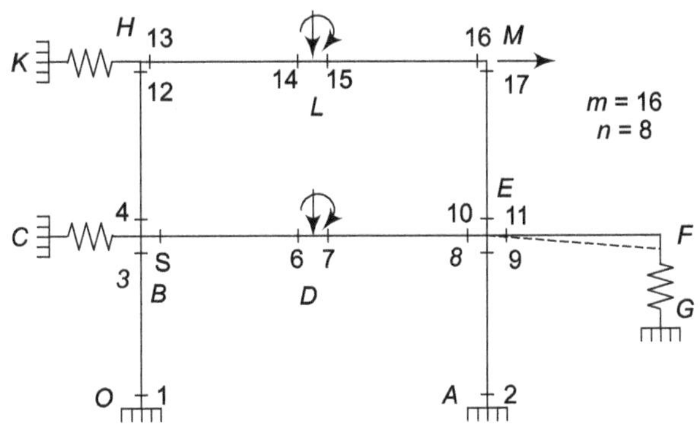

Fig. 4.24 *Spring Support at Cantilever End*

Number of independent equations of equilibrium = 11.

The six joint mechanisms, two beam mechanisms and two sway mechanisms for the two storeys totaling to ten are obvious. And these do not include the constraining force due to the spring FG. The mechanism for the eleventh equation of equilibrium is shown dotted, which consists of the moment q_{11} and the constraining orce due to the spring FG.

The formulation of the equilibrium equation

$$B_{rX\,(m+e)}\, q_{(m+e)X1} = P \qquad \text{...(a)}$$

is illustrated with the help of an example, Fig. Ex. 4.41, at the end of the Article.

4.10.2 Energy Equations

For the evaluation of q_i, consisting of the moments, and reactions in the case of spring constraints, in addition to the independent equations of equilibrium, we need to formulate n equations based on energy principles. We use the principle of minimum complementary total potential, Chapter I and Summary, page 238. Based on the principle of minimum complementary total potential,

$$\frac{\partial \pi}{\partial R_i} = \frac{\partial U}{\partial R_i} - \delta_i = 0 \qquad \text{.....(a)}$$

for $i = 1$ to n. The forces R_i as redundants are included in q_i, refer Art. 1.6.6 and page 238 Eq. (b).

Robinson has shown an automatic generation of λ matrix. The procedure of generation consists of:

(i) As explained in Article 2.4, the matrix B is linearly transformed, using Jordan elimination procedure, which should result into r number of columns having unity in each of these columns and no two unities appear in one row. All other elements of these r columns are zeros. At the end of this step the remaining n columns, not normalized, are isolated as redundancy columns. These redundancy columns are attributed to R_1 , R_2 , ..., R_n in the sequence of their appearance from left to right. Let us designate the matrix consisting of the redundancy columns by R. At this stage, we also know which of the q are treated as R_1, R_2, ..., R_n.

(ii) Obtain B^0 matrix by replacing the columns of B corresponding to R_1, R_2, ..., R_n with zero. Premultiply B^0 matrix by R^t and then the resulting product is multiplied by -1. Let

$$S = -\,R^t\,B^0$$

(iii) The n number of columns, corresponding to the redundants, of the S matrix are null columns. In each of these null columns place 1 in specified locations, that is, the first null column will have unity in the first row, the second null column will have unity in the second row and so on, till the n-th null column has unity placed in the n-th row. Such a resulting matrix may be verified to be the same as λ appearing in Eq. (b2).

$$\frac{\partial U}{\partial R_1} = \frac{\partial U}{\partial q_1}\frac{\partial q_1}{\partial R_1} + \frac{\partial U}{\partial q_2}\frac{\partial q_2}{\partial R_1} + \cdots + \frac{\partial U}{\partial R_1} + \cdots + \frac{\partial U}{\partial q_{r+n}}\frac{\partial q_{r+n}}{\partial R_1}$$

$$\vdots$$

$$\frac{\partial U}{\partial R_k} = \frac{\partial U}{\partial q_1}\frac{\partial q_1}{\partial R_k} + \frac{\partial U}{\partial q_2}\frac{\partial q_2}{\partial R_k} + \cdots + \frac{\partial U}{\partial R_1}\frac{\partial R_1}{\partial R_k} + \cdots + \frac{\partial U_1}{\partial R_k} + \cdots + \frac{\partial U}{\partial q_{r+n}}\frac{\partial q_{r+n}}{\partial R_k}$$

$$\vdots$$

$$\frac{\partial U}{\partial R_n} = \frac{\partial U}{\partial q_1}\frac{\partial q_1}{\partial R_n} + \frac{\partial U}{\partial q_2}\frac{\partial q_2}{\partial R_n} + \cdots + \frac{\partial U}{\partial R_1}\frac{\partial R_1}{\partial R_n} + \cdots + \frac{\partial U_1}{\partial R_n} + \cdots + \frac{\partial U}{\partial q_{r+n}}\frac{\partial q_{r+n}}{\partial R_n}$$

When written in matrix form,
Redundancy column

$$\begin{bmatrix} \dfrac{\partial U}{\partial R_1} \\ \vdots \\ \dfrac{\partial U}{\partial R_k} \\ \vdots \\ \dfrac{\partial U}{\partial R_n} \end{bmatrix} = \begin{bmatrix} \dfrac{\partial q_1}{\partial R_1}\dfrac{\partial q_2}{\partial R_1}\ldots 1 \ldots \dfrac{\partial R_k}{\partial R_1}\cdots\dfrac{\partial R_n}{\partial R_1}\cdots\dfrac{\partial q_{r+n}}{\partial R_1} \\ \vdots \\ \dfrac{\partial q_1}{\partial R_k}\dfrac{\partial q_2}{\partial R_k}\cdots\dfrac{\partial R_1}{\partial R_k}\ldots 1 \ldots\dfrac{\partial q_{r+n}}{\partial R_k} \\ \vdots \\ \dfrac{\partial q_1}{\partial R_n}\dfrac{\partial q_2}{\partial R_n}\cdots\dfrac{\partial R_1}{\partial R_r}\cdots\dfrac{\partial R_k}{\partial R_n}\ldots 1 \ldots\dfrac{\partial q_{r+n}}{\partial R_n} \end{bmatrix} \begin{bmatrix} \dfrac{\partial U}{\partial q_1} \\ \dfrac{\partial U}{\partial q_2} \\ \vdots \\ \dfrac{\partial U}{\partial R_1} \\ \vdots \\ \dfrac{\partial U}{\partial q_{r+n}} \end{bmatrix} \quad \text{...(b1)}$$

We may write the equation ($b1$) in the contracted form as

$$\frac{\partial U}{\partial R_k} = \lambda \frac{\partial U}{\partial q_1} \quad \text{...(b2)}$$

SUMMARY OF ANALYSIS USING ENERGY PRINCIPLES

Virtual Work Principle

1. Real force F_i and q_i ride virtual deformations δ_i
2. The principle is used for satisfying equilibrium
3. Minimum total potential π

   $$\frac{\partial \pi}{\partial \delta_i} = 0$$

   π = strain energy U + potential V due to loads

Complementary Virtual Work Principle

1. Virtual forces F_i and q_i ride real deformations δ_i
2. The Principle is used for computing real deformations like deflections along virtual or dummy loads, flexibilities f_{ij} and u_i^0
3. Minimum total complementary potential π^*

   $$\frac{\partial \pi^*}{\partial q_i} = \frac{\partial U^*}{\partial q_i} + \frac{\partial V}{\partial q_i} = 0$$

   π^* =Complementary strain-energy U^* + potential V due to loads

Note for Unified Force Method

In the case of elastic systems $U = U^*$

If the loads F_i are specified, and there are no relative deformations along the unspecified redundanats, then for elastic systems.

$$\frac{\partial V}{\partial R_i} = 0 \text{ and as } R_i \text{ are included in } q_i \text{ set,}$$

$$\frac{\partial \pi}{\partial q_i} = \frac{\partial U}{\partial q_i} = 0 \text{ (Minimum Strain Energy Principle)}$$

And if there are settlements, lack of fit and temperature effects, and these give rise to relative deformations δ_i along the releases corresponding to redundants R_i then for elastic systems

$$\frac{\partial V}{\partial R_i} = \frac{-\partial}{\partial R_i}\left(\sum R_i\,\delta_i\right) = -\,\delta_i$$

$$\therefore \qquad \frac{\partial \pi}{\partial R_i} = \frac{\partial U}{\partial R_i} - \delta_i = 0$$

Jordan elimination can be used such that ill conditioning is avoided, and R_i among q_i are chosen automatically without additional effort.

From Eqs (a) and (b2) we obtain the following Eq. (c) in contracted form

$$\frac{\partial \pi}{\partial R_k} = \lambda\frac{\partial U}{\partial q_i} - \delta_i = 0 \qquad \text{...(c)}$$

Eq. (c) results in n equations, but we have not yet explicitly introduced q_i in Eq. (c). To achieve this, we know,

$$U = \frac{1}{2}[q_1\, q_2 \cdots q_{r+n}]\, F_m [q_1\, q_2 \cdots q_{r+n}]^t$$

Where $F_m = [(f_m)_a, (f_m)_b, \cdots (f_m)_1 \cdots (f_m)_2]$

Where s is the number of members including extensible members as springs.

For inextensible members

$$f_m = \begin{bmatrix} \frac{L}{3EI} & \frac{L}{6EI} \\ \frac{L}{6EI} & \frac{L}{3EI} \end{bmatrix} \text{ and}$$

for springs $f_m = \frac{1}{K}$ where k the spring constant.

And $$\frac{\partial U}{\partial R_k} = F_m\, q^t$$

Substituting in Eq. (c)

$$\lambda F_m \begin{bmatrix} q_1 \\ q_2 \\ \vdots \\ q_{r+n} \end{bmatrix} - \begin{bmatrix} \delta_1 \\ \delta_2 \\ \vdots \\ \delta_n \end{bmatrix} = 0 \qquad \text{...(d)}$$

Thus Eq. (d) together with Eq. (a) of previous Article. 4.10.1 provide as many equations as the number of internal forces q_i. Combining the two equations.

$$\begin{bmatrix} B \\ \cdots \\ \lambda F_m \end{bmatrix} \begin{bmatrix} q_1 \\ q_2 \\ \vdots \\ R_1 \\ \vdots \\ R_k \\ \vdots \\ R_n \\ \vdots \\ q_{r+n} \end{bmatrix} = \begin{bmatrix} P_1 \\ P_2 \\ \vdots \\ P_r \\ \cdots \\ \delta_1 \\ \delta_2 \\ \vdots \\ \delta_n \end{bmatrix} \qquad \text{...(e}_1\text{)}$$

Or in the contracted form

$$\begin{bmatrix} B \\ \cdots\cdots \\ \lambda F_m \end{bmatrix}_q = \begin{bmatrix} P \\ \cdots \\ \delta \end{bmatrix} \qquad \text{...(e}_2\text{)}$$

And $\{\delta\} = \{0\}$, if there is no temperature effect/lack of fit/settlement. The Jordan elimination is further continued till the $\begin{bmatrix} B \\ \cdots\cdots \\ \lambda F_m \end{bmatrix}$ matrix is reduced to 'scrambled' unit matrix and the corresponding right hand column matrix of the equation gives the elements of q matrix, that is, the internal forces in the scrambled order.

In the following, some typical illustrations are analysed by the above proposed to elucidate the unified force method. The reader would infer that

(i) the unified method overcomes the difficulty of choosing proper releases as experienced by the conventional flexibility method,

(ii) the m_0, m_i diagrams are not required to be drawn and instead write r, independent equations of equilibrium, and

(iii) additional n equations are obtained automatically from B matrix and the flexibility matrix F_m.

The above presentation of the 'energy equations' is more elaborately explained through the illustrations for comprehending the theory under lining the 'Unified force method' – Statics and Energy Principles.

Illustration 4.41. Analyse the frame, Fig. Ex. 4.21, by the unified force method. $KL^3/EI = 1$

Equilibrium Equations, Art. 3.10.1

$$M = 6,\ e = 1 \text{ and } n = 2 \text{ Fig. 3.41(b)},$$

Number of independent equations of equilibrium,

$$R = m - n + e = 5.$$

Corresponding five basic mechanisms for writing independent equations of equilibrium are:

Three joint mechanisms (Joint B, F and C)

One beam mechanism (Beam BFC)

One sway mechanism.

Based upon the basic mechanism, equilibrium equations are written in the matrix form, Eq. (i). The internal forces,

Their number equal to $r + n = 7$, are $q_1, q_2, \ldots, q_7$, Fig. Ex.4.21 (b)

$$\begin{array}{rc}
 & \begin{matrix}1 & \;\;2 & \;\;3 & \;\;4 & \;\;5 & \;\;6 & \;\;7\end{matrix} \\
\begin{array}{r}\text{Joint Mechanism}\left\{\begin{array}{c} \\ \\ \\ \end{array}\right. \\ \text{Beam Mechanism} \\ \text{Sway Mechanism}\end{array} &
\begin{bmatrix}
1 & -1 & 0 & 0 & 0 & 0 & 0 \\
0 & 0 & 1 & -1 & 0 & 0 & 0 \\
0 & 0 & 0 & 0 & 1 & -1 & 0 \\
0 & -1 & 2 & 0 & -1 & 0 & 0 \\
\frac{2}{3} & 0 & 0 & 0 & 0 & -1 & L
\end{bmatrix}
\end{array}
\begin{bmatrix} q_1 \\ q_2 \\ q_3 \\ q_4 \\ q_5 \\ q_6 \\ q_7 \end{bmatrix}
=
\begin{bmatrix} 0 \\ 0 \\ 0 \\ \frac{PL}{2} \\ \frac{PL}{5} \end{bmatrix}$$

$$B \qquad\qquad q \qquad = \qquad p \qquad\qquad \text{...(ia)}$$

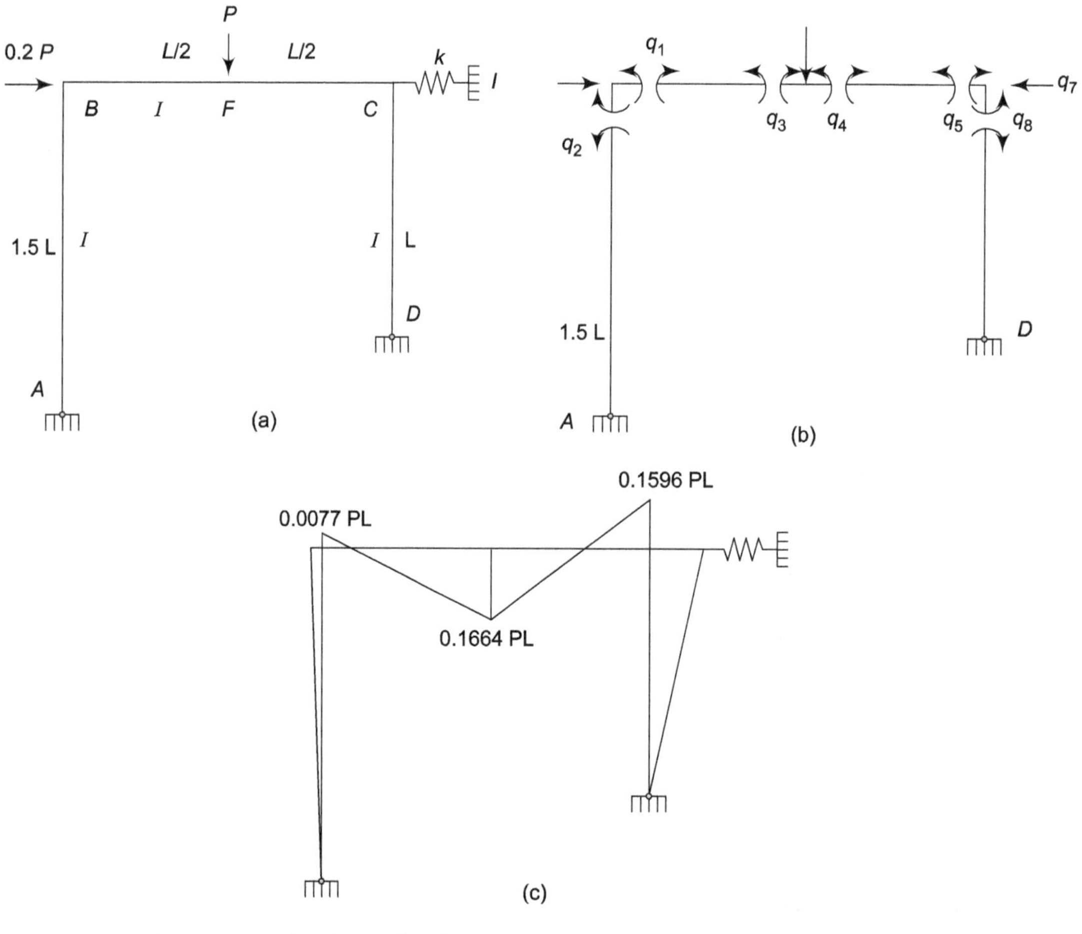

Fig. Ex 4.21 *Single Portal with Hinge Supports and Lateral Spring at Rigid Joint*

Normalizing the columns 2, 3, 4, 5 and 6, we obtain

$$\begin{array}{ccccccc} R_1 & 2 & 3 & 4 & 5 & 6 & R_2 \end{array}$$

$$\begin{bmatrix} -1.00 & \boxed{1} & 0 & 0 & 0 & 0 & 0 \\ -.833 & 0 & 0 & \boxed{1} & 0 & 0 & -.5L \\ -.667 & 0 & 0 & 0 & \boxed{1} & 0 & -L \\ -.833 & 0 & \boxed{1} & 0 & 0 & 0 & -.5L \\ -.667 & 0 & 0 & 0 & 0 & \boxed{1} & -L \end{bmatrix} \begin{bmatrix} q_1 \\ q_2 \\ q_3 \\ q_4 \\ q_5 \\ q_6 \\ q_7 \end{bmatrix} = \begin{bmatrix} 0 \\ .15PL \\ -.2PL \\ .15\,PL \\ -.2PL \end{bmatrix}$$

Or $\qquad \overline{B} \qquad q = \overline{P} \qquad$ ...(ib)

In contrast to B in (*ia*), when normalized it is designated $\overline{B}$

$$R^T = \begin{bmatrix} -1.0 & -.833 & -.667 & -.833 & -.667 \\ 0 & -.5L & -L & -.5L & -L \end{bmatrix}$$

$$B^0 = \begin{bmatrix} 0 & \boxed{1} & 0 & 0 & 0 & 0 & 0 \\ 0 & 0 & 0 & \boxed{1} & 0 & 0 & 0 \\ 0 & 0 & 0 & 0 & \boxed{1} & 0 & 0 \\ 0 & 0 & \boxed{1} & 0 & 0 & 0 & 0 \\ 0 & 0 & 0 & 0 & 0 & \boxed{1} & 0 \end{bmatrix}$$

$$S = R^T B^0 = \begin{bmatrix} 0 & 1 & .833 & .833 & .667 & .667 & 0 \\ 0 & 0 & .500 & .500 & 1.00 & 1.00 & 0 \end{bmatrix}$$

λ is the same as S except that we make $S_{11} = 1$ and $S_{27} = 1$ in place of each being equal to zero in the S matrix.

$$F_m = \frac{L}{6EI}\left[3, \begin{bmatrix} 1 & 1/2 \\ 1/2 & 1 \end{bmatrix}, \begin{bmatrix} 1 & 1/2 \\ 1/2 & 1 \end{bmatrix}, 2, 6L^2 \right]$$

$$\lambda F_m = \begin{bmatrix} 1 & .472 & .444 & .388 & .361 & 0.449 & 0 \\ 0 & 0.42 & .083 & .167 & .210 & 0.334 & L \end{bmatrix}$$

From Eq. (e_2), Article 4.10.2

$$\begin{bmatrix} \overline{B} \\ \lambda F_m \end{bmatrix} q = \begin{bmatrix} \overline{P} \\ Q \end{bmatrix} \qquad \text{...(ii)}$$

Normalizing the columns 1 and 7, and rows 6 and 7 of the coefficient matrix of q, Eq. (ii), that is continuing with the Jordan elimination procedure, we get

$$\begin{bmatrix} 0 & 1 & 0 & 0 & 0 & 0 & 0 \\ 0 & 0 & 0 & 1 & 0 & 0 & 0 \\ 0 & 0 & 0 & 0 & 1 & 0 & 0 \\ 0 & 0 & 1 & 0 & 0 & 0 & 0 \\ 0 & 0 & 0 & 0 & 0 & 1 & 0 \\ \boxed{1} & 0 & 0 & 0 & 0 & 0 & 0 \\ 0 & 0 & 0 & 0 & 0 & 0 & \boxed{1} \end{bmatrix} \begin{bmatrix} q_1 \\ q_2 \\ q_3 \\ q_4 \\ q_5 \\ q_6 \\ q_7 \end{bmatrix} = \begin{bmatrix} -.0077 \\ .1664 \\ -.1596 \\ .1664 \\ -.1596 \\ -.0077 \\ \left(\dfrac{.0455}{L}\right) \end{bmatrix} PL$$

$$q^t = PL\,[-.0077, -.0077, .1654, .1664, -.1596, -.1596, .0455/\text{L}]$$

The elements of q define completely the bending moment ordinates, Fig. Ex. 4.21(c).

Illustration 4.22. Analyse the frame, Fig. Ex. 4.22 (a)

The internal moments, $qt = [q_1, q_2, \ldots, q_7]$ are shown in Fig. 9(b).

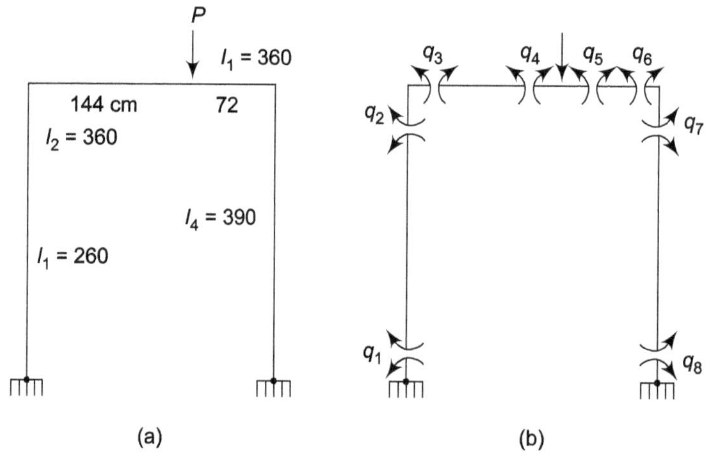

Fig. Ex. 4.22

Number of independent equations of equilibrium, $r = 5$. Based on joint mechanisms (= 3), beam mechanism (= 1) and sway mechanism (= 1), we obtain the following equilibrium equation.

$$\begin{bmatrix} 0 & 1 & -1 & 0 & 0 & 0 & 0 & 0 \\ 0 & 0 & 0 & 1 & -1 & 0 & 0 & 0 \\ 0 & 0 & 0 & 0 & 0 & 1 & -1 & 0 \\ 0 & 0 & -1 & 3 & 0 & -2 & 0 & 0 \\ 0 & -1 & 0 & 0 & 0 & 0 & 1 & -1 \end{bmatrix} \begin{bmatrix} q_1 \\ q_2 \\ q_3 \\ q_4 \\ q_5 \\ q_6 \\ q_7 \\ q_8 \end{bmatrix} = \begin{bmatrix} 0 \\ 0 \\ 0 \\ 0 \\ 144P \\ 0 \end{bmatrix}$$

choosing first, seventh and eight as redundancy column

$$\begin{matrix} 1 & 2 & 3 & 4 & 5 & 6 & 7 & 8 \end{matrix}$$

$$\underbrace{\begin{bmatrix} -1 & \boxed{1} & 0 & 0 & 0 & 0 & -1 & 1 \\ -\frac{1}{3} & 0 & 0 & \boxed{1} & 0 & 0 & -1 & \frac{1}{3} \\ 0 & 0 & 0 & 0 & 0 & \boxed{1} & -1 & 0 \\ -\frac{1}{3} & 0 & 0 & 0 & \boxed{1} & 0 & -1 & \frac{1}{3} \\ 1 & 0 & -1 & 0 & 0 & 0 & 1 & -1 \end{bmatrix}}_{B} q = \begin{bmatrix} 0 \\ 48P \\ 0 \\ 48P \\ 0 \end{bmatrix}$$

$$q = \bar{P} \qquad \text{...(i)}$$

$$S = -R^T B^0 = \begin{bmatrix} 0 & 1 & 1 & \frac{1}{3} & \frac{1}{3} & 0 & 0 & 0 \\ 0 & 1 & 1 & 1 & 1 & 1 & 0 & 0 \\ 0 & -1 & -1 & -\frac{1}{3} & -\frac{1}{3} & 0 & 0 & 0 \end{bmatrix}$$

$$\lambda F_m = \begin{bmatrix} 0.6 & 0.6 & .155 & 0.11 & 0.022 & 0.011 & 0 & 0 \\ 0.2 & 0.4 & .200 & 0.20 & 0.100 & 0.100 & .266 & .133 \\ -.2 & -.4 & -.155 & -.11 & -.022 & -.011 & .133 & .266 \end{bmatrix}$$

and solving the equation, $\begin{bmatrix} \bar{B} \\ \lambda F_m \end{bmatrix} q = \begin{bmatrix} \bar{P} \\ 0 \end{bmatrix}$, we get

$$q^t = P[7.682, -11.862, -11.852, 34.218, 34.218, -14.768, -14.768, 4.775]$$

Angle discontinuities, $\hat{\theta}$ caused by linear settlement Δ: On the merit of the release, we chose Hinge releases and basic mechanisms for writing the equations of equilibrium. As a consequence of the choice of releases, we need to compute the angle discontinuities due to linear settlements. Consider frame Fig. 4.25. The degree of indeterminacy of the Frame being considered is two. Given: Support E is a hinge.

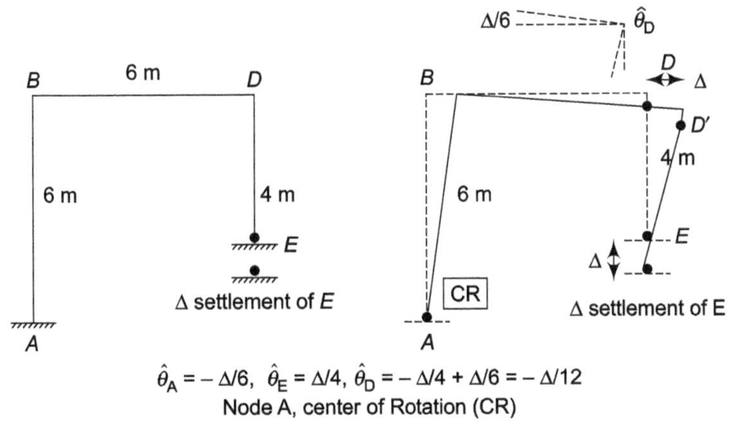

Fig. 4.25 *Illustration for Angle discontinuities, $\hat{\theta}$ due to settlement Δ*

The structure is made determinate by choosing the hinge releases at A and D. As a result of settlement of support E, as shown results in angle discontinuities at the hinges. The evaluation of the angle discontinuities are explained in the Figure. The reader is aware that the determinate structures do not develop internal stresses due to settlement or any other 'geometric loading'. If the angle discontinuity is resulting in tension inside, it is assumed positive. Similar approach we used in Chapter 3.

$$\text{The angle discontinuities at the hinge releases, } \delta = \begin{pmatrix} -\dfrac{\Delta}{12} \\ -\dfrac{\Delta}{6} \end{pmatrix}$$

The details of analysis of the plane frame for settlement of support are illustrated by the following illustration, Ex. 4.23.

Illustration 4.23. Analyse the frame, Ex. 4.23, by unified force method. The support E settles by Δ. Support E is a hinge

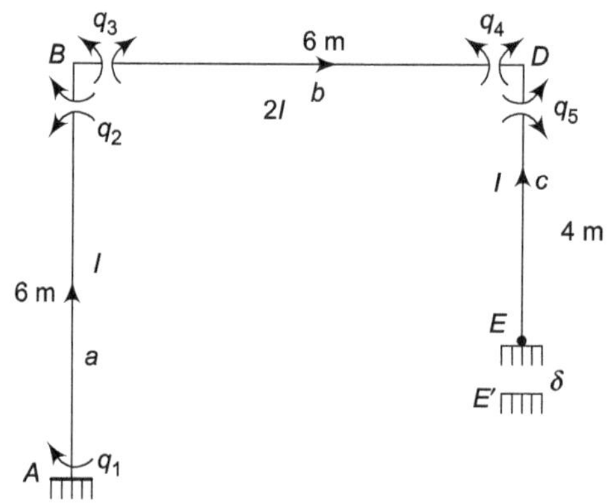

Fig. Ex. 4.23

The internal forces q_1 to q_5 for the numbers, a, b, c, are shown in Fig. Ex. 4.23. For the member c, end 1 being hinged.

$$(f_m)_c = (L/3EI)_c$$

$m = 5$, $n = 2$, $r = 3$. The three independent equations are obtained from two joint mechanisms and one sway mechanism.

$$\begin{array}{ccc} B & q & = 0 \end{array}$$

$$\begin{bmatrix} 0 & 1 & \boxed{-1} & 0 & 0 \\ 0 & 0 & 0 & \boxed{1} & -1 \\ -1 & \boxed{1} & 0 & 0 & -1.5 \end{bmatrix} \begin{bmatrix} q_1 \\ q_2 \\ q_3 \\ q_4 \\ q_5 \end{bmatrix} = \begin{bmatrix} 0 \\ 0 \\ 0 \end{bmatrix}$$

Choosing q_1 and q_5 as redundant and normalizing,

$$\Rightarrow \qquad \overline{B} = \begin{bmatrix} 1 & \vdots & 0 & \boxed{-1} & 0 & \vdots & 1.5 \\ 0 & \vdots & 0 & 0 & \boxed{1} & \vdots & -1 \\ -1 & \vdots & \boxed{1} & 0 & 0 & \vdots & -1.5 \end{bmatrix}$$

$$\lambda = \begin{bmatrix} 1 & 1 & 1 & 0 & 0 \\ 0 & 1.5 & 1.5 & 1 & 1 \end{bmatrix}$$

$$\lambda F_m = \begin{bmatrix} 1.5 & 1.5 & 1 & 0.5 & 0 \\ 0.75 & 1.5 & 2 & 1.75 & 1.333 \end{bmatrix} \frac{1}{EI}$$

Continuing with the Jordan elimination procedure for the equation.

$$\begin{bmatrix} \overline{B} \\ \lambda\, F_m \end{bmatrix}_q = \begin{bmatrix} 0 \\ \delta \end{bmatrix}, \text{ where } \delta = \begin{bmatrix} -\dfrac{\Delta}{12} \\ -\dfrac{\Delta}{6} \end{bmatrix}$$

$$\Rightarrow \begin{bmatrix} 0 & 0 & -1 & 0 & 0 \\ 0 & 0 & 0 & 1 & 0 \\ 0 & 1 & 0 & 0 & 0 \\ \boxed{1} & 0 & 0 & 0 & 0 \\ 0 & 0 & 0 & 0 & \boxed{1} \end{bmatrix} q = \begin{bmatrix} 0.0311 \\ 0.0246 \\ -0.0311 \\ -0.068 \\ 0.0246 \end{bmatrix} EI\delta$$

$$q_1 = -0.068 \text{ EI } \Delta$$

$$q_2 = q_3 = -0.0311 \text{ EI } \Delta$$

$$q_{4\backslash} = q_5 = 0.0246 \text{ EI } \Delta$$

4.11 APPLICATION OF FORCE METHOD TO BRIDGE DECKS

4.11.1 General

Bridge decks are supported on parallel abutments and piers if any. Un like bridge decks, the building floor decks are generally on the four sides.

It is an application of force method for the analysis to bridge decks. A bridge deck consists of: slab, longitudinal girders, and transverse girders. The transverse girders are required to distribute heavy live loads acting on the bridge decks on to the longitudinal girders. A simplification is achieved by merging stiffnesses of transverse diaphragms into the slab stiffness. The idealization achieves equivalent stiffness of uniform transverse medium connecting the longitudinal girders of the bridge decks. The longitudinal girders are assumed to be of zero torsional stiffness. The concentrated loads are represented by Harmonic series in the longitudinal direction. It helps to reduce the surface problem to a simple beam on spring supports. If the loads are expressed as harmonic series in the longitudinal direction, the effect of the longitudinal girders with zero torsion stiffness may be replaced by discrete springs. The slab is thus reduced to a continuous beam on flexible supports. The force method is more suitable than the displacement method for the idealized bridge deck. The structural idealization of a uniform transverse medium and harmonic expansion of loading is used by Hendry and Jaeger, but they use displacement method of analysis. The analysis based on force method results in reduced computations and quick solution is possible without having to use tables and graphs. The proposed method is equally suited for preparing charts and design graphs for load distribution, with much less effort than by the Hendry-Jaeger method.

4.11.2 Proposed Method and Assumptions

Refer Fig. 4.26. A reasonable structural idealization replaces the actual transverse medium consisting of slab and diaphragms by a uniform slab of equivalent stiffness, while the longitudinal girders are retained as they are in actual structure.

The treatment is greatly simplified by harmonic analysis as proposed by Hendry and Jaeger*. They utilize the property that in a system of interconnected beams (grids), if one of the longitudinal is loaded in such a way that the load intensity is proportional to the deflection at that point, all the longitudinal in the system are assumed to deflect in the same form. This condition is satisfied by replacing the load term by the corresponding Fourier series.

Hendry and Jaeger have prepared extensive design tables, charts and graphs based on this structural idealization, which can be readily used for any practical design.

Hendry and Jaeger have used displacement method of analysis of bridge deck. It is observed that the analysis can be made much more direct and simplified if we use the force method as proposed herein, reducing the computational efforts to a great extent.

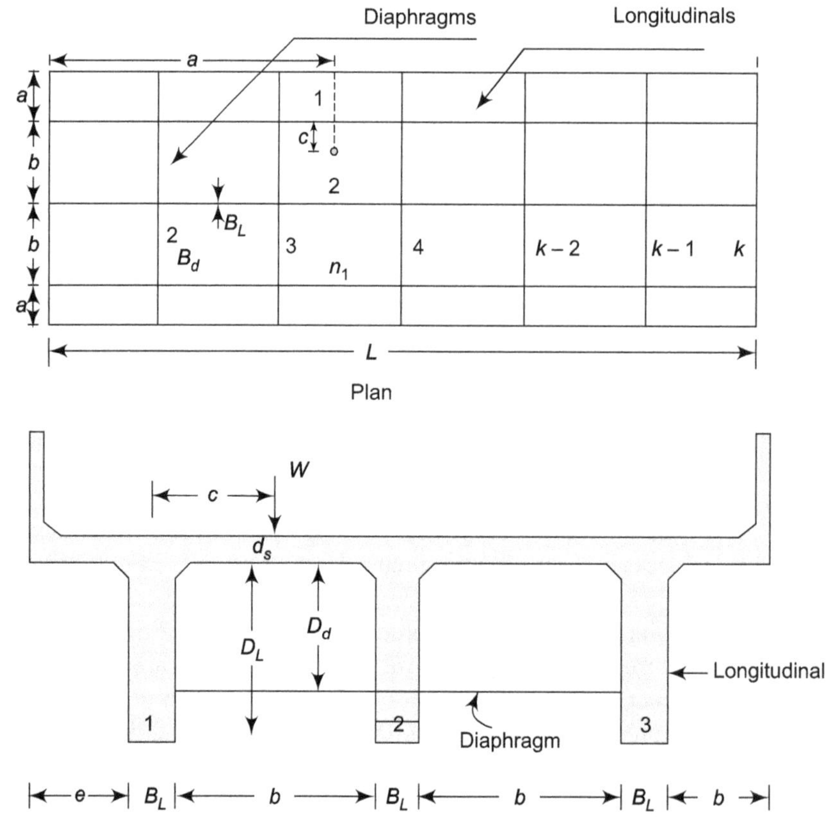

For given example b = 2.5 m, e = 1.25 m, C = 1.25 m, L = 24 m, a = 12 m, d_s = 0.20 m, D_L = 1.20 m, D_d = 0.80 m, B_L = 0.40 m, F_d = 0.30t

Fig. 4.26 *Bridge Deck*

*Hendry and Jaeger, "The Analysis of grid frame works and related structures", Chatto & Windus Ltd., London, WC-2.

The important assumptions are:

(i) The transverse medium of slab and diaphragms, is replaced by a continuous medium of the same total transverse moment of inertia.

(ii) Torsion stiffness of the transverse girders is neglected.

(iii) Torsion stiffness of the longitudinal beams is also neglected.

Torsion of the transverse medium is usually negligible and even if only five transverse diaphragms are used, this assumption gives negligible error. The transverse diaphragms are also referred herein as transversals.

Torsional stiffness of the longitudinal can also be neglected in a practical design problem due to following reasons:

(a) In a practical case, the actual loading on the deck consists of a train of loads sufficiently wide so that net eccentricity of load system, and hence the torsion on the structure is greatly reduced. The dispersion of load in the transverse direction, through wearing coat and deck slab, further reduces these effects.

(b) The longitudinal are sufficiently deep as compared to width. The overall torsional stiffness, therefore, is negligible as compared to their flexural stiffness.

4.11.3 Idealizations and Reduction of Surface to Linearity

A general bridge deck with n longitudinal and k transversals is considered. Let I_L be the moment of inertia of a longitudinal girder and $I_{t1}, I_{t2}..., I_{tk}$ be the moments of inertial of first, second, ... and kth transversals. T-beam action is considered while computing the moments of inertia. I_s is the moment of inertia of the remaining portion of the slab. If ΣI_t is total moment of inertia of transverse medium then

$$\Sigma I_t = I_{t1} + I_{t2} + \ldots + I_{tk} + I_s \qquad \text{...(1)}$$

The moment of inertia of equivalent uniform slab is

$I_t = \Sigma I_t / L$ where L is span of bridge (Fig. 4.26).

Let us consider the effect of a concentrated load of magnitude W acting at a distance "a" from the support on the slab in between the first and second girder. This forms the most general case of loading.

Let us consider any longitudinal girder acted upon by a load W_i on it at a distance "a" from the left support. Harmonics are used in the following to reduce two dimensional variation to a linear beam idealization.

This can be expressed in the form of Fourier series, Eq. (4.3.2).

$$(q_x)_i = \sum_{m=1}^{\infty} \frac{2W_i}{L} \sin\frac{m\pi a}{L} \sin\frac{m\pi x}{L} \qquad \text{...(2)}$$

Each Fourier term can be taken independently for the analysis. The result is obtained by adding up the contributions due to all harmonic components. Consider the effect of mth harmonic term of the load series, Eq. (2).

$$(q_{mx})_i \frac{2W_i}{L} \sin\frac{m\pi a}{L} \sin\frac{m\pi x}{L} = (\bar{q}_m)_i \sin\frac{m\pi x}{L} \qquad \text{...(3)}$$

Where $(\bar{q}_m)_i = \dfrac{2W_i}{L} \sin \dfrac{m\pi a}{L}$

The differential equation for the deflection of girder is

$$\left(EI \frac{d^4 y}{dx^4} \right)_i = (q_{mx})_i \qquad \text{...(4)}$$

Integrating equation (4) four times we get the deflection from mth harmonic,

$$(y_{mx})_i = \frac{L^4}{m^4 \pi^4} \left(\frac{\bar{q}_m}{EI_L} \right)_i \sin \frac{m\pi x}{L} = (\bar{y}_m)_i \sin \frac{m\pi x}{L} \qquad \text{...(5)}$$

Flexural stiffness $(K_m)_i$ of longitudinal, which is the same as load per unit deflection on the girder i, is

$$(K_m)_i = \frac{(q_{mx})_i}{(y_{mx})_i} = \frac{(\bar{q}_{mx})_i}{(\bar{y}_{mx})_i} = \frac{m^4 \pi^4 (EI_L)_i}{L^4} \qquad \text{...(6)}$$

The longitudinal girder can now be replaced by a spring with stiffness equal to K_m. Thus the analytical model of the bridge deck is simplified to the beam on elastic supports as shown in Fig. 4.27.

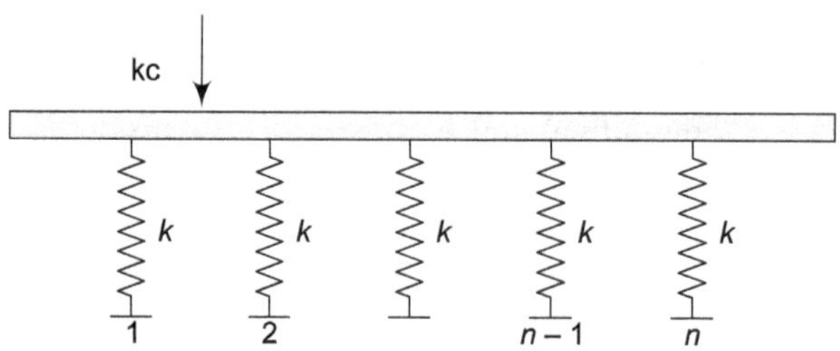

Fig. 4.27

This statically indeterminate structure can be converted into a determinate structure by introducing a suitable release system. The releases in the proposed method consist of hinges at the joints of slab with girders 2, 3 … to $n - 1$. The degree of indeterminacy in this case is $n - 2$ where n is number of longitudinal girders. Hendry and Jaeger use slope deflection equations (displacement method) and the unknown are $(n - 2)$ rotations and $(n - 1)$ relative displacements. Thus total unknowns are $(2n - 3)$ as against $(n - 2)$ in the proposed method. The reduced number of unknowns is the distinct advantage of the proposed method.

Furthermore, in Hendry-Jaeger method, the loads placed between the girders are required to be transformed into equivalent loads at girder positions treating the slab as continuous over non-sinking supports. No such transformation is necessary in the proposed method.

The unknown forces in the proposed method will be the moments at the hinges to balance the effect of the release system.

The required $(n - 2)$ equations to determine these $(n - 2)$ unknown moments can be obtained from the compatibility of slopes in the slabs at these release hinges. The flexibility coefficients can be computed as follows (Fig. 4.28):

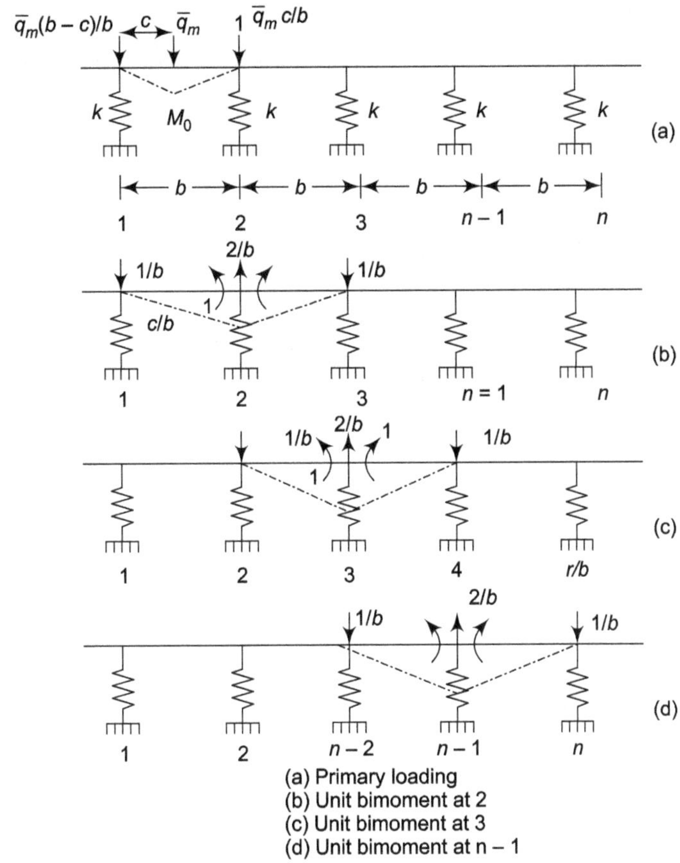

Fig. 4.28 *Flexibility Coefficients*

We define rotation θ_{ij} as angle discontinuity at hinge i due to a unit bimoment at j, and θ_{i0} as angle discontinuity at hinge i due to primary loading. The equation of compatibility of rotation of hinge i can be written as

$$m_2\theta_{i2} + m_3\theta_{i3} + m_4\theta_{i4} + \ldots m_{n-1}\theta_{in-1} + \theta_{i0} = 0 \qquad \ldots(7)$$

Similar expressions can be written for all hinges for $i = 2$ to $i = n - 1$, Eq. (8a)

$$m_2\theta_{22} + m_3\theta_{23} + m_4\theta_{24} + \ldots m_{n-1}\theta_{2,n-1} + \theta_{20} = 0$$

$$m_2\theta_{32} + m_3\theta_{33} + m_4\theta_{34} + \ldots m_{n-1}\,\theta_{3,n-1} + \theta_{30} = 0$$

....

....

....

$$m_2\theta_{n-1.2} + m_3\theta_{n-1.3} + m_4\theta_{n-1.4} + \ldots m_{n-1}\,\theta_{n-1,n-1} + \theta_{n-1.0} = 0 \qquad \ldots(A.8a)$$

or $$[\theta_{ij}]\,[m_i] = -\,[\theta_{io}] \qquad \ldots(A.8b)$$

We may use flexibilities f_{ij} in place of θ_{ij}. Flexibilities are deformations, and it is being used in place of figure to reassert that flexibilities are deformations.

Derivation of Coefficients

(1) θ_{io}. The principle of complementary virtual work is used to derive these coefficient directly. From Fig. 4.28 (a) and (b), we get

$$\theta_{20} = \left[\int_o^c \frac{q_m}{EI_t}.\frac{b-c}{b}.x.\frac{x}{b}dx + \int_o^{b-c}\frac{q_m}{EI_t}.\frac{c}{b}.\frac{b-x}{b}dx + \frac{q_m}{k_m}.\frac{b-c}{b}.\frac{1}{b} - \frac{q_m}{k_m}.\frac{c}{b}.\frac{2}{b}\right]$$

$$= \frac{q_m}{6b\,EI_t}\left[b^2 - c^2\right] + \frac{q_m}{b^2\,k_m}\left[b - 3c\right] = 0 \quad \text{...(9)}$$

For the case considered $\theta_{30} = \theta_{40} = \ldots\ \theta_{n-1.0} = 0$

However in a general case when free bending moment for primary loading is present in all spans $\theta_{30}, \theta_{40}, \ldots\ \theta_{n-1.0}$ can be computed likewise.

(2) θ_{ij} Similarly by principle of complementary virtual work we get.

$$\theta_{22} = \left[2\int_o^b \frac{1}{b}.x.\frac{1}{b}.x\,dx + \frac{1}{b^2k_m} + \frac{2}{bk_m}.\frac{2}{b} + \frac{1}{b^2k_m}\right]$$

$$\theta_{22} = \left[\frac{2}{3EI_t} + \frac{6}{b^2\,k_m}\right] \quad \text{...(10)}$$

$$\theta_{23} = \left[2\int_o^b \frac{1}{b}.x.\frac{1}{b}.(b-x).d.c - \frac{2}{bk_m}.\frac{1}{b} - \frac{1}{bk_m}.\frac{2}{b}\right]$$

$$= \left[\frac{b}{6EI_t} - \frac{4}{b^2k_m}\right] \quad \text{...(11a)}$$

And finally

$$\left.\begin{aligned}\theta_{n-1,\,n-1} &= \left[\frac{2b}{3EI_t} - \frac{6}{b^2k_m}\right]\\ \theta_{n-2,\,n-1} &= \left[\frac{b}{6EI_t} - \frac{4}{b^2k_m}\right]\end{aligned}\right\} \quad \text{...(11b)}$$

Thus after computing these flexibility coefficient, we can obtain the values of unknown moments m_2, $m_3, \ldots\ m_{n-1}$ from the Eq. (8). The reactions R_n on the girder due to given loading and the redundants are computed from statics, Eq. (12).

$$R_{n,\,m} = R_{n,\,0} + R_{n,\,1} + R_{n,\,2} + \ldots + R_{n,\,n-1} \quad \text{...(12)}$$

The moments on the longitudinal girder, say, nth girder will be

$$M_{n,\,m} = \frac{R_nL^2}{m^2\pi^2}$$

For the mth harmonic. Final comments can be computed by summing up the values due to all harmonics.

$$M_n = M_{n,\,1} + M_{n,\,2} + \cdots + R_{n,\,m} \quad \text{...(13)}$$

The ratio of the moment M to the free bending moment qL^2/π^2 is called the moment distribution coefficient of the girder.

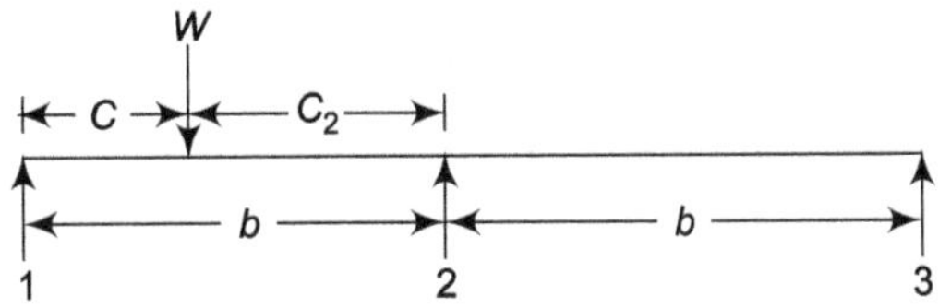

Fig. 4.11.4

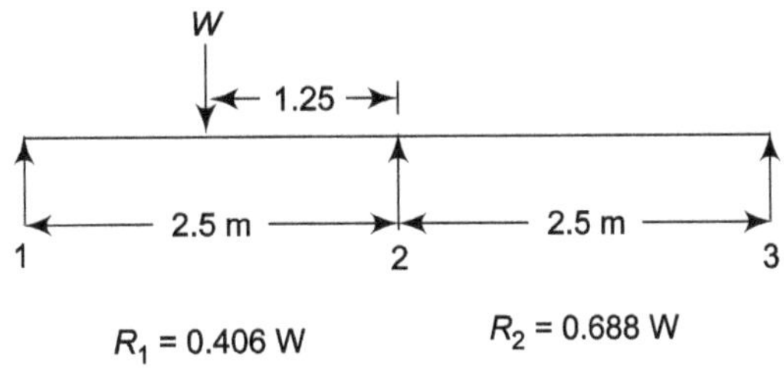

Fig. 4.11.5 *Three Girder Bridge*

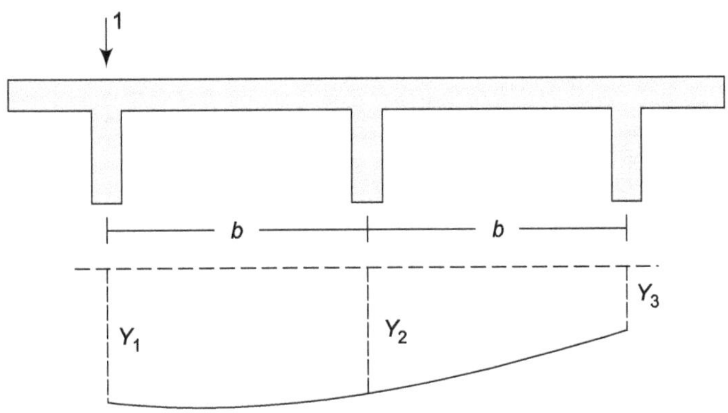

Fig. 4.11.6

4.11.4 Analysis of 3-girder Bridge by Proposed Method

The three girder bridge (Fig. 4.26) is analysed by the force method for the following cases.

Case A	Number of Diaphragm	5
Case B	"	3
Case C	"	2
Case D	"	Nil

Longitudinal and transverse stiffnesses

The longitudinal girders and diaphragm are cast monolithically with the slab, hence they will be treated as T beams. The effective widths of the resulting T or L beams are computed on the basis of I.S. 456-1964 and are used in Table A-1.

For a point load W at mid-span the Fourier series is

$$q_x = \sum_{m=1}^{\infty} \frac{2W}{L} \sin\frac{m\pi}{2} \sin\frac{m\pi x}{L}$$

$$= \frac{2W}{L} \sum_{m=1}^{\infty} \sin\frac{m\pi}{2} \sin\frac{m\pi x}{L} \qquad \text{...(a)}$$

From Eq. (A-6) the spring constant of the longitudinal girder is

$$K_m = \frac{m^4\pi^4 EI_L}{L^4} = 0.6316 m^4 E \times 10^{-4} \qquad \text{...(b)}$$

Flexibility coefficients

The load is acting midway between girders 1 and 2, hence $c = b/2$. From Eq. (9)

$$\theta_{20} = \frac{\bar{q}_m c(b^2 - c^2)}{6bEI_t} + \frac{\bar{q}_m (b - 3c)}{b^2 k_m}$$

$$= \left[\frac{\bar{q}_m b^2}{16EI_t} - \frac{\bar{q}_m}{2bk_m}\right] \qquad \text{...(c)}$$

Add from Eq. (A.10)

$$\theta_{22} = \left[\frac{2b^2}{3EI_t} + \frac{6}{b^2 k_m}\right] \qquad \text{...(d)}$$

Eq. (8) yields

$$m_2\theta_{22} + \theta_{20} = 0$$

(In our Ch 2 notation $fx + u_0 = 0$)

Or

$$m_2 = -\frac{\theta_{20}}{\theta_{22}}$$

$$= -\frac{\left[\frac{\bar{q}_m b^2}{16EI_t} - \frac{\bar{q}_m}{2bk_m}\right]}{\left[\frac{2b^2}{3EI_t} - \frac{6}{b^2 k_m}\right]}$$

$$= -\bar{q}_m \frac{\left[\frac{1}{2bk_m} - \frac{b^2}{16EI_t}\right]}{\left[\frac{6}{b^2 k_m} - \frac{2b}{3EI_t}\right]} \qquad \text{...(e)}$$

Reactions at the girders

$$\left.\begin{aligned} R_1 &= R_{10} + R_{12} = \bar{q}_m/2 + m_2/b \\ R_2 &= R_{20} + R_{22} = \bar{q}_m/2 - 2m_2/b \\ R_1 &= R_{10} + R_{12} = \quad +m_2/b \end{aligned}\right\} \qquad \text{...(f)}$$

The calculations for first nine harmonics for all the four cases considered are reproduced in a tabular form in Tables A-1 to A-3.

Girder moments

The moment from these individual girder loads R_1, R_2 and R_3 are obtained by integration, i.e.,

$$M_{1m} = \frac{R_1 L^2}{m^2 \pi^2},\ M_{2m} = \frac{R_2 L^2}{m^2 \pi^2},\ M_{3m} = \frac{R_3 L^2}{m^3 \pi^2}$$

The final values of the moments are obtained by superimposing the contribution from each harmonic term. The values of the girder moments for the cases considered are computed for first-five terms and also for first nine terms of the series, and are given in Table A-3.

If we consider only finite terms of the series, say, the first five terms, we observe that the moment series has not converged to the true values of the moment and if M is the free bending moment then the error δM equal to $\{M - (M_1 + M_2 + M_3\}$ will not be equal to zero.

Error $\qquad \delta M = \{M - (M_1 + M_2 + M_3)\}$ ...(g)

It is proposed that the error δM be distributed on the girders adjoining the load under consideration in the inverse ration of their distances from the load, i.e., if W is at a distance c_1 from girder 1 and c_2 from girder 2 (Fig. 4.11.4), then

$$\delta M_1 = \delta M \frac{C_2}{b} \text{ and } \delta M_1 = \delta M \frac{C_1}{b} \qquad \text{...(h)}$$

And the corrected girder moments are

$$M_1^{\ c} = M_1 + \delta M_1, \text{ and } M_2^{\ c} = M_2 + \delta M_2 \qquad \text{...(i)}$$

It is observed that equation (i) yields results fairly close to those obtained by Hendry and Jaeger (Table A-4).

If however a more accurate assessment of moment is required, it can be achieved by analyzing the deck slab as continuous over longitudinal girders and computing the reactions R_1, R_2 and R_3 of the girders. The error δM is now distributed to the girders in proportion to their reactions, i.e.,

$$\delta M_1 = \delta M \times \frac{R_1}{W},\ M_2 = \delta M \times \frac{R_2}{W},\ \delta M_3 = \delta M \times \frac{R_3}{W} \qquad \text{...(j)}$$

Eq. (j) gives the values of moments, which are the same as given by Hendry and Jaeger's method (Table A-4).

Table A-1

CASE	I_L	I_T	$\frac{b^2}{16I}$	$(2/3)(b/I_1 t)$
A	0.2156	0.01095	35.60	152.10
B	0.216	0.00742	53.96	230.30
C	0.2156	0.00412	94.80	403.20
D	0.2156	0.00067	583.00	2487.80

Table A-2

No.	Parameter		m = 1	m = 3	m=5	m=7	m=9
1	K_m/E		0.6316×10^{-4}	$0.6136 \times 3^4 \times 10^{-4}$	$0.6316 \times 7^4 \times 10^{-4}$	$0.6316 \times 7^4 \times 10^{-4}$	$0.6316 \times 9^4 \times 10^{-4}$
2	$E/(2bK_m)$		3166.50	39.10	5.07	1.32	0.48
3	$6E/(b^2K_m)$		15190	187.53	24.30	6.30	2.30
4	$\frac{E}{2bKm} - \frac{b^2}{16I}$	A	3130.8	+3.44	-30.59	-34.35	-35.18
		B	3112.5	-14.86	-48.89	-52.64	-53.48
		C	3071.70	-55.70	-89.70	-93.50	-94.30
		D	2583.5	-543.90	-577.90	-581.70	-589.50
5	$\frac{6E}{b^2Km} - \frac{2b}{3I_t}$	A	15342.1	339.63	176.4	158.4	154.4
		B	15420.3	417.83	254.5	236.6	232.6
		C	15593.2	590.73	427.5	409.5	405.5
		D	17677.8	2675.33	2512.1	2494.1	2490.1
6	$m_2 = \frac{\text{Row 4}}{\text{Row 5}}$	A	0.204	+0.0101	-0.1734	-0.2168	-0.2279
		B	0.202	-0.0356	-0.1920	-0.2225	-0.2299
		C	0.197	-0.0943	-0.2098	-0.2283	-0.2325
		D	0.1465	-0.2033	-0.2300	-0.2332	-0.2351
7	$R_3 = R_{12} = \frac{m_2}{b}$	A	0.08160	+0.00404	-0.06936	-0.08672	-0.09116
		B	0.08080	-0.01424	-0.07680	-0.08900	-0.09196
		C	0.07886	-0.03722	-0.0392	-0.09132	-0.09300
		D	0.05860	-0.08132	-0.09200	-0.09328	-0.09404
8	$R_1 = R_{10} + R_{12}$	A	0.58160	+0.50404	+0.43064	+0.41328	+0.40884
		B	0.58080	+0.46576	+0.42320	+0.41106	+0.40804
		C	0.57830	+0.46228	+0.41608	+0.40868	+0.40700
		D	0.55860	+0.41868	+0.40800	+0.40672	+0.40596
9	$R_1 = R_{10} + R_{22}$	A	0.33680	+0.49192	+0.63872	+0.67344	+0.68238
		B	0.33840	+0.53848	+0.65360	+0.67800	+0.68392
		C	0.34240	+0.57544	+0.66784	+0.68264	+0.68600
		D	0.38280	+0.66264	+0.68400	+0.68656	+0.68808

Table A-3 *Moments in longitudinal girders*

Moments		m = 1	m = 3	m = 5	m = 7	m = 9	(1+3+5)	(1+3+5+9)
Girder-1 $M_1\left[\frac{2W}{L}\right]\times$	A	0.58160	+0.05600	+0.01723	+0.00843	+0.00505	0.65483	0.66831
	B	0.58080	+0.05397	+0.01693	+0.00839	+0.00504	0.65470	0.66513
	C	0.57880	+0.05136	+0.01664	+0.00834	+0.00502	0.64680	0.66018
	D	0.55860	+0.04652	+0.01632	+0.00829	+0.00501	0.62144	0.63471
Girder-2 $\overline{M_2} = \left[\frac{2W}{L}\right]\times$	A	0.33680	+0.05466	+0.02555	+0.01375	+0.00842	0.41701	0.43918
	B	0.33840	+0.05872	+0.02614	+0.01384	+0.00844	0.42326	0.44554
	C	0.34240	+0.06394	+0.02671	+0.01393	+0.00847	0.43305	0.45545
	D	0.38280	+0.07363	+0.02766	+0.01401	+0.00849	0.48379	0.50529
Girder-3 $\overline{M_3} = \left[\frac{2W}{L}\right]\times$	A	+0.08160	+0.00045	-0.00277	-0.00177	-0.00113	-0.07928	-0.07638
	B	+0.08080	-0.00158	-0.00307	-0.00182	-0.00115	-0.07615	-0.07318
	C	+0.07880	-0.00419	-0.00336	-0.00186	-0.00115	-0.07125	-0.06824
	D	+0.05860	-0.00904	-0.00368	-0.00368	-0.00116	-0.04588	-0.04287

Table A-4 *Case of 3-girder bridge deck with five diaphragms final girder moments (common multiplier = $2WL/\pi^2$)*

Moment	*By Proposed Method*	*Hendry and Jaeger's Method*	
		By static corrections	By coefficients only
M_1	**0.65483**		0.65350
Corrected by first approach	0.69612		
Corrected by second approach	0.68836	0.68830	
M_2	**0.41701**		0.41710
Corrected by first approach	0.45830		
Corrected by second approach	0.47392	0.47390	
M_3	**0.07928**		0.07920
Corrected by first approach	0.07928		
Corrected by second approach	0.07152	0.7140	

4.11.5 Comments

1. The structural idealization of transverse medium and harmonic analysis, which reduces the problem to a simple beam on flexible supports, are common with the proposed method and Hendry and Jaeger's method. However, the proposed force method has following advantages over the Hendry and Jaeger's method which is a displacement method.
   (a) Given loading system need not be converted to equivalent girder loads.
   (b) The number of redundants is $n - 2$ in the proposed method as against $2n - 3$ unknown deformations in Hendry and Jaeger's method.
2. In Hendry and Jaeger's method the convergence of series is improved by computing the loaded girder moments which are obtained by subtracting the sum of the unloaded girder moments from the free moment. The procedure is applicable even when the number of harmonics is limited to five.
   The same results can be obtained in the proposed method by applying corrections. A further simplification has been proposed which avoids computations of equivalent girder loads. The error in the latter approach is 1.13% for heaviest loaded girder of the illustrated example.
3. The computation being simple and direct, for the reasons stated above, direct analysis may be recommended in place of the use of charts and graphs. However the proposed method can also be used conveniently to prepare the design aids.
4. The results presented in Table A-3 confirm that the number of transversals when more than two (which are provided one at each end) do not significantly improved the load distribution. On the other hand, the interior diaphragms increase the dead loads.

Chapter 5: Approximate Methods of Analysis of Tall Building Frames

Chapter

5

Approximate Methods of Analysis of Tall Building Frames

5.1 SCOPE

The approximate methods presented herein are a good preparation for Ch-9 pertaining to tall buildings. The approximate methods presented herein have been extensively used by the building designers for tall skeletal frames (without shear walls), having equal storey heights along the entire height. The reasons are the simplicity in the analysis and generally conservative nature of the results provided by the approximate methods. The results of the numerical examples solved at the end of the chapter are compared with the 'exact' methods, which justify the continued use of the approximate methods. Even the so-called 'exact' methods are limited by structural idealizations, uncertainty about material properties and other approximations which may not be fully realized. Any 'exact' analysis requires apriori decision regarding material properties and other approximations which may not be realized. Any 'exact' analysis requires prior decision about initial geometrical properties like cross-sectional areas, moments of inertia, foot print of the building complex, types of foundation, etc.

A more analysis-oriented designer may use the approximate methods for initial estimate of cross-sectional dimensions and for deciding the type of foundation suitable for the soil conditions. A computer oriented designer by using these method, is provided with a feel of the order of intermediate important numerical results, which is generally denied to him while using computers. Even perhaps the computer results may require to be checked by these approximate methods. It is not the intention of this Chapter to design the buildings with irregular heights, of course limited to skeletal frames. The reader should appreciate the bases for different modeling, one for gravity loads and altogether different idealization/assumptions for the case of lateral loadings. With the advent of computers and easy access to the software, it is not unusual to end up with different set of results. The approximate methods provide the feel of the results and numbers, and enable us in the decision making. The computer based design of tall buildings with and without shear walls are discussed separately in chapter 8, under the title Tall Buildings.

The approximate methods are presented separately for vertical load and horizontal loads. The separate analysis facilitates the critical design combinations based on the appropriate code provisions. The important reason for proposing different approximate methods for gravity and horizontal loads is that in the case of gravity loads; the moments and shears experienced by any one girder load are localized. However a lateral load, say, at the top level causes moments and shears in many storeys and the largest moments and shears develop in the bottom storey. The wind loads control the design consideration for high rise buildings.

The tedious nature of computations may have been experienced by the reader, while analyzing even moderately tall frames (Ex. 2.28, Chapter 2). In the absence of the computer, the approximate methods are useful for analyzing such frames and even perhaps much taller frames. It has been observed that the cantilever and portal methods of analysis for lateral loads give good results for frames as tall as 25 and 35 storeys respectively. The methods assume reasonable symmetry of frame-geometry. In this chapter approximate analysis is presented for lateral loads only. In the case of gravity loads, only structural idealizations (substitute frames) are discussed for obvious reasons given in the following articles.

5.2 APPROXIMATE METHOD FOR FRAMES UNDER VERTICAL LOADS

The approximation is due to the structural idealization, when a complex frame is reduced to a simple frame. For this simple frame, which we prefer to call as "substitute frame", any of the exact methods of analysis may be used. Based on live load positions, critical for maximum moments (and sometimes shears) at selected points, the substitute frames are introduced. The degrees of indeterminacy, required for force method of analysis, are stated against each of these substitute frames, but it is more convenient to analyze these substitute frames by displacement methods like moment distribution or Kani's iteration, Chapter 8. The vertical loads acting on any girder are transferred to the foundation by developing essentially axial stresses in columns. The flexural stresses, however, are limited to the members surrounding the girder under the vertical loads. Reasonably good estimate of vertical reactions at the bottom ends can be made for the gravity loads for the design of footings. It is this localized nature of flexural stresses (moment and shears) that justifies the use of substitute frames.

5.2.1 Substitute Frames Based on Qualitative Influence Lines

The critical design moments occur at mid-span and at ends of girders and also at the ends of columns. For critical moments at these points due to live loads, influence lines are required to be drawn for design moments and shears in girder and moments and thrusts in columns. The Muller-Breslau's principle* is extremely useful to sketch the influence lines. For our purpose, the shapes of influence lines are sufficient to decide critical position of live loads. The influence lines for a general building frame are drawn (Fig. 5.1). In the Fig. 5.1, the qualitative influence lines drawn, is for a typical and mid-span moment at point x, load all the spans which are lifted up. It may be noticed that deformation shapes of all other girders and columns are half waves without exceptions. For reasons of localized nature of moments and shears, only span AB is laded and all other members adjoining the nodes A and B (defined by the girder ends) are retained.

The far-ends (designated by O) of all the adjoining members are considered hinged, so as to result in larger moment at x, for reasons of safe design [Fig. 5.1(b), Substitute Frame]. The degree of indeterminacy, n, of the substitute frame is nine.

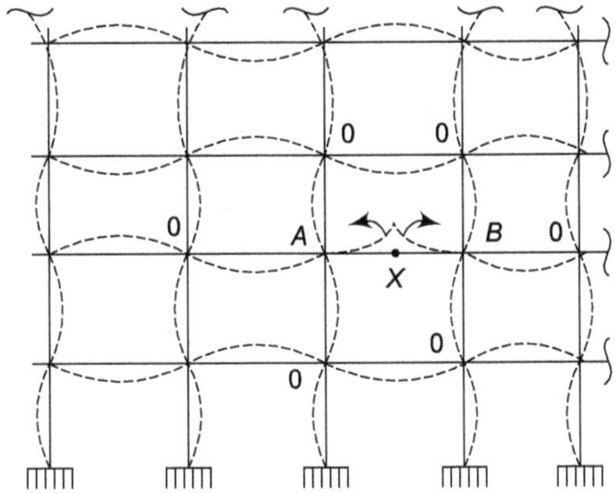

(a) Frame

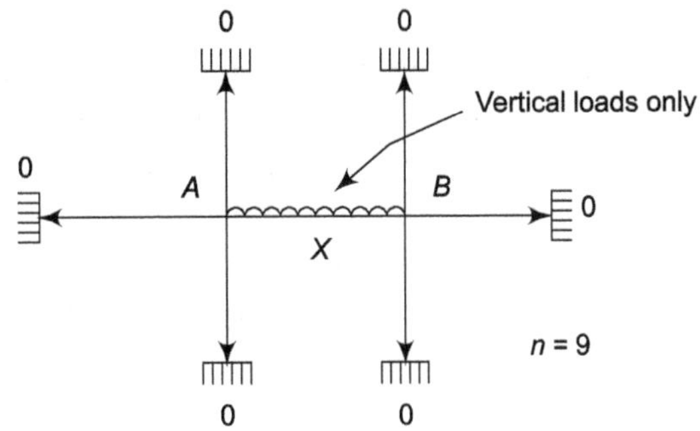

(b) Substitute frame for mid-span moments

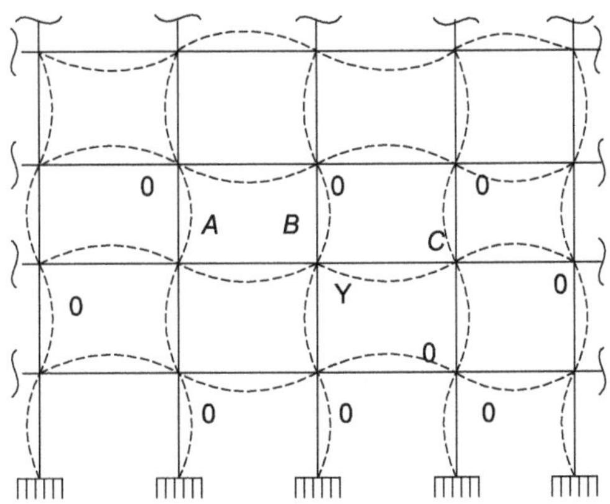

(c) Frame

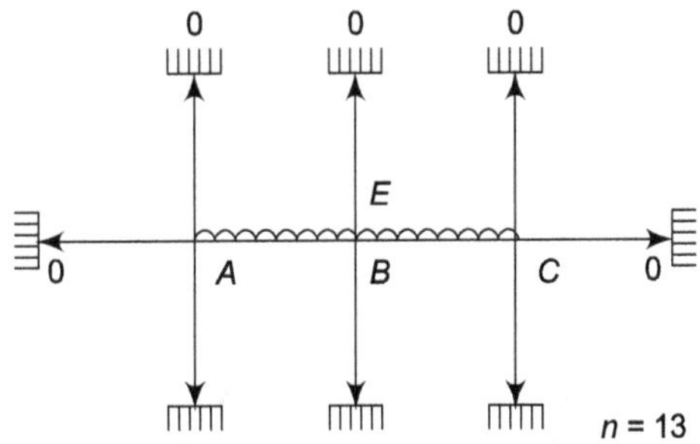

(d) Substitute frames for end moment of span BC

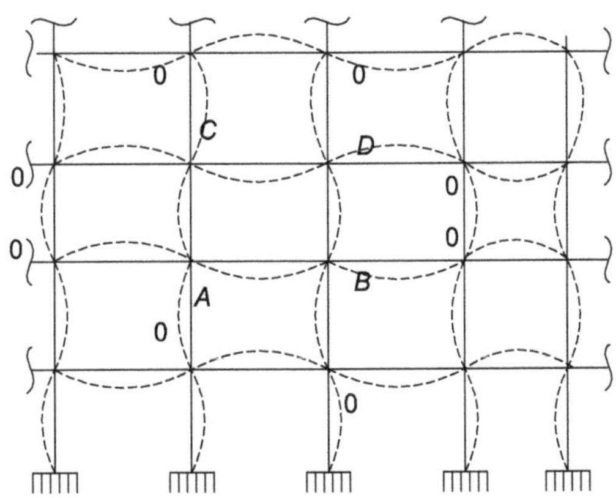

(e) Frame

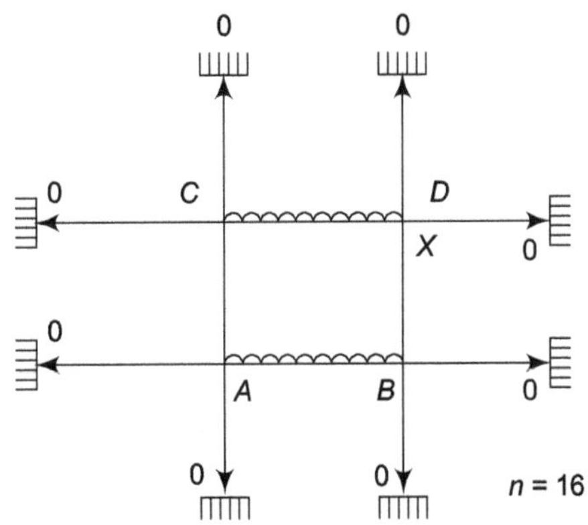

(f) Substitute frame for column moment at X,

Fig. 5.1 *Qualitative Influence Lines for Moments at X and Substitute Frame under Vertical Live Loads*

Consider the influence line, Fig. 5.1(*c*), for maximum moment at *x*, left end of girder *BC*. In this case all column member-deformation-shapes are half waves. In the case of girders, however, other girders within the bay of *BC* have kinks which interrupt half waves. For maximum moment at *x*, we need to load all the spans which undergo upward deflection similar to the span containing the point *x*. However because of the localized nature we load the girders *AB* and *BC* only and for substitute frame retain the adjoining members, Fig. 5.1(*d*). The members with kinks are not relevant to the substitute frame. The degree of indeterminacy is thirteen.

From similar considerations, the substitute frame, Fig. 5.1(*f*), is for maximum moment at *x*, at the top end of column *BD*, Fig. 5.1(*e*). It does not require to emphasize that for maximum axial load for the column *BD*, all the girder members above the girder *CD* require to be loaded. The degree of indeterminacy is sixteen.

The same substitute frame maybe used for vertical dead loads also. If the frame is not symmetrical the complete frame analysis for no sway, will show horizontal constraining forces. These horizontal constraining forces, which are generally ignored, may be merged with the given horizontal loads due to wind or earthquake and analysed separately, Article 5.3.

The above discussion is valid for the building frames with rigid joints. In the case of steel frames without welded joints and under vertical loads, it is sufficient to consider the particular girder (under load) with suitable constraining coefficients for computing end moments.

5.3 APPROXIMATE METHODS FOR FRAMES UNDER LATERAL LOADS

Only two approximate methods, (1) Cantilever method and (2) Portal method, are described. Both the methods may be expected to give similar results, say upto 15 storeys. The cantilever methods, however, is expected to give generally satisfactory results even upto 35 storeys. The portal method does not require to specify flexural rigidities (*EI*) and axial stiffnesses (*EA*). It may be recalled that in the case of force method (Chapter 2) we generally ignored extensibility of members (except for arches and similar structures), consequently *EA* values were not required to be specified. The cantilever method does not require to specify *EI* values and it is not sensitive to *EA* values. The methods are based on visualization of the deflected shape of the frames due to lateral loads at floor levels. Consider a typical frame under lateral loads, Fig. 5.2(*a*). It has been found from the nature of deformations of such frames that the points of inflection occur at points close to mid-heights of columns and mid-spans of girders for lateral loads. If N and C are the numbers of storeys and columns respectively, the number of releases introduced = number of hinge releases = $(2C - 1)N$.

For the frame considered, Fig. 5.2(*a*), the number of hinge releases = $(2 \times 4 - 1)\,6 = 42$.

The degree of indeterminacy, $n = 3\,(C - 1)\,N$.

For the frame, $n = 54$

The residual indeterminacy = $3(C - 1)\,N - (2C - 1)\,N = (C - 2)\,N$

Or residual indeterminacy per storey = $C - 2$

The above residual indeterminacy is removed by the additional assumptions made in the two methods.

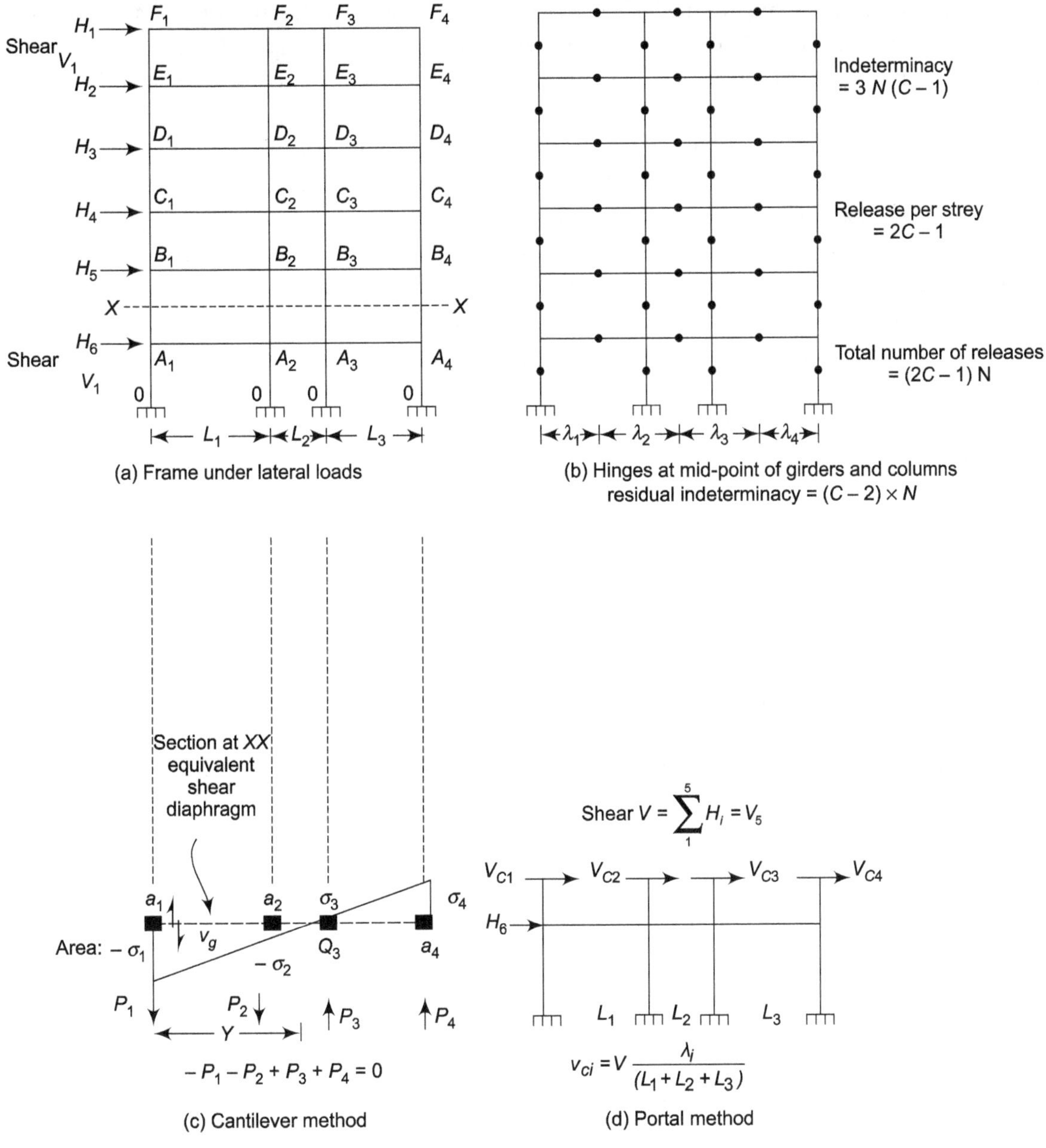

Fig. 5.2 *Frame Under Lateral Loads*

Cantilever Method: As the name suggests, the frame is idealized as a cantilever fixed at the base. The cross-sectional area of the cantilever is assumed to consist of individual column area with the girders providing an equivalent shear diaphragm assuming plane section remaining plane, Fig. 4.2(*c*). For locating the centroid, the columns are assumed of equal cross-sectional area. The location of Zero strain or stress at centroid and linear variation of deformations ensures vertical equilibrium. The linear variation of deformation is used to express the column force $p_2, p_3, \ldots, p_c$ in terms of p_1. The evaluation of p_1, say at the section *XX*, Fig. 5.2(a), is achieved by satisfying moment equilibrium of the free body above the section considered. The section *XX*, is taken at mid-height of the columns, because at any other position column moments also appear in the moment equilibrium equation. While specifying the axial force of all the columns of a storey, we have used two equations of equilibrium (vertical force equilibrium and moment equilibrium) and thus we have exactly eliminated the residual indeterminacy.

The axial forces, P_i are evaluated starting from the top floor. With the axial forces evaluated and considering the free body above the inflexion points of the top storey columns, girder shears V_g are found. Having specified the points of inflexion of girders, the end moments are $(V_g L_g)/2$, where L_g is the girder span.

In the case of an interior storey we take a corresponding free body defined by the points of inflexion of columns at top and bottom of the floor of storey. The remaining procedure of findings shears and drawing girder moments is the same. Here we are required to consider axial forces acting at both the ends of individual columns for finding shears.

With the girder moments known, we find the column moments by satisfying equilibrium of joints starting from the left.

Before we pass on to the portal methods, let us solve a numerical example based on the cantilever method.

Ex. 5.1. Analyse the frame, Fig. Ex. 5.1(a).

Solution: Ex. 5.1 (Cantilever method):

The assumed points of inflexion are shown in Fig. (b). For the idealized cantilever, normal stresses acting at a cross-section *XX* are shown in Fig. (c). Based on equal area A of the columns, the centroid.

$$\bar{Y} = \frac{AY_i}{3A} = 10 \text{ m}$$

From the consideration of linear variation of the stress on an idealized cross-section, Fig. (c),

$$P_2 = 0.2\,P_1 \text{ and } P_3 = 0.8\,P_1 \quad \text{...(i)}$$

P_1 is tension and P_2, P_3 are compressions. The above relations between P_i are valid at all the storey levels.

The cantilever moment at level $X_3 - X_3$

$$M_{x3} = 3 \times 2 = 6 \text{ tm} \quad \text{...(ii)}$$

Moment of resistance due to the column forces cut by section *XX*, $M_x = 18\,P_3 + 12P_2$.

Substituting the values of P_2 and P_3 from the relations *(i)*,

$$M_x = 18(0.8)\,P_1 + 12\,(0.2)\,P_1 = 16.8\,P_1 \quad \text{...(iii)}$$

Equating M_{x3} *to* M_x given by the Eq. *(iii)*, we get axial forces of columns at level $X_3 - X_3$.

$$P_1 = 6/16.8 = 0.357\,t \text{ and from Eq. (i)}$$

$$P_2 = 0.071\,t \text{ and } P_3 = 0.286\,t$$

Or symbolically we may write

$$\{P_i\}_{\text{level }3} = [-0.357, 0.071, 0.286]$$

Positive sign is used for compression

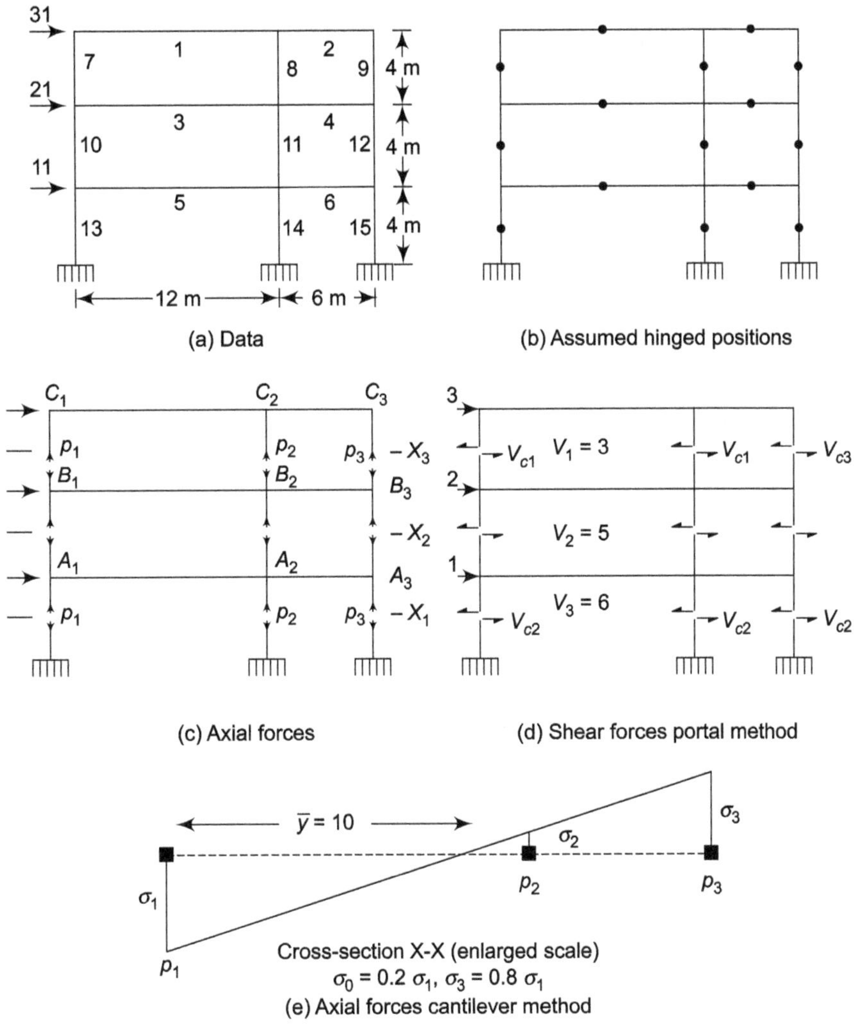

Fig. Ex. 5.1 *Approximate Methods, Cantilever Method*

Similarly $\quad M_{x2} = 3 \times 6 + 2 \times 2 = 22$ tm

and $\quad P_1 = 22/16.8 = 1.310$ t

and $\quad \{P_i\}_{\text{level }2} = [-1.310. 0.262, 1.048]$

Finally it can be shown that

$$\{P_i\}_{\text{level }1} = [-2.620. 0.524. 2.095]$$

Girder moments:

Procedure

Consider the top girder C_1, C_2, C_3 and $\{P_i\}_{\text{level } 3}$

The girder shears are

$$V_{g1} = 0.357,\ V_{g2} = -0.286 \text{ t}$$
$$m_{g1} = (V_{g1}\, L_{g2})/2 = 2.142 \text{ tm}$$
$$m_{g2} = (V_{g2}\, L_{g2})/2 = 0.858 \text{ tm}$$

Consider the girder $B_1B_2B_3$ and $\{P_i\}$ at the level 2 and 3

The girder shears are

$$V_{g1} = 0.953,\ V_{g2} = -0.762\ t$$
$$m_{g1} = 5.718 \text{ tm}$$
$$m_{g2} = 2.286 \text{ tm}$$

Consider the girder $A_1A_2A_3$, and $\{P_1\}$ at the levels 1 and 2

$$V_{g1} = 1.310,\ V_{g2} = 1.048 \text{ t}$$
$$m_{g1} = 7.860 \text{ tm}$$
$$m_{g2} = 3.171 \text{ tm}$$

Column moments:

Columns end moments are found from moment equilibrium of joints.

Consider moment equilibrium of the joints C_1, C_2 and m_{e3}.

$$m_{e1} = 2.142 \text{ tm},\ m_{e2} = 3.00 \text{ tm},$$
$$m_{e3} = 0.858 \text{ tm, or symbolically}$$
$$\{m\}_{\text{level } C} = [2.142, 3.00, 0.858]$$

Transfer the moment to the other ends of the columns of storey 3 and considering moment equilibrium of the joints B_1, B_2, B_3 we get

$$\{m\}_{\text{level } B} = [3.576, 4.464, 1.428]$$

Transfer the moments to the other ends of the columns of storey 2.

Considering the moment equilibrium of the joints A_1, A_2, A_3, we get

$$\{m\}_{\text{level } A} = [4.264, 6.537, 1.718]$$

Complete the m-diagram for the columns under the level A.

It is now possible to draw the final bending moment diagram, Fig. Ex. 5.1. For the sake of comparison of values obtained by approximate methods and an 'exact' method is given in Table 5.1. The results based on approximate methods are reasonably in agreement with the 'exact' results. The same example is solved below by the portal method.

Portal method: We had already seen (Article 5.3.1) that after introducing the points of inflexion, it is required to specify at least (C-2) unknown forces per storey. To achieve this, while analyzing by portal method, the additional assumption specifies column shear as a percentage of total horizontal shear. The shear resisted by a column is assumed to be proportional to the width of aisle the column supports. The width of aisle λ_i, is given by the mean of the span-lengths adjoining the columns, Fig. 5.2(d). With the column shear V_{ci} evaluated, the procedure of finding column moments is the same as evaluation of girder moments from girder shears. The column-end-moment is equal to $V_{ci}l_c / 2$, where l_c is the column length. The value of V_{ci} can be evaluated form equilibrium of horizontal forces.

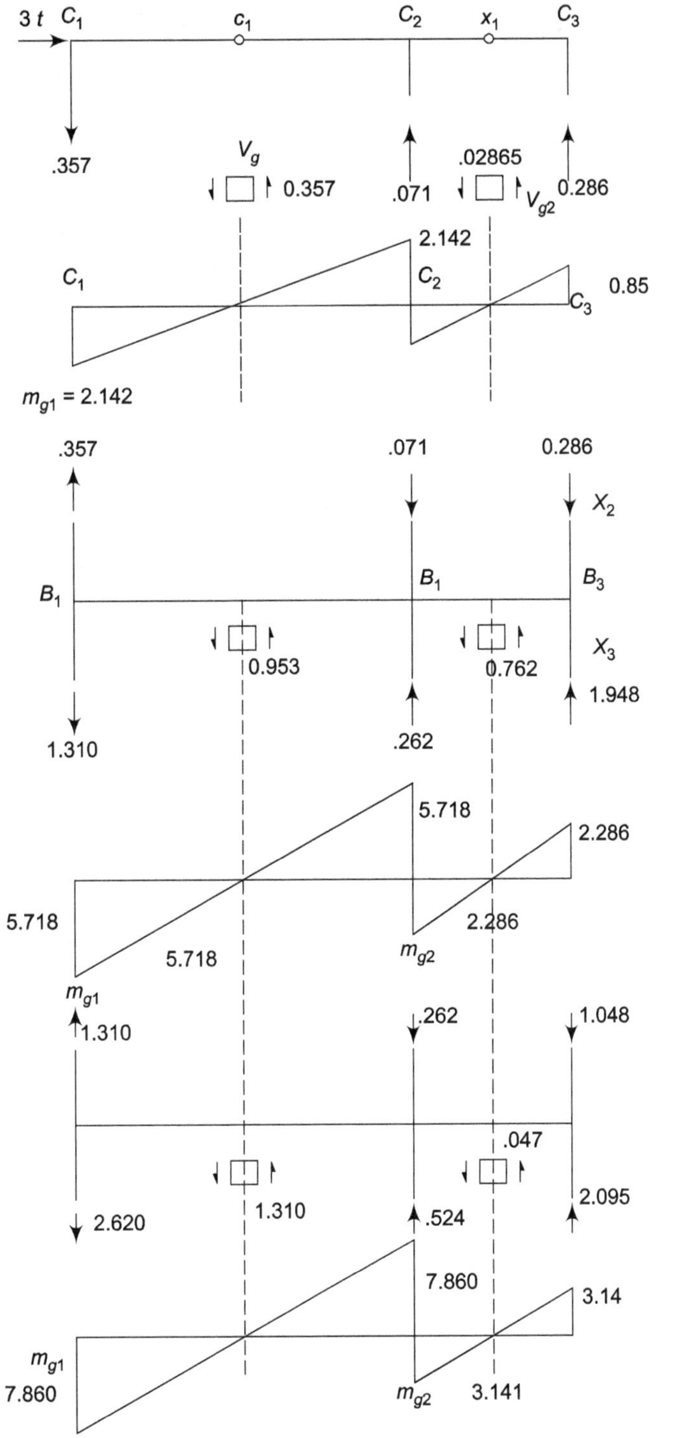

Fig. Ex. 5.1(b) *Cantilever Method-Girder Moments*

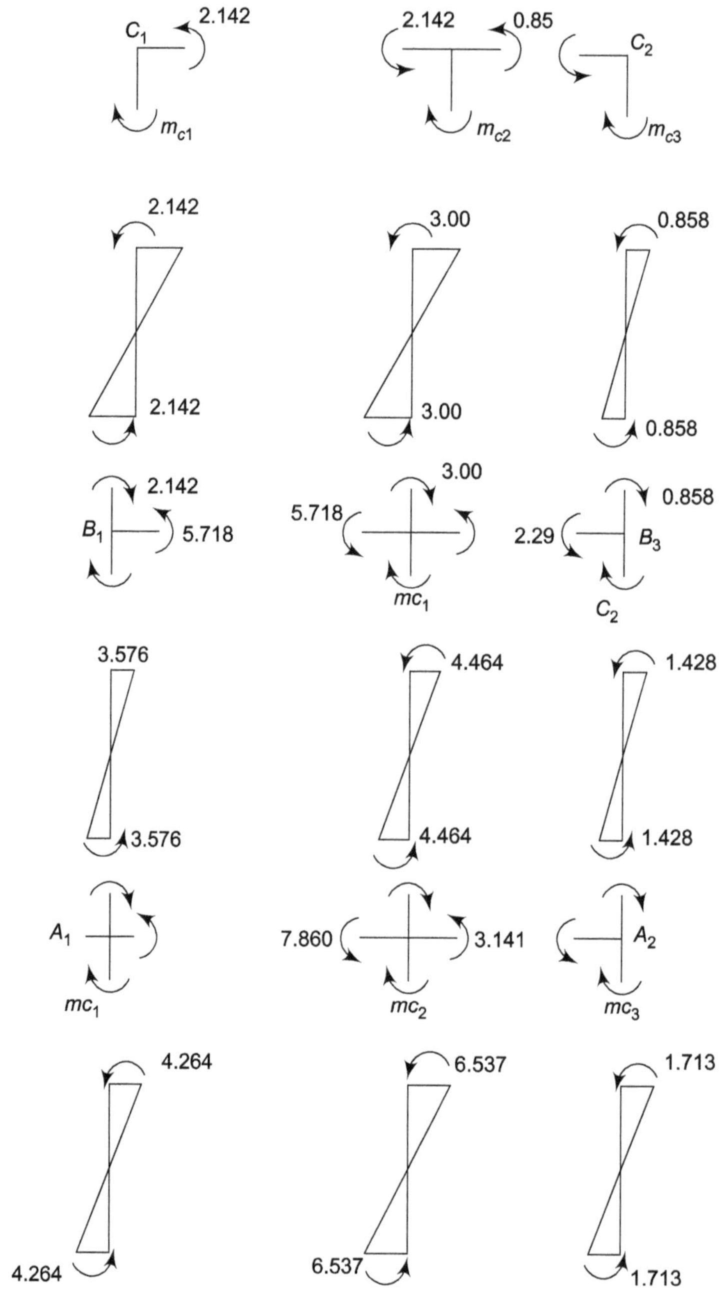

Fig. Ex. 5.1(c) *Cantilever Method-Column Moments*

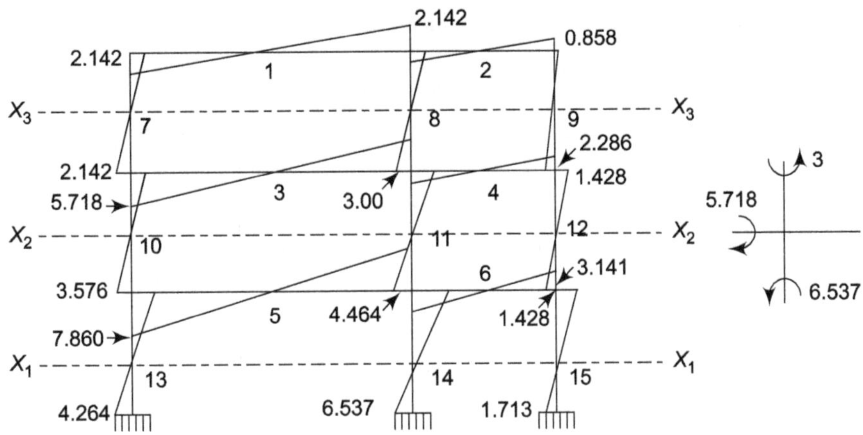

Fig. Ex. 5.1(d) *Final M-Diagram (Cantilever Method)*

$$V_{ci} = k_i V$$

Where shear distribution factor,

$$\therefore \qquad K_i = \frac{\lambda_i}{\Sigma \lambda_i}$$

And V is the total shear at a section x where V_{ci} are evaluated. The girder end-moments are evaluated from the equilibrium of the joints. The sequence of the joints is considered from left to right and top to bottom (similar to cantilever method for finding column-end-moments). While specifying the column shears, we have used equation of equilibrium for horizontal forces. Vertical force equilibrium is automatically satisfied by the nature of shears developed due to numerically equal end moments.

Solution: Ex. 5.2, Portal method:

Width of aisles are:

$$\lambda_1 = 6, \lambda_2 = 6 + 3 = 9 \text{ and } \lambda_3 = 3$$

Shear distribution factors are

$$k_1 = \frac{6}{18} = \frac{1}{3}$$

$$k_2 = \frac{9}{18} = \frac{1}{2}$$

$$k_3 = \frac{3}{18} = \frac{1}{6}$$

The members are designated by the numbers shown in Fig. Ex. 5.1(a) and the moment diagrams may be drawn.

Table 5.1 *Comparison of moments by different methods, ex. 5.1*

Methods	*'Cantilever'*	*Portal*	**' Exact '*	
Members			*Left end*	*Right end*
1	2.142	2.00	2.191	1.707
2	0.858	1.00	1.096	1.580
3	5.718	5.37	5.252	4.386
4	2.286	2.266	2.626	3.493
5	7.860	7.340	6.773	5.446
6	3.141	3.660	3.387	4.714
			Bottom end	Top end
7	2.142	2.000	1.809	2.191
8	3.000	3.000	2.548	2.803
9	0.858	1.000	1.070	1.580
10	3.576	3.340	3.223	3.444
11	4.464	5.000	4.397	4.464
12	1.428	1.660	2.049	2.423
13	4.264	4.000	4.450	3.550
14	6.537	6.000	4.893	4.435
15	1.713	2.000	4.008	2.665

* Flexural rigidities of columns, girders of 6 m span and girders of 12 m span equal to *EI*, 1.5 *EI* and 3 *EI* respectively were used while solving by Kani's method, 13 (cycles). In the case of the bottom end of the base column (member 15) with the smallest aisle width the moment values is considerably underestimated, and this is compensated by an overestimation of interior base column, Fig. Ex. 5.1.

The total shear V of different storeys are:

At top storey i.e. at level x_3x_3, Fig. Ex. 5.1(c),

$$V_{x3} = 3t$$

Similarly at x_2x_2 $\quad V_{x2} = 3 + 2 = 5t$, and

at x_1x_1 $\quad V_{x1} = 3 + 2 + 1 = 6t$.

Consider the level x_3x_3,

$$V_1 = k_1\, V_3 = 1$$

$$V_2 = k_2\, V_3 = 3/2$$

and $\quad V_3 = k_3\, V_3 = 1/2$

or symbolically

$$V_{x3} = (1, 3/2, 1/2)$$

Similarly the column shears at $x_2 - x_2$ and $x_1 - x_1$ are

$$V_{x2} = (5/3, 5/2, 5/6) \text{ and}$$

$$V_{x1} = (2, 3, 1)$$

With the evaluation of column shears, finding column-end moments and girder-end moments is simple and direct, and the computations are left to the reader to complete. The final results are given in Table 5.1.

5.4 LATERAL DEFLECTIONS

With increasing height of frames, the deflections at upper storeys may become critical. With the column moments specified, starting from bottom fixed end, it is possible to compute deflections of a column member which should be equal to frame deflections at floor levels. These computations for deflections do not use girder forces. It may be observed from the Table 5.1 that the error in moment values of the bottom column is relatively more. The deflections at higher floor levels are sensitive to the errors in moments near the fixed end and the deflection values so computed are unreliable. In the following a more reliable method based on virtual work considerations, is presented.

5.4.1 Deflection of Frames by Virtual Work Method

We have already used extensively the virtual work method for computing deflections, where the force or statical system is virtual system and the real structure is the deformation or kinematic system. The application for the frame analyzed by approximate method is shown with the help of same example (Ex. 5.1) considered in the preceding Articles. Let us consider the evaluation of relative lateral deflection, δ of the frame between the levels x_2x_2 and x_3x_3, Fig. Ex. 5.1(d) and Fig. 5.3. The moment m_i within the segment may be based on cantilever or portal method. Consider the results of the cantilever method, Fig. 5.1(d).

It is assumed that all columns at a section deflect by the same value Δ.

The statical system is drawn for unit shear V at levels x_3x_3 and x_2x_2. The statical moment diagram, $\bar{m}$, is drawn for $V = 1$, similar to the procedure used in portal method. The numerical values are shown in Fig. 5.3. The relative deflection Δ is given by relation

$$\Delta = \int \frac{m}{EI} \bar{m}\, ds \qquad \text{...(i)}$$

As m and $\bar{m}$ diagrams are simple triangles and if the end ordinates of the triangles are m_e and $\bar{m}_e$ for any segment i, the $\int \frac{m}{EI} \bar{m}\, ds$ for the triangular segments is given by

$L_i/6EI\,(2m_e\text{–}(\bar{m}_e)) = (m_e\, \bar{m}_e\, l_i)/3EI$

Where l_i is the length of segment covered by the triangular m and $\bar{m}$ diagram, Fig. 5.3.

Therefore,

$$\Delta = \frac{1}{3E} \sum_{i=1}^{n} \frac{m_e\, m_e\, l_i}{I_i} \qquad \text{...(ii)}$$

Where n is the number of segments. For the frame, Figure 5.3, the number of segments is ten, that is, six column segments and four girder segments (each girder has two triangular segments). The contribution to the right side of Eq. (ii), due to the six columns segments is equal to (for the problem let the columns be of the same moment of inertia equal to I):

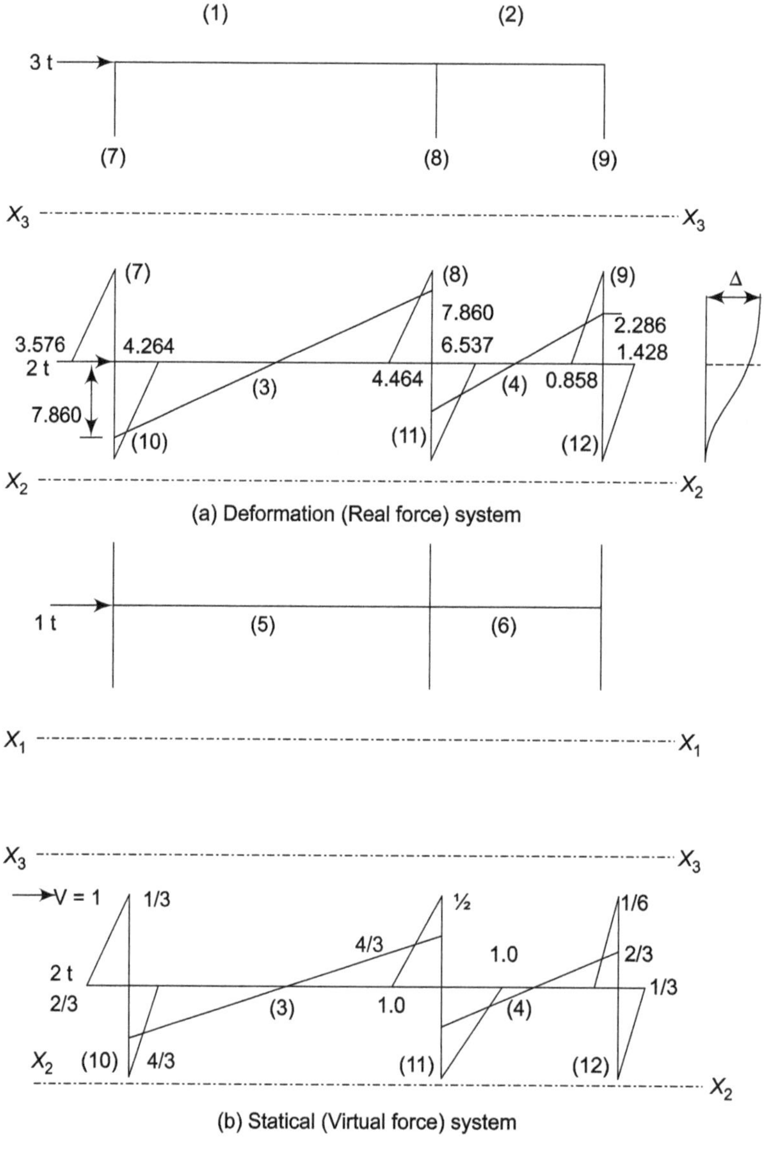

(a) Deformation (Real force) system

(b) Statical (Virtual force) system

Fig. 5.3 *Deflection by Virtual Work Mehod*

$$\frac{(l_i = 2)}{3EI}\left[(2.142 + 3.576)\frac{2}{3} + (3.576 + 4.464)\,1.0 + (0.858 + 1.428)\frac{1}{3}\right]$$

$$= 8.02/EI$$

The contribution to the right side of the Eq. (ii), due to the four girder segments is equal to (let the left and right girders be of moment of inertia equal to 31 and 1.51 respectively)

$$\frac{6}{3E(3I)}\left[2(5.718)\frac{4}{3}\right]+\frac{6}{3E(1.5I)}\left[2(2.286)\frac{2}{3}\right]$$

$$= \frac{1}{EI}(14.23)$$

$$\therefore \qquad \Delta = \frac{1}{EI}(8.02+14.23)$$

$$\Delta = 22.25/EI$$

Similarly the relative deflections between $x_1 - x_1$ and $x_2 - x_2$, and x_3x_3 and top can be evaluated. The deflection between the bottom 0 and x_1x_1 is actual deflection at x_1. Al these deflections when summed up give the deflection of the top (roof).

The deflections, Δ, are sensitive to the *EI* values which are altogether ignored while computing moments in the approximate methods. Had we considered the girder flexural rigidities equal to 2*EI* and *EI*, then the deflection $\Delta = 29.37/EI$.

However, the values of the moments given by exact analysis are not sensitive to the similar changes in *EI* values and are reasonably close to the values given by approximate methods.

Because of wall paneling, the stiffness of the frames are increased considerably and therefore the actual deflections are accordingly reduced. The accuracy with which the deflections for the approximate methods are computed should be adequate.

Ex. 5.3. Analyse the frame of Ex. 2.28 of Chapter II by an approximate method, Fig. Ex. 5.3(a).

In the case of single bay multistory frame, the cantilever and portal methods give identical results. The final bending moment diagram is given, Fig. Ex. 5.2(b). The 'exact' values for the frame are given in the alternative solution of the Ex. 2.2 of Chapter II. The reader may like to compute deflections using virtual work method for the deformation system (m diagram) given by the approximate solution and the exact solution and check the reliability of deflections computed by the approximate method. The statical (virtual) system will be similar to Fig. 5.3(b). In this case, the computations are much simpler.

The results given in Fig. Ex. 5.2(b) were used for the iteration of the simultaneous equations (alternative solution) based on flexibility method for the Ex. 2.27, Chapter 2.

Concluding Remarks: While solving the problems by the approximate methods, flexural rigidities (*EI*) and axial stiffnesses (*EA*) are not used. Several modifications for locating the points of inflection for improving the accuracy of the methods have been suggested in the available literature, which have not been included here for two reasons, (1) to retain the methods in the simplest form as first order solutions and (2) for improved accuracy, it is more advisable to use displacement methods like moment distribution and Kani's iteration. The errors observed in the base columns may be attributed to the actual inflection points not occurring at mid-heights. The relative stiffnesses of girder to column have considerable effect on location of inflection points for the base columns. The moment at the bottom of a base column is always larger than its top moment. Sometimes when the floor stiffnesses are relatively low, the point of inflection may not occur at all in the lowest storey. The floor stiffnesses are low when only slabs or slabs with shallow beams of long spans are used. In the case of relatively rigid flooring system, while analysing such frames by portal methods, the column shears may be proportioned in the ratio of lateral stiffnesses ($12\,EI_i/h^3$) instead of aisle lengths λ_i. Adaptation of these methods for building frames with shear walls is not included.

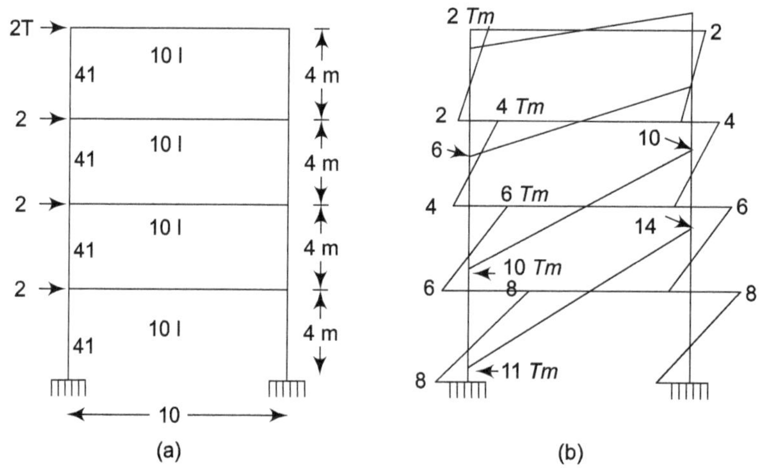

Fig. Ex. 5.3 *Same as the Frame Considered in Chapter 2*

PART B

6. Displacement Method
7. Displacement Method-Illustrations
8. Iterative Techniques
9. Introduction to Applied Dynamics and Design of Tall Buildings

Chapter 6: Displacement Method

Chapter

6

Displacement Method

It is advised to go through the following brief section, 'General'; and for a proper appreciation of the Displacement method, the reader is advised to come back to the same section 'General' after completing the study of the entire Chapter 6. **It is reiterated that the theory developed assumes small deformations.** The same assumption was used while sketching collapse mechanisms while presenting Plastic analysis of Plane frames. In the text even while sketching mechanism of a truss small deformations are assumed. Large deformations, if considered result in geometric non linearity, and it is excluded from the scope. Sometimes the same numerical example is considered for illustrating different approaches of the analysis. The examples included in this chapter are essential to the understanding of the theory. The numerical examples are selectively chosen as illustrations to complement the theory. In the computer era, it is felt that the large number of numerical examples deluges the reader. It is believed that mostly every serious reader with necessary preparations has the potentialities of innovation as well as creation. Every structure designed has scope for innovation. In order to achieve such skills for innovation all along, be it theory or illustrations, concepts are more relevant.

Illustrations and a popular structural system known as "Mero System", are included in Chapter 7. The two chapters, six and seven are interrelated.

6.1 GENERAL

The title suggests that the unknowns are the displacements and the method is known as **displacement method**. It is converse of **force method**. Because Equilibrium conditions are used for arriving at the Global Stiffness equation, the displacement method is also known as **equilibrium method**. The member properties are used while writing the member stiffnesses in contrast to the member flexibilities in the

Force method, which also uses member properties). The method is also known as **Stiffness method**, because the stiffnesses are used at member level (step 2 of the basic steps), and more importantly, we write the structural stiffness at the global level.

In as much as choice of release system is important for the force method, while using the Displacement method, the modeling of a given structure before the analysis in respect of nodes and joints need to be done with care with the aim to keep the degrees of freedom to a small number. The degrees of freedom equal to the unknown displacements depends on the number of joints and locations of the nodes at locations other than the joints. The illustrations considered are mostly straight prismatic skeletal members, and these include trusses, beams, plane frames, grids, and space frames. The degrees of freedom are also governed by approximation of inextensibility, that is, negligible effect of axial deformations on bending moments and shears. For manual calculations particularly the assumption of inextensibility reduces the computational effort quite significantly. Wherever inextensibility is used, the load matrix and the corresponding degrees of freedom need to be understood properly. All these aspects have been discussed while formulating the equations for the unknown displacements. The slope deflection method is illustrated as a particular case of displacement method applied to small frames with the assumed inextensibility. In the case of continuous beams, however, axial deformations get delinked from flexural deformations. While discussing trusses with hinged joints, the reader is cautioned against likely mechanisms. Sketching of such mechanisms of unstable trusses is also included. Sometimes a hybrid approach, combined force and displacement methods are computationally very efficient. The types of examples suitable for the mixed methods and associated problems are illustrated.

It will be obvious from the following that Equilibrium is the consideration in the formulation of equations (step 3 or final step of the basic steps) for the unknown DISPLACEMENTS. These are the numerical values of displacements pertaining to the joints. Each numerical value is associated with the corresponding deformed configuration. With every numerical value of displacement, it is important to visualize the associated deflected shapes. In the following, the sketching of deflected shapes are emphasized. For a proper conceptualization, the deflected shapes and likely inflection points are the skills which are important for any designer. While presenting the mathematical contents, an attempt has been made to make the reading smooth and maximum of conceptualization.

These are also referred as GLOBAL DISPLACEMENTS. The global displacements are used differently for (1) Connection Matrix approach and (2) Direct Stiffness approach. Both the approaches belong to the Displacement Method.

6.1.1 Basic Steps

Analogous to the **force method**, conversely the basic steps are:

Step 1, Compatibility: It essentially describes the compatibility relationship of member-end deformations and the global deformations. (Displacements and deformations are used synonymously.) The member axes of a skeletal-straight-prismatic member of a plane frame are designated as (x, y) where the x-axis is a centroidal line (line obtained by connecting the centroid of the cross sections). The end-1 pertains to the commencement of the x-axis, and end-2 to the end of the member axis.

Step 2, Force-deformation relationship: The relationship is given by the member stiffness, k_m

A plane frame with rigid joints is considered for explaining the theory. The details are developed under Direct Stiffness Approach.

$P_m = k_m d_m$, where $P_m = \{P_{m1}\ P_{m2}\}$ and $P_{mi} = \{P_{mx}\ P_{my}\ m\}_i$. The index i takes the values 1 or 2.

Corresponding to $P_m = \{P_{m1}\ P_{m2}\}$, $D_m = \{d_{m1}\ d_{m2}\}$ and $d_{mi} = \{\delta_{mx}\ \delta_{my}\ \theta\}_i$

The sign conventions for the member-end forces are given in Figs 6.1 and 6.2.

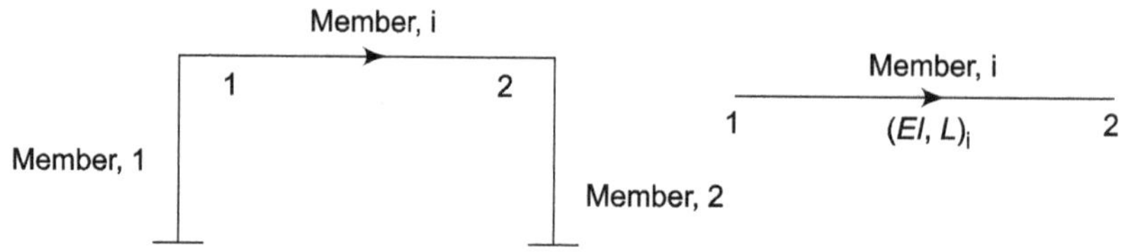

Fig. 6.1 *Plane Frame with a Typical Member i*

Stiffness of a member of plane frame, Eq. (i),

$$K_m = \begin{bmatrix} K_{11} & K_{12} \\ K_{21} & K_{22} \end{bmatrix} \text{ and}$$

$$K_{11} = \begin{bmatrix} \frac{EA}{L} & 0 & 0 \\ 0 & \frac{12\,EI}{L^3} & \frac{6\,EI}{L^2} \\ 0 & \frac{6\,EI}{L^2} & \frac{4\,EI}{L} \end{bmatrix}$$

$$K_{12} = \begin{bmatrix} -\frac{EA}{L} & 0 & 0 \\ 0 & -\frac{12\,EI}{L^3} & \frac{6\,EI}{L^2} \\ 0 & -\frac{6\,EI}{L^2} & \frac{4\,EI}{L} \end{bmatrix} \text{and } K_{21} = K_{12}^T$$

$$K_{22} = \begin{bmatrix} \frac{EA}{L} & 0 & 0 \\ 0 & \frac{12\,EI}{L^3} & -\frac{6\,EI}{L^2} \\ 0 & -\frac{6\,EI}{L^2} & \frac{4\,EI}{L} \end{bmatrix} \qquad \text{...(i)}$$

The above member stiffnesses, Eq. (i) are derived in Article 6.4.1.

In the case of member of a space frame, the corresponding stiffnesses K_{11}, $K_{12} = K_{21}$, and K_{22} are given, in the following and Fig. 6.3.

$$
K_{11} = \begin{array}{c} \\ \end{array}
\begin{array}{cccccc}
\delta_{x1} & \delta_{y1} & \delta_{z1} & \theta_{x1} & \theta_{y1} & \theta_{z1}
\end{array}
$$

$$
K_{11} = \left[\begin{array}{ccc|ccc}
\frac{EA}{L} & 0 & 0 & 0 & 0 & 0 \\
0 & \frac{12\,EI_z}{L^3} & 0 & 0 & 0 & \frac{6\,EI_z}{L^2} \\
0 & 0 & \frac{12\,EI_y}{L^3} & 0 & -\frac{6\,EI_z}{L^2} & 0 \\
0 & 0 & 0 & \frac{GJ}{L} & 0 & 0 \\
0 & 0 & -\frac{6\,EI_y}{L^2} & 0 & \frac{4\,EI_y}{L^3} & 0 \\
0 & \frac{6\,EI_z}{L^2} & 0 & 0 & 0 & \frac{4\,EI_z}{L^2}
\end{array}\right]
$$

$$
\begin{array}{cccccc}
\delta_{x1} & \delta_{y1} & \delta_{z1} & \theta_{x1} & \theta_{y1} & \theta_{z1}
\end{array}
$$

$$
K_{21} = [K_{12}]^t = \left[\begin{array}{ccc|ccc}
-\frac{EA}{L} & 0 & 0 & 0 & 0 & 0 \\
0 & -\frac{12\,EI_z}{L^3} & 0 & 0 & 0 & -\frac{6\,EI_z}{L^2} \\
0 & 0 & -\frac{12\,EI_y}{L^3} & 0 & \frac{6\,EI_y}{L^2} & 0 \\
0 & 0 & 0 & -\frac{GJ}{L} & 0 & 0 \\
0 & 0 & -\frac{6\,EI_y}{L^2} & 0 & \frac{4\,EI_y}{L^3} & 0 \\
0 & \frac{6\,EI_z}{L^2} & 0 & 0 & 0 & \frac{2\,EI_z}{L}
\end{array}\right]
$$

K_{22} has off diagonal elements which are shears at the opposite ends and therefore take signs different from the off diagonals of $k_{11.}$ The member-end forces and deformation are shown in Fig. 6.3.

$$
\begin{array}{cccccc}
\delta_{x1} & \delta_{y1} & \delta_{z1} & \theta_{x1} & \theta_{y1} & \theta_{z1}
\end{array}
$$

$$
K_{22} = \left[\begin{array}{ccc|ccc}
0 & 0 & 0 & 0 & 0 & 0 \\
0 & 0 & 0 & 0 & 0 & \frac{-6\,EI_z}{L^2} \\
0 & 0 & 0 & 0 & \frac{6\,EI_y}{L^2} & 0 \\
0 & 0 & 0 & 0 & 0 & 0 \\
0 & 0 & \frac{6\,EI_y}{L^2} & 0 & 0 & 0 \\
0 & \frac{-6\,EI_z}{L^2} & 0 & 0 & 0 & \frac{4\,EI_z}{L}
\end{array}\right]
$$

Step-3, Equilibrium Conditions: The equilibrium of appropriately chosen free bodies or from virtual work considerations establishes the equations for the unknown displacements. The unknown displacements are also referred as degrees of freedom. In fact, when we are describing the joint coordinates, we are in a way referring to the unknown displacements pertaining to the joints. The number of the unknowns is given by the number of joints multiplied by multipliers depending upon the properties of the joints.

6.1.2 Degrees of Freedom

The multipliers to the number of joints depend on the type of structure. The degrees of freedom for the different types of structures are:

- Plane Truss (Hinged joints): two translations per joint
- Space truss (Hinged joints): three translations per joint
- Plane frame (Rigid jointed and inextensibility not assumed): three per joint (two translations and one rotation)
  If inextensibility is assumed: 3(Number of joints) - Number of inextensible members
- Space frame (Rigid jointed and inextensibility not assumed): six per joint (three translations and three rotations)
- Tall Building frames: The degrees of freedom considering the number of Floor slabs, and inextensible girders are discussed in Chapter 9

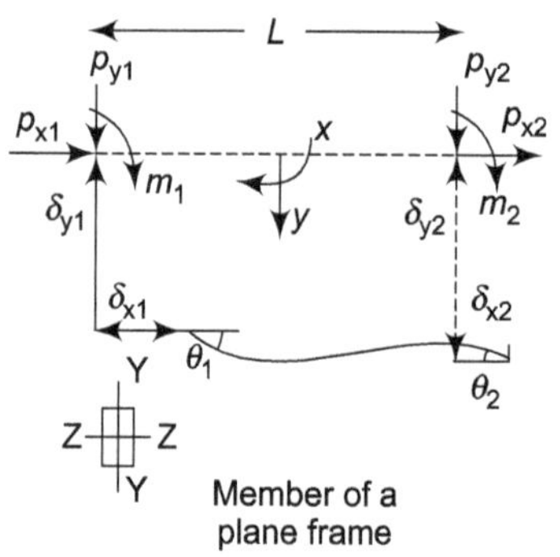

Fig. 6.2 *Member of a Plane Frame*

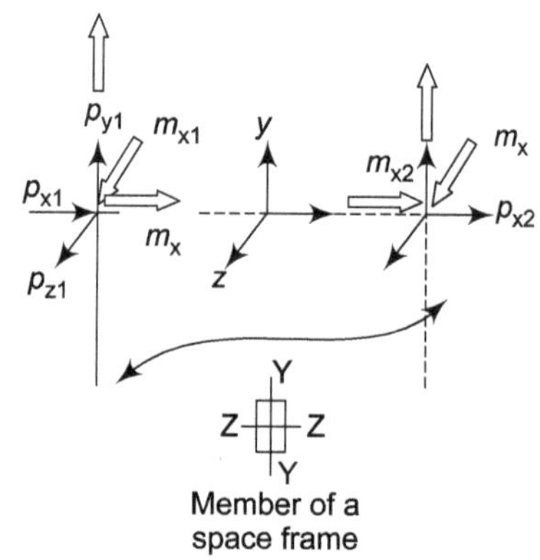

Fig. 6.3 *Member of a Space Frame*

6.1.3 Plane Hinged Truss

Consider a Plane Truss in the following with joints hinged. The members are under axial forces, p_{x1} and p_{x2}, and the corresponding member stiffness is given by Eq. (ii).

$$\begin{pmatrix} p_{x1} \\ p_{x2} \end{pmatrix} = \begin{pmatrix} \frac{EA}{L} & -\frac{EA}{L} \\ -\frac{EA}{L} & \frac{EA}{L} \end{pmatrix} \begin{pmatrix} d_{x1} \\ d_{x2} \end{pmatrix} \quad \ldots\text{(ii)}$$

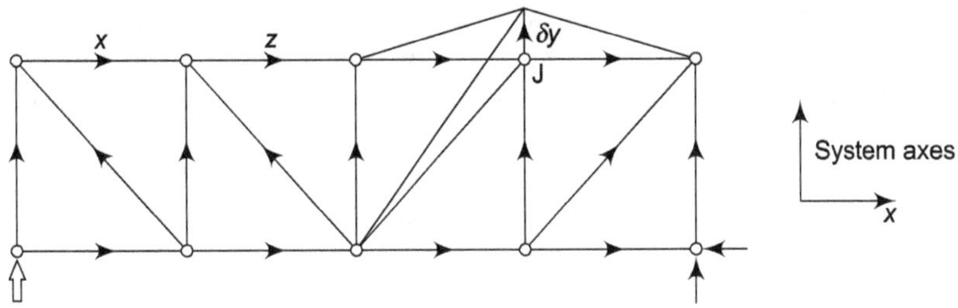

Members effected by the joint displacement δy

Fig. 6.4(a) *Typical Vertical Joint Displacement of a Plane Truss*

6.2 DEFLECTED SHAPES

6.2.1 Deflected Shape of Hinge Jointed Truss

(a) Displacement shapes of a stable truss: The values of the displacements are the ordinates of the displacement shapes. A shape is associated with each values of displacement. The deflected shape associated with δy of a truss is shown, Fig. 6.4(a). Correlate the shape with the assumption of small deformation and the hinged joints of the truss.

Appreciate the statement conforming to the assumption that the vertical member and the diagonal meeting the joint are stressed (stretched or affected); whereas the horizontals meeting, the joint *J*, Fig. 6.4, are not stressed. Why? Think. Referring to the horizontals, in the case of small deformations, $\sin \theta = \tan \theta = \theta$.

6.2.2 Continuous Beam

A typical rotation θ_c of a continuous beam on rigid support is shown in Fig. 6.4(b).

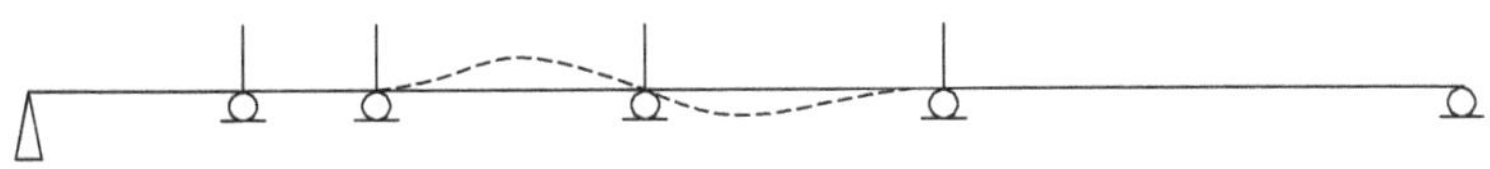

Rotation of a typical joint at support C of a Continuous Beam

Fig. 6.4 (b) *Displaced Configuration of Stable Structure – Continuous Beam*

6.2.3 Sketching of Deflected Shape of a Typical Plane Frame

Small deformations and Inextensibility.

Similar exercises are undertaken for beams, plane frames with inextensibility and without inextensibility. Refer the Fig. 6.4. Member affected are identified by .... It is a good preparation for writing the connection matrices or Compatibility matrices, that establish the relationship between (a) the member-end deformations of all the members of the frame with (b) the Joint or global deformations.

Even though we are assuming "small deformations, the displacements (deformations) are magnified while sketching the deflected shapes". The scale used for sketching the displacements is considerably magnified. The plastic analysis presented in Chapter 3 applied to plain frames also assumed "small deformations", while magnifying the mechanisms.

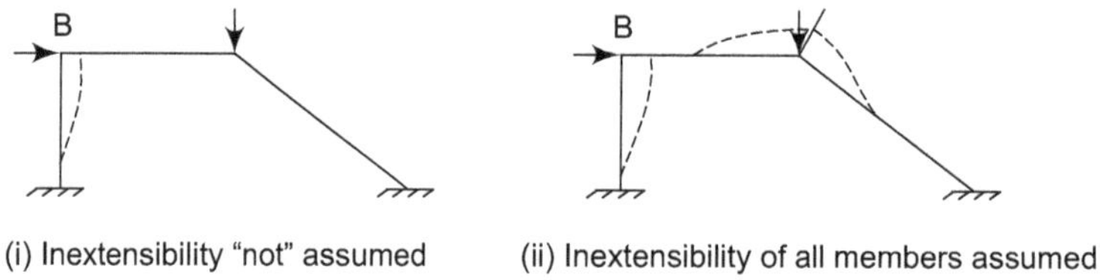

Fig. 6.4(c) *Lateral Displacement of Joint B: Two Cases of (i) Axial Deformations Allowed, and (ii) Inextensibility*

For the case of the members assumed inextensible, the degrees of freedom = 3
θ_A, θ_B are also sketched in the following:
Corresponding load elements are:

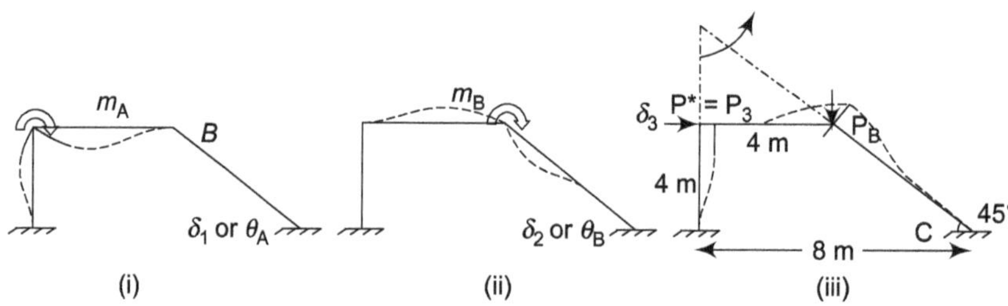

Fig. 6.4(d) *Sketching of Three Degrees of Freedom of Plane Frame*

P^* corresponding to P_3 cannot be visualized and it is equal to $\boldsymbol{P_A \times 1 - P_B \times 1}$ refer Figure.

6.3 DISPLACED SHAPE OF UNSTABLE STRUCTURES

While discussing plastic analysis of plane frames, several mechanisms of unstable plane frames were sketched. For a change, let us consider a hinged truss. If the diagonal meeting at the joint *J*, is removed, you can sketch a displaced configuration without stretching any member. Such a truss is a 'Mechanism'; it is being displaced without any load acting on the truss.

The degrees of freedom are uniquely fixed. Whereas in the case of Force Method after finding out the degree of indeterminacy the choice of release system is not unique. The sketching of mechanism is recalled in the following, Fig. 6.5.

Mechanism Explained

- The truss is made unstable by removing the diagonal *CL*, Fig. 6.5,
- *ABCD* and *FGHL* are rigid bodies, and their centers of rotation are '*A*' and '*H*' respectively.
- Assume θ rotation. The new position of *ABCD*, is shown as *AB′C′D′* (*A*, the hinge is centre of rotation),
- Taking a clue from *D′*, the node *L* can move vertically only. *H* is centre of rotation of *FGHI*,
- Taking a clue from *C*, inextensibility is ensured, by moving *F* upward,

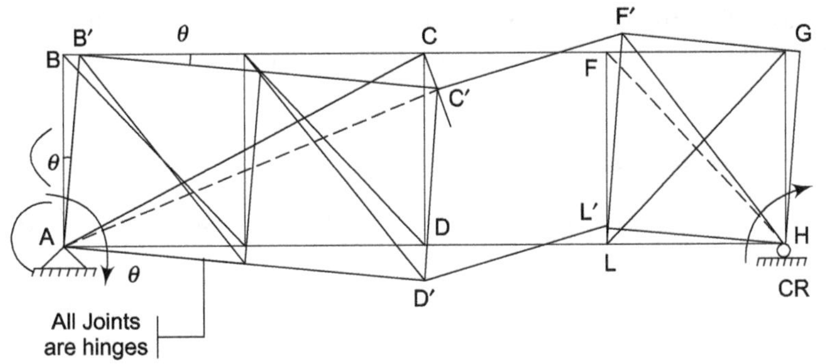

Fig. 6.5 *Mechanism Caused by Missing Diagonal Explained*

- Based on property of roller at *H*, and *L* moving vertically upward, *H* is a Centre of Rotation, and
- The displaced configuration of the diagonals is shown dotted. The dotted lines *AC* and *HF* represent the rigid bodies *ABCD* and *FGHL* respectively.

The topic of the analysis of trusses is comprehensively discussed in Article 6.7 with emphasis on fabrication of trusses for large spans.

6.4 TWO APPROACHES OF DISPLACEMENT METHOD

The two approaches are: (1) Direct Stiffness Approach, and (2) Connection Matrix Approach.

If the 'inextensibility' of the members constituting a joint is not assumed, the equilibrium of the joints is provided by the members meeting at each joint. The Direct Stiffness Approach (DSM) does not assume inextensibility. A minimum of two members constitute a joint or node. If there are, say, four members meet at a joint; the equilibrium is governed by those four members only. The approach for different cases is illustrated by the illustrations. Obviously the DSM is preferred if the inextensibility of the members is not valid. But there are some cases like tall buildings, the floor slab and girders are practically inextensible and it limits the degrees of freedom for the analysis due to the earthquake effects. In such cases the equilibrium equations are revised and the property of inextensibility is incorporated subsequently considering 'master' and 'slave' joints.

In the case of 'Connection Matrix Approach (CMA), it connects the member deformations (d_m) with the joint deformations (d_j). The compatibility matrix '*A*' in the relation, Eq. (6.1) is based on the inextensibilities assumed.

$$\mathbf{d}_m = \mathbf{A}\,\mathbf{d}_j \qquad \text{...(6.1)}$$

If may be pointed out here that the load element corresponding to the independent joint displacement is obtained from the principle of 'contra gradience',

$$\mathbf{P}_j = \mathbf{A}^T\mathbf{p}_m \qquad \text{...(6.1a)}$$

The CMA does not require the transformations of member forces and member deformations from the member axes to the global axes. In the case of DSM the member force $\mathbf{p}_m$ in member coordinates are required to the transformed to the member force $\mathbf{p}'_m$ in flobal or system coordinates for assembling the equilibrium matrix,

$$\mathbf{P}'_m = \mathbf{K}'_j\,\mathbf{d}'_j \qquad \text{...(6.1b)}$$

The systematization for satisfying compatibility, a substitute to the connection matrix for computer coding is provided by the 'Code Number Technique'. It being a technique, the code numbers are explained through typical illustrations in the following.

6.4.1 Direct Stiffness Approach

The structures considered are **(i) Plane Frame, (ii) Grids, (iii) Space Frame,** and **(iv) Space Truss (MERO).**

Plane Frame: Matrix representation of the Displacement Method. To begin with 2D frames with in-plane loading will be considered first. Exact analysis without omitting axial deformation will be considered first, to be followed by approximate analysis and matrix condensation. In the case of matrix approach, it is more convenient to express a given loading system by an equivalent joint load and member-end forces can then be described in terms of end deformations only. These aspects, particular to matrix approach will be discussed elsewhere.

Step 1: The compatibility conditions have already been discussed, $d_m = A\, d_j$.

Step 2 (a): Force deformation relations in member coordinates (2-D frame member, in-plane loading).

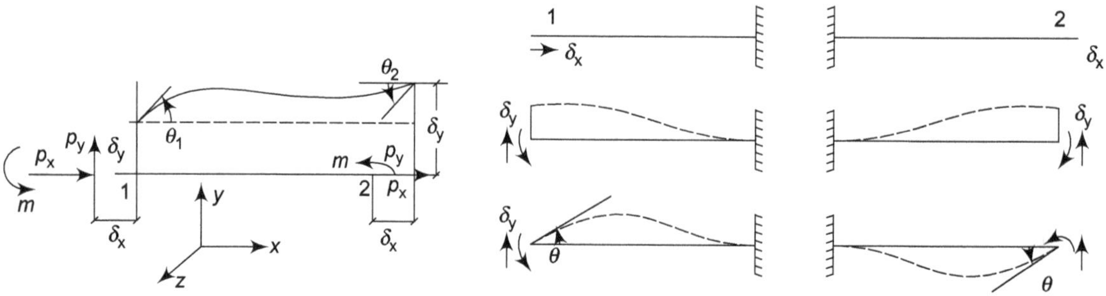

(a) (b) Super position

Fig. 6.6 *End Deformations*

$$P_1 = \begin{bmatrix} P_X \\ P_Y \\ m_1 \end{bmatrix}, P_2 = \begin{bmatrix} P_X \\ P_Y \\ m_2 \end{bmatrix}$$

It can be shown from Fig. 6.5 that the end forces P_1 and P_2 due to d_1 alone are:

Due to $(\delta_x)_1 = 1$; $(P_x)_1 = \dfrac{EA}{L}$, $(P_y)_1 = 0$; $m_1 = 0$; $(P_x)_2 = -\dfrac{EA}{L}$, $(P_y)_2 = 0$; $m_2 = 0$;

Due to $(\delta_y)_1 = 1$; $(P_x)_1 = 0$, $(P_y)_1 = \dfrac{12\,EI}{L^3}$, $m_1 = \dfrac{6\,EI}{L^2}$; $(P_x)_2 = 0$; $(P_y)_2 = \dfrac{-12\,EI}{L^3}$, $m_2 = \dfrac{6\,EI}{L^2}$

Due to $\theta_1 = 1$; $(P_x)_1 = 0$, $(P_y)_1 = \dfrac{6\,EI}{L^2}$, $m_1 = \dfrac{4\,EI}{L}$; $(P_x)_2 = 0$; $(P_y)_2 = \dfrac{6\,EI}{L^2}$, $m = \dfrac{2\,EI}{L}$

When the above relations are expressed in matrix form

$$\underbrace{\begin{bmatrix} P_x \\ P_y \\ m \end{bmatrix}_1}_{P_1} = \underbrace{\begin{bmatrix} \dfrac{EA}{L} & 0 & 0 \\ 0 & \dfrac{12EI}{L^3} & \dfrac{6EI}{L^2} \\ 0 & \dfrac{6EI}{L^2} & \dfrac{4EI}{L} \end{bmatrix}}_{K_{11}} \underbrace{\begin{bmatrix} \delta_x \\ \delta_y \\ \theta \end{bmatrix}_1}_{d_1} ; \underbrace{\begin{bmatrix} P_x \\ P_y \\ m \end{bmatrix}_2}_{P_2} = \underbrace{\begin{bmatrix} -\dfrac{EA}{L} & 0 & 0 \\ 0 & \dfrac{-12EI}{L^3} & \dfrac{-6EI}{L^2} \\ 0 & \dfrac{6EI}{L^2} & \dfrac{2EI}{L} \end{bmatrix}}_{K_{21}} \underbrace{\begin{bmatrix} \delta_x \\ \delta_y \\ \theta \end{bmatrix}_1}_{d_1} \quad \text{...(i)}$$

Alternatively we may write due to d_1 alone,

$$P_1 = K_{11}\, d_1 \quad \text{and} \quad P_2 = K_{21}\, d_1 \quad \text{(ii)}$$

Where K_{11} and K_{21} are defined as per the equalities.
Similarly, due to deformation d_2 alone we can obtain the relations,

$$P_1 = K_{12}\, d_2 \text{ and } P_2 = K_{22}\, d_2 \quad \text{...(iii)}$$

Where

$$K_{12} = \begin{bmatrix} \dfrac{-EA}{L} & 0 & 0 \\ 0 & \dfrac{-12EI}{L^3} & \dfrac{6EI}{L^2} \\ 0 & \dfrac{-6EI}{L^2} & \dfrac{2EI}{L} \end{bmatrix} \text{ and } K_{22} = \begin{bmatrix} \dfrac{EA}{L} & 0 & 0 \\ 0 & \dfrac{12EI}{L^3} & \dfrac{-6EI}{L^2} \\ 0 & \dfrac{-6EI}{L^2} & \dfrac{4EI}{L} \end{bmatrix} \quad \text{...(iv)}$$

Superposing the results we write

$$\left.\begin{aligned} &P_1 = K_{11}\, d_1 + K_{12}\, d_2 \text{ and} \\ &P_2 = K_{21}\, d_1 + K_{22}\, d_2 \\ &\text{Or when written in the matrix form} \\ &\begin{bmatrix} P_1 \\ P_2 \end{bmatrix} = \begin{bmatrix} K_{11} & K_{12} \\ K_{21} & K_{22} \end{bmatrix} \begin{bmatrix} d_1 \\ d_2 \end{bmatrix} \text{ member} \end{aligned}\right\} \quad \text{...(6.2)}$$

Where each sub matrix (K_{11}, K_{12}, K_{22}) is of size 3 × 3. It may be verified from Eq. 6.2 that

$$K_{12} = (K_{21})^t.$$

The development of the theory for grids, space truss and space frame is given separately, Article 6.12.

6.4.2 Connection Matrix Approach

In the following all the Members are assumed inextensible. The method is illustrated with the help of a typical plane frame.

6.4.2.1 Illustration 1 – Two Bay Plane Frames

Two Bay Plane Frames: After computing the nodal loads from the fixed edge forces (nodes clamped). The number of nodal displacements under the action of the nodal loads, depends upon the assumption regarding inextensibility of member and type of the frame. In this case of plane frames, the number of unknown displacements or degrees of freedom is equal to $2n$-m. As an illustration, consider plane frame. The Examples are so chosen that these are amenable to hand computations, essential for conceptualization. The code number technique is explained with the help of the illustration. The master and slave nodes are related to the 'code number' technique. The frame has only one translation, and the other three are rotations.

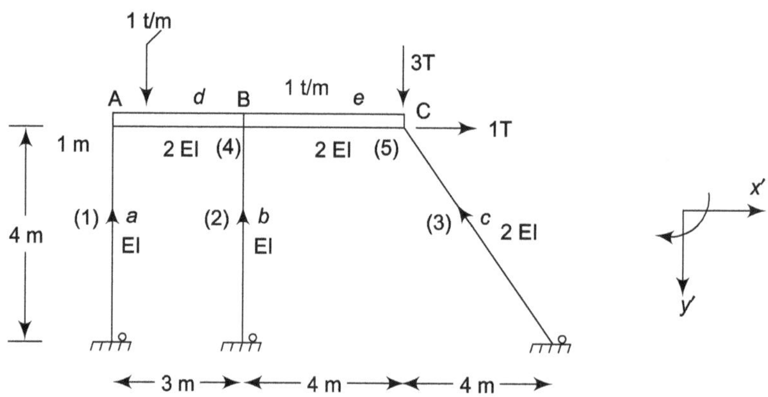

Fig. 6.7 *Illustration 1a–Data*

Displacement Analysis Part I: Analysis for the clamped nodes

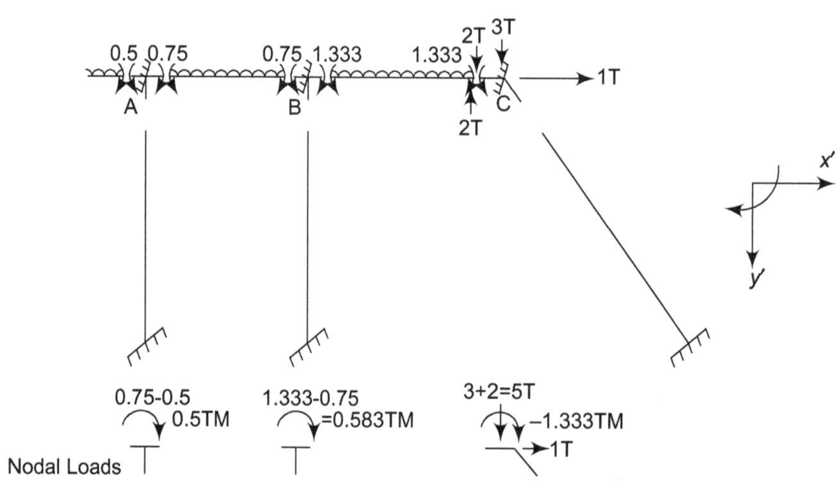

Fig. 6.7 *Illustration 1b–Fixed End Forces*

Displacement Analysis Part II: Inextensible Members

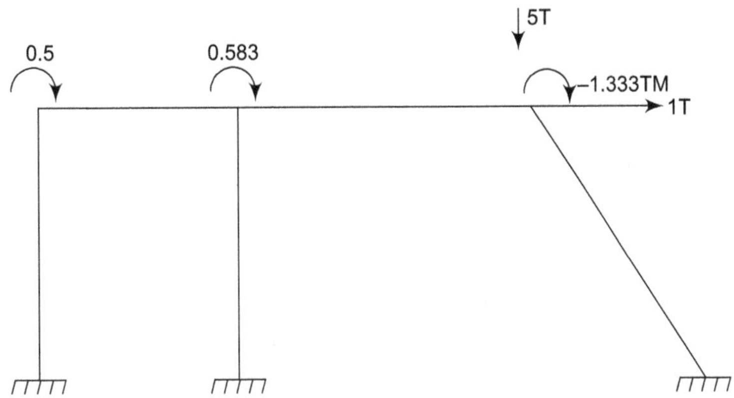

Fig. 6.7 *Illustration 1c–Final Nodal Loads*

Number of displacements = $[3n - m]$
Number of nodes, $n = 3$,
Number of inextensible members, $m = 5$(all members assumed inextensible)
Number of the unknown displacements = 4 and they are [θ_1, θ_2, θ_3 and δ_4] which are sketched in following:

Step 1: Compatibility, $d_m = A\ d_j$ (Eq. 6.1 reproduced) ...(i)

Step 2: Generalized stress strain relation, $p_m = K_m\ d_m$ ...(ii)

Displacement θ_1, members affected *a* and *d*

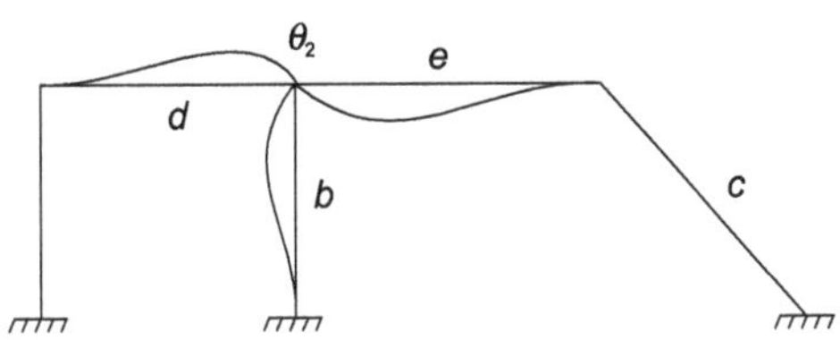

Displacement θ_2, members affected *b*, *d*, *e*

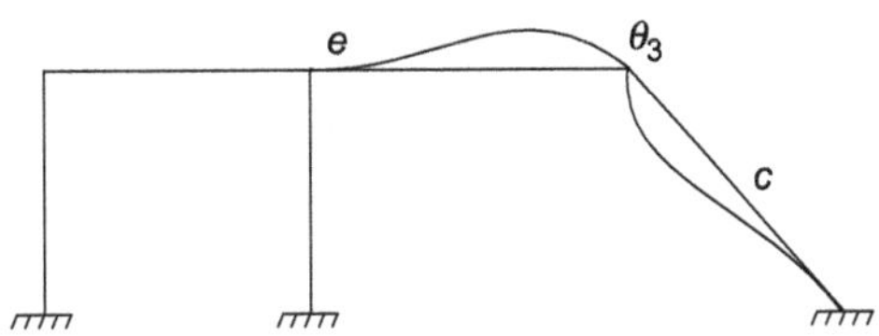

Displacement θ_3, members affected *c*, *e*

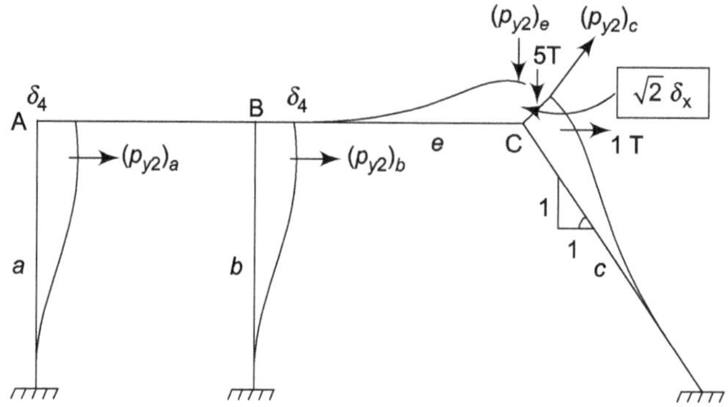

Slave Nodes B and C Dependent on the δ_4 of the node A

Fig. 6.8 *Illustration 1–Degrees of freedom*

It may be noted that for δ_4, 'A' is the 'master' node and B, C are the 'slave' nodes. For writing equilibrium equations, virtual displacements diagrams will be used (virtual work method). The shape of the four virtual displacement diagrams will be chosen to be the same as the true unknown displacement diagrams, Fig. 6.7.

$$\delta_1{}^* = (\theta_1 = 1),\ \delta_2{}^* = (\theta_2 = 1),\ \delta_3{}^* = (\theta_3 = 1),\text{ and } \delta_4{}^* = (\delta_4 = 1)$$

And the equation based on virtual work principle are

$$M_1\ \delta_1{}^* = (m_2)_a\ \delta_1{}^* + (m_1)_b\ \delta_1{}^*$$

$$m_1 = 0.5$$

$$M_1 = (m_2)_a + (m_1)_b \qquad \text{...(i)}$$

$$M_2 = (m_2)_b + (m_2)_d + (m_1)_e \qquad \text{...(ii)}$$

$$M_2 = 0.583$$

$$M_3 = (m_2)_c + (m_2)_e \qquad \text{...(iii)}$$

$$M_3 = -1.333$$

$$1 \times \delta_1{}^* - 5\ \delta_4{}^* = (p_{y2})_a\ \delta_4{}^* + (p_{y2})_b\ \delta_4{}^* + (p_{y2})_c\ \sqrt{2}\ \delta_4{}^* + (p_{y2})_e\ \delta_4{}^*$$

The shears p_{y2} for the relevant members are shown on the Fig. 6.8

Taking $\delta_4{}^* = 1$

$$P_4 = (p_{y2})_a + (p_{y2})_b + \sqrt{2}(p_{y2})_c + (p_{y2})_e \qquad \text{...(iv)}$$

$$P_4 = -4$$

The equilibrium equations (i) to (iv) give

$$M_1 = K_{11}\theta_1 + K_{12}\theta_2 + K_{13}\theta_3 + K_{14}\theta_4$$
$$M_2 = K_{21}\theta_1 + K_{22}\theta_2 + K_{23}\theta_3 + K_{24}\delta_4$$
$$M_3 = K_{31}\theta_1 + K_{32}\theta_2 + K_{33}\theta_3 + K_{34}\delta_4$$
$$P_4 = K_{41}\theta_1 + K_{42}\theta_2 + K_{43}\theta_3 + K_4\delta_4$$

The contribution to k_{ij} in the above equation are be computed member use in the following table:

Table 6.1 *Correlation of member codes and Global stiffness coefficients Initialize $k_{ij} = 0$ Members a, b, c, d, e are digitized and designated 1, 2, 3, 4, 5.*

Member	*Code No. Member Code*	*Co-efficient* k_{ij}
1	(0; 1, 4)	$K_{11} = 0 + \left(\frac{4EI}{L}\right)_1$ $K_{44} = 0 + \left(\frac{12EI}{L^3}\right)_1$ $K_{14} = 0 + \left(-\frac{6EI}{L^2}\right)_1$
2	(0; 2, 4)	$K_{22} = 0 + \left(\frac{4EI}{L}\right)_2$ $K_{44} = \left[\frac{12EI}{L^3}\right]_1 + \left(\frac{12EI}{L^3}\right)_2$ $K_{24} = 0 + \left(-\frac{6EI}{L^2}\right)_1$
3	(0; 3, 4)	$K_{33} = 0 + \left(\frac{4EI}{L}\right)_3$ $K_{44} = \left[\left(\frac{12EI}{L^3}\right)_1 + \left(\frac{12EI}{L^3}\right)_2\right] + \left(\sqrt{2} \times \frac{12EI}{L^3}\right)_3 \sqrt{2}$ $K_{34} = 0 + \left(-\frac{6EI}{L^2}\right)_3 \sqrt{2}$
4	(1; 2, 0)	$K_{11} = \left[\left(\frac{4EI}{L}\right)_1\right] + \left(\frac{4EI}{L}\right)_4$ $K_{22} = \left[\left(\frac{4EI}{L}\right)_2\right] + \left(\frac{4EI}{L}\right)_4$ $K_{12} = 0 + \left(\frac{2EI}{L}\right)_4$

(*Contd.*)

(Contd.)

Member	*Code No. Member Code*	*Co-efficient* k_{ij}
5	(2; 3, 4)	$K_{22} = \left[\left(\frac{4EI}{L}\right)_2 + \left(\frac{4EI}{L}\right)_4\right] + \left(\frac{4EI}{L}\right)_5$ $K_{33} = \left(\frac{4EI}{L}\right)_3 + \left(\frac{4EI}{L}\right)_5$ $K_{44} = \left[\left(\frac{12EI}{L^3}\right)_1 + \left(\frac{12EI}{L^3}\right)_2 + \left(\frac{24EI}{L^3}\sqrt{2}\right)_3\right] + \left(\frac{12EI}{L^3}\right)_5$ $K_{23} = 0 + \left(\frac{2EI}{L}\right)_5$ $K_{24} = \left[\left(-\frac{6EI}{L^2}\right)_1\right] + \left(\frac{6EI}{L^2}\right)_5$ $K_{34} = \left[\left(-\frac{6EI}{L^2}\right)_3 \sqrt{2}\right] + \left(\frac{6EI}{L^2}\right)_5$

The modified co-efficient $K_{13} = 0$

Substituting the values of EI and L values of each member

$$3.667\,\theta_1 + 1.333\,\theta_2 + 0\,\theta_3 - .375\,\delta_4 = \frac{0.25}{EI}$$

$$1.333\,\theta_1 + 5.666\,\theta_2 + 0\,\theta_3 - .375\,\delta_4 = \frac{0.583}{EI}$$

$$0\,\theta_1 + \theta_2 + 3.415\,\theta_3 + .219\,\delta_4 = -\frac{1.333}{EI}$$

$$-.375\,\theta_1 + 0.375\,\theta_2 + .219\,\theta_3 + 1.015\,\delta_4 = -\frac{4}{EI} \quad \text{...(v)}$$

Alternative:

Alternatively if P_4 is the load corresponding to δ_4, taking $\delta_4^* = 1$

Work done by the equivalent load P_4 is equal to $P_4\,\delta_4^*$ ...(a)

Work done the applied loading 1T & 5T (Part – II)

$$= 1 \times \delta_4^* - 5 \times \text{(virtual vertical displacement of the joint C)}$$

$$= 1 \times \delta_4^* - 5 \times \delta_4^* \quad \text{...(b)}$$

Equating *(a)* and *(b)* and cancelling the virtual displacement δ_4^*

$$P_4 = -4$$

Step 3: Equation of equilibrium: The equations are obtained from virtual work considerations.

From contragradiance, that is, if $d_m = A\,dj$,

$$p_j = A^t\, p_m, \text{ substituting for } p_m$$

$$= A^t\, K_m\, \mathrm{A}\, d_j$$

or $$p_j = K_j\, d_j \text{ where } K_j = A^t\, K_m\, A$$

It may be verified that $p_j = K_j\, d_j$ is the same as Eq. (v).

The inverse of the coefficient matrix

$$\begin{bmatrix} 0.3200 & -.0878 & .0163 & .1470 \\ -.0878 & +.2140 & -.0562 & -.0992 \\ .0163 & -.0562 & .3120 & .0406 \\ .0470 & -.0992 & -.0406 & 1.0800 \end{bmatrix}$$

$$\theta_1 = \frac{0.32 \times 0.25 - .0878 \times .583 - .0163 \times 1.333 - .147 \times 4}{EI} = \frac{-0.581}{EI}$$

$$\theta_2 = \frac{-.0878 \times 0.25 + .214 \times .583 + .0562 \times 1.333 + .0992 \times 4}{EI}$$

$$= -.022 + .125 + .075 + .3968 = \frac{.575}{EI}$$

$$\theta_3 = \frac{.0163 \times 0.25 - .0562 \times .583 - .312 \times 1.333 + .0406 \times 4}{EI}$$

$$= .0041 - .033 - .416 + .1624 = -\frac{.282}{EI}$$

$$\delta_4 = \frac{.01427 \times 0.25 - .0992 \times .583 - .0406 \times 1.333 - 1.08 \times 4}{EI}$$

$$= .0037 - .058 + .054 - 4.320 = -\frac{4.287}{EI}$$

$$\theta_1 = -\frac{.581}{EI}$$

$$\theta_2 = -\frac{.575}{EI}$$

$$\theta_3 = -\frac{.282}{EI}$$

$$\delta_4 = -\frac{4.287}{EI}$$

Table 6.2 *Member Forces*

Member	$\left(\frac{2EI}{L}\right)_i \theta$	$\left(\frac{4EI}{L}\right)_i \theta$	$\left(\frac{6EI}{L^2}\right)_i \delta$	*M*	*Fem*	*Final Moment*
(1) M_1	- .290	0	1.608	1.318	0	1.318
M_2	0	- .580	1.608	1.028	0	1.028
(2) M_1	+ .287	0	1.608	1.895	0	1.895
M_2	0	+ .575	1.608	2.183	0	2.183
(3) M_1	- .199	0	2.270	-2.071	0	2.071
M_2	0	- .398	2.270	1.872	0	1.872
(4) M_1	+ .787	- 1.549	0	-.782	- .75	- 1.532
M_2	- .774	- 1.354	0	+.760	+ .75	+ 1.510
(5) M_1	- .278	1.15	3.215	-2.348	-1.333	- 3.681
M_2	+ .575	- .546	3.215	-3.201	+1.333	- 1.868

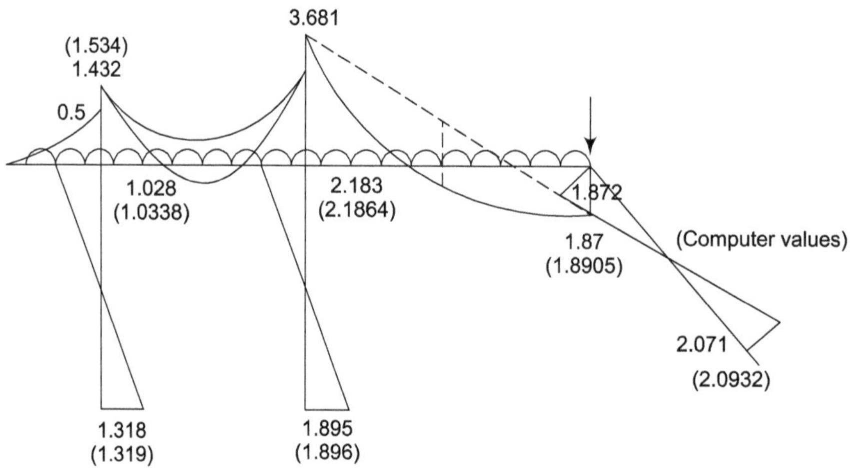

Fig. 6.9 *Illustration 1–Final Bending Moment Diagram (over)*

Master and Slave Nodes: Lateral deflections of the nodes *A*, *B*, *C*, with the *EI* values substituted, δ_A = –.3648, –.3651, –.3660. The Cantilever tip δ_x = –.03648.

It justifies the assumption of node *A* as master and the rest slaves. This aspect is discussed in detail elsewhere, tall buildings. In spite of the lateral load at node *C* in the *x*-direction, the frame deflects to the left. Why?

6.4.2.2 Illustration 2: Gallery

Portal Frame with gallery roof: Generalization of the translational degrees of freedom is presented in Chapter 7, Plane Frames.

Sketching of deflected shapes, particularly for translation and with inextensible members, is a skill. The attempt herein is to familiarize the readers with this skill.

The solution is no different from a Portal frame. But the visualization of the translational displacement is interesting, when the members are assumed inextensible. The displacements of the nodes are:

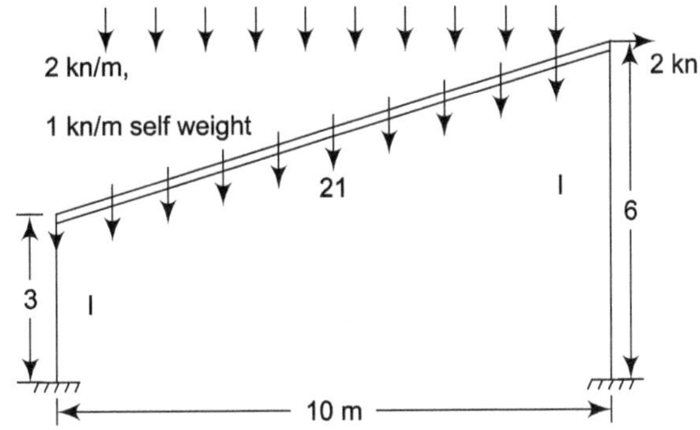

Fig. 6.10 *Illustration 2–Data*

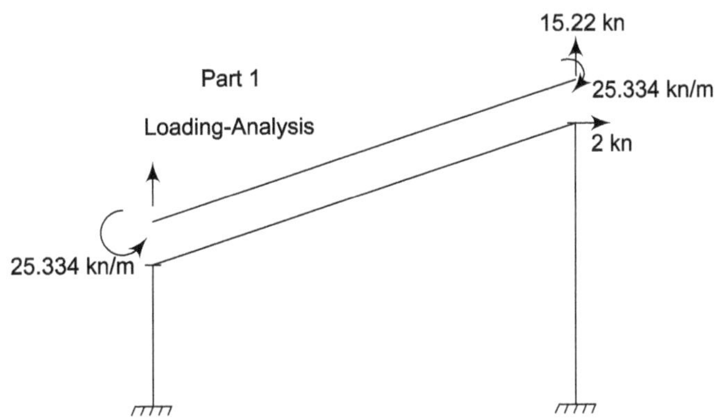

Fig. (i) *Illustration 2–Part I of Analysis–Fixed and Moments and Shears*

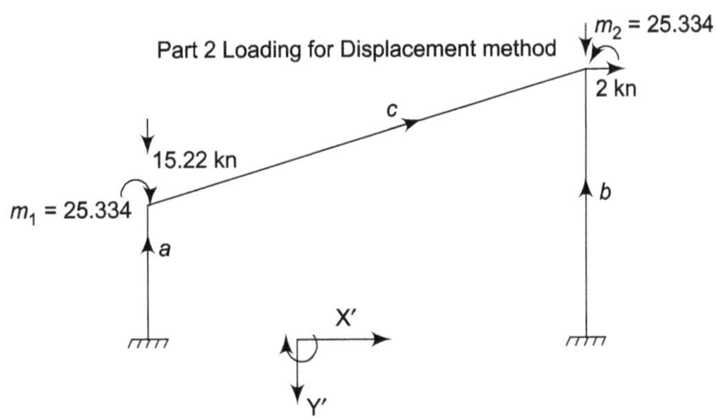

Fig. (ii) *Illustration 2–Part II of Analysis–Final Nodal Loads*

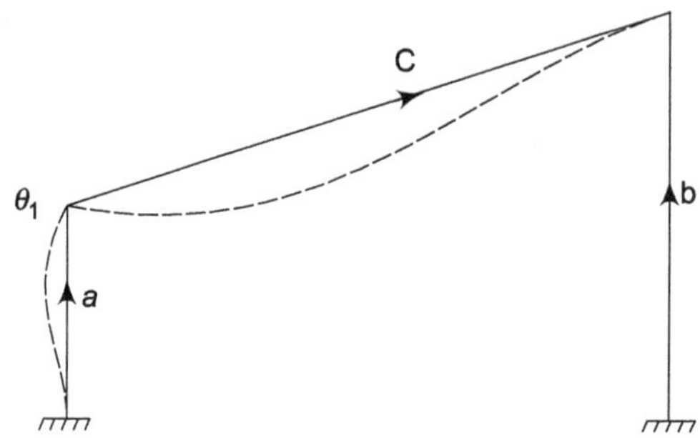

Fig. (iii) *Part II–Displacement θ_1*

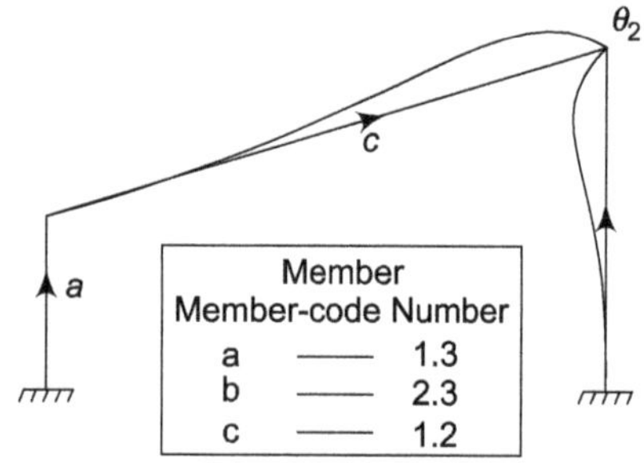

Fig. (iv) *Part II–Displacement θ_2*

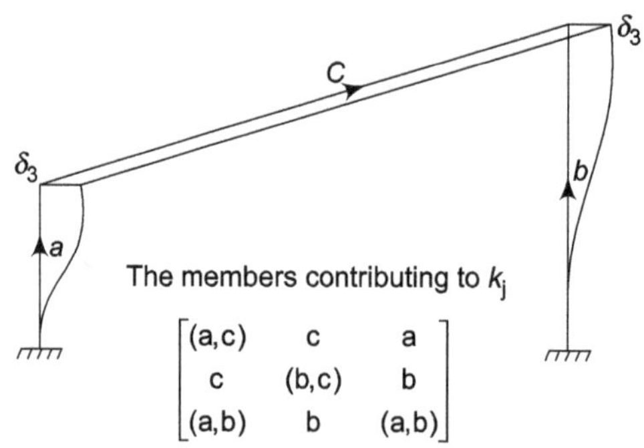

Fig. (v) *Part II–Displacement δ_3*

Illustration2: Degrees of Freedom (θ_1, θ_2, δ_3)
And the values of Matrix K_j; K_{11}, K_{12}, K_{13}, K_{22}, K_{23}, K_{33}

May be written with the help of the free bodies of the nodes 1 and 2.

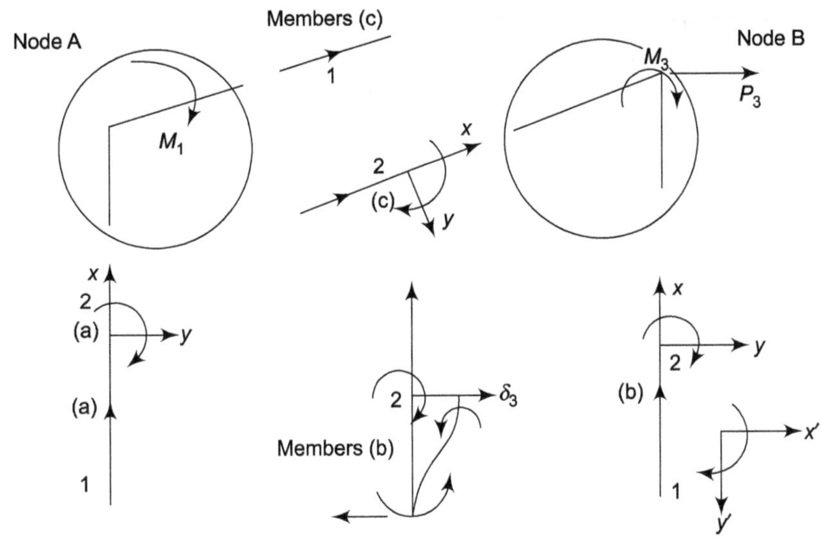

$$K_{11} = \left(\frac{4\,EI}{L}\right)_a + \left(\frac{4\,EI}{L}\right)_c \qquad K_{12} = +\left(\frac{2\,EI}{L}\right)_c \qquad K_{13} = -\left(\frac{6\,EI}{L^2}\right)_a$$

$$K_{22} = \left(\frac{4\,EI}{L}\right)_b + \left(\frac{4\,EI}{L}\right)_c \qquad K_{23} = -\left(\frac{6\,EI}{L^2}\right)_b$$

$$K_{33} = \left(\frac{12\,EI}{L^3}\right)_a + \left(\frac{12\,EI}{L^3}\right)_b$$

Illustration: Part II, Free Bodies and Global Stiffnesses

Substituting the values of I and L of the member a, b, c
$K_{11} = 2.1EI = K_{12} = .383EI$, $K_{13} = .667EI$; $K_{22} = 1.433EI$, $K_{23} = -0.167EI$ and $K_{33} = 0.5\,EI$. The other elements follow from symmetry.

Alternative:

Connection Matrix Approach: Alternatively we may get the K_j using the connection matrix approach.

$$K_j = \Sigma_{a,b,c}\, A^t\, K_m A \text{ Let us illustrate for the member } a\text{: } A_a = \begin{bmatrix} 0 & 0 & 1 \\ 0 & 0 & 0 \\ 1 & 0 & 0 \end{bmatrix}$$

$$(A^t\, K_m A)_a = \begin{bmatrix} \left(\frac{4\,EI}{L}\right)_a & 0 & -\left(\frac{6\,EI}{L^2}\right)_a \\ 0 & 0 & 0 \\ -\left(\frac{6\,EI}{L^2}\right)_a & 0 & \left(\frac{12\,EI}{L^3}\right)_a \end{bmatrix}$$

Similar contributions from other member (b) and (c) are added
Solving the Equilibrium Equation, $P_j = K_j\, d_j$

For the unknowns displacements, d_j, we get $EI\, d_j = \begin{Bmatrix} 25.929 \\ -21.001 \\ 31.561 \end{Bmatrix}$

Part 2 LOADING the Bending Moment Diagram is:

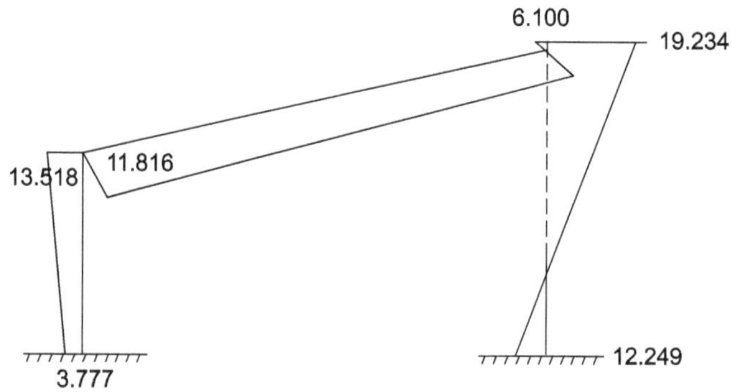

Fig. *Illustration Part II of Analysis–Bending Moment Diagram*

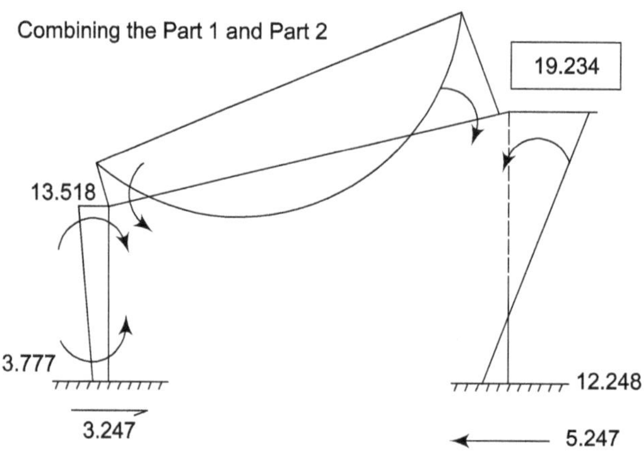

Fig. *Illustration 2–Superposition of Part I and II–Final B.M. Diagram*

6.4.2.3 Illustration 3: 'A' Frame

In the following the deflected shapes and degrees of freedom are given. The detailed analysis for Part I and Part II may be taken up as exercises. For the inextensibility assumed, there is only one Center of Rotation (CR), and therefore only on translation in addition to the two rotations at *A* and *B*.

Single bay portal frame with inclined legs. The inclined legs result in an interesting deflected shape caused by the translation with the inextensible members.

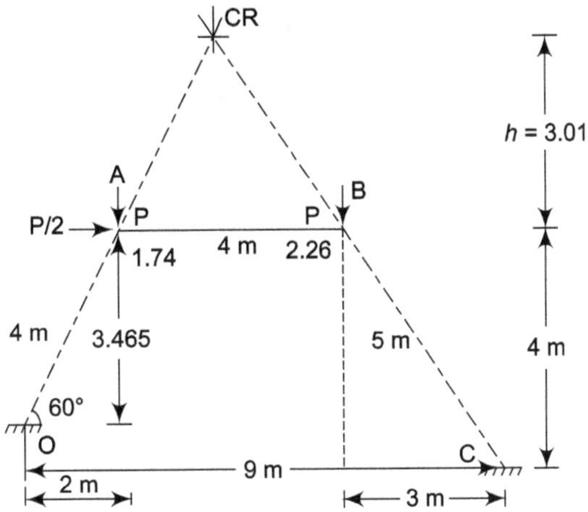

Fig. 6.11 *Illustration 3–Data*

Rotations: The deflected shapes corresponding to the rotations θ_A and θ_B are sketched separately. The degrees of freedom corresponding to the translations for the plane frame are:
2 × (number of nodes) – number of inextensible members. The supports are assumed fixed.
No. of rotations = 2 ; θ_A and θ_B
No of translations = 2 (number of movable joints) – number of inextensible members
Number of translations = 2 (2) – 3 = 1, δ_3
The deflected shape for the translation is sketched in the following.

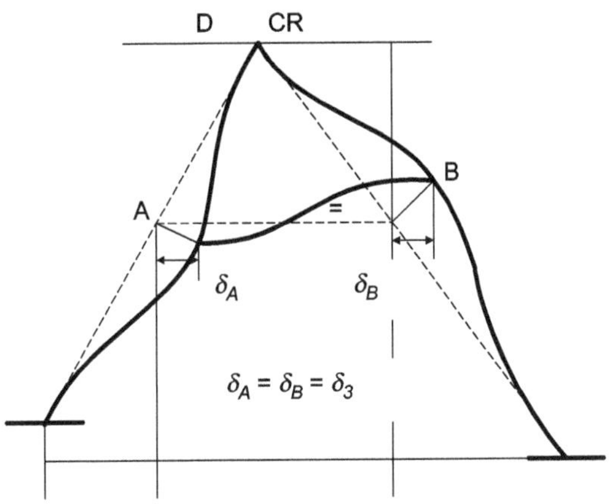

Geometry of Displacement δ_3

Translation and free body *B* for writing the equilibrium equation corresponding to the translation.
The unknown are

$$\theta_1 = \theta_A$$
$$\theta_2 = \theta_B$$
$$\delta_3 = \delta_A = \delta_B$$

Member	Member code number
a	1, 3
b	2, 3
c	1, 2, 3

Illustration: "*A*" Frame

The rotation of the joints *A* and *B* about the center of rotation, *CR* is given by $\frac{\delta}{h} = \frac{\delta RA}{AD} = \frac{\delta RB}{BD}$

Explanation:

Consider a "*A*" frame *OADEC*. It may be verified that if the five members *a, b, c, d, e* of the plane frame are assumed inextensible, the degrees of freedom, that is, the

Number of unknown displacements is:

$$3 \times (\text{number of movable nodes}) - \text{number of inextensible members}$$

$$3 \times (3) - 5 = 4$$

The three unknown displacements are three rotations θ_A, θ_B, θ_C, one translation δ. The translation δ is shown in the figure. The joints *A* and *B* rotate about the node D that is centre of rotation.

6.4.2.4 *Illustration 4: Portal Frame with Inclined Legs*

Portal Frame with an inclined leg used as an illustration. Even though, it is the case of one bay only, the number of translations remains one only. The inclined legs give rise to increased stiffness.

Direct Stiffness method for a plane frame results in three degrees of freedom per node. Their number is equal to six. In the following the members *a, b, c* are assumed inextensible.

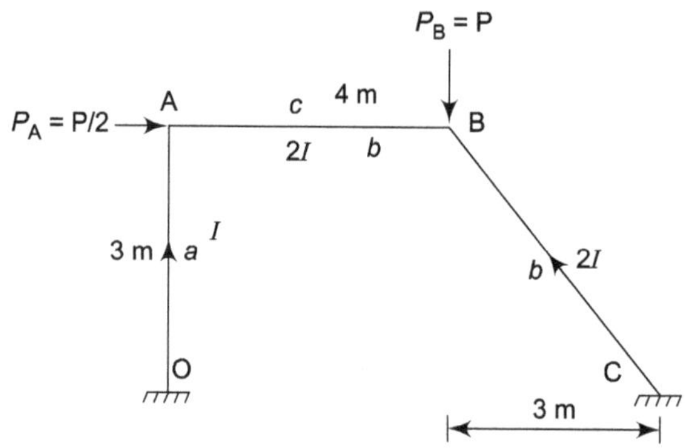

Fig. 6.12 *Illustration 4–Under Nodal Loads Only*

Joint loads at joints *A, B*

Between the joints, no member loads

Degrees of freedom for plane

$$\text{Frames} = 3 \times \text{number of nodes} - \text{Number of inextensible members}$$

It follows that the degrees of freedom = 3

$$\delta_1 = \theta_A,\ \delta_2 = \theta_B \text{ and } \delta_3.$$

These are sketched in the following:

Corresponding load displacements. Computation of P_3, Virtual work considerations.

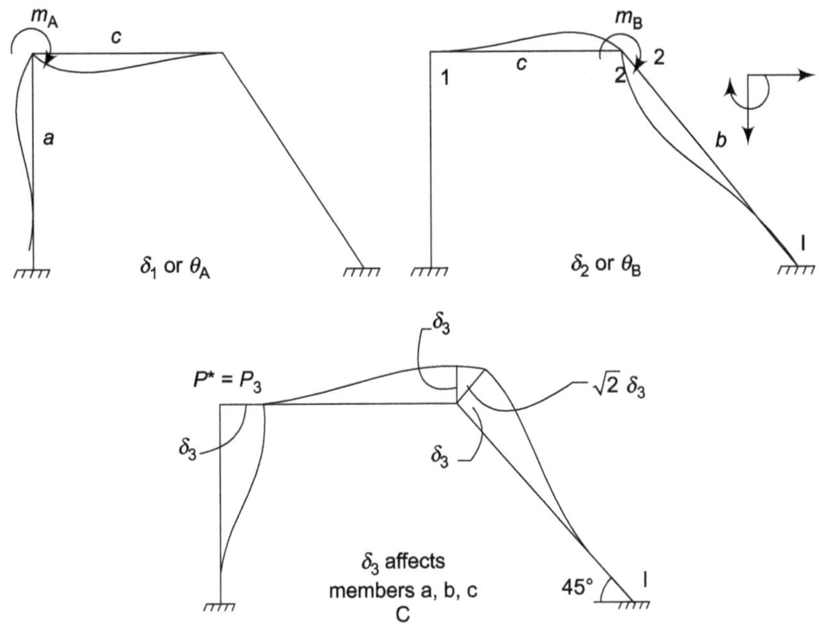

Fig. (a) *Illustration 4–Deflected Shapes*

P* corresponding to δ_3 cannot be visualized and it is equal to $-P/2$ obtained from $P_A \times 1 - P_B \times 1$, refer the figure.

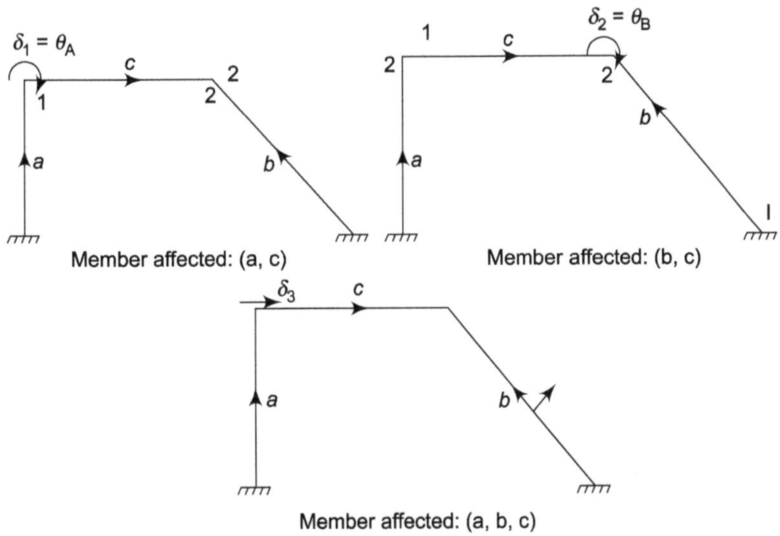

Fig. (b) *Illustration 4–Members Affected by the Nodal Displacements*

These virtual displacements will be used while writing the global stiffness elements based on the code numbers. The code numbers define the compatibility of the member deformations and joint deformations.

$$\begin{pmatrix}\delta_{y2}\\ \theta_1\\ \theta_2\end{pmatrix}_i \quad \overbrace{\begin{pmatrix}0\\0\\1\end{pmatrix}_a,\begin{pmatrix}0\\0\\0\end{pmatrix}_b\begin{pmatrix}0\\1\\0\end{pmatrix}_c}^{\theta_A} \quad \overbrace{\begin{pmatrix}0\\0\\0\end{pmatrix}_a,\begin{pmatrix}0\\0\\1\end{pmatrix}_b,\begin{pmatrix}0\\0\\1\end{pmatrix}_c}^{\theta_B} \quad \overbrace{\begin{pmatrix}1\\0\\0\end{pmatrix}_a,\begin{pmatrix}\sqrt{2}\\0\\0\end{pmatrix}_b,\begin{pmatrix}-1\\0\\0\end{pmatrix}_c}^{\delta_3}$$

The member wise connection matrices are given as per the deformations. In fact, the code numbers of the members are given by the joint deformations and connection / compatibility A are given by

$$A_b\begin{pmatrix}0&0&1\\0&0&0\\1&0&0\end{pmatrix}_a,\ A_b=\begin{pmatrix}0&0&\sqrt{2}\\0&0&0\\0&1&0\end{pmatrix}_a,\ A_c=\begin{pmatrix}0&0&-1\\1&0&0\\0&1&0\end{pmatrix}_c$$

Contribution of the member "a" to the global Matrix K is given by the code number $(1, 3)_a$ leading to K_{11}, K_{13} using the connection matrix approach, the contribution by the member "a" is given by;

$$A_a^t\,(K_m)_a\,A_a=\begin{pmatrix}0&0&1\\0&0&0\\1&0&0\end{pmatrix}_a^t,\begin{pmatrix}\frac{12\,EI}{L^3}&-\frac{6\,EI}{L^2}&-\frac{6\,EI}{L^2}\\ *&\frac{4\,EI}{L}&\frac{2\,EI}{L}\\ *&*&\frac{4\,EI}{L}\end{pmatrix}_a,\begin{pmatrix}0&0&1\\0&0&0\\1&0&0\end{pmatrix}_a$$

$$K_a=\begin{pmatrix}\frac{4\,EI}{L}&0&-\frac{6\,EI}{L^2}\\0&0&0\\-\frac{6\,EI}{L^2}&0&\frac{12\,EI}{L^3}\end{pmatrix}_a \tag{i}$$

Similarly, the member codes of member, b is $(1, 3)_b$ leading to K_{22}, K_{23}, K_{32}, K_{33} of the global stiffness matrix

$$K_b=(A^t)_b\,(K_m)_b\,(A)_b=\begin{pmatrix}0&0&-\frac{6EI}{L^2}\\0&\frac{4\,EI}{L}&\frac{2EI}{L}\\0&\frac{-6\sqrt{2}\,EI}{L^2}&\frac{12\left(\sqrt{2}\right)\left(\sqrt{2}\right)EI}{L^3}\end{pmatrix}_b \tag{ii}$$

The member code number of member "C" is $(1, 2, 3)_a$ leading to $K_{11}, K_{12}, K_{13}, K_{22}, K_{23}, K_{33}$. The other elements of the global K are written from symmetry considerations. Member "C" contribution to the global

$$K_c = \begin{pmatrix} \frac{4\,EI}{L} & \frac{2\,EI}{L} & \frac{6EI}{L^2} \\ * & \frac{4\,EI}{L} & \frac{6EI}{L^2} \\ * & * & \frac{12\,EI}{L^3} \end{pmatrix}_c \tag{iii}$$

The global stiffness matrix is obtained by adding (i), (ii), (iii)

$$K = \begin{pmatrix} \left(\frac{4\,EI}{L}\right)_a + \left(\frac{4\,EI}{L}\right)_c & \left(\frac{2\,EI}{L}\right)_c & \left(-\frac{6\,EI}{L^2}\right)_a + \left(\frac{6\,EI}{L^2}\right)_c \\ * & \left(\frac{4\,EI}{L}\right)_b + \left(\frac{4\,EI}{L}\right)_c & \left(\frac{-6\sqrt{2}\,EI}{L^2}\right)_b + \left(\frac{6\,EI}{L^2}\right)_c \\ * & * & \left(\frac{12\,EI}{L^3}\right)_a + \left(\frac{24\,EI}{L^3}\right)_b + \left(\frac{12\,EI}{L^3}\right)_c \end{pmatrix}$$

$$A_a = \begin{bmatrix} 0 & 0 & 1 \\ 0 & 0 & 0 \\ 1 & 0 & 0 \end{bmatrix}; A_b = \begin{bmatrix} 0 & 0 & \sqrt{2} \\ 0 & 0 & 0 \\ 0 & 1 & 0 \end{bmatrix}; A_c = \begin{bmatrix} 0 & 0 & -1 \\ 1 & 0 & 0 \\ 0 & 1 & 0 \end{bmatrix}$$

In the preceding the global stiffness matrix was obtained from

$$K = \sum \left(A^t\ K\ A\right)_{a,b,c}$$

Substituting the member properties: $I_a = I$ and $I_b = I_c = 2I$

$$K = EI\begin{pmatrix} 3.333 & 1 & 0.083 \\ * & 3.886 & -0.193 \\ * & * & 1.448 \end{pmatrix} \text{ and } P_j = \begin{Bmatrix} 0 \\ 0 \\ -\frac{P}{2} \end{Bmatrix}$$

The equilibrium equation:

$P_j = K\,d$ is solved for d by iteration, 3 cycles only

$$\delta_1 = \theta_A = \frac{0.0152}{EI}, \delta_2 = \theta_2 = \frac{-0.02156}{EI} \text{ and } \delta_3 = \frac{-0.3485}{EI}$$

Using the member stiffness $P_m = k_m d_m$, the member end forces are determined. The moment values are given in the following.

The hand computed values are given within parenthesis.

The small variations in the following member end moments at each joint are because of the approximation of the solution by three iterative cycles only.

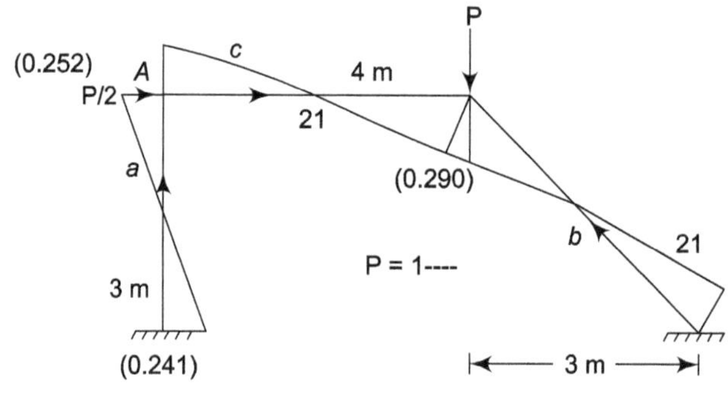

Fig. (c) *Illustration 4–Final Bending Moment Diagram*

M_{1a} = 0.2424 KNm

M_{2a} = 0.2526

M_{1b} = 0.3052

M_{2b} = 0.2879

M_{1c} = –0.2525

M_{2c} = –0.2893

6.4.2.5 'A-Frame' Generalized

There is only one *CR*, and it helps in sketching the deflected shapes for the two levels. It is interesting to sketch the deflected shapes caused by translation.

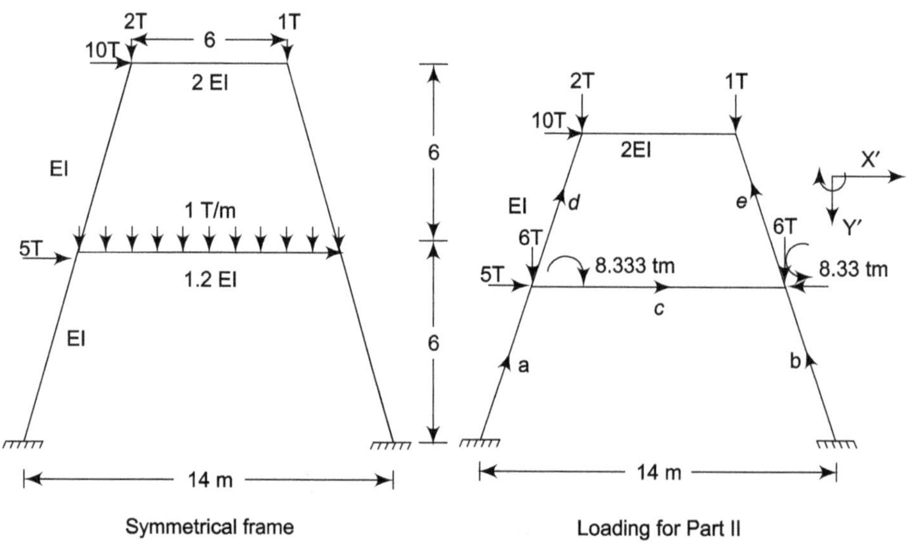

Fig. 6.13 *Generalization*

Two storey – single bay with inclined legs: Illustration 5:

Member Code and Joint Code: A two storey frame (single bay), an illustration and not an merely example. The numericals are slidelined with intention. It helps in assembling global stiffness matrix by inspection, essential for building the concepts and also for our understanding.

Rotational degrees of freedom θ_A, θ_B, θ_C, θ_D may be sketched by the reader.

Translational degrees of freedom δ_5 and δ_6 are sketched in the following:

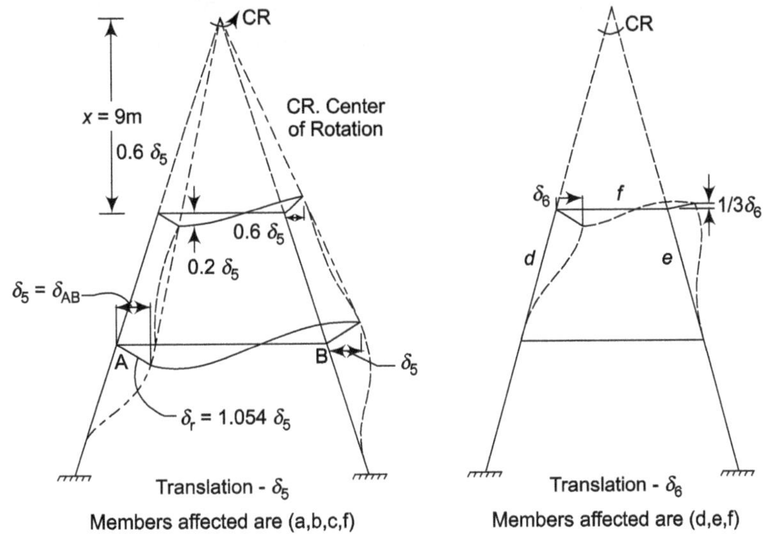

(a) Deflected Shapes and Members Affected

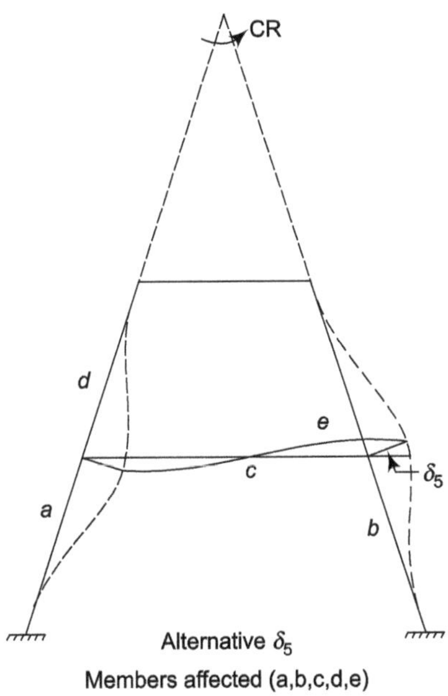

(b) Alternative Translation

The members of the two storey frame are assumed inextensible. A joint code establishes relation of the joint deformation with the connected members. The translational degrees of freedom are included within the joint-degree of freedom. It may be noted that δ_5 results in translation of the girder *CD* also. The translation, δ_6, however does not give rise to the deformation of girder *AB* and other members of the lower storey. The translation δ_5 may be assumed differently. Refer Fig. Alternative δ_5.

Member Code: A member code of, say member '*a*' helps obtain contribution of member '*a*' towards the global stiffness matrix K_j, Similarly consideration of all the members helps us obtaining the total K_j. Similarly consideration of all the members helps us obtaining the total K_j

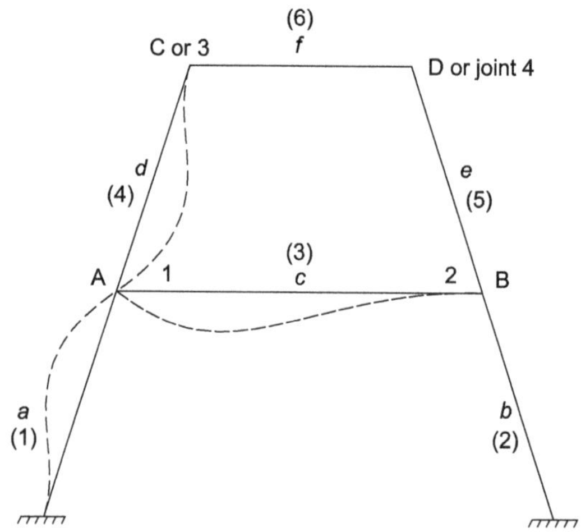

(c) A Typical Joint Rotation, θ_A or θ_1

The member affected are '*a, c,* and *d*'. Accordingly if all the Degrees of Freedom are considered, the member codes are:

Member	*Member Code*
'a' or 1	(1,5)
'b' or 2	(2,5)
'c' or 3	(1,2,5)
'd' or 4	(1,3,5,6)
'e' or 5	(2,4,5,6)
'f' or 6	(3,4,5,6)

Inferences: If is good to remind our selves that while we are using 'member iodes', we are assembling the global stiffness from each member, and when all the members are exhausted, the global stiffness is complete. It is similar to assembling $(A^T k_m A)_i$ of each member '*i*'. It may populate the entire K_j. In the following, we consider the use of 'joint codes'. The approach is some as considering each free body and complete the contributions from the members meeting the joint. Each 'free body-equilibrium' completes the row of the global stiffness, K_j, we are assembling row-wise.

Joint Codes: The typical joint rotation, θ_1

Affects the members {1, 3, 4}

Similarly, θ_2 affects the members {2, 3, 5}

θ_3 affects the members {4, 6}

θ_4 affects the members {5, 6}

θ_5 affects the members {1, 2, 3, 4, 5, 6}. The displacement δ_6 results in the associated translations of the nodes *A*, *B* as well as *C*, *D*. Finally.

δ_6 affects the members {4, 5, 6}

From the above, we may write the final **joint codes.** And these are

Joint	*Joint Code*	*Associated Displacements* θ_1 *to* δ_6
'A' or 1	1, 3, 4	[1, 2, 3; 5, 6]
'B' or 2	2, 3, 5	[1, 2, 4; 5, 6]
'C' or 3	4, 6	[1, 3, 4; 5, 6]
'D' or 4	5, 6	[2, 3, 4; 5, 6]

6.5 SUMMARY: 'MEMBER CODE' AND 'JOINT CODE

As seen from the above examples, the codes facilitates Visual Assembly of Global Stiffness Matrix refer particularly the Illustration 4.5. The code number techniques are used for assembling the stiffness matrix while using the Connection matrix Approach. It is a substitute to writing equilibrium equation in the case of Direct Stiffness method.

$K_j = \sum_{a,b,c,d,e,f} K_i$. The coding technique is summarized in the following:

Member Code	*Joint Code*
A member code, say a member '*i*' helps obtain contribution of the member towards the Global Stiffness Matrix K_j.	The joint codes enable to obtain row-wise the Global Stiffness Matrix K_j.
Consideration of all the members results in obtaining the K_j.	Joint code of a joint '*i*' enables to obtain the ith row. Consideration of all the joints results in K_j.
$K_j = \sum_{1\ to\ m} K_i$	$K_j = \sum_{j=1\ to\ n} K_j$
Where '*m*' is the number of members.	Where '*n*' is the number of joints.

The approach for the space frames is similar. In the case of trusses, however, the number of joints being relatively fewer as compared to the number of members and there is no inextensibility, the approach using the 'joint Codes' may be advisable for the assembly of the Global Stiffness matrix. The global Stiffness matrix is in fact an Equilibrium matrix.

6.5.1 Illustration 6: Unsymmetrical Gable Frame; Application of Center of Rotation

The frame being geometrically unsymmetrical, it is not possible to superpose symmetrical and antisymmetrical solutions. The total degrees of freedom equal to 5, three rotations and two translations are required to be evaluated. In the case of Force Method, the unknowns are three redundant only. If inextensibility is not assumed the number of unknown displacements are increased from 5 to 9. The first part of the analysis is fixed edge analysis, which gives the nodal loads. The second part is the Displacement method. For the translational displacements the centers of rotations for the members, '*c*' and '*d*' are used. Corresponding to the two centers of rotation there are two independent translations. Number of movable joints = 3, and four inextensible members.

Number of independent translations = $2 \times 3 - 4 = 2$

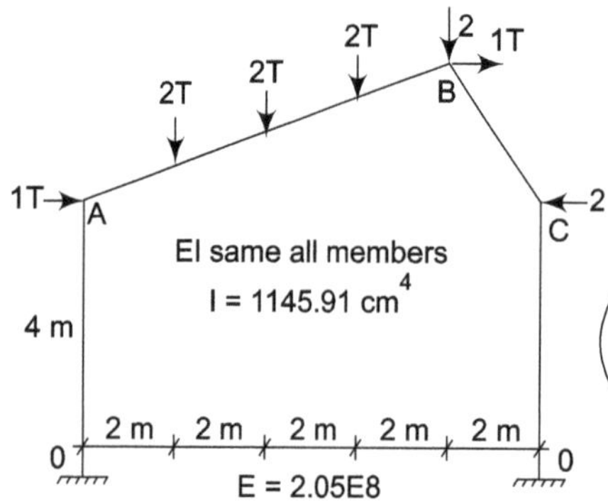

Fig. 6.14 *Illustration 6–Given Loading and Geometry*

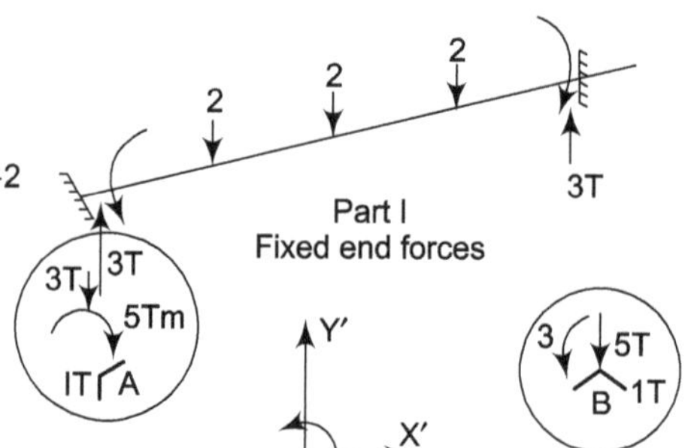

Fig. *Illustration 1 Part I*

The second translation, δ_2 may be taken differently.

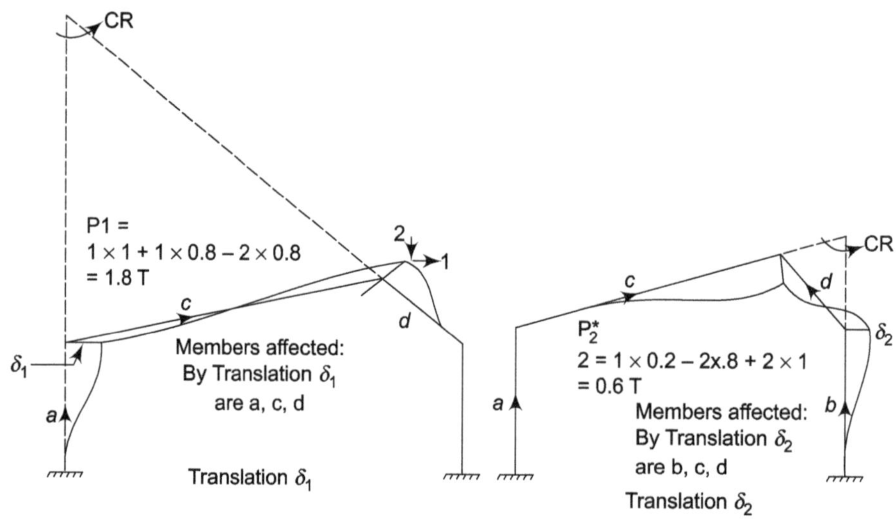

Fig. (a) *Illustration 6–Deflected Shapes*

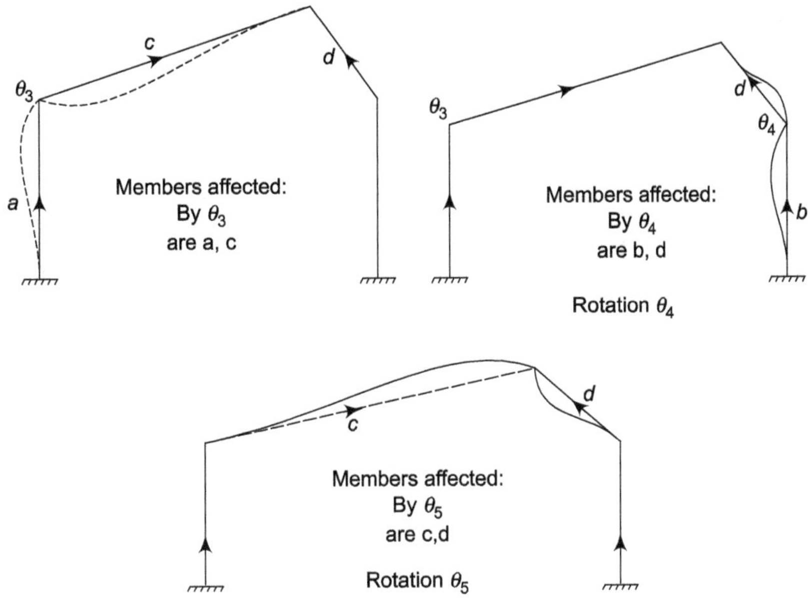

Fig. (b) *Illustration 6–Deflected Shapes*

Member	Member Code
'a'	(1,3)
'b'	(2,4)
'c'	(1,2,3,5)
'd'	(1,2,4,5)

Members affected by the displacement 1,2,3,4,5 give the code numbers. The member codes enable us to directly generate the global stiffness matrix. The Equilibrium equations by mapping the table:

$$P'_j = \begin{array}{c} \begin{array}{ccccc} \delta_1 & \delta_2 & \theta_3 & \theta_4 & \theta_5 \end{array} \\ \begin{bmatrix} a,c,d & c & a,c & * & c \\ * & b,c,d & c,d & b & c,d \\ * & * & a,c & * & d \\ * & * & * & b,d & d \\ * & * & * & * & c,d \end{bmatrix} \end{array} d'$$

It is simple and direct to write the member-k-matrices.

From the part I of the Analysis $P'_j = \{1.8, 0.6, -5, 0\}$ member a, b, c, d $\rightarrow$ obtain member forces, p_m member forces $= K_m\, d_m = K_m\, A\, d_j$

$$d_j = \{0.86651.279, -.0045, -.0040, -.0017\}$$

Draw BM Diagram

Result of (Part) I:

Fixed End Analysis:

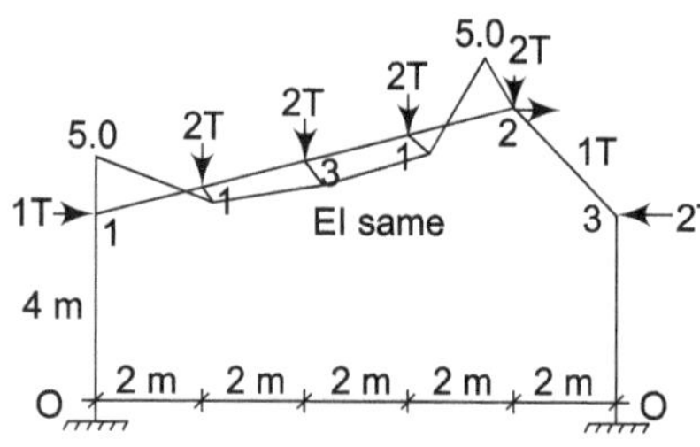

Fig. (c) *Part I-Fixed End Analysis*

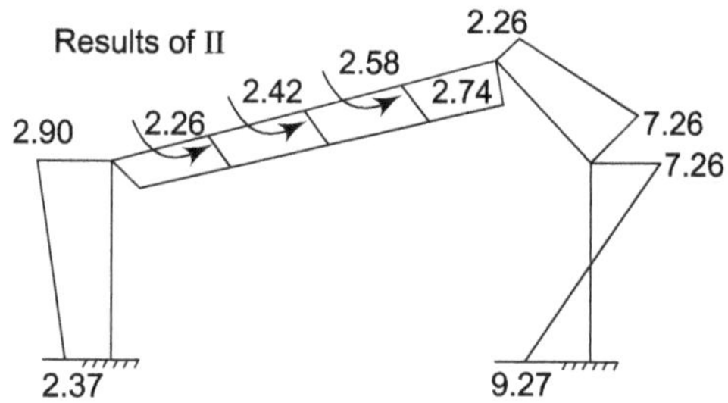

Fig. (d) *Part II-Displacement Analysis*

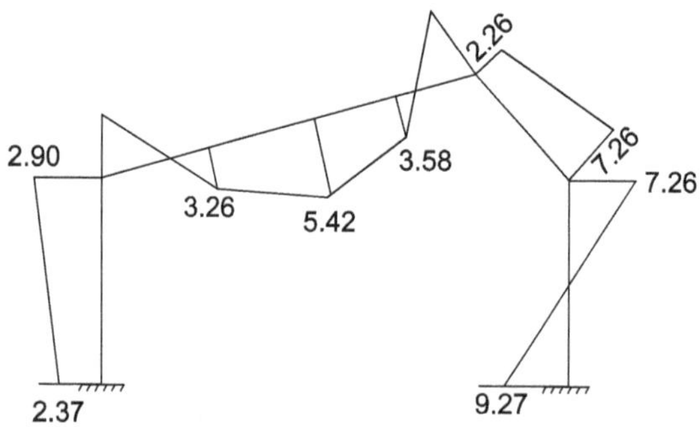

Fig. (e) *Superposition of I and II*

Illustration 6–*Final Bending Moment Diagram*

Results without assuming inextensibility:

Joint	δ_x	δ_x	θ
1	0.8665	– .002	– .0045
2	0.9486	– .3363	.004
3	1.2795	– .0027	– .0017

The illustration justifies the assumption of inextensibility. The number of the unknown displacements gets reduced from nine to F_{ii}.

6.5.2 A Preparation for Tall Buildings: Two Bay-two Storey Plane Frame

Illustration-7: The illustration of the plane frame with two bays and two storey has been chosen as an introduction to the analysis of Tall Buildings. In the land mark technical paper by Clough et al. Plane

Frame is chosen as an element for the total analysis of a Tall Building System. The numbering is done from roof to the ground supports. Here the roof is the top girder with the adjoining nodes. The floor below is a typical floor line but in this case two girders with the three nodes, and the supports. The designations and loadings with the cross sectional properties are shown, Fig. 1a.

Part–I of the analysis consists for the fixed-edges. The bending moment diagram for arriving at the nodal moments is shown, Fig. 1(b). The members are assumed in-extensible, the load matrix referring to the global axes,

$$P_j = \{-53.333, 53.333, \quad -53.333, 40, 13.333, 20, 20\}$$

'Master node' and 'Slave nodes': The lateral loads are clubbed together at the end as p'_6 and p'_7. In a building system δ'_6 is the lateral load at the master node and the lateral displacements of the other nodes in the same floor are the slave nodes and their displacement are governed by the master node. The floor below undergoes δ'_7 and the other two nodes at the same floor are the slave nodes, governed by the assumption of inextensibility.

The seven degrees of the freedom are sketched, Fig. (iii). The stiffness matrix, k_m for the chosen member axes is also written for ready reference. The members affected by each degree of freedom helps us to write the Table of the member-codes.

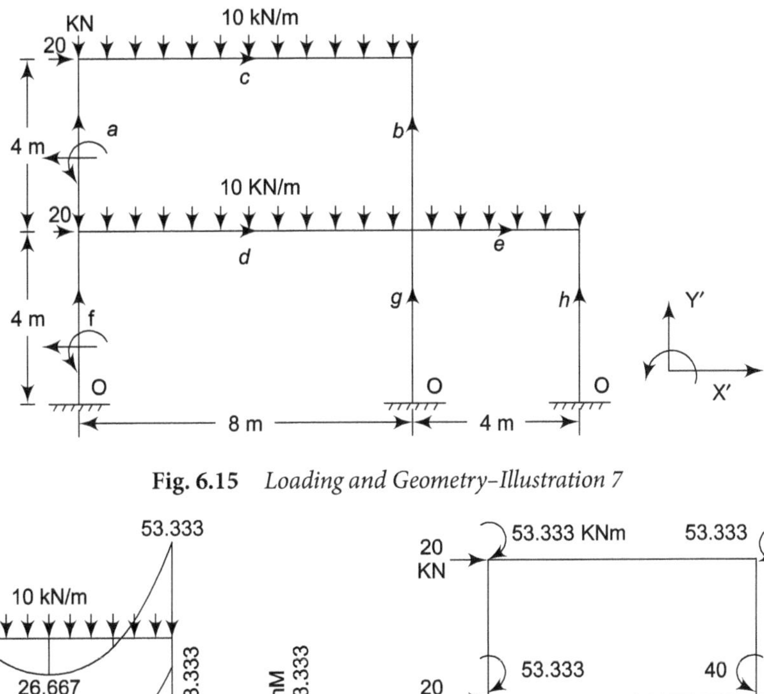

Fig. 6.15 *Loading and Geometry–Illustration 7*

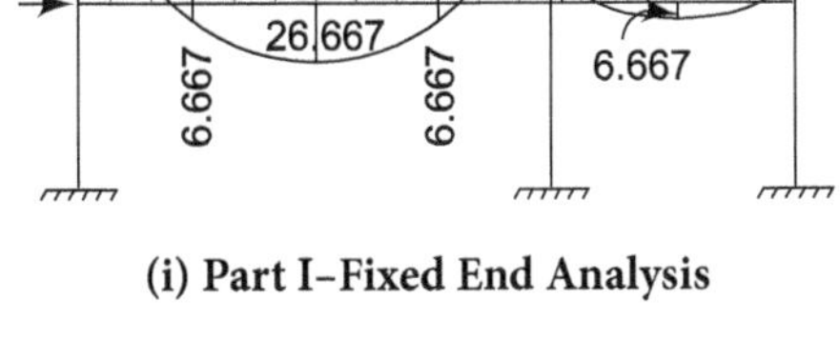

(i) Part I–Fixed End Analysis

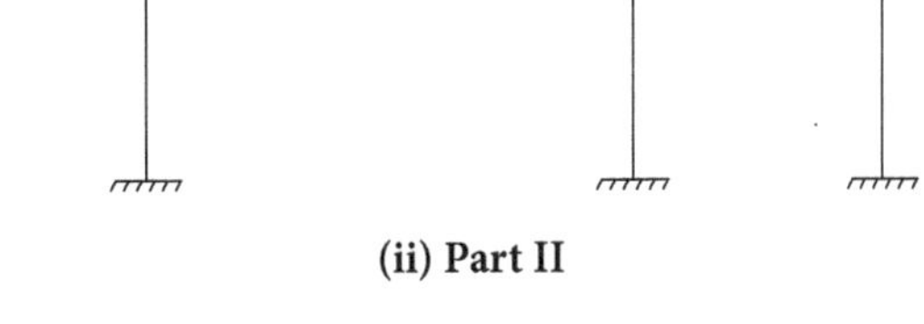

(ii) Part II

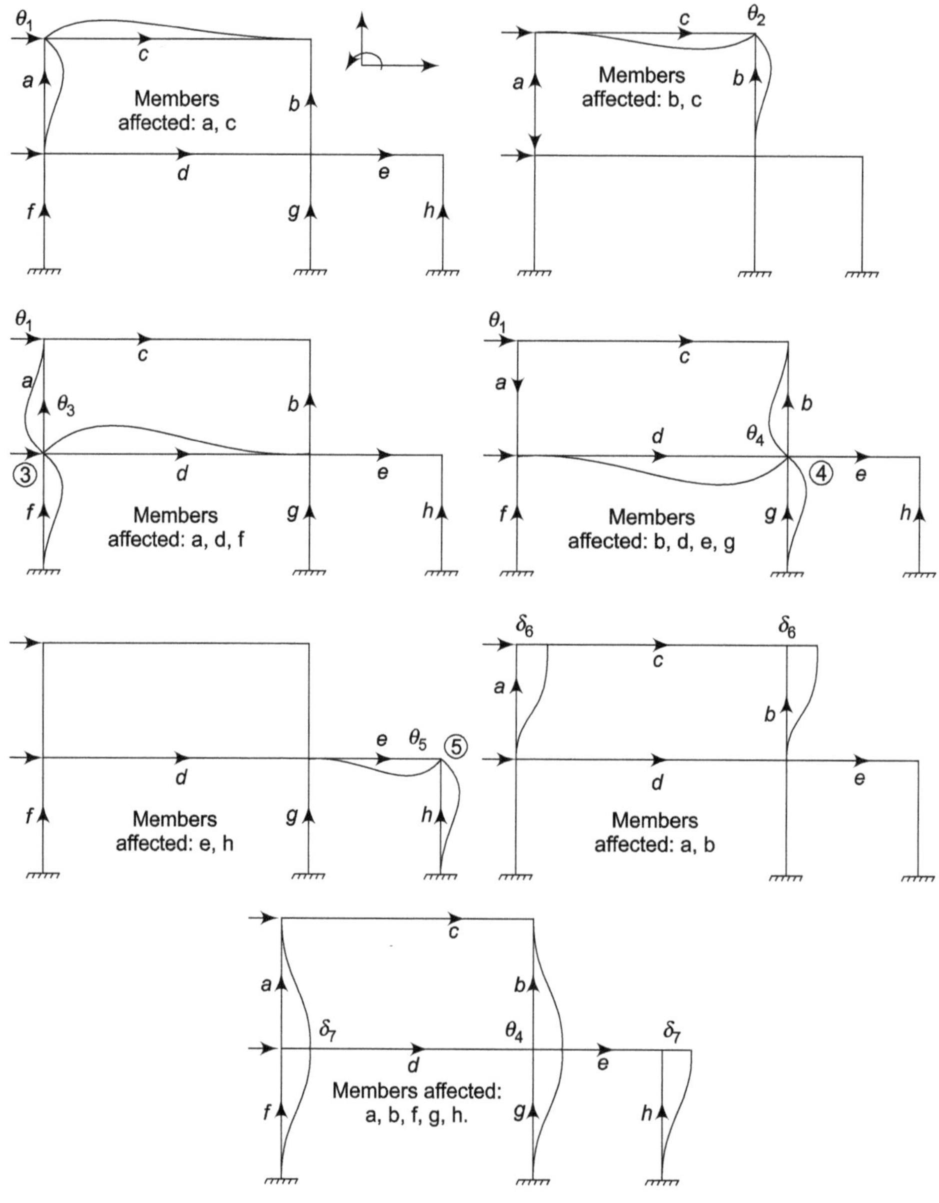

(iii) Deflected Shapes of Two Bay–Two Storey Frame

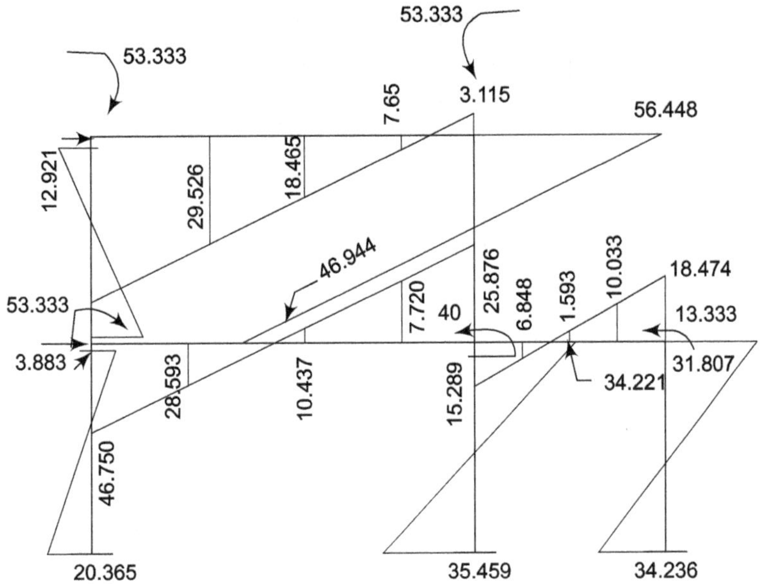

(iv) Part II - Bending Moment Diagram for the Nodal Loads

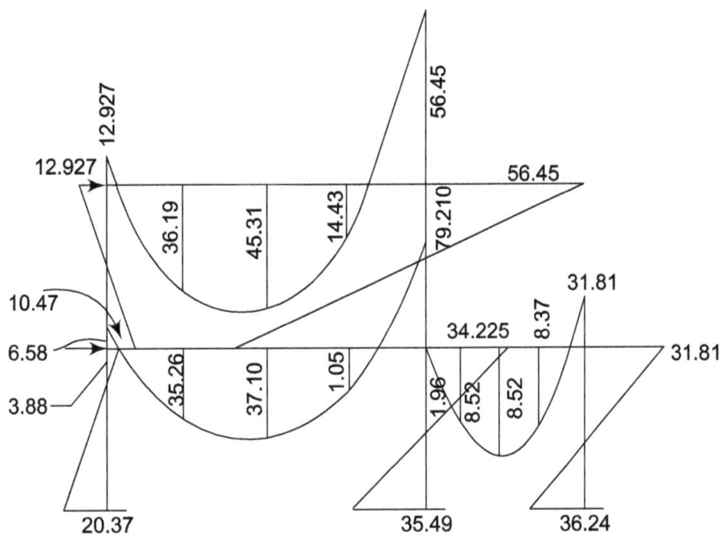

(v) Final Bending Moment Diagram from Superposition of Part-I and Part-II

Table *Member Codes*

Member	*Member Code*
a	1, 3, 6, 7
b	2, 4, 6, 7
c	1, 2
d	3, 4
e	4, 5
f	3, 7
g	4, 7
h	5, 7

Generation of the global Stiffness matrix, K_j':

Contribution of each member towards the global matrix, K_j'. With some practice, it is possible to generate the entire K_j' by inspection. Alternatively we may use Connection matrix approach.

$$K_j' = \sum_{a \text{ to } h} A^T k_m A$$

Consider member 'a'. Its contribution is $\left[A^T k_m A\right]_a$

$$A_a = \begin{bmatrix} 0 & 0 & 0 & 0 & 0 & -1 & 1 \\ 1 & 0 & 0 & 0 & 0 & 0 & 0 \\ 0 & 1 & 0 & 0 & 0 & 0 & 0 \end{bmatrix}$$

And the property of symmetry of the global stiffness matrix is used for checking the accuracy. The procedure is repeated for all the member and the summation gives the global stiffness matrix, $K_j' = \sum_{a \text{ to } h} A^T k_m A$

$$\left[A^T k_m A\right]_a = \begin{bmatrix} \frac{4EI}{L} & 0 & \frac{2EI}{L} & 0 & 0 & \frac{6EI}{L^2} & -\frac{6EI}{L^2} \\ 0 & 0 & 0 & 0 & 0 & 0 & 0 \\ \frac{2EI}{L} & 0 & \frac{4EI}{L} & 0 & 0 & \frac{6EI}{L^2} & -\frac{6EI}{L^2} \\ 0 & 0 & 0 & 0 & 0 & 0 & 0 \\ 0 & 0 & 0 & 0 & 0 & 0 & 0 \\ \frac{6EI}{L^2} & 0 & \frac{6EI}{L^2} & 0 & 0 & \frac{12EI}{L^3} & -\frac{12EI}{L^3} \\ -\frac{6EI}{L^2} & 0 & -\frac{6EI}{L^2} & 0 & 0 & -\frac{12EI}{L^3} & \frac{12EI}{L^3} \end{bmatrix}_a$$

Let us try by inspection: The member stiffness matrix has four typical values.

$$\frac{12EI}{L^3}, \frac{6EI}{L^2}, \frac{4EI}{L}, \frac{2EI}{L}$$

Consider member code 1, 3, 6, 7 of member 'a'.

The four typical values occupy positions 1,1; 3,3; 6,6; 7,7

Along the diagonal, the global stiffnesses, $k'_{1,1} = k'_{3,3}$ correspond to the Rotations θ_1, θ_3. Therefore $k'_{1,1} = k'_{3,3} = \frac{4EI}{L}$. Similarly

$$k'_{1,3} = k'_{3,1} = \frac{2EI}{L}$$

Corresponding to lateral translation, $k'_{6,6} = k'_{7,7} = \frac{12EI}{L^3}$

And as the shears at the other end of a member is opposite, $k'_{6,7} = k'_{7,6} = -\frac{12EI}{L^3}$

As δ_6 pertains to the upper floor, $k'_{1,6} = \frac{6EI}{L^2}$ and the moment at the bottom, $k'_{1,7} = \frac{6EI}{L^2}$ and the moment at the bottom $k'_{1.7} = -\frac{6EI}{L^2}$

Similar are the stiffness $k'_{3,6}$ and $k'_{3,7}$. It may be of interest to note that θ_2, θ_4, θ_5, do not affect the member '*a*', and therefore second, fourth and fifth are all zero. So also are the corresponding columns of the global stiffness. Because the global matrix has to be symmetrical, the property is used for accuracy of the elements.

The above inspection and connection matrix approach pertains to the chosen member '*a*'. the same is repeated for the contributions from the other members '*b*', '*c*', '*d*', '*e*', '*f*', '*g*', '*h*'. The final global, stiffness matrix is obtained by the summation,

$$K'_j = \sum_{a \text{ to } h} A^T k_m A \qquad \text{...(6.3)}$$

Unknown Displacements: The nodal displacements take place at the joints from the part II. The inversion of the global matrix is used for computing the unknown displacements.

$$d'_j = \{K'_j\}^{-1} p'_j$$

Ans: d'_j = {–.0089, .0039, –.0078, –.0006, –.0011, 4.922, 2.316} cm

A good analysis relies on the inspection for accuracy of the computer result.

Computations subsequent to the Displacement Method.

- The post analysis consists of computations of member displacements,
- Compute member forces, $p_m = k_m d_m$ for all the members, and
- Finally draw Bending Moment Diagram

The bending moment diagram based on the Part-II, Displacement method is plotted, Fig. (iv).

The final result is obtained by superposing the results obtained from the Part-I and Part-II, Fig. (v).

6.6 FORCE-DEFORMATION RELATION IN SYSTEM COORDINATES

It was stated in Art. 6.4 that the direct stiffness method requires transformation from p_m to p'_m. Transformation matrices for different cases:

Before writing the equations of equilibrium for each node, it is necessary that the end forces of different members meeting at the node should refer to a single coordinate system. It is in this connection that we have to fix system coordinates.

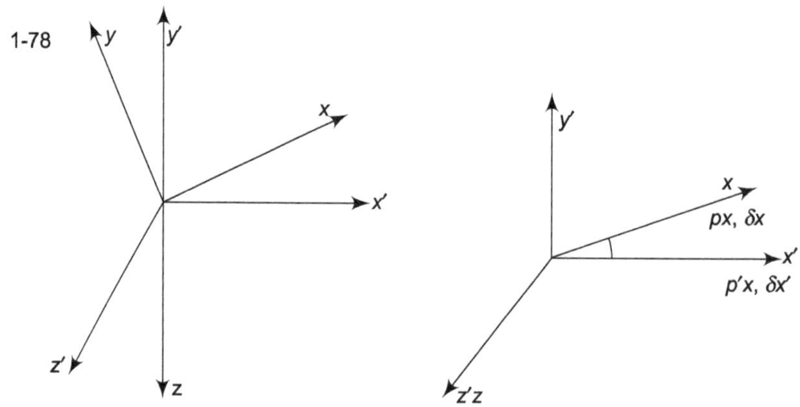

(a) 3D – Frame general loading (b) 2D – Frame in plane loading

Fig. 6.16 *Member and System Coordinates*

Referring to Fig. 6.16 where there is only rotational transformation at a point:

$$P'_m = T\,P_m \qquad \ldots(6.4)$$

$$\begin{bmatrix} P'_x \\ P'_y \\ P'_z \\ m'_x \\ m'_y \\ m'_z \end{bmatrix} = \begin{bmatrix} \cos(x'_x) & \cos(P'_y) & \cos(P'_z) & 0 & 0 & 0 \\ \cos(y'_x) & \cos(y'_y) & \cos(y'_z) & 0 & 0 & 0 \\ \cos(z'_x) & \cos(z'_y) & \cos(z'_z) & 0 & 0 & 0 \\ 0 & 0 & 0 & \cos(x'_x) & \cos(x'_y) & \cos(x'_z) \\ 0 & 0 & 0 & \cos(y'_x) & \cos(y'_y) & \cos(y'_z) \\ 0 & 0 & 0 & \cos(z'_x) & \cos(z'_y) & \cos(z_z) \end{bmatrix} \begin{bmatrix} P_x \\ P_y \\ P_z \\ m_x \\ m_y \\ m_z \end{bmatrix}$$

or simply

$$P' = T\,P \qquad \ldots(6.5)$$

and as the direction cosine matrices are orthogonal

$$T^{-1} = T^t \text{ and } d = T^t\, d'$$

∴ The matrix relations

$P_1 = K_{11}\, d_1 + K_{12}\, d_2$ and $P_2 = K_{21}\, d_1 + K_{22}\, d_2$ may now be expressed with the help of above transformation in system coordinates.

$P'_1 = K'_{11}\, d'_1 + K'_{12}\, d'_2$ and $P'_2 = K'_{21}\, d'_1 + K'_{22}\, d'_2$ where

$K'_{11} = T\,K_{11}\,T^t$ or in general $K_{ij'} = TK_{ij}T^t$, i and j take values 1 and/or 2 only. ...(6.6a)

In the case of 2D-frames with inplane loading, it is advantageous to coincide z and z' axes as we have components b_x, p_y, and m only the transformation matrix Eq. 6.5 gets simplified to

$$\begin{pmatrix} p'_j \\ p'_y \\ m' \end{pmatrix} = \begin{pmatrix} c & -s & 0 \\ s & c & 0 \\ 0 & 0 & 1 \end{pmatrix} \begin{pmatrix} p_x \\ p_y \\ m \end{pmatrix} \quad \ldots(6.6b)$$

K' matrices for the above kind of transformations are given in Table,
where $c = \cos\alpha = \cos(x', x)$ and $s = \sin\alpha$

Assembling of the member stiffnesses for obtaining the global matrix of a structure is explained considering a plane frame. The procedure for assembling the member stiffnesses for the space frame remains unchanged.

6.7 TRANSFORMATION MATRICES, ASSEMBLY FOR PLANE FRAME

The assembly of stiffness matrix, we recall (DSM), is based on equations of equilibrium of the nodes as the free bodies. Let us consider a typical node, E and the equivalent load on the node be P_E. We have from Fig. 6.17(b)

$$P'_E = (P'_2)_g + (P'_2)_i + (P'_1)_j + (P'_1)_1 + (P'_1)_m$$

And also that

$$(P'_2)_g = (K'_{21})_g\,[(d'_1)_g = d'_B] + (K'_{22})_g\,[(d'_2)_g = d'_E]$$

$$(P'_2)_i = (K'_{21})_i\, d'_D + (K'_{22})_i\, d'_E$$

$$(P'_1)_j = (K'_{11})_j\, d'_E + (K'_{12})_j\, d'_F$$

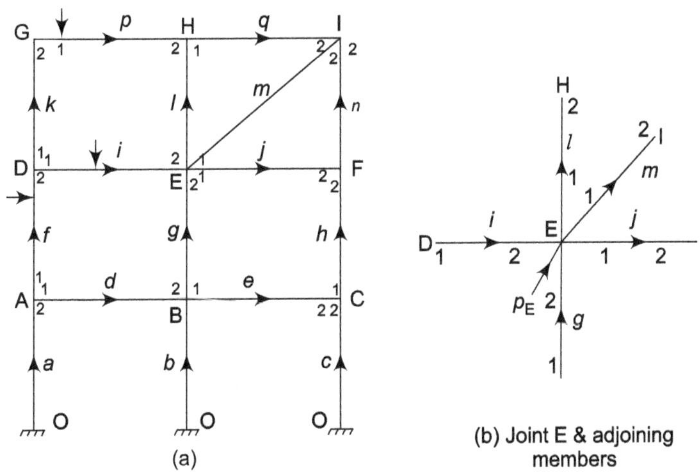

Fig. 6.17 *Illustration for Assembling Stiffness Matrix*

Coefficients of the K′ matrices
(s = sin α, c = cos α)

$$K'_{11} = \begin{pmatrix} c^2\dfrac{EA}{L}+s^2\dfrac{12EI}{L^3} & sc\left(\dfrac{EA}{L}-\dfrac{12EI}{L^3}\right) & -s\dfrac{6EI}{L^2} \\ sc\left(\dfrac{EA}{L}-\dfrac{12EI}{L^3}\right) & s^2\dfrac{EA}{L}+c^2\dfrac{12EI}{L^3} & c\dfrac{6EI}{L^2} \\ -s\dfrac{6EI}{L^2} & c\dfrac{6EI}{L^2} & \dfrac{4EI}{L} \end{pmatrix}$$

$$K'_{12} = \begin{pmatrix} -\left(c^2\dfrac{EA}{L}+s^2\dfrac{12EI}{L^3}\right) & -sc\left(\dfrac{EA}{L}-\dfrac{12EI}{L^3}\right) & -c\dfrac{6EI}{L^2} \\ -sc\left(\dfrac{EA}{L}-\dfrac{12EI}{L^3}\right) & -\left(s^2\dfrac{EA}{L}+c^2\dfrac{12EI}{L^3}\right) & c\dfrac{6EI}{L^2} \\ +s\dfrac{6EI}{L^2} & -c\dfrac{6EI}{L^2} & \dfrac{2EI}{L} \end{pmatrix}$$

$$K'_{21} = \begin{pmatrix} -\left(c^2\dfrac{EA}{L}+s^2\dfrac{12EI}{L^3}\right) & -sc\left(\dfrac{EA}{L}-\dfrac{12EI}{L^3}\right) & s\dfrac{6EI}{L^2} \\ -sc\left(\dfrac{EA}{L}-\dfrac{12EI}{L^3}\right) & -\left(s^2\dfrac{EA}{L}+C^2\dfrac{12EI}{L^3}\right) & -c\dfrac{6EI}{L^2} \\ -s\dfrac{6EI}{L^2} & c\dfrac{6EI}{L^2} & \dfrac{2EI}{L} \end{pmatrix}$$

$$K'_{21} = \begin{pmatrix} \left(c^2\dfrac{EA}{L}+s^2\dfrac{12EI}{L^3}\right) & sc\left(\dfrac{EA}{L}-\dfrac{12EI}{L^3}\right) & s\dfrac{6EI}{L^2} \\ sc\left(\dfrac{EA}{L}-\dfrac{12EI}{L^3}\right) & s^2\dfrac{EA}{L}+c^2\dfrac{12EI}{L^3} & -c\dfrac{6EI}{L^2} \\ s\dfrac{6EI}{L^2} & -c\dfrac{6EI}{L^2} & \dfrac{4EI}{L} \end{pmatrix}$$

$$(P'_1)_1 = (K'_{11})_1\, d'_E + (K'_{12})_l\, d'_H$$
$$(P'_1)_m = (K'_{11})_m\, d'_E + (K'_{12})_m\, d'_I$$

The above relations when written in the matrix form, we get Eq. 6.7 for equilibrium of node *E*.

$$P'_I = \left[0, (k'_{21})_g, 0, (k'_{21})_i, (k'_{22})_g + (k'_{22})_i + (k'_{11})_j + (k'_{11})_l + (k'_{11})_m, (k'_{12})_j, 0, (k'_{12})_l, (k'_{12})_m\right] \begin{bmatrix} d'_A \\ d'_B \\ d'_C \\ d'_D \\ d'_E \\ d'_F \\ d'_G \\ d'_H \\ d'_I \end{bmatrix} \quad \ldots(6.7)$$

Diagonal matrix consists of summation of the submatrices.

Similar equations of equilibrium can be established for all other nodes, and when all such matrix equations are assembled we get the stiffness matrix for the entire structure. And the final stiffness matrix equation of a typical row is given by Eq. (6.7), given above.:

When all the rows are assembled, the following complete global matrix is obtained, Eq. (6.8).

$$\begin{bmatrix} P'_A \\ P'_B \\ P'_C \\ P'_D \\ P'_E \\ P'_F \\ P'_G \\ P'_H \\ P'_I \end{bmatrix} = \begin{bmatrix}
(K'_{22})_a + (K'_{11})_d + (K'_{11})_f & (K'_{12})_d & 0 & (K'_{12})_f & 0 & 0 & 0 & 0 & 0 \\
(K'_{21})_d & (K'_{22})_a + (K'_{11})_d + (K'_{11})_e + (K'_{11})_f & (K'_{12})_e & 0 & (K'_{12})_g & 0 & 0 & 0 & 0 \\
0 & (K'_{21})_e & (K'_{22})_c + (K'_{22})_e + (K'_{11})_h & 0 & 0 & (K'_{12})_g & 0 & 0 & 0 \\
(K'_{21})_f & 0 & 0 & (K'_{22})_f + (K'_{11})_i + (K'_{11})_k & (K'_{12})_i & 0 & (K'_{12})_k & 0 & 0 \\
0 & (K'_{21})_g & 0 & (K'_{21})_i & (K'_{22})_g + (K'_{11})_i + (K'_{11})_g + (K'_{11})_i + (K'_{11})_{n1} & (K'_{12})_j & 0 & (K'_{12})_l & (K'_{12})_m \\
0 & 0 & (K'_{21})_h & 0 & (K'_{21})_j & (K'_{22})_h + (K'_{22})_i + (K'_{11})_k & 0 & 0 & (K'_{12})_n \\
0 & 0 & 0 & (K'_{21})_k & 0 & 0 & (K'_{22})_k + (K'_{11})_P & (K'_{12})_P & 0 \\
0 & 0 & 0 & 0 & (K'_{21})_i & 0 & (K'_{21})_P & (K'_{22})_l + (K'_{22})_P + (K'_{11})_q & (K'_{12})_q \\
0 & 0 & 0 & 0 & (K'_{21})_m & (K'_{21})_n & 0 & (K'_{21})_q & (K'_{22})_m + (K'_{22})_n + (K'_{22})_q
\end{bmatrix} \begin{bmatrix} d_A \\ d_B \\ d_C \\ d_D \\ d_E \\ d_F \\ d_G \\ d_H \\ d_I \end{bmatrix}$$

$$P' = K' \, d' \qquad \ldots(6.8)$$

The solution of Eq. (6.6.4) gives the unknown displacements d_A, d_B to d_I. For a complete analysis member end forces are also required to be found. The procedure for the evaluation of member end forces is explained by considering a typical member I, Fig. 6.18.

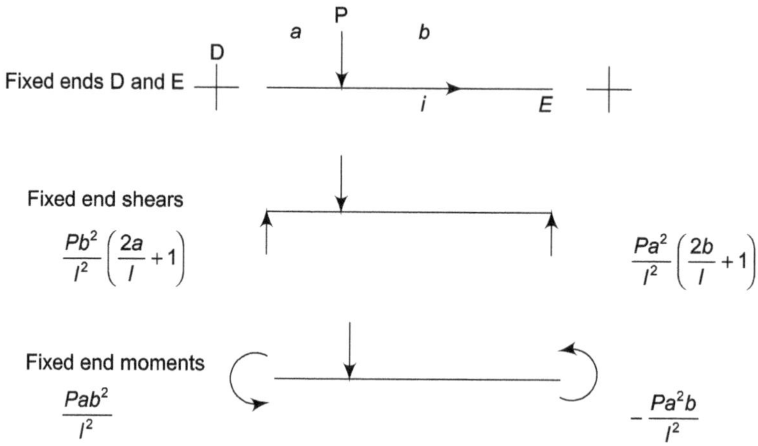

Fig. 6.18 *Evaluation of Member Forces–Fixed-end forces*

$$\underbrace{\begin{bmatrix} p_1 \\ p_2 \end{bmatrix}_i}_{\text{Net- Forces}} = \underbrace{\begin{bmatrix} K_{11} & K_{12} \\ K_{21} & K_{22} \end{bmatrix}_i}_{\text{Due to end deformations}} \begin{bmatrix} d_1 \\ d_2 \end{bmatrix}_i + \underbrace{\begin{bmatrix} p_1 \\ p_2 \end{bmatrix}_i}_{\text{caused by fixed end forces and moments}}$$

Where $$\begin{bmatrix} d_1 \\ d_2 \end{bmatrix}_i = \begin{bmatrix} d'_D \\ d'_E \end{bmatrix}_i \quad \ldots(6.9)$$

All the quantities to the right of Eq. (6.9) being known the net (p_1, p_2) values for a member are evaluated.

The above analysis is further elaborated with the help of numerical examples. The concept of matrix condensation for the negligible effect of axial deformations on moments and shears is explained in entirety using the same examples.

Example 6.1

(A) Exact solution without neglecting axial deformations for their effect on moments of shears. Consider, Fig. Ex. 1

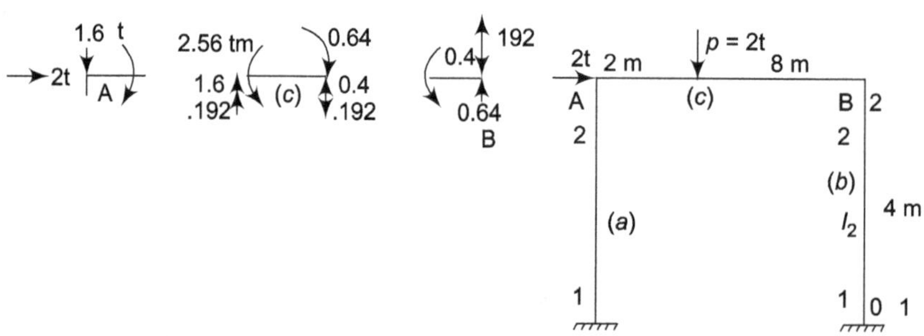

Fig. Ex. 6.1

Load matrices are given by

$$P'_A = \begin{bmatrix} 2 \\ -1.6-1.92 \\ -2.56 \end{bmatrix}$$

$$P'_B = \begin{bmatrix} 1 \\ -0.4+1.92 \\ 0.64 \end{bmatrix}$$

The assembly of stiffness matrix for the entire structure Fig. (a) is given by the Eq. (6.10).

$$\begin{bmatrix} P'_A \\ P'_B \end{bmatrix} = \begin{bmatrix} (K'_{22})_a + (K'_{11})_c & (K'_{12})_c \\ (K'_{21})_c & (K'_{22})_c + (K'_{22})_b \end{bmatrix} \begin{bmatrix} d'_A \\ d'_B \end{bmatrix} \qquad \ldots(6.10)$$

Member stiffness matrices in system coordinates are:

At end 1:

$$(K'_{11})_c = \begin{bmatrix} \left(\frac{EA}{L}\right)_c & 0 & 0 \\ 0 & \left(\frac{12EI}{L^2}\right)_c & \left(\frac{6EI}{L^2}\right)_c \\ 0 & \left(\frac{6EI}{L^2}\right)_c & \left(\frac{4EI}{L}\right)_c \end{bmatrix}$$

At end 2:

$$(K'_{22})_c = \begin{bmatrix} \left(\frac{EA}{L}\right)_c & 0 & 0 \\ 0 & \left(\frac{12EI}{L^3}\right)_c & -\left(\frac{6EI}{L^2}\right)_c \\ 0 & -\left(\frac{6EI}{L^2}\right)_c & +\left(\frac{4EI}{L}\right)_c \end{bmatrix}$$

and

$$(K'_{12})_c = \begin{bmatrix} \left(\frac{EA}{L}\right)_c & 0 & 0 \\ 0 & -\left(\frac{12EI}{L^3}\right)_c & \left(\frac{6EI}{L^2}\right)_c \\ 0 & -\left(\frac{6EI}{L^2}\right)_c & +\left(\frac{2EI}{L}\right)_c \end{bmatrix}$$

Similarly directly by inspection it is possible to write the stiffness matrices for vertical members.

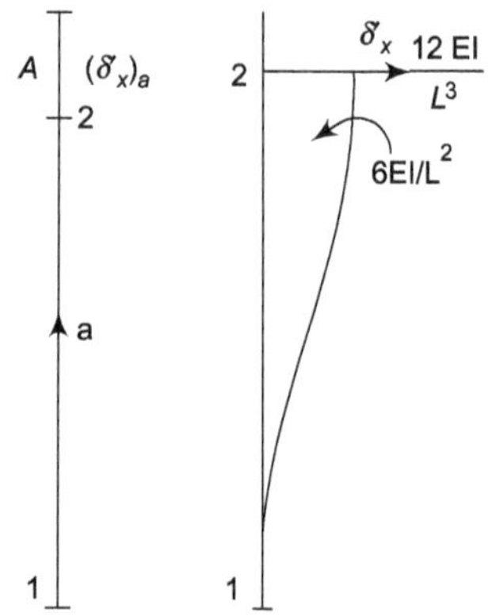

(i) Vertical member 'a'

$$
(K'_{22})_a = \begin{bmatrix} \left(\dfrac{12EI}{L^3}\right)_a & 0 & \left(\dfrac{6EI}{L^2}\right)_a \\ 0 & \left(\dfrac{EI}{L}\right)_a & 0 \\ \left(\dfrac{6EI}{L^2}\right)_a & 0 & \left(\dfrac{4\,EI}{L}\right)_a \end{bmatrix}
$$

Assembling the individual matrices of the entire structure we write:

$$
\begin{bmatrix} \begin{pmatrix} P'_x \\ P'_y \\ m' \end{pmatrix}_A \\ \begin{pmatrix} P'_x \\ P'_y \\ m' \end{pmatrix}_B \end{bmatrix} = \left(\begin{vmatrix}
\left(\dfrac{EA}{L}\right)_c + \left(\dfrac{12EI}{L^3}\right)_a & 0 & \left(\dfrac{6EI}{L^2}\right)_a & -\left(\dfrac{EA}{L}\right)_c & 0 & 0 \\
0 & \left(\dfrac{EA}{L}\right)_c + \left(\dfrac{12EI}{L^3}\right)_c & \left(\dfrac{6EI}{L^2}\right)_c & 0 & -\left(\dfrac{12EI}{L^3}\right)_b & \left(\dfrac{6EI}{L^2}\right)_c \\
\left(\dfrac{6EI}{L^2}\right)_a & \left(\dfrac{6EI}{L^2}\right)_c & \left(\dfrac{4EI}{L}\right)_c + \left(\dfrac{4EI}{L}\right)_a & 0 & -\left(\dfrac{6EI}{L^2}\right)_c & \left(\dfrac{2EI}{L^2}\right)_c \\
-\left(\dfrac{EA}{L}\right)_c & 0 & 0 & \left(\dfrac{EA}{L}\right)_c + \left(\dfrac{12EI}{L^3}\right)_b & 0 & \left(\dfrac{6EI}{L^2}\right)_b \\
0 & -\left(\dfrac{12EI}{L^3}\right)_c & -\left(\dfrac{6EI}{L^2}\right)_c & 0 & \left(\dfrac{EA}{L}\right)_b + \left(\dfrac{12EI}{L^3}\right)_c & \left(\dfrac{6EI}{L^2}\right)_c \\
0 & 6\left(\dfrac{EI}{L^2}\right)_c & 2\left(\dfrac{EI}{L^2}\right)_c & 2\left(\dfrac{EI}{L^2}\right)_b & -\left(\dfrac{6EI}{L^2}\right)_c & \left(\dfrac{EA}{L}\right)_b + \left(\dfrac{12EI}{L^3}\right)_c
\end{vmatrix} \right) \begin{bmatrix} \begin{pmatrix} \delta'_x \\ \delta'_y \\ \theta \end{pmatrix}_A \\ \begin{pmatrix} \delta'_x \\ \delta'_y \\ \theta \end{pmatrix}_B \end{bmatrix} \qquad \ldots(6.11)
$$

(B) Axial deformations neglected for their effect on moments and shears.

In most cases of rigid frames with vertical & and horizontal members of moderately tall (say up to 10 storeys, in a multi-storeyed building), the effect of axial deformations on bending moments and shears is negligible, i.e., if we are to evaluate $(P'_x, m')_A$ and $(P'_x, m')_B$, the effect of $(\delta'_y)_A$ and $(\delta'_y)_B$ being negligible, in anticipation, we eliminate the second and fifth column of the stiffness matrix. And as we are justified in neglecting the axial deformations for the shears and moments only and not for P'_y, we exclude the rows corresponding to $(P'_y)_A$ and $(P'_y)_B$. The matrix equation, therefore, gets condensed to

$$\begin{bmatrix} \begin{pmatrix} P'_x \\ m' \end{pmatrix}_A \\ \begin{pmatrix} P'_x \\ m' \end{pmatrix}_B \end{bmatrix} = \begin{bmatrix} \left(\frac{12EI}{L^3}\right)_a + \left(\frac{EA}{L}\right)_c & \left(\frac{6EI}{L^2}\right)_a & 0 & 0 \\ \left(\frac{6EI}{L^2}\right)_a & \left(\frac{4EI}{L}\right)_c + \left(\frac{4EI}{L}\right)_a & 0 & \left(\frac{2EI}{L}\right)_c \\ -\left(\frac{EA}{L}\right)_c & 0 & \left(\frac{12EI}{L^3}\right)_b + \left(\frac{EA}{L}\right)_c & \left(\frac{6EI}{L^2}\right)_b \\ 0 & \left(\frac{2EI}{L}\right)_c & \left(\frac{6EI}{L^2}\right)_b & \left(\frac{4EI}{L}\right)_c + \left(\frac{4EI}{L}\right)_b \end{bmatrix} \begin{bmatrix} \begin{pmatrix} \delta'_x \\ \theta' \end{pmatrix}_A \\ \begin{pmatrix} \delta'_x \\ \theta' \end{pmatrix}_B \end{bmatrix}$$

We add the first and third rows. The terms $\frac{EA}{L}$ get cancelled $c, a \rightarrow$ add for members c and a.

$$\begin{bmatrix} (P'_x)_A + (P'_x)_B \\ (m')_A \\ (m')_B \end{bmatrix} = \begin{bmatrix} \left(\frac{12EI}{L^3}\right)_a & \left(\frac{6EI}{L^2}\right)_a & \left(\frac{12EI}{L^3}\right)_b & \left(\frac{6EI}{L^2}\right)_b \\ \left(\frac{6EI}{L^2}\right)_a & \left(\frac{4EI}{L}\right)_{c,a} & 0 & \frac{2EI}{L} \\ 0 & \frac{2EI}{L} & \left(\frac{6EI}{L^2}\right)_b & \left(\frac{4EI}{L}\right)_{b,c} \end{bmatrix} \begin{bmatrix} \begin{pmatrix} \delta'_x \\ \theta' \end{pmatrix}_A \\ \begin{pmatrix} \delta'_x \\ \theta' \end{pmatrix}_B \end{bmatrix}$$

Taking advantage of $(\delta'_x)_A = (\delta'_x)_B = \delta'_x$ and adding First and Third columns of the above matrix, we get the final Equilibrium Matrix.

$$\begin{pmatrix} P'_{xA} + P'_{xB} \\ m'_A \\ m'_B \end{pmatrix} = \begin{pmatrix} \left(\frac{12EI}{L^3}\right)_a + \left(\frac{12EI}{L^3}\right)_b & \left(\frac{6EI}{L^2}\right)_a & \left(\frac{6EI}{L^2}\right)_b \\ \left(\frac{6EI}{L^2}\right)_a & \left(\frac{4EI}{L^3}\right)_c + \left(\frac{4EI}{L}\right)_a & \left(\frac{2EI}{L}\right)_c \\ \left(\frac{6EI}{L^2}\right)_b & \left(\frac{2EI}{L}\right)_c & \left(\frac{4EI}{L}\right)_c + \left(\frac{4EI}{L}\right)_b \end{pmatrix} \begin{pmatrix} \delta'_x \\ \theta'_A \\ \theta'_B \end{pmatrix} \quad \text{...(6.12)}$$

$(\delta'_x)_A = (\delta'_x)_B$

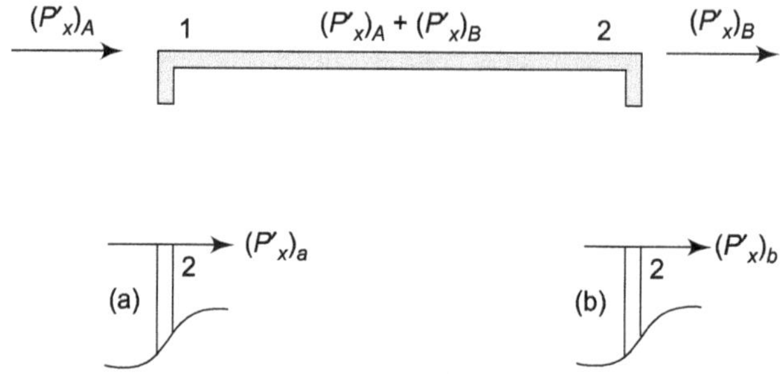

Girder Free body

After some familiarity, it is possible to write the above stiffness matrix purely from inspection. The elements of the first row give the horizontal shear equilibrium at girder level, Free body Fig. (ii). Equilibrium Eq: $(P'_x)_A + (P'_x)_B = (P'_x)_a + (P'_x)_b$ eliminates the axial forces in girder '*c*'.

6.7.1 Gable Frame: The Solution is Explained with the Help of An Illustration–(DSM) Without Making the Approximation of Member in Extensibility–Illustration 8

Direct Stiffness Method (DSM)for the analysis of a gable frame. The frame is particularly chosen as it entails all the concepts and at the same time the size of the problem is manageable for hand computations. For this example also, we consider the two cases: (a) Without the approximation of member inextensibility, and (b) With inextensibility.

(i) Without neglecting axial deformations

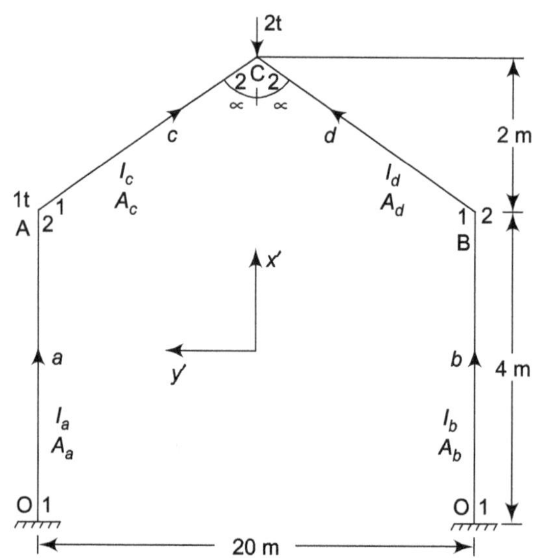

Fig. 6.19 *Gable Frame Illustration–8*

$$\begin{pmatrix} P'_A \\ P'_B \\ P'_C \end{pmatrix} \begin{pmatrix} (K'_{22})_a + (K'_{11})_c & 0 & (K'_{12})_c \\ 0 & (K'_{22})_b + (K'_{11})_d & (K'_{12})_d \\ (K'_{21})_c & (K'_{21})_d & (K'_{22})_c + (K'_{22})_d \end{pmatrix} \begin{pmatrix} d'_A \\ d'_B \\ d'_C \end{pmatrix} \quad \ldots(6.12)$$

The sub matrices of the Eq. (6.12) are defined in the following, Eqs (i) to (v).

$$(K'_{22})_{a,b} = \begin{bmatrix} \left(\frac{EI}{L}\right)_{a.b} & 0 & 0 \\ 0 & \left(\frac{12EI}{L^3}\right)_{a.b} & \left(\frac{-6EI}{L^2}\right)_{a.b} \\ 0 & \left(\frac{-6EI}{L^2}\right)_{a.b} & \left(\frac{4EI}{L}\right)_{a.b} \end{bmatrix} \quad \ldots(i)$$

$a \cdot b \rightarrow a \text{ or } b$

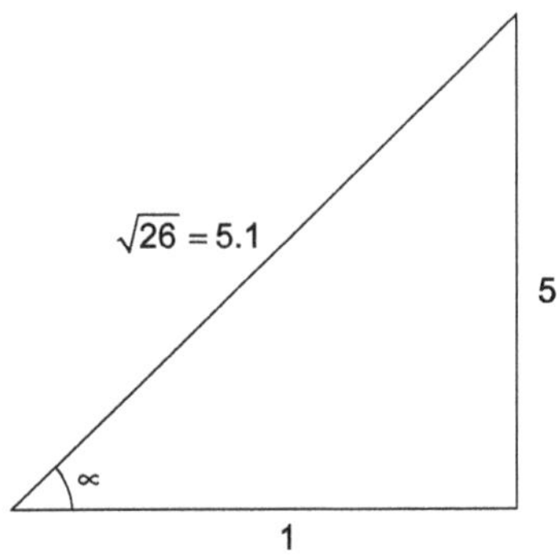

Angle Designation

For $(K_{ij}')_c$ and $(K_{ij}')_d$ the values are

$$\alpha = \tan^{-1} 5, \quad \sin \alpha = \frac{5}{\sqrt{26}} = 0.980$$

$$\cos \alpha = 0.196$$

For member (C),

$$C_C = 0.196 \text{ and } S_C = -0.980$$

For member (d),

$$C_d = 0.196 \quad \text{and} \quad S_d = +0.980$$

Let $C_C = C_d = C, S_d = -S_C = S$

$$(K'_{11})_c = \begin{bmatrix} \left(C^2 \dfrac{EA}{L} + s^2 \dfrac{12\,EI}{L^3}\right)_c; & SC\left(\dfrac{EA}{L} - \dfrac{12EI}{L^3}\right)_c; & -\left(S\dfrac{6EI}{L^2}\right)_c \\ SC\left(\dfrac{EA}{L} - \dfrac{12EI}{L^3}\right)_e; & \left(S^2 \dfrac{EA}{L} + C^2 \dfrac{12\,EI}{L^3}\right)_e; & \left(C\dfrac{6EI}{L^2}\right)_e \\ -\left(S\dfrac{6EI}{L^2}\right); & \left(C\dfrac{6EI}{L^2}\right)_c & \left(\dfrac{4EI}{L}\right)_c \end{bmatrix} \quad \ldots \text{(ii)}$$

$$(K'_{22})_c = \begin{bmatrix} \left(C^2 \dfrac{EA}{L} + S^2 \dfrac{12EI}{L^3}\right)_c; & SC\left(\dfrac{EA}{L} - \dfrac{12EI}{L^3}\right)_c; & \left(S\dfrac{6EI}{L^2}\right)_c \\ SC\left(\dfrac{EA}{L} - \dfrac{12EI}{L^3}\right)_c; & \left(S^2 \dfrac{EA}{L} + C^2 \dfrac{12EI}{L^3}\right)_c; & -\left(C\dfrac{6EI}{L^2}\right)_c \\ \left(S\dfrac{6EI}{L^2}\right)_c; & -\left(C\dfrac{6EI}{L^2}\right)_c & \left(\dfrac{4EI}{L}\right)_c \end{bmatrix} \quad \ldots \text{(iii)}$$

$$(K'_{12})_c = \begin{bmatrix} -\left(C^2 \dfrac{EA}{L} + S^2 \dfrac{12EI}{L^3}\right)_c; & -SC\left(\dfrac{EA}{L} - \dfrac{12EI}{L^3}\right)_c; & -\left(S\dfrac{6EI}{L^2}\right)_c \\ -\left\{SC\left(\dfrac{EA}{L} - \dfrac{12EI}{L^3}\right)_c\right\}; & -\left(S^2 \dfrac{EA}{L} + C^2 \dfrac{12EI}{L^3}\right)_c; & \left(C\dfrac{6EI}{L^2}\right)_c \\ \left(S\dfrac{6EI}{L^2}\right)_c; & -\left(C\dfrac{6EI}{L^2}\right)_c; & \left(\dfrac{2EI}{L}\right)_c \end{bmatrix} \quad \ldots \text{(iv)}$$

The stiffness matrix $(K_{ij}')_d$ is obtained by substituting appropriate values of A, E, I, L and S, C for the member 'a' to 'd', in Eqs (ii) to (iv). And when these submatrices are entered in the global K, Eq. (6.12) we obtain a nine by nine global stiffness matrix K. We have all the information, i.e., load matrix, Eq. (v) and K matrix for the evaluation of the Unknown nine displacement for the problem chosen as an illustration.

Load matrix is given by

$$P'A = \begin{bmatrix} 0 \\ -1 \\ 0 \end{bmatrix} \quad P'B = \begin{bmatrix} 0 \\ 0 \\ 0 \end{bmatrix} \quad P'C = \begin{bmatrix} -2 \\ 0 \\ 0 \end{bmatrix} \quad \ldots \text{(v)}$$

Substituting the values of submatrices defined by Eq. (i) to (iv) into the Eq. (6.12) we get the total global matrix of size (9×9) and the solution of the matrix Eq. (6.12) and using the given load matrix Eq. (v), we get the 'exact' solution after solving for the unknown displacements from the equation.

$$\begin{pmatrix} \delta X' \\ \delta Y' \\ \theta \end{pmatrix}_A, \begin{pmatrix} \delta X' \\ \delta Y' \\ \theta \end{pmatrix}_B, \begin{pmatrix} \delta X' \\ \delta Y' \\ \theta \end{pmatrix}_C$$

The deformations are found from compatibility

$$(d_1')_a = (d_1')_b = 0; \quad (d_2')_a = (d_1')_C = (d_A')$$

Or $$\begin{pmatrix}\delta X_1'\\ \delta Y_1'\\ \theta_1\end{pmatrix}_a, \begin{pmatrix}\delta X_1'\\ \delta Y_1'\\ \theta_1\end{pmatrix}_b = \begin{pmatrix}0\\ 0\\ 0\end{pmatrix}; \begin{pmatrix}\delta X_2'\\ \delta Y_2'\\ \theta_2'\end{pmatrix}_a = \begin{pmatrix}\delta X_1'\\ \delta Y_1'\\ \theta_1'\end{pmatrix}_c = \begin{pmatrix}\delta X_A\\ \delta X_A\\ \theta_A\end{pmatrix}$$

$$(d_1')_d = (d_2')_b = (d_B')_d, \; (d_2')_c = (d_2')_d = (d')_c$$

$$\begin{pmatrix}\delta X_1'\\ \delta Y_1'\\ \theta_1'\end{pmatrix}_d = \begin{pmatrix}\delta X_2'\\ \delta Y_2'\\ \theta_2'\end{pmatrix}_b = \begin{pmatrix}\delta X_B'\\ \delta Y_B'\\ \theta_B\end{pmatrix}_d ; \begin{pmatrix}\delta X_2'\\ \delta Y_2'\\ \theta_2\end{pmatrix}_c = \begin{pmatrix}\delta X_2'\\ \delta Y_2'\\ \theta_2'\end{pmatrix}_d = \begin{pmatrix}\delta X_C'\\ \delta Y_C'\\ \theta_C'\end{pmatrix}$$

and the member end stress resultant are found by the force deformation relationship e.g., the following are the relations for a typical member, say,

$$(p_1')_c = (K'_{11})_C\,(d_1')_C + (K'_{12})_C\,(d_2')_C$$

and $$(p_2')_c = (K'_{12})_C\,(d_1')_C + (K'_{22})_C\,(d_2')_C$$

6.7.2 Code Number Technique Applied to Sub Matrices

Member codes using the sub matrices for directly assembling the global matrix: In the following the global matrix is generated with out having to assemble global matrix 9 × 9. The approach is suitable for the machine assembling of the matrix equation.

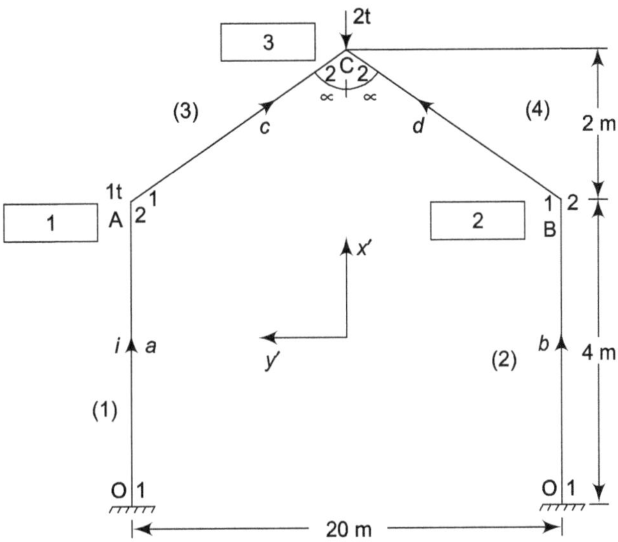

Fig. 6.20 *Digitization of the Members and Joints*

The code number techniques were discussed using the scalar stiffness, Article 6.5. In the following, the sub matrices are used in place of the scalars. It is known by now that the code number techniques facilitate the computer coding. The Fig. 6.19 is redrawn with the digitized member designations as well as the digitized joint designation. The member numbers are shown within parenthesis ( ), and the joints are designated within the squares, Fig. 6.20.

The coding of the members is explained in the following. The member designations use the sub matrices K_{ij} of the size (3×3). It will be interesting going through the table.

Member (a)	*Digitized (1)*	*Joints (O, A)*	*Member code Digitized [0, 1]*	*Member to Global Contribution $K'_{11} \rightarrow K'_{11}$*
(b)	(2)	(O, B)	[0, 2]	$K'_{22} \rightarrow K'_{22}$
(c)	(3)	(A, C)	[1, 3]	$K'_{11} \rightarrow K'_{11}$ $K'_{12} \rightarrow K'_{13}$ $K'_{22} \rightarrow K'_{33}$
(d)	(4)	(B, C)	[2, 3]	$K'_{11} \rightarrow K'_{22}$ $K'_{12} \rightarrow K'_{23}$ $K'_{22} \rightarrow K'_{33}$

If the global contributions are appropriately placed according to the table above, we get:

$$\begin{pmatrix} (K'_{22})_a + (K'_{11})_c & 0 & (K'_{12})_c \\ * & (K'_{22})_b + (K'_{11})_d & (K'_{12})_d \\ * & * & (K'_{22})_c + (K'_{22})_d \end{pmatrix}_{9 \times 9} \qquad \ldots(6.12a)$$

Taking note of symmetry of the Structural stiffness of Stable Structures, it may be seen that the Eq. (6.12) is identically same as Eq. (6.12a) stated for scalar approach. The alternative approach of assembling the Global stiffness matrix has the advantage of machine programming in the digitized form.

6.7.3 Solution Assuming Inextensibility

It is a significant assumption, useful not only for small rigidly jointed structures but also for tall buildings with rigid floor diaphragms (inextensible) and inextensible girders. Obtaining the solution for the case of inextensibility, making use of the Eq. (6.12) and then choosing appropriately the load matrices is possible. It requires proper choice of free bodies for the reduced number of independent displacements, recognizing master and slave joints. However the effort involved does not justify the condensation of a size (9×9) for the example of the gable to a reduced size $(5 \times 5)^*$. The case (ii) assuming inextensibility is covered elsewhere, Chapter 7 and 8, using connection matrix approach and the Iterative techniques.

6.8 Revision, Member and System Axes

Member and System Axes (Direct Stiffness Approach), and Transformation matrices necessitated due to change of axes, member to system

Recapitulation prismatic straight member:

* Refer Ex. 6.1. The matrix K_j of size 6×6 was reduced to size 3×3 by using the approximation of inextensibility.

6.8.1 Straight Prismatic Member of a Space Frame

In the case of a 3D space frames the load and deformations matrices are of size 6 × 1 each.

$$\theta = \begin{bmatrix} P_x & P_y & P_z & m_x & m_y & m_z \end{bmatrix}^t \text{ and}$$

$$d = \begin{bmatrix} \delta_x & \delta_y & \delta_z & \theta_x & \theta_y & \theta_z \end{bmatrix}^t$$

The sub matrices K_{11}, K_{12}, and K_{22} are of size 6 × 6 each and are described by the following relations.

$$K_{11} = \begin{bmatrix} \frac{EA}{L} & 0 & 0 & 0 & 0 & 0 \\ 0 & \frac{12EI_z}{L^3} & 0 & 0 & 0 & \frac{6EI_z}{L^3} \\ 0 & 0 & \frac{12EI_y}{L^3} & 0 & -\frac{6EI_y}{L^2} & 0 \\ 0 & 0 & 0 & \frac{GJ}{L} & 0 & 0 \\ 0 & 0 & -\frac{6EI_y}{L^2} & 0 & \frac{4EI_y}{L} & 0 \\ 0 & \frac{6EI_z}{L^2} & 0 & 0 & 0 & \frac{4EI_z}{L} \end{bmatrix}$$

GJ/L is torsional stiffness ...(6.13)

K_{22}, and $K_{21} = (K_{12})^t$ have already been given.

It can be easily shown that the results are for the particular cases of Eq. (6.13). Further a grid problem where the loading is perpendicular to the plane of the grid is also a particular case of Eq. (6.13) and *p, k, d* matrices are given by the relations given by Eq. (4.9)

$$P = \begin{Bmatrix} p_y \\ m_x \\ m_z \end{Bmatrix}, d = \begin{Bmatrix} \delta_y \\ \theta_x \\ \theta_z \end{Bmatrix}, K_{11} = \begin{bmatrix} \frac{12EI_z}{L^3} & 0 & \frac{6EI_z}{L^2} \\ 0 & \frac{GJ}{L} & 0 \\ \frac{6EI_z}{L^2} & 0 & \frac{4EI_z}{L} \end{bmatrix}$$

$$K_{12} = (K_{21})^t = \begin{bmatrix} \frac{-12EI_z}{L^3} & 0 & \frac{6EI_z}{L^2} \\ 0 & -\frac{GJ}{L} & 0 \\ -\frac{6EI_z}{L^2} & 0 & \frac{2EI_z}{L} \end{bmatrix}$$

and
$$K_{22} = \begin{bmatrix} +\frac{12EI_z}{L^3} & 0 & -\frac{6EI_z}{L^2} \\ 0 & \frac{GJ}{L} & 0 \\ -\frac{6EI_z}{L^2} & 0 & \frac{4EI_z}{L} \end{bmatrix}$$

We will take up additional problem of grid analysis at the end of the section for the application of the theory. Continuing with the member stiffness of a plane frame and continuous beams, at the end supports, it is quite frequent to provide a hinge. In such a case, considering modified stiffness is useful, particularly when hand calculations are envisaged.

Modified matrices for one of the ends of member hinged can be used to eliminate the unknown deformations at hinged ends.

6.9 CONSOLIDATION OF THEORY

Before taking up solution of some illustrations, it is good to visualize the coverage so far on the Displacement Method. It is for no purpose that we discussed at great detail symmetry and anti symmetry applied to Grids. We are going to use symmetry and antisymmetry with advantage for arriving at the member stiffnesses.

6.9.1 Effect of the Assumption of Inextensibility and Corresponding Displacements

The number of unknown displacements depends upon the assumption made regarding inextensibility of the members constituting the frame. If inextensibility is not assumed in the case of plane frame with rigid joints, the additional unknowns are equal to the number of inextensible members. The following are the typical illustrations.

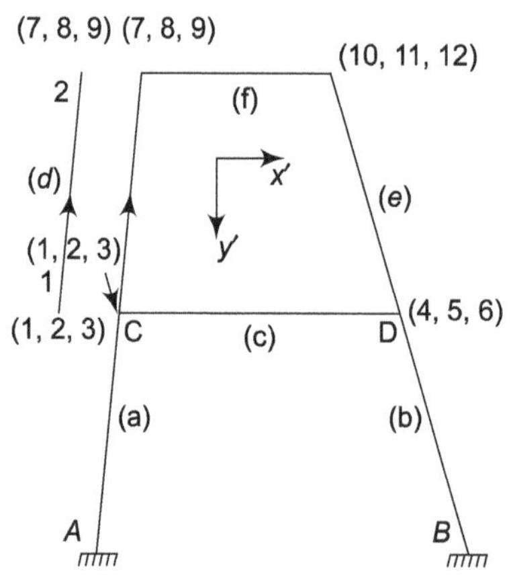

No. of deformations = 12

(a) Frame with rigid joints

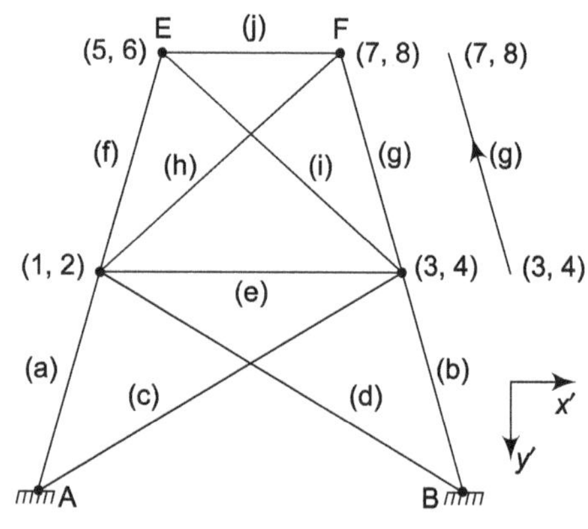

No. of deformations = 8

(b) Truss with Hinge joints

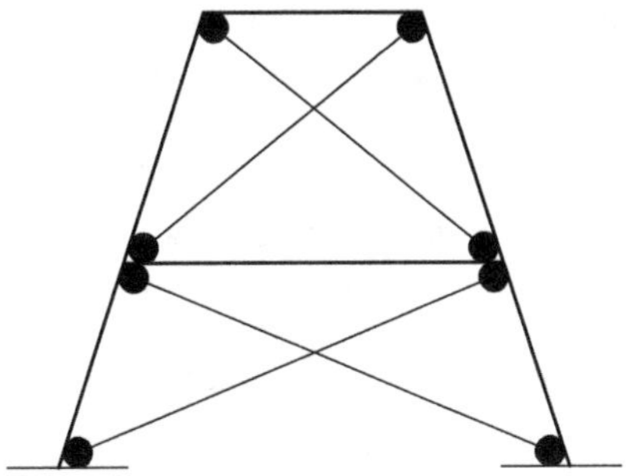

(c) Trussed Frame. Diagonals hinged at their ends.

Fig. 6.21 *Unknown Displacements*

Referring to Fig. 6.21(a), there are two unknown translations and one rotation per node and the total number of unknowns = $3n$ (=12 for the frame Fig. 6.21(a) where n is the number of the nodes undergoing displacements). In the case of plane hinged trusses, the total number of unknown displacement is equal to $2n$, (=8 for the frame, Fig. 6.21(b)). In the case of truss members axial deformation is the only member deformation.

Referring to the rigid plane frames, if all the members are assumed to be inextensible, the total number of unknowns gets reduced by the number of members (m). No. of unknown displacements = $3n - m$.

If only some of the members are inextensible then m will be the number of the inextensible members only. Referring to the Fig. 6.21(a), $n = 4$, $m = 6$.

If all the members are inextensible.

No. of unknown displacements = $3 \times 4 - 6 = 6$

Consider the rigid frame connected by the diagonals which are pinned at the ends. It is the case of a "Trussed frame". If it is a case of a Trussed Frame, the assumption of inextensibility of the frame members may not be realized. The assumption is a suspect, Fig. 6.21(c).

If only girders *CD* and *EF* are assumed to be inextensible, $m = 2$, the number of unknowns of the original frame is ten. It is sensitive to the assumption regarding inextensibility.

It is interesting to find (i) degree of indeterminacy by Force Method, and (ii) DOF by Displacement Method. While analysing by the Displacement method, part I of the analysis, which is a primary analysis, it is done for the nodes clamped. From the fixed edge forces from the primary analysis nodal loads are computed. The nodes are allowed to under displacements caused by the nodal loads. The displacement method evaluates the nodal deformations (displacements and rotations) caused by the computed nodal loads.

The three basic steps are: (1) Compatibility, (2) "Force-deformation" relations or member stiffnesses and (3) Equilibrium. But in the following we try to look at from a different perspective.

6.9.2 Basic Steps for Member Stiffnesses from a Different Perspective: Superposition of Symmetry and Antisymmetry

Step 1 ***Compatibility and code numbers:*** Refer Fig. 6.5(a) for the member *(d)*. From compatibility conditions the deformations of the member *(d)* are completely defined by the deformation of the nodes C, (1,2,3) and D, (7,8,9). Therefore the deformations relevant to the member *d* are (1,2,3) and (7,8,9). Therefore, we say, the code number of member *d*, [1,2,3:7,8,9] defines the compatibility of the member. Similarly we write the code numbers of the members *q,b,c,e,f are:*

[0,0,0;1,2,3], [0,0,0;4,5,6], [1,2,3;4,5,6], [4,5,6;10,11,12], and finally [7,8,9;10,11,12] respectively.

Another alternative way of writing compatibility is to write code number of the joints. The member numbers meeting at a joint is used to write the joint code number. Referring to the Fig. 6.5 (a), the code number of joint C is given by the members *a, c,* and *d* meeting the joint, and is equal to [*a, c, d*; 2,1,1]. The code numbers are used to generate the compatibility matrix.

Step 2 For a change, we look at force deformation relations as superposition of Symmetry and Antisymmetry. Let the member axis (i) be the undeformed position of the member. The elastic deformations are with respect to the axis (iii). The corresponding rigid body translation are δ, and rigid body rotation ψ, Fig. (a). Consider flexure only. The end moments are M_1 and M_2 at the ends 1 and 2 respectively. And θ_1 and θ_2 be the corresponding end rotations, shown in Fig. (b).

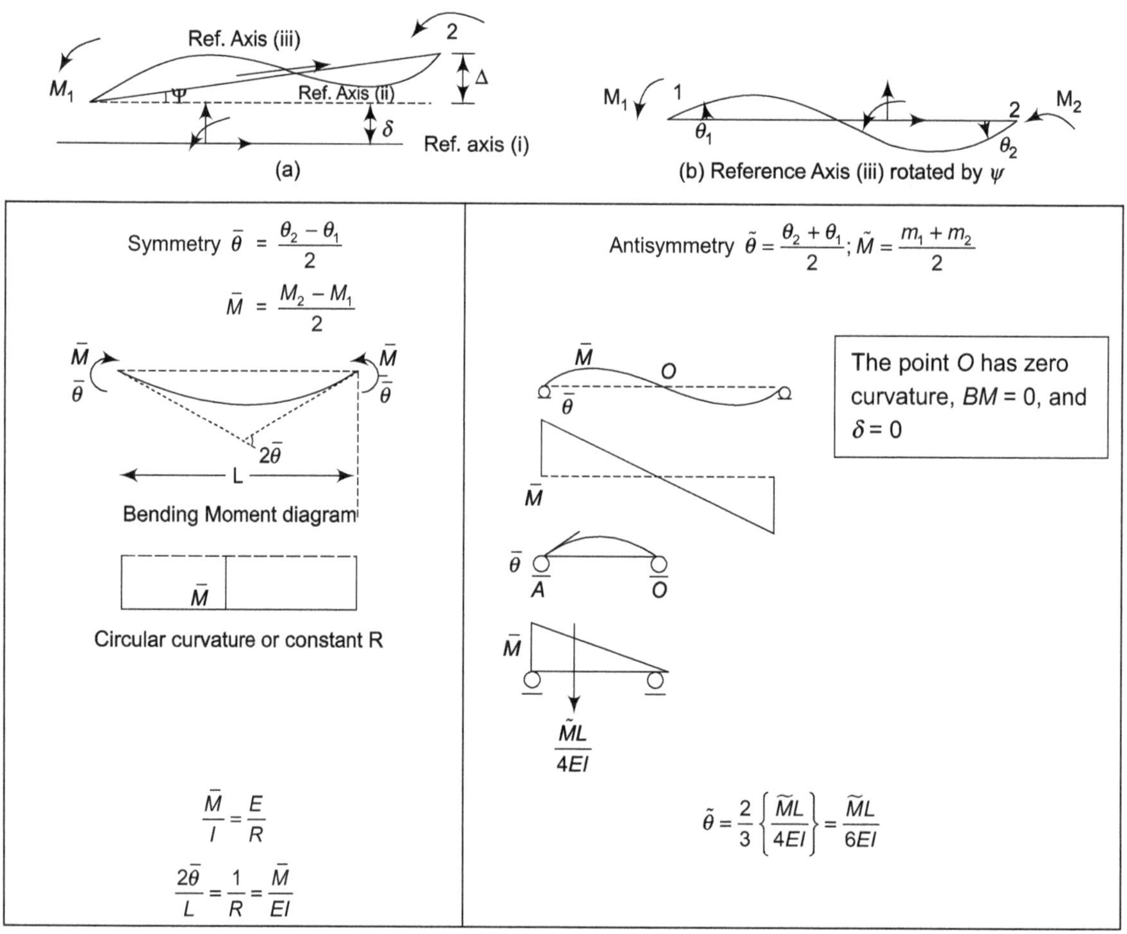

(c) Symmetry and Antisymmetry for Axis (iii)

Fig. 6.22 *Member Axes and Flexural Stiffness*

Substituting for $\tilde{\theta} = \dfrac{\theta_2 - \theta_1}{2}$

and $\bar{M} = (M_2 - M_1)/2$

Substituting for $\tilde{\theta}$ and $\widetilde{M}$

$$\tilde{\theta} = \frac{(\theta_2 + \theta_1)}{2} \text{ and } \widetilde{M} = (M_2 + M_1)/2$$

We obtain the results:*

$$\left.\begin{aligned} M_1 &= 4EI/L\,\theta_1 + 2EI/L\,\theta_2 \\ M_2 &= 2EI/L\,\theta_1 + 4EI/L\,\theta_2 \end{aligned}\right\} \quad \text{or} \quad \begin{pmatrix} M_1 \\ M_2 \end{pmatrix} = \overbrace{\begin{bmatrix} 4EI/L & 2EI/L \\ 2EI/L & 4EI/L \end{bmatrix}}^{K_m} \begin{bmatrix} \theta_1 \\ \theta_2 \end{bmatrix} \quad \ldots\text{(i)}$$

For a more inquisitive reader, there is yet another alternative of using natural (dimensionless) coordinates, (ξ, η) with origin at the midpoint of the beam member. Refer Article 6.15.

The additional component δ can be shown as equivalent to rotations equal to $\psi = \Delta L$. Therefore the moments M_1 and M_2 due to Δ only are obtained by substituting $\theta_1 = \theta_2 = -\Delta/L$

$$\left.\begin{aligned} M_1 &= \frac{4EI}{L}\left(\frac{-\Delta}{L}\right) + \frac{2EI}{L}\left(\frac{-\Delta}{L}\right) = -\frac{6EI\Delta}{L^2} \\ \text{Similarly,} \quad M_2 &= -\frac{6EI}{L^2}\delta \end{aligned}\right\}$$

Combining the effect of δ, Eq. (3) with the moments obtained for θ_1 and θ_2 (Eq. 2), we obtain the moments for the reference axis Fig. 9(a) (ii).

$$\left.\begin{aligned} M_1 &= \frac{4EI}{L}\theta_1 + \frac{2EI}{L}\theta_2 - \frac{6EI}{L^2}\Delta \\ \text{and} \quad M_2 &= \frac{2EI}{L}\theta_1 + \frac{4EI}{L}\theta_1 - \frac{6EI}{L^2}\Delta \end{aligned}\right\} \quad \ldots\text{(ii)}$$

If the reference x axis is taken along tangent at end-1, in this case also we eliminate rigid body deformations and the member deformation d_m is defined by two components only, that is,

$$d_m = (\delta_2, \theta_2)$$

$$\left.\begin{aligned} \text{and} \quad M_1 &= \frac{2EI}{L}\theta_2 - \frac{6EI}{L^2}\delta \\ \text{and} \quad M_2 &= \frac{4EI}{L}\theta_2 - \frac{6EI}{L^2}\delta \end{aligned}\right\} \quad \ldots\text{(iii)}$$

$$P_{y2} = -\frac{6EI}{L^2}\theta_2 + \frac{12EI}{L^3}\delta, \text{and } K_m \text{ is given by:} \begin{pmatrix} P_{y2} \\ m_2 \end{pmatrix} = \begin{pmatrix} \dfrac{12EI}{L^3} & -\dfrac{6EI}{L^2} \\ -\dfrac{6EI}{L^2} & \dfrac{4EI}{L} \end{pmatrix} \begin{pmatrix} \delta_2 \\ \theta_2 \end{pmatrix}$$

Step 3 Free body approach for writing moment equations of equilibrium is simple. But writing the equations of equilibrium for the assumption of inextensibility requires proper choice of free bodies

* We may obtain K_m by inverting the flexibility matrix, Art. 4.4. The approach presented and also Art. 6.15 derive the stiffness matrices, K_m directly.

for eliminating axial forces (Centers of rotation). Alternatively we can use virtual work method which is definitely more appropriate. The approach is better explained while solving the typical structural problems. This is done in Chapter 7 Displacement Method continued.

In the case of hinged trusses, the nodal loads directly give the load elements and the number of unknowns and load elements are $2n$ each, where n is the number of nodes.

6.10 MIXED METHOD

Sometimes it is more convenient to use Mixed Method of analysis which results in number of unknowns less than those encountered when analyzed by force method or displacement method. As an example, consider the frame, Fig. 6.23. The number of unknowns for the different methods is

Force Method = 18

Displacement Method = 7 + 7 = 14, when inextensibility is assumed.

Mixed Method = 3 + 7 = 10

The unknowns, 3 redundants X_1, X_2, X_3 and 7 deformations θ_4, θ_5, θ_6, θ_7, θ_8, θ_9, δ_{10} are shown.

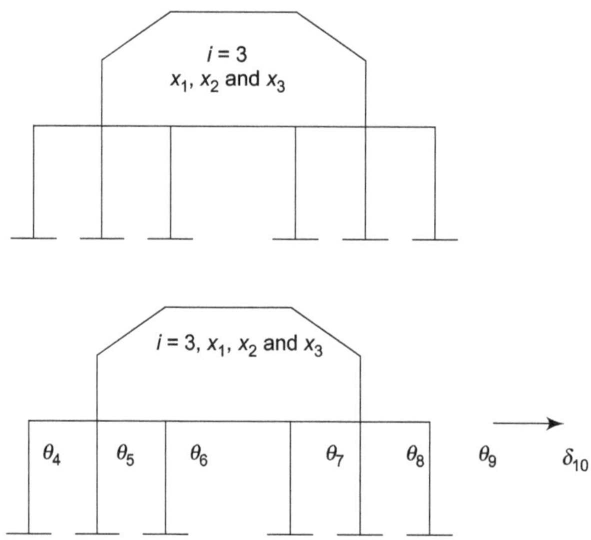

Fig. 6.23 *Mixed Method*

Caution: The mixed method results in a coefficient matrix which is not totally symmetric.

The reader is advised to solve a numerical example by the mixed method and notice lack of symmetry.

6.11 DIRECT STIFFNESS METHOD—MORE CASES

6.11.1 Plane Frames

Plane Frames have been discussed in Article 6.4.

6.11.2 Direct Stiffness Method: Grillage as Illustration

It is a step to Generalization. While presenting grid under the Force Method, we discussed its degree of indeterminacy depending upon the torsion stiffness of its members. The grid members provide an efficient system for transferring live loads to the supports. In the absence of the grid members, the slabs become uneconomically thick. There comes a stage the self weight due to the increased thickness of the slabs compensate the corresponding enhancements in stiffness/strength. The Box sections give higher strength to weight ratio. The building floor grids are generally supported on the four sides.

Like in the case of continuous beams, inextensibility is not relevant.

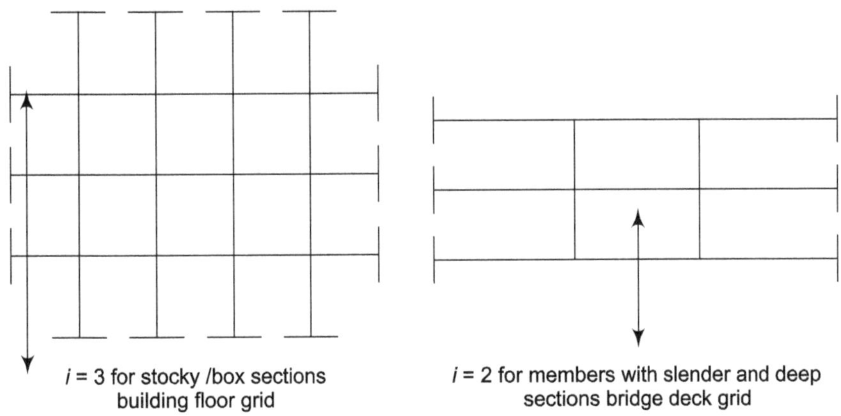

Fig. 6.24 *Types of Grids*

The bridge grids, because of the functional requirements, are supported on the two parallel sides. The members parallel to the supports are also known as diaphragms. These serve the purpose of dispersing the wheel loads on the bridge decks. In the case of grids, they are geometrically symmetrical at least in one direction. Elsewhere, superposition was used for splitting the loads with an aim to reduce the number of unknowns to be evaluated. The loads are perpendicular to the plane of the grids.

The number of internal forces, was discussed in the context of evaluating the number of unknown redundant. In the case of Displacement method, however, the number of unknown displacements (degree of freedom) is not sensitive to the value of '*i*'. Consider a simple grid. It is chosen to enable manual computations.

6.12 GRID ILLUSTRATIONS

6.12.1 Illustration: 1

Concrete Member Properties:

$$35 \text{ cm} \times 15 \text{ cm}$$

$$I_z = 53593 \text{ cm}^4$$

$$J = I_x = 2877.63$$

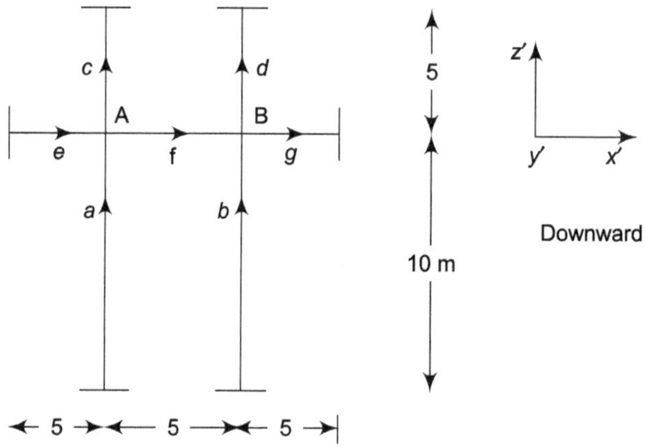

Fig. *Illustration 1(a)–Plan of a Building Floor*

$P_A = 40$ kN, vertical upward

Perpendicular to the plane of grid

$$P_B = 20 \text{ kN}$$

$$P_J' = \{[M_x, P'_Y, M_Z]_A{}' \; [M_x, P'_Y, M_Z]_B\}$$

$$P_J' = \{0, -40, 0, 0, -20, 0\}$$

It is required to find $d_J' \to d_m$

And finally p_m of all the members

Member stiffness matrix:

$$\begin{Bmatrix} p_1 \\ p_2 \end{Bmatrix} = \begin{bmatrix} K_{11} & K_{12} \\ K_{21} & K_{22} \end{bmatrix} \begin{Bmatrix} d_1 \\ d_2 \end{Bmatrix}$$

$$K_{11} = \begin{bmatrix} \frac{GJ}{L} & 0 & 0 \\ 0 & \frac{12EI_z}{L^3} & \frac{6EI_z}{L^2} \\ 0 & \frac{6EI_z}{L^2} & \frac{4EI_z}{L} \end{bmatrix}, K_{22} = \begin{bmatrix} \frac{GJ}{L} & 0 & 0 \\ 0 & \frac{12EI_z}{L^3} & -\frac{6EI_z}{L^2} \\ 0 & -\frac{6EI_z}{L^2} & \frac{4EI_z}{L} \end{bmatrix}$$

$$K_{12} = \begin{bmatrix} -\frac{GJ}{L} & 0 & 0 \\ 0 & -\frac{12EI_z}{L^3} & \frac{6EI_z}{L^2} \\ 0 & \frac{6EI_z}{L^2} & \frac{4EI_z}{L} \end{bmatrix} [K_{12}]^{\mathrm{T}} = K_{21}$$

There is no transformation required for the members '*e*', '*f*', '*g*'.

For the members '*a*', '*b*', '*c*', '*d*', however

$$P_1 = K_{11}\, d_1 + K_{12}\, d_2$$

$$P_2 = K_{21}\, d_1 + K_{22}\, d_2$$

Because of the change of the reference axes, the matrices appearing in the equation above need transformations. And

$$P_1' = TK_{11}\, T^t\, d_1 + TK_{12}\, T^t\, d_2$$

$$P_2' = TK_{21}\, T^t\, d_1 + TK_{22}\, T^t\, d_2$$

Where

$$T = \begin{bmatrix} 0 & 0 & -1 \\ 0 & 1 & 0 \\ 1 & 0 & 0 \end{bmatrix}$$ By now this must be an obvious step of the equation:

Similarly $$\begin{Bmatrix} m_x' \\ p_y' \\ m_z' \end{Bmatrix} = \begin{bmatrix} 0 & 0 & -1 \\ 0 & 1 & 0 \\ 1 & 0 & 0 \end{bmatrix} \begin{Bmatrix} m_x \\ p_y \\ m_z \end{Bmatrix}$$

$$\begin{Bmatrix} \theta_x \\ \delta_y \\ \theta_z \end{Bmatrix} = \begin{bmatrix} 0 & 0 & 1 \\ 0 & 1 & 0 \\ -1 & 0 & 0 \end{bmatrix} \begin{Bmatrix} \theta_x' \\ \delta_y' \\ \theta_z' \end{Bmatrix}$$

You are familiar of the property: TT^t = identity or unit matrix

$$K_j' = \begin{bmatrix} (K_{22})_a + (K_{22})_e + (K_{11})_f + (K_{11})_c & ;(K_{12})_f \\ * & ;(K_{22}')_b + (K_{22}')_f + (K_{11}')_g + (K_{11}')_d \end{bmatrix}$$

The equilibrium equation, $P_j' = K_j' = d_j'$

The inversion of K_j' and the given p_j', gives $d_j' = \left[K_j'\right]^{-1} P_j'$

$$d_j' = \{.002 \quad 1.231 \quad .022 \quad ; \quad .0016 \quad .8673 \quad -.0025\}\text{cm and radians.}$$

d_m and p_m are the results obtained from the "Post Displacement Analysis".

Similar B M Diagrams may be drawn for the members, '*a*, *c*' and '*b*, *d*'

Alongside torsion in the members, '*e*, *f*, *g*', Fig. 6.25.

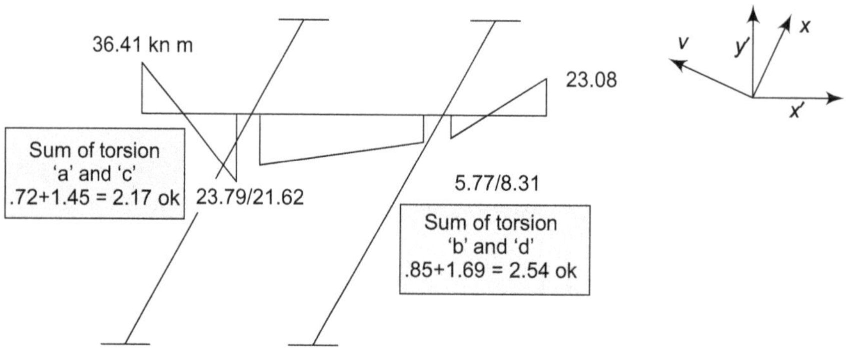

Fig. 6.25 *Bending Moment Diagram of Members 'e, f, g' Alongside Torques on Perpendicular Members*

6.12.2 Illustration 2

Transverse Plane Grids (Continued), superposition of symmetry and antisymmetry

The case of grids transverse loaded is illustrated. Because the grid is symmetrical, advantage is taken of the cases of Symmetry and Antisymmetry. The problem is amenable to hand computations.

Grid is also an important structure. The grids are transversely loaded. The grids are used for (a) building floors, and (bridge Decks). In the case of the bridge decks the longitudinal girders are deep. The deep girders have negligible torsion resistance. The typical cases are, therefore the girders with and without torsion-stiffness. In the case of force method there is an attraction to make the approximation of zero torsion: the degree of indeterminacy gets significantly reduced. It may be recalled Art 4.11 that the example of a bridge deck was reduced to a beam-problem by judiciously choosing the Harmonic analysis. In the case of the Displacement method for solving the grid problems, the number of unknown displacements do not dependent on the assumption pertaining to the torsion stiffness.

Ex. Typical building Floor.

(Solution of a plane grid by the stiffness method)

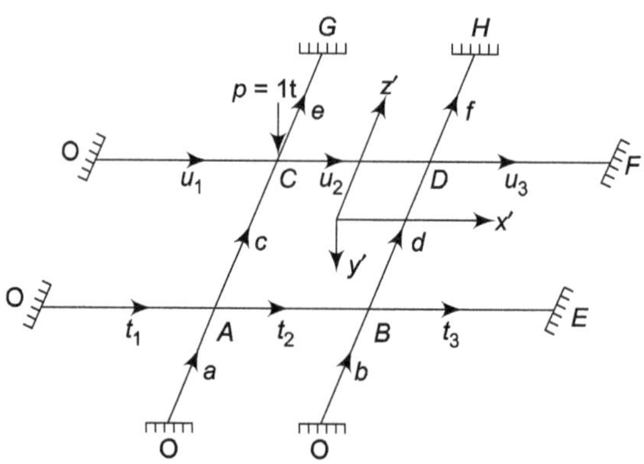

All members are circular in section $GJ/L = 0.72\ EI/L$

Symmetrical Grid frame

The problem is divided into four cases as shown below.

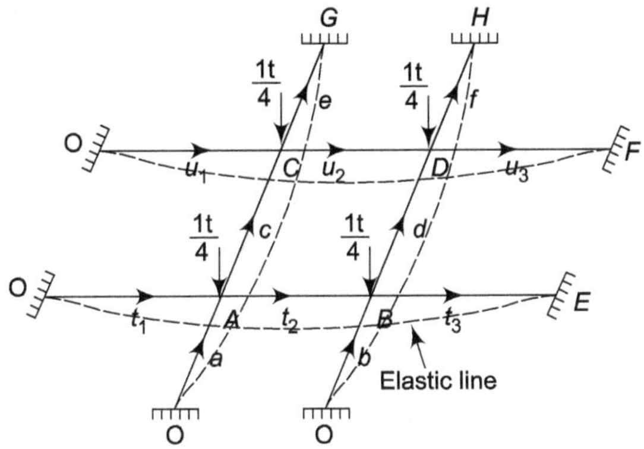

I. Symmetrical-Symmetrical case

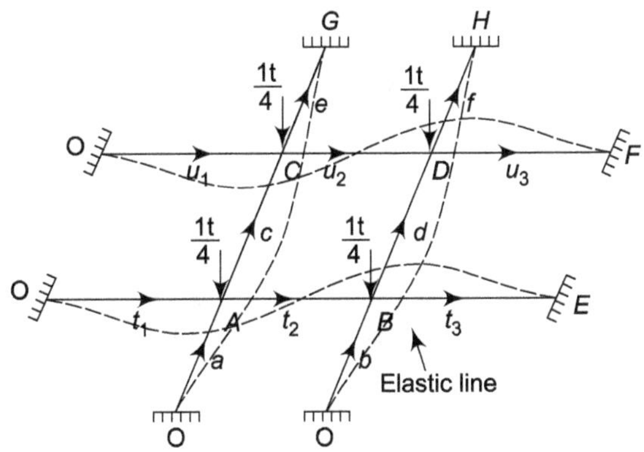

II. Symmetrical-Antisymmetrical case

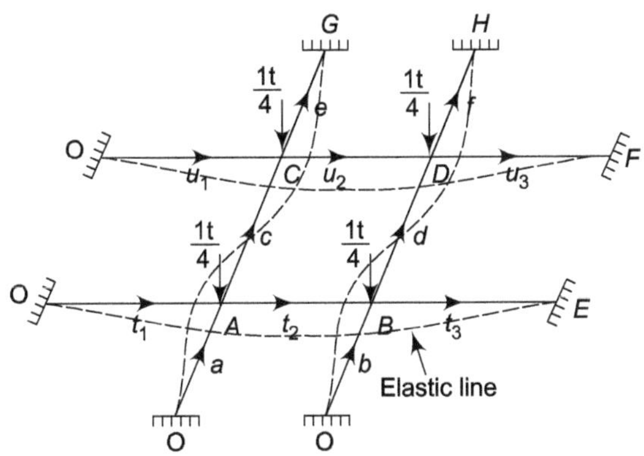

III. Antisymmetrical-Symmetrical case

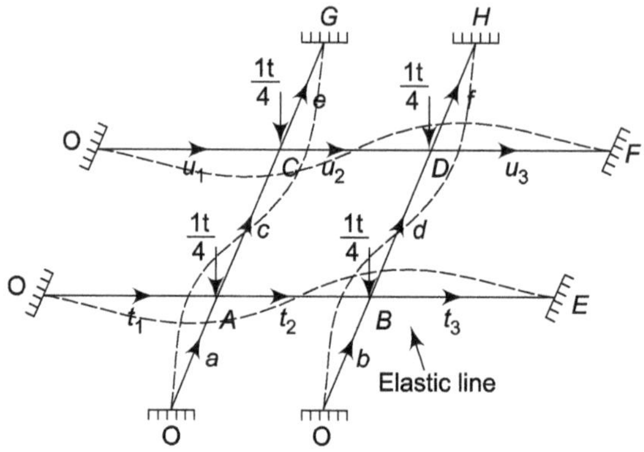

IV. Symmetrical-Antisymmetrical case

Considering any joint, say, the joint A, the load and deformation matrix can be connected as,

$$P_A = \begin{array}{c} \begin{matrix} d_A & \quad d_B & \quad d_C & d_D \end{matrix} \\ \begin{bmatrix} (K'_{22})_{t1,a} & (K'_{12})_{t2} & (K'_{12})_c & 0 \\ & + & & \\ & (K'_{11})_{t2,c} & & \\ & + & & \\ & (K'_{11})_c & & \end{bmatrix} \end{array} \begin{bmatrix} d_A \\ d_B \\ d_C \\ d_D \end{bmatrix}$$

Where $(K'_{11})_{t1,\,a} = (K'_{11})_{t1} = (K'_{11})_a$ and $(K'_{11})_{t2,\,C} = (K'_{11})_{t2} + (K'_{11})_C$

The stiffness matrix for a member which is parallel to X' – axis does not require any transformation and is written unaltered, Force at the end 1 of the member.

$$K_{11} = K'_{11}$$

$$\begin{pmatrix} P'_Y \\ m'_X \\ m'_2 \end{pmatrix}_1 = \begin{pmatrix} \dfrac{12\,EI}{L^3} & 0 & \dfrac{6EI}{L^2} \\ 0 & \dfrac{GJ}{L} & 0 \\ \dfrac{6EI}{L^2} & 0 & \dfrac{4EI}{L} \end{pmatrix} \begin{pmatrix} \delta'_Y \\ \theta'_X \\ \theta'_Z \end{pmatrix}_{2+}$$

$$K_{12} = K'_{12}$$

$$\begin{pmatrix} -\dfrac{12\,EI}{L^3} & 0 & \dfrac{6EI}{L^2} \\ 0 & -\dfrac{GJ}{L} & 0 \\ -\dfrac{6EI}{L^2} & 0 & \dfrac{4EI}{L} \end{pmatrix} \begin{pmatrix} \delta'_Y \\ \theta'_X \\ \theta'_Z \end{pmatrix}_2$$

At the end of Ex. 2 of the member.

$$K_{21} = K'_{21}$$

$$\begin{pmatrix} P'_Y \\ m'_X \\ m'_2 \end{pmatrix}_2 = \begin{pmatrix} -\frac{12\,EI}{L^3} & 0 & \frac{6EI}{L^2} \\ 0 & -\frac{GJ}{L} & 0 \\ \frac{6EI}{L^2} & 0 & \frac{2EI}{L} \end{pmatrix} \begin{pmatrix} \delta'_Y \\ \theta'_X \\ \theta'_Z \end{pmatrix}_{1+}$$

$$K_{22} = K'_{22}$$

$$\begin{pmatrix} \frac{12\,EI}{L^3} & 0 & -\frac{6EI}{L^2} \\ 0 & \frac{GJ}{L} & 0 \\ -\frac{6EI}{L^2} & 0 & \frac{4EI}{L} \end{pmatrix} \begin{pmatrix} \delta'_Y \\ \theta'_X \\ \theta'_Z \end{pmatrix}_{1+}$$

If the member lies parallel to the Z- axis, Force at the and 1 of the member

$$K'_{11}$$

$$\begin{pmatrix} P'_Y \\ m'_X \\ m'_2 \end{pmatrix}_1 = \begin{pmatrix} \frac{12\,EI}{L^3} & +\frac{6EI}{L^2} & 0 \\ +\frac{6EI}{L^2} & \frac{4EI}{L} & 0 \\ 0 & 0 & \frac{GJ}{L} \end{pmatrix} \begin{pmatrix} \delta'_Y \\ \theta'_X \\ \theta'_Z \end{pmatrix}_{1+}$$

$$K'_{12}$$

$$\begin{pmatrix} -\frac{12\,EI}{L^3} & \frac{6EI}{L^2} & 0 \\ \frac{6EI}{L^2} & \frac{2EI}{L} & 0 \\ 0 & 0 & -\frac{GJ}{L} \end{pmatrix} \begin{pmatrix} \delta'_Y \\ \theta'_X \\ \theta'_Z \end{pmatrix}_2$$

At the end of 2 of the members

$$\begin{pmatrix} P'_Y \\ m'_X \\ m'_Z \end{pmatrix}_2 = \overbrace{\begin{pmatrix} -\frac{12\,EI}{L^3} & -\frac{6EI}{L^2} & 0 \\ \frac{6EI}{L^2} & \frac{2EI}{L} & 0 \\ 0 & 0 & -\frac{GJ}{L} \end{pmatrix}}^{K'_{21}} \begin{pmatrix} \delta'_Y \\ \theta'_X \\ \theta'_Z \end{pmatrix}_{1+}$$

$$\overbrace{\begin{pmatrix} \frac{12\,EI}{L^3} & -\frac{6EI}{L^2} & 0 \\ -\frac{6EI}{L^2} & \frac{4EI}{L} & 0 \\ 0 & 0 & \frac{GJ}{L} \end{pmatrix}}^{K'_{22}} \begin{pmatrix} \delta'_Y \\ \theta'_X \\ \theta'_Z \end{pmatrix}_{1+}$$

Writing $d_B = z_b\, d_A$ and $d_C = z_C\, d_A$ and $d_D = z_d\, d_A$

Substituting these in the above matrix we get,

$$[P'_A] = [K'_{22}]_2 + (K'_{22})_{t1} + (K'_{11})_{t2} + (K'_{11})_C + (K'_{12})_{t2}\, Z_b + (K'_{12})_C Z_c]\,[d'_A]$$

Where Z_b matrix is such that, $d_B = z_b\, d_A$

Case I (Symmetrical – Symmetrical)

$$\begin{pmatrix} \delta'_Y \\ \theta'_X \\ \theta'_Z \end{pmatrix}_B = \overbrace{\begin{pmatrix} 0 & 0 & 0 \\ 0 & 1 & 0 \\ 0 & 0 & -1 \end{pmatrix}}^{z_b} \begin{pmatrix} \delta'_Y \\ \theta'_X \\ \theta'_Z \end{pmatrix}_A$$

$$d_C = z_C\, d_A$$

$$\overbrace{\begin{pmatrix} \delta'_Y \\ \theta'_X \\ \theta'_Z \end{pmatrix}_C}^{d_C} = \overbrace{\begin{pmatrix} 0 & 0 & 0 \\ 0 & -1 & 0 \\ 0 & 0 & +1 \end{pmatrix}}^{z_C} \overbrace{\begin{pmatrix} \delta'_Y \\ \theta'_X \\ \theta'_Z \end{pmatrix}_A}^{d_A}$$

$$d_D = z_d\, d_A$$

$$\begin{matrix} d_D & & z_d & & d_A \end{matrix}$$

$$\begin{pmatrix} \delta'_Y \\ \theta'_X \\ \theta'_Z \end{pmatrix}_D = \begin{pmatrix} 0 & 0 & 0 \\ 0 & -1 & 0 \\ 0 & 0 & -1 \end{pmatrix} \begin{pmatrix} \delta'_Y \\ \theta'_X \\ \theta'_Z \end{pmatrix}_A$$

$$(K'_{12})t_2\, Z_b = \begin{pmatrix} -\dfrac{12\,EI}{L^3} & 0 & \dfrac{6}{L^2} \\ 0 & -\dfrac{GJ}{L} & 0 \\ -\dfrac{6\,EI}{L^2} & 0 & \dfrac{2EI}{L} \end{pmatrix} \begin{pmatrix} 1 & 0 & 0 \\ 0 & 1 & 0 \\ 0 & 0 & -1 \end{pmatrix} = \begin{pmatrix} -\dfrac{12\,EI}{L^3} & 0 & -\dfrac{6\,EI}{L^2} \\ 0 & -\dfrac{GJ}{L} & 0 \\ -\dfrac{6\,EI}{L^2} & 0 & -\dfrac{2\,EI}{L} \end{pmatrix}$$

and

$$(K'_{12})_C\, Z_C = \begin{pmatrix} -\dfrac{12\,EI}{L^3} & \dfrac{6\,EI}{L^2} & 0 \\ 0 & -\dfrac{GJ}{L} & 0 \\ 0 & 0 & -\dfrac{CJ}{L} \end{pmatrix} \begin{pmatrix} 1 & 0 & 0 \\ 0 & -1 & 0 \\ 0 & 0 & 1 \end{pmatrix} = \begin{pmatrix} -\dfrac{12\,EI}{L^3} & -\dfrac{6\,EI}{L^2} & 0 \\ -\dfrac{6\,EI}{L^2} & -\dfrac{2\,EI}{L} & 0 \\ 0 & 0 & -\dfrac{CJ}{L} \end{pmatrix}$$

$$[P_A'] = \begin{bmatrix} +0.25 \\ 0 \\ 0 \end{bmatrix} = \begin{pmatrix} \dfrac{24\,EI}{L^3} & -\dfrac{6\,EI}{L^2} & -\dfrac{6\,EI}{L^2} \\ -\dfrac{6\,EI}{L^2} & \left(\dfrac{6\,EI}{L^2}\right)+\left(\dfrac{GJ}{L}\right) & 0 \\ -\dfrac{6\,EI}{L^2} & 0 & \left(\dfrac{6\,EI}{L}\right)+\left(\dfrac{GJ}{L}\right) \end{pmatrix} \begin{bmatrix} \delta'_Y \\ \theta'_X \\ \theta'_Z \end{bmatrix}_A$$

Since $CJ/2 = 0.72\ EI/L$, substituting the value of GJ/L

$$\begin{bmatrix} +0.25 \\ 0 \\ 0 \end{bmatrix} = \begin{pmatrix} \dfrac{24\,EI}{L^3} & -\dfrac{6\,EI}{L^2} & -\dfrac{6\,EI}{L^2} \\ -\dfrac{6\,EI}{L^2} & \dfrac{6.72\,EI}{L} & 0 \\ -\dfrac{6\,EI}{L^2} & 0 & \dfrac{6.72\,EI}{L} \end{pmatrix} \begin{bmatrix} \delta'_Y \\ \theta'_X \\ \theta'_Z \end{bmatrix}_A$$

The determinant of the K-matrix is,

$$\Delta = \left(\frac{24EI}{L^3}\right)\left[\left(\frac{6.72EI}{L^3}\right)\right] - \left(\frac{6EI}{L}\right)\left[\left(\frac{6EI}{L}\right)\times\left(\frac{6.72EI}{L}\right)\right] - \left(\frac{6EI}{L^2}\right)\left[\left(\frac{6EI}{L^2}\right)\times\left(\frac{6.72EI}{L}\right)\right]$$

Inverting the K-Matrix,

$$\begin{bmatrix}\delta'_Y\\ \theta'_X\\ \theta'_Z\end{bmatrix}_A = \left(\frac{L^5}{600}(EI)^3\right)\begin{bmatrix}\frac{45.16(EI)^2}{L^2} & \wedge & \wedge\\ \frac{40.32(EI)^2}{L^3} & \wedge & \wedge\\ \frac{40.32(EI)^2}{L^3} & \wedge & \wedge\end{bmatrix}\begin{bmatrix}+0.25\\ 0\\ 0\end{bmatrix} = \frac{1}{100\,EI}\begin{bmatrix}1.88L^2\\ 1.68L^2\\ 1.68L^2\end{bmatrix}$$

Elements marked "∧" are not required for the final results.

Since $d_B = z_d\, d_A$

$$D_B = \left(\frac{1}{100EI}\right)\begin{pmatrix}0 & 0 & 0\\ 0 & -1 & 0\\ 0 & 0 & -1\end{pmatrix}\begin{pmatrix}1.88L^3\\ 1.68L^2\\ 1.68L^2\end{pmatrix} = \left(\frac{1}{100EI}\right)\begin{pmatrix}+1.88L^3\\ +1.68L^2\\ -1.68L^2\end{pmatrix}$$

Similarly $d_C = z_C\, d_A$

$$\therefore \qquad d_C = \frac{1}{100EI}\begin{pmatrix}+1.88L^3\\ -1.68L^2\\ 1.68L^2\end{pmatrix}$$

And

$$d_D = \left(\frac{1}{100EI}\right)\begin{pmatrix}1.88L^3\\ -1.68L^2\\ -1.68L^2\end{pmatrix}$$

Case II (Symmetrical – Antisymmetrical)

$$\begin{pmatrix}\delta'_Y\\ \theta'_X\\ \theta'_Z\end{pmatrix}_B = \overset{z_b}{\begin{pmatrix}-1 & 0 & 0\\ 0 & -1 & 0\\ 0 & 0 & 1\end{pmatrix}}\begin{pmatrix}\delta'_Y\\ \theta'_X\\ \theta'_Z\end{pmatrix}_A$$

$$z_C$$

$$\begin{pmatrix} \delta'_Y \\ 0'_X \\ \theta'_Z \end{pmatrix}_C = \begin{pmatrix} 1 & 0 & 0 \\ 0 & -1 & 0 \\ 0 & 0 & 1 \end{pmatrix} \begin{pmatrix} \delta'_Y \\ 0'_X \\ \theta'_Z \end{pmatrix}_A$$

$$z_d$$

$$\begin{pmatrix} \delta'_Y \\ \theta'_X \\ \theta'_Z \end{pmatrix}_d = \begin{pmatrix} -1 & 0 & 0 \\ 0 & 1 & 0 \\ 0 & 0 & 1 \end{pmatrix} \begin{pmatrix} \delta'_Y \\ \theta'_X \\ \theta'_Z \end{pmatrix}_A$$

$$(K_{12}')t_2 Z_b = \begin{pmatrix} \frac{12\,EI}{L^3} & \frac{6\,EI}{L^2} & \frac{6\,EI}{L^2} \\ 0 & \frac{GJ}{L} & 0 \\ \frac{6\,EI}{L^2} & 0 & \frac{2\,EI}{L} \end{pmatrix}$$

$$(K_{12}')_C Z_C = \begin{pmatrix} -\frac{12\,EI}{L^3} & -\frac{6\,EI}{L^2} & 0 \\ -\frac{6\,EI}{L^2} & -\frac{2\,EI}{L} & 0 \\ 0 & 0 & -\frac{GJ}{L} \end{pmatrix}$$

The determinant of the global K matrix,

$$\Delta = \frac{46EI}{L^3}\left[\left(\frac{8.16\,EI}{L}\right)\times\left(\frac{10.72\,EI}{L}\right)\right] - \frac{6EI}{L^2}$$

$$\left[\left(\frac{6EI}{L^2}\right)\times\left(\frac{10.72\,EI}{L}\right)\right] - \frac{6EI}{L^2}\left[\left(\frac{8.16\,EI}{L}\right)\times\frac{6EI}{L^2}\right]$$

$$= 3520\frac{(EI)^3}{} L^5$$

$$\begin{bmatrix} \delta'_Y \\ \theta'_X \\ \theta'_Z \end{bmatrix}_A = \left(\frac{L^5}{3520}(EI)^2\right) \begin{bmatrix} \frac{45.16(EI)^2}{L^2} & \wedge & \wedge \\ \frac{40.32(EI)^2}{L^3} & \wedge & \wedge \\ \frac{40.32(EI)^2}{L^3} & \wedge & \wedge \end{bmatrix} \begin{bmatrix} +0.25 \\ 0 \\ 0 \end{bmatrix} = \left(\frac{1}{1000EI}\right) \begin{bmatrix} +6.22L^3 \\ +4.56L^2 \\ -3.48L^2 \end{bmatrix}$$

$$d_B = \left(\frac{1}{1000EI}\right)\begin{bmatrix} -6.22L^3 \\ -4.56L^2 \\ -3.48L^2 \end{bmatrix} \quad \text{and} \quad d_C = \left(\frac{1}{1000EI}\right)\begin{bmatrix} +6.22L^3 \\ -4.56L^2 \\ -3.48L^2 \end{bmatrix}$$

and
$$d_D = \left(\frac{1}{1000EI}\right)\begin{bmatrix} -6.22L^3 \\ +4.56L^2 \\ -3.48L^2 \end{bmatrix}$$

Case III (Antisymmetrical - Symmertical)

$$\begin{pmatrix} \delta'_Y \\ \theta'_X \\ \theta'_Z \end{pmatrix}_B = \overset{z_b}{\begin{pmatrix} 1 & 0 & 0 \\ 0 & 1 & 0 \\ 0 & 0 & -1 \end{pmatrix}} \begin{pmatrix} \delta'_Y \\ \theta'_X \\ \theta'_Z \end{pmatrix}_A$$

$$\begin{pmatrix} \delta'_Y \\ \theta'_X \\ \theta'_Z \end{pmatrix}_C = \overset{z_C}{\begin{pmatrix} -1 & 0 & 0 \\ 0 & 1 & 0 \\ 0 & 0 & -1 \end{pmatrix}} \begin{pmatrix} \delta'_Y \\ \theta'_X \\ \theta'_Z \end{pmatrix}_A$$

$$\begin{pmatrix} \delta'_Y \\ \theta'_X \\ \theta'_Z \end{pmatrix}_D = \overset{z_d}{\begin{pmatrix} -1 & 0 & 0 \\ 0 & 1 & 0 \\ 0 & 0 & 1 \end{pmatrix}} \begin{pmatrix} \delta'_Y \\ \theta'_X \\ \theta'_Z \end{pmatrix}_A$$

$$(K'_{12})_C Z_C = \begin{pmatrix} \frac{12EI}{L^3} & \frac{6EI}{L^2} & 0 \\ \frac{6EI}{L^2} & \frac{2EI}{L} & 0 \\ 0 & 0 & \frac{GJ}{L} \end{pmatrix}$$

$$(K'_{12})_{t2} Z_b = \begin{pmatrix} -\frac{12EI}{L^3} & 0 & -\frac{6EI}{L^2} \\ 0 & \frac{CJ}{L} & 0 \\ -\frac{6EI}{L^2} & 0 & -\frac{2EI}{L} \end{pmatrix}$$

$$[P_A] = \begin{bmatrix} -0.25 \\ 0 \\ 0 \end{bmatrix} = \begin{bmatrix} \dfrac{48\,EI}{L^3} & \dfrac{6\,EI}{L^2} & -\dfrac{6\,EI}{L^2} \\ \dfrac{6\,EI}{L^2} & \dfrac{10.72\,EI}{L} & 0 \\ -\dfrac{6\,EI}{L^2} & 0 & -\dfrac{8.16\,EI}{L} \end{bmatrix} \begin{bmatrix} \delta'_Y \\ \theta'_X \\ \theta'_Z \end{bmatrix}_A$$

The determinant of the K-matrix is,

$$\Delta = \frac{48EI}{L^3}\left[\left(\frac{10.72\,EI}{L}\right)\times\left(\frac{8.16\,EI}{L}\right)\right] - \frac{6EI}{L^2}\left[\left(\frac{6EI}{L^2}\right)\times\left(\frac{8.16\,EI}{L}\right)\right] - \frac{6EI}{L^2}\left[\left(\frac{6EI}{L^2}\right)\times\frac{(10.72\,EI)}{L}\right]$$

$$= 3520\,\frac{(EI)^3}{L^5}$$

$$\begin{bmatrix} \delta'_Y \\ \theta'_X \\ \theta'_Z \end{bmatrix}_A = \left(\frac{L^5}{3520}(EI)^3\right) \begin{bmatrix} \dfrac{87.50(EI)^2}{L^2} & \wedge & \wedge \\ -\dfrac{48.96(EI)^2}{L^3} & \wedge & \wedge \\ \dfrac{64.32(EI)^2}{L^3} & \wedge & \wedge \end{bmatrix} \begin{bmatrix} -0.25 \\ 0 \\ 0 \end{bmatrix}$$

$$= \left(\frac{1}{1000\,EI}\right)\begin{bmatrix} +6.22L^3 \\ +3.48L^2 \\ -4.56L^2 \end{bmatrix}$$

$$d_B = \left(\frac{1}{1000\,EI}\right)\begin{bmatrix} -6.22L^3 \\ +3.48L^2 \\ +4.56L^2 \end{bmatrix},\ d_C = \left(\frac{1}{1000\,EI}\right)\begin{bmatrix} +6.22L^3 \\ +3.48L^2 \\ +4.56L^2 \end{bmatrix}$$

$$d_D = \left(\frac{1}{1000\,EI}\right)\begin{bmatrix} +6.22L^3 \\ +3.48L^2 \\ -4.56L^2 \end{bmatrix}$$

Case IV (Antisymmetrical – Antisymmetrical)

$$z_b$$

$$\begin{pmatrix} \delta'_Y \\ \theta'_X \\ \theta'_Z \end{pmatrix}_B = \begin{pmatrix} -1 & 0 & 0 \\ 0 & -1 & 0 \\ 0 & 0 & 1 \end{pmatrix} \begin{pmatrix} \delta'_Y \\ \theta'_X \\ \theta'_Z \end{pmatrix}_A$$

$$z_C$$

$$\begin{pmatrix} \delta'_Y \\ \theta'_X \\ \theta'_Z \end{pmatrix}_C = \begin{pmatrix} -1 & 0 & 0 \\ 0 & 1 & 0 \\ 0 & 0 & -1 \end{pmatrix} \begin{pmatrix} \delta'_Y \\ \theta'_X \\ \theta'_Z \end{pmatrix}_A$$

$$z_d$$

$$\begin{pmatrix} \delta'_Y \\ \theta'_X \\ \theta'_Z \end{pmatrix}_D = \begin{pmatrix} 1 & 0 & 0 \\ 0 & -1 & 0 \\ 0 & 0 & -1 \end{pmatrix} \begin{pmatrix} \delta'_Y \\ \theta'_X \\ \theta'_Z \end{pmatrix}_A$$

$$(K'_{12})_{t2}\, Z_b = \begin{pmatrix} \dfrac{12\,EI}{L^3} & 0 & \dfrac{6\,EI}{L^2} \\ 0 & \dfrac{GJ}{L} & 0 \\ \dfrac{6\,EI}{L^2} & 0 & \dfrac{2\,EI}{L} \end{pmatrix},$$

$$(K'_{12})_C\, Z_b = \begin{pmatrix} \dfrac{12\,EI}{L^3} & \dfrac{6\,EI}{L^2} & -\dfrac{6\,EI}{L^2} \\ \dfrac{6\,EI}{L^2} & \dfrac{2\,EI}{L} & 0 \\ 0 & 0 & \dfrac{GJ}{L} \end{pmatrix}$$

$$[P_A] = \begin{bmatrix} -0.25 \\ 0 \\ 0 \end{bmatrix} = \begin{bmatrix} \dfrac{72\,EI}{L^3} & \dfrac{6\,EI}{L^2} & \dfrac{6\,EI}{L^2} \\ \dfrac{6\,EI}{L^3} & \dfrac{12.16\,EI}{L} & 0 \\ -\dfrac{6\,EI}{L^2} & 0 & \dfrac{12.16\,EI}{L} \end{bmatrix} \begin{bmatrix} \delta'_Y \\ \theta'_X \\ \theta'_Z \end{bmatrix}_A$$

The determinant of the K-matrix is,

$$\Delta = \frac{72EI}{L^3}\left[\frac{(12.16)^2(EI)^2}{L^2}\right]-\left[\frac{6EI}{L^2}\times\left(\frac{12.16EI}{L}\right)\right]-\frac{6EI}{L^2}\left[\left(\frac{6EI}{L^2}\right)\times\left(\frac{12.16\,EI}{L}\right)\right]$$

$$= 9766\,\frac{(EI)^3}{L^5}$$

$$[d_A] = \begin{bmatrix}\delta'_Y\\ \theta'_X\\ \theta'_Z\end{bmatrix}_A = \left(\frac{L^5}{9766}(EI)^3\right)\begin{bmatrix}\dfrac{147.8(EI)^2}{L^2} & \wedge & \wedge\\ -\dfrac{72.96(EI)^2}{L^3} & \wedge & \wedge\\ -\dfrac{72.96(EI)^2}{L^3} & \wedge & \wedge\end{bmatrix}\begin{bmatrix}-0.25\\ 0\\ 0\end{bmatrix}$$

$$= \left(\frac{1}{1000EI}\right)\begin{bmatrix}+3.78\,L^3\\ -1.86\,L^2\\ -1.86\,L^2\end{bmatrix}$$

$$d_B = \left(\frac{1}{1000\,EI}\right)\begin{bmatrix}+3.78\,L^3\\ -1.86\,L^2\\ +1.86\,L^2\end{bmatrix};\; d_C = \left(\frac{1}{1000\,EI}\right)\begin{bmatrix}+3.78\,L^3\\ +1.86\,L^2\\ -1.86\,L^2\end{bmatrix};$$

$$d_C = \left(\frac{1}{1000\,EI}\right)\begin{bmatrix}-3.78\,L^3\\ -1.86\,L^2\\ -1.86\,L^2\end{bmatrix};$$

At the joints A, B, C and D summing up the corresponding deformations of the four cases.

$$d_A = \begin{bmatrix}+\dfrac{0.015L^3}{EI}\\ +\dfrac{0.0268L^2}{EI}\\ +\dfrac{0.0106L^2}{EI}\end{bmatrix};\; d_B = \begin{bmatrix}+0.01045\,\dfrac{L^3}{EI}\\ +0.01386\,\dfrac{L^2}{EI}\\ -0.01386\,\dfrac{L^2}{EI}\end{bmatrix}$$

$$d_C = \begin{bmatrix}+0.0352\,\dfrac{L^3}{EI}\\ -0.016\,\dfrac{L^2}{EI}\\ +0.016\,\dfrac{L^2}{EI}\end{bmatrix};\; d_C = \begin{bmatrix}+0.015\,\dfrac{L^3}{EI}\\ -0.016\,\dfrac{L^2}{EI}\\ +0.0268\,\dfrac{L^2}{EI}\end{bmatrix}$$

Evaluation of the member end forces:

From compatibility,

$$d_A = (d_2')_a = (d_2')_{t1} = (d_1')_c = (d_1')_{t2},$$

and $$d_B = (d_2')_b = (d_2')_{t2} = (d_1')_{t3} = (d_1')_d,$$

and $$d_c = (d_2')_c = (d_2')_{u1} = (d_1')_{u2} = (d_1')_e,$$

and $$d_D = (d_2')_d = (d_2')_{u2} = (d_1')_{u3} = (d_1')_f,$$

Consider Joint A:

$$\begin{bmatrix}(p_y)_1\\(m_x')_1\\(m_z')_1\end{bmatrix}_{t1} = \begin{pmatrix}\frac{12\,EI}{L^3} & 0 & \frac{6\,EI}{L^2}\\ 0 & \frac{0.72EI}{L} & 0\\ \frac{6EI}{L^2} & 0 & \frac{4EI}{L}\end{pmatrix}\begin{pmatrix}0\\0\\0\end{pmatrix}_1 + \begin{pmatrix}-\frac{12\,EI}{L^3} & 0 & \frac{6\,EI}{L^2}\\ 0 & -\frac{0.72EI}{L} & 0\\ -\frac{6EI}{L^2} & 0 & \frac{4EI}{L}\end{pmatrix}\begin{pmatrix}-\frac{0.015L^3}{EI}\\ +\frac{0.0268L^2}{EI}\\ -\frac{0.0156L^2}{EI}\end{pmatrix}_2$$

$$\begin{bmatrix}(p_y)_1\\(m_x')_1\\(m_z')_1\end{bmatrix}_{t1} = \begin{pmatrix}+0.1164\\-0.0192L\\+0.0688L\end{pmatrix}_{1.t1}$$

$$\begin{bmatrix}(p_y)_2\\(m_x')_2\\(m_z')_2\end{bmatrix}_{t1} = \begin{pmatrix}\frac{12\,EI}{L^3} & 0 & -\frac{6\,EI}{L^2}\\ 0 & \frac{0.72EI}{L} & 0\\ -\frac{6EI}{L^2} & 0 & \frac{4EI}{L}\end{pmatrix}\begin{pmatrix}-\frac{0.015L^3}{L^2}\\ +\frac{0.0268L^2}{EI}\\ -\frac{0.0156L^2}{EI}\end{pmatrix}_2$$

$$(P_2)_{t1} = \begin{pmatrix}-0.1164\\+0.0192L\\+0.0476L\end{pmatrix}$$

Similarly other end force can be found out.

6.13 DIRECT STIFFNESS METHOD FOR SPACE FRAME

The six degrees of freedom per node are too many for the evaluation of the unknown displacements for manual computations. A computer algorithm is attached for the Tall Building Systems. In the case of a tall building system, taking advantage of master and slave nodes in a floor slab the degrees of freedom per floor are: 3 times the number of columns + three rigid body displacement of the master node. The total number of unknowns is equal to:

ns * {3 times column lines + 3} This is a too brief an introduction for Space Frames.

The reader is referred to the land mark papers stated in Chapter 9.

An interesting case of a Space Truss is given in Chapter 7 as an illustration.

The matrix force method applied to space frames was presented in Chapter 4. The staircase (dog lagged) required six redundants only to be applied. For the example considered, the adoption of Force Method is attractive. However, the solution of staircase problems by displacement method is presented in the following:

6.13.1 Illustration I of Space Frames

3-Member Space Frame: M-30, Fe 415

Members (1) and (3) require axes transformation

$$p_1' = k'_{11} d_1' + k'_{12} d_2' \quad \text{where } (k'_{ig})_{(i)} = T_{(i)} k_{ij} T_{(i)} \text{ for a member } i$$
$$p_2' = k'_{21} d_1' + k'_{22} d_2'$$

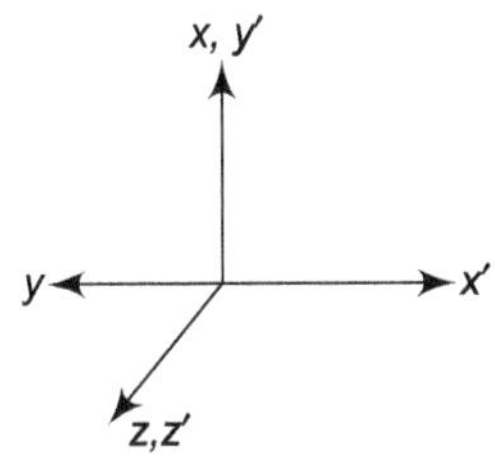

Member 1

The transformation matrices for the members (1) and (3) are given in the following:

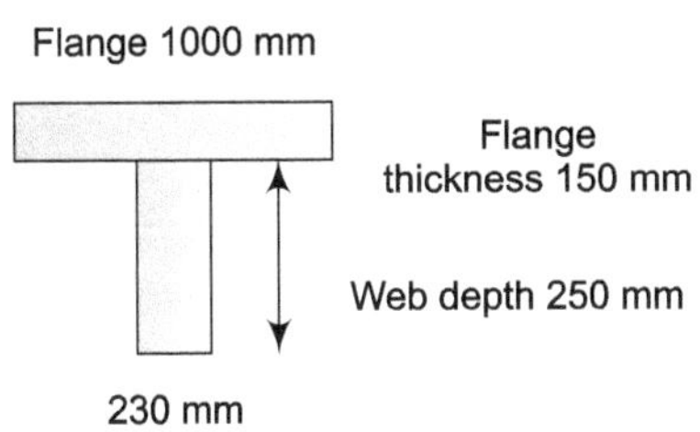

Cross Section: Section

Cross beams at mid-span of each member ensures total flange contributing to the T-beam action

$$\text{LL} = 3\text{KN}/m^2$$

Check Statics	$\Sigma x = \Sigma z = 0$	Support reaction
Σy	12.53	←3.19 KN
	+20.47	(.)12.53
	33.00 KN	↓0.48
	(4+4+3) × 3 = 33 KN ok	

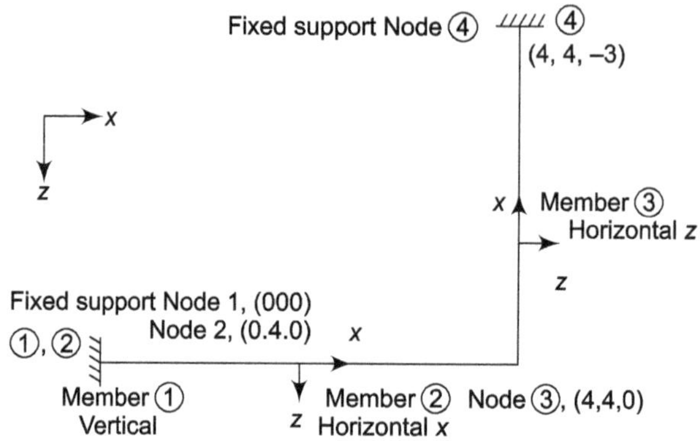

Fig. *Illustration 1(a)–Space Frame, Plan*

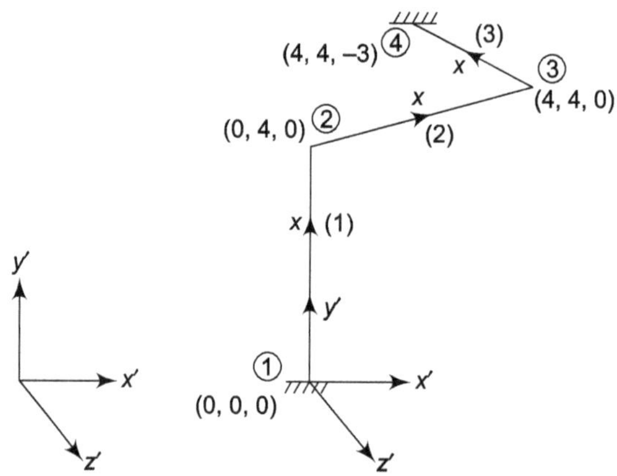

Fig. *Illustration 1(b)–Space Frame, View*

Support reactions

Supports	*1*	*4*
Rx	→3.19	←3.19 KN
Ry	20.47 upward	12.53 upward
Rz	0.48 ↑	↓0.48

Table 6.3 Joint Deflections under Live Load, GY 3KN/m along the member axes. Rotations are not given

Joint	*(1)*	*(2)*	*(3)*	*(4)*
Coordinates	0, 0, 0 m	0, 4, 0	4, 4, 0	4, 4, -3
Deflections				
δ_x	0	.0074	.0071	0
δ_y	0	.0013	.1130	0
δ_z	0	.0082	0.0	0

Transformation Matrices: The members (1) and (3) require K_{ij} to be transformed

$$T_{(i)} = \begin{bmatrix} 0 & -1 & 0 & \vdots & & 0_{3\times3} & \\ 1 & 0 & 0 & \vdots & & & \\ 0 & 0 & 1 & \vdots & & & \\ \cdots & \cdots & \cdots & \vdots & \cdots & \cdots & \cdots \\ & & & \vdots & 0 & -1 & 0 \\ & 0_{3\times3} & & \vdots & 1 & 0 & 0 \\ & & & \vdots & 0 & 0 & 1 \end{bmatrix}_{(1)}$$

Transformation
$$T_{(3)} = \begin{bmatrix} 0 & 0 & 1 & \vdots & & 0_{3\times3} & \\ 0 & 1 & 0 & \vdots & & & \\ -1 & 0 & 0 & \vdots & & & \\ \cdots & \cdots & \cdots & \vdots & \cdots & \cdots & \cdots \\ & & & \vdots & 0 & 0 & 1 \\ & 0_{3\times3} & & \vdots & 0 & 1 & 0 \\ & & & \vdots & -1 & 0 & 0 \end{bmatrix}_{(3)}$$

Assembling of K_j' is done in the following equilibrium equation. There are twelve unknown displacements, six at node 2 and six at node 3. The translations are shown in the Table 1. The member forces are given in Table 2. It is reiterated that if the problem is analysed by the Force Method, the number of unknown redundant is six only. This is further emphasized in the staircase problem in the following.

6.13.2 Illustration 2 : Five Member Space Frame, Staircase without Supports Under Intermediate Nodes

Five Member space frame: The geometry member and node designations, and loading are shown in Fig. Illustration 2a. It is a typical example of a space frame without a support under the landing. The coordinates of the nodes are given, Fig. Illustration 2b. Based on the geometry, the global stiffness matrix is given in the following.

$$\begin{bmatrix} p_2' \\ p_3' \\ p_4' \\ p_5' \end{bmatrix} = \begin{bmatrix} (K_{22}')_1 + (K_{11}')_2 & (K_{12}')_2 & 0 & 0 \\ (K_{21}')_2 & (K_{22}')_2 + (K_{11}')_3 & (K_{12}')_3 & 0 \\ 0 & (K_{21}')_3 & (K_{22}')_3 + (K_{11}')_4 & (K_{12}')_4 \\ 0 & 0 & (K_{21}')_4 & (K_{22}')_4 + (K_{11}')_5 \end{bmatrix}$$

Load case :
Member load, Live Load,
GY =3 KN/m along the member axes.

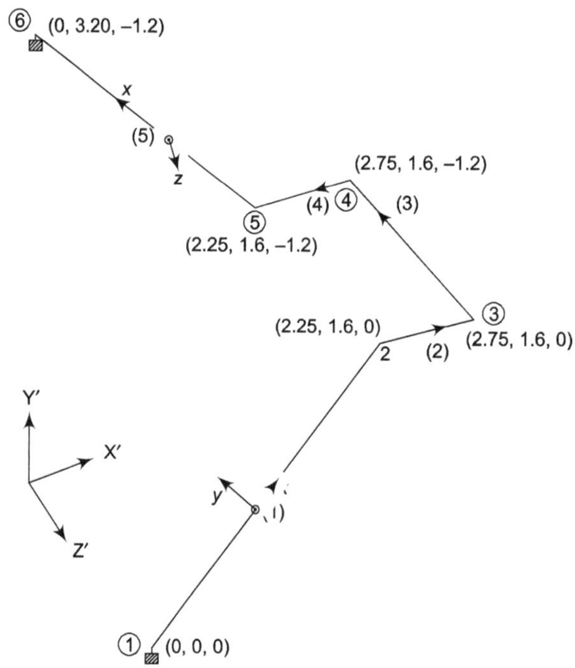

Fig. *Illustration 2a–View of Staircase*

The member stiffnesses K_{ij} of 1, 3, 4, and 5 need to be transformed for obtaining the global matrices,

$$K'_{ij} = T\,K_{ij}\,T^t$$

The Transformation matrix T of the members is the appropriate direction cosines.

In contrast to the Force Method presented in Chapter 2, the Displacement Method of the Five Member staircase requires the evaluation of four nodal displacements. The support points are assumed fixed. In view of the large number of deformations equal to twenty four, the illustration has been analysed by software (Staad Pro).

The inverse of the $(K_j)_{24\times 24}$ gives the deformations d'_2, d'_3, d'_4, d'_5.

The compatibility equation, $d_m = A\,d_j$.

The member forces are then determined from the member stiffness equations.

$$p_m = k_m\,d_m$$

The member forces are given in Table 6.5

Force Method: It requires evaluation of six redundants only.

Results:

Support Reactions (KN):

Joint	R_x	R_y	R_z	M_x	M_y	M_z
1	8.00	11.58	0.00	2.39	−4.86	5.06
6	−8.10	11.58	0	−2.39	−4.86	5.06

Table 6.4 *Joint Translations caused by the given UDL GY –3.00 KN/m along the member length.*

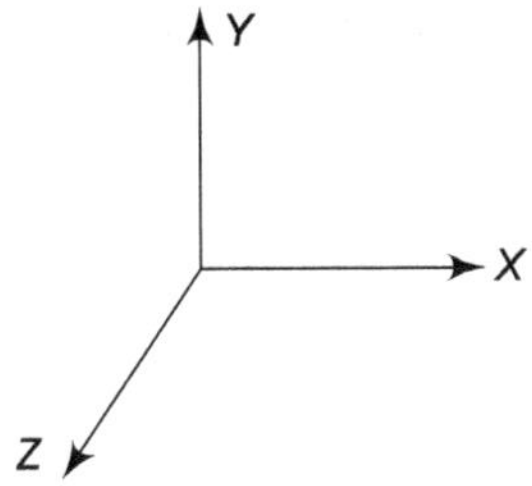

Joint.	*2*			*3*			*4*			*5*		
δ_x cm	.0096			.0095			–.0095			– .0096		
δ_y cm		–.0147			–.0195			– .0195			– .0147	
δ_{xz} cm			–.0074			–.0149			– .0149			–.0074

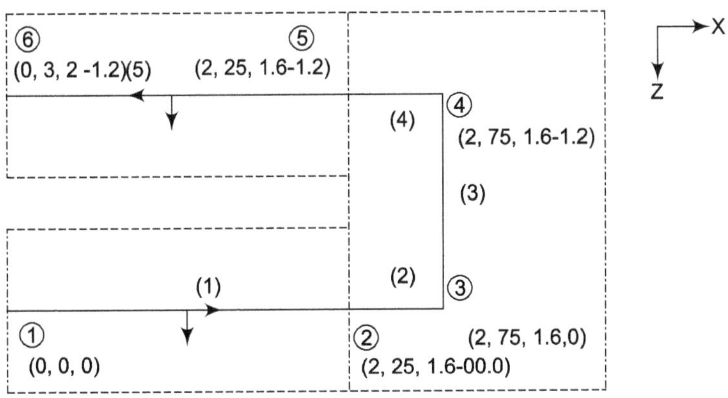

Fig. *Illustration 2b – Plan of Staircase*

Table 6.5 *Member end forces space structure, staircase*

Member Load Case	*Member JT*	*Load Case*	*AXIAL*	*SHEAR-Y*	*SHEAR-Z*	*TORSION*	*MOM-Y*	*MOM-Z*	*Remarks*
Load (1)	1	1	23.02	8.21	0.00	–1.50	–9.25	8.75	
	2		–14.72	3.46	0.00	1.50	9.25	–2.20	
Load (2)	1	2	13.31	4.75	0.00	-0.86	–5.35	5.06	
	2		–8.51	2.00	0.00	0.86	5.35	–1.28	
Combined (3)	1	3	36.33	12.95	0.00	–2.36	–14.59	13.81	
	2		–23.23	5.47	0.00	2.36	14.59	–3.48	
Node (2)	2	1	14.00	5.71	0.00	4.17	–8.40	2.20	
lc -1	3		–14.00	–3.11	0.00	–4.14	8.40	0.00	
lc -2	2	2	8.10	3.30	0.00	2.39	–4.86	1.28	

(*Contd.*)

(Contd.)

Member Load Case	Member JT	Load Case	AXIAL	SHEAR-Y	SHEAR-Z	TORSION	MOM-Y	MOM-Z	Remarks
lc -2	3		-8.10	-1.80	0.00	-2.39	4.86	0.00	
lc -3	2	3	22.10	9.01	0.00	6.53	-13.26	3.48	
	3		-22.10	-4.91	0.00	-6.53	13.26	0.00	
Node (3)	3	1	0.00	3.11	14.00	0.00	-8.40	4.14	
lc -1	4		0.00	3.11	-14.00	0.00	-8.40	-4.14	
	3	2	0.00	1.80	8.10	0.00	-4.86	2.39	
lc -2	4		0.00	1.80	-8.10	0.00	-4.86	-2.39	
lc -3	3	3	0.00	4.91	22.10	0.00	-13.26	6.53	
	4		0.00	4.91	-22.10	0.00	-13.26	-6.53	
Node (4)	4	1	-14.00	-3.11	0.00	-4.14	8.40	0.00	
lc -1	5		14.00	5.71	0.00	4.14	-8.40	-2.20	
lc -2	4	2	-8.10	-1.80	0.00	-2.39	4.86	0.00	
	5		8.10	3.30	0.00	2.39	-4.86	-1.28	
lc -3	4	3	-22.10	-4.91	0.00	-6.53	13.26	0.00	
	5		22.10	9.01	0.00	6.53	-13.26	3.48	
Node (5)	5	1	-14.72	3.46	0.00	1.50	9.25	2.20	
lc -1	6		23.02	8.21	0.00	-1.50	-9.25	-8.75	
	5	2	-8.51	2.00	0.00	0.86	5.35	1.28	
lc -2	6		13.31	4.75	0.00	-0.86	-5.35	-5.06	
lc -3	5	3	-23.23	5.47	0.00	2.36	14.59	3.48	
	6		36.33	12.95	0.00	-2.36	-14.59	-13.81	

Table 6.6 *Support reactions, stair case*

Support	Load	Force X, KN	Force Y	Force Z	Moment x	Moment y	Moment z
1	LL 30 KN/m^2	8.10	11.58	0	2.39	-4.86	5.06
	(Self Weight)	(14.00)	(20.03)	0	(4.14)	(-8.4)	(8.75)
6	LL	-8.10	11.58	0	-2.39	-4.86	5.06
	(Self Weight)	(-14.00)	(20.03)	0	(-4.14)	(-8.40)	(8.75)

Check Statics: For LL
Sum of vertical reactions = 11.58 × 2 = 23.16 KN
Inclined length = 2.761 m, Total length = 2.761 + 0.5 + 1.2 + 0.5 + 2.761 = 7.722
Load GY = 7.722 × 3 = 23.165 KN Statics Satisfied.

The values of δ and support reactions caused by self weight are given in parentheses ( ). In concrete structures, self weight effect is significantly higher than those caused by the live loads.

The values of deflections caused by the self weight (not shown in the table) are considerably higher than the Live Load. This is generally the case with concrete structures.

6.14 OBSERVATIONS: USE OF SHAPE FUNCTIONS

The results like: $M_2 = M_1 = \frac{6EI}{L^2}$, $Py_1 = -Py_2 = \frac{12EI}{L^3}$ caused by $(\delta_y)_1 = 1$, and similar results due to $\theta_1 = 1$ acting at the end-1 are generally shown to be obtained by the Force Method using "Moment Area" theorems. These aspects have been elaborated in the Chapter 3. The same results for d_1 as well as d_2 may be obtained directly without having to use the Force method. The alternative approach is based upon the displacement fields or best known as shape functions. The approach is general and a similar approach is used for arriving at stiffnesses for the "Finite Elements". Keeping this objective in mind, the use of shape functions and their applications for the skeletal members is shown in the following. As stated earlier, the distributed loading acting on a member is converted to the nodal loads based on the shape functions. It is the nodal loads that are used for the Part-2 of the Displacement Analysis with the joins released. The shape functions are also used for dynamic analysis converting the distributed mass as the nodal masses. The nodal or joint masses particular importance while performing the Dynamic Analysis, Chapter 9. In fact the Earthquake analysis/design which is a pseudo dynamic analysis, is also known as analysis for the mass loading. The earthquake analysis is necessitated primarily due to mass loading.

The mass matrix consistent with stiffness matrix for the cases, (a) Prismatic, and (b) Nonprismatic are presented. The shape functions used for arriving at the member stiffnesses are used for obtaining the 'consistent mass' matrices. The mass matrices are assembled for the 'global mass matrix' on the same way as assembling the member stiffnesses. An example at the end helps in comprehending theses aspects.

6.15 NONPRISMATIC STRAIGHT SKELETAL MEMBERS

Shape functions: The same shape functions are consistently used in the following for stiffness matrix, K as well as 'mass matrix', M. The mass matrix is used in the dynamic analysis of skeletal structures, particularly the grids used as bridge-decks.

When a curved nonprismatic arch under coplanar loading was considered for anlaysis, the arch was divided into number of prismatic segments. The stiffness matrix for nonprismatic members enables us to consider each segment without having to be approximated as prismatic segment. The advantage is gained in the form of the reduced number of segments, and therefore the reduced degree of freedom. In the design of box culverts for the bridges, the heavy live loads are assumed to be disbursed trapezoidally. The abutment segment, connecting the slab deck above to the raft below is assumed as a trapezoidal member. A numerical example of a trapezoidal member is discussed.

6.15.1 Stiffness Matrix for Nonprismatic Straight Skeletal Members

Normally force method (moment area) is used for arriving at the flexibility coefficients. Herein, the Shape Functions are used, and formulation for K and M matrices are developed for the Nonprismatic skeletal members. It is direct and simple to use but it approximates the deflected shapes (elastic line) by the shape functions assumed.

Rigid bodies – Moments of inertia of an idealized rigid members from kinematic considerations is developed under applied Dynamics.

Our discussion is pertaining to the elastically deforming skeletal members. With the mass matrices developed similar to stiffness matrices, the beam problem is reduced to mass less elastically deforming straight member.

Degrees of Freedom of the ends of a straight member:

Consider the ends of a member of a plane frame

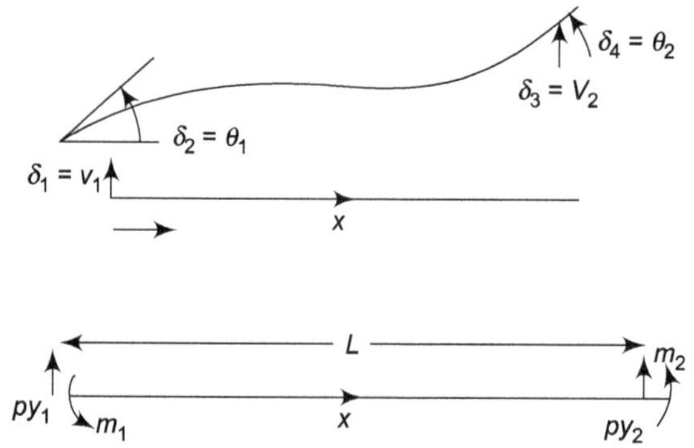

Fig. 6.26(i) *Member Stiffness of a Nonprismatic Member*

Member end forces are:

$$P_1 = \begin{bmatrix} P_{y1} \\ m_1 \end{bmatrix} \quad P_2 = \begin{bmatrix} P_{y2} \\ m_2 \end{bmatrix}$$

Shape Functions:

Deformed shapes corresponding to the individual degree of freedom:

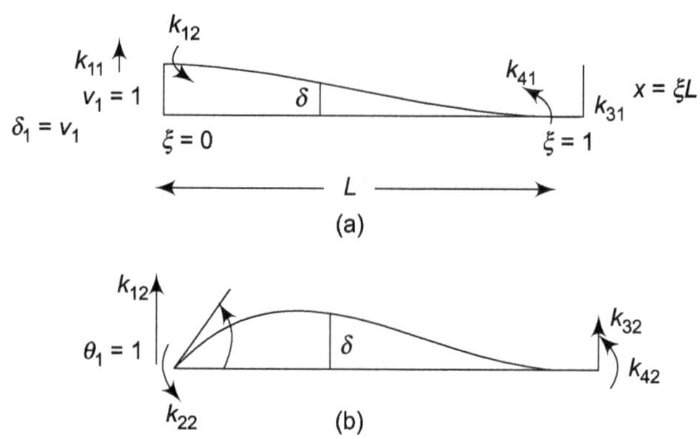

$\delta_2 = \theta_1$

$$v \text{ due to } d_1 = (n_1\ n_2) \begin{bmatrix} v \\ \theta \end{bmatrix}_1 = N_1\, d_1$$

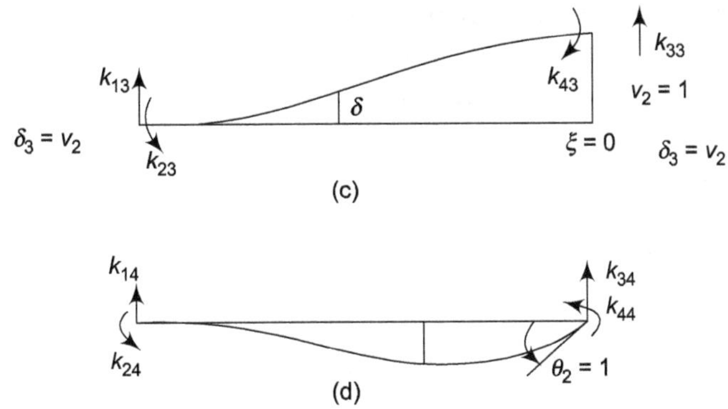

Fig. 6.26(ii)

$$\delta_4 = \theta_2$$

$$d_2 = \begin{bmatrix} v_2 \\ \theta_2 \end{bmatrix}$$

$$d = (1 - \xi)\, \xi^2 L \quad n_4 = -\delta = -(1 - \xi)\, \xi^2 L$$

$$v \text{ due to } d_2 = N_2\, d_2 \quad N_2 = [n_3, n_4] \text{ and } d_2 = \begin{bmatrix} V \\ \theta \end{bmatrix}_2$$

Summing up the results of the Eqs (i) and (ii)

$$v = (n_1\, n_2\, n_3\, n_4) \begin{bmatrix} \delta_1 \\ \delta_2 \\ \delta_3 \\ \delta_4 \end{bmatrix}$$

or
$$v = (N_1\, N_2) \begin{bmatrix} d_1 \\ d_2 \end{bmatrix} \qquad \text{...(6.13)}$$

N_1, N_2 are referred as the shape functions or displacement-fields corresponding to the deformation d_1 and d_2 respectively.

Member Stiffness Matrix, K_m

$$P = K_m\, d$$

where
$$P = \begin{bmatrix} p_1 \\ p_2 \end{bmatrix} \text{ and } d = \begin{bmatrix} d_1 \\ d_2 \end{bmatrix}$$

Based on the principle of virtual work, we treat the real system to be the force system. Flexural moments only are considered. The real force system consists end forces (p_{y1}, m_1, p_{y2}, m_2) and internal flexural moment, $m = EI\,v''$ where $v'' = \dfrac{d^2 v''}{dn^2}$ and v defined by the Eq. (1).

Choose the virtual deformation system from deformed shapes corresponding to the individual degrees of freedom. Let us consider shape function n_1 which gives

$d^* = (1, 0, 0, 0)$ and deformation or generated strain

$$\left(\frac{1}{R}\right)^* = n_1''$$

$$d^* \times p = \int \left(\frac{1}{R}\right)^* m\, dx$$

From virtual work principle where $m = EI\, v''$ (internal flexural moment)

$$1 \times p_{y1} = \frac{1}{L^2} \int n_1'' EI\, v''\, dx$$

$$\frac{1}{R} = \frac{n_1''}{L^2}$$

Substituting for $v = (n_1\, n_2\, n_3\, n_4)\, d$

$$p_{y1} = \frac{1}{L^4} \left(\int n_1'' EI \left(n_1'', n_2'', n_3'', n_4''\right) dx \right) d \qquad \ldots\text{(i)}$$

$d = [\delta_1, \delta_2, \delta_3, \delta_4]$ being independent of x it is taken outside the integral

Similarly choosing virtual deformation corresponding to the shape function n_2, it can be shown that

$$m_1 = \left(\int n_2'' EI \left(n_1'', n_2'', n_3'', n_4''\right) dx \right) d \qquad \ldots\text{(ii)}$$

Similarly choosing virtual deformation corresponding to n_3'' and n_4''

$$P_{y2} = \left(\int n_3'' EI \left(n_1'', n_2'', n_3'', n_4''\right) dx \right) d \qquad \ldots\text{(iii)}$$

$$m_2 = \left(\int n_4'' EI \left(n_1'', n_2'', n_3'', n_4''\right) dx \right) d \qquad \ldots\text{(iv)}$$

Combining the Eqs (i) to (iv) and using the rotations N_1 and N_2 as per equation (i) and (ii)

$$\begin{bmatrix} p_1 \\ p_2 \end{bmatrix} = \left(\int N_1'', N_2'' \right)^t EI \left(N_1'', N_2''\right) dx \begin{bmatrix} d_1 \\ d_2 \end{bmatrix}$$

Thus

$$P \quad = \quad K \quad d \qquad \ldots\text{(6.14a)}$$

Where

$$K = \left(\int N_1'', N_2'' \right)^t EI \left(N_1'', N_2''\right) d_2 \qquad \ldots\text{(6.14b)}$$

When written in the expanded form

$$K = \left(\int \left(n_1'', n_2'', n_3'', n_4''\right)^t EI \left(n_1'', n_2'', n_3'', n_4''\right) dx \right)$$

$$K = \begin{bmatrix} (n_1'')^2 & (n_1'', n_2'') & (n_1'', n_3'') & (n_1'', n_4'') \\ * & (n_2'')^2 & (n_2'', n_3'') & (n_2'', n_4'') \\ * & * & (n_3'')^2 & (n_3'', n_4'') \\ * & * & * & (n_4'')^2 \end{bmatrix}$$

If EI is varying with x, i.e., $EI = EI(x)$
Each element of the stiffness matrix K,

$$K_{ij} = \int EI(x)\, n_1'' n_2''\ \mathrm{dx} \qquad \ldots(6.15)$$

Example 1 Compute K_{24} for a prismatic beam of length L, (EI is constant)

$$n_2 = \xi\,(1 - \xi)^2\, L$$

$n_4 = -\,(1 - \xi)\,\xi^2\, L$ using the relation $d/dx = 1/(Ld\,\xi)$

$$n_2'' = \frac{1}{L}(-\,4 + 6\,\xi)$$

$$n_4'' = \frac{1}{L}(-\,2 + 6\,\xi)$$

$$n_2' = 2(1 - \xi)\,\xi + (1 - \xi)^2$$

$$n_2'' = 2(1 - 2\,\xi) - 2(1 - \xi)$$

$$= (-4 + 6\,\xi)$$

$K_{24} = \dfrac{EI}{L}\displaystyle\int_0^1 (n_2'' n_4'')\, d\xi$ using the relation $d_x = Ld\,\xi$

$$K_{24} = 2EI/L$$

Similarly the other elements x_{ij} can be compute and following is the function K matrix for the prismatic beam. The element K_{24} as compared is shown within. The reader also is required to verify K_{33} discussed at the end of the article.

$$K = \begin{bmatrix} \dfrac{12EI}{L^3} & \dfrac{6EI}{L^2} & -\dfrac{12EI}{L^3} & \dfrac{6EI}{L^2} \\ * & \dfrac{4EI}{L} & -\dfrac{6EI}{L^2} & \boxed{\dfrac{2EI}{L}} \\ * & * & \dfrac{12EI}{L^3} & \dfrac{6EI}{L^2} \\ * & * & * & \dfrac{4EI}{L} \end{bmatrix}$$

Example 2 Compute K_{24} for the nonprismatic beam, Fig. Ex. 2 the values of n_2'', n_4'' and the integrand I at the stations are also given

[The reader should take note of the approximation of using the same function even for the nonprismatic member. With the kind of nonprismatic members, we come across, the engineering accuracy is achieved, and the reader need not be unduly disturbed.]

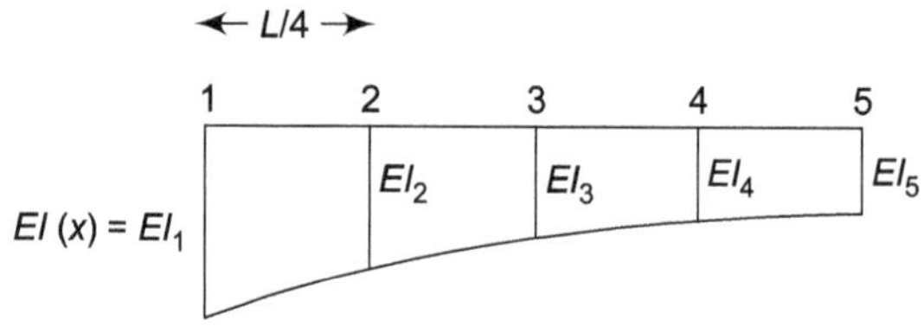

Fig. Ex. 2

Computations for evaluating K_{24}

$$n_2'' = \frac{1}{L}\times \quad -4, \quad -2.5, \quad -1, \quad 0.5, \quad +2$$

$$n_4'' = \frac{1}{L}\times \quad -2, \quad -0.5, \quad 1, \quad 2.5, \quad +4$$

$$I = \frac{1}{L^2} \quad 8EI_1, \quad 1.25EI_2, \quad -EI_3, \quad 1.25EI_4, \quad +8EI_5$$

Using Simpsons rule for quadrature

$$K_{24} = \frac{1}{L^2}\left[\frac{L}{12}(8EI_1 + 5EI_2 - 2EI_3 + 5EI_4 + 8EI_5)\right]$$

Check when $EI_i = EI$, $K_{24} = 2EI/L$ as computed in Ex. 1

Details of calculations for computing K_{ij}

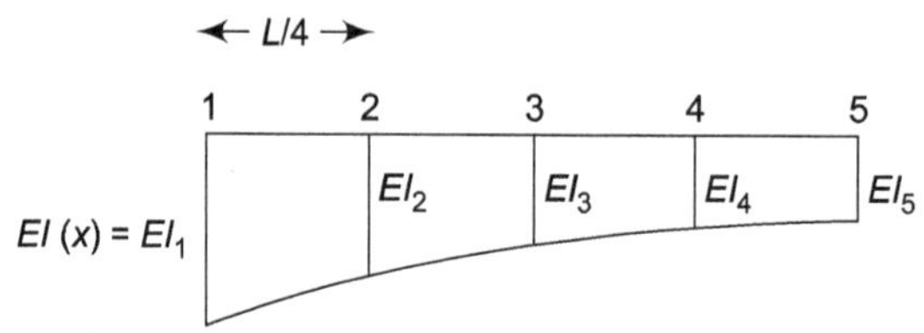

n_1''	$1/L^2$	-6,	-3,	0,	3,	6
n_2''	$1/L$	-5,	$-5/2$,	-1,	$1/2$,	2
n_3''	$1/L^2$	6,	3,	0,	-3,	-6
n_4''	$1/L$	-2,	$-1/2$,	1,	$5/2$,	4
$EI\,(n_1'')^2$	$\frac{E}{L^4}\times$	$36I_1$,	$9I_2$,	0,	$9I_4$,	$36I_5$
$EI\,n_1''n_2''$	$\frac{E}{L^3}$	$24I_1$,	$\frac{15}{2}I_2$,	0,	$\frac{3}{2I_4}$,	$12I_5$

(Contd.)

(*Contd.*)

$EI\,n_1''n_3''$	$\frac{E}{L^4}$	$-36I_1$,	$-9I_2$,	0,	$-9I_4$,	$-36I_5$,
$EI\,n_1''n_4''$	$\frac{E}{L^3}$	$12I_1$,	$\frac{3}{2I_2}$,	0,	$\frac{15}{2I_4}$,	$24I_5$,
$EI\,(n_2'')^2$	$\frac{E}{L^2}$	$16I_1$,	$\frac{25}{4I_2}$,	I_3,	$\frac{1}{4}I_4$,	$4I_5$,
$EI\,n_2''n_3''$	$\frac{E}{L^2}$	$-24I_1$,	$-\frac{15}{2}I_2$,	0,	$-\frac{3}{2I_4}$,	$-12I_5$,
$EI\,n_1''n_4''$	$\frac{E}{L^2}$	$8I_1$,	$\frac{5}{4I_2}$,	$-I_3$,	$\frac{5}{4}I_4$,	$8I_5$,
$EI\,(n_3'')^2$	$\frac{E}{L^4}$	$36I_1$,	$9I_2$,	0,	$9I_4$,	$36I_5$,
$EI\,n_3''n_4''$	$\frac{E}{L^3}$	$-12I_1$,	$-\frac{3}{2}I_2$,	0,	$-\frac{15}{2I_4}$,	$-24I_5$,
$EI\,(n_4'')^2$	$\frac{E}{L^2}$	$4I_1$,	$\frac{1}{4}I_2$,	I_3,	$\frac{25}{4}I_4$,	$16EI_5$,

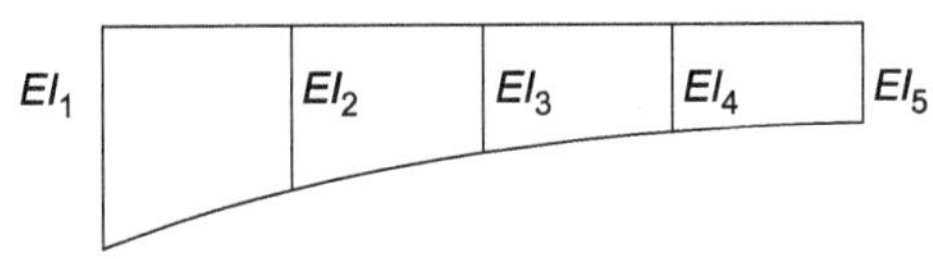

Summary of n_i''

$$n_1'' = (-6 + 12\xi) / L^3$$
$$n_2'' = (-4 + 6\xi) / L$$
$$n_3'' = (6 - 12\xi) / L^2$$
$$n_4'' = (-2 + 6\xi) / L$$

And for the case of nonprismatic beam of Ex. 2, the stiffness coefficients are given below. The computations of the integrands are shown in the table above.

$$K_{11} = \int EI(x)(n_1'')^2\,dx = \frac{3}{L^3}(EI_1 + EI_2 + EI_4 + EI_5)$$

$$K_{12} = \int EI(x)n_1''n_2''\,dx = \frac{1}{2L^2}(4EI_1 + 5EI_2 + EI_4 + 2EI_5)$$

$$K_{13} = \int EI(x)n_1''n_3''\,dx = -\frac{3}{L^3}(EI_1 + EI_2 + EI_4 + EI_5)$$

$$K_{14} = \int EI(x)n_1''n_4''\,dx = \frac{1}{2L^2}(2EI_1 + EI_2 + 5EI_4 + 4EI_5)$$

$$K_{22} = \int EI(x)(n_2'')^2\,dx = \frac{1}{12L}(16EI_1 + 25EI_2 + 2EI_3 + EI_4 + 4EI_5)$$

$$K_{23} = \int EI(x)\, n_2''\, n_3''\, dx = -\frac{1}{2L^2}\,(4EI_1 + 5EI_2 + EI_4 + 2EI_5)$$

$$K_{24} = \int EI(x)\, n_2''\, n_4''\, dx = \frac{1}{12L}\,(8EI_1 + 5EI_2 + 2EI_3 + 5EI_4 + 8EI_5)$$

$$K_{33} = \int EI(x)\, (n_3'')^2\, dx = \frac{3}{L^3}\,(EI_1 + EI_2 + EI_4 + EI_5)$$

$$K_{34} = \int EI(x)\, n_3''\, n_4''\, dx = -\frac{1}{2L^2}\,(2EI_1 + EI_2 + 5EI_4 + 4EI_5)$$

$$K_{44} = \int EI(x)\, (n_4'')^2\, dx = \frac{1}{12L}\,(4EI_1 + EI_2 + 2EI_3 + 25EI_4 + 16EI_5)$$

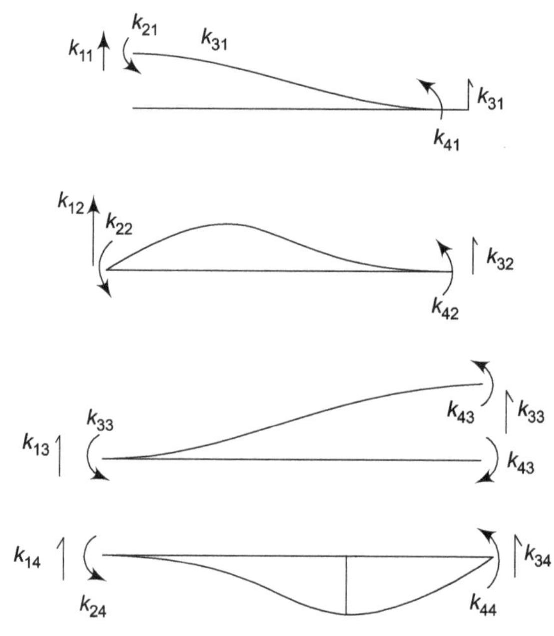

6.15.2 Consistent Mass Matrix of a Straight Member

The distributed loads acting on a member, in the case of static analysis, were converted to nodal loads, by clamping the nodes, as part-I of the analysis. The displacement method is for part-II of the analysis for the unbalanced forces at the nodes. If we are required to do dynamic analysis, the locational effects are important. In the following, the distributed mass is converted to an equivalent **Nodal consistent mass matrix**. The shape functions used for arriving at member stiffness will be used for obtaining the consistent mass matrices. These consistent mass matrices of the members are assembled in the same way we assembled the member stiffnesses for obtaining the global stiffness matrix of an entire structure.

$$m = \text{mass percent length} = \rho A$$

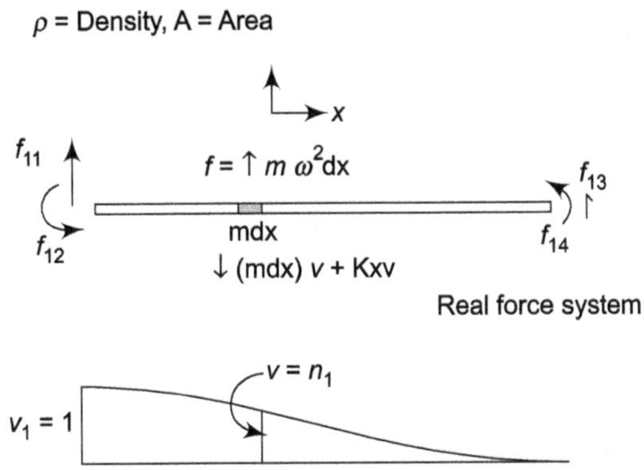

Deformation system for computing mass-coefficients m_{11}, m_{12}, m_{13}, m_{14}

Fig. 6.27 *Consistent Mass Coefficients*

$$m\ddot{v} + k\,v = p \qquad\qquad v = (n_1\, n_2\, n_3\, n_4)\begin{bmatrix}\delta_1\\ \delta_2\\ \delta_3\\ \delta_4\end{bmatrix}$$

$$v = (N_1\; N_2)\; d$$

For the cases of consistent mass and finite degrees of freedom

$$M\,\ddot{d} + K\,d = P$$

P is the load matrix consisting of the nodal loads

$$F_I + f_s = P$$

P, Equivalent nodal loads for distributed loads is given by

$$P_i = \int n_i\; p(x,t)\,dx$$, see the solved example of at the end.

Assembling of the consistent mass matrices of the members is done to obtain the final global mass matrix (consistent mass matrix). The global mass matrix (consistent mass matrix) will be achieved following the same procedure as used for assembling the stiffness matrix. In the following consistent mass matrix of a member is derived.

Consistent Mass Matrix [M] of a Member

We treat the real system to be the force system, when the inertial forces at the ends of a member are:

$f_I = [f_{I1}\, f_{I2}\, f_{I3}\, f_{I4}]^T$ and the inertial force in the interior is given by

$$f = -(m\;dx)\;\ddot{v}$$

or

$$f = -m\;\ddot{v}\;dx$$

If the virtual deformation chosen is $\delta^* = (1, 0, 0, 0)$ and virtual internal deformation corresponding to f

$$v^* = n_1$$

from the virtual work principle

$$f_{I1} = \int n_1 \,(m)\, \ddot{v}\, dx$$

Substituting for $v = (n_1\, n_2\, n_3\, n_4)$

$$f_{I1} = \left(\int n_1 (m) (n_1, n_2, n_3, n_4)\, dx\right) \ddot{d} \qquad \ldots\text{(i)}$$

similarly choosing n_2, n_3, n_4 in succession as virtual deformation systems,

$$f_I = \left(\int m \begin{pmatrix} n_1 \\ n_2 \\ n_3 \\ n_4 \end{pmatrix} (n_1, n_2, n_3, n_4)\, dx\right) \ddot{d}$$

and

$$f_I = M\, (\ddot{d}) \qquad \ldots\text{(ii)}$$

Where M is given by

$$M = \int m \begin{bmatrix} n_1^2 & (n_1\, n_2) & (n_1\, n_3) & (n_1\, n_4) \\ * & n_2^2 & (n_2\, n_3) & (n_2\, n_4) \\ * & * & n_3^2 & (n_3\, n_4) \\ * & * & * & n_4^2 \end{bmatrix} dx \qquad \ldots(6.16)$$

Example 3 Compute element m_{24} for a beam of length L with a constant mass per unit length.

We consider shape functions are arrived at based on the end conditions.

$$n_2 = \xi\,(1 - \xi)^2\, L$$

$$n_4 = -\,(1 - \xi)\, \xi^2\, L$$

$$n_2\, n_4 = -\,\xi^3\,(1 - \xi)^3\, L^2$$

$$m_{24} = -\,mL^3 \int_0^1 \xi^3\,(1 - \xi)^3\, d\xi$$

Use $d = L\, d\xi$

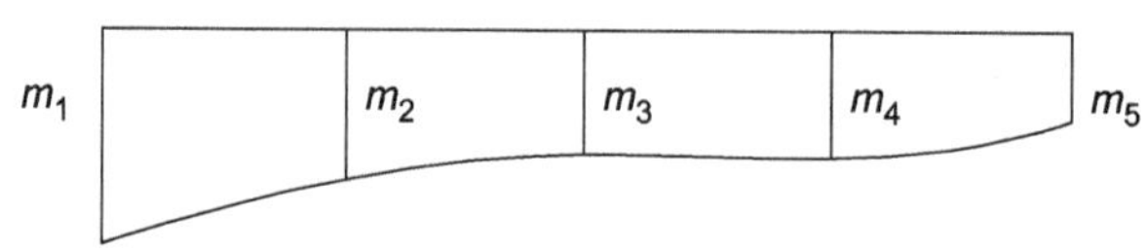

$$n_2 = L\times \quad 0 \qquad 9/64 \qquad 1/8 \qquad 3/64 \qquad 0$$

$$n_4 = L\times \quad 0 \qquad -3/64 \qquad -1/8 \qquad -2/64 \qquad 0$$

$$I = L^2\times \quad 0 \qquad \frac{-27}{64\times 64} m_2, \qquad \frac{-1}{64} m_3, \qquad \frac{-27}{64\times 64} m_4, \qquad 0$$

$$m_{24} = L^3\left[\frac{L}{12}\left(\frac{-4\times 27}{64\times 64}m_2 + \frac{-2}{64}m_3 + \frac{-4\times 27}{64\times 64}m_4\right)\right]$$

$$= \frac{L^3}{12}\left[\left(\frac{-27}{64\times 16}m_2 + \frac{-32}{64\times 16}m_3 - \frac{27}{64\times 16}m_4\right)\right]$$

$$m_{24} = \frac{-L^3}{12\times 16\times 64}[+27m_2 + 32\,m_3 + 27\,m_4]$$

Illustration: Compute m_{24} for a prismatic beam of length = L

Mass per unit length = constant= m

$$n_2 = \xi\,(1-\xi)^2\,L$$

$$n_4 = -\,(1-\xi)\,\xi^2\,L$$

$$n_2\,n_4 = \xi^3\,(1-\xi)^3\,L^2$$

$$m_{24} = -mL^3\int_0^1 \xi^2(1-\xi)^3\,d\xi \quad \text{use } dx = Ld\xi$$

$$m_{24} = -mL^3/140$$

Similarly the other elements $m_2\,j$ can be computed. The following is the consisted mass matrix for the beam of uniform mass. The value of m_{24} computed may be verified in the following.

$$M = mL^3\begin{bmatrix} \frac{13}{35L^3} & \frac{11}{20L} & \frac{9}{70L^2} & -\frac{13}{420\,L} \\ * & \frac{1}{105} & \frac{13}{420\,L} & -\frac{1}{140} \\ * & * & \frac{13}{35\,L^2} & -\frac{11}{210L} \\ * & * & * & \frac{1}{105} \end{bmatrix}$$

M matrix is symmetrical and positive definite.*

Example 4 Compute M_{24} for a beam of uniform mass. Use the above result of varying mass along its length,

$$M_{24} = \frac{-L^3}{12\times 16\times 64}\left[27m_2 + 32m_3 + 27m_4\right]$$

for computation details obtained already

Check when $m_1 = m_2 = m_3 = m_4 = m$

For uniform mass beam

* Preliminary back ground about Matrices is given in Appendix A.

$$M_{24} = \frac{-86}{12\times16\times64} mL^3 \approx \frac{mL^3}{143}$$ as against $-mL^3/140$
Computed *m Ex. 3*
Error 2:14 %

Details of calculations for computing m_{ij}

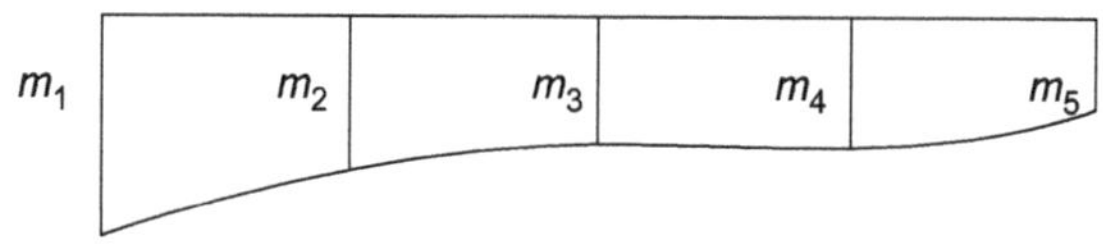

$n_1 =$	1	$\frac{27}{32}$	$\frac{1}{2}$	$\frac{5}{32}$	0	
$n_2 =$	0	$\frac{9}{64}$	$\frac{1}{8}$	$\frac{3}{64}$	0	$\times$L
$n_3 =$	0	$\frac{10}{64}$	$\frac{1}{2}$	$\frac{54}{64}$	1	
$n_4 =$	0	$-\frac{3}{64}$	$-\frac{1}{8}$	$-\frac{9}{64}$	0	$\times L$
mn12 =	m1	$\frac{729}{32\times32}m_2$	$\frac{1}{4}m_2$	$\frac{25}{32\times32}m_4$	0	
$mn_1\ n_2 =$	0	$\frac{243}{32\times64}m_2$	$\frac{1}{16}m_3$	$\frac{15}{32\times62}m_4$		$\times L$
$mn_1n_3 =$	0	$\frac{270}{32\times62}m_2$	$\frac{1}{4}m_3$	$\frac{270}{32\times62}m_4$	0	
$mn_1n_4 =$	0	$\frac{81}{62\times32}m_2$	$-\frac{1}{16}m_3$	$-\frac{45}{64\times32}m_4$	0	$\times L$
$mn_2^2 =$	0	$\frac{81}{64\times64}m_2$	$\frac{1}{16}m_3$	$\frac{9}{64\times64}m_4$	0	$\times L$
$mn_2n_3 =$	0	$\frac{90}{64\times64}m_2$	$\frac{1}{16}m_3$	$\frac{162}{64\times64}m_4$	0	$\times L$
$mn_2n_4 =$	0	$-\frac{27}{64\times64}m_2$	$-\frac{1}{16}m_3$	$-\frac{27}{64\times64}m_4$	0	$\times L^2$
$mn_3^2 =$	0	$\frac{100}{64\times64}m_2$	$\frac{1}{4}m_3$	$\frac{2916}{64\times64}m_4$	m_5	
$mn_3n_4 =$	0	$-\frac{30}{64\times64}m_2$	$-\frac{1}{16}m_3$	$-\frac{486}{64\times64}m_4$	0	$\times L$
$mn_4^2 =$	0	$\frac{9}{64\times64}m_2$	$\frac{1}{64}m_3$	$\frac{81}{64\times64}m_2$	0	$\times L^2$

Summary of m_{ij}

The use of the entire mass-coefficient for nonuniform mass beam are given for ready reference. See table given elsewhere for computation details:

Summary and comparison with direct derivation.

Consider $m_i = m$

	Shape Functions	Direct derivation
$m_{11} = \frac{L}{12}\left(m_1 + \frac{729}{256}m_2 + \frac{1}{2}m_3 + \frac{25}{256}m_4\right)$	= 1138 mL/3072,	13/35 mL
$m_{12} = \frac{L^2}{12}\left(\frac{243}{513}m_2 + \frac{1}{8}m_3 + \frac{15}{512}m_4\right)$	= 161 mL^2/3072,	11/210 mL^2
$m_{13} = \frac{L}{12}\left(\frac{135}{256}m_2 + \frac{1}{2}m_3 + \frac{135}{256}m_4\right)$	= 191 mL/1536,	9/70 mL
$m_{14} = \frac{L^2}{12}\left(\frac{81}{512}m_2 + \frac{1}{8}m_3 + \frac{45}{512}m_4\right)$	= – 95 mL^2/3072,	– 13/420 mL^2
$m_{22} = \frac{L^3}{12}\left(\frac{81}{1024}m_2 + \frac{1}{32}m_3 + \frac{9}{1024}m_4\right)$	= 61 mL^3/6144,	1/105 mL^3
$m_{23} = \frac{L^2}{12}\left(\frac{45}{512}m_1 + \frac{1}{8}m_3 + \frac{81}{512}m_4\right)$	= 95 mL^2/3072,	13/420 mL^2
$m_{24} = \frac{-L^3}{12\times16\times64}\left(27\,m_2 + 32\,m_3 + 27\,m_4\right)$	= – 43 mL^3/6144,	– 1/140 mL^3
$m_{33} = \frac{L}{12}\left(\frac{25}{256}m_2 + \frac{1}{2}m_3 + \frac{729}{256}m_4 + m_5\right)$	= 1138 mL/3072,	13/35 mL
$m_{34} = \frac{L^2}{12\times16\times64}(30\,m_2 + 128\,m_3 + 486\,m_4)$	= –161 mL^2/3072,	–11/210 mL^2
$m_{44} = \frac{L^3}{12\times16\times64}(9\,m_2 + 32\,m_3 + 81\,m_4)$		

Where m_i = 61 mL^3/6144, 1/105 mL^3

One obvious approximation in the above is that when the beam is nonprismatic (or varying mass), the deflected shapes given by the shape functions n_2 is retained the same as considered for the prismatic members. To this extent, the shape function is not unique and evaluation of K_{ij} or m_{ij} are approximately correct values. It was already reiterated.

To illustrate the approximation and its effect on the results, consider the stiffness of a nonprismatic beam element.

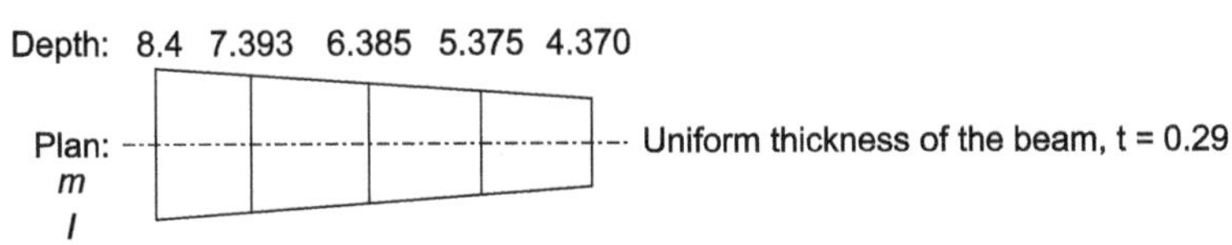

Details of typical Computations, of K_{22}, K_{42}, K_{23}, K_{44}, K_{34} and comparison are shown below.

EI_1	EI_2	EI_3	EI_4	EI_5
$16 \times EI_1 = 6.9648E8$	$25 \times E_8 = 9.57775$	$2\,x = .6618E8$	.2787 E8	$4x = .9060$ E8
$8 \times E_8 = 3.4824$	$5 \times E_8 = 1.9155$	$-2\,x = -.6618E8$	$5\,x\,EI_4 = 1.3945$ E8	$8x = 1.8120$ E8
$4 \times E_8 = 1.7412$ E8	$5 \times E_8 = 1.9155$	0	.2787 E8	$2\,x = .4530$ E8
$K_{22} = 6.9648$ E8 9.5775 .6618 .9060 18.1101/48	$K_{42} = 3.4824$ 1.9155 1.3945 1.8120 8.6044 - .6618	$K_{32} = 1.7412$ 1.9155 .2787 .4530 .1371	$K_{44} = 1.7418$ .3831 .6618 6.9725 3.6240 .2788	$K_{34} = 0.8706$ .3831 1.3945 .9060 -.1111
$K_{22} = 0.3773$ E8	$K_{42} = 0\,.1655$	$K_{23} = K_{32} = -.1371$ E8	$K_{44} = .2788E8$	$K_{34} = .1111E8$
Values from Moment area: 0.3755 E8	Values from Moment area 0.1579 E8	Value from moment area 0.1334 E8	Value from Moment area 0.2711 E8	Value from Moment area 0.1072 E8

In the following WS rotation in moment area method is same as K in Shape function approach.

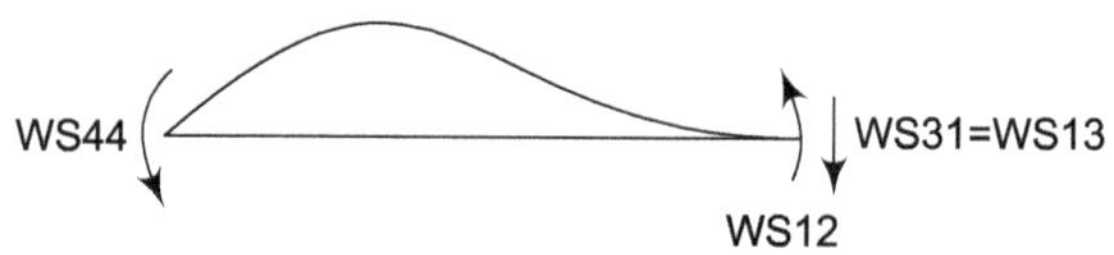

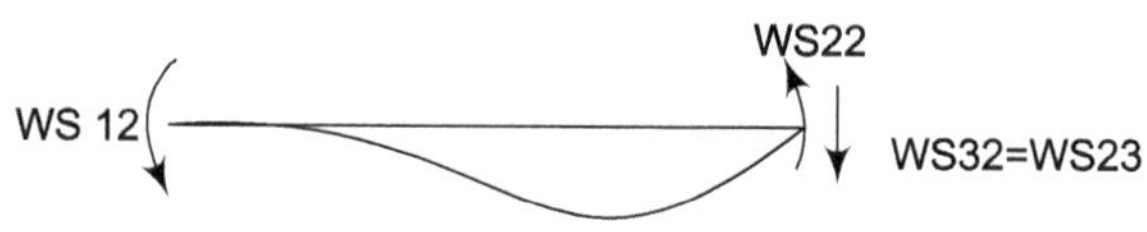

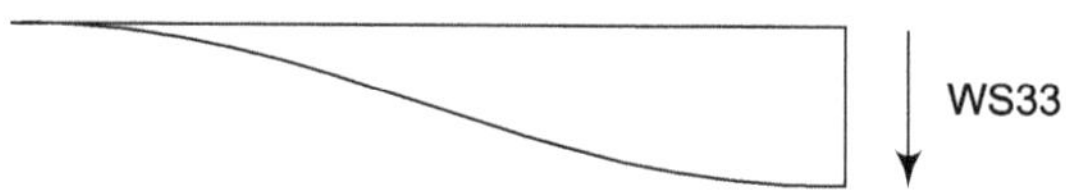

$$\begin{bmatrix} .3755\,E8 & .1579\,E8 & .1334\,E8 \\ * & .2711\,E8 & .1072\,E8 \\ * & * & ? \end{bmatrix}$$

The reader may compute K_{33} and compare his result with the WS_{33} from the shape function considerations.

6.15.3 Member Stiffnesses Based on Shape Functions

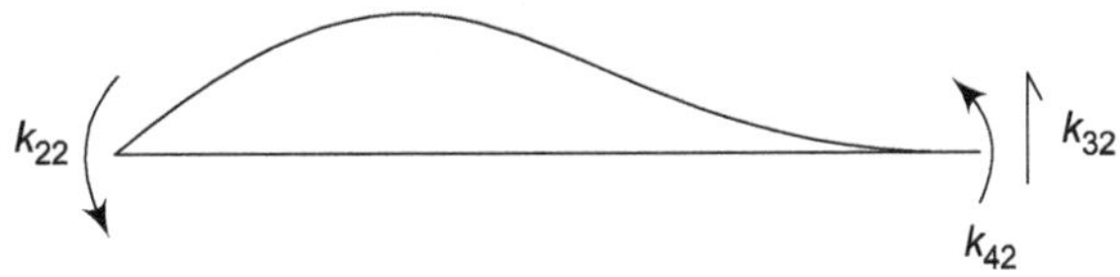

$$WS_{11} = K_{22}$$

$$WS_{12} = -K_{24}$$

$$K_{22} = \frac{1}{12L}(16EI_1 + 25EI_2 + 2EI_3 + EI_4 + 4EI_5)$$

$$K_{42} = \frac{1}{12L}(8EI_1 + 5EI_2 + 2EI_3 + 5EI_4 + 8EI_5)$$

$$K_{23} = \frac{-1}{2L^2}(4EI_1 + 5EI_2 + EI_4 + 2EI_5)$$

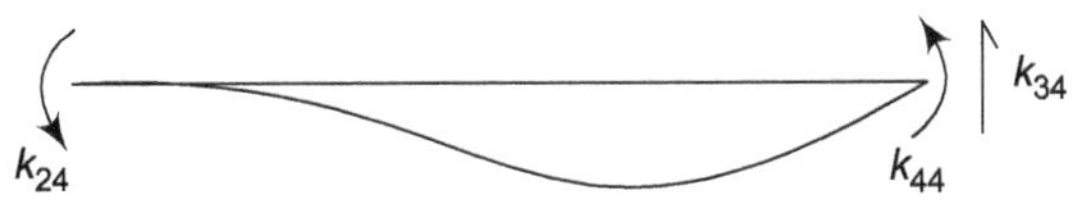

$$WS_{12} = K_{24}$$

$$WS_{22} = K_{44}$$

$$WS_{23} = -K_{34}$$

$$K_{44} = \frac{1}{12L}(4EI_1 + EI_2 + 2EI_3 + 25EI_4 + 16EI_5)$$

$$K_{34} = \frac{-1}{12L^2}(2EI_1 + EI_2 + 5EI_4 + 4EI_5)$$

$$WS_{33} = K_{33}$$

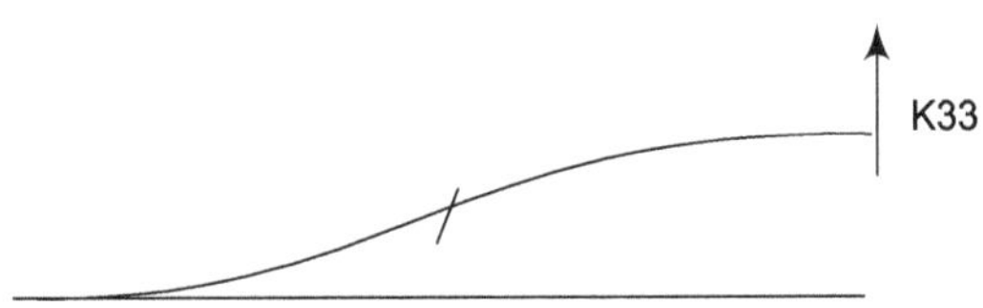

$$K_{33} = \frac{3}{L^3}(EI_1 + EI_2 + EI_4 + EI_5)$$

$$K_{33} = \frac{3}{L^2}\ (0.4353 + 0.3831 + 0.3309 + 0.2787 + 0.2265)\ E_8 = 0.07755$$

Evaluation of K_{22}, K_{42}, and K_{44}

Table 6.4 *Comparison*

	Shape function	*Moment Area*	*% Error*
K_{22}	0.3773 E8	0.3755 E8	0.53
K_{42}	0.1655 E8	0.1579 E8	4.81
K_{23}	-0.1371 E8	-0.1334 E8	2.77
K_{44}	0.2788 E8	0.2711 E8	2.84
K_{34}	-0.1111 E8	-0.1072 E8	3.64
K_{33}	0.0776 E8	Not computed- to be completed as an exercise	

*An approximation is implied while assuming shape functions. The errors are computed, treating the moment area values as the basic values. Similar procedures are used for assembling the member mass matrices (consistent mass matrices) for obtaining the global mass matrix for the entire structure. Such a Global mass matrix is also known as consistent mass matrix. Its use in dynamic analysis, say in the case of grids, gives better result than the lumped masses. In the case of Tall Buildings, however, the approximation of lumped masses is adequate.

6.15.4 Illustration: Portal Frame with Different Combinations of Girder–Floor–Masses with the Columns (Skeletal) Masses, Three Cases

(a) $m_f = m_c$,
(b) $m_f = 5\ m_c$, and
(c) $m_f = 10\ m_c$.

Assume inextensibility of the members. Obtain global matrix.

The reader's conclusions are important regarding lumping of the masses at the floor level.

Refer the below Illustration for Consistent Mass Matrix.

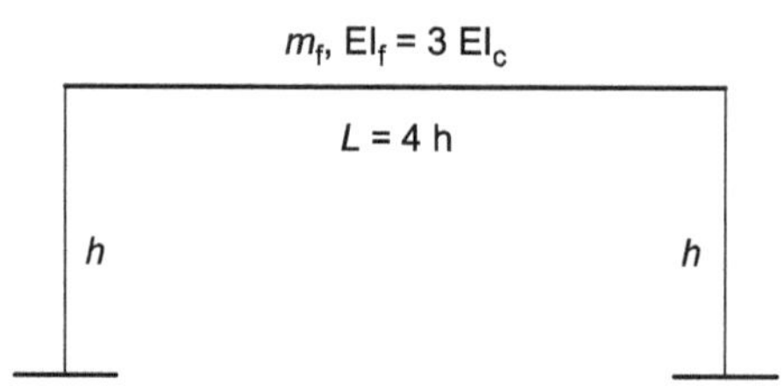

Fig. *Illustration for Consistent Mass Matrix*

Chapter 7: Displacement Method—Illustrations

Chapter

7

Displacement Method-Illustrations

Displacement method is continued in the following with the important aspect of the member stiffnesses having been covered in Chapter 6, the emphasis is mainly on the illustrations.

This chapter is devoted mainly for highlighting the concepts through the illustrative examples. The illustrations are not numerical but are to be treated as case studies to explain the theory and related concepts. Both the approaches, Direct Stiffness approach and connection matrices for satisfying compatibility of the Displacement method are presented. The appropriateness of the assumption of the inextensibility of the members, which influences the degrees of freedom, is explained while analyzing the illustrations using the connection matrix approach. The sketching of the deflected shapes corresponding to the inextensibilities is an intelligent exercise, and this aspect is conspicuously missing in the Direct Stiffness method.

In the case of trusses also, we may use connection matrix approach but there being no such assumption of inextensibility, the unknown displacements are unaltered. For the sake of directness and simplicity, direct stiffness approach is recommended. In the case of small frames and also tall buildings, the assumption of inextensibility is attractive, and therefore the preferences in favor of connection matrix approach for the Displacement method.

It is good to recall the Force method by way of comparison with the Displacement method.

In the following, we are comparing the two methods, Force and Displacement Methods, before we proceed to the Displacement method.

7.1 COMPARISON OF FORCE AND DISPLACEMENT METHODS

Force Method	*Displacement Method*
1. Solution for forces	1. Solution for deformation at chosennodes
2. Generally complete geometry is included while analyzing for chosen determinate structure.	2. Generally simple straight prismatic members considered.
3. The method is suitable when degree of Indeterminacy is not large say. 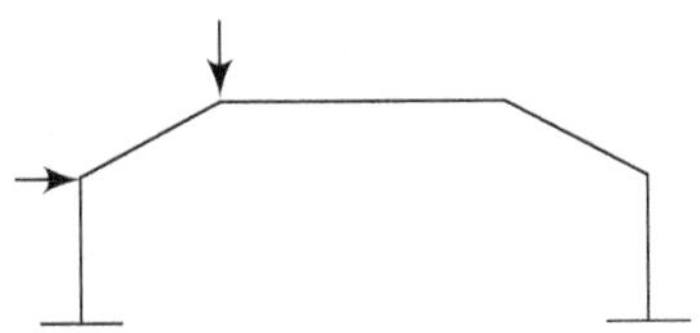 **Fig. 7.1** *(a) Indeterminacy is small, n = 3*	3. Suitable where number of nodes is not large, say: 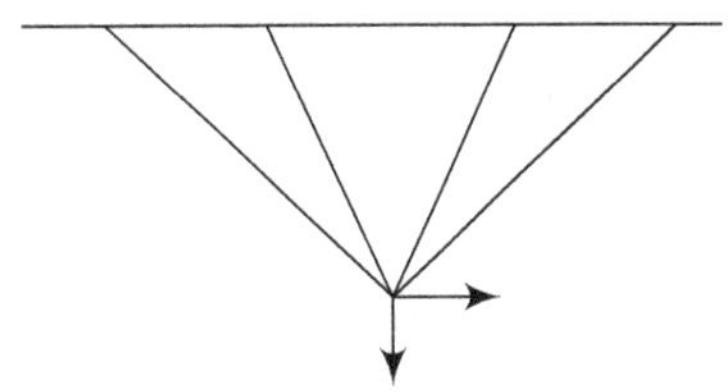 **Fig. 7.1** *(b) Degrees of Freedom, n = 2*
4. If the chosen releases are not properly chosen, the flexibility matrix may be ill conditioned.	4. Generally the matrix is well conditioned.
5. The banded form is not obvious. We had attempted for some particular types leading diagonal of structures like tall buildings.	5. Non zero elements are along diagonals. It is possible to express K matrix in tridiagonal form (Chapter 9) whereby the unknowns can be solved without resorting to complete inversion (Successive elimination and back substitution). Suitable for multistorey buildings.
6. Not well suited for dynamic analysis. The global flexibility F is useful for evaluating frequencies.	6. Well suited for dynamic problems for evaluating modes of vibrations and corresponding frequencies. We evaluate say the flexibility matrix F for the tall buildings for the chosen degrees of freedom and obtain frequencies.
7. Suitable for elastic – plastic analysis of frames and evaluating capacity of Structures needed for 'Push over' - Design against the earthquakes.	7. The degrees of freedom increase with the development of plastic hinges.

Recapitulation: Generation of Equilibrium Matrix $[B_0\ B]$ of force method and Compatibility Matrix A of connection Matrix approach of Displacement method:

While using the force method, the member forces based on equilibrium populating the $[B_0\ B]$, take into account the different orientation of the member axes. The choice of release system and formulation of the $[B_0\ B]$ renders the method to be highly conceptual but the computer algorithm may not be direct.

In the case of "Direct Stiffness Method", the equilibrium is satisfied joint wise. The method, therefore, is suitable for the machine programming and the axes transformation is accomplished by the 'direction cosine matrices', that is, by transformation matrix, T.

Like the $[B_0\ B]$ in the Force method and the connection matrix A of the Displacement method, axes transformations are accounted for while writing these matrices, the $[B_0\ B]$ and $[A]$ respectively. The assembly for obtaining global matrix, $K = \Sigma\, A^t\, k_m\, A$, is assembled member wise, where as the "Direct Stiffness" method assembles joint wise forming a row of the global stiffness.

The final step of the compatibility in the Force Method is accomplished by **"Contragradience"- the principle of complementary virtual work,** and

The final step of the Equilibrium in the connection matrix approach (of displacement method) is accomplished by the **principle of virtual work**. For facilitating the computer programming, the connection matrix A is generated by the code numbers, the essence of the compatibility condition.

The computers are used extensively in all walks of life. In this direction, the matrix methods of Structural Analysis based on Force method was introduced. The learning of the subject of 'Structural analysis and design' needs to be oriented keeping in mind the use of computers. More important than using commercial software and manually solving a large number of numerical problems are to (a) understand conceptual as well as (b) in-depth knowledge orienting ourselves to make our learning amenable to writing simple software and also appreciate the scope and limitations of the commercially available software.

In this direction, we have learnt the subject of structural analysis, (i) fairly comprehensively about the Force method and also (ii) the displacement method. The displacement method has come a long way in its use for developing the computer software.

Our strategy* in the following was to take up first the simple case of truss (i = 1). We follow it up by considering continuous beams (i = 2) and rigid jointed structures like plane frames, it may be appropriate to introduce some more aspects of computer logic and its blending with computer programming. For the analysis of plane frames, we continue to use the gable frame as a 'reference structure'. There is a significant advantage in using the connection matrix approach in preference to the direct stiffness approach for a structure like gable frame where inextensibility is assumed with advantage. For the manual computations, our attempt is directed toward reducing the number of unknowns to be evaluated. For the rigid jointed frames, the assumption of inextensibility in conjunction with the connection matrix approach make the analysis more attractive for the hand computations.We commence with the trusses before taking up the continuous beams and frames.

Truss: Consider a simple truss with hinged joints, Figs 7.2(a), (b), (c).

All members ISLT50

Force method: Link (2) is a roller support.

It is a determinate truss.

Four unknown displacements.

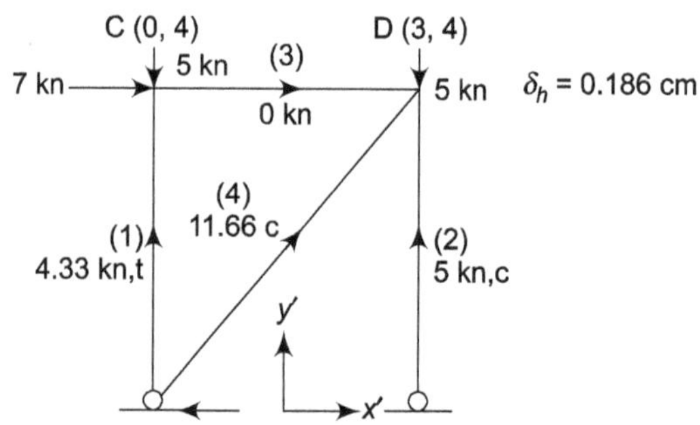

Fig. 7.2(a)

* By now the reader has acquired sufficient background in Structural Analysis. In fact, without having to go thru from the beginning of the Ch. 7, he may pick up any illustration of his choice. Initially the presentation may seem too elementary we pick up as we move on.

Case 1

$$P'_C = (7, 5)\ kn, P'_D = (0, 5)\ kn$$

All joints including the support are hinged
All members T –section of same area ISLT50

Force method:

It is a determinate truss. Member forces are shown along side.

Displacement method: 2 dof per node.

Four unknown displacements.

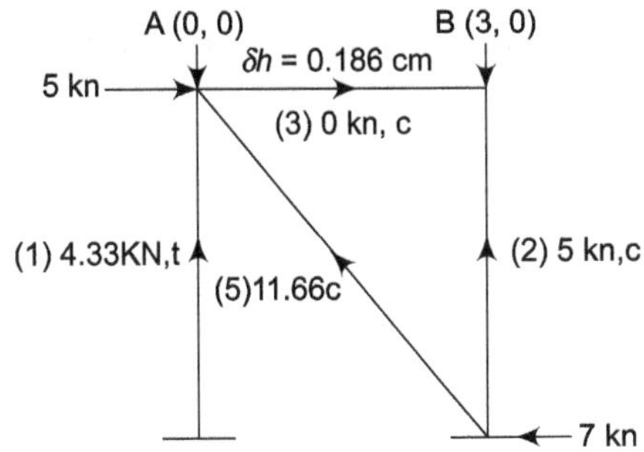

Fig. 7.2(b)

Case 2, Fig. 7.2(c)

$$P'_c = (7, 5)\ kn, P'_d = (0, 5)\ kn$$

All joints including the support are hinged
All members T –section of same area
ISLT50
Notation: c , member is in compression
t, member in tension

Force method: It is a indeterminate truss, n=1
Member forces are shown along side.

Displacement method: 2 dof per node.

Four unknown displacements.

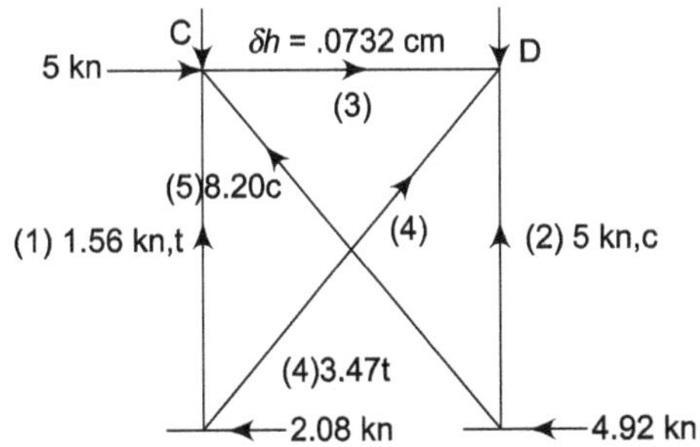

Fig. 7.2(c)

Case 3

A simple truss with hinged joints, Figs 7.2(a) to (c)

Inferences: As long as the nodes are unaltered, as is the case above, the number of unknown displacements are equal to four.

The first case is determinate, and one could write the member forces practically by inspection. The cross-sectional properties do not influence the member forces.

Case 1: Even though, the longest member is in tension, the vertical link is in compression =14.33

Case 2: Not only that the maximum horizontal deflection is considerably reduced (=.0732 cm), the member forces are also moderated. When analyzed by the Force method, degree of indeterminacy is equal to one. Such examples have been discussed in the preceding chapters.

For the case 3, formulation of the global stiffness is derived in the following.

7.1.1 Direct stiffness approach for Plane Truss with Hinged Joints

Refer Fig. 7.3

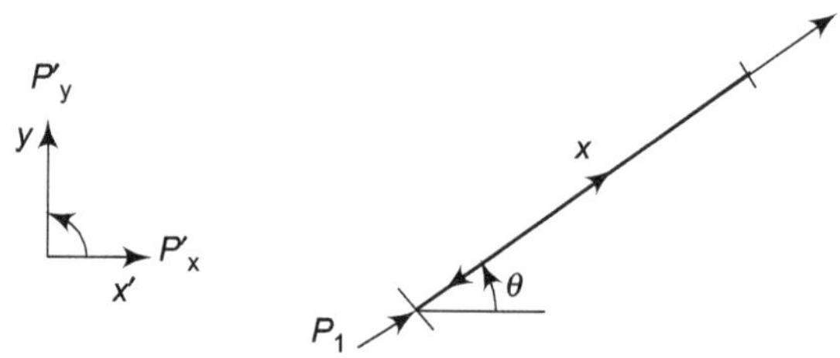

Fig. 7.3 *Member Stiffness of a Member of Truss*

$$p_1 = K\,d_1 - K\,d_2$$

$$p_2 = -K\,d_1 + K\,d_2, \text{ where } K = EA/L$$

Or in the matrix form, the member stiffness is given by

$$\begin{pmatrix} p_1 \\ p_2 \end{pmatrix} = \begin{pmatrix} K, -K \\ -K, +K \end{pmatrix} \begin{pmatrix} d_1 \\ d_2 \end{pmatrix} \text{ the forces } p \text{ and } d \text{ are along member axis } x$$

Referring to the global axis, $p'_x = p\cos\theta$ and,

$$p'_y = p\sin\theta$$

Or in the matrix form, $p' = T\,p$, where $T = \begin{bmatrix} \cos\theta \\ \sin\theta \end{bmatrix}$; $p' = \begin{bmatrix} p'_x \\ p'_y \end{bmatrix}$

From contragradience, $d = T^t\,p'$

Making the above substitution: $p'_1 = K\,d'_1 - K\,d'_2$

$$p'_1 = (T\,K\,T^t)\,d'_1 - (T\,K\,T^t)\,d'_2$$

$$p'_2 = -(T\,K\,T^t)\,d'_1 + (T\,K\,T^t)\,d'_2$$

or

$$p'_1 = K'\,d'_1 - K'\,d'_2$$

$$p'_2 = -K' d'_1 + K' d'_2, \text{ where } K' = (T K T^t)$$

and

$$K' = \begin{pmatrix} \cos^2\theta, \sin\theta\cos\theta \\ \sin\theta\cos\theta, \sin^2\theta \end{pmatrix} \frac{EA}{L}$$

Writing the equations of equilibrium for the free body of the node C:

$$P_C' = p_2'{}_{(1)} + p_1'{}_{(3)} + p_2'{}_{(5)}$$
$$P_D' = p_2'{}_{(2)} + p_2'{}_{(3)} + p_2'{}_{(4)}$$

Substituting the member forces in the above,

$$\begin{pmatrix} P_C' \\ P_D' \end{pmatrix} = \begin{pmatrix} K'_{(1)} + K'_{(3)} + k'_{(5)} & -K'_{(3)} \\ -K'_{(3)} & K'_{(2)} + K'_{(3)} + k'_{(4)} \end{pmatrix} \begin{pmatrix} d_C' \\ d_D' \end{pmatrix}$$

From the above equilibrium equation, we solve for the unknown displacements,

- From the joint displacements, we find member end deformations from compatibility
- From the member stiffnesses, we find the member end forces

For the final results, refer the figure for case 2.

7.1.2 Connection Matrix Approach Applied to Plane Truss with Hinged Joints

The assumption of approximation pertaining to the inextensibility of the members is applicable for structures which deform mainly due to flexure, bending moment and shear. Even though such an assumption is not valid in the case of trusses, where axial force is the only internal force, the method is applicable to the Trusses and does not imply any approximation. Instead of using the member axis, in the undeformed position as done in the preceding illustration, when taken in the deformed position, we are considering the relative deformation with respect to the end-1 and therefore the member stiffness is simplified to $k_m = EA/L$. Thereby, the size of the connection matrix "A" is considerably reduced.

Truss: Consider a prismatic member of a truss (with all joints hinged). Because the axial force does not change along its length, the total length is considered in place of an infinitesimal length, Fig. 1. The member stiffness is given by $f_m = EA/L$. Irrespective of its being a plane truss or space truss the member stiffness remains unaltered. The method is illustrated by solving a plane truss, Fig. 7.4. The hinged nodes A and B are the supports and the nodes C and D are the free nodes. With each node having two degrees of freedom each.

$$P_m = P_1 = P_2, \quad \text{and} \quad \delta_m = \delta_2 - \delta_1$$

Member stiffness

$$k_m = EA/L$$

Member and Global Axes

Each member, (1) to (5), has its own member axis x. The direction of the member axis fixes the terminal points (1 or 2) of the individual members. As these member axes are acting in the direction depending upon their orientations. For the total structure, there is required to be specified global axes 'X' and 'Y' which pertain to the nodes, where members in different directions meet.

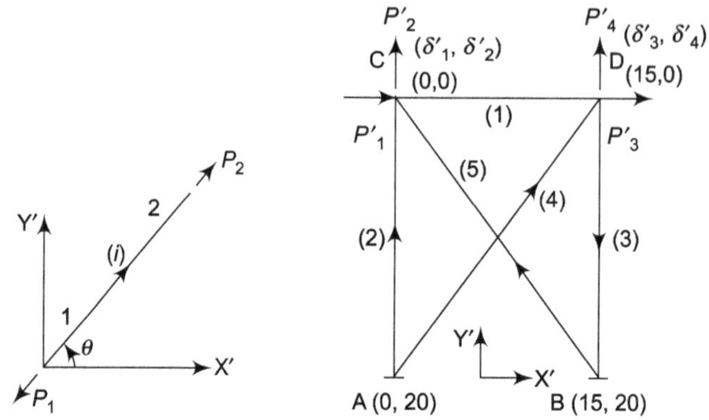

(a) A Typical inclined member (i) (b)

Fig. 7.4 *Member Stiffness for Truss Member*

Member Stiffness K_m, in the relation $P_m = K_m\, \delta_m$ is given by

$(k_m)_i = (EA)_i/L_i$, and

Member forces $P_m = P_1 = P_2$ and corresponding member deformation $\delta_m = \delta_2 - \delta_1$ Fig. 1.

Referring to the global axes X', Y', we are required to write the equilibrium $P' = K'\, d'$ for the evaluation of the unknown displacements d'. In the following, the displacement method is developed which essentially consists of obtaining the global matrix K'.

Preliminaries: It being a plane truss ; the free nodes C and D have two degrees of freedom each. The four unknown displacements ds, are:

Node C: $[\delta'_1\ \delta'_2]$, and Node D: $[\delta'_3\ \delta'_4]$

Corresponding Load Vector P'_j: $[P'_1\ P'_2\ P'_3\ P'_4]$

Basis Steps

Step 1. Compatibility using connection approach

Connection matrices A_i are defined by

$$(\delta_m)_i = A_i\, [\delta'_1\ \delta'_2\ \delta'_3\ \delta'_4]^t$$

These being members of truss, $(\delta_2 - \delta_1)_i = A_i\, [\delta'_1\ \delta'_2\ \delta'_3\ \delta'_4]^t$

Referring to Fig. 7.4(a):

$$A_i = [c, s; c, s], \text{ where } c = \cos\theta, \text{ and } s = \sin\theta$$

$$(\delta_m)_1 = (\delta_2 - \delta_1)_1 = [0, 0, 1, 0]\begin{bmatrix}\delta'_1\\ \delta'_2\\ \delta'_3\\ \delta'_4\end{bmatrix}$$

Similarly the connection matrices for the members (2), (3), and (4) are:

$$A_2 = [0\quad 0\quad 1\quad 0],$$
$$A_3 = [0\quad 0\quad 0\quad 1],$$
$$A_4 = [0\quad 0\quad c\quad s],$$
$$A_5 = [0\quad 0\quad -c\quad s],$$

Global stiffness matrix is obtained from equilibrium condition. Centrogredience is used for satisfying the equilibrium.

$$P' = K\,d'_j, \text{ where}$$
$$K = \Sigma\,(A^t\,k_m\,A)_i$$

$(A^t\,k_m\,A)_i$ is the contribution of the member i towards the global stiffness, K.

In the above member, contributions towards the global stiffness K_j have been computed. From each member separately. The connection matrix A_i of each member achieves transformation, that is, no separate transformations are required. In the case of direct stiffness approach, the transformation of each member stiffness is required to be done before assembling the global stiffness matrix.

Post Processing:

- Invert K_j and compute Joint Displacements d_j
- Compute member deformations from $d_m = A_i\,d_j$
- Compute member forces $P_m = K_m\,d_m$

We are now at a stage to take up the formal design.

Having recalled the Direct stiffness and connection matrix approaches for a simple case of a truss, we a simple case of a truss discuss the computerization for the displacement method. The direct stiffness formulations for assembling the Global Stiffness matrix are relatively simpler. Computerization is being presented for the connection matrix approach.

7.2 COMPUTERIZATION-LOGISTICS

Because the question of inextensibility doesnot arise for the trusses, the joints are chosen as the free bodies for writing the equations of equilibrium. The members connected to a joint are the only members influenced by the joint deformation. Similarly a member is influenced by the adjoining joints only. We either give (a) Ncode and VNcode for the joints being considered or (b) Ncode and VNcode for the members.

7.2.1 Plane Truss

The presentation of the analysis using, (i) Scalar approach, and (ii) Matrix approach. In the following presentation is directed towards computer orientation.

Write the coordinates of all the nodes, Fig. 7.5(a), referring to the system axes.

The node A is chosen as origin Coordinates (0, 0).

B	(15, 0)
C	(0, 15)
D	(15, 15)
E	(– 15, 20)
F	(0, 20)
G	(15, 20)
H	(30, 20)

Degrees of freedom arise due to the displacements of the nodes A, B, C, and D only.

And the corresponding deformations are eight degrees of freedom = $[\delta_1, \delta_2, \delta_3, \delta_4, \delta_5, \delta_6, \delta_7, \delta_8]$ Fig. 7.5(a). The lengths, inclinations and direction cosines of the members are entered in a tabular form. The values of a typical member (4) are shown as an illustration.

The inclination of the member is measured with respect to the positive x' axis of the coordinate system.

Table 7.1 *Direction cosines of a typical member (4)*

Members	Coordinates		$\Delta X = x'_2 - x'_1$	$\Delta Y_2 = y'_2 - y'_1$	$l = [\Delta x^2 + \Delta y^2]^{1/2}$	$cx' = \frac{\Delta x}{l}$	$cy' = \frac{\Delta y}{l}$
	at end 1 (x'_1, y'_1)	at end 2 (x'_2, y'_2)					
1.							
2.							
3.							
4.	(0, 0)	(15, 20)	15	20	25	.6	.8
5.							
6.							
7.							
8.							
9.							
10.							
11.							
12.							

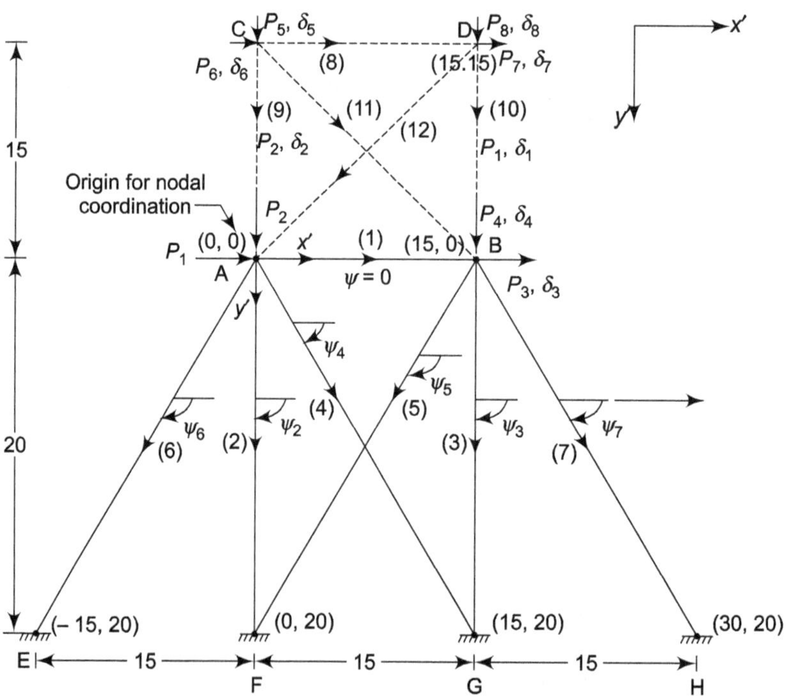

(a) Plane Hinged Truss

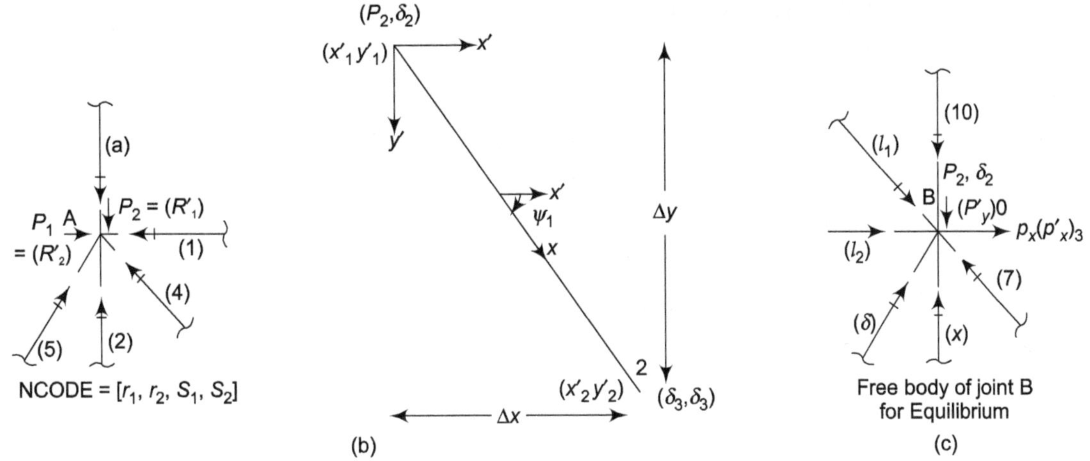

Hinged Plane Truss and Free Bodies: Member and System Axes

Fig. 7.5 *Plane Truss—Geometry and Member Designations*

Consider a typical member i of a plane truss Fig. 7.5(b)

Assumed positive directions are shown

$$\cos(x, x') = \frac{\Delta x}{l}, \ \cos(x, y') = \frac{\Delta y}{l}, \ l = [(\Delta x)^2 + (\Delta y)^2]^{1/2}$$

Let the direction cosines cos (x, x'), cos (x, y') & cos (x, z') be represented by $C_{x'}$, $C_{y'}$ and $C_{z'}$ respectively

And $\psi_i = (x, x')_i$

Step 1: Member deformations

$$(\delta_x)_i = (\delta x_2 - \delta x_1)_i \tag{1)a}$$

Where δx_1 and δx_2 are the relative member end deformation at ends 1 and 2 respectively measured along member x-axis.

From compatibility

$$\delta x_1 = (\delta' x)_1 \, cx' + (\delta' y)_1 \, cy'$$
$$\delta x_2 = (\delta' x)_2 \, cx' + (\delta' y)_2 \, cy' \tag{1)b}$$

Where $(\delta' x)_1$ & $(\delta' y)_1$ are the deformation components of the end-1 measured along x', y' and similary $(\delta'_x)_2$ and $(\delta' y)_2$ pertain to end-2.

Let the ends 1 and 2 be represented by r and s (as per system numbering) respectively.

If the nodal deformation $(\delta r_1, \delta r_2)$ represent $(\delta' x_1, \delta' y_1)$ and $(\delta s_1, \delta s_2)$ represent $(\delta' x_2, \delta' y_2)$ from equation (1) and substituting for $\delta' x_1, \delta' y_1, \delta' x_2, \delta' y_2$

$$(\delta' x)_i = -(c' x)_i \, \delta r_1 - (c' y)_i \, \delta r_2 + (c' x)_i \, \delta s_1 + (c' x)_i \, \delta s_2 \tag{I}$$

Equation (I) is the compatibility equation which can be easily generated by knowing r_1, r_2 at the end 1, and s_1, s_2 at the and the direction cosines. We designate these by

$$\text{NCODE} = [r_1, r_2; s_1, s_2]$$
$$\text{VCODE} = [-c'x, -c'y; c'x, c'y]$$

For the frame, Fig. 7.5(a)

Say for member (4), $(\text{NCODE})_4$ is defined by $r_1 = 1, r_2 = 2, s_1 = 0$ and $s_2 = 0$

Or $(\text{NCODE})_4 = [1, 2, 0, 0]$, and $(\text{VNCODE})_4 = [-.6, -.8, .6, .8]_4$

Similarly, say, for member 11

$$(\text{NCøDE} = [5, 6, 3, 4] \quad \text{and}$$
$$(\text{VNCøDE} = [-.707, -.707, .707, .707]$$

Step 2: Member force – deformation relations

$$(p_m)_i = (k_m)_i\,(\delta_m)_i$$

Where for hinged trusses $(\delta_m)_i = (\delta_x)_i$ (II)

$$(p_m)_i = (p_x)_i,\ (k_m)_i = \frac{EA}{L}$$

Or combining with Eq. (I)

$$(p_x)_i = (p_m)_i = \left(\frac{EA}{L}\right)_2 [-(cx')\,\delta r_1 - (cy')\,\delta r_2 + (cx')\,\delta s_1 - (cy')\,\delta s_1]$$

Similarly force expressions are written for all the members.

Step 3: Equilibrium

For a planar truss two equations of equilibrium for (i), horizontal and (ii) vertical components of forces are written for each joint. Consider free body of joint A, Fig. 7.5(c). for the joint, if n is the number of members meeting at a joint, $n = 5$ and the five members are [1, 2, 4, 6, 9]

And it is designated as the JOINT CODE of the joint A..

From the equilibrium of the joint A,

$$(p'_x)_A = p_1 = \sum_{i=1.5} (p'_x)_i$$

Where $(p'_x)_n$ is the horizontal component of the force in a member, n and

$$(p'_x)_1 = (p_m)_1\,(c'_x)_1,$$
$$(p'_x)_2 = (p_m)_2\,(c'_x)_2,$$
$$(p'_x)_3 = (p_m)_4\,(c'_x)_4, \text{ and}$$
$$(p'_x)_4 = (p_m)_6\,(c'_x)_6,$$
$$(p'_x)_5 = (p_m)_9\,(c'_x)_9,$$

Similarly

$$(p'_y)_A = p_2 = \sum_{i=1,s} (p'_y)_i$$
$$(p'_y)_1 = (p_m)_1,\,(c'_y)_1,$$
$$(p'_y)_2 = (p_m)_2,\,(c'_y)_2,$$
$$(p'_y)_3 = (p_m)_4,\,(c'_y)_4, \text{ and}$$

$$(p'_y)_4 = (p_m)_6, (c'_y)_6,$$
$$(p'_y)_5 = (p_m)_9, (c'_y)_9,$$

Quite often it is expedient to consider each member separately and write its contribution to the different nodal equation of equilibrium. As an illustration, for the member $i = 4$, $NCODE = [1, 2, ,0,0]$, it contributes to the equilibrium equation corresponding to the nodal forces p_1 and deformations δ_1, δ_2 based on which the element of *NCODE* are defined.

And the contributions are

For $(p'_x)_A = P_1 = -(c'_x)_i$ and

For $(p'_y)_A = P_2 = -(c'_y)_i$

And contributions at the node *E* (in this case zero displacement) are

For $(p'_x)_E =$ Horizontal reaction at $E = (c'_x)_i$ and similarly

For $(p'_y)_E =$ vertical reaction at $E = (c'_y)_i$

Tabulating the contributions from different members meeting at the joints, A, B, C, and D we have, 7.2:

Table 7.2 *Contributions from different members meeting at the joints, A, B, C, and D*

Equilb Equation	*Nodal Load, P*	*Member designation*											
		(1)	*(2)*	*(3)*	*(4)*	*(5)*	*(6)*	*(7)*	*(8)*	*(9)*	*(10)*	*(11)*	*(12)*
1		$-(c'_x)$	$-c'_x = 0$	-	$-(c'_x)$	-	$-c'_x$	-	-	c'_x	-	-	c'_x
2	p_2	$-(c'_y) = 0$	$-c'_y$	-	$-(c'_y)$	-	$-c'_y$	-	-	c'_y	-	-	c'_y
3	p_2	$(c'_x) = 1$	-	$-c'_x = 0$	-	$-c'_x$	-	$-c'_x$	-	-	$-c'_x$	c'_x	-
4	p_4	$(c'_y) = 0$	-	$-c'_y$	-	$-c'_y$	-	$-c'_y$	-	-	$-c'_y$	c'_y	-
5	p_5	-	-	-	-	-	-	-	$-c'_x$	$-c'_x$	-	$-c'_x$	-
6	p_6	-	-	-	-	-	-	-	$-c'_y$	$-c'_y$	-	$-c'_y$	-
7	p_7	-	-	-	-	-	-	-	c'_x	-	$-c'_x$	-	$-c'_x$
8	p_8	-	-	-	-	-	-	-	c'_y	-	$-c'_y$	-	$-c'_y$
Not for evaluating displacement	$(R'_x)_E$	-	-	-	-	-	c'_x	-	-	-	-	-	-
	$(R'_y)_E$	-	-	-	-	-	c'_y	-	-	-	-	-	-
	$(R'_x)_F$	-	$c'_x = 0$	-	-	-	-	-	-	-	-	-	-
	$(R'_y)_F$	-	$c'_y = 1$	-	-	-	-	-	-	-	-	-	-
	$(R'_x)_G$	-	-	$c'_x = 0$	c'_x	c'_x	-	-	-	-	-	-	-
	$(R'_y)_G$	-	-	$c'_y = 1$	c'_y	c'_y	-	-	-	-	-	-	-
	$(R'_x)_H$	-	-	-	-	-	-	c'_x	-	-	-	-	-
	$(R'_y)_H$	-	-	-	-	-	-	c'_y	-	-	-	-	-

Thus from the table, consider the column corresponding to the member $i = 4$. The member contributes to the equilibrium for the nodal loads p_1, p_2 and $(p'_x)_E$, $(p'_y)_E$ and the contribution are

$$\begin{matrix} p_1 \\ p_2 \\ (p'_x)_E \\ (p'_y)_E \end{matrix} \begin{bmatrix} -c'_x \\ -c'_y \\ c'_x \\ c'_y \end{bmatrix}_4^{(p_m)_i} \quad ; i = 4 \qquad \text{(III)}$$

It may be noted that the direction cosines are the same as given in VNCODE for the member $i = 4$ as defined while writing compatibility (Step-1). The property known as centrogredience will be discussed while presenting the matrix displacement method. The column corresponding to $i = 4$ is marked red and it defines $[\text{VNCODE}]_4$ for the member.

As the equilibrium equations for evaluating the unknown displacements are written for the nodes A and B, C and D only. We have 8 equations corresponding to the 8 unknowns (δ_1, δ_2), (δ_3, δ_4), (δ_5, δ_6) and (δ_7, δ_8), for the nodes A, B, C and D respectively. Thus in the system of equations,

$$\left.\begin{aligned} p_1 &= k_{11}\delta_1 + k_{12}\delta_2 + k_{13}\delta_3 + k_{14}\delta_4 + k_{15}\delta_5 + k_{16}\delta_6 + \dots \\ p_2 &= k_{21}\delta_1 + k_{22}\delta_2 + k_{23}\delta_3 + k_{24}\delta_4 + k_{25}\delta_5 + k_{26}\delta_6 + \dots \\ p_3 &= k_{31}\delta_1 + k_{32}\delta_2 + k_{33}\delta_3 + k_{34}\delta_4 + k_{35}\delta_5 + k_{36}\delta_6 + \dots \\ p_4 &= k_{41}\delta_1 + k_{42}\delta_2 + k_{43}\delta_3 + k_{44}\delta_4 + k_{45}\delta_5 + k_{46}\delta_6 + \dots \end{aligned}\right\} \qquad (3)$$

$$p_y =$$

The member $i = 4$, NCODE (1, 2, 0, 0) contributes to the coefficient δ_1 only δ_2. The corresponding stiffness coefficient following the NCODE values are k_{11}, k_{12}, and k_{22} for which the member 4 contributes. From Eq. III

Due to member 4

$$p_1 = -(c'_x)_4 (p_m)_4 \qquad (4)$$

$$\left.\begin{aligned} p_2 &= -(c'_y)_4 (p_m)_4 \\ \text{Substitute for } (\delta_2)_4 (p_m)_4 &= (k_m)_4 (\delta_x)_4 \end{aligned}\right\}$$

$$\text{VNCODE} = [-c'_x, -c'_y; c'_x, c'_y]$$

The 3rd and 4th elements c'_x, c'_y are not relevant because of the deformation at the end 2 of the member 4, are equal to zero.

Relevant VNCODE $= [-c'_x, -c'_y]$ and

From Eq (I),

$$(\delta_x)_4 = (c'_x)_4 \cdot \delta_1 - (c'_y)_4 \cdot \delta_2$$

For Step-2

$$(p_m)_4 = (k_m)_4 [-(c'_x)_4 \, \delta_1 - (c'_y)_4 \, \delta_2]$$

Substituting equation (4)

$$p_1 = -(c'_x)_4 (k_m)_4 [-(c'_x)_4 \, \delta_1 - (c'_y)_4 \, \delta_2]$$

$$p_2 = -(c'_y)_4 (k_m)_4 [-(c'_x)_4 \, \delta_1 - (c'_y)_4 \, \delta_2]$$

Expanding

$$p_1 = k_{11}\delta_1 + k_{22}\delta_2; \quad p_2 = k_{21}\delta_1 + k_{22}\delta_2 \text{ where}$$

$$k_{11} = +(c'_x)_4\,(k'_m)_4\,(c'_x)_4 = (c'_x)^2_4\,(k'_m)^2_4;\; k_{12} = k_{21},\, k_{22}$$

Similarly contributions to the stiffnesses are found from all the other members, $i = 1$ to 12. It is a bit rough going but it is hoped the readerslike going through with concentration.

7.2.2 Space Trusses

Similarly extend the theory to the 3D – Truss.

Coordinates $(x, y, z)_i$, $(x, y, z)_i$

$$l_i = (\Delta x^2 + \Delta y^2 + \Delta z^2)^{1/2}$$

$$cx' = \Delta x/l$$

$$cy' = \Delta y/l$$

$$cz' = \Delta z/l$$

$$\delta_{x1} = (\delta x')_1\, cx' + (\delta y')_1\, cy' + (\delta z')_1\, cz'$$

$$\delta_{x2} = \ldots$$

$$\text{NCODE} = [r_1, r_2, r_3; s_1, s_2, s_3]$$

$$\text{VNCODE} = [-c'_x, -c'_y, -c'_z, c'_x, c'_y, c'_z]$$

Equilibrium equations are obtained from the joint free bodies.

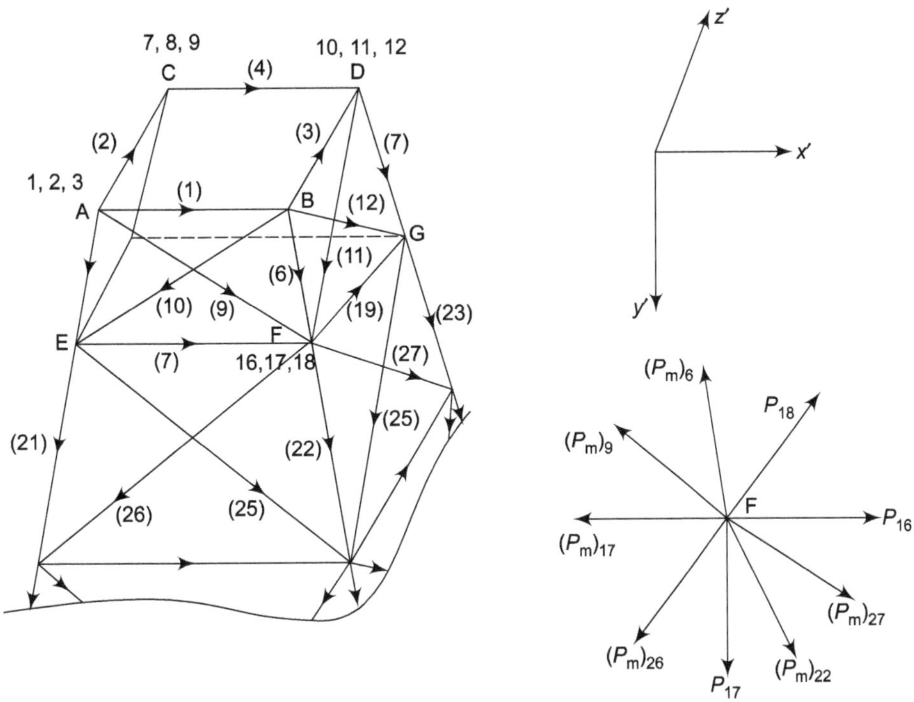

Fig. 7.6 *A Typical Space Tower*

The global stiffness matrix, $k_j = \Sigma(A^t K_m A)_i$ where A_i is the compatibility matrix of member i defined by $(d_m)_i = (A)_i\, d_j$ and $P_j = k_j d_j$

$$(p'_x)_I = p_1 = \sum_I (p'_x)_i$$

$$(p'_y)_I = (p'_y) = p_2 = \Sigma_{i=1}$$

$$(p'_z)_I = p_3 = \sum_I (p'_z)_i$$

Assembly of stiffness matrix based on equilibrium follows from contragradience consideration. *i.e.*, memberwise. Alternatively consider a member '*i*', and define A_i – differently i.e., $p_j = A_i\,(p_m)_i$

$$p_j = \begin{bmatrix} -c'_x \\ -c'_y \\ -c'_z \\ c'_x \\ c'_y \\ c'_z \end{bmatrix}_i (p_m)_i \text{ and } d_m = (A^T)_i\, d_j$$

And for the case consider $k_j = \Sigma(A\, k_m\, A^t)_i$

Mero System: It is an example of space truss. The assembly of the truss is achieved by pipe members and nodes. The system is described in detail at the end of the Chapter, 7.16.

7.2.3 Continuous Beams

Once you chose a coordinate system, we do not change it till the solution is complete. The member axes are marked. No axes transformation is required.The geometrical properties not being symmetrical, we cannot split into symmetry and symmetry. In the following we solve the problem using the more familiar Slope Deflection method. To begin with consider a simple case of continuous beam on non-sinking supports. It will be followed by a general case of the beams using the matrix approach Illustration-1:

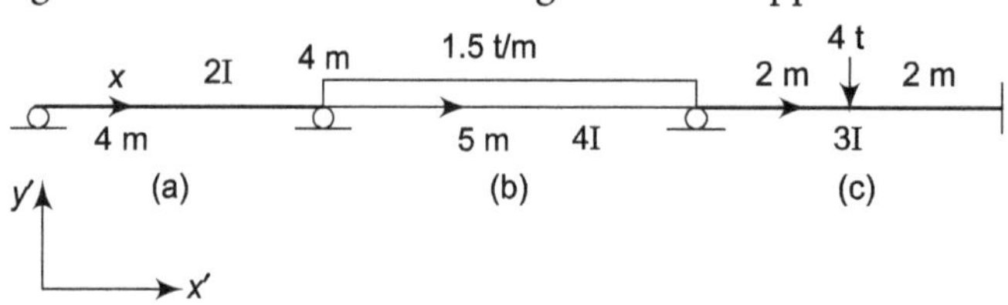

Fig. 7.7 *Continuous Beam*

And repeat the solution based on matrix approach.

Slope Deflection method

Member force deformation relations

For a member (i).

Recall: $(m_1)_i = (4EI/L)\,(\theta_1)_i + (2EI/L)\,(\theta 2_1)_i$

$(m_2)_i = (2EI/L)\,(\theta_1)_i + (4EI/L)\,(\theta 2_1)_i$

Even though we are not encountering relative lateral displacement, but the end moments can be obtained by changing the reference axis. Thereby the transverse relative displacement are simplified to the end rotations only. The example 2 falls under this category.

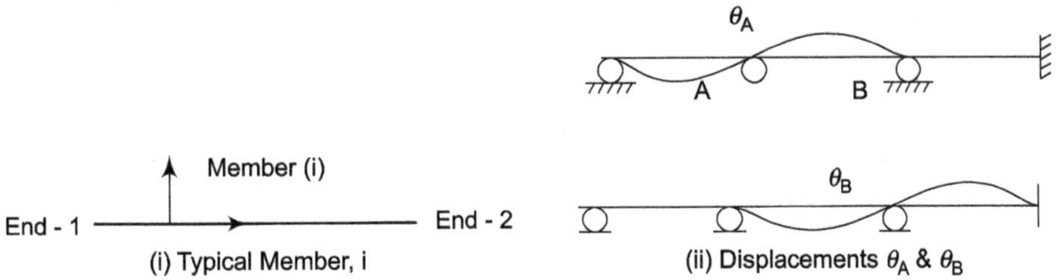

Fig. 7.8 *Degrees of Freedom θ_A and θ_B*

Step 1: Compatibility at the joints:

$$(\theta_2)_a = (\theta_1)_b = \theta_A \,;\, (\theta_2)_b = (\theta_1)_c = \theta_B$$

There are only two independence deformations θ_A and θ_B

Step 2: Force-Deformation relations:

Considering the modified stiffness of the member "*a*",

$$(m_2)_a = 3\left(\frac{EI}{L}\right)_a (\theta_2)_a + (fem)_{2a}$$

Using compatibility conditions at joints:

$$(m_2)_a = \frac{6EI}{4}\theta_A + 0$$

Similarly,

$$\left.\begin{aligned} (m_1)_b &= \frac{8EI}{5}(2\theta_A + \theta_B) + 3.125 \\ (m_2)_b &= \frac{8EI}{5}(\theta_A + 2\theta_B) - 3.125 \\ (m_1)_c &= \frac{6EI}{4}2\theta_B + 2 \end{aligned}\right\}$$

$$(m_2)_c = \frac{6EI}{4}\theta_B - 2$$

Step 3: Equilibrium Conditions:

Moment equilibrium of the joint *A* gives: $(m_2)_a + (m_1)_b = 0$

Similarly from moment equilibrium of the joint $B = (m_2)_b + (m_1)_c = 0$

Substituting the values of the member moments from the step-2,

$$4.7\,\theta_A + 1.6\,\theta_B = -\,3.125/EI$$

$$1.6\,\theta_A + 6.2\,\theta_C = 1.125/EI$$

Solving the above equations:

$$\theta_A = -\,0.8/EI\,;\, \theta_B = 0.386/EI$$

Substituting the values of the "force deformation" relations, step 2

$$(m_2)_a = -\,(m_1)_b = -\,1.2 \text{ tm};\ (m_2)_b = -\,(m_1)_c = -\,3.165 \text{ tm};\ (m_2)_c = -\,1.42 \text{ tm}$$

B.M. Diagram

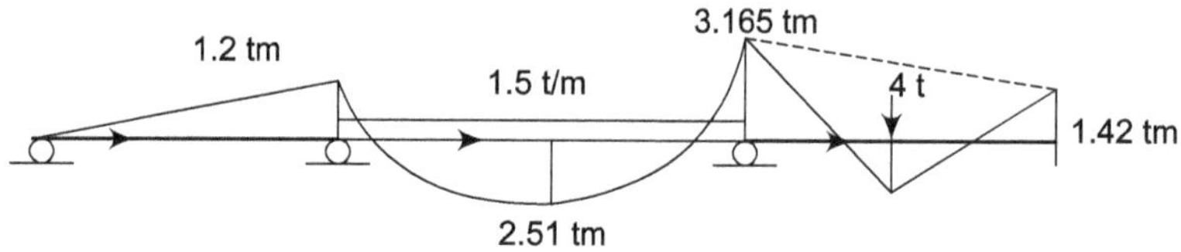

Fig. 7.9 *Final BM Diagram of a Typical Continuous Beam*

Alternative solution following the matrix approach:

Member	*Code number*	*Stiffness coefficients*
A	1	$(K'_{11})_a = (3EI/L)_a$
B	1, 2	$(K'_{11})_b = (4EI/L)_b$ $(K'_{12})_b = (2EI/L)_b$ $(K'_{22})_b = (4EI/L)_b$
C	2	$(K'_{22})_c = (3EI/L)_c$

$$\begin{pmatrix} (K'_{11})_i & (K'_{12})_i \\ \bullet & (K'_{22})_i \end{pmatrix}$$

We write the above table by inspection. We write the equilibrium equations for the solution of the unknown rotations, θ_A, and θ_B. The stiffness coefficient $(K'_{21}) = (K'_{12})^T$

Illustration 2: Continuous beam with a flexible support. Use modified stiffness for the end span.

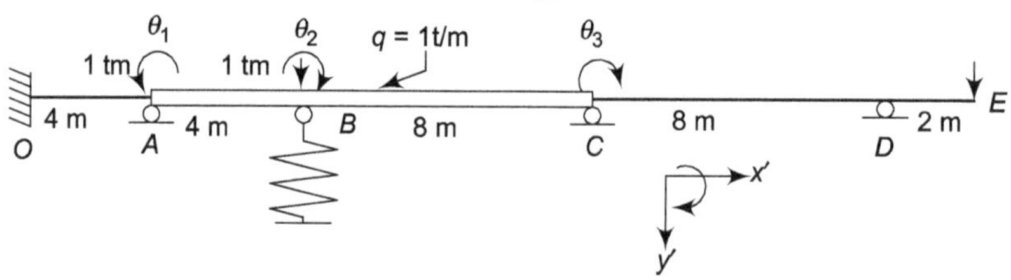

Fig. 7.10 *Continuous Beam with Flexible Support*

The number of unknowns are four, θ_1, θ_2, θ_3, corresponding to M'_1, M'_2, M'_3 and P'_4.

Part I: The nodal loads are computed for the fixed nodes, and these are added to the given distributed loads.

Deflected Shape for **Part I** "no nodal –displacement- Analysis".

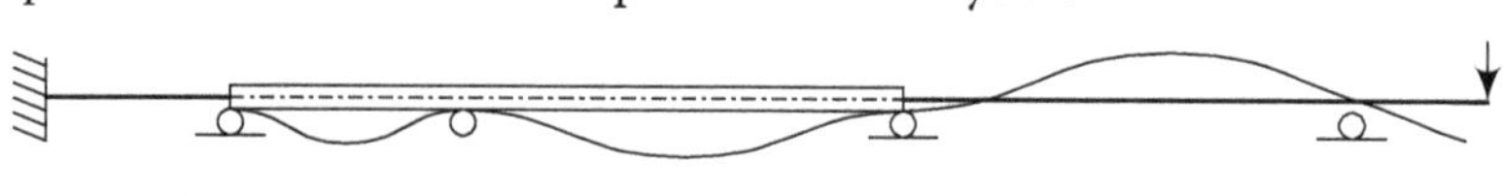

$M'_1 = 1/3M'_2 = 5\ t_m\ M'_3 = -\ 22/3\ t_m$

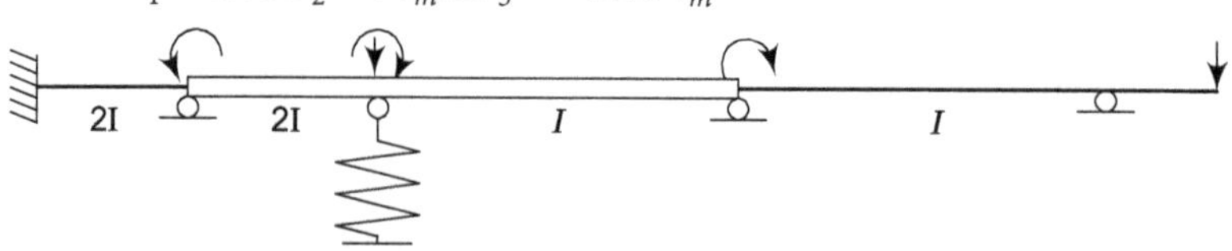

The corresponding nodal loads for **Part II,** Displacement method of analysis are

$M'_1 = 1/3\ t_m$, $M'_2 = 5\ t_m$, $M'_3 = -\ 22/3$, and $P'_4 = 6\ t$.

A typical Nodal Load M'_1 **and corresponding θ1 of Part II**

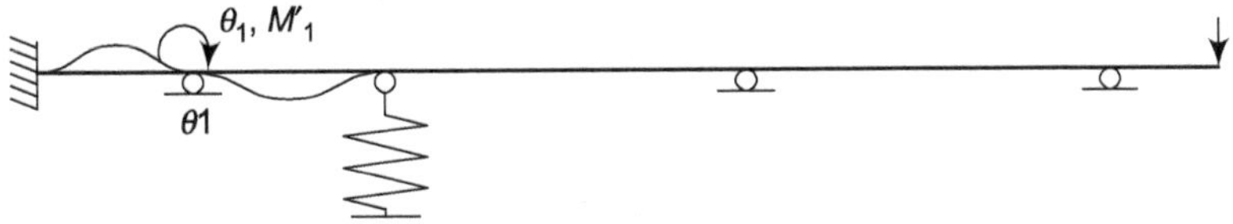

The code numbers while establishing the compatibility of the member–end and nodal deformations, help in writing the final equilibrium equations.

Sketching of deflected shape is used to know the members effected by each degree of freedom

A typical Nodal Load M'_2 **and corresponding** θ_2 **of Part II**

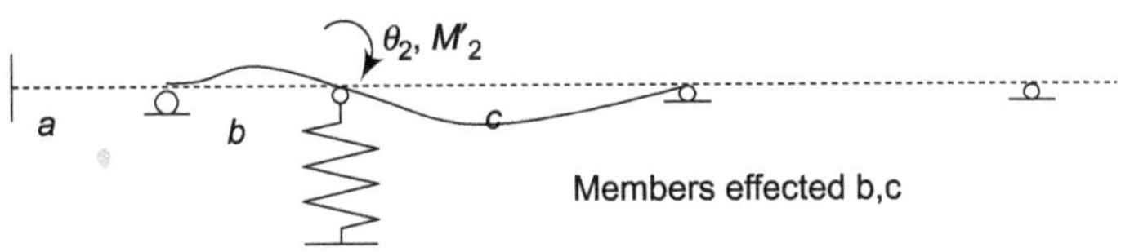

A typical Nodal Load M'_3 **and corresponding** θ_3 **of Part II**

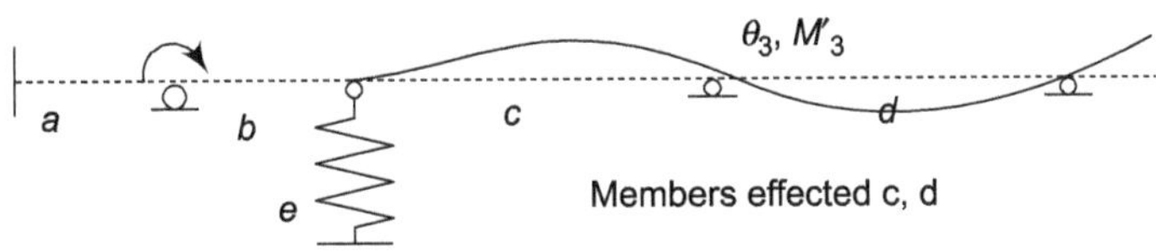

A typical Nodal Load M'_3 **and corresponding** δ_4 **of Part II**

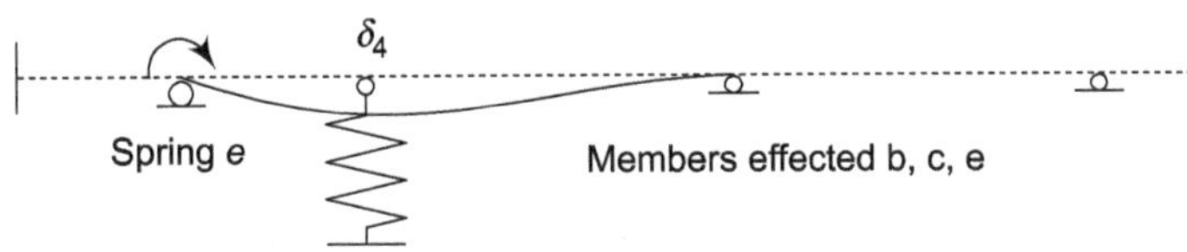

Table 7.3 *Code numbers and stiffness coefficients*

Member	*Code No.*	*Stiffness Coefficients*
A	1	$K'_{11} = \frac{(4EI)_a}{L} = \frac{4E(2I)}{4} = 2EI$
B	1, 2, 4	$K'_{11} = \frac{2EI + (4EI)_b}{4} + \frac{4E(2I)}{4} = 4EI$
		$K'_{12} = \frac{(2EI)_b}{L} = \frac{-6E(2I)}{(4)^2} = \frac{3}{4}EI$
		$K'_{14} = \frac{(-6EI)_b}{L^2} = \frac{4E(2I)}{4} = 2EI$
		$K'_{22} = \frac{(4EI)_b}{L^2} = \frac{4E(2I)}{4} = 2EI$

(*Contd.*)

(Contd.)

Member	Code No.	Stiffness Coefficients
		$K'_{24} = \frac{(-6EI)_b}{L^2} = -\frac{3}{4}EI$
		$K'_{44} = \frac{(12EI)_b}{L^3} = \frac{12E(2I)}{(4)^3} = \frac{3}{8}EI$
C	2, 3, 4	$K'_{22} = 2EI + \frac{(4EI)_c}{L} = 2EI + \frac{4EI}{8} = \frac{5}{2}EI$
		$K'_{23} = \frac{(2EI)_c}{L} = \frac{2EI}{8} = \frac{EI}{4}$
		$K'_{24} = \frac{(6EI)_c}{L^2} + \frac{(-3EI)}{4} = -\frac{3}{4}EI + \frac{6EI}{8^2} = -\frac{21}{32}EI$
		$K'_{33} = \frac{(4EI)_c}{L} = \frac{4EI}{8} = \frac{EI}{2}$
		$K'_{34} = \frac{(6EI)_c}{L^2} = \frac{3EI}{32}$
		$K'_{44} = \frac{(12EI)_c}{L^3} + \frac{(12EI)}{8^3} = \frac{3}{128}EI + \frac{3EI}{8} = \frac{48}{128}EI + \frac{3}{8}EI = \frac{51}{128}EI$
D	3	$K'_{33} = \frac{EI}{2} + \frac{(3EI)_d}{L} = \frac{EI}{2} + \frac{3EI}{8} = \frac{7EI}{8}$
E	4	

Conceptualization: We visualize the part I of the analysis, as if the nodes are fixed externally, and thereby **Part I** of the analysis is made possible. When we release, so to say that external fixity of the nodes is released to deform, resulting in the nodal and associated **Displacements**. The evaluation of the displacements is the **Part II** of the analysis for which we are using the Displacement method. This conceptualization is true irrespective of the type of structure, Truss or Continuous beams or Plane Frame or Grid or Space frame, etc. Modified stiffnesses are used for the end span with simple end support.

Forming the equilibrium equations:

$$K_{11}'\theta_1 + K_{12}'\theta_2 + K_{13}'\theta_3 + K_{14}'\delta_4 = M_1' \quad \text{...(I)}$$

$$K_{21}'\theta_1 + K_{22}'\theta_2 + K_{23}'\theta_3 + K_{24}'\delta_4 = M_2' \quad \text{...(II)}$$

$$K_{31}'\theta_1 + K_{32}'\theta_2 + K_{33}'\theta_3 + K_{34}'\delta_4 = M_3' \quad \text{...(III)}$$

$$K_{41}'\theta_1 + K_{42}'\theta_2 + K_{43}'\theta_3 + K_{44}'\delta_4 = P_4' \quad \text{...(IV)}$$

Substituting the values we get

$$4EI\,\theta_1 + EI\,\theta_2 + 0\,\theta_3 + \frac{(-3EI)}{4}\delta_4 = \frac{1}{3} \quad \text{...(V)}$$

$$EI\,\theta_1 + \frac{5EI}{2}\theta_2 + \frac{EI}{4}\theta_3 + \frac{(-21EI)}{32}\delta_4 = 5 \qquad \ldots(\text{VI})$$

$$0\,\theta_1 + \frac{EI}{2}\theta_2 + \frac{7EI}{8}\theta_3 + \frac{(3EI)}{32}\delta_4 = -\frac{22}{3} \qquad \ldots(\text{VII})$$

$$\left(-\frac{3}{4}EI\right)\theta_1 + \frac{(-21EI)}{32}\theta_2 + \frac{(3EI)}{32}\theta_3 + \frac{51}{128}(EI + K_e)\,\delta_4 = -\frac{22}{3} \qquad \ldots(\text{VIII})$$

The solution gives the deformations

Using slope deflection equations (*i.e.*, member stiffness), member forces are evaluated.

Even though, we use the code number technique for the continuous beam, its suitability will be more obvious when assumptions like inextensibility is used. It is a preparation to the 'code number technique'.

7.2.4 Rigid Jointed Plane Frames and Inextensibility

The displacements at every node consist of rotations and translations. The translations are linked to inextensibility, and they are, therefore separated from rotations.

Inextensibility associated translations are governed by bending Moments

The following structures are in this category of translations.

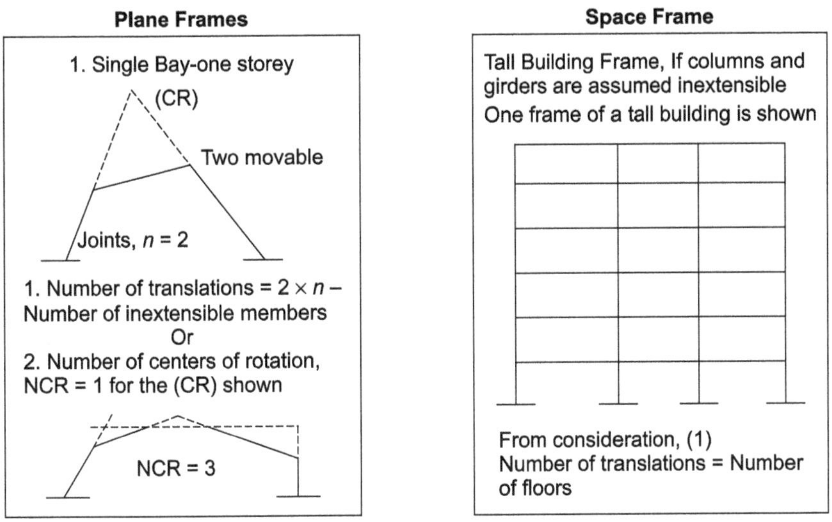

Fig. 7.11 *Continuous Beam with Flexible Support*

The concept of the center of rotation, while deciding, the number of independent translations:

The consideration (2) is used for writing the independent equations of equilibrium corresponding to the number of unknown translations. The 'center of rotation', eliminates the axial forces when the moments are taken about the CR.

Refer text for 'two bays', two storey, and 'A' frame for illustrating the consideration (1). Space frames are discussed separately in the text.

7.3 CONNECTION MATRIX APPROACH FOR RIGID JOINTED FRAMES

A gable frame has been chosen for illustrating the inextensibility. For enabling better familiarization; the same example was used for the illustrations elsewhere also.

Illustration 3: CONNECTION MATRIX METHOD Applied to Gable Frame

Step-1 of the Displacement method. Compatibility Connection Matrix Approach is preferred.

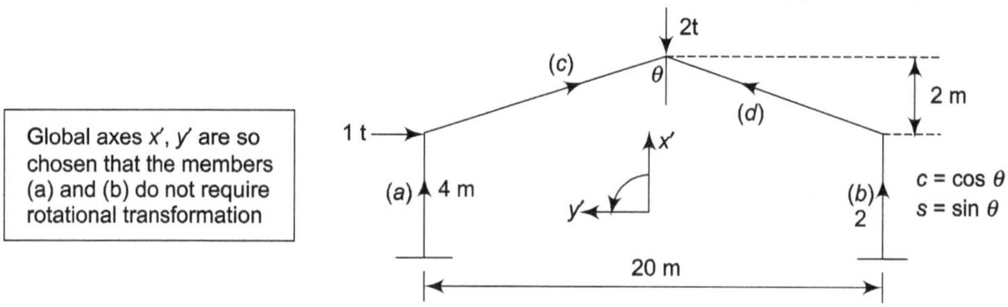

Fig. 7.12 *Illustration-3*

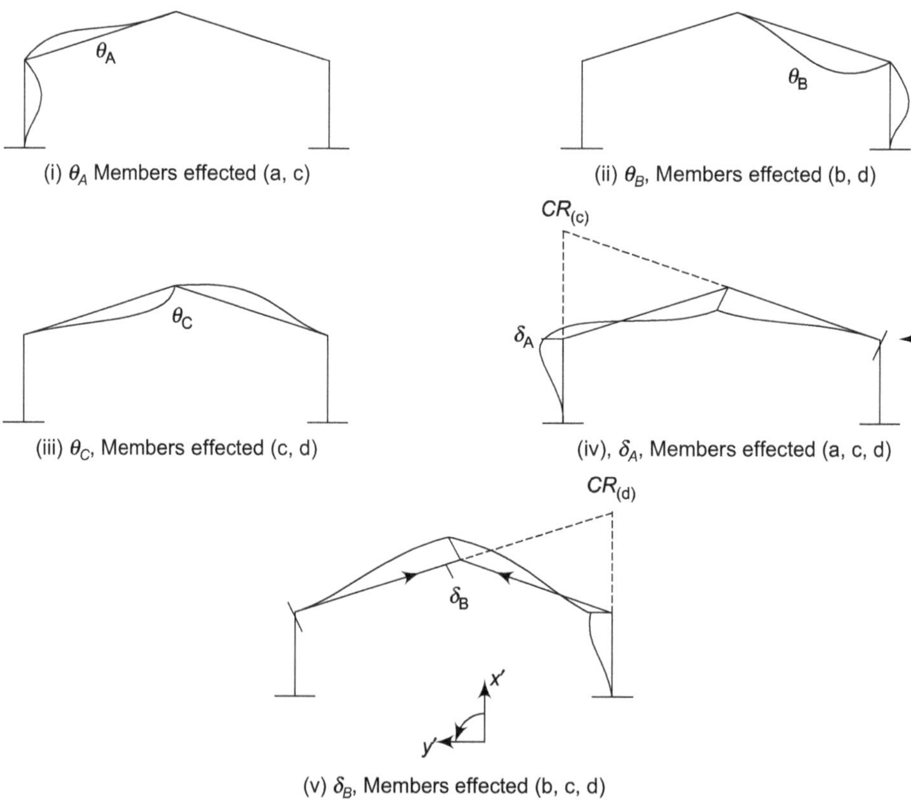

Fig. 7.13 *Illustration 3-(Virtual Displacement)*

The five unknown displacements, three rotations and two translations are sketched in the following. The sketching helps us to know the members affected by each displacement, an important prerequisite for writing the connection matrix. Consider two cases: (1) Uniform I/L, and (2) Uniform I

I. Compatibility Equation

$$\underbrace{\begin{bmatrix} \begin{pmatrix} \theta_1 \\ \delta_1 \end{pmatrix}_a \\ \begin{pmatrix} \theta_2 \\ \delta_2 \end{pmatrix}_b \\ \begin{pmatrix} \theta_1 \\ \theta_2 \\ \delta_Z \end{pmatrix}_c \\ \begin{pmatrix} \theta_1 \\ \theta_2 \\ \delta_Z \end{pmatrix}_d \end{bmatrix}}_{d_m} = \underbrace{\begin{bmatrix} 1 & 0 & 0 & 0 & 0 \\ 0 & 0 & 0 & 1 & 0 \\ 0 & 1 & 0 & 0 & 0 \\ 1 & 0 & 0 & 0 & 1 \\ 1 & 0 & 0 & 0 & 0 \\ 0 & 0 & 1 & 0 & 0 \\ 0 & 0 & 0 & -\dfrac{1}{2C} & +\dfrac{1}{2C} \\ 0 & 1 & 0 & 0 & 0 \\ 0 & 0 & 1 & 0 & 0 \\ 0 & 0 & 0 & \dfrac{1}{2C} & \dfrac{1}{2C} \end{bmatrix}}_{A} \underbrace{\begin{bmatrix} \theta_A \\ \theta_B \\ \theta_C \\ \delta_A \\ \delta_B \end{bmatrix}}_{d_j}$$

II. Force Deformation Relationship: $d_m = k_m\, d_m$

$$\begin{bmatrix} \begin{pmatrix} m_2 \\ P_{y2} \end{pmatrix}_a \\ \begin{pmatrix} m \\ P_{y2} \end{pmatrix}_b \\ \begin{pmatrix} m_1 \\ m_2 \\ P_{y2} \end{pmatrix}_c \\ \begin{pmatrix} m_1 \\ m_2 \\ P_{y2} \end{pmatrix}_d \end{bmatrix} = \begin{bmatrix} \dfrac{4EI}{L}, \dfrac{-6EI}{L^2} & & & \\ \dfrac{-6EI}{L^2} + \dfrac{12EI}{L^3} & & & \\ & \dfrac{4EI}{L}, \dfrac{-6EI}{L^2} & & \\ & \dfrac{-6EI}{L^2}, + \dfrac{12EI}{L^3} & & \\ & & \dfrac{4EI}{L}, \dfrac{2EI}{L}, \dfrac{-6EI}{L^2} & \\ & & \dfrac{2EI}{L}, \dfrac{4EI}{L}, \dfrac{-6EI}{L^2} & \\ & & \dfrac{-6EI}{L^2}, \dfrac{-6EI}{L^2}, \dfrac{+12EI}{L^3} & \\ & & & \dfrac{4EI}{L}, \dfrac{2EI}{L}, \dfrac{-6EI}{L^2} \\ & & & \dfrac{2EI}{L}, \dfrac{4EI}{L}, \dfrac{-6EI}{L^2} \\ & & & \dfrac{-6EI}{L^2}, \dfrac{-6EI}{L^2}, \dfrac{+12EI}{L^3} \end{bmatrix} \begin{bmatrix} \begin{pmatrix} \theta_2 \\ \delta_2 \end{pmatrix}_a \\ \begin{pmatrix} \theta_2 \\ \delta_2 \end{pmatrix}_b \\ \begin{pmatrix} \theta_1 \\ \theta_2 \\ \delta_2 \end{pmatrix}_c \\ \begin{pmatrix} \theta_1 \\ \theta_2 \\ \delta_2 \end{pmatrix}_d \end{bmatrix}$$

III. Equation of equilibrium are implied in the relation $K_j = A^t K_m A$ K_j is stiffness in system coordinate for the entire structure

Procedure Chart:

Note: has $I = I_a$, that is, I corresponds to the member subscripted with L

$$
k_m A = \begin{bmatrix}
\begin{matrix} \frac{4EI}{L_a}, \frac{-6EI}{L_a^2} \\ \frac{-6EI}{L_a^2} + \frac{12EI}{L_a^3} \end{matrix} & & & \\
& \begin{matrix} \frac{4EI}{L_b}, \frac{-6EI}{L_b^2} \\ \frac{-6EI}{L_b^2} + \frac{12EI}{L_b^3} \end{matrix} & & \\
& & \begin{matrix} \frac{4EI}{L_c}, \frac{2EI}{L_c}, \frac{-6EI}{L_c^2} \\ \frac{2EI}{L_c}, \frac{4EI}{L_c}, \frac{-6EI}{L_c^2} \\ \frac{-6EI}{L_c^2}, \frac{-6EI}{L_c^2}, \frac{+12EI}{L_c^3} \end{matrix} & \\
& & & \begin{matrix} \frac{4EI}{L_d}, \frac{2EI}{L_d}, \frac{-6EI}{L_d^2} \\ \frac{2EI}{L_d^2}, \frac{4EI}{L_d^2}, \frac{-6EI}{L_d^2} \\ \frac{-6EI}{L_d^2}, \frac{-6EI}{L_d^2}, \frac{+12EI}{L_d^3} \end{matrix}
\end{bmatrix}_{10 \times 10}
\begin{bmatrix}
1\,0\,0 & 0 & 0 \\
0\,0\,0 & 1 & 0 \\
0\,1\,0 & 0 & 0 \\
0\,0\,0 & 0 & 1 \\
1\,0\,0 & 0 & 0 \\
0\,0\,1 & 0 & 0 \\
0\,0\,0 & -\frac{1}{2C} & \frac{1}{2C} \\
0\,1\,0 & 0 & 0 \\
0\,0\,1 & 0 & 0 \\
0\,0\,0 & \frac{1}{2C} & -\frac{1}{2C}
\end{bmatrix}
(A)\ 10 \times 5
$$

The above stiffness matrix is not the same as the stiffness matrices appearing in the other examples. It is discussed to obtain equivalent or generalized joint loads from the member fixed end forces. In the case of the "connection matrix approach the load matrix corresponding to the above stiffness matrix is given by

$$P_f = -A^t P_m$$

The minus sign is due to the joint ends developing opposite forces to that of the member ends meeting the joint. This follows from the law of centragradience. For the example considered

$$P'_j = \left\{ (m_2)_a^* + (m_1)_c^*, (m_2)_b^* + (m_1)_d^*, (m_2)_c^* + (m_2)_d^*, (p_{y2})_a^* - \frac{1}{2c}(p_{y2})_c^* + \frac{1}{2c}(p_{y2})_d^*, \right.$$

$$\left. (p_{y2})^*{}_b + \frac{1}{2c}(p_{y2})^*{}_c - \frac{1}{2c}(p_{y2})^*{}_d \right\}$$

Where m^* and P^* are the fixed end forces (moment and shears) comparing with the load matrix, the variations are in the fourth and fifth row only.

$$A^T K_m A = \begin{bmatrix} 1 & 0 & 0 & 0 & 1 & 0 & 0 & 0 & 0 & 0 \\ 0 & 0 & 1 & 0 & 0 & 0 & 0 & 1 & 0 & 0 \\ 0 & 0 & 0 & 0 & 0 & 1 & 0 & 0 & 1 & 0 \\ 0 & 1 & 0 & 0 & 0 & 0 & -\frac{1}{2c} & 0 & 0 & \frac{1}{2c} \\ 0 & 0 & 0 & 1 & 0 & 0 & \frac{1}{2c} & 0 & 0 & -\frac{1}{2c} \end{bmatrix} \begin{bmatrix} \frac{4EI}{L_a} & 0 & 0 & \frac{6EI}{L^2_d} & 0 \\ -\frac{6EI}{L^2_a} & 0 & 0 & \frac{12EI}{L^3_a} & 0 \\ 0 & \frac{4EI}{L_b} & 0 & 0 & -\frac{6EI}{L^2_c} \\ 0 & -\frac{6EI}{L^2_b} & 0 & 0 & \frac{12EI}{L^3_c} \\ \frac{4EI}{L_c} & 0 & \frac{2EI}{L_c} & \frac{3EI}{L^2_c c} & -\frac{3EI}{L^2_c c} \\ \frac{2EI}{L_c} & 0 & \frac{2EI}{L_c} & \frac{3EI}{L^2_c c} & -\frac{3EI}{L^2_c c} \\ -\frac{6EI}{L^2_c} & 0 & -\frac{6EI}{L^2_c} & -\frac{6EI}{L^3_c c} & +\frac{6EI}{L^2_c c} \\ 0 & \frac{4EI}{L_d} & \frac{2EI}{L_d} & -\frac{3EI}{L^2_d c} & \frac{3EI}{L^2_d c} \\ 0 & \frac{2EI}{L_d} & \frac{4EI}{L_d} & -\frac{3EI}{L^2_d c} & \frac{3EI}{L^2_d c} \\ 0 & -\frac{6EI}{L^2_d} & -\frac{6EI}{L^2_d} & \frac{6EI}{L^3_d c} & -\frac{6EI}{L^2_d c} \end{bmatrix}$$

$(A^t)\ 5 \times 10$ $\qquad$ $(K_m \times A)\ 10 \times 5$

$$[K_j]_{5\times 5} = \begin{bmatrix} \frac{4EI}{L_a} + \frac{4EI}{L_c} & 0 & \frac{2EI}{L_c} & \frac{-6EI}{L^2_a} - \frac{3EI}{L^3 c} & -\frac{3EI}{L^2_c c} \\ 0 & \frac{4EI}{L_b} + \frac{4EI}{L_d} & \frac{2EI}{L_d} & -\frac{3EI}{L^2_d c} & \frac{-6EI}{L^2_b} - \frac{3EI}{L^3_d c} \\ \frac{2EI}{L_c} & \frac{2EI}{L_d} & \frac{4EI}{L_c} + \frac{4EI}{L_d} & \frac{3EI}{L^2_c c} - \frac{3EI}{L^2_d c} & -\frac{3EI}{L^2_c c} - \frac{3EI}{L^2_d c} \\ \frac{6EI}{L^2_a} + \frac{3EI}{L^2_a c} & -\frac{3EI}{L^2_d c} & \frac{3EI}{L^2_c c} - \frac{3EI}{L^2_d c} & \frac{12EI}{L^3_a} + \frac{3EI}{L^3_c c^2} + \frac{3EI}{L^2_d c^2} & -\frac{3EI}{L^3_c c^2} - \frac{3EI}{L^3_d c^2} \\ -\frac{3EI}{L^2_c c} & \frac{-6EI}{L^2_d} - \frac{3EI}{L^2_d c} & -\frac{3EI}{L^3_c c} + \frac{3EI}{L^2_d c} & -\frac{3EI}{L^3_c c^2} - \frac{3EI}{L^2_d c} & \frac{12EI}{L^3_b} + \frac{3EI}{L^3_c c^2} + \frac{3EI}{L^3_d c^2} \end{bmatrix}$$

Normally, we do the analysis in two parts: The first part pertains to Fixed-end analysis and freeze the fixed end force to be added to the results of the second part. The first part is used to compute the nodal loads for performing the part II.

The problem given consists of the nodal loads only.

The fourth load element is computed using the [Fig. 7.13 (iv)] for the virtual displacements, and the given loading and the fourth load element, P'_4 as the real system.

$$P'_4 \times \delta^* = -1 \times \delta^* + 2 \times \delta^*/4 \times 10, \text{ giving } P'_4 = 4\,t$$

Using the Fig(e) for the virtual displacements, and the given loading and the fifth load element, P_5' as the real system.

$$P'_5 \times \delta^* = 2 \times \delta^*/4 \times 10, \text{ giving } P'_4 = 5\,t.$$

The final Bending Moment Diagrams for the Two Cases are given, Fig. 7.14 for the two cases

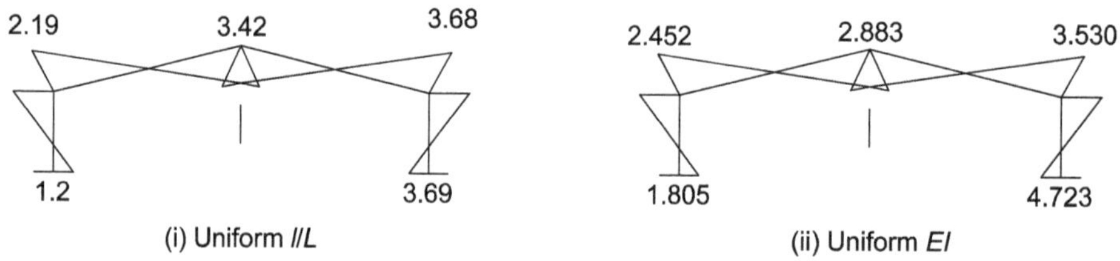

Fig. 7.14 *Bending Moment Diagrams for Cases (i) and (ii)*

7.4 MATRIX CONDENSATION

The approximate analysis neglecting the effect of axial deformations on moments and shears condenses the stiffness matrix significantly. In the above approach, matrix condensation is implied while writing the connection matrix. The condensation using free body diagrams from first principles was shown with the help of the Eq. 6.12 of portal frame and 6.12(a) gable frame. The same results can be obtained by matrix operations. As the final result excludes the "EA/L" terms, to simplify the procedure, we may drop these terms. For the approximateanalysis, let $\bar{d}$ be the displacement matrix which consists of independent displacement components. For the example of the gable.

$$d_j = \{(\delta'_x, \delta'_y, \theta)_A, (\delta'_x, \delta'_y, \theta)_B, (\delta'_x, \delta'_y, \theta)_C\};\ P_j = K_d\, d_j$$

$$\tilde{d} = \{\theta_A, \theta_B, \theta_C, (\delta'_y)_A, (\delta'_y)_B\};\ \tilde{d} = \tilde{K}\,\tilde{d}$$

And the relation between them is given by Z matrix in the equation below:

Where

$$Z = \begin{bmatrix} 0 & 0 & 0 & 0 & 0 \\ 0 & 0 & 0 & 1 & 0 \\ 1 & 0 & 0 & 0 & 0 \\ 0 & 0 & 0 & 0 & 0 \\ 0 & 0 & 0 & 0 & 1 \\ 0 & 1 & 0 & 0 & 0 \\ 0 & 0 & 0 & -S/2C & S/2C \\ 0 & 0 & 0 & 1/2 & 1/2 \end{bmatrix}$$

Similarly from the free body diagrams (Figs. 7.13(iv) & (v)) we can establish the relation.

$$\tilde{P} = Q P_J$$

Where is the load matrix for the approximate analysis.

For the gable considered

$$\begin{bmatrix} m_A \\ m_B \\ m_C \\ \tilde{P}_4 \\ \tilde{P}_5 \end{bmatrix} = \begin{bmatrix} 0 & 0 & 1 & 0 & 0 & 0 & 0 & 0 & 0 \\ 0 & 0 & 0 & 0 & 0 & 1 & 0 & 0 & 0 \\ 0 & 0 & 0 & 0 & 0 & 0 & 0 & 0 & 0 \\ 0 & 1 & 0 & 0 & 1 & 0 & 0 & 0 & 0 \\ 0 & 2cl & 1 & 0 & 0 & 0 & sli - cli & 1 & 1 \end{bmatrix} \begin{bmatrix} \begin{pmatrix} P'_x \\ P'_y \\ m \end{pmatrix}_A \\ \begin{pmatrix} P'_x \\ P'_y \\ m \end{pmatrix}_B \\ \begin{pmatrix} P'_x \\ P'_y \\ m \end{pmatrix}_C \end{bmatrix} \quad \ldots(7)$$

$$\tilde{P} = Q P_J$$

The stiffness matrix is given by $[\tilde{K} = Q\, K_j\, Z]$

The results for the two cases of (a) uniform I/L: The member (2) is critically stressed, and the Bending moment values are, end-1, 3.64 and end-2, 3.66t m,

(b) Constant moment of inertia, that is, all members of uniform I value; The member (2) is critically stressed, and the Bending moment values are, 4.71 and 3.53 tm.

7.5 GABLE FRAME BY 'SLOPE-DEFLECTION APPROACH'

7.5.1 Centers of Rotation for Deflected Shapes and Writing Equations of Equilibrium

I. **Compatibility conditions:**

$$\left.\begin{aligned} &(\theta_2)_a = (\theta_1)_c = \theta_A \\ &(\theta_2)_c = (\theta_2)_d = \theta_C;\ (\theta_2)_b = (\theta_1)_d = \theta_B \\ &(\delta_2)_a = (\delta'_Y)_A;\ (\delta_2)_c = -\frac{1}{2c}(\delta'_Y)_A + \frac{1}{2}(\delta'_Y)_B; \\ &(\delta_2)_b = (\delta'_Y)_B;\ (\delta_2)_d = +\frac{1}{2c}(\delta'_Y)_A - \frac{1}{2}(\delta'_Y)_B; \end{aligned}\right\} \quad \ldots(\text{I})$$

Where $c = \cos\theta,\ s = \sin\theta$

II. **Force deformation relations:**

$$
\left.
\begin{aligned}
(m_1)_a &= 2\left(\frac{EI}{L}\right)_a (\theta_2)_a - 6\left(\frac{EI}{L^2}\right)_a (\delta_2)_a + (f.e.m. = 0)\\
(m_2)_a &= 4\left(\frac{EI}{L}\right)_a (\theta_2)_a - 6\left(\frac{EI}{L^2}\right)_a (\delta_2)_a + (f.e.m. = 0)\\
(m_1)_c &= 4\left(\frac{EI}{L}\right)_c (\theta_1)_c + 2\left(\frac{EI}{L}\right)_c (\theta_2)_c - 6\left(\frac{EI}{L^2}\right)_c (\delta_2)_a + (f.e.m. = 0)\\
(m_2)_c &= 2\left(\frac{EI}{L}\right)_c (\theta_1)_c + 4\left(\frac{EI}{L}\right)_c (\theta_2)_c - 6\left(\frac{EI}{L^2}\right)_c (\delta_2)_c + (f.e.m. = 0)\\
(m_1)_d &= 4\left(\frac{EI}{L}\right)_d (\theta_1)_d + 2\left(\frac{EI}{L}\right)_d (\theta_2)_d - 6\left(\frac{EI}{L^2}\right)_d (\delta_2)_d + (f.e.m. = 0)\\
(m_2)_d &= 4\left(\frac{EI}{L}\right)_d (\theta_1)_d + 2\left(\frac{EI}{L}\right)_d (\theta_2)_d - 6\left(\frac{EI}{L^2}\right)_d (\delta_2)_d + (f.e.m. = 0)\\
(m_2)_b &= 4\left(\frac{EI}{L}\right)_d (\theta_2)_b - 6\left(\frac{EI}{L^2}\right)_b (\delta_2)_b + (f.e.m. = 0)
\end{aligned}
\right\} \quad \ldots\text{(II)}
$$

III. **Equilibrium equations:** Expressing the member deformations (member coordinates x, y, z) in terms of joint deformations (system coordinates x', y', z') we write the following moments equations of equilibrium.

$$(m_2)_a + (m_1)c = 0; \quad (m_1)_d + (m_2)_b = 0$$

$$(m_2)_c + (m_2)_d = 0$$

The force equations of equilibrium will be written for the free bodies Fig. d and e.

$$(P'_y)_A + (P'_y)_B + (P'_y)_C = [(P_y)_2]_a + [(P_y)_2]_b \text{ or}$$

$$-1 = [-(m_1 + m_2)/L]_a + [-(m_1 + m_2)/L]_b$$

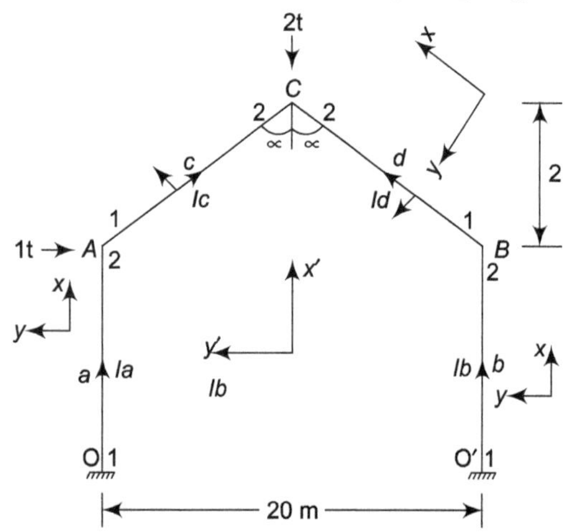

(a) Loadings and Coordinate System of a Gable Frame

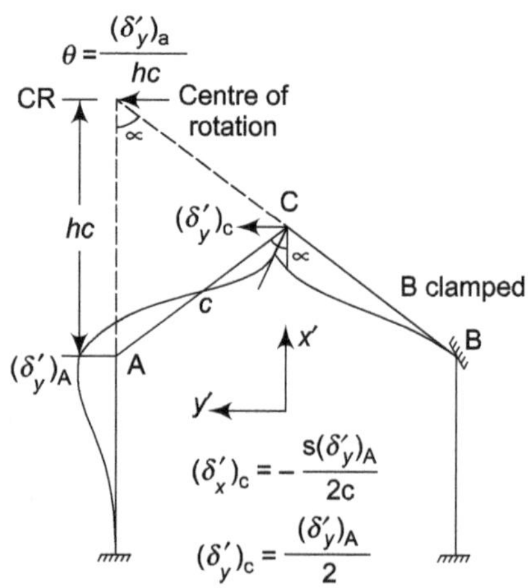

(b) Center of Rotationof Member c

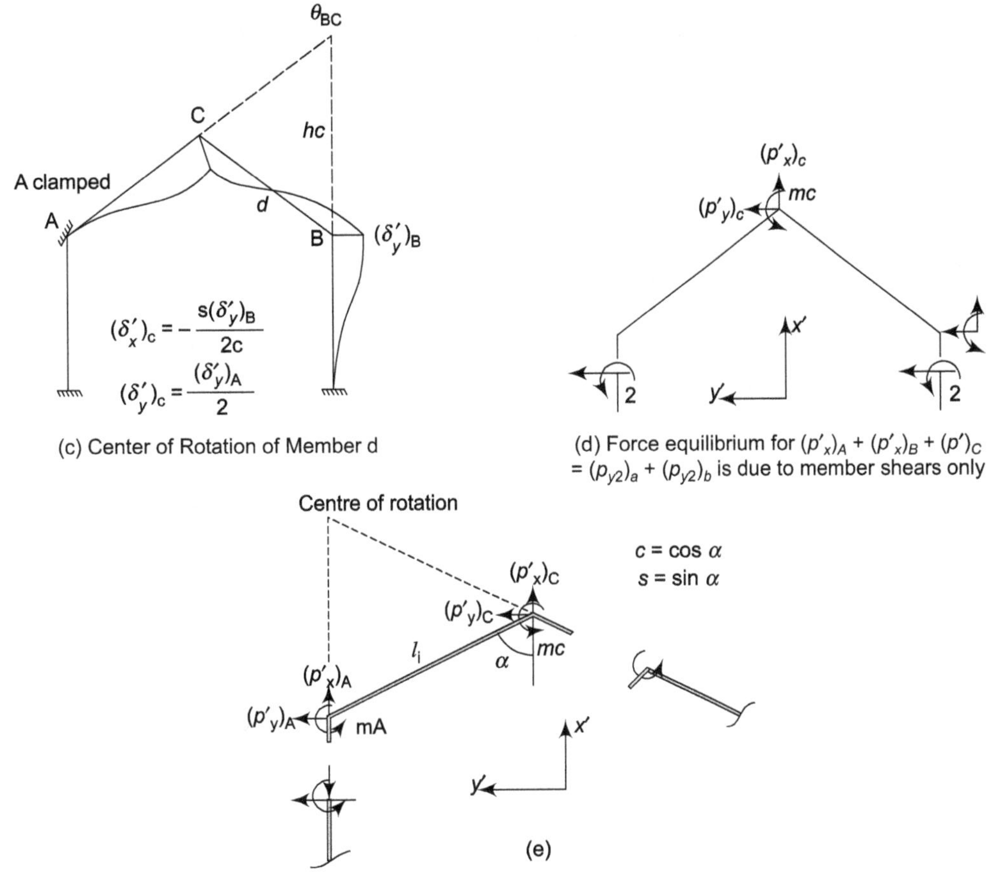

Fig. 7.15 *Gable Frame with Inextensible Members: Free Bodies Corresponding to the Two Independent Displacements*

The members are assumed to behave as rigid links because the axial deformations are neglected. Fig. (*e*). Equilibrium for the free body as shown.

$$m_A - (P'_y)_A\,(2cl_i) - (P'_y)_c\,cl_i + (P'_x)_c\,sl_i + m_c = -\,[(P'_y)_2]_a\,2cl_i + (m_2)_a - l_i\,[(P'_y)_2]_d + \text{ or}$$

$$-\,24 = -\,2cl_i\left\{-\left(\frac{m_1+m_2}{L}\right)_a\right\}-\left\{-\left(\frac{m_1+m_2}{L}\right)_d\right\}+(m_2)_a+(m_2)_d$$

Substituting the values Eq. (II) in the equations of equilibrium III

$$\left\{\left(\frac{4EI}{L}\right)_a+\left(\frac{4EI}{L}\right)_c\right\}\theta_A+0\theta_B+\left(\frac{2EI}{L}\right)_c\theta_C+\left[\left(\frac{-6EI}{L^2}\right)_a+\frac{1}{2c}\left(\frac{6EI}{L^2}\right)_c\right](\delta'_y)_A-\frac{1}{2c}\left(\frac{6EI}{L^2}\right)_c(\delta'_y)_B=0$$

$$0\theta_A+\left\{\left(\frac{4EI}{L}\right)_b+\left(\frac{4EI}{L}\right)_d\right\}\theta_B+\left(\frac{2EI}{L}\right)_d\theta_C-\frac{1}{2c}\left(\frac{6EI}{L^2}\right)_d(\delta'_y)_A+\left[\left(\frac{-6EI}{L^2}\right)_b+\frac{1}{2c}\left(\frac{6EI}{L^2}\right)_d\right](\delta'_y)_B=0$$

$$\left(\frac{2EI}{L}\right)_c \theta_A + \left(\frac{2EI}{L}\right)_c \theta_B + \left\{\left(\frac{4EI}{L}\right)_c + \left(\frac{4EI}{L}\right)_d\right\}\theta_c + \left[\frac{1}{2c}\left(\frac{-6EI}{L^2}\right)_c - \frac{1}{2c}\left(\frac{6EI}{L^2}\right)_d\right](\delta'_y)_A$$

$$+\left[\frac{1}{2c}\left(\frac{6EI}{L^2}\right)_c + \frac{1}{2c}\left(\frac{6EI}{L^2}\right)_d\right](\delta'_y)_B = 0$$

$$-\left(\frac{6EI}{L^2}\right)_a \theta_A - \left(\frac{6EI}{L^2}\right)_B \theta_B + 0\theta_C + \left(\frac{12EI}{L^3}\right)_a (\delta'_y)_A + \left(\frac{12EI}{L^3}\right)_b (\delta'_y)_B = -1$$

$$\left\{2cli\left(\frac{6EI}{L^2}\right)_a + \left(\frac{4EI}{L}\right)_a \theta_A\right\} + \left(\frac{8EI}{L^2}\right)_d \theta_B + \left(\frac{10EI}{L}\right)_d \theta_C + \left[-(1+4c)\left(\frac{6EI}{L^2}\right)_a - \frac{3}{2c}\left(\frac{6EI}{L^2}\right)_d\right](\delta'_y)_A$$

$$+\left[\frac{3}{2c}\left(\frac{6EI}{L^2}\right)_d\right](\delta'_y)_B = -16$$

The solution of the above simultaneous equation gives values of θ_A, θ_B, θ_C, $(\delta'_y)_A$, and which in turn based upon compatibility conditions, Step I, gives member deformation. Making use of Eq. (II) the member "end moments and end shear" can be evaluated.

7.5.2 Illustration: Examples of Portal Frames with Inextensible Members

The following are some typical Frames and the translational Degrees of Freedom:
No. of translational Degrees of Freedom = 1
$\Delta_A = \Delta_B$ With and without extensibility evaluated, Ex. 6.1.
Art. 6.7. Compared with and without extensibility compared.

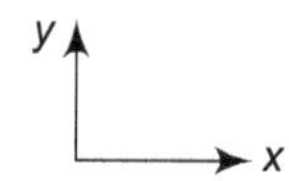

No. of translational Degrees of Freedom = 1, Illustration-4, Art. 6.4.5.2
$(\Delta_x)_A = (\Delta_x)_B$

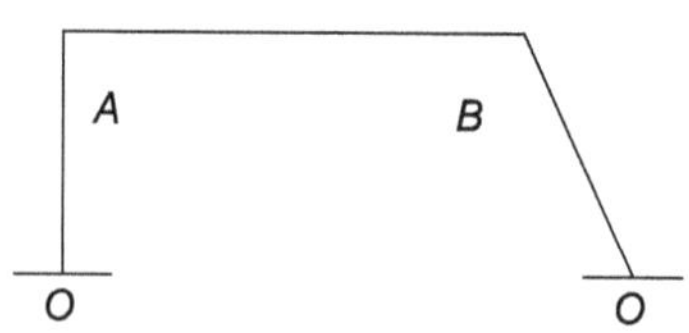

No. of translational Degrees of Freedom = 1
$(\Delta_x)_A$ is not equal to $(\Delta_x)_B$

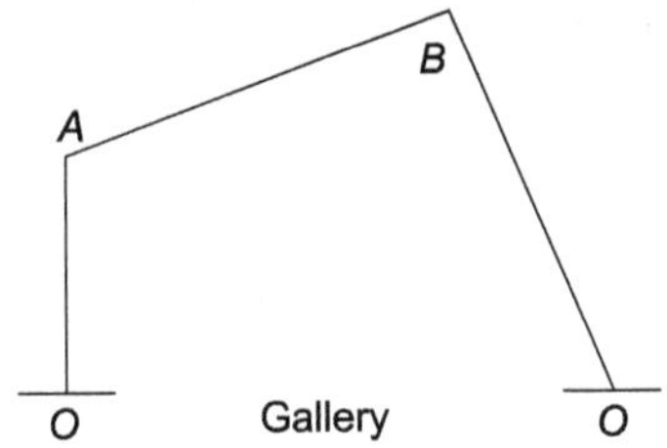

No. of translational Degrees of Freedom = 1
$(\Delta_x)_A = (\Delta_x)_B$ Illustration-2 Art. 6.4.5.2

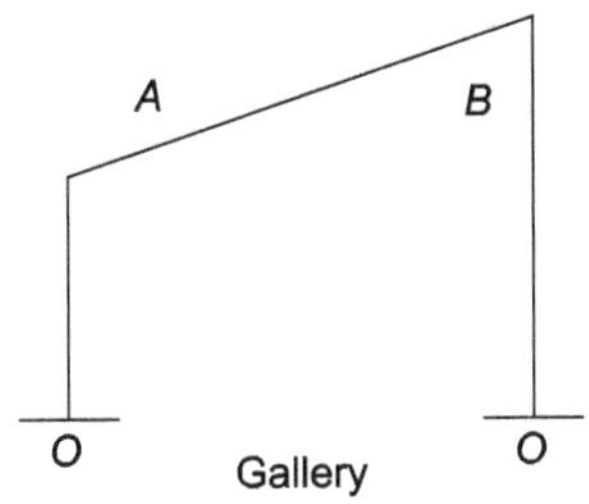

No. of translational Degrees of Freedom = 2
$(\Delta_x)_A$ is not equal to $(\Delta_x)_B$, With and without inextensibility compared
Illustration-3 Art. 7.3 and Art. 7.7

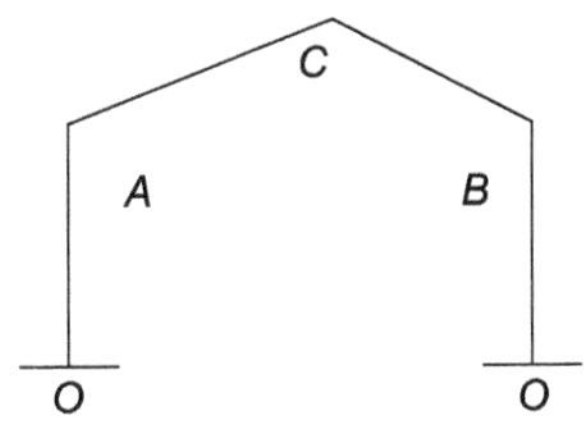

Translational Degrees of Freedom = 2 (No of free joints) – No. of members
$(\Delta_x)_A$, $(\Delta_x)_B$, $(\Delta_x)_C$, $(\Delta x)_D$ Only any three translations are independent.

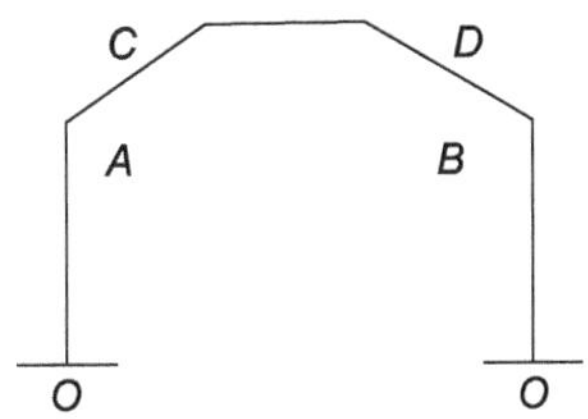

7.6 DIRECT STIFFNESS METHOD

Article 6.71 Analyis of Gable frame by DSM was discussed.
In the following the gable frame analyzed by the direct stiffness method (DSM) for the case without neglecting axial deformations is reproduced in the following.

The geometry and other data are reproduced in the following.

$$\begin{bmatrix} P'_A \\ P'_B \\ P'_C \end{bmatrix} = \begin{bmatrix} (k'_{22})_a + (k'_{11})_c & 0 & (k'_{11})_c \\ 0 & (k'_{22})_b + (k'_{11})_d & (k'_{12})_d \\ (k'_{21})_c & (k'_{21})_d & (k'_{22})_c + (k'_{11}) \end{bmatrix} \quad (1)$$

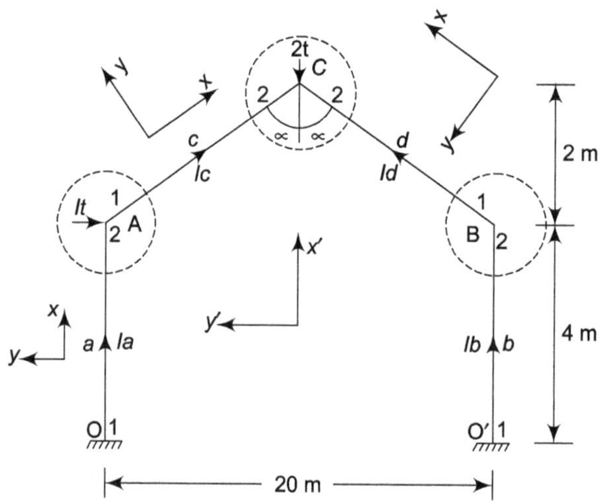

Data Gable Frame

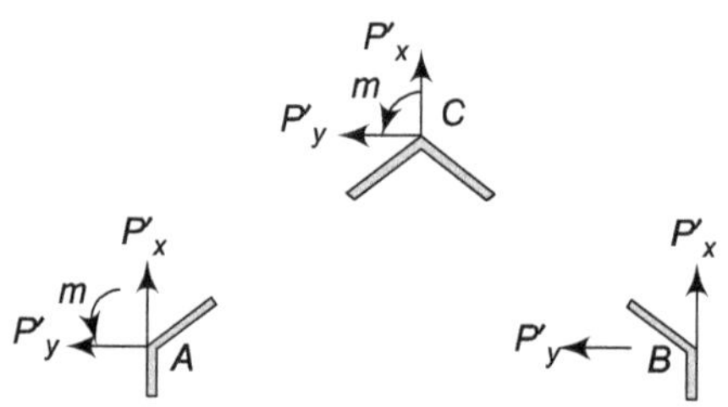

Free bodies for moment equilibrium
m_A, m_B, m_C corresponding to θ_A, θ_B, θ_C

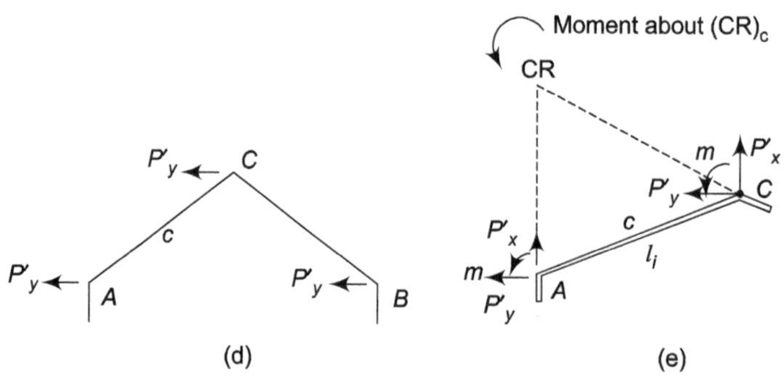

Free Bodies Corresponding to δ_4 and δ_5

Fig. 7.16

The Eq. 6.12(a) developed in the Art. 6.7.1 when written in the expanded form

$$\begin{bmatrix} \begin{pmatrix} (p'_x) \\ (p'_y) \\ m \end{pmatrix}_A \\ \begin{pmatrix} p'_x \\ p'_y \\ m \end{pmatrix}_B \\ \begin{pmatrix} p_x \\ p_y \\ m \end{pmatrix}_C \end{bmatrix} = \begin{bmatrix} k_{11} & k_{12} & k_{13} & - & - & - & k_{17} & k_{18} & k_{19} \\ k_{21} & k_{22} & k_{23} & - & - & - & k_{27} & k_{28} & k_{21} \\ k_{31} & k_{32} & k_{33} & k_{34} & k_{35} & k_{36} & k_{37} & k_{38} & k_{39} \\ k_{41} & - & - & - & - & - & - & - & - \\ k_{51} & - & - & - & - & - & - & - & - \\ k_{61} & - & - & - & - & - & - & - & - \\ k_{71} & k_{72} & k_{73} & k_{74} & k_{75} & k_{76} & k_{77} & k_{78} & k_{79} \\ k_{81} & k_{82} & k_{83} & k_{84} & k_{85} & k_{86} & k_{87} & k_{88} & k_{89} \\ k_{91} & k_{92} & k_{93} & k_{94} & k_{95} & k_{96} & k_{97} & k_{98} & k_{99} \end{bmatrix} \begin{bmatrix} \begin{pmatrix} (\delta'_x) \\ (\delta'_y) \\ \theta \end{pmatrix}_A \\ \begin{pmatrix} \delta'_x \\ \delta'_y \\ \theta \end{pmatrix}_B \\ \begin{pmatrix} \delta_x \\ \delta_y \\ \theta \end{pmatrix}_C \end{bmatrix} \quad \text{...(6.12(a))}$$

7.7 CONDENSATION OF GLOBAL STIFFNESS OF GABLE FRAME

Direct Stiffness Approach was considered while analyzing the Ex. 6.1. Assuming inextensibility, the size of the global stiffness matrix 6×6 was condensed to the size 3×3. Similarly, the Illustration-3, Art. 7.3 analyzed a gable frame used connection matrix approach assuming inextensibility for all the members. The important aspect to be noted is the use of 'contragradience' for arriving at the load elements corresponding to the translations. The same gable frame when analyzed by slope deflection method, we used suitable free bodies for writing the equations of equilibrium. In the following, the condensation of global matrix (size 9x9) based on Direct Stiffness approach is being condensed by assuming inextensibility. The load elements corresponding to the two unknown translations are obtained by considering suitable free bodies for writing corresponding equilibrium equations. If all the members of the frame are assumed inextensible, the condensed gJobal stiffness matrix is of size 5×5. The analysis is similar to Ex. 6.1.

The equation developed in Chapter 6, Eq. 6.12(a) is the equilibrium equation from DSM, which does not assume inextensibility. But our degrees of freedom based on the inextensibility are:

$$[\theta_A \quad \theta_B \quad \theta_C \quad (\delta'_y)_A \quad (\delta'_y)_B]^T$$

We use the joints A, B, C as the free bodies for the moment equilibrium, and use the free bodies :

Fig. 7.15(d) for the force equilibrium with the gable as the free body corresponding to the unknown translation, $(\delta'_y)_A$. And

Fig. 7.15(e) for the moment equilibrium of the forces acting on the free body, moments taken about the Center of Rotation, $(CR)_C$ as shown for the unknown translation, $\delta_5 = (\delta'_y)_B$.

All the free bodies are reproduced, Fig. 7.16.

From the Eq. 6.12(a) we pick the three rows corresponding to the moment loads m_A, m_B, m_C.

The load elements corresponding to the unknown translations $(\delta'_y)_A$ and $(\delta'_y)_B$ are arrived from the the free bodies from the free bodies, Figs 7.16(d) and (e).

Fig. 7.16(d) is the free body of the 'gable roof and the load element is given by: $(P'_x)_A + (P'_x)_B + (P'_x)_C$, and substituting the values of loads at A, B, C of the gable roof, $(P'_4) = -1 + 0 + 0 = -1$.

Similarly the load element from the free body, Fig. 7.15(e), taking the moments of all the end forces about the Center of Rotation of element (c), we get

$$(P_5') = m_A - 2\,c\,I_i\,(P_y')_A - c\,I_i\,(P_y')_C + s\,I_i\,(P_x')_C + m_c$$

Where l_i is length of member 'c' and $s = \sin\alpha$ and $c = \cos\alpha$.

The load matrix for the condensed equation,

$$P_s = [m_A, m_B, m_C, P_4' = \sum_A^C (p_y'), (P_5') = m_A - 2cI_i(P_y')_A - cI_i(P_y')_C + sI_i(P_x')_C + m_C]^T$$

According to the load elements :

$$P'_4 = \sum_A^C (p_y'),\ (P_5') = m_A - 2\,c f|\,(P_y')_A - cl_i\,(P_y')_c + sl_i\,(P_x')_c + m_c]^T$$

Similar transformations* are to be performed on the rows and columns of the coefficient matrix (9x9) guided by the load elements.

The condensed coefficient matrix K_s is given in the following equation, Eq. 6.12(b) and (c). While using the free body approach for the equilibrium, the resulting Ks may not be symmetrical. The **principle of contragradience** used for the Illustration-3, The coefficient matrix Ks is always symmetrical.

7.8 NON-LINEARITY DUE TO LARGE DISPLACEMENTS

"The displacement method has been used in the analysis of rigidly jointed frames where changes of geometry effects are taken into account. The effects of change of geometry on the analysis have been grouped into three categories:

1. Change of member lateral stiffness due to large v, Fig. 7.17
2. Finite deflection of joints, Fig. 7.17(b), and
3. Change in member length due to bowing, Fig. 7.17(c).

The non linearity due to the second category only is to give rise to an additional moment as shown in Fig. 13(b) (i). The use of elastic stability functions tends to restrict its use to problems in which the material characteristics remain elastic. It also lends to complicate the iteration technique because the Equilibrium for $[M_A - (P'_y)_A\, z\, cl_i + (P'_y)_C\, cl_i + (P'_x)_C\, sl_i + m_c]$ is due to member shear and member moments only and not due to member axial forces

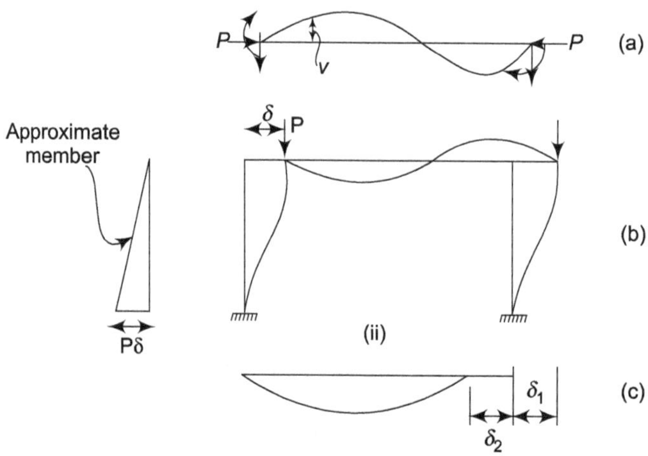

Fig. 7.17 Three cases of Non-linearity

* The reader may like to read the Appendix A pertaining to the linear transformations of a matrix equation.

$$\begin{bmatrix} m'_A = 0 \\ m'_B = 0 \\ m'_C = 0 \\ (P'_y)_A + (P'_y)_B + (P'_y)_C = -1 \\ \left\{ \begin{matrix} m_A - 2cl_i (P'_y)_A - cl_i (P'_y)_C \\ + sl_i \, (P'_x)_C + m_c = -16 \end{matrix} \right\} \end{bmatrix} = P_s \quad \text{and} \quad d_s = \begin{bmatrix} \theta_A \\ \theta_B \\ \theta_C \\ \delta_4 \\ \delta_5 \end{bmatrix}; \begin{matrix} P_s = K_s \, d_s \\ \delta_4 = (\delta'_y)_A \, ; \delta_5 = (\delta'_y)_B \end{matrix}$$

$$P_s = \begin{bmatrix} K_{33} & K_{36} & K_{39} & K_{32} - \frac{3}{2c} K_{37} + \frac{1}{2} K_{38} & K_{33} - \frac{3}{2c} K_{37} + \frac{1}{2} K_{38} \\ K_{63} & K_{66} & K_{69} & K_{62} - \frac{3}{2c} K_{67} + \frac{1}{2} K_{68} & K_{65} - \frac{3}{2c} K_{67} + \frac{1}{2} K_{68} \\ K_{93} & K_{96} & K_{99} & K_{92} - \frac{3}{2c} K_{97} + \frac{1}{2} K_{98} & K_{95} - \frac{3}{2c} K_{97} + \frac{1}{2} K_{98} \\ K_{23} + K_{53} + K_{83} & K_{26} + K_{56} + K_{86} & K_{29} + K_{59} + K_{89} & \begin{matrix} K_{22} - \frac{3}{2c} K_{27} + \frac{1}{2} K_{28} \\ K_{52} - \frac{3}{2c} K_{57} + \frac{1}{2} K_{58} \\ K_{82} - \frac{3}{2c} K_{87} + \frac{1}{2} K_{88} \end{matrix} & \begin{matrix} K_{25} - \frac{3}{2c} K_{27} + \frac{1}{2} K_{28} \\ K_{55} - \frac{3}{2c} K_{57} + \frac{1}{2} K_{58} \\ K_{85} - \frac{3}{2c} K_{87} + \frac{1}{2} K_{88} \end{matrix} \\ \begin{matrix} K_{33} \\ -2cl_i \, K_{23} \\ -cl_i \, K_{83} \\ +sl_i \, K_{73} + K_{93} \end{matrix} & \begin{matrix} K_{36} \\ 2cl_i \, K_{26} \\ cl_i \, K_{86} \\ sl_i \, K_{76} + K_{96} \end{matrix} & \begin{matrix} K_{39} \\ 2cl_i \, K_{29} \\ cl_i \, K_{89} \\ sl_i \, K_{79} + K_{99} \end{matrix} & \begin{matrix} K_{32} - \frac{3}{2c} K_{37} + \frac{1}{2} K_{38} \\ -2cl_i \left(K_{22} - \frac{3}{2c} K_{27} + \frac{1}{2} K_{38} \right) \\ -2cl_i \left(K_{82} - \frac{3}{2c} K_{87} + \frac{1}{2} K_{88} \right) \\ -sl_i \left(K_{72} - \frac{3}{2c} K_{77} + \frac{1}{2} K_{78} \right) \\ +K_{92} - \frac{3}{2c} K_{97} + \frac{1}{2} K_{88} \end{matrix} & \begin{matrix} K_{35} - \frac{3}{2c} K_{37} + \frac{1}{2} K_{38} \\ -2cl_i \left(K_{25} - \frac{3}{2c} K_{27} + \frac{1}{2} K_{38} \right) \\ -2cl_i \left(K_{85} - \frac{3}{2c} K_{87} + \frac{1}{2} K_{88} \right) \\ -sl_i \left(K_{75} - \frac{3}{2c} K_{77} + \frac{1}{2} K_{78} \right) \\ +K_{95} - \frac{3}{2c} K_{97} + \frac{1}{2} K_{98} \end{matrix} \end{bmatrix} \begin{bmatrix} \theta_A \\ \theta_B \\ \theta_C \\ (\delta'_y)_A \\ (\delta'_y)_B \end{bmatrix} \quad \text{...Eq. 7.12(b)}$$

Neglecting EA/L terms and substituting the values of in the above equation, we get

$$P_s = \begin{bmatrix} \left(\frac{4EI}{L}\right)_a + \left(\frac{4EI}{L}\right)_c & 0 & \left(\frac{2EI}{L}\right)_c & -\left(\frac{6EI}{L^2}\right)_c + \frac{1}{2c}\left(\frac{6EI}{L^2}\right)_c & -\frac{1}{2c}\left(\frac{6EI}{L^2}\right)_c \\ 0 & \left(\frac{4EI}{L}\right)_b + \left(\frac{4EI}{L}\right)_d & \left(\frac{2EI}{L}\right)_d & -\frac{1}{2c}\left(\frac{6EI}{L^2}\right)_d & -\left(\frac{6EI}{L^2}\right)_b + \frac{1}{2c}\left(\frac{6EI}{L^2}\right)_d \\ +\left(\frac{2EI}{L}\right)_c & +\left(\frac{2EI}{L}\right)_d & \left(\frac{4EI}{L}\right)_c + \left(\frac{4EI}{L}\right)_d & \frac{1}{2c}\left(\frac{6EI}{L^2}\right)_c - \frac{1}{2c}\left(\frac{6EI}{L^2}\right)_d & -\frac{1}{2c}\left(\frac{6EI}{L^2}\right)_c + \frac{1}{2c}\left(\frac{6EI}{L^2}\right)_d \\ -\left(\frac{6EI}{L^2}\right)_a & -\left(\frac{6EI}{L}\right)_b & 0 & \left(\frac{12EI}{L^2}\right)_a & \left(\frac{12EI}{L^3}\right)_b \\ \left(\frac{4EI}{L}\right)_a + 2cl_i\left(\frac{6EI}{L^2}\right)_a & 2\left(\frac{4EI}{L}\right)_d & 10\left(\frac{EI}{L}\right)_d & -(1+4c)\left(\frac{4EI}{L^2}\right)_a - \frac{3}{2c}\left(\frac{6EI}{L^2}\right)_d & \frac{3}{2c}\left(\frac{6EI}{L^2}\right)_d \end{bmatrix} \begin{bmatrix} \theta_A \\ \theta_B \\ \theta_C \\ (\delta'_y)_A \\ (\delta'_y)_B \end{bmatrix}$$

7.9 REITERATION OF INEXTENSIBILITY AND FREE BODIES

The test case of 'Gable Frame' is continued.

Fig. 7.18(b) and (c). The member assumed to behave as rigid link because axial deformations are neglected.

Loading and Coordinate System

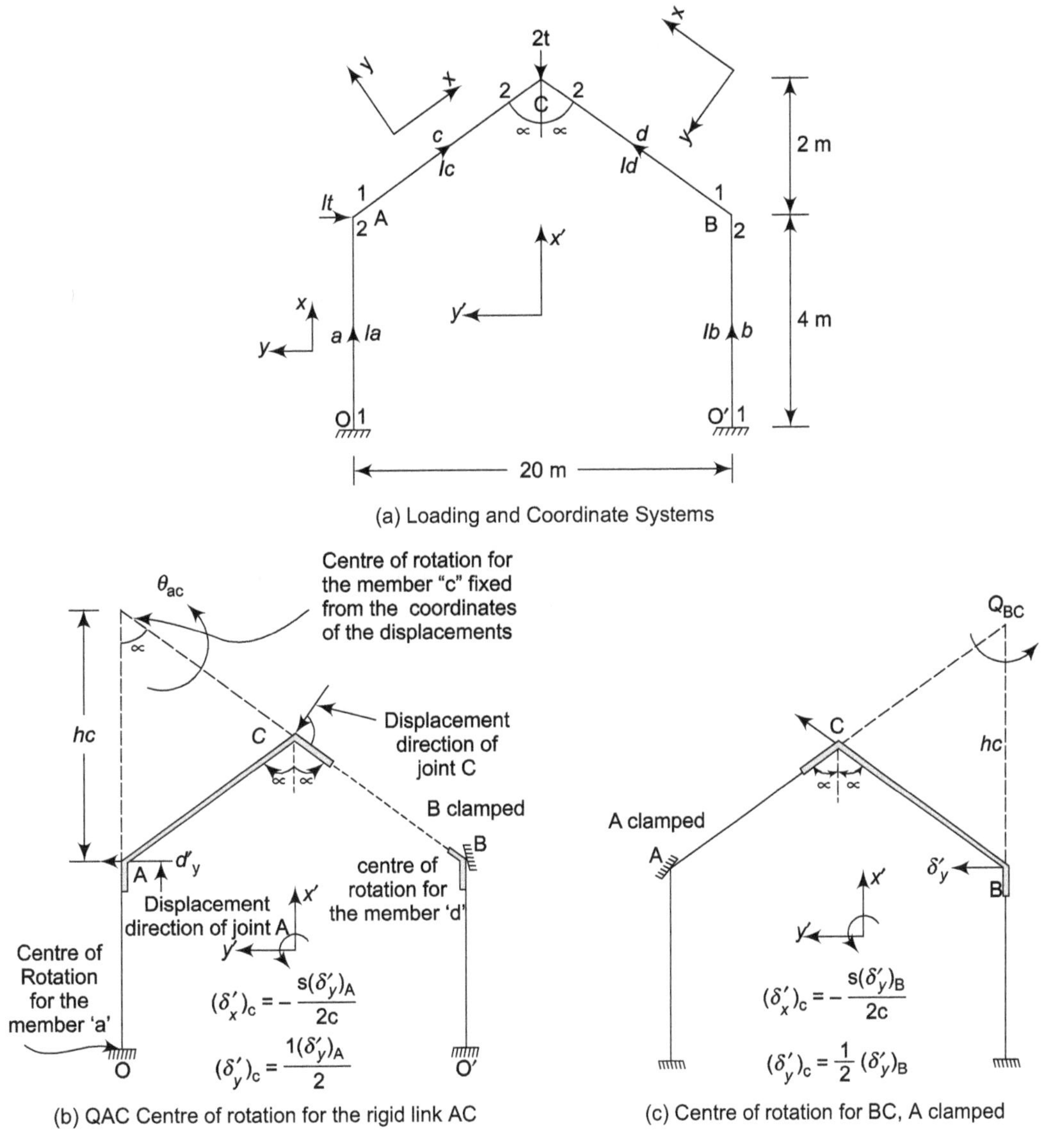

Fig. 7.18

The assumption of inextensibility gives rise to two independent translational displacements. We may choose any two of the above three free bodies. However, it is preferable to use free bodies (b) and (c). The connection matrix approach using the principle of contragrandience may be preferred over approach using, the equations written based on free bodies (b) and (c).

7.10 GENERALIZATION

In the case of the frame, Fig. 7.19, we have six nodes and seven members

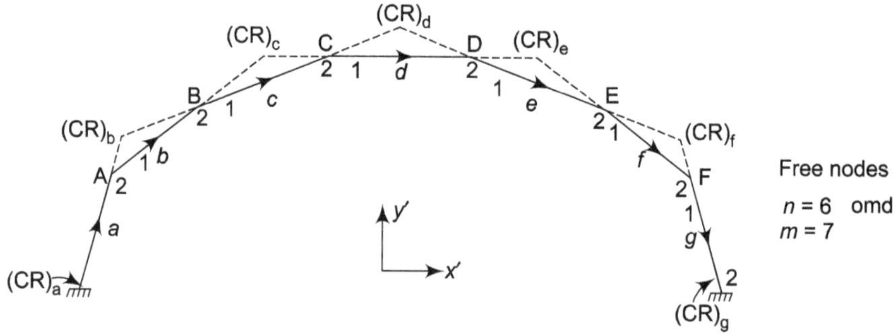

Fig. 7.19

The number of independent deformations for the approximate analysis is equal to $(3n - m)$ =

$$3 \times 6 - 7 = 11$$

i.e., θ_A, θ_B, θ_C, θ_D, θ_E, θ_F and the twelve displacement components of the nodes (two per node *i.e.*, (δ_x') are connected by seven axial inextensibility conditions equal to seven members. Thus we have six independent rotations and five (*i.e.*, 12 – 7 = 5) independent translations, and the total number of deformation components are eleven (*i.e.*, 6 + 5 = 11). The equations of equilibrium are written for the six joint free bodies and the five free bodies corresponding to the five instantaneous centers of rotation, $(CR)_b$, $(CR)_c$ ... $(CR)_r$. Each free body pertaining to its centre of rotation will be considered in the same way as we used for the equilibrium equation of the gable frame. If may also be noted, in the case of the gable frame, that we could have taken the free body corresponding to the centre of rotate, as many free bodies as the centers of rotation.

Due to large number of unknowns the displacement method for the frame given in Fig. 7.19 and also the gable frame is not recommended. The Force Method is definitely more convenient and it should be preferred for such frames.

7.11 MISCELLANEOUS STRUCTURES

7.11.1 Typical Illustrations of Introducing Symmetry

An unsymmetrical structure is made symmetrical to enable applying superposition of Symmetry and antisymmetry. This is achieved by a hypothetical cantilever extension beyond C.

(*i*) Symmetrical case:

$$\theta_B = -\theta_C,\ \delta_{BC} = 0$$

$$(m_1)a = +.833 \text{ tm};\ (m_1)_b = -.833 \text{ tm}$$

$$(m_2)a = +1.667 \text{ tm};\ (m_2)_b = -1.667 \text{ tm}$$

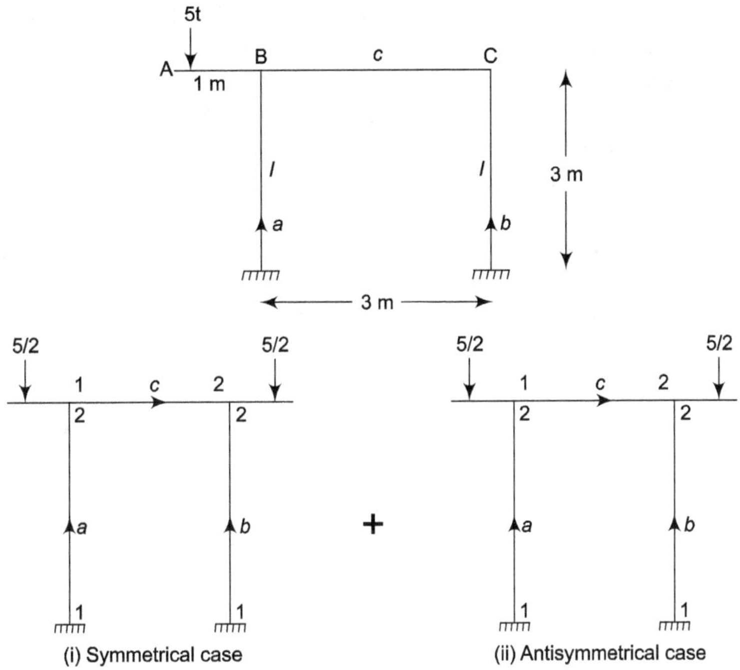

Fig. 7.20 *Device for Validating Symmetry and Antisymmetry*

(*ii*) Antisymmetrical case:

$$\theta_B = \theta_C, \delta_{BC} \text{ to be evaluated}$$

$$(m_1)a = -.356 \text{ tm}; (m_1)_b = -.356 \text{ tm}$$

$$(m_2)a = +.358 \text{ tm}; (m_2)_b = +.358 \text{ tm}$$

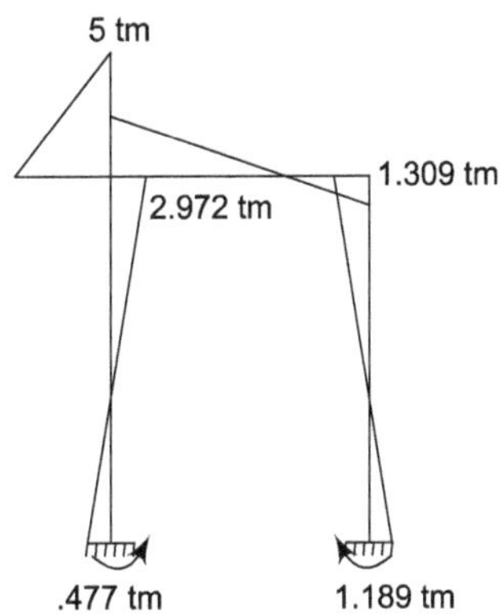

Fig. 7.21 *Final Bending Moment Diagram*

7.11.2 Multi Bay Frame

The connection matrices establish compatibility and it is not required to do transformations from member axes to system axes seperately. It is particularly useful when the approximation like inextensibility is used.

In fact, even for large structural systems like Tall buildings, the connection matrices are used with advantage for arriving at the 'storey stiffness matrix'. In such a case, the storey is being treated as a member. While arriving at the stiffness matrix; the approximation of inextensibility of the girders is assumed.

The essential steps or parts of analysis remain the same as discussed in the preceding. The variations will be obvious while discussing the details and each part of analysis. When the effects of axial deformations are to be neglected this approach does not call for matrix condensation separately.

Compatibility conditions: Compatibility conditions which establish the connection between member-end and joint deformations, demand the deformation of nodes, d_i, are written directly in system coordinates and for the entire structure. For example consider the two bay frame of the Fig. 7.22. Treat as a revision exercise. A more general case with inclined leg was illustrated in Ch 6-Illustration.

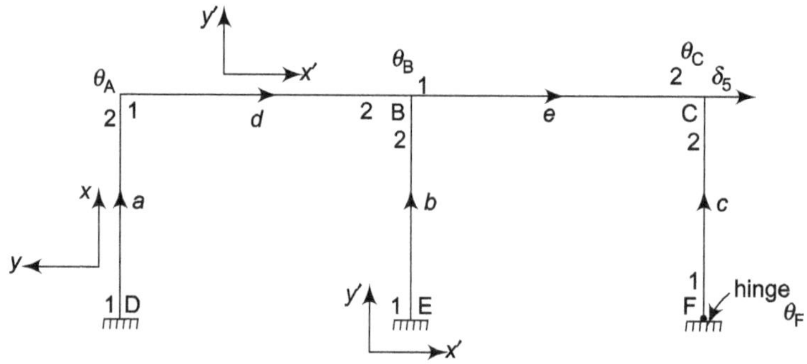

Fig. 7.22 *Two Bay Frame*

Step 1: Compatibility

The independent joint deformations are:

$$d_j = \{\theta_A, \theta_B, \theta_C, \theta_F, \delta_y]^T$$

Directly by inspection the following connection matrix may be written. Modified stiffness for member 'C' is not considered

$$\underset{d_m}{\begin{bmatrix} \begin{pmatrix} \theta_2 \\ \delta_2 \end{pmatrix}_a \\ \begin{pmatrix} \theta_2 \\ \delta_2 \end{pmatrix}_b \\ \begin{pmatrix} \theta_1 \\ \theta_2 \\ \theta_3 \end{pmatrix}_c \\ \begin{pmatrix} \theta_2 \\ \delta_2 \end{pmatrix}_d \\ \begin{pmatrix} \theta_1 \\ \delta_2 \end{pmatrix}_e \end{bmatrix}} = \underset{A}{\begin{bmatrix} 1 & 0 & 0 & 0 & 0 \\ 0 & 0 & 0 & 0 & -1 \\ 0 & 1 & 0 & 0 & 0 \\ 0 & 0 & 0 & 0 & -1 \\ 0 & 0 & 0 & 1 & 0 \\ 0 & 0 & 1 & 0 & 0 \\ 0 & 0 & 0 & 0 & -1 \\ 1 & 0 & 0 & 0 & 0 \\ 0 & 1 & 0 & 0 & 0 \\ 0 & 1 & 0 & 0 & 0 \\ 0 & 0 & 1 & 0 & 0 \end{bmatrix}} \underset{d_j}{\begin{bmatrix} \theta_A \\ \theta_B \\ \theta_C \\ \delta_D \end{bmatrix}} \qquad \ldots(4)$$

or simply, $d_m = A.\, d_j$ where A is the connection matrix which establishes the compatibility condition.

Step 2: Member stiffnesses or Force deformations relations corresponding to elements a, b, c, d, and e may be written as follows

$$\begin{bmatrix} \begin{pmatrix} m_2 \\ (p_y)_2 \end{pmatrix}_a \\ \begin{pmatrix} m_2 \\ (p_y)_2 \end{pmatrix}_b \\ \begin{pmatrix} m_1 \\ m_2 \\ (p_y)_2 \end{pmatrix}_c \\ \begin{pmatrix} m_1 \\ m_2 \end{pmatrix}_d \\ \begin{pmatrix} m_1 \\ m_2 \end{pmatrix}_e \end{bmatrix} = \begin{bmatrix} \begin{pmatrix} \frac{4EI}{L}, -\frac{6EI}{L^2} \\ -\frac{6EI}{L^2}, +\frac{12EI}{L^3} \end{pmatrix}_a & & & & \\ & \begin{pmatrix} \frac{4EI}{L}, -\frac{6EI}{L^2} \\ -\frac{6EI}{L^2}, +\frac{12EI}{L^3} \end{pmatrix}_b & & & \\ & & \begin{pmatrix} \frac{4EI}{L}, +\frac{2EI}{L}, -\frac{6EI}{L^2} \\ \frac{2EI}{L}, +\frac{4EI}{L}, -\frac{6EI}{L^2} \\ -\frac{6EI}{L^2}, -\frac{6EI}{L^2}, +\frac{12EI}{L^3} \end{pmatrix}_c & & \\ & & & \begin{pmatrix} \frac{4EI}{L}, +\frac{2EI}{L} \\ \frac{2EI}{L}, +\frac{4EI}{L} \end{pmatrix}_d & \\ & & & & \begin{pmatrix} \frac{4EI}{L}, +\frac{2EI}{L} \\ \frac{2EI}{L}, +\frac{4EI}{L} \end{pmatrix}_e \end{bmatrix} \begin{bmatrix} \begin{pmatrix} \theta_2 \\ \delta_2 \end{pmatrix}_a \\ \begin{pmatrix} \theta_2 \\ \delta_2 \end{pmatrix}_b \\ \begin{pmatrix} \theta_1 \\ \theta_2 \\ \delta_2 \end{pmatrix}_c \\ \begin{pmatrix} \theta_1 \\ \theta_2 \end{pmatrix}_d \\ \begin{pmatrix} \theta_1 \\ \theta_2 \end{pmatrix}_e \end{bmatrix}$$

Step 3: Equilibrium condition

Principle of virtual work may be used conveniently which is same as satisfying equation of equilibrium. The work done by the forces p_m at the ends of members while riding through the deformation d_m should be equal to the work done p_m, p_j while riding through the corresponding deformations, d_j (d_m and d_j deformations are compatible). Let us recall contragradience in the following.

$$d^t{}_m\, p_m = d^t{}_j\, p_j \qquad \ldots\text{(i)}$$

But the connection matrix gave the relation $d_m = A\, d_j$

or $\quad d^t{}_m = d^t{}_j A^t$ ...(ii)

From (*i*) and (*ii*), $d^t{}_j A^t p_m = d^t{}_j \cdot p_j$

or simply $\quad p_j = A^t p_m$ ...(iii)

We know from member "force deformation"

$$p_m = k_m d_m$$

and substituting for pmin Eq. (iii)

$$P_j = A^t k_m d_m$$

But as $\quad d_m = A\, d_j$ (Compatibility)

$$P_j = \underbrace{A^t k_m A d_i}_{K} \qquad \text{...(iv)}$$

The final stiffness matrix $K = A^t k_m A$, for any structure can be obtained by direct multiplication of the matrices.

Let us carry through the steps to obtain K matrix in the relation Eq. (iv) for the chosen structure,

Let $\quad (I/L)_a = (I/L)_b = (I/L)_c = (I/L)_1$ and $(I/L)_d = (I/L)_c = (I/L)_2$

$$K_m A = \begin{bmatrix} \left(\frac{4EI}{L}\right)_1 & 0 & 0 & 0 & \left(\frac{6EI}{L^2}\right)_1 \\ -\left(\frac{6EI}{L^2}\right)_1 & 0 & 0 & 0 & -\left(\frac{12EI}{L^3}\right)_1 \\ 0 & \left(\frac{4EI}{L}\right)_1 & 0 & 0 & \left(\frac{6EI}{L^2}\right)_1 \\ 0 & -\left(\frac{6EI}{L^2}\right)_1 & \left(\frac{6EI}{L^2}\right)_1 & 0 & -\left(\frac{12EI}{L^3}\right)_1 \\ 0 & 0 & \left(\frac{2EI}{L}\right)_1 & \left(\frac{4EI}{L}\right)_1 & \left(\frac{6EI}{L^2}\right)_1 \\ 0 & 0 & \left(\frac{4EI}{L}\right)_1 & \left(\frac{2EI}{L}\right)_1 & \left(\frac{6EI}{L^2}\right)_1 \\ 0 & 0 & -\left(\frac{6EI}{L^2}\right)_1 & -\left(\frac{6EI}{L^2}\right)_1 & -\left(\frac{12EI}{L^3}\right)_1 \\ \left(\frac{4EI}{L}\right)_2 & \left(\frac{2EI}{L}\right)_2 & 0 & 0 & 0 \\ \left(\frac{2EI}{L}\right)_2 & \left(\frac{4EI}{L}\right)_2 & 0 & 0 & 0 \\ 0 & \left(\frac{4EI}{L}\right)_2 & \left(\frac{2EI}{L}\right)_2 & 0 & 0 \\ 0 & \left(\frac{2EI}{L}\right)_2 & \left(\frac{4EI}{L}\right)_2 & 0 & 0 \end{bmatrix}$$

And $$K = A^t k_m A = \begin{bmatrix} \left(\frac{4EI}{L}\right)_2 + \left(\frac{4EI}{L}\right)_1 & \left(\frac{2EI}{L}\right)_2 & 0 & 0 & \left(\frac{6EI}{L^2}\right)_1 \\ \left(\frac{2EI}{L}\right)_1 & \Sigma\left(\frac{4EI}{L}\right)_{1,2,2} & \left(\frac{2EI}{L}\right)_2 & 0 & \left(\frac{6EI}{L^2}\right)_1 \\ 0 & \left(\frac{2EI}{L}\right)_2 & \Sigma\left(\frac{4EI}{L}\right)_{2,1} & \left(\frac{2EI}{L}\right)_1 & \left(\frac{6EI}{L^2}\right)_1 \\ 0 & 0 & \left(\frac{2EI}{L}\right)_1 & \left(\frac{4EI}{L}\right)_1 & \left(\frac{6EI}{L^2}\right)_1 \\ \left(\frac{6EI}{L^2}\right)_1 & \left(\frac{6EI}{L^2}\right)_1 & \left(\frac{6EI}{L^2}\right)_1 & \left(\frac{6EI}{L^2}\right)_1 & \Sigma\left(\frac{12EI}{L^3}\right)_{1,1,1} \end{bmatrix}$$

The procedure following the evaluation of K matrix is the same as discussed already Eq. (iv)

7.12 BOX FRAME

Box frames are commonly used for various applications. Either case, (1) it issupported over the piers/ abutments or (2) the bottom member is supported by a soil bed. In the latter case, it is required to obtain stiffness matrix of a member directly resting on soil bed. The idealization of the soil bed uses Winkler's model.

7.12.1 Case 1: Analysis of 'Box Frame'

The box is supported by hinged supports.
Using connection matrix method.

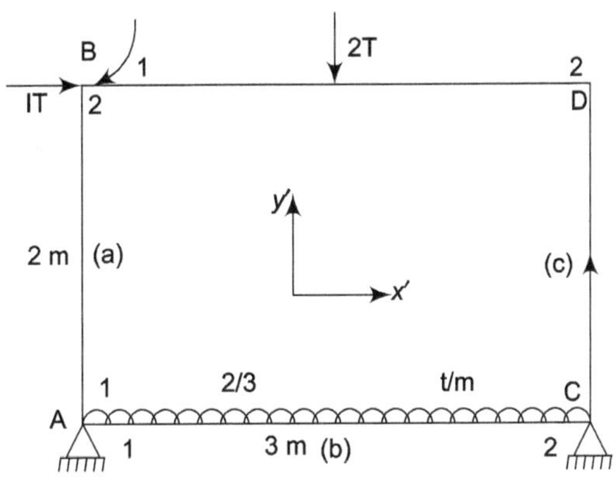

(a) Box Frame

The problem is first solved by connection matrix approach

Where $K' = A^t K_f A$

A. Connection matrix (Compatibility)
B. K_f Stiffness matrix of all members in member coordinates.

The same result (k_j) is obtained (alternatively) simply by inspection.
"A" matrix in the equation

$$\delta_m = A\delta_j$$

$$\begin{bmatrix} \begin{pmatrix} \theta_1 \\ \theta_Z \\ \delta_Y \end{pmatrix}_a \\ \begin{pmatrix} \theta_1 \\ \theta_2 \end{pmatrix}_b \\ \begin{pmatrix} \theta_1 \\ \theta_Z \\ \delta_Y \end{pmatrix}_c \\ \begin{pmatrix} \theta_1 \\ \theta_2 \end{pmatrix}_d \end{bmatrix} = \begin{bmatrix} 1 & 0 & 0 & 0 & 0 & 0 \\ 0 & 0 & 0 & 0 & 0 & 0 \\ 0 & 0 & 0 & 0 & -1 & 0 \\ 1 & 0 & 0 & 0 & 0 & 0 \\ 0 & 0 & 1 & 0 & 0 & 0 \\ 0 & 0 & 0 & 0 & 0 & 0 \\ 0 & 0 & 0 & 1 & 0 & 0 \\ 0 & 0 & 0 & 0 & 0 & -1 \\ 0 & 1 & 0 & 0 & 0 & 0 \\ 0 & 0 & 0 & 1 & 0 & 0 \end{bmatrix} \begin{bmatrix} \theta_A \\ \theta_B \\ \theta_C \\ \theta_D \\ \delta x'_B \\ \delta x'_D \end{bmatrix}$$

$$\underset{\text{Member}}{\delta_m} = A \underset{\text{Joint}}{\delta_J}$$

Step 2: Member force deformation relations:

$$\begin{bmatrix} M_1 \\ M_2 \\ S \end{bmatrix} = \begin{bmatrix} 4EI/L & 2EI/L & -6EI/L^2 \\ 2EI/L & 4EI/L & -6EI/L^2 \\ -6EI/L^2 & -6EI/L^2 & 12EI/L^2 \end{bmatrix} \begin{bmatrix} \theta_1 \\ \theta_2 \\ \delta_y \end{bmatrix}$$

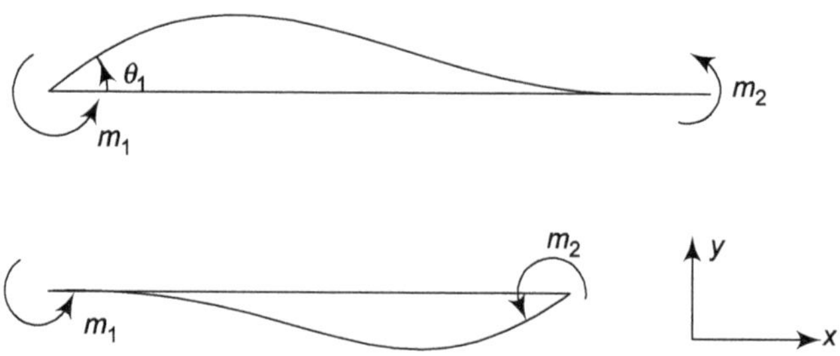

The above are the independent deformations

(b) Stiffness for a prismatic member

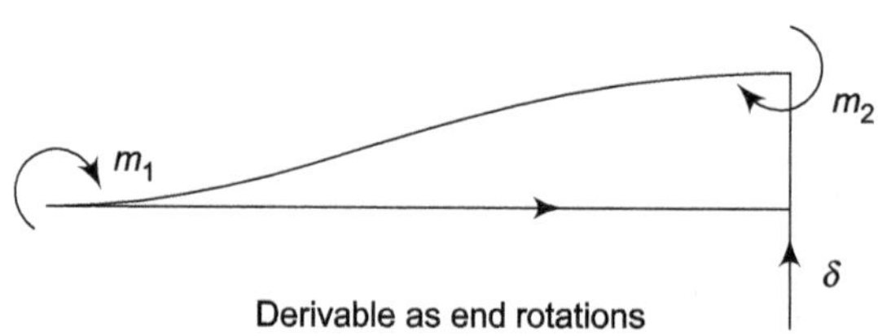

Derivable as end rotations

The above are the independent deformations. The results for a beam on elastic bed are given in the following.

The above member deformation is derivable as end rotations

The stiffness coefficients:

$$K_f A = \begin{bmatrix} \frac{4EI}{L_a} & \frac{2EI}{L_a} & 0 & 0 & \frac{4EI}{L_1^2} & 0 \\ \frac{2EI}{L} & \frac{4EI}{L_a} & 0 & 0 & \frac{6EI}{L_1^2} & 0 \\ \frac{-6EI}{L^2} & \frac{-6EI}{L_a^2} & 0 & 0 & \frac{-12EI}{L_1^3} & 0 \\ \frac{4EI}{L_b} & 0 & \frac{2EI}{L_b} & 0 & 0 & 0 \\ \frac{2EI}{L_b} & 0 & \frac{4EI}{L_b} & 0 & 0 & 0 \\ 0 & 0 & \frac{4EI}{L_b} & \frac{2EI}{L_c} & 0 & 0 \\ 0 & 0 & \frac{2EI}{L_c} & \frac{4EI}{L_c} & 0 & \frac{6EI}{L_c^2} \\ 0 & 0 & \frac{-6EI}{L_c^2} & \frac{-6EI}{L_c^2} & 0 & \frac{-12EI}{L_c^3} \\ 0 & \frac{4EI}{L_d} & 0 & \frac{2EI}{L_d} & 0 & 0 \\ 0 & \frac{2EI}{L_d} & 0 & \frac{4EI}{L_d} & 0 & 0 \end{bmatrix} \quad \ldots(4)$$

1	2	3	4	5	6	7	8	9	10
$\begin{pmatrix} \frac{4EI}{L} & \frac{2EI}{L} & \frac{6EI}{L^2} \\ \frac{2EI}{L} & \frac{4EI}{L} & \frac{-6EI}{L^2} \\ \frac{-6EI}{L^2} & \frac{-6EI}{L^2} & \frac{12EI}{L^3} \end{pmatrix}_a$									
			$\begin{pmatrix} \frac{4EI}{L} & \frac{2EI}{L} \\ \frac{2EI}{L} & \frac{4EI}{L} \end{pmatrix}_b$						

1	2	3	4	5	6	7	8	9	10
					$\begin{pmatrix} \frac{4EI}{L} & \frac{2EI}{L} & \frac{-6EI}{L^2} \\ \frac{2EI}{L} & \frac{4EI}{L} & \frac{-6EI}{L^2} \\ \frac{-6EI}{L^2} & \frac{-6EI}{L^2} & \frac{12EI}{L^3} \end{pmatrix}_c$				
								$\begin{pmatrix} \frac{4EI}{L} & \frac{2EI}{L} \\ \frac{2EI}{L} & \frac{4EI}{L} \end{pmatrix}_d$	

Finally substituting for

$$L_a = 2 \text{ m}, \quad L_b = 3 \text{ m}, \quad L_c = 2 \text{ m}$$

$$A^t K_f A = \begin{bmatrix} \frac{10}{3} & 1 & \frac{2}{3} & 0 & \frac{3}{2} & 0 \\ 1 & \frac{10}{3} & 0 & \frac{2}{3} & -\frac{3}{2} & 0 \\ \frac{2}{3} & 0 & \frac{10}{3} & 1 & 0 & \frac{3}{2} \\ 0 & \frac{2}{3} & 1 & \frac{10}{3} & 0 & \frac{3}{2} \\ \frac{3}{2} & \frac{3}{2} & 0 & 0 & \frac{3}{2} & 0 \\ 0 & 0 & \frac{3}{2} & \frac{3}{2} & 0 & \frac{3}{2} \end{bmatrix} EI = K' \qquad \ldots(5)$$

And $\quad P_j = K'_j,\ \delta_j,\ \{P_j\} = \{m_A \quad m_B \quad m_C \quad m_D \quad P'_{XB} \quad P'_{XD}\}$

$$\{\delta_j\} = \{\theta_A \quad \theta_B \quad \theta_C \quad \theta_D \quad \delta'_{XB} \quad \delta'_{XD}\}$$

The above matrix can be condensed to 5 × 5 because δ'_{XB} $\delta'_{XB} = \delta'_{X'D}$.

The solution for δ_j can be substantially simplified by considering symmetrical and antisymmetrical analysis.

Case 1:

Symmetrical Analysis :

$$\theta_C = -\theta_A \quad \text{and} \quad \theta_D = -\theta_B \quad \text{and} \quad \delta'_y = 0$$

Solving for

$$\theta_A = -0.94/EI$$

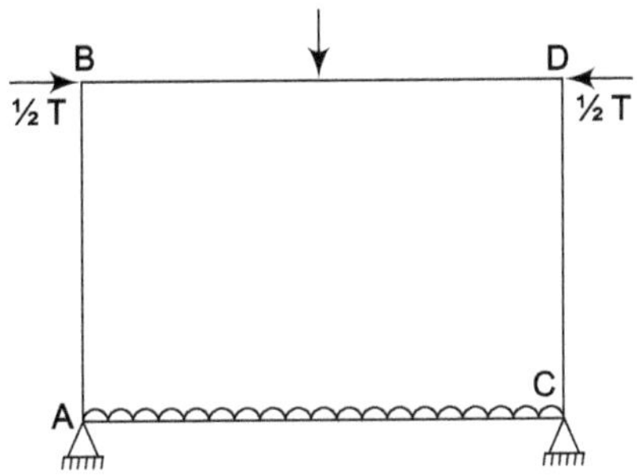

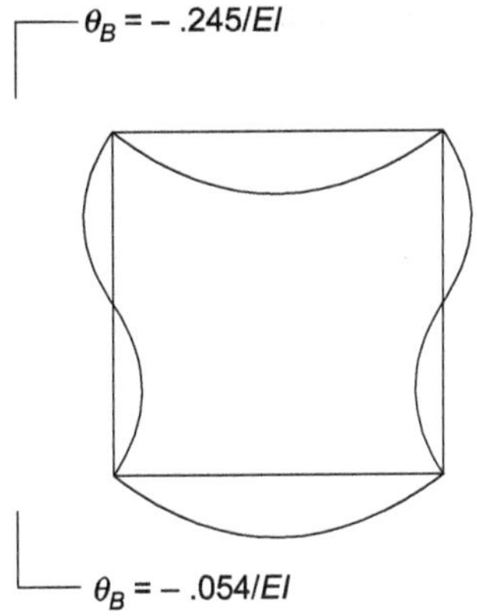

$$\theta_B = -.245/EI$$

$$M_{AB} = \frac{4EI}{L_a}\left(\frac{-0.94}{EI}\right) + \frac{2E}{L_a}\left(\frac{-.245}{EI}\right)$$

$M_{AB} = .434$ and similarly $M_{BA} = -.586$ tm

Check: $M_{AC} = -M_{AB}$ and $M_{BD} = M_{BA}$

$$M_{BC} = \frac{2EI}{L_c} \times \left(\frac{-246}{EI}\right) + (Fem = 5/4) = -0.537$$

Note: Modified stiffness for Symmetrical rotation $\frac{2EI}{L}$

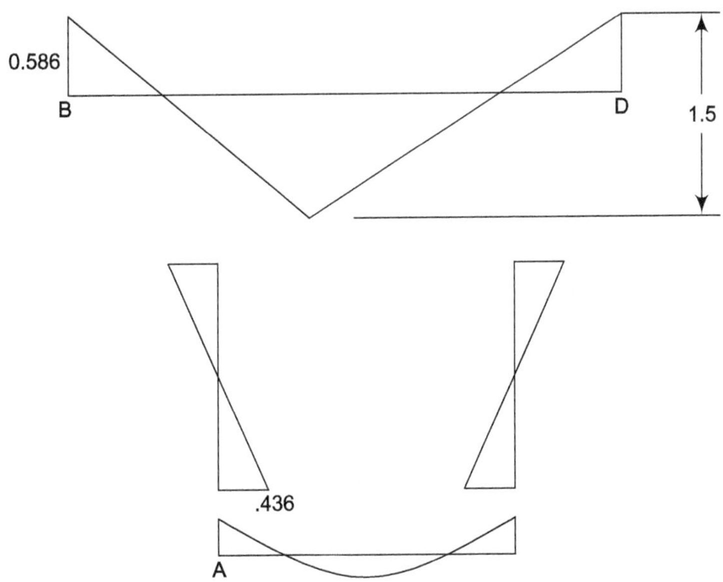

BM Diagram for Symmetrical Loading

Case II: Antisymmetric loading

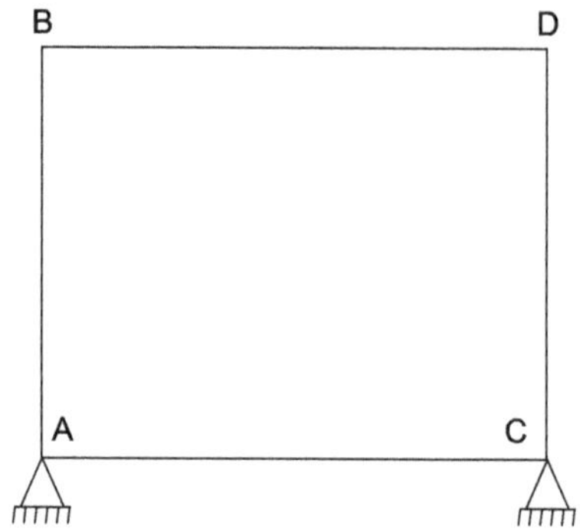

$$\theta_A = \theta_C,\ \theta_B = \theta_D \quad \text{and} \quad \delta_{XB} = \delta_{XD}$$

$$\begin{bmatrix} M_A = 0 \\ M_B = 0 \\ P_{Bx} = 0 \end{bmatrix} = \begin{bmatrix} \frac{10}{3}+\frac{2}{3} & 1 & \frac{3}{2} \\ 1 & \frac{10}{3}+\frac{2}{3} & \frac{3}{2} \\ \frac{3}{2} & \frac{3}{2} & \frac{3}{2} \end{bmatrix} = \begin{bmatrix} \theta_A \\ \theta_B \\ \delta_{XB} \end{bmatrix} \qquad \text{...(5b)}$$

Solving for δ_j:

$$\theta_A = \theta_B = -\frac{1}{4EI} + \delta_{XB} = \frac{5}{6EI}$$

There is no fixed end moment contribution in this case and $M_{AB} = \ = M_{AC} = ½$

The B.M. diagram is given below.

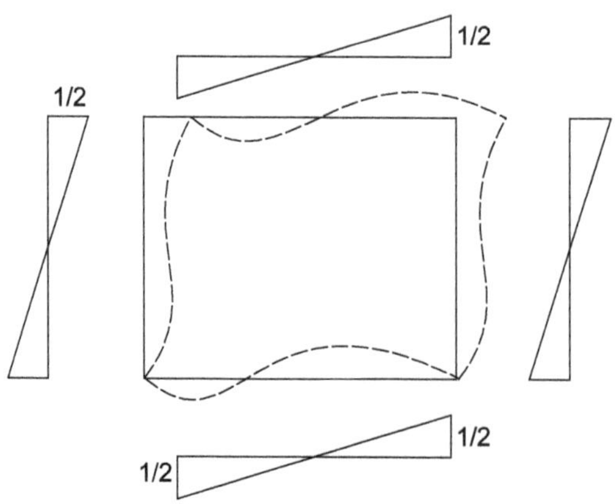

B.M. Diagram for Antisymmetrical Loading Superposition of Case I and II

Alternatively

$$\begin{bmatrix} P_A \\ P_B \\ P_C \\ P_D \end{bmatrix} = \begin{bmatrix} (K_{11})_a + (K_{11})_b & (K_{12})_a & (K_{12})_b & 0 \\ * & (K_{22})_a + (K_{11})_d & 0 & (K'_{12})_d \\ * & * & (K'_{22})_b + (K_{11})_c & (K'_{12})_c \\ * & * & (K'_{21})_c & (K'_{22})_b \end{bmatrix} \begin{bmatrix} \delta_A \\ \delta_B \\ \delta_C \\ \delta_d \end{bmatrix} \quad \ldots(6)$$

Directly by inspection without intermediate steps.

The Eqs. (5) and (7) are identical

Foundation Beam on Winkler's bed

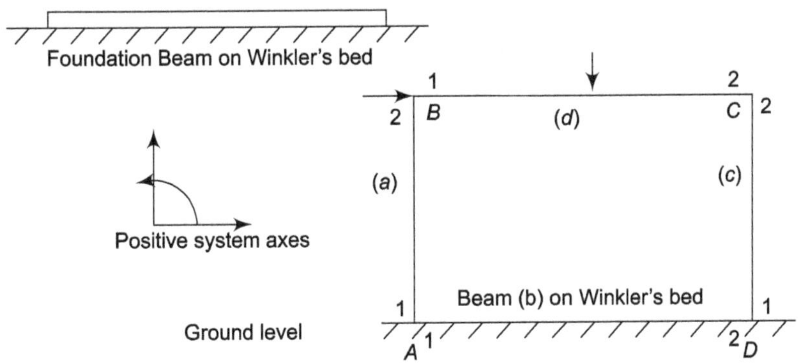

7.13 THE BOTTOM MEMBER IS SUPPORTED BY A SOIL BED: STIFFNESS MATRIX OF FOUNDATION BEAM

In this case, it is required to obtain stiffness matrix of a member directly resting on soil bed. One of the idealizations of the soil bed is that it uses Winkler's model. In the following, the stiffness matrix of a beam on elastic foundation (Winkler's Bed) is given without proof. References for the derivation are given.

Refer the attached figure for the member stiffness. Where,

$$y_1 = \sinh \lambda x \sin \lambda x; \; y_2 = \cosh \lambda x \sin \lambda x + \sinh \lambda x$$

$$y_3 = \cosh \lambda x \sin \lambda x - \sinh \lambda x; \; y_4 = \cosh \lambda x \cos \lambda x$$

$\lambda = 4$, Where K is the foundation modulus

The following is the stiffness matrix of a finite beam on winkler's bed.

$$\begin{Bmatrix} \begin{pmatrix} P_y \\ m \end{pmatrix}_1 \\ \begin{pmatrix} P_y \\ m \end{pmatrix}_2 \end{Bmatrix} = \frac{2\lambda EI}{2y_{1^2}} \begin{pmatrix} S_{11} & S_{12} & S_{13} & S_{14} \\ S_{12} & S_{22} & S_{14} & S_{24} \\ S_{13} & S_{14} & S_{11} & -S_{12} \\ S_{14} & S_{24} & -S_{12} & S_{22} \end{pmatrix} \begin{Bmatrix} \begin{pmatrix} P_y \\ m \end{pmatrix}_1 \\ \begin{pmatrix} P_y \\ m \end{pmatrix}_2 \end{Bmatrix}$$

Fig. 7. *Stiffness Matrix of Foundation Beam*

The typical values are:

$$S_{11} = 2\lambda^2 (y_2 y_4 + y_1 y_3)$$

$$S_{12} = ½\,\lambda\,(y_2^{\,2} + y_3^{\,2})$$
$$S_{13} = -\,2\,\lambda^2\,y_2$$
$$S_{14} = 2\,\lambda\,y_1$$
$$S_{22} = (y_1\,y_2 - y_3\,y_4)$$
$$S_{24} = y_3$$

We don't consider uniformly distributed load on foundation beam resting on Winkler's Bed.

7.14 SINGLE BAY TWO STOREY FRAME

Elastic Line: In the following, it is attempted to familiarize with the sketching of elastic lines (deflected shapes) with reference to typical member axes, viz, (a) tangent at end-1 of member in deformed position, and (b) line joining the end points in deformed position. The columns are inextensible. The deformation are shown exaggerated.

'A Single Bay Two Storey Frame'

The following frame, single bay and two storeys is a good preparation for the analysis of multi storey frame, multiple bays and multiple storey. Ignore the axial deformations of girders.

Two cases are considered in the following:

Case 1: All members are assumed inextensible, and

Case 2: Only the two girders are assumed inextensible. The lateral deflections δ'_x of the nodes C and D are assumed equal. So also the δ'_x of the nodes A & B.

Member coordinates with two independent relative deformations.

Fig. 7.23 (a)The relative deformations are measured wrt the tangent at end-1 of each member.

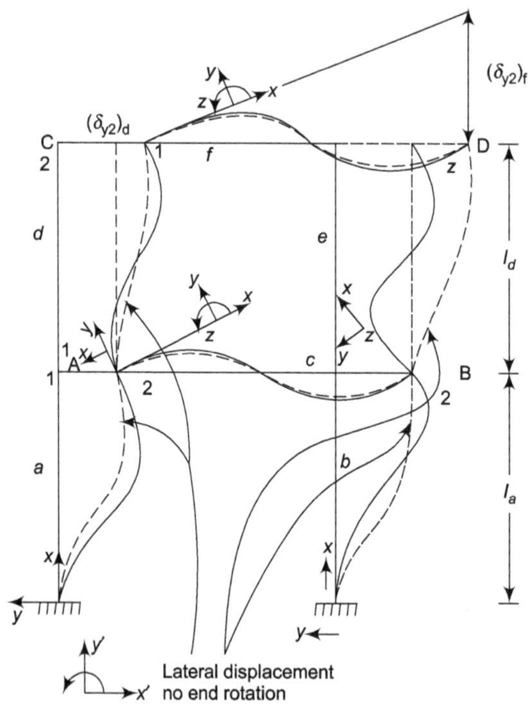

Fig. 7.23 *(a) Reference Axis : Tangent at End-1*

7.14.1 Case 1: All Members Inextensible

Reduced size of [A] matrix: If inextensibility is not assumed (let us call it, DSM analysis), number of rows of a connection matrix is equal to 6 times the number of members constituting a frame. If the axial deformations are neglected for their effect on moments and shears, we need to know two end forces, moment and shear at the end-2. There are corresponding two independent deformations, rotations and lateral displacement with respect to the tangent at the end-1. Member reference lines with tangents drawn at one end and relative deformation exaggerated are shown, Fig. 7.23(a).

Alternative member reference line the two independent end deformations are measured w.r.t. the line joining the ends in deformed position. The end deformations are the end rotations θ_1 and θ_2 measured with respect to the line joining end points in the deformed position. The end forces for such a system of two end deformations are the moment m_1 and m_2. Such member reference lines are shown in Fig. 7.23(b).

With any one of the above member reference lines the size of connection matrix (rows) gets reduced to 2 times the number of members constituting a frame. The number of columns of the connection matrix is governed by the number of independent displacements. The number of deforming joints $n = 4$. The number of in extensible members, $m = 6$. The number of independent displacements $3n - m = 6$. The size of connection matrix for the case and chosen reference lines is $(2m) \times$ Degrees of freedom. For the example considered size $= 12 \times 6$.

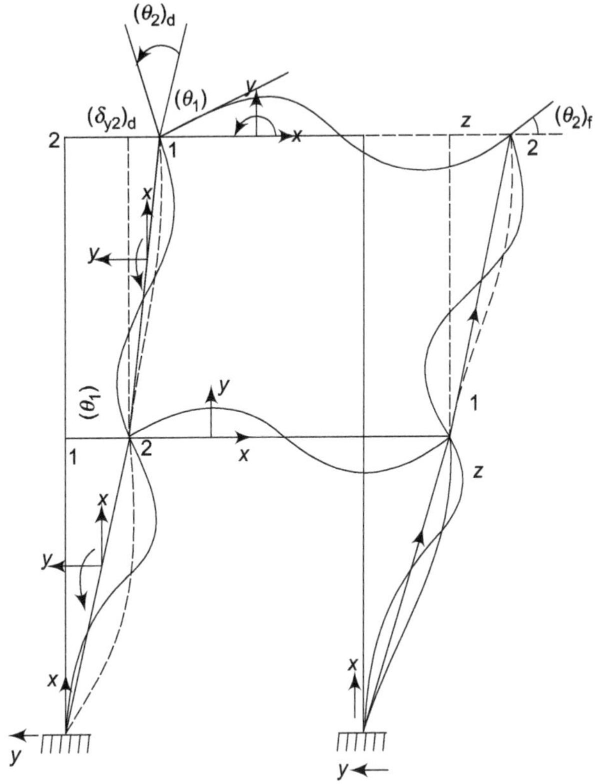

Fig. 7.23 *(b) Reference Axis : Line Joining the End Points in Deformed Position*

The member stiffness matrices for the above choice of member reference lines are:

(a) Relative deformations $(\delta_Y)_2$, δ_2 with respect to tangent at end.

$$k_m = \begin{bmatrix} \dfrac{12EI}{L^3} & -\dfrac{6EI}{L^2} \\ -\dfrac{6EI}{L^2} & \dfrac{4EI}{L} \end{bmatrix}$$

(b) End rotations and with respect to the line joining deformed position of ends 1 and 2.

$$k_m = \begin{bmatrix} \dfrac{4EI}{L} & \dfrac{2EI}{L} \\ \dfrac{2EI}{L} & \dfrac{4EI}{L} \end{bmatrix}$$

Case (a) with km of the members as stated. $K_m = \lceil (k_m)_a, (k_m)_b, (k_m)_c, (k_m)_d, (k_m)_e, (k_m)_f \rfloor$

$A^t k_m =$

$\left(\frac{-12EI}{L^3}\right)_a$	$\left(\frac{6EI}{L^2}\right)_a$	$\left(\frac{-12EI}{L^3}\right)_b$	$\left(\frac{6EI}{L^2}\right)_b$	0	0	$\left(\frac{12EI}{L^3}\right)_d$	$\left(\frac{6EI}{L^2}\right)_d$	$\left(\frac{12EI}{L^3}\right)_e$	$\left(\frac{-6EI}{L^2}\right)_d$	0	0
$\left(\frac{-6EI}{L^2}\right)_a$	$\left(\frac{4EI}{L}\right)_a$	0	0	$\left(\frac{-6EI}{L^2}\right)_c$	$\left(\frac{2EI}{L}\right)_c$	$\left(\frac{-6EI}{L^2}\right)_d$	$\left(\frac{2EI}{L}\right)_d$	0	0	0	0
0	0	$\left(\frac{6EI}{L^2}\right)_b$	$\left(\frac{4EI}{L}\right)_b$	$\left(\frac{-6EI}{L^2}\right)_c$	$\left(\frac{4EI}{L}\right)_c$	0	0	$\left(\frac{-6EI}{L^2}\right)_c$	$\left(\frac{2EI}{L}\right)_d$	0	0
0	0	0	0	0	0	$\left(\frac{12EI}{L^3}\right)_d$	$\left(\frac{6EI}{L^2}\right)_d$	$\left(-\frac{12EI}{L^3}\right)_a$	$\left(\frac{6EI}{L^2}\right)_a$	0	0
0	0	0	0	0	0	$\left(\frac{-6EI}{L^2}\right)_d$	$\left(\frac{4EI}{L}\right)_d$	0	0	$\left(\frac{-6EI}{L^2}\right)_e$	$\left(\frac{2EI}{L}\right)_e$
0	0	0	0	0	0	0	0	$\left(\frac{-6EI}{L^2}\right)_a$	$\left(\frac{4EI}{L}\right)_a$	$\left(\frac{-6EI}{L^2}\right)_f$	$\left(\frac{4EI}{L}\right)_f$

In the relation $d_m = Ad_j$; 'A' matrix is defined in the following

$A_{12\times 6} =$

−1	0	0	0	0	0
0	1	0	0	0	0
−1	0	0	0	0	0
0	0	1	0	0	0
0	$-l_c$	0	0	0	0
0	−1	1	0	0	0
1	$-l_d$	0	−1	0	0
0	−1	0	0	1	0
1	0	$-l_e$	−1	0	0
0	0	−1	0	0	1
0	0	0	0	$-l_f$	0
0	0	0	0	−1	0

$$K_J = \begin{bmatrix}
\left(\frac{12EI}{L^3}\right)_a + \left(\frac{12EI}{L^3}\right)_b + \left(\frac{12EI}{L^3}\right)_d + \left(\frac{12EI}{L^3}\right)_e & \left(\frac{6EI}{L^2}\right)_a + \left(-\frac{6EI}{L^2}\right)_d & \left(\frac{6EI}{L^2}\right)_d + \left(-\frac{6EI}{L^2}\right)_e & \left(-\frac{12EI}{L^3}\right)_d + \left(-\frac{12EI}{L^3}\right)_e & \left(-\frac{6EI}{L^2}\right)_d & \left(-\frac{6EI}{L^2}\right)_d \\
\left(\frac{6EI}{L^2}\right)_a + \left(-\frac{6EI}{L^2}\right)_d & \left(\frac{4EI}{L}\right)_a + \left(\frac{4EI}{L}\right)_c + \left(\frac{4EI}{L}\right)_d & \left(\frac{2EI}{L}\right)_c & \left(\frac{6EI}{L^2}\right)_d & \left(\frac{2EI}{L}\right)_d & 0 \\
\left(\frac{6EI}{L^2}\right)_b + \left(\frac{6EI}{L^2}\right)_e & \left(\frac{2EI}{L}\right)_d & \left(\frac{4EI}{L}\right)_b + \left(\frac{4EI}{L}\right)_c + \left(\frac{4EI}{L}\right)_e & \left(\frac{6EI}{L^2}\right)_d & 0 & \left(\frac{2EI}{L}\right)_a \\
\left(-\frac{12EI}{L^3}\right)_d + \left(-\frac{12EI}{L^3}\right)_e & \left(\frac{6EI}{L^2}\right)_d + & \left(\frac{6EI}{L^2}\right)_e & \left(\frac{12EI}{L^3}\right)_d + \left(+\frac{12EI}{L^3}\right)_e & \left(\frac{6EI}{L^2}\right)_d & \left(\frac{6EI}{L^2}\right)_e \\
\left(-\frac{6EI}{L^2}\right)_d & \left(\frac{2EI}{L}\right)_d & 0 & \left(\frac{6EI}{L^2}\right)_d & \left(\frac{4EI}{L}\right)_d + +\left(\frac{4EI}{L}\right)_f & \left(\frac{2EI}{L}\right)_f \\
-\left(\frac{6EI}{L^2}\right)_c & 0 & \left(\frac{2EI}{L}\right)_e & \left(\frac{6EI}{L^2}\right)_e & \left(\frac{2EI}{L}\right)_f & \left(\frac{4EI}{L}\right)_e + \left(\frac{4EI}{L}\right)_f
\end{bmatrix}$$

Example 1. (a) Write connection matrix for the frame Fig. 7.23(a). Neglect axial deformation of columns and girders for their effect on moments and shear.

(b) Obtain K_j matrix

(c) Obtain load matrix for the frame, and indicate the unknown deformations.

Solution:

$$A = \begin{bmatrix} -1 & 0 & 0 & 0 & 0 & 0 \\ 0 & 1 & 0 & 0 & 0 & 0 \\ -1 & 0 & 0 & 0 & 0 & 0 \\ 0 & 0 & 1 & 0 & 0 & 0 \\ 0 & -L_c & 0 & 0 & 0 & 0 \\ 0 & -1 & 1 & 0 & 0 & 0 \\ 1 & -L_d & 0 & -1 & 0 & 0 \\ 0 & -1 & 0 & 0 & 1 & 0 \\ 1 & 0 & -L_e & -1 & 0 & 0 \\ 0 & 0 & -1 & 0 & 0 & 1 \\ 0 & 0 & 0 & 0 & -L_f & 0 \\ 0 & 0 & 0 & 0 & -1 & 0 \end{bmatrix}; P_m = \begin{bmatrix} \begin{pmatrix} P_Y \\ m \end{pmatrix}_a \\ \begin{pmatrix} P_Y \\ m \end{pmatrix}_b \\ \begin{pmatrix} P_Y \\ m \end{pmatrix}_c \\ \begin{pmatrix} P_Y \\ m \end{pmatrix}_d \\ \begin{pmatrix} P_Y \\ m \end{pmatrix}_e \\ \begin{pmatrix} P_Y \\ m \end{pmatrix}_f \end{bmatrix}; d_j = \begin{bmatrix} d'_A = d'_B \\ \theta_A \\ \theta_B \\ d'_C = d'_D \\ \theta_C \\ \theta_D \end{bmatrix}$$

(b) $K_j = A^t K_m A$

$$K_m = \begin{bmatrix} (K_m)_a & 0 & 0 & 0 & 0 \\ 0 & (K_m)_b & 0 & 0 & 0 \\ 0 & 0 & (K_m)_c & 0 & 0 \\ 0 & 0 & 0 & (K_m)_d & 0 \\ 0 & 0 & 0 & 0 & (K_m)_e \\ 0 & 0 & 0 & 0 & 0 \end{bmatrix}$$

For a typical girder member $k_m = \begin{bmatrix} \left(\dfrac{12EI}{L^3}\right)_f & \left(\dfrac{-6EI}{L^2}\right)_f \\ \left(\dfrac{-6EI}{L^2}\right)_f & \left(\dfrac{4EI}{L^2}\right)_f \end{bmatrix}$

(c) P_j matrix is obtained from the fixed or no displacement considerations directly

The following free bodies form the bases for computing the load elements.

Theory for computing the load elements corresponding to the unknown Displacements.

- Moment loads and vertical loads pertain to the joints. Therefore, transfer Fixed end forces, shears and moments to the joints.
- Because the girders are assumed inextensible, we take the free bodies, AB and CD (case 2).
- If the columns are assumed inextensible, case 1, the fixed edge shears of the girder-ends get directly transferred to the base. The axial forces in the columns do not appear because of the assumed column inextensibility.

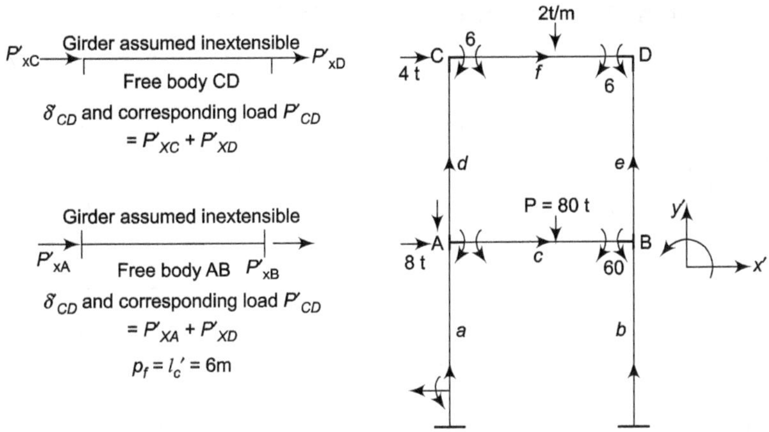

$$P_j = \begin{bmatrix} 8 \\ -40 \\ -60 \\ 4 \\ -6 \\ 6 \end{bmatrix}$$

Fig. (i) *Example-1*

7.14.2 Girders Only are Assumed Inextensible

Example 2: Write connection matrix for the chosen frame. Neglect axial deformations of the girders only for their effect on moments and shears.

(*a*) Obtain K_j matrix

(*b*) Obtain load matrix for the frame Fig. (ii). and indicate the unknown deformations to be evaluated.

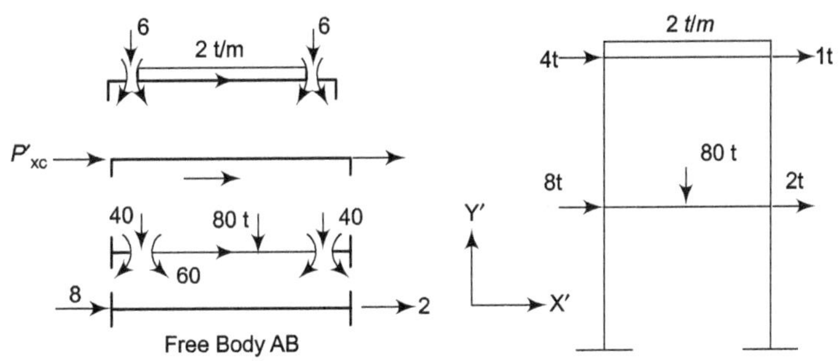

Fig. (ii) *Example-2*

$$P'_J = \begin{bmatrix} 10 \\ \hline -40 \\ \hline -60 \\ \hline -40 \\ 60 \\ \hline 5 \\ -6 \\ -6 \\ \hline -6 \\ 6 \end{bmatrix}$$

(a) Consider the tangent at end 1 to be the reference line, Fig. 7.23 (a)

$$
\begin{bmatrix}
\begin{pmatrix} (\delta_x)_2 \\ (\delta_y)_2 \\ \theta_2 \end{pmatrix}_a \\
\begin{pmatrix} (\delta_x)_2 \\ (\delta_y)_2 \\ \theta_2 \end{pmatrix}_b \\
\begin{pmatrix} (\delta_Y)_2 \\ \theta_2 \end{pmatrix}_c \\
\begin{pmatrix} (\delta_x)_2 \\ (\delta_y)_2 \\ \theta_2 \end{pmatrix}_d \\
\begin{pmatrix} (\delta_x)_2 \\ (\delta_y)_2 \\ \theta_2 \end{pmatrix}_e \\
\begin{pmatrix} (\delta_Y)_2 \\ \theta_2 \end{pmatrix}_f
\end{bmatrix}
=
\begin{bmatrix}
0 & +1 & 0 & 0 & 0 & 0 & 0 & 0 & 0 & 0 \\
-1 & 0 & 0 & 0 & 0 & 0 & 0 & 0 & 0 & 0 \\
0 & 0 & 1 & 0 & 0 & 0 & 0 & 0 & 0 & 0 \\
0 & 0 & 0 & +1 & 0 & 0 & 0 & 0 & 0 & 0 \\
-1 & 0 & 0 & 0 & 0 & 0 & 0 & 0 & 0 & 0 \\
0 & 0 & 0 & 0 & 1 & 0 & 0 & 0 & 0 & 0 \\
0 & -1 & -L_e & +1 & 0 & 0 & 0 & 0 & 0 & 0 \\
0 & 0 & -1 & 0 & 1 & 0 & 0 & 0 & 0 & 0 \\
0 & -1 & 0 & +1 & 0 & 0 & 0 & 0 & 0 & 0 \\
1 & 0 & -L_d & 0 & 0 & -1 & 0 & 0 & 0 & 0 \\
0 & 0 & -1 & 0 & 0 & 0 & 0 & 1 & 0 & 0 \\
0 & 0 & 0 & -1 & 0 & 0 & 0 & 0 & 1 & 0 \\
1 & 0 & 0 & 0 & -L_e & -1 & 0 & 0 & 0 & 0 \\
0 & 0 & 0 & 0 & -1 & 0 & 0 & 0 & 0 & 1 \\
0 & 0 & 0 & 0 & 0 & 0 & -1 & -L_f & 1 & 0 \\
0 & 0 & 0 & 0 & 0 & 0 & 0 & -1 & 0 & +1
\end{bmatrix}
\begin{bmatrix}
(\delta'_X)_{AB} \\
\begin{pmatrix} \delta'_y \\ \theta \end{pmatrix}_A \\
\begin{pmatrix} \delta'_y \\ \theta \end{pmatrix}_B \\
(\delta'_X)_{CD} \\
\begin{pmatrix} \delta'_y \\ \theta \end{pmatrix}_C \\
\begin{pmatrix} \delta'_y \\ \theta \end{pmatrix}_D
\end{bmatrix}
$$

(b) $$K_J = A^t K_m A$$

K_m matrix for the members of the frame Fig. 7.23 (a) is given by

$$
K_m = \begin{bmatrix}
(K_m)_a & & & & & \\
& (K_m)_b & & & & \\
& & (K_m)_c & & & \\
& & & (K_m)_d & & \\
& & & & (K_m)_e & \\
& & & & & (K_m)_f
\end{bmatrix}
$$

For a typical girder $(K_m)_f = \begin{pmatrix} \left(\dfrac{12EI}{L^3}\right)_f & \left(\dfrac{-6EI}{L^2}\right)_f \\ \left(\dfrac{-6EI}{L^2}\right)_f & \left(\dfrac{4EI}{L}\right)_f \end{pmatrix}$

For a typical column $(K_m)_d = \begin{pmatrix} \dfrac{EA}{L} & 0 & 0 \\ 0 & \dfrac{12EI}{L^3} & \dfrac{-6EI}{L^2} \\ 0 & \dfrac{-6EI}{L^2} & \dfrac{4EI}{L} \end{pmatrix}_d$

$$
A^Tk_m = \left[\begin{array}{cccccccccccccccc}
0 & \left(\frac{-12EI}{L^2}\right)_a & \left(\frac{6EI}{L^2}\right)_a & 0 & \left(\frac{-12EI}{L^3}\right)_b & \left(\frac{-6EI}{L^2}\right)_e & 0 & 0 & 0 & \left(\frac{12EI}{L^2}\right)_d & -\left(\frac{6EI}{L^2}\right)_d & 0 & -\left(\frac{2EI}{L^2}\right)_d & -\left(\frac{6EI}{L^2}\right)_a & 0 & 0 \\
\left(\frac{EA}{L}\right)_a & 0 & 0 & 0 & 0 & 0 & \left(\frac{-12EI}{L^2}\right)_e & \left(\frac{6EI}{L^2}\right)_a & \left(\frac{-EA}{L}\right)_d & 0 & 0 & 0 & 0 & 0 & 0 & 0 \\
0 & \left(\frac{-6EI}{L^2}\right)_a & \left(\frac{4EI}{L}\right)_a & 0 & 0 & 0 & \left(\frac{-6EI}{L^2}\right)_c & 2\left(\frac{EI}{L}\right)_c & 0 & \left(\frac{-6EI}{L^2}\right)_d & \left(\frac{2EI}{L}\right)_d & 0 & 0 & 0 & 0 & 0 \\
0 & 0 & 0 & \left(\frac{EA}{L}\right)_b & 0 & 0 & \left(\frac{12EI}{L^2}\right)_c & \left(\frac{-6EI}{L^2}\right)_c & \left(\frac{EI}{L}\right)_d & 0 & 0 & -\left(\frac{EA}{L}\right)_a & 0 & 0 & 0 & 0 \\
0 & 0 & 0 & 0 & \left(\frac{-6EI}{L^2}\right)_b & \left(\frac{4EI}{L}\right)_b & \left(\frac{-6EI}{L^2}\right)_c & \left(\frac{4EI}{L}\right)_c & 0 & 0 & 0 & 0 & \left(-\frac{6EI}{L^2}\right)_c & \left(\frac{2EI}{L}\right)_c & 0 & 0 \\
0 & 0 & 0 & 0 & 0 & 0 & 0 & 0 & 0 & \left(\frac{-12EI}{L^2}\right)_d & \left(\frac{6EI}{L^2}\right)_d & 0 & \left(-\frac{12EI}{L^2}\right)_c & \left(\frac{6EI}{L^2}\right)_c & 0 & 0 \\
0 & 0 & 0 & 0 & 0 & 0 & 0 & 0 & 0 & 0 & 0 & 0 & 0 & 0 & \left(-\frac{12EI}{L^3}\right)_f & \left(\frac{6EI}{L^2}\right)_f \\
0 & 0 & 0 & 0 & 0 & 0 & 0 & 0 & 0 & \left(\frac{-6EI}{L^2}\right)_d & \left(\frac{4EI}{L}\right)_d & 0 & 0 & 0 & \left(-\frac{6EI}{L^2}\right)_f & 2\left(\frac{EI}{L}\right)_f \\
0 & 0 & 0 & 0 & 0 & 0 & 0 & 0 & 0 & 0 & 0 & \left(\frac{EA}{L}\right)_e & 0 & 0 & \left(\frac{12EI}{L^3}\right)_f & \left(-\frac{6EI}{L^2}\right)_f \\
0 & 0 & 0 & 0 & 0 & 0 & 0 & 0 & 0 & 0 & 0 & 0 & \left(-\frac{6EI}{L^2}\right)_e & \left(\frac{4EI}{L}\right)_e & \left(-\frac{6EI}{L^2}\right)_f & 4\left(\frac{EI}{L}\right)_f
\end{array}\right]
$$

$d_m = A\, d_j$ where;

$$A = \begin{pmatrix} 0 & 1 & 0 & 0 & 0 & 0 & 0 & 0 & 0 & 0 \\ -1 & 0 & 0 & 0 & 0 & 0 & 0 & 0 & 0 & 0 \\ 0 & 0 & 1 & 0 & 0 & 0 & 0 & 0 & 0 & 0 \\ 0 & 0 & 0 & 1 & 0 & 0 & 0 & 0 & 0 & 0 \\ -1 & 0 & 0 & 0 & 0 & 0 & 0 & 0 & 0 & 0 \\ 0 & 0 & 0 & 0 & 1 & 0 & 0 & 0 & 0 & 0 \\ 0 & -1 & -l_c & 1 & 0 & 0 & 0 & 0 & 0 & 0 \\ 0 & 0 & -1 & 0 & 1 & 0 & 0 & 0 & 0 & 0 \\ 0 & -1 & 0 & 1 & 0 & 0 & 0 & 0 & 0 & 0 \\ 1 & 0 & -l_d & 0 & 0 & -1 & 0 & 0 & 0 & 0 \\ 0 & 0 & -1 & 0 & 0 & 0 & 0 & 1 & 0 & 0 \\ 0 & 0 & 0 & -1 & 0 & 0 & 0 & 0 & 1 & 0 \\ -1 & 0 & 0 & 0 & -l_c & -1 & 0 & 0 & 0 & 0 \\ 0 & 0 & 0 & 0 & -1 & 0 & 0 & 0 & 0 & 1 \\ 0 & 0 & 0 & 0 & 0 & 0 & -1 & -l_f & 1 & 0 \\ 0 & 0 & 0 & 0 & 0 & 0 & 0 & -1 & 0 & 1 \end{pmatrix}$$

Comments: The bottom supports are considered fixed. The number of members, $m = 6$ which consist of four column members and two girder members. The number of deforming nodes, $n = 4$.

For the Example-1, all members inextensible, the independent degrees of freedom or number of independent displacements is given by $3n - m = 3 \times 4 - 6 = 6$. The size of the global stiffness matrix, K_s is six by six and the matrix has been evaluated and the load matrix, P'_j shown in Fig. (i). the global deformations are obtained by inverting K_s and $\mathbf{d}'_j = [\mathbf{K}_s]^{-1} P'_j$

If the girders alone are assumed inextensible, Example-2, The number of inextensible members, $m = 2$, and the degrees of freedom are given by $3n - 2 = 10$. The global matrix K_s of size ten by ten has been evaluated and the corresponding load matrix shown in Fig. (ii). Evaluate $\mathbf{d}'_j$.

The post processing consists of:

1. Compute member deformations, $\mathbf{d_m = A\, d'_j}$
2. Compute member forces of all the members, $(\mathbf{p_m = k_m\, d_m})_\mathbf{I}$ **I = 1 to m**
3. Compute the member forces at the end-1 for all the members.

We are ready to proceed to design if the global deformations and the member forces are acceptable.

The reader may have found Article 1.3 to Article 1.6 to be interesting but these articles require a good amount of concentration. In the following it is compensated by the Article 7.17 interesting but fight to go thru.

$K_j = A^t k_m A$ where;

$K_j =$

$\sum_{a,b,d,e} \frac{12EI}{L^3}$	0	$\left(\frac{6EI}{L^2}\right)_a + \left(\frac{6EI}{L^2}\right)_d$	0	–	$-\left(\frac{12EI}{L^2}\right)_d + -\left(\frac{12EI}{L^2}\right)_e$	0	$-\left(\frac{6EI}{L^2}\right)_d$	0	$-\left(\frac{6EI}{L^2}\right)_e$
0	$\left(\frac{EA}{L}\right)_a + \left(\frac{12EI}{L^3}\right)_b + \left(\frac{EA}{L}\right)_d +$	$\left(\frac{6EI}{L^2}\right)_a$	$-\left(\frac{12EI}{L^3}\right)_e + -\left(\frac{EA}{L}\right)_d$	$\left(\frac{-6EI}{L^2}\right)_c$	0	0	0	0	0
$\left(\frac{6EI}{L^2}\right)_a -\left(\frac{6EI}{L^2}\right)_d$	$\left(\frac{6EI}{L^2}\right)_c$	$\sum_{a,c,d} \frac{4EI}{L}$	$-\left(\frac{6EI}{L^2}\right)_c$	$\left(\frac{2EI}{L}\right)_c$	$\left(\frac{6EI}{L^2}\right)_d$	0	$2\left(\frac{EI}{L}\right)_d$	0	0
0	$-\left(\frac{12EI}{L^3}\right)_c -\left(\frac{EA}{L}\right)_d$	$-\left(\frac{6EI}{L^2}\right)_c$	$\sum_{b,d} \frac{EA}{L} + \left(\frac{12EI}{L^2}\right)_c$	$-\left(\frac{6EI}{L^2}\right)_c$	0	0	0	0	0
$\left(\frac{6EI}{L^2}\right)_b + -\left(\frac{6EI}{L^2}\right)_e$	$\left(\frac{6EI}{L^2}\right)_c$	$\left(\frac{2EI}{L}\right)_c$	$-\left(\frac{6EI}{L^2}\right)_c$	$\sum_{b,c,e} \frac{4EI}{L}$	$\left(\frac{6EI}{L^2}\right)_e$	0	0	0	$\left(\frac{2EI}{L}\right)_e$

(Contd)

(Contd)

$-\left(\frac{12EI}{L^3}\right)_d + -\left(\frac{12EI}{L^3}\right)_e$	0	$\left(\frac{6EI}{L^2}\right)_d$	0	$\left(\frac{6EI}{L^2}\right)_e$	$\left(\frac{12EI}{L^3}\right)_d + \left(\frac{12EI}{L^3}\right)_e$	0	$\left(\frac{6EI}{L^2}\right)_d$	0	$\left(\frac{6EI}{L^2}\right)_e$
0	0	0	0	0	0	$\left(\frac{12EI}{L^3}\right)_f$	$\left(\frac{6EI}{L^2}\right)_f$	$-\left(\frac{12EI}{L^2}\right)_f$	$\left(\frac{6EI}{L^2}\right)_f$
$-\left(\frac{6EI}{L^2}\right)_d$	0	$\left(\frac{2EI}{L}\right)_d$	0	0	$\left(\frac{6EI}{L^2}\right)_d$	$\left(\frac{6EI}{L^2}\right)_f$	$\sum_{d,f}\frac{4EI}{L}$	$-\left(\frac{6EI}{L^2}\right)_f$	$\left(\frac{2EI}{L}\right)_f$
0	0	0	$-\left(\frac{EA}{L}\right)_e$	0	0	$-\left(\frac{12EI}{L^3}\right)_f$	$-\left(\frac{6EI}{L^2}\right)_f$	$\left(\frac{EA}{L}\right)_e + +\left(\frac{12EI}{L^2}\right)_f$	$-\left(\frac{6EI}{L^2}\right)_f$
$-\left(\frac{6EI}{L^2}\right)_e$	0	0	0	$\left(\frac{2EI}{L}\right)_e$	$\left(\frac{6EI}{L^2}\right)_e$	$-\left(\frac{6EI}{L^2}\right)_f$	$\left(\frac{2EI}{L}\right)_f$	$-\left(\frac{6EI}{L^2}\right)_f$	$\sum_{e,f}\frac{4EI}{L}$

7.15 ANALYSIS OF 'A-FRAME'

A typical structure an 'A-Frame' is discussed for its peculiarities in the following, utilizing this approach. Consider two cases: (1) All members in-extensible, and (2) Girder alone in extensible.

Case 1: We are using **'connection' for compatibility, the equilibrium** is satisfied by writing virtual work equation. The centrogredience enables us to use the compatibility matrix **A** for arriving at the Load vector corresponding to the reduced number of independent displacements. Consider the bases to be fixed. The other nodes are three.

The frame has three nodes. The number of unknowns by Direct Stiffness Approach:

Number of nodes × Degrees of freedom per node = 3 × 3 = 9. Inextensibility is not assumed.

Case 1: The number of unknowns by the Connection Matrix Approach assuming all members inextensible:

Refer Art. 6.4.5.5 A simple version is recalled in the following.

Nine as arrived in the case of Direct Stiffness Approach less the number of members giving rise to revised dof due to the assumption of inextensibility conditions = 9 – 5 = 4.

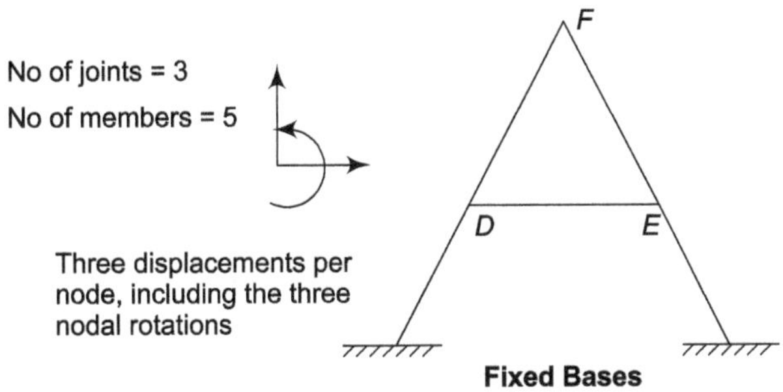

Fig. 7.24 *'A' Frame: All Members Mextensible*

The four unknowns consist of three rotations at the nodes D, E, and F; one translation, say at the node D, if the node D is treated as the Master node. The translational displacement only is sketched in the following. The numerical examples are given elsewhere.

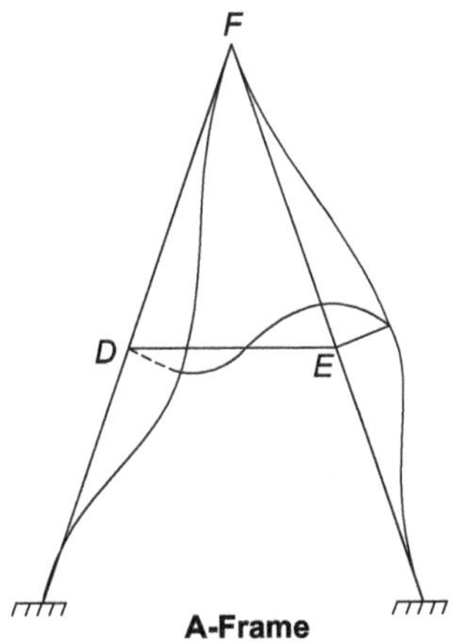

Fig. 7.25 *Deflected Shape-For All Members Inextensible, the Top Node does not Deflect (Stationary)*

If the members are assumed inextensible, there is only one translation as shown.

The top node of the 'A' frame is stationary.

Case 2: Girder '*c*' of the '*A*' frame is assumed inextensible. In addition to the three rotations at the nodes *A*, *B*, and *C*, the number of translations is equal to $2n - 1 = 5$. The five translations are sketched in the following:

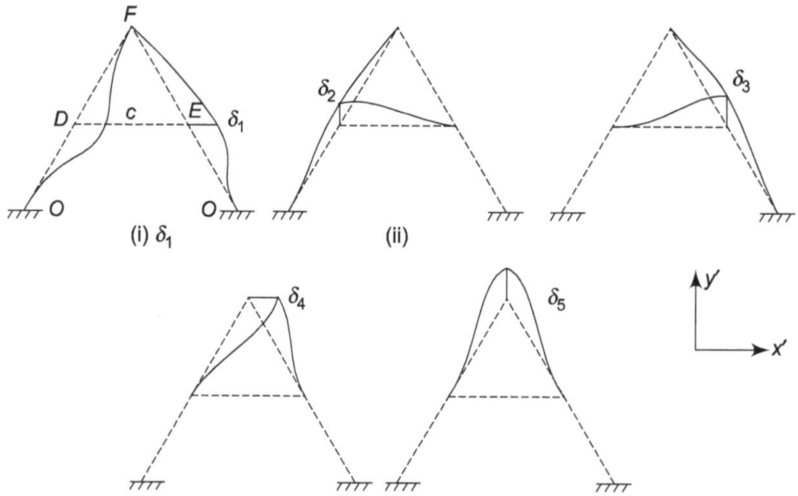

'A' Frame: Girder Alone is Inextensible: Five Translations

In the case of only one inextensible member, it is a fit case for formulating the stiffness matrix by the Direct stiffness approach and formulate corresponding global matrix *K*, of size 9×9, and then simplify the global stiffness accounting for the one member inextensibility, that is, the global stiffness matrix of size 9×9 is condensed to the size 8×8. The eight unknowns are three rotations and five translations – the node *E* is slave to the node *D*, Fig. (i), caused by inextensibility.

7.16 PLANE AND SPACE TRUSSES, AND 'MERO' SPACE TRUSS SYSTEM

The topic of the analysis of trusses is comprehensively discussed herein with emphasis on fabrication of trusses for large spans.

Trusses : 'Mero' System

Simple plane trusses are discussed with the objective of introducing to Mero Space Truss system. The Mero Space Truss system has been used extensively for large span structures, particularly the Airports structures in India.

The beams and Frames are essentially flexural members. The rolled Steel Joists for bending only and H sections for the columns which combine flexure with axial forces are the familiar choices of the designers. In the case of concrete, T-beams for girders and circular/near-square columns are the most common choices. The compound (or built up) steel sections for larger spans and Box sections are the preferred as skeletals. In the case of concrete, corresponding choices are prestressed concrete 'I' and boxes of various configurations are quite common for girders and frames.

For longer spans and large areas to be covered with minimal number of columns, 2-dimensional and 3-dimensional trusses are the designers' choice. The trussed girders are preferred over plate girders and compound/built-up sections because direct stresses/axial forces mobilize the entire section efficiently. For larger areas to be covered Mero's structural system has been extensively used. The ideal cross section for such assemblies is the pipes for their uniform stiffness. These structures are idealized as hinged joints. The Mero's structural system is essentially a double membrane. The degrees of freedom are:

Plane Trusses: Two unknown displacements per joint. And for the **Space trusses:** The displacements are three per joint.

While presenting the Force method, it has been amply demonstrated that the number of unknown redundant is relatively small. The analysis of the determinate structure requires significant computational effort. And because the formulation of the equilibrium equations for the matrix is significantly simpler the Displacement method is attractive. It is left to the ingenuity of the engineer for using Displacement method for the Hybrid structures. The trussed beam discussed under the Force method is one such hybrid structure.

Tension structures, Cable domes, and Tensegrity: These structures are not within the scope of this book.

The topic of the truss is covered under, (a) Plane Trusses, and (b) Space Trusses.

The beams and Frames are essentially flexural members. The rolled Steel Joists for bending only and H sections for the columns which combine flexure with axial forces are the familiar choices of the designers. In the case of concrete, T-beams for girders and circular/near-square columns are the most common choices. The compound (or built up) steel sections for larger spans and Box sections are preferred as skeletal. In the case of concrete corresponding choices are pre-stressed concrete 'I' and **boxes** of various configuration for girders and frames.

For longer spans and large areas to be covered with minimal number of columns, 2-dimensional and 3-dimensional trusses are the designers' choice. The trussed girders are preferred over plate girders and compound/built-up sections because direct stresses/axial forces mobilize the entire section efficiently. For larger areas to be covered Mero's structural system has been extensively used. The ideal cross section for such assemblies is the pipes for their uniform stiffness and easiness for fabrication. These structures are idealized as hinged joints. The Mero's structural system is essentially a double membrane truss. The degrees of freedom are:

The topic of the truss is covered under, (a) Plane Trusses, and (b) Space Trusses. The plane truss is being considered in the following for familiarization purpose before taking up Mero Space truss.

Plane Trusses:

Consider a plane truss, Fig. 1. All the members of the truss are taken as equal. It facilitates fabrication. In the following Pipes, ST PIPS15, are assumed for the members. The pipe sections are suitable as compression members also.

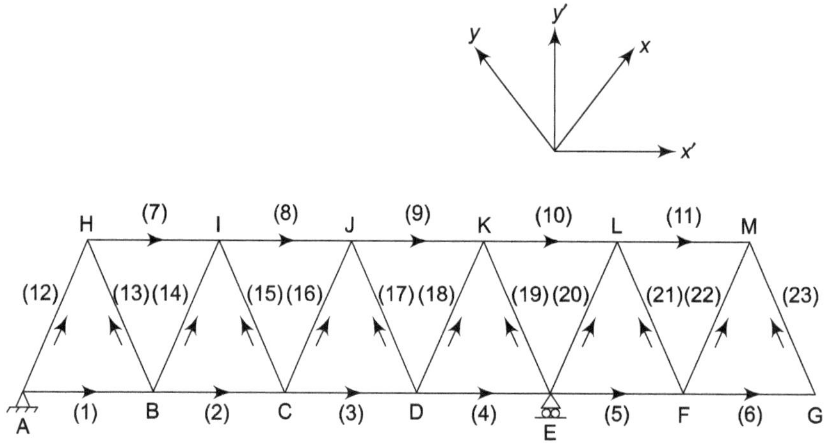

Fig. 1 *(Plane Truss)*

Nodes B, C, D, F, G, H, I, J, K, L, M loaded:

$$P'_y = -20 \text{ kN}$$

Member axes, x, are shown. The starting and end points of the member axes are the ends 1 and 2 of each member. The axes of the members 1 to 11 are in the same direction as the system axes. The other members (12) to (23) need transformations to describe the member stiffnesses referring to the system axes.

$$p_{m1} = K_{11}d_1 + K_{12}d_2$$

$$p_{m2} = K_{21}d_1 + K_{22}d_2$$

The stiffnesses $K_{11}, K_{12}, K_{21}, K_{22}$ * need to be transformed.

$$P' = Tp$$

$$T = \begin{bmatrix} \cos x'x & \cos x'y \\ \cos y'x & \cos y'y \end{bmatrix} \quad \text{and} \quad K' = T\,k\,T^t$$

Referring to the global axes, the member stiffness as given in the following.

$$p'_{m1} = K'_{11}d'_1 + K'_{12}d'_2$$

$$p_{m2} = K'_{21}d'_1 + K'_{22}d'_2 \text{ where } K' = T\,k\,T^t$$

The Direct Stiffness approach is used to write

$$p'_j = k'_j d'_j$$

consists of the nodal loads. The values of p'_x, p'_y of the node A and p'_y of the node E are the unknown reactions. Corresponding to these unknown reactions the known displacements are each equal to 0. In the example chosen the structure being determinate, the unknown reactions may be evaluated and these are:

0.0, 45.0; 0.0, 175 KN

We eliminate corresponding rows and columns and the reduced K'_j is inverted and thus find out the unknown displacements.

The unknown displacements are:

Node	δ_x' *cm*	δ_y' *cm*
1	0	0
2	.049	- 0.470
3	.1311	- 0.580
4	.1584	- 0.347
5	.0437	0
6	- .0874	- .8296
7	- .1093	- 1.6214
8	.2841	- .2760
9	.2049	- .5757
10	.1313	- .4968
11	.1721	-.1656
12	.3906	- .4022
13	.4562	- 1.2444

With the nodal deformations evaluated, the next steps are to find member deformations from compatibility and thereby enables us to find the member forces using the member-stiffnesses.

More than anything else, the illustration establishes that for the chosen example, the FORCE METHOD is the obvious choice. The structure being determinate from statics alone the member forces are evaluated.

For example referring to the node A,

$$(p_m)_{12} \cos 30 = 45 \text{ and } (p_m)_{12} = 51.964\text{C}$$

$$(p_m)_{23} \cos 30 = 20 \text{ and } (p_m)_{23} = 23.094\text{T}$$

and so on

Some typical member forces are given in the following:

Node	*Member Forces KN*
1	25.980T
2	43.300T
3	14.430T
4	60.620C
5	69.280C
6	11.550C
7	51.960C
8	48.420C
9	34.640T
10	109.700C
11	92.380C
12	23.090T

Having familiarized with a plane hinged truss, we move to Mero Space Truss in the following:

Space truss considered is the double membrane Mero System

Consider 4 × 3 modules. Member length 2 m same for all the members it is generally a requirement.

For enabling the requirement, $h = \sqrt{2\left(\frac{a}{2}\right)^2} = 0.707a$

Total number of the members = 96

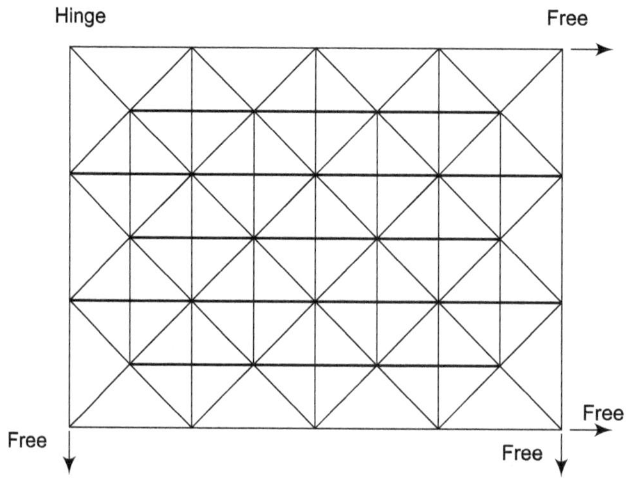

No. of Top nodes = 20

Top members = 31

No. of bottom nodes = 12

= No. of modules

Total nodes = 32

No. of bottom member = 17

No. of Diagonal member = 38

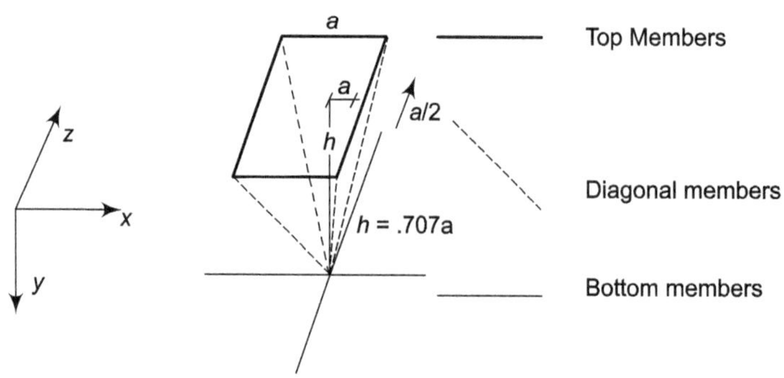

Fig. 1(a) *A Typical Module*

$$\Sigma r + \Sigma i = 104$$

No. of free bodies = no. of nodes = 32

Mero System:

Each joint consists of spherical nodes, number of bolts equal to the number of members, and cones welded to the ends of pipe members meeting at the joint.

A lever cum-pin across the hole in the sleeve removes the slackness in the members.

Illustration: Consider Mero 6 × 5

Number of Nodes = 72, All members of equal length = 2 m, h = 1.414 m

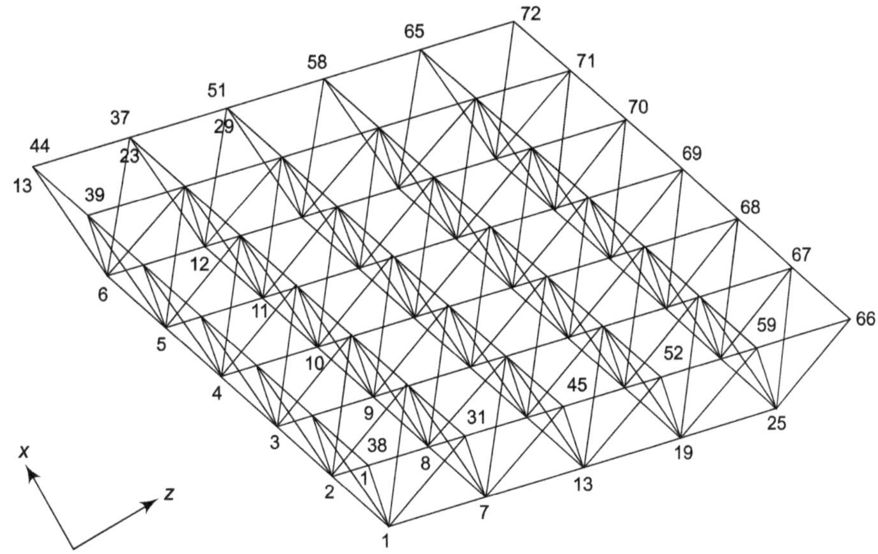

Fig. 1 *(b) Mero 6 × 5*

------ PIPS 20, Members 10, 11, 40, 41, A = 903 cm^2, 00 = 6.033 cm, TW = 0.391 cm

——— The other members PIPS 15, A = 5.155 cm^2, 00 = 4.826 cm

Top nodes:

38	39	40	41	42	43	44
31	32	33	34	35	36	37
45	46	47	48	49	50	51
52	53	54	55	56	57	58
59	60	61	62	63	64	65
66	67	68	69	70	71	72

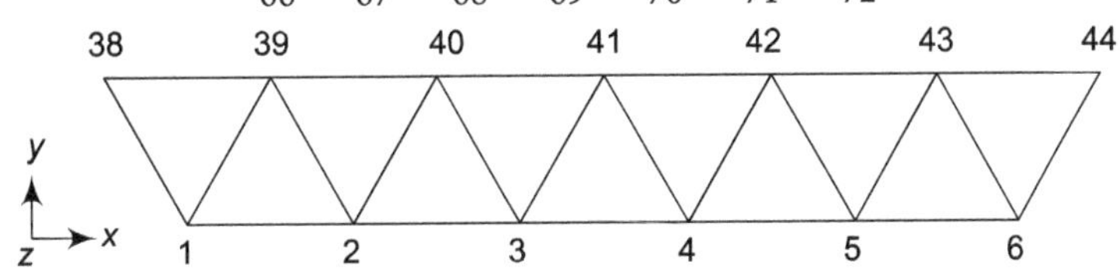

Bottom nodes:

1	2	3	4	5	6
7	8	9	10	11	12
13	14	15	16	17	18
19	20	21	22	23	24
25	26	27	28	29	30

Fig. 2 *(a) Top and Bottom Nodes*

If equal lengths is the length of the diagonal members, and H is the height of the diagonal members,

$$l_d^2 = \left(\frac{a}{2}\right)^2 + \left(\frac{a}{2}\right)^2 + H^2$$

$$l_d^2 = \left(\frac{a^2}{2}\right) + H^2$$

72	65	58	51	37	44
71	64	57	50	36	43
70	63	56	49	35	42
69	62	55	48	34	41
68	61	54	47	33	40
67	60	53	46	32	39
66	59	52	45	31	38

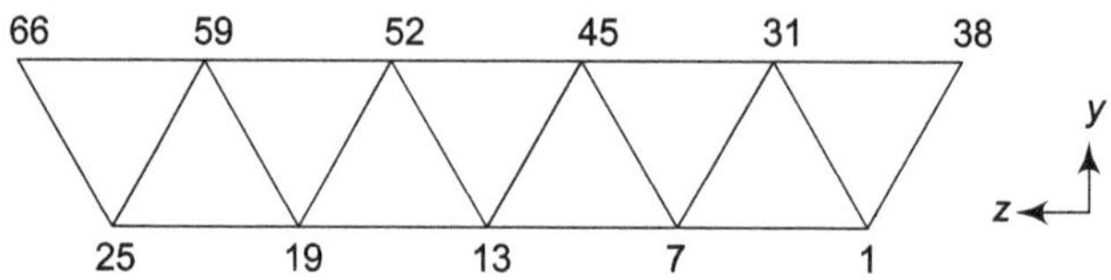

25	19	13	7	1
26	20	14	8	2
27	21	15	9	3
28	22	16	10	4
29	23	17	11	5
30	24	18	12	6

Fig. 2 *(b) End view with Top and Bottom Nodes*

Equal Lengths:

If l_d is to be kept equal to "a",

$$H = \frac{a}{\sqrt{2}} = 0.707a$$

For $H = 0.707$a, the length of the top, bottom, and diagonal members are all equal.

Number of bolts equal to the number of members, with cones welded to the ends of members meeting at the joint. A lever across the hole in the sleeve helps remove the slackness in the members by turning the bolts.

In the real life structural system, the number of the nodes runs into thousands.

The displacement method is used for the Analysis.

The unknown degrees of freedom for this simple structure consisting of mere 12 modules is equal to 96 – 8 = 88.

The slope of the roof cladding is achieved by the vertical members screwed to the top of each node supporting the cladding.

It is not always possible to have equal lengths of the pipe members.

It is a Double Membrane system.

All members of length = 2 m, covered area, 12 m × 10 m, h = 1.414 m

Number of top nodes = 42, node numbers 31 to 72, Fig. 2(a).

And number of modules = number of bottom nodes = 30, 1 to 30, Fig. 2(a) and (b).

Loading: (a) Roof loading

Sheeting, purlins and fixtures 0.55 KN/Sqm

Live Load 0.75

Nodal loads, sheeting etc. = 0.55 (2 * 2) = 2.20 KN

LL = 0.75 (2 * 2) = 3.0 KN

} Joint Load = 5.20 KN, on nodes 31 to 72

EQ load As per IS Code, top nodal load = 2.2 + 25% 3.0 = 2.95 KN, Joints 31 to 72

Similarly bottom nodal load = 0.50 (2 * 2) = 2.0 KN applied at the nodes 1 to 30.

Provide PIPES 15 (Area = 0.80 or 5.15, OD = 1.9 in, t = 0.14 in and E = 205 KN/mm^2) These are is sections

Based on the analysis results, the support reactions are:

Node 1: R_y = 69.6 kN

6, 30, and 25: R_y = 69.6 kN

Table 7.4 *Deflections of typical nodes*

Joint designation**	34	48	55	69	54
Vertical downward	0.7038	0.7101	0.7101	0.7063	0.835

The nodes are marked in Fig. 1(b)

Table 7.5 *Member forces of typical members*

Member designation**	3	97	193	196	214	215	10,11,40,41
Member Axial force	44.86T	37.46T	29.44C	29.53C	29.44C	29.53C	33.048C

The members above are marked in Fig. 1(b)

Discussions: Particularly when we have not considered the self weight, and also that the members, 10, 11, 40, and 41 are in compression, we may provide heavier section,

PIPS 20 (area = 1.07 in^2)

Table 7.6 *Revised deflections of typical nodes*

Joint designation	34	48	55	69	54
Vertical downward	0.6881	0.6983	0.6983	0.6841	0.636

Table 7.7 *Revised member forces of typical members*

Member designation*	3	97	193	196	214	215	10,11,40,41
Member Axial force	45.26T	37.73T	29.22C	30.30C	28.73C	30.30C	34.77C

There is a marginal increase in the member force 33.048 to 34.77 C but the ratio of areas = 1.07/.80 = 1.3375, that is, an increase of 33.75% OK. The revision is recommended.

The LIC-six storey complex at Connaught Place was the first Mero-Structure erected in India. The author was associated in the designing and testing process of the structure. In addition to the advantage of the ease of erection at great heights without any welding and props/or supports, the system allows the added advantage of grouping of the members depending upon the magnitude of the axial forces. From any standards, the illustration chosen (with 72 nodes, 240 members) can only be termed as a modest structure. The critically stressed compression members at the top were chosen as a group of a different pipe-size. In a real life structure, however, there may be a number of groupings running into dozens.

CHAPTER 8: ITERATIVE TECHNIQUES

Chapter

8

Iterative Techniques

8.1 ITERATIVE TECHNIQUES—DISPLACEMENT METHOD

The 'Moment Distribution' proposed by Hardy Cross is an incremental iterative technique for computing the unknown internal moments at the ends of the members meeting the joints based on equilibrium considerations, a Displacement Method. It was in general use from 1935–1959. From the year 1960, the computer based solutions for large structural systems gradually superseded the moment Distribution. Its wide and continued popular acceptability till then for small plane frames and continuous beams are a tribute to Hardy Cross.

8.2 PREDICTION AND CORRECTION—CONCEPTUALIZATION

The first part of the analysis consists of predicting the end moments of all the members by assuming their ends to be fixed (or clamped) at the joints. These predicted moments are corrected by satisfying equilibrium of the joints.The fixed end moments correspond to the *'no displacement'* (of joints) analysis. It is highly conceptual in the sense that the clamped joints have to be released and correcting moments are computed from Equilibrium considerations. All these constitute the **Part I** , a "no displacement" analysis of the Moment Distribution.

In the **Part II** of the analysis the correcting moments are distributed between the ends of the members meeting at each joint. The releasing allows for displacements, rotation and translation; and then again clamping of the joints of the frames are done sequentially. Consequent to the sequential *release* of the joints, at the end of each release, the sum of the end moments at each joint being released is *balanced*, in proportion to the *relative stiffnesses* of the members meeting at the joint. The joints taken up in a sequence, are selected by the designer, achieving a *zero unbalance*. In the process of allowing for the displacement at each release, we are computing the correcting moments based on the relative stiffnesses of the members. The balancing by introducing correcting moment at the end of each member meeting the joint while keeping the far end clamped, is the process of *correction*. Having introduced the correcting moments at each member end meeting at the released joint, we need to carry over the effect of balancing to the far end. Whenever a joint is released all other joints remain clamped. The effect is *carried over* to the far end of the members by a *factor "Half"* applicable to the *prismatic members*. This operation is performed for all the members meeting at the joints. Following a *sequence,* all the joints are released in a sequence, corrections applied as balancing moments, and while balancing the moments, the effect is carried over to the far ends. The cycles of operations are continued till convergences is achieved.

All the joints are clamped while doing Part I of the analysis, enabling the computations of the fixed end moments, as predicted moments. The fixed end moments correspond to the *'no displacement'* of the joints. You may have started to feel that it is a highly conceptual visualization, in the sense that the clamped joints are released and correcting moments are computed based on relative stiffnesses for satisfying Equilibrium conditions. The releasing of a joint results in undergoing Displacements, Fig. 8.1.

The releasing and clamping of the joints of the frames being done sequentially is a structural process. Consequent to the sequential *release* of the joints, at the end of each release, the sum of the end moments at the joint being released is *balanced,* in proportion to the *relative stiffnesses* of the members meeting at the joint constitute a structural process. Thus the joints taken up in a sequence, to be decided by the designer, achieves *zero unbalance.* The carried over moments disturb the equilibrium, necessitating the next cycle of corrections of the Moment Distribution. The process gives the feel of the structural behavior.

In the above, we introduced the terminologies, shown in the italics, like, no displacement of joints, clamping of joints, fixed end moments consequent to clamping, releasing of the joints, unbalanced moments, distribution in proportion to the relative stiffnesses, zero-unbalance, carrying over to the far end of each member by a factor 'half' for the prismatic members. These operations are explained in the following, with the help of illustrations.

Consider a Joint *O*, see Fig. 8.1, being released and the 'balancing-moments' are determined based on the relative stiffnesses of the four members. The far ends of the members continue to remain clamped. As a result, the four far ends *A*, *B*, *C*, and *D* receive the carried over moments. Thus we continue the process till all the unbalanced joints of a structure require minimal correcting moments. The sum of fixed end moments, correcting (incremental) moments of each member, give the final bending moment. The 'incremental iterative technique' is also known as relaxation technique.

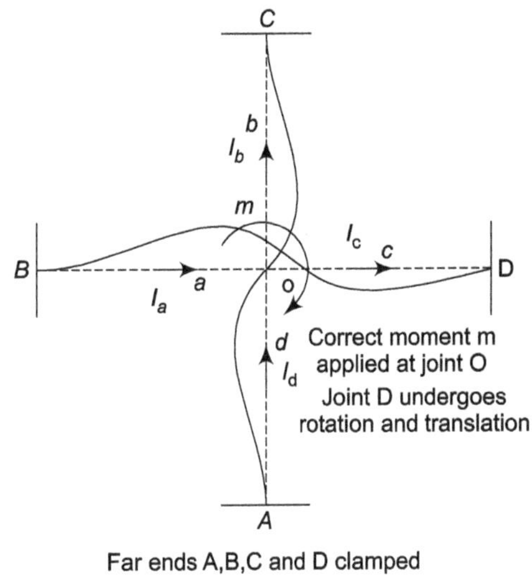

Fig. 8.1 *Typical Joint of a Frame Undergoing Displacements Caused by Correcting Moment m-Joint O Released*

Even though the design offices are becoming increasingly dependent on the computer software based on Force or Displacement methods, the designers are well advised to solve the problems manually, wherever possible, by making use of the Hardy Cross-moment Distribution. In fact many feel, Hardy Cross paper (1949) deserved recognition no less than a Nobel Prize (instituted in 1895/ first award 1901).

'Moment Distribution' and Kani's method: The mathematical presentation of the moment distribution is the "incremental iterative technique".The Kani's method is an iterative technique. It is nothing but a tabular presentation of the technique for computing rotations. Even though there is not much of structural concept associated with the Kani's technique, it has certain advantages like convergence and minimization of errors. Both the methods applicable to the rigid jointed structures only are presented in the following. We will see more of these elucidated in the presentation to follow.

8.3 GENERAL CONSIDERATIONS OF THE METHODS, INEXTENSIBILITY AND RAPID CONVERGENCE

Iterative procedures, which include incremental iterative technique, assume the members to be inextensible. Both the techniques use the displacement method. The simultaneous equations based on the stiffness or displacement method are "well conditioned." The leading diagonal elements of the stiffness matrix being large, the solution of the unknowns by iterative methods are most suitable due to rapid convergence. The method, proposed by G. Kani, is an iterative-displacement-method and it solves for the unknown displacements. The "Kani's Method" is sometimes preferred over Hardy Cross-moment distribution, even though the latter method based on "relaxation" of joint for moment equilibrium, is more conceptual. Every step of the moment distribution method, as we have seen, can be interpreted in physical terms. In the case of rigid frames and particularly multi-story frames, several translational displacements (degrees of sway) are the unknowns in addition to the joint rotations. Complete sets of moment distributions (which satisfy moment equilibrium of joints) are required to be performed, (i) for the given loading. It constraints against sway, while allowing for the joint rotations, and (ii) releasing of the sway constraints. The fixed end moments are computed for each case of "sway", based on no-joint-rotation. Then each sway case is considered, allowing for the joint rotations only. For the entire frame it is required to perform two steps of the moment distributions (i) for given loading and (ii) each independent auxiliary sway displacement individually which satisfy moment equilibrium only. The "force equilibrium" conditions are used to obtain the actual values of the displacements used for the unit sway corrections. The evaluation of the independent linear displacements requires solution of simultaneous equations for the unknown sways in addition to several operations of the moment distributions for (i) and (ii). The illustrations complement the concept outlined above. The "Kani's-iterative method" to solve for the unknown deformations (rotations and linear displacements) has significant advantages over the moment distributions method for the reason that separate iterations are not required for sway corrections for the force equilibrium. Because of the assumption of 'inextensibility', the axial forces remain undetermined. The axial forces are computed separately using appropriate free bodies with flexural forces acting on them, computed by the iterative techniques.

8.4 MATHEMATICAL ITERATIVE METHODS

Before presenting the iterative method for the structural frames a brief description of two standard mathematical iterative methods, (1) Jacobi method and (2) Gauss-Seidel method, are given in the following. The "Kani's iterative method" is essentially the Gauss-Seidel method of solution of the displacement equations of the structural frames. The Article 8.5 presents the moment distribution as a particular case of "relaxation method" for the solution of simultaneous equations obtained by satisfying equilibrium of the joints of a continuous beam.

8.4.1 Jacobi Method of Solution of Simultaneous Equations

Let us consider the following equations as an illustration:

$$\left.\begin{aligned} a_{11}x_1 + a_{12}x_2 + a_{13}x_3 + a_{14}x_4 &= c_1 \\ a_{21}x_1 + a_{22}x_2 + a_2x_3 + a_{24}x_4 &= c_2 \\ a_{31}x_1 + a_{32}x_2 + a_{33}x_3 + a_{34}x_4 &= c_3 \\ a_{41}x_1 + a_{42}x_2 + a_{43}x_3 + a_{44}x_4 &= c_4 \end{aligned}\right\}$$

For rapid convergence, the pivotal or diagonal elements a_{11}, a_{22}, a_{33} and a_{44} should be large. In the case of the Displacement methods the diagonal elements are dominant. The unknowns associated with the pivotal elements are expressed in terms of the other unknowns.

$$\begin{aligned} x_1 &= \frac{1}{a_{11}}(c_1 - a_{12}x_2 - a_{13}x_3 - a_{14}x_4) \\ x_2 &= \frac{1}{a_{22}}(c_2 - a_{21}x_1 - a_{23}x_3 - a_{24}x_4) \\ x_3 &= \frac{1}{a_{33}}(c_3 - a_{31}x_1 - a_{32}x_2 - a_{34}x_4) \\ x_4 &= \frac{1}{a_{44}}(c_4 - a_{41}x_1 - a_{42}x_2 - a_{43}x_4) \end{aligned} \qquad \text{...(a)}$$

The first cycle of the iterative procedure consists of solving for the unknown appearing on the left hand side by considering c_1, c_2, c_3, and c_4 only. In this cycle the x_1, x_2, x_3, and x_4 values appearing in the right hand side are taken equal to zero. Thus we obtain

$$x_1^{(1)} = c_1/a_{11} \quad x_2^{(1)} = c_2/a_{22} \quad x_3^{(1)} = c_3/a_{33} \quad \text{and} \quad x_4^{(1)} = c_4/a_{44}$$

The superscript (1) of x_1 indicates the first cycle value in general, $x_1^{(n)}$, means the value of x_1 obtained in the nth cycle of the iteration.

The first set of values is used in the second cycle to obtain $x_1^{(2)}$. The procedure is continued till $x_1^{(n)} \approx x_1^{(n-1)}$ i.e., we stop iterations at the nth cycle when the difference between the x_1 values of two consecutive cycles $(n-1)$ and (n) is within prescribed acceptable variation.

8.4.2 Gauss-Seidel Method of Solution of Simultaneous Equations

We start with equation (*b*) of the Jacabi method. The approximate values of x_1 of the first cycle are computed as given below.

$x_1^{(1)} = \dfrac{c_1}{a_{11}}$, that is, in the first equation we assume

$x_1 = x_2 = x_3 = x_4 = 0$

Substitute the approximate values of x_1, i.e., $x_1 = x_1^{(1)}$, $x_3 = x_4 = 0$ into the second equation of *(b)* to obtain

$$x_2^{(1)} = \frac{1}{a_{22}}\left(c_2 - a_{21}x_1^{(1)}\right)$$

Similarly,

$$x_1^{(1)} = \frac{1}{a_{33}}\left(c_3 - a_{31}x_1^{(1)} - a_{32}x_1^{(1)}\right)$$

$$x_4^{(1)} = \frac{1}{a_{44}}\left(c_4 - a_{41}x_1^{(1)} - a_{42}x_2^{(1)} - a_{43}x_3^{(1)}\right)$$

The above evaluations of x_1 complete one cycle.
The set of equations for the second cycle are:

$$x_1^{(2)} = \frac{1}{a_{11}}\left(c_1 - a_{12}x_2^{(1)} - a_{13}x_3^{(1)} - a_{14}x_4^{(1)}\right)$$

$$x_2^{(2)} = \frac{1}{a_{22}}\left(c_2 - a_{21} - x_2^{(2)} - a_{23}x_3^{(1)} - a_{24}x_4^{(1)}\right)$$

$$x_3^{(2)} = \frac{1}{a_{33}}\left(c_3 - a_{31}x_1^{(2)} - a_{32}x_2^{(2)} - a_{34}x_4^{(2)}\right)$$

and

$$x_4^{(2)} = \frac{1}{a_{44}}\left(c_4 - a_{41}x_1^{(2)} - a_{42}x_2^{(2)} - a_{43}x_3^{(2)}\right)$$

The x_i values of the nth cycle are given by:

$$x_1^{(n)} = \frac{1}{a_{11}}\left(c_1 - a_{12}x_2^{(n-1)} - a_{13}x_3^{(n-1)} - a_{14}x_4^{(n-1)}\right)$$

$$x_2^{(n-1)} = \frac{1}{a_{22}}\left(c_2 - a_{21}x_1^{(n)} - a_{23}x_3^{(n-1)} - a_{24}x_4^{(n-1)}\right)$$

$$x_3^{(n)} = \frac{1}{a_{33}}\left(c_3 - a_{31}x_2^{(n-1)} - a_{33}x_3^{(n-1)} - a_{34}x_4^{(n-1)}\right)$$

$$x_4^{(n)} = \frac{1}{a_{44}}\left(c_4 - a_{41}x_1^{(n)} - a_{42}x_2^{(n)} - a_{43}x_3^{(n)}\right)$$

It is obvious that the Gauss-Seidel method is superior to the Jacobi in the sense that the latest value of x_i on the right hand side of the system are used to compute the new values. Incremental iteration is explained while solving continuous beam Art. 8.7.

It may be inferred from the solution Art. 8.7 that the moment distribution and the incremental relaxation method of the solution Art. 8.2 are similar. The moment distribution procedure also balances the joints incrementally. In the case of frames with sway, the moment distribution algorithm, as it satisfies moment equilibrium only, requires several sway corrections depending on the independent number of joint translations. Whereas in the case of the relaxation method, the elements of A matrix in the relation $AX = C$ can be written directly which incorporates the effect of sway (i.e., the Moment as well as Force Equilibrium equations). The matrix C is the generalized load matrix obtained from the given loading and chosen free bodies for the equilibrium equations. The relaxation procedure presented gives all the unknowns, which are used to find the member end moments. The relaxation algorithm based on mathematical or numerical technique as compared to the moment distribution is more convenient to write a computer program.

Conceptually, however, we are not visualizing the structure while doing the iterative operations which is a great advantage in favor of the moment distribution. Both the incremental iterative techniques, viz, (a)Relaxation numerical algorithm and (b) Moment Distribution have been used for solving a Continuous Beam problem Tables 8.7 and 8.8 are given at the end of the section pertaining to Moment Distribution. The reader may like to assess the advantages of Moment distribution over the mathematical algorithms. It is a good preparation before we move over to Kani's technique applied to Continuous Beams.

8.4.3 Elimination and Back Substitution

The algorithm of elimination and back substitution is explained for the solution of simultaneous equations obtained from the equilibrium consideration while analyzing a gable frame. There is no iteration implied, but the presentation is aimed at comparison. It is used with advantage for the analysis of tall Buildings. Mathematically speaking the augmented matrix is reduced to a triangular matrix using the process of *elimination*. After finding the displacement from the last equation, back substitution is used to find the other displacements by the *back substitution*.The technique is used in the following for the analysis of Gable Frame. The technique is simple, and it is useful. It competes well with the relaxation and iterative solutions. We had a glimpse of the mathematical solution when we discussed unified energy principle in Chapter 4. The elimination and back substitution algorithm is 'exact', where as the iterative procedures are basically 'approximate'.

The numerical techniques are presented in the Appendix A, Mathematical preliminaries.

8.5 ROTATION AND SWAY OF FRAMES

The rotations and translation are represented by the model, Fig. 8.1. The moment load applied at the joint O, in general, results in not only rotation of the members meeting at the joint but also in translations. The continuous beams on non-sinking roller supports allow rotations only at the roller or hinged joint. The joints of the frames and also beams on spring supports undergo rotation as well as translations. The algorithm of Moment distribution, however, is based on rotations and the moment equilibrium. The sways or translations are treated as unknowns and separate moment distributions are done for each case of the sways, and equations are formed for the evaluation of the unknown sways or translational displacements. The aspect of sways or translations is demonstrated with the help of illustrations. The only shortcoming of the model, Fig. 8.1, is that it allows axial deformations along with rotations. Our iterative techniques and for that matter all our frames assume inextensibility. In the case of beams and grids, however, axial effect is delinked from the flexural effect. We used the assumption of inextensibility for the analysis of frames discussed in Chapters 6 and 7.

8.6 ANALYSIS OF GABLE FRAME USING ELIMINATION AND BACK SUBSTATION

The equations are formed from the equilibrium considerations. The geometry and loading data are given in Fig. 8.2(a). Relative stiffnesses are assumed to be the same for all the members. The globel axes are shown. The member axes are marked on the members. The degrees of freedom are three rotations corresponding to the three joints *A*, *B*, and *C*, and two translations. The rotations and translations are grouped separately, Fig. 5.2 (a) and (b).

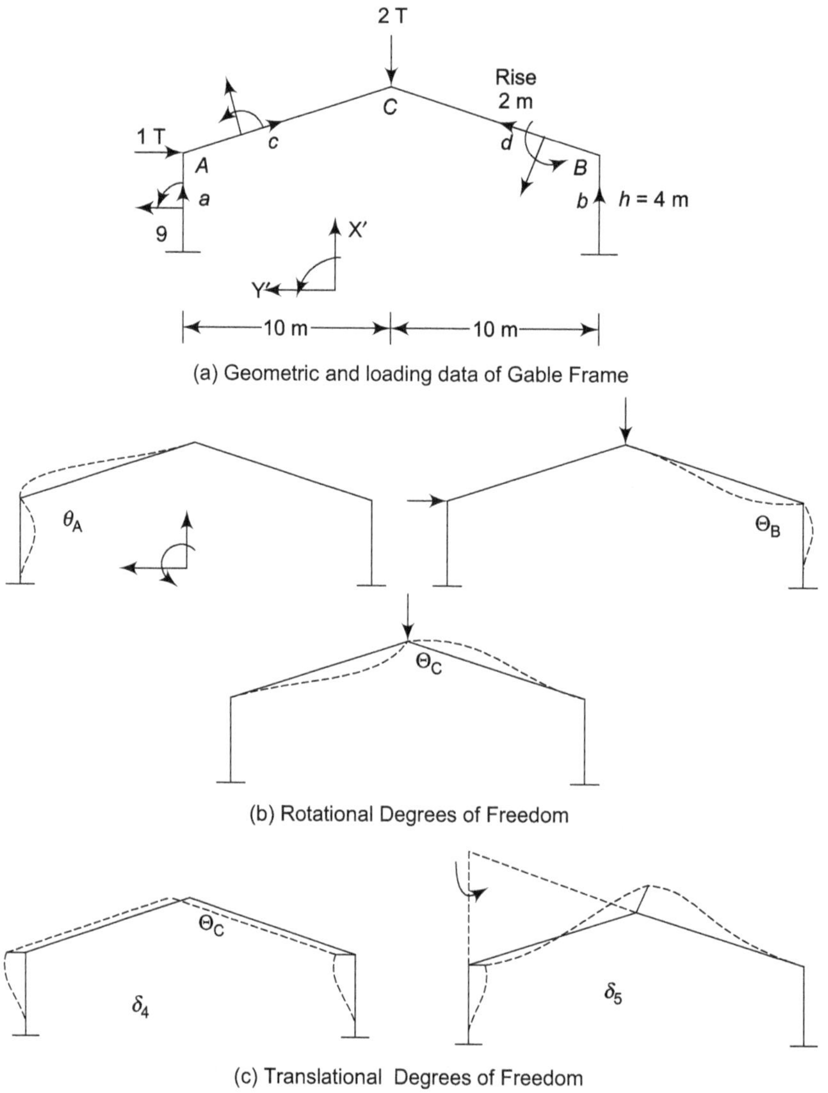

Fig. 8.2 *Symmetrical Gable Frame and Degrees of Freedom for Unsymmetrical Loading*

The geometry and loading are shown in Fig 8.2(a). The member and global axes are shown. The global axes are so chosen that the members 'a' and 'b' do not require any axes transformation. The relative stiffness of all the members assumed unity. The problem will be solved by (a) Slope deflection method (relative stiffnesses equal), where the five simultaneous equations are solved by the method of elimination and back substitution (*EI* values same), described already, Article 8.3.2 , and (b) by "Moment Distribution". For moment distribution, it will be solved for the members of the same *EI* values. The sensitivity is presented at the end.

The equations I to V of the example, based on slope deflection method are given in the following matrix form. The relative stiffness of all the members is taken equal to unity.

Free bodies for equilibrium equations

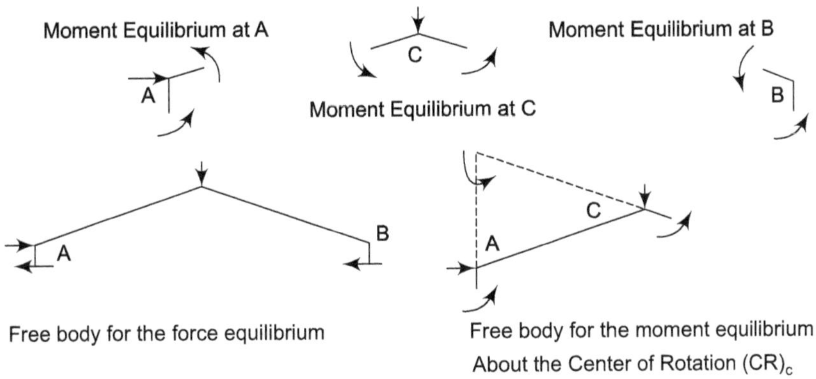

Fig. 8.3 *Free Bodies corresponding to the Degrees of Freedom*

$$\begin{bmatrix} 8 & 0 & 2 & 0 & -1.5 \\ 0 & 8 & 2 & -1.5 & 0 \\ 2 & 2 & 8 & 0 & 0 \\ -1.5 & -1.5 & 0 & 0.75 & 0.75 \\ 10 & 8 & 10 & -9 & 4.5 \end{bmatrix} \begin{bmatrix} \theta_A \\ \theta_B \\ \theta_C \\ (\delta'_y)_A \\ (\delta'_y)_B \end{bmatrix} = \begin{bmatrix} 0 \\ 0 \\ 0 \\ -1 \\ -16 \end{bmatrix}$$

The (*EI*) values are merged with the deformations. The relaxation method for the above system may not be suitable unless we do some linear combinations. The linear combinations are necessary to make the leading diagonal elements dominant. In the present case, the elimination procedure (presented while discussing rank of a matrix and stable frames) is used.

Table 8.1 *Elimination and back substitution*

Equation No.	X_1	X_2	X_3	X_4	X_5	C	*Remarks*
(1) (i) (ii)	8	0	2	0	–1.5	0	Choose the row for normalizing the first
	1	0	.25		.1875	0	column(Over).
(2) (i) (ii)	0	8	2	–1.5	0	0	Choose Eq(2) for normalizing the second
	0	1	.25	–.1875	0	0	column. Eq(2)/8.
							Element being nonzero, normalize. By dividing the Eq. 2, Eq. 2/8 (Over)
(3) (i) (ii)	2	2	8	0	0	0	Reduce the second column element = 0.Eq.
	0	2	7.5	0	.375	0	3- 2*Eq.1(ii)
		0	7	.375	.375	0	Reduce the third column element=0. Eq.
			1.	.0536	.0536		3-2*Eq.2(ii)
							Divide Eq3 by 7
(4)	-1.5	-1.5	0	0.75	0.75	-1	Similarly reduce the Eq.4, so that fourth
				1	1	-2.330	column element is normalized.
							We get the fourth row modified to
(5)	10	8	10	-9	4.5	-16	Finally, when the four column elements of
	0	0	0	0	13.5	33.30	Eq. 5 are reduced to zero, we get

From the last row of Table 8.1,

We get, $13.50\,(\delta' y)_B = -33.3$

$\therefore \quad (\delta' y)_B = -0.247$ and from the fourth row

$$(\delta' y)_A + (\delta' y)_B = -2.330$$

$\therefore \quad (\delta' y)_A = +\,0.137$

Elimination is complete and we start back substitution.

Similarly, the third, second and first rows and the values of θ_C, θ_B and θ_A respectively

$$\theta_C = -0.0536\,(-2.47 + 0.14) = +.125$$

$$\theta_B = -.25(.125) + .1875\,(+\,0.14) = -.005$$

$$\theta_A = -.25(.125) + .1875\,(-2.47) = -.495$$

Member Moments:

$$(m_1)_A = -1.20\,;\ (m_2)_a = -2.19$$

$$(m_2)_B = +\,3.69\,;\ (m_2)_b = +3.68$$

$$(m_1)_C = 4(-.125) + 2(-.495) + 1.5(0.14 + 2.47) = +3.420$$

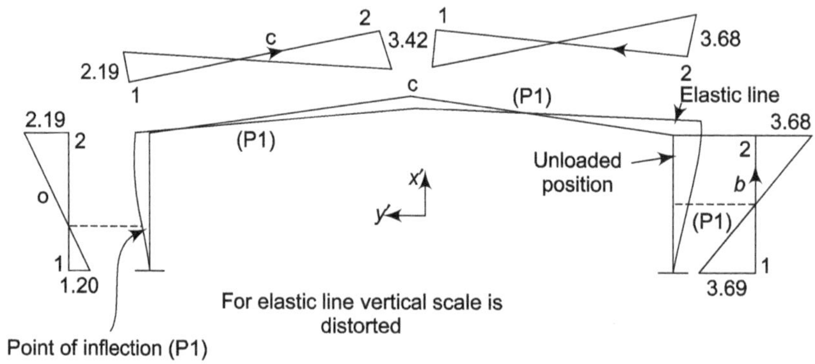

Fig. 8.4 *Final Bending Moment Diagram and Elastic line*

8.7 SOLUTION OF GABLE FRAME BY MOMENT DISTRIBUTION

All members are considered to have Equal *EI* values. The reader is advised to compare the results. In the following, the problem is solved without taking advantage of symmetrical and antisymmetrical loading. Primary moment distribution is performed for joint moments computed from the Part I. The joints are allowed to rotate. The moment distribution values corresponding to the two sways also allow joint rotations in the two swayed positions.

Consider lateral displacement, $(\delta' y)_A$ only

Rotations due to lateral end displacements of different members

$$(\psi)_a = \frac{(\delta' y)_A}{4}, (\psi)_b = 0, (\psi)_c = -\frac{(\delta' y)_A}{4}, (\psi)_d = (\delta' y)_A$$

Fixed end moments = $(\text{FEM})_a = 6\,(EI/L)_a\,x = (\psi)_a = -\,1.5\,(\delta' y)_A\,EI$

$$(\text{FEM})_c = 1.5\,(\delta' y)_A\,EI,\ (\text{FEM})_d = -\,1.5\,(\delta' y)_A\,EI$$

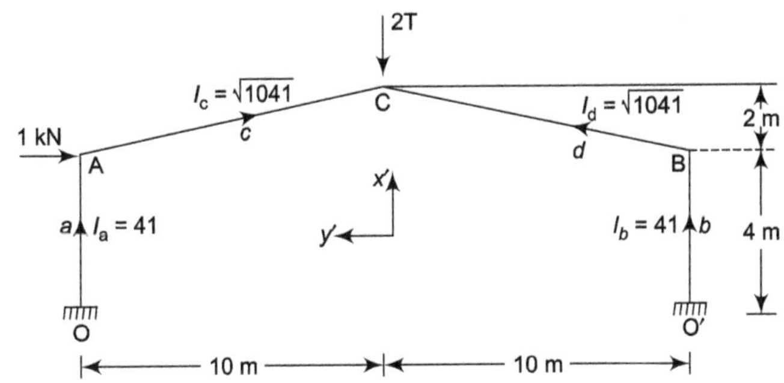

Fig. 8.5 *Data-equal EI for Moment Distribution*

Symmetrical and antisymmetrical loading on the symmetrical gable: In the following, the above problem is solved by taking advantage of the geometric symmetry of the gable. The loading is split into symmetrical and antisymmetrical loading. And the results of the two cases are superposed to obtain the final result.

8.7.1 Case 1: Symmetrical Loading

It may be verified that the lateral sway of member "*c*" = $\delta/2$, 2 m is the distance of the (CR)c from the joint *A*, where δ is the later displacement. Rotation of '*c*' is $\delta/2$ clockwise.

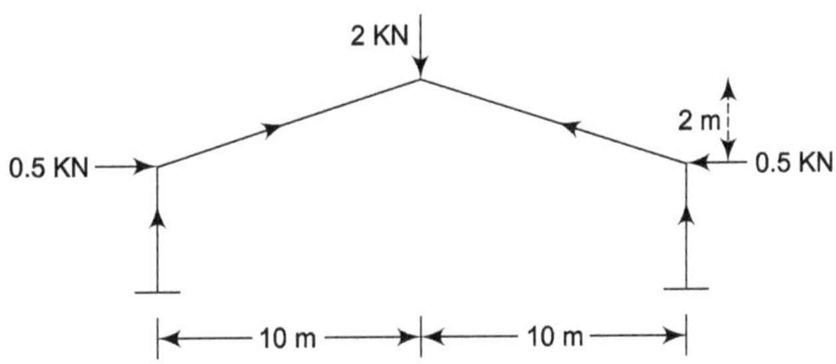

Symmetrical case

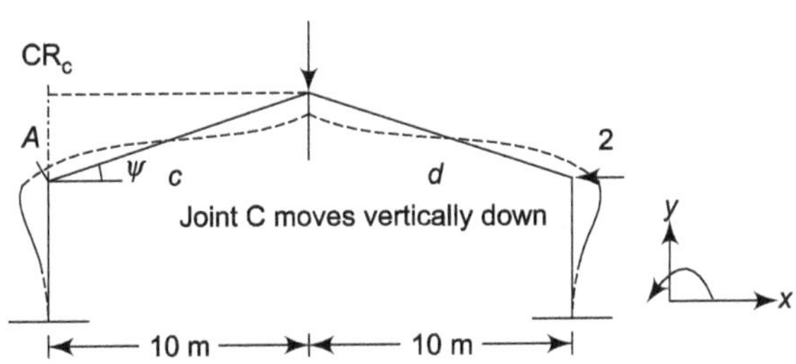

Elastic line - Symmetrical case 1

Numerical computations required for the moment distribution:

$$(\text{FEM})_c = (6\,EI\delta/L^2\,)_c = 6EI(0.5\,\delta)/\sqrt{104} = 0.2942\,\delta EI$$

Similarly $(\text{FEM})_a = (-0.3750)\,\delta EI$, the unbalanced moment at joint $A = 0.2942 - 0.3750 = -0.0808\,\delta EI$

Relative stiffnesses: $1/\sqrt{104}$: ¼, 0.2817: 0.7183, Common multiplier = δEI

The moment distribution for the symmetrical case is shown below.

Table 8.2 *Moment Distribution*

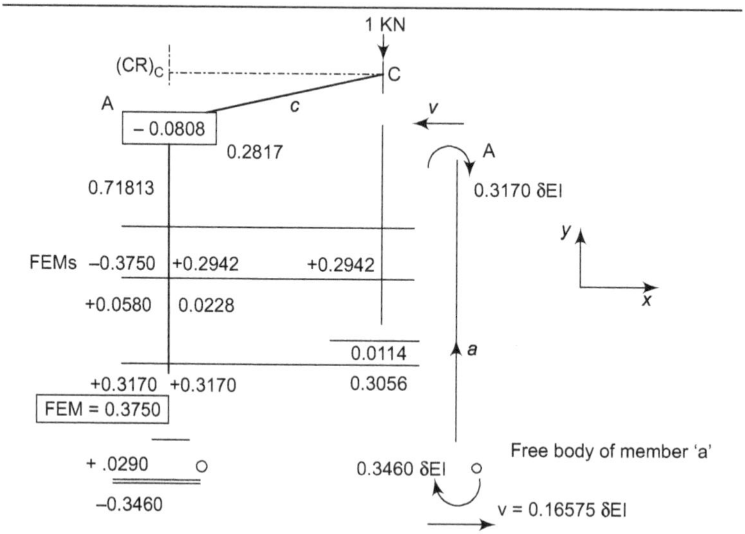

For the free body OAC, take moments about the $(CR)_c$
$6\,V - 0.3460\ \delta EI + .3056\ \delta EI + 0.5 \times 2 - 1 \times 10 = 0,$
The common multiplier, $\delta EI = 9.4340$
The Bending Moment diagram for the symmetrical case:

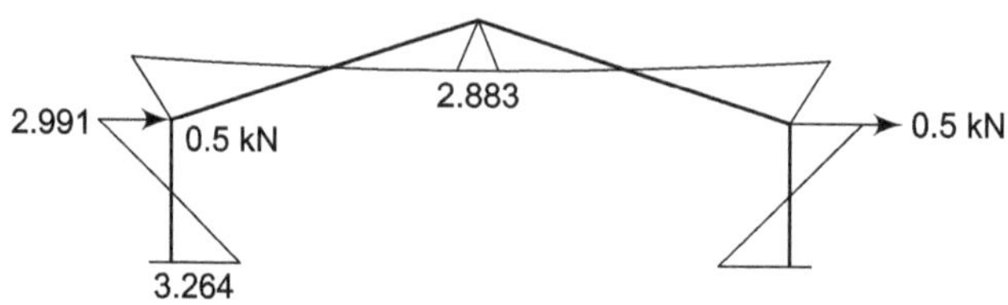

Fig. 8.6 *Bending Moment Diagram for Symmetrical Case*

8.7.2 Case 2: Antisymmetrical Loading

It is good to sketch elastic line in anticipation of the analysis. Guess the inflexion points. At the symmetry point bending moment is zero, that is, the end 2 of member 'c' has zero moment. We modify the stiffness by a factor ¾.

Relative stiffnesses $(1/\sqrt{104}) \times 3/4 : 1/4$ (same EI)

0.227 : 0.773 These values are entered at the appropriate ends of the members '*a*' and '*c*'.

For this case, we may treat the entire structure as a free body. Using the statical condition of sum of the horizontal forces = 0.0

$(0.0788 + 0.0788)\ \delta EI - 2 \times 0.5 = 0.0$

$$\delta EI = 1/0.1576 = 6.345$$

With the unknown δEI evaluated, we obtain the bending moments for the antisymmetrical loading.

Superposing the two cases, symmetrical loading and antisymmetrical loading, the following is the final Bending Moment Diagram.

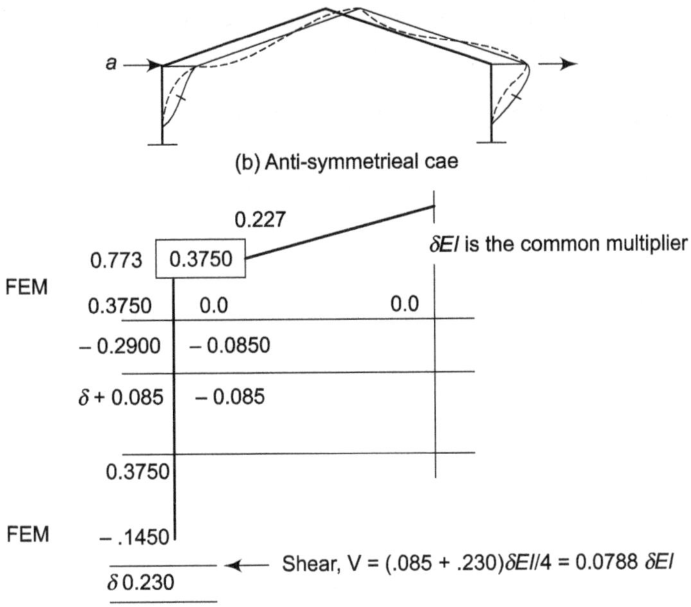

Fig. 8.7 *Antisymmetrical Loading, Elastic Line and Table of Moment Distribution*

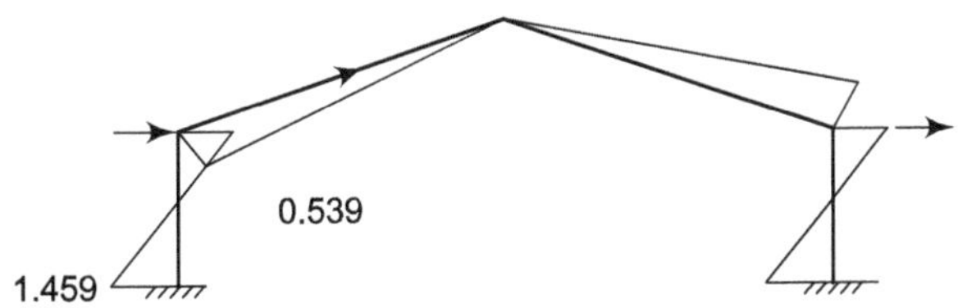

Fig. 8.8 *Bending Moment Diagram for the Antisymmetrical Loading*

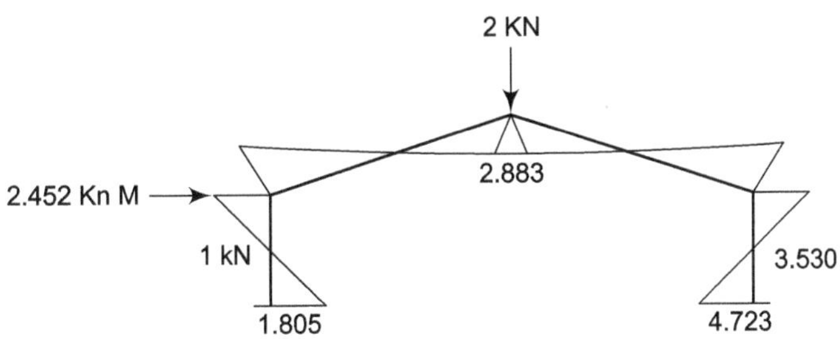

Fig. 8.9 *Final Bending Moment Diagram*

Comment on:

1. Effect of the change in the cross-sectional properties on the behavior of the structure.
2. Why larger moment on the members on the right of the crown.
3. Location of the inflexion points.
4. Your preference of the method? Mathematical elimination and back substitution.

The frame being geometrically symmetrical, we may or may not consider the analysis for symmetrical and antisymmetrical loading. (Your opinion)

Manual computations if not made a prerequisite; it should at least be used along with the computer results from the analysis using the commercial software, for a proper appreciation of the behavior of the structures.

8.8 MOMENT DISTRIBUTION APPLIED TO CONTINUOUS BEAMS

The application of continuous Beams on non-sinking supports is an example of a structure, where joints undergo rotation only, Fig. 8.10.

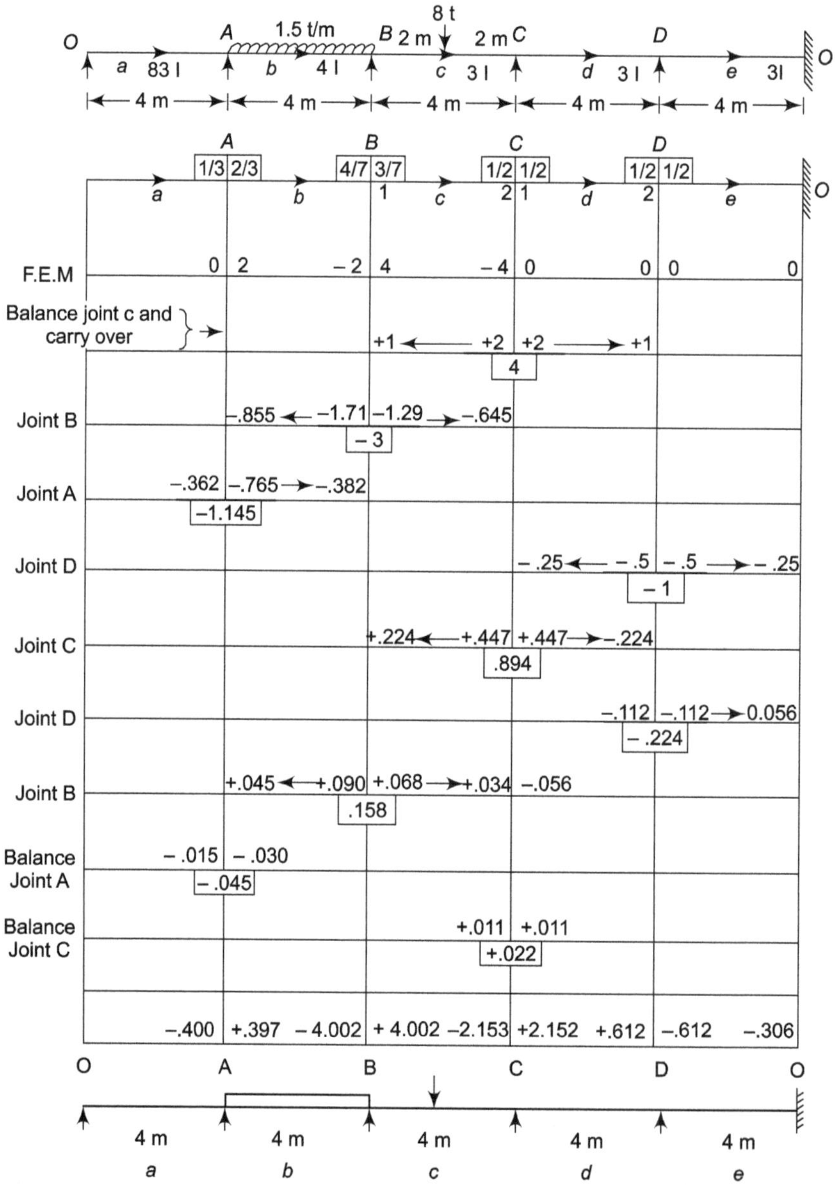

Fig. 8.10 *Continuous Beam on Non-Sinking Roller Supports*

In the following, the same problem of continuous beam is anatlyzed mathematically using relaxation technique. The values are compared in the Table 8.3.

Table 8.3 *Implementation of relaxation technique*

$\bar{\theta} = EI/L.\theta$

$\bar{\theta}_A$	$\bar{\theta}_B$	$\bar{\theta}_C$	$\bar{\theta}_D$				$\bar{\theta}_C$ −4/24	$\varepsilon^{(1)}$	$\bar{\theta}_B$ −3/28	$\varepsilon^{(2)}$	$\bar{\theta}_A$ 1.143\24	$\varepsilon^{(3)}$	$\bar{\theta}_D$ −1/24	$\varepsilon^{(4)}$	$\bar{\theta}_C$ .893\24	$\varepsilon^{(5)}$	$\bar{\theta}_D$ −.223\24	$\varepsilon^{(6)}$	$\bar{\theta}_B$ −.158\28	$\varepsilon^{(7)}$	$\bar{\theta}_A$	ERROR $\varepsilon^{(8)}$ Negl-ected
24	8	0	0	$\bar{\theta}_A$		$m_A = -2$	0	−2	−.857	−1.143	−1.143	0	0	0	0	0	0	0	−0.452	−04.5	−04.5	0
8	28	6	0	$\bar{\theta}_B$		$m_B = -2$	1	−3	−3	0	−.381	+.381	0	−.361	.223	.158	0	+.158	.158	0	−.015	+.015
0	6	24	6	$\bar{\theta}_C$	=	$m_C = 4$	4	−0	−6.43	6.45	0	6.43	−.25	.893	.893	0	−.056	+.056	.0348	.022	0	+.922
0	0	6	24	$\bar{\theta}_D$		$m_D = 0$	1	−1	0	−1	0	−1.0	−1.0	0	.223	−..223	−..223	0	0	0	0	0
	A			X	=	C	$C^{(1)}$		$C^{(2)}$		$C^{(3)}$		$C^{(4)}$		$C^{(5)}$		$C^{(6)}$		$C^{(7)}$		$C^{(8)}$	

Adding the value of θ_A

$$\bar{\theta}_A = -1.143/24 - 0.45/24 = -1/24(1.188) = -.0495,$$

similarly

$$\bar{\theta}_B = -1/28\,(3 + .158) = -.1015$$

$$\bar{\theta}_C = -1/24\,(4 + .893) = +.2040$$

$$\bar{\theta}_D = -1/24\,(1 + .223) = -.0510$$

$$(m_2)_a = 3(EI/L)_a\,\theta_A = 3 \times 8/3\,\bar{\theta}_A = 8 \times (-0.0495) = -.0396 \text{ tm},$$

similarly,

$$(m_1)_b = 4(EI/L)_b\,\theta_A + 2 \times (EI/L)_b\,\theta_B + 2 = .0395 \text{ tm}$$

$$(m_2)_b = 4(EI/L)_b\,\theta_B + 2 \times (EI/L)_b\,\theta_A - 2 = 4.0210 \text{ tm}$$

$$(m_1)_c = 12\bar{\theta}_B + 6\bar{\theta}_C + 4 = 4.0060 \text{ tm}$$

$$(m_2)_c = 6\bar{\theta}_B + 12\bar{\theta}_C - 4 = -2.1610 \text{ tm}$$

$$(m_1)_d = 12\bar{\theta}_C + 6\bar{\theta}_D = -2.1420 \text{ tm}$$

$$(m_2)_d = 12\bar{\theta}_D + 6\bar{\theta}_C = +0.612 \text{ tm}$$

Table 8.4 *Comparison of moment values computed by moment distribution (nine cycles) and Relaxation Technique. The values given are in tm.*

Technique		*a*	*b*	*c*	*d*	*e*	*Remarks*
Moment Distribution	Ends						All values to be
	1	0	.397	4.002	2.152	−.612	Multiplied by
	2	−.400	−4.002	−2.152	0.612	−.306	*EI*
Relaxation	1	0	0.395	4.006	−2.142	−0.612	
	2	−.396	4.01	−2.161	0.612	−0.306	
$(EI/L)_i$	$8/3I$		$4I$	$3I$	$3I$		

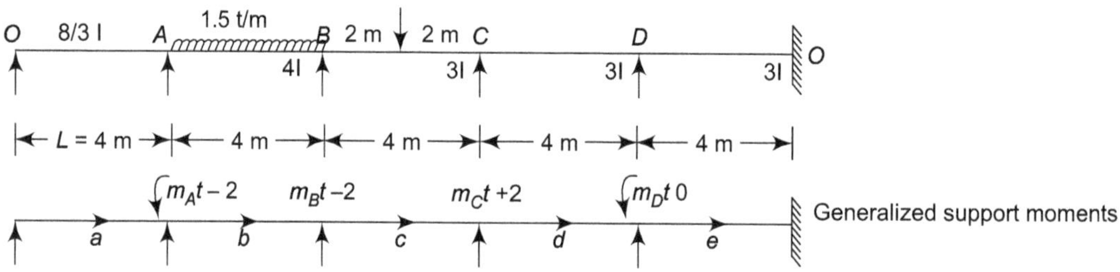

Fig. 8.11(a) *Continuous Beam*

8.9 ELIMINATION AND BACK SUBSTITUTION APPLIED TO BEAMS

Before we move to the Kanis technique, it is desirableto discuss the procedure of 'elimination and back substitution', introduced already, Article 8.4.3. Even though Kanis is not an iterative technique, the awareness in its totality is aimed at.

In the case of continuous beams, the members do not require coordinate transformation for the member stiffnesses. The member coordinates are the same as the system coordinates. In the case of the stiffness method, the supports are generally designated by capital letters starting from extreme left and, "O". The loading is transformed into the equivalent moment loads at the supports. The moment loads are shared by the member ends meeting at the supports and whereas the shear loads get directly transferred to the supports, Fig. 8.11.

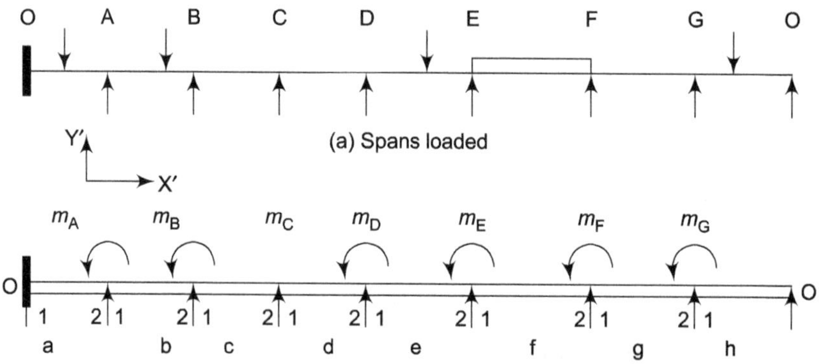

(b) Equivalent nodal loads m_I

Fig. 8.11(b) *Continuous Beam of Eight Spans*

8.9.1 Continuous Beam of Seven Interior Nodes

With the objective of limiting the size of the global stiffness matrix. In the Fig. 8.11 the number of spans is restricted to eight, Fig. 8.11 (a) to (h). Unlike Tall Buildings, where the number of levels could be large, it is rare to have number of spans more than eight from considerations of expansion joints. Accordingly, the number of nodes considered is seven. The end supports, do not add to the degrees of freedom. If it is a simple support, we have been considering modified member stiffness.

Following the notation of the "stiffness method", the matrix equation Eq. (i) is obtained. In this equation, observe particularly the "Tridiagonalization" in Eq. (i) obtained for non-sinking supports. The Tridiagonal form introduced in the presentation is a preparation for the analysis of Tall Buildings, Chapter 9. Instead of scalars, we will be using matrices for storey stiffnesses of Tall Buildings.

In the case of continuous beams axial deformations are delinked from the 'flexural deformations'. In the case of frames, however, we assume delinking. The global stiffness matrix for the continuous beam considered, $P = K\,d'$, K is given by Eq. (i).

$$\begin{bmatrix} k_{22a}+k_{11b} & k_{12b} & 0 & 0 & 0 & 0 & 0 \\ k_{21b} & k_{22b}+k_{11c} & k_{12c} & 0 & 0 & 0 & 0 \\ 0 & k_{21c} & k_{22c}+k_{11d} & k_{12d} & 0 & 0 & 0 \\ 0 & 0 & k_{21d} & k_{22d}+k_{11e} & k_{12e} & 0 & 0 \\ 0 & 0 & 0 & k_{21e} & k_{22e}+k_{11f} & k_{12f} & 0 \\ 0 & 0 & 0 & 0 & k_{21f} & k_{22f}+k_{11g} & k_{12g} \\ 0 & 0 & 0 & 0 & 0 & k_{21g} & k_{22g}+k_{11h} \end{bmatrix} \quad \text{(i)}$$

8.9.2 Generalized Algorithm for Elimination and Back Substitution

The generalization attempted is useful for many types of structures, particularly the tall buildings. And it will lead to an algorithm suitable for the numerical computations. Having presented the method for seven interior nodes, the generalization is given in the following.

Generalized Continuous Beam

We write the following equilibrium equations, Fig. 8.11:

If $\{\mathbf{P}\} = \{m_A, m_B, m_C, m_D, \ldots\, m_H, m_I, m_J \ldots, m_M\}$, and

$\{\mathbf{d'}\} = \{\theta_A, \theta_B, \theta_C, \theta_D, \ldots\, \theta_H, \theta_I, \theta_j \ldots, \theta_M\}$; modified stiffness is considered.

We may write the equilibrium equation as,

$$P' = K\,d'$$

The first three rows of the matrix , K is given by

	θ_A	θ_B	θ_C	θ_D
	$2(k_a + k_b)$,		k_b,	0
	k_a,	$2(k_b + k_c)$,		k_c,
	$0\ k_c$,	$2(k_c + k_d)$,		k_d
	θ_H	θ_I	θ_J	
Interior rows are:	k_h,	$2(k_h + k_i)$,	ki,	
	k_i,	$2(k_i + k_j)$,	k_j,	

The last row corresponding to θ_M

	θ_{M-1}	θ_M	
is given by:	k_m,	$2(k_m + k_n)$	(ii)

From the above equations we perform the process of elimination starting from the top and proceed up to the last mth equation.

After the process of elimination from thelast equation, we evaluate θ_M, and back substitution is used to evaluate $\theta_{M-1}, \theta_{M-2}, \ldots \theta_2$ and θ_1. Following back substitution we find $\theta_{M-1}, \ldots, \theta_A$.

8.10 KANI'S TECHNIQUE

Kanis and Moment Distribution algorithms are essentially for Rigid Jointed Structures, continous Beams and Plane Frames.

Kanis iterative method of analysis of continuous beams resting on rigid supports is developed in the following.

8.10.1 Kani's Method Applied to Continuous Beams

Consider Fig. 8.12 in the following

(a) Joint and "member–ends" designated following "stiffness method".

(b) Equivalent moment load: $M_I = -(\text{FEM}_2)_I + (\text{FEM}_1)_J$

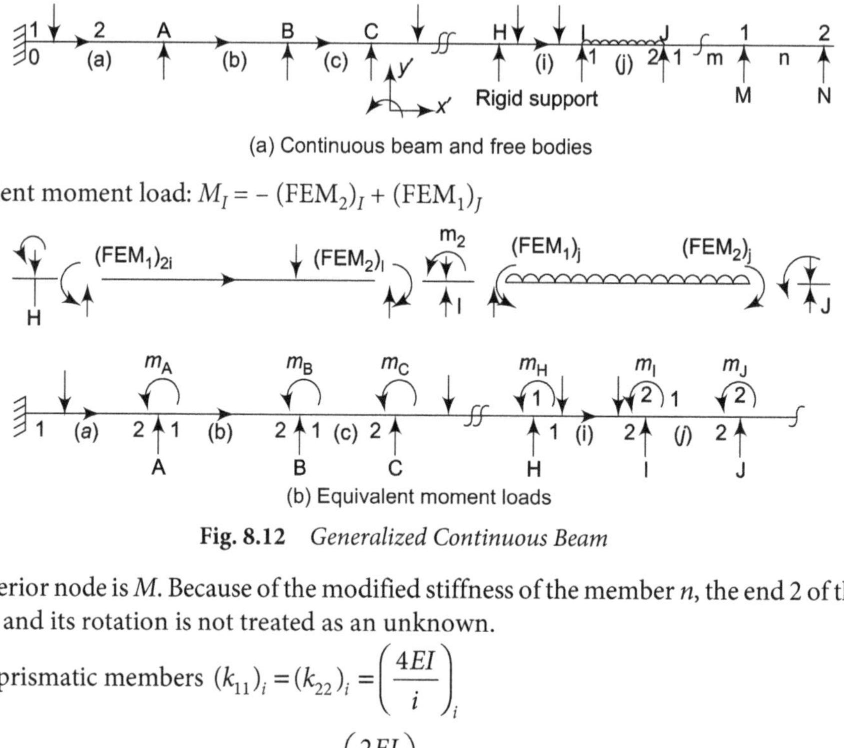

Fig. 8.12 *Generalized Continuous Beam*

The last interior node is M. Because of the modified stiffness of the member n, the end 2 of the member is the node N, and its rotation is not treated as an unknown.

Where for prismatic members $(k_{11})_i = (k_{22})_i = \left(\dfrac{4EI}{i}\right)_i$ ...(1)

If $$k_i = (k_{12})_i = (k_{21})_i = \left(\frac{2EI}{L}\right)_i \qquad ...(2)$$

Considering any typical row, say, corresponding *to* m_I,

$$m_i = (k_{21})_i\theta_H + [(k_{22})_i + (k_{11})_j]\theta_I + (k_{12})_j\theta_J \qquad ...(3)$$

For a given structure under loads, m_I is known and from relation, Eq. 2,

$$m_I = k_i\theta_H + 2(k_i + k_j)\ \theta_I + (k_j)\theta_J \qquad ...(4)$$

Following the above notations, and the representation of members and Joints shown below,

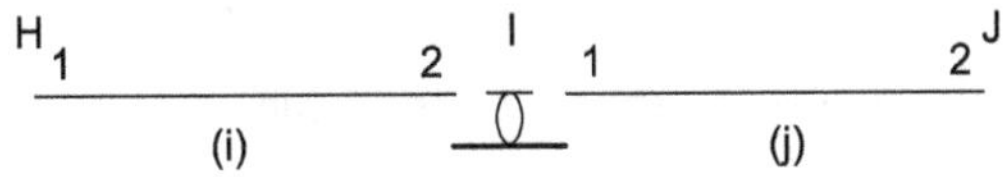

Fig. 8.13 *Member and Joint Designation—A Typical Joint I is Considered*

In the following, we use the Kani's algorithm for the evaluation of

Let $k_i\theta_I = (\overline{M_2})_i$ i.e., rotation to the right of member i is multiplied by k_i

$k_i\theta_J = (\overline{M_1})_J$ i.e., rotation to the left of member j is multiplied by k ...(5)

By rotation of indices

$$k_j\theta_J = (\overline{M_2})_J \text{ and } k_i\theta_H = (\overline{M_1})_i$$

From Eqs. (4) and (5):

$$\theta_I = \frac{-1}{2(k_i + k_j)}\left[M_1 + (\overline{M_1})_i + (\overline{M_2})_j\right]$$

Or

$$k_i\,\theta_I = (\overline{M_2})_i = \frac{-K_i}{K_i + K_j}\left[-m_I + (\overline{M_1})_i + (\overline{M_2})_j\right]$$

$$k_J\,\theta_I = (\overline{M_1})_j = \frac{-K_J}{K_i + K_j}\left[-m_I + (\overline{M_1})_i + (\overline{M_2})_j\right]$$

$$\text{and } \underbrace{(M_1)_i}_{\text{Final moment}} = \underbrace{2(\overline{M_1})_i}_{\text{Distribution moment}} + (\overline{M_2})_i + (\text{FEM})_{1i};\ (M_2)_i = 2(\overline{M_2})_i + (\overline{M_1})_i + (\text{FEM})_{2i} \qquad \text{...(6)}$$

The Eq. (6) is the basis of the iteration method as proposed by Kani. If a member i, has one of its ends hinged then k_1 requires to be modified to $\frac{3}{4}\left(\frac{2.EI}{L}\right)_i = \left(\frac{1.5EI}{L}\right)_i$

8.10.2 Illustration 1: Analysis of Continuous Beam on Non-sinking Supports

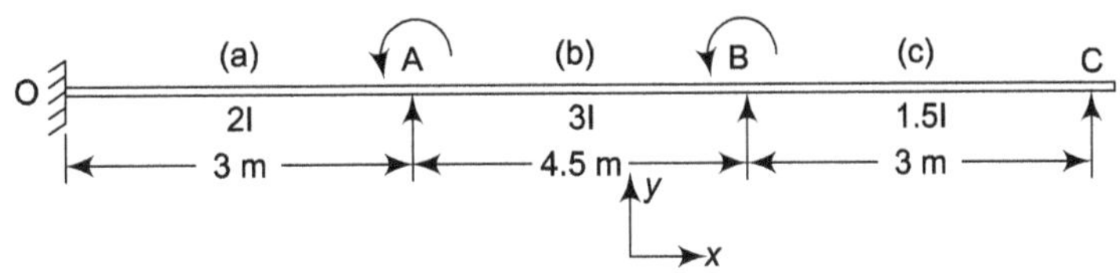

Fig. 8.13 *Illustration (1) $m_A = m_B = 1\ t_m$*

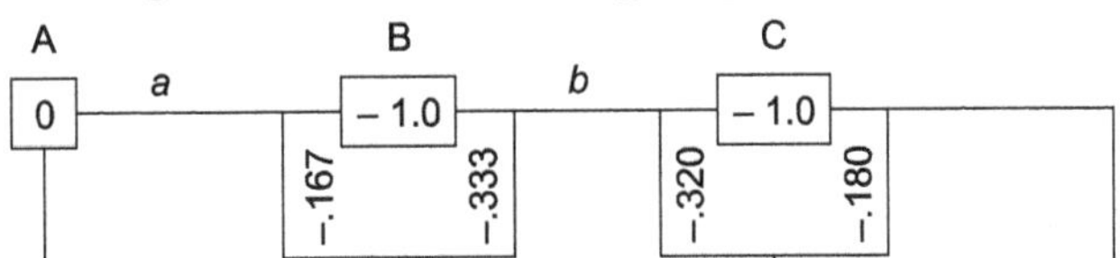

	A	B		C		
I Cycle	0	0	0	0	0	0
II	0	–.167	–.333	–.215	–.120	0
III		–.131	–.262	–.236	–.133	
IV		–.127	–.255	–.238	–.134	
		–.127	–.254	–.239	–.134	

Operation Table

The final end moments are computed in the following table:

Table 8.5 *Computation of final end moments*

Moment at	*Fixed end moments*	*Distribution moments*	*Carry over moments*	*Final moments*
(1)	(2)	(3)	(4)	(5)
$(M_1)_a$	0	0	−.127	−.127
$(M_2)_a$	0	2(−.127) = −.254	0	−.254 tm
$(M_1)_b$	0	2(−.234) =−.508	−.239	−.747 tm
$(M_2)_b$	0	2(−.239) = −.478	−.254	−.732 tm
$(M_1)_c$	0	2(−.134) = −.268	0	−.268 tm

$$k_a = \frac{2EI}{3}, k_b = \frac{2(3EI)}{4.5}, k_c = \frac{1.5(1.5EI)}{3} = \frac{2.25\ EI}{3}$$

$$k_a = 0.67\ \text{EI};\ k_b = 1.33\ \text{EI};\ k_c = 0.75\ \text{EI}$$

At Support A: Distribution Factors

$$\frac{-k_i}{2(k_i + k_j)} = \frac{-k_a}{2(k_a + k_b)} = \frac{-0.67}{2(0.67 - 1.33)} = \frac{0.67}{4} = -0.167$$

$$\frac{+k_i}{2(k_i - k_j)} = \frac{-k_b}{2(k_b + k_c)} = \frac{-1.33}{4} = -0.333$$

At support B

$$\frac{-k_i}{2(k_i + k_j)} = \frac{-k_b}{2(k_b + k_c)} = \frac{-0.89}{2(1.33 + .75)} = \frac{-1.33}{4.16} - 0.32$$

$$\frac{-k_i}{2(k_j + k_j)} = \frac{-k_c}{2(k_b + k_c)} = \frac{-0.75}{4.16} = -0.18$$

I cycle:

Based on Eq. (6), Support A.

$$(\overline{M}_2)_a = -0.167[-m_A + (\overline{M}_1)_a + (\overline{M}_2)_b]$$

$$(\overline{M}_1)_b = -0.333[-m_A + (\overline{M}_1)_a + (\overline{M}_2)_b]$$

and for the first approximation, $(\overline{M}_1)_a = (\overline{M}_2)_b = 0$

$$\therefore \quad (\overline{M}_2)_a = -0.167 \times 1 = -0.167 \text{ and } (\overline{M}_1)_b = -0.333$$

and similarly, based on Eq. (6), Support *B*

$$(\overline{M}_2)_a = \frac{-k_b}{2(k_b + k_c)}\left[-m_B + (\overline{M}_1)_b + (\overline{M}_2)_c\right]$$

$$(\overline{M}_1)_a = \frac{-k_c}{2(k_b + k_c)}\left[-m_B + (\overline{M}_1)_b + (\overline{M}_2)_c\right]$$

Where $m_B = -1$, $(\overline{M}_1)_b$ as obtained in the previous step is equal to –0.333 and $(\overline{M}_2)_c = \left(\frac{EI\,\theta_2}{L}\right)_c$ is not to be considered for the reason that modified stiffness has been considered.

$$(\overline{M}_2)_b = -0.320\,[+1 + (-0.333)] = -0.320(0.667)$$

$$(\overline{M}_2)_b = 0.213$$

$$(\overline{M}_1)_c = -0.180 \times 0.667 = -0.120$$

The above values are now used to get new set of values of $\overline{M}$ values. Based on Eq. (6), applied to Support A:

$$\text{as } (\theta_1)_a = 0,\ (\overline{M}_1)_a = \left(\frac{2EI}{L}\right)(\theta_1)_a = 0$$

$$(\overline{M}_2)_a = -0.167\,[+1 + 0 + (-.213)] = -.167 \times (.787) = -.131$$

$$(\overline{M}_1)_b = -0.333\,[+1 + 0 + (-.213)] = -.333 \times (.787) = -.262$$

Support B

$$(\overline{M}_2)_b = -0.320\,[+1 - .262] = -.320 \times .787 = -.236$$

$$(\overline{M}_1)_c = -.180 \times .738 = -.133$$

Similarly the values of III and IV cycle are given in the operation table. The final end moments are given in tabular form, Table 8.6.

Observation: Beams are the simplest of the rigid jointed structures. The illustrations are going to be covered in the sequence: continuous beams including sway due to sinking of the supports, Portal Frame, and finally a multi bay and multi storey plane frame.

8.10.3 Comparison of Convergence Compared to Moment Distribution

A simple beam problem is chosen.

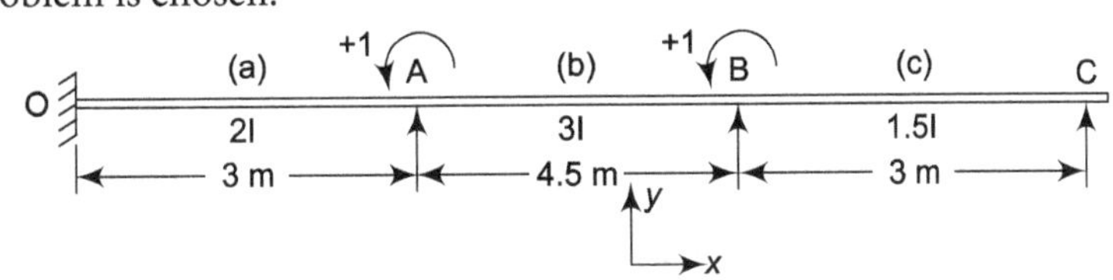

Fig. *Illustration 2*

Convergence of the Iterative Methods (Continued)

(a) By Kani's algorithm
The Example of Illustration 1 is also solved in the following forComparing convergence with moment distribution for the continuous beam.

(b) The extreme end conditions are changed to simple supports.

Illustration 3 The extreme end conditions are changed to simple supports.

(a) Kani's Method

	A		B		C	
	−.214	−.286	−.25	−.25	−.167	−.333
	−.417		−.083		+.33	
	+.089	+.119	−.009	+.009	−.061	−.091
	+.091	+.122	+.0055	+.0055	−.064	−.0955
	+.088	+.118	+.007	+.007	−.064	−.096
	+.088	+.117	+.007	+.097	−.054	−.096
Final Moments	−.575	+.575	−.201	+.201	−.370	+.370

(c) Moment Distribution

	.428	.572	.5	.5	.4	.6
	.750	+.333	+.333	+.250	.250	+.563
	+.178	+.239	+.0415	+.0415	.125	+.183
		+.0207	+.1195	.0625	+.0207	
	+.0088	.01 19	.0285	.0285	.0083	.0124
		.0142	.0059	.0041	.0142	
	+.0061	+.0081	+.005	+.005	+.0057	+.0085
		+.0025	+.005	+.003	+.0025	
	.001 1	.0014	.0035	.0035	.0010	.0015
Final Moments	−.576	+.576	.201	+.201	.370	+.370

Comments:

It may be inferred that

1. Kani's method based on the numerical iterative technique converges faster than Moment Distribution.
2. But for the cases like gable frames, beams on flexible supports, the method is not recommended, Article 8.10.
3. An algorithm is presented, and it is left to the reader todecide the suitability.
4. In the case of Moment Distribution, any error made gets *embedded*. In the case of Kani's algorithm it gets *corrected* in the subsequent cycles.

Analysis by Moment Distribution for the case of extreme ends simply supported:

8.10.4 Prescribed Settlement of a Support

Illustration 4A

The support B settles by a given value, say, equal to 1 cm. Given $EI = 100$ tm^2

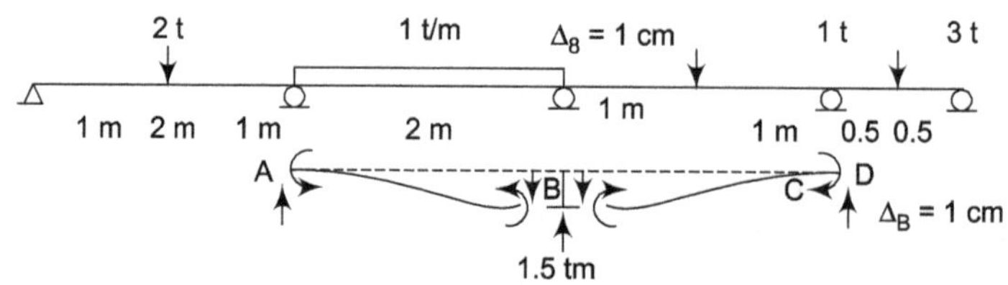

Fig. 8.12 *Illustration 4A: Prescribed Settlement of a Support of a Continuous Beam*

Based on the fixed-end-conditions for the prescribed settlement, the total nodal loads are shown in Fig. (ii).

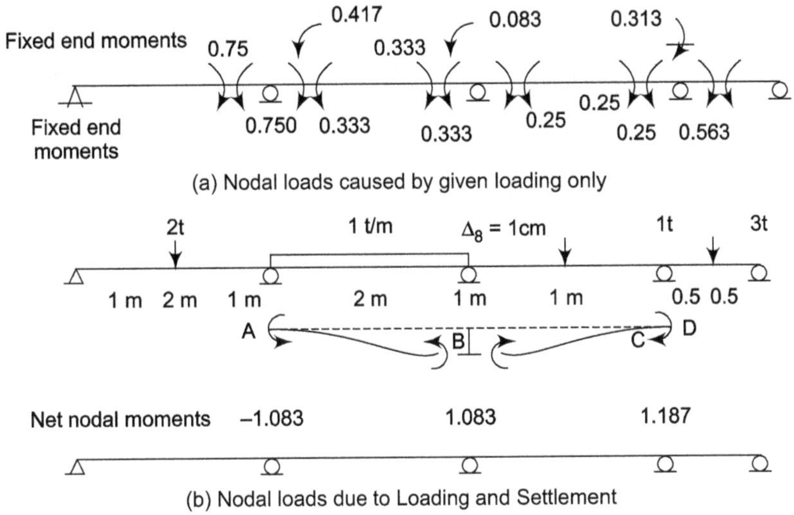

Fig. 8.13 *Prescribed Settlement as Loading Combined with Given Loads*

Illustration 4B

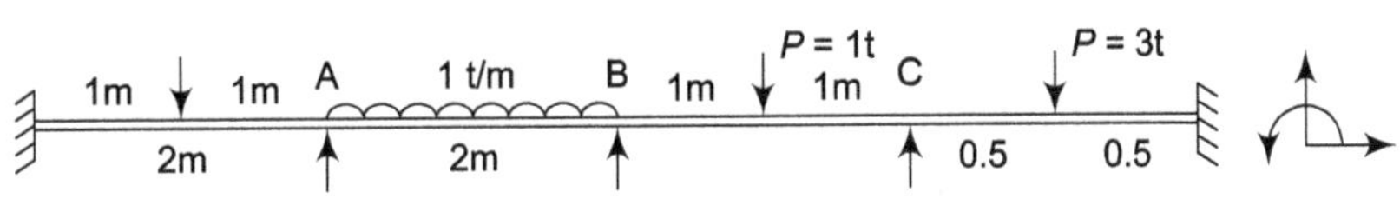

(a) Kani's Method:

	A		B		C	
Rotation factor / FEM sum	−.25	−.167	−.25	−.083	−.167	+.125
	+.0417	+.0417	−.0103	−.0103	−.0225	−.045
	+.038	+.038	+.0168	+.0168	−.0236	−.0472
	+.0373	+.0375	+.0170	+.017	−.0236	−.0472
Final Moments	+.5375 −.4245	+.4245	−.262	+.262	−.422 +.280	+.280

(b) Moment Distribution:

	.5	.5	.5	.5	.333	.667
	+.500 −.500	+.333	+.333	+.250	−.250	+.375
	+.0835	+.0835	+.041	+.0415	−.042	.083
	.0417	+.0207	+.0417	+.021	+.207	.415
	−.0103	+.0103	−.0103	+.0103	-.0069	-.0138
	−.005	−.005	−.005	−.0034	.0069	.0069
	+.0025	+.0025	−.0042	+.0042	+.0016	+.0032
	.0012	+.0012	+.0012	+.0008	+.0021	.0016
	−.001	−.001	−.006	−.0004	−.0007	−.0014
Final Moments	+.5379 - .425	+.425	−.260	+.260	−.280 +.280	−.422

Ex. Effect of settlement of supports of a continuous beam

In the case of sinking support, say support I, Sinks by $(-\Delta_I)$, it gives rise to support moments

	A		B		C	
Method	.214	+1.085	.25	+.083	.20	+.1.187
f.e.m	−.75	+.1.833	+1.167	−.125	−1.25	−.563
I Cycle	−232	+.310	−.098	+.098	−.218	+.327
II Cycle	+.252	+.337	+.054	+.054	−.225	−.339
III Cycle	−.243	+.325	+.045	−.045	+.229	+.343
IV Cycle	−.242	−.322	+.045	−.045	−.228	−.343
Final Moments	−1.234	+1.234	+.935	−.932	−.1249	+.1.249

$$m_h = -\left(\frac{6EI\Delta I}{L^2}\right)_1$$

$$m_l = \left(\frac{6EI}{L^2}\right)_j \Delta - \left(\frac{6EI}{L^2}\right)_i \Delta \text{ and}$$

$$m_j = \left(\frac{6EI\Delta_I}{L^2}\right)_J$$

Which should be added to the support moment load (Ex.1)? The support moments may be due to applied loads, if any, on the beam members. The fixed end moments of members due to sinking of support, should appropriately be added in the column 4 of the table, Fig. 6, for getting the final end moments. The above procedure in essence means that we find generalized or equivalent load with the settlement deformation incorporated and thereafter Kani's method issued for joint rotations only. In the case of spring supports the settlements are given by the structural response.

8.10.5 Method of Successive Corrections

Analysis by successive corrections: if convergence is fast it is a very useful technique. We in the following apply the technique to continuous beams with spring support. In fact the technique can be used for more than one spring. The aspect may become clear at the end of the solution by using the technique.

Illustration: Consider the following beam

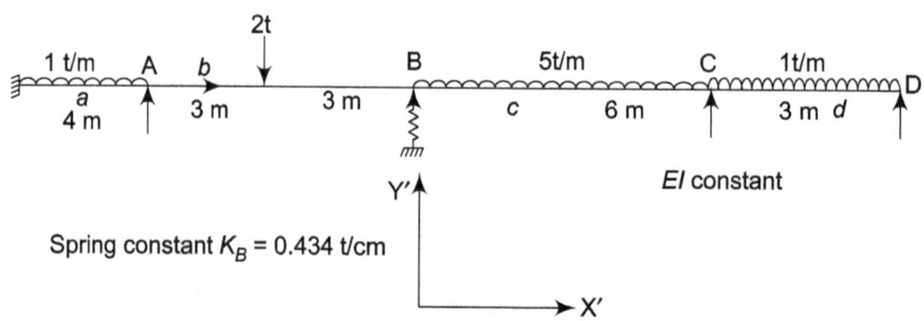

Fig. 8.14 *Spring Support Treated as Prescribed Settlement and Successive Correction*

Step 1. Compute R_B for the situation $\Delta_B = 0$

The results for all supports assumed non sinking are:

Ans. $(m_1)_a = 1.287$, $(m_2)_a = -1.426$, $(M_2)_b = 1.554$, $(M_2)_C = -1.357$

From the following free body, R_B due to end moments = .021 + .033 = .054

R_B due to fixed end shears= 1 +1.5 = 2.5

Total R_B =2.554 t

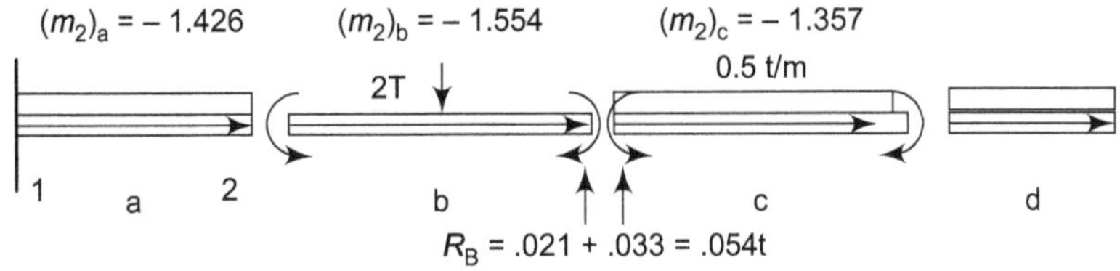

Fig. 8.15 *Free Body for Combined Reaction at Spring Support*

Revised $\Delta_B = 2.554/0.434 = 5.885$. We may round off to a value equal to 6 cm

Step 2. Compute $\Delta B \approx 6$ cm for the computed R_B of step1, and evaluate fixed end computes $6EI\Delta/L_2$ = 1 tm

Analyze the beam similar to the solution of the case with prescribed settlement.

Step 3. Computed new value of R_B and compare with the result of step 1. If different (within permissible limits), repeat the cycle.

Ans. R_B of second cycle = 2.086

R_B of third cycle = 2.191

R_B of fourth cycle = 2.167

The final moments are

$$(m_1)_a = 1.031,\ (m_2)_a = -1.92,\ (m_2)_b = -.89,\ (m_2)_c = 1.855 \text{ tm.}$$

The analysis by the 'successive correction' can certainly be used for more than one spring support.

The same Illustration above is solved by a modified algorithm which avoids the above process of successive correction.

8.11 EXTENSION OF KANI'S ALGORITHM TO CONTINUOUS BEAM WITH SPRING SUPPORT

Let us consider sinking being caused due to the elastic support. The elastic support considered as a spring. In the case of spring support, equivalent support loads are also to be considered, and the load in the stiffness matrix will be shared by the shears at ends of members meeting at the support and the supporting spring below, Fig. 8.11. It will be a good preparation for the theory pertaining to multi-storey frames. It is a good exercise in formulating a numerical algorithm.

From moment equilibrium it can be shown that

$$m_I = K_i\theta_H + 2\left[k_i + k_j\right]\theta_I + k_j\theta_J + \overline{\overline{M}}_i + \overline{\overline{M}}_j \qquad \ldots\text{(i)}$$

Or

$$(\overline{M}_2)_i = \frac{-k_i}{2(k_i + k_j)}\left[-m_I + (\overline{M}_1)_i + (\overline{M}_2)_j + \overline{\overline{M}}_i + \overline{\overline{M}}_j\right]$$

$$(\overline{M_1})_j = \frac{-k_i}{2(k_i+k_j)}\left[-m_I+(\overline{M_1})_i+(\overline{M_2})_j+\overline{\overline{M}}_i+\overline{\overline{M}}_j\right] \qquad \ldots\text{(ii)}$$

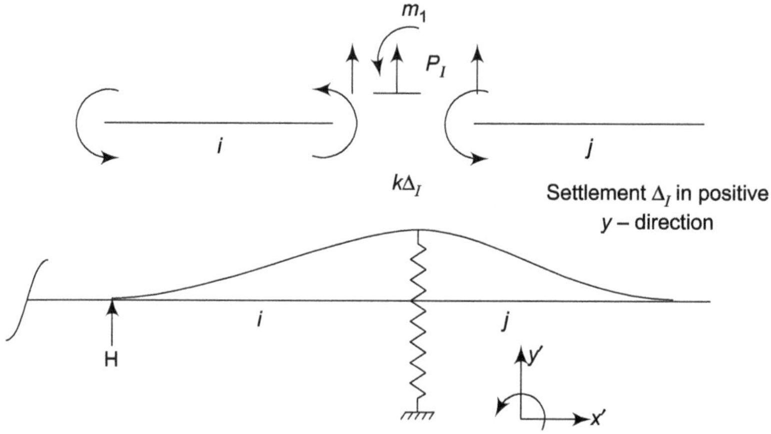

$$\overline{\overline{M}}_i = \frac{-6\,E\,|\Delta|}{L_i^2} \text{ and } \overline{\overline{M}}_j = \frac{-6\,E\,|\Delta|}{L_j^2}$$

Fig. 8.16 *Generalized Forces at a Spring Support of Continuous Beam*

From vertical force equilibrium we have

$$[p'y]_I = -\left[\frac{M_1+M_2}{L}\right]_i + \left[\frac{M_1+M_2}{L}\right]_j$$

$$U = -(P'\,y)_I\,L_i = 1.5 \qquad \ldots\text{(iii)}$$

Where due to end deformations

$$M_1 = \frac{4EI\theta_1}{L}+\frac{2EI\theta_2}{L}-\frac{6EI\Delta_I}{L^2}$$

And, therefore $$M_2 = \frac{2EI\theta_1}{L}+\frac{4EI\theta_2}{L}-\frac{6EI\Delta_I}{L^2}$$

$$(M_1+M_2)_i = \frac{6EI}{L}\theta_1+\frac{6EI}{L}\theta_2-\frac{12EI}{L^2}\Delta_I \qquad \ldots\text{(iv)}$$

But if end 1 of member *i* is hinged, we are familiar with the modified stiffness. In the following example gives an illustration of the above.

$$(M_1+M_2)_i = (M_2)_i = \frac{3EI}{I}\theta_2-\frac{3EI}{L^2}\Delta_I \qquad \ldots\text{(va)}$$

Similar equation can be written for the number *J*.

$$(P_y')_I = -\frac{1}{L_i}\left[3(\overline{M_1}+\overline{M_2})_1-\left(\frac{12EI}{L^2}\right)_i\Delta_I\right]$$

$$+\frac{1}{L_j}\left[3(\overline{M_1}+\overline{M_2})_J+\left(\frac{12EI}{L^2}\right)_i\Delta_I\right]+K\Delta_I \qquad \ldots\text{(vb)}$$

And if end 1 of I is hinged,

$$(P'_y)_I = -\frac{1}{L_i}\left[2\overline{M}_2 - \left(\frac{3EI}{L^2}\right); \Delta_I\right] + \frac{1}{L_j}\left[3\left(\overline{M}_1+\overline{M}_2\right)_I + \left(\frac{12EI}{L^2}\right)_J \Delta_J\right] + K\,\Delta_I$$

Or

$$(P'_y)_I = +3\left[-\left(\frac{\overline{M}_1+\overline{M}_2}{L}\right)_i + \left(\frac{\overline{M}_1+\overline{M}_2}{L}\right)_j\right] + \left[\left(\frac{12EI}{L^3}\right)_i + \left(\frac{12EI}{L^3}\right)_j + \underline{K}\right]\Delta_I = 0$$

Or

$$\Delta_I = \frac{1}{\left[\left(\frac{12EI}{L^3}\right)_i + \left(\frac{12EI}{L^3}\right)_j + \underline{K}\right]}\left[-(p'_y)_i + 3\left\{\left(-\frac{\overline{M}_1+\overline{M}_2}{L}\right)_i + \left(\frac{\overline{M}_1+\overline{M}_2}{L}\right)_j\right\}\right]. \qquad \text{...(vi)}$$

Or

$$\overline{\overline{M}}_i = \left(\frac{-6EI}{L^2}\right)_i \Delta_I$$

$$\overline{\overline{M}}_i = \frac{3}{2}\frac{\left(\frac{-6EI}{L^2}\right)_i}{\left[+\left(\frac{6EI}{L^2}\right)_i + r\left(\frac{6EI}{L^2}\right)j + \frac{1}{2}\underline{KL_i}\right]}\left[-\frac{1}{3}\times U - \left(\overline{M}_1+\overline{M}_2\right)_i + r\left(\overline{M}_1+\overline{M}_2\right)_j\right] \qquad \text{...(vii)}$$

Where $U = -(p'_y)_I L_i$

And $r = \dfrac{L_i}{L_j}$

Similarly

$$\overline{\overline{M}}_j = -\frac{3}{2}\frac{\left(\frac{6EI}{L^2}\right)_j}{\left[\left(\frac{6EI}{L^2}\right)_i + r\left(\frac{6EI}{L^2}\right)_j + \frac{1}{2}\underline{KL_i}\right]}\times\left[-\frac{1}{3}U - \left(\overline{M}_1+\overline{M}_2\right)_i + r\left(\overline{M}_1+\overline{M}_2\right)_j\right] \qquad \text{...(viia)}$$

If the end 1 of member is hinged

$$\overline{\overline{M}}_i = \frac{3}{2}\frac{\left(\frac{3EI}{L^2}\right)_j}{\left[\left(\frac{1.5EI}{L^2}\right)_i + r\left(\frac{6EI}{L^2}\right)_j + \frac{1}{2}\underline{KL_i}\right]}\times\left[-\frac{1}{3}U - \frac{2}{3}\times\left(\overline{M}_2\right)_i + r\left(\overline{M}_1+\overline{M}_2\right)_j\right] \qquad \text{...(viib)}$$

and corresponding change in the expression for $\overline{\overline{M}}_j$. It is only in parenthetic values of *i*, that is, $(-6EI/L^2)$ in place of $(3EI/L^2)$ in the numerator.

The use of the above pivotal equations, Eqs (ii), and (viia) (viib) is explained by the illustrations below.

Illustration for Continuous Beam

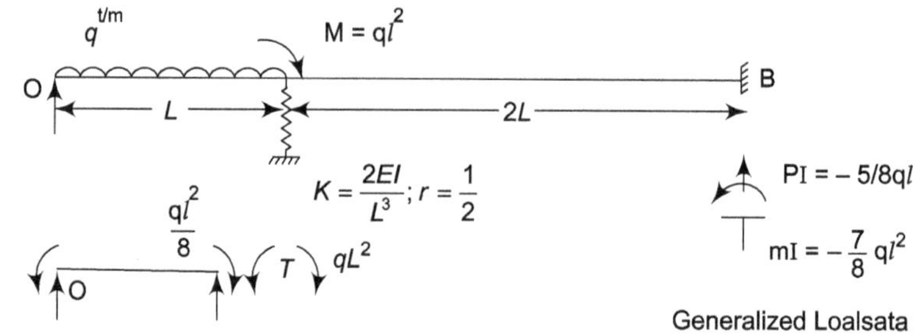

Fig. 8.17 *Continuous Beam with a Spring Support*

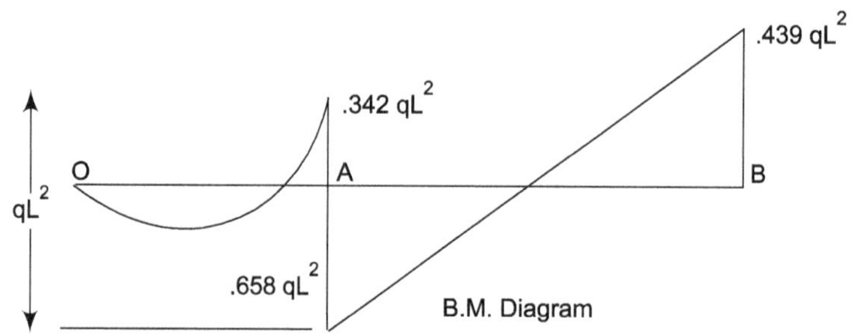

Fig. 8.18 *Illustration of a Spring Support of a Continuous Beam*

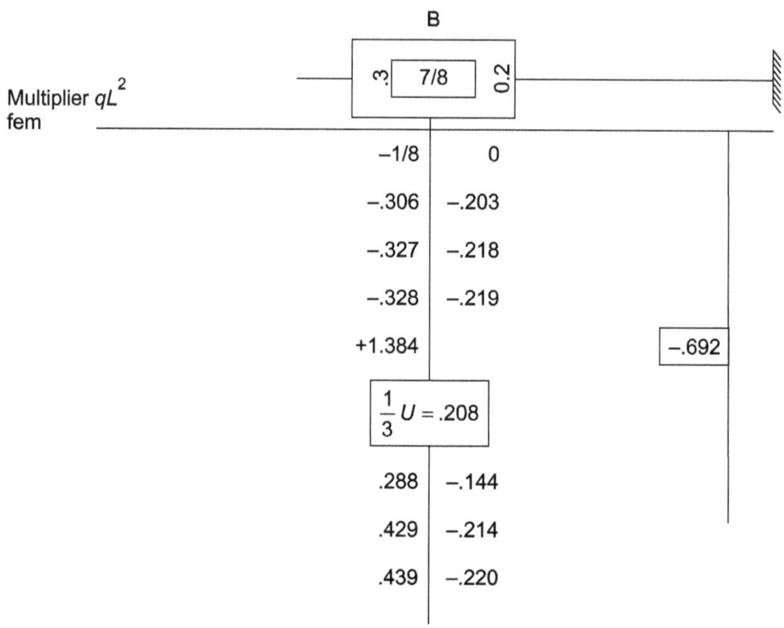

Results.

$$(M_1)_a = 0$$

$$(M_2)_a = .125 + 2(-.328) + .439 = -.342$$

$$(M_1)_b = 0 + 2\,(-.219) -.220 = -.439$$

$$(M_2)_b = 0 - .219 - .220 = -.439$$

Explanation of computations.

Cycle I: Start with the pivotal equation (vii).

$$\frac{1}{3}U = \frac{1}{3}\left(-\left(P'_y\right)_A L\right) = \frac{1}{3} \times \left(\frac{5}{8}qL\right)L = .208qL^2$$

Let $\overline{M}$ values be equal to zero

$$\overline{\overline{M}}_a = (.208)\ 1.384 = .288$$

$$\overline{\overline{M}}_b = (.208)\ (-0.692) = -.144$$

$$(\overline{M}_2)_a = (7/8 + .288 - .144)\ (-.3) = -.306, \text{ and}$$

$$(\overline{M}_1)_b = (1.019)\ (-.2) = -.203$$

Cycle II

$$\frac{1}{3}U = \frac{2}{3}\left(\overline{M}_2\right)_a + 2\left(\overline{M}_1 + \overline{M}_2\right)_b = +.310\ qL^2$$

$$\overline{\overline{M}}_a = .310 \times 1.384 = +.429 \quad \text{and} \quad \overline{\overline{M}}_b = .310(-.692) = -.214$$

$$\left(\overline{M}_2\right)_a = -.327 \quad \text{and} \quad \left(\overline{M}_1\right)_b = -.218$$

Cycle III

$$\frac{1}{3}U - \frac{2}{3} - \left(\overline{M}_2\right)_a + r\left(\overline{M}_1 + \overline{M}_2\right)_b = .317\ qL^2$$

$$\overline{\overline{M}}_a = .317 \times 1.38 = .439, \overline{\overline{M}}_b = -.220, (\overline{M}_2)_a = -.328,$$

$$(\overline{M}_1)_b = -.220$$

From $(\overline{M}_2)_a, (\overline{\overline{M}})_a, (\overline{M}_1)_b, (\overline{M}_2)_b$ and $(\overline{\overline{M}})_b$ compute end moments of members a and b

Refer the figure for BM Diagram

In the following a simple portal frame is chosen as an illustration.

Illustration 1: Member end moments for the portal frame. We are moving away from the spring support.

Solution:

From the joint free bodies given in the figure, the equivalent or generalized joint loads computed from the fixed end moments of the members are,

$$m_A = -1.041\ tm. \text{ and } m_B = +1.041\ tm.$$

The values of $(-m_A)$ and $(-m_B)$ are entered within the inner squares of the corresponding joints, A and B. The (p'_x) values are not used directly, instead $(1/3)U$ values for the different storeys are required to be computed. In the present example, we have single storey.

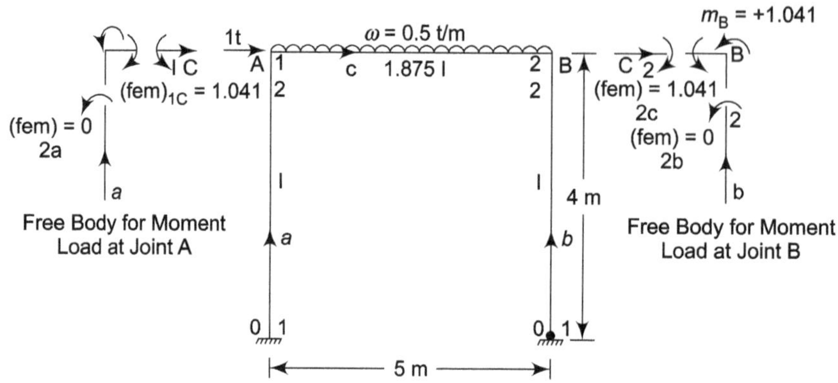

Fig. 8.10 *Illustration 1–Portal Frame*

one such value to be computed

$$(1/3)U = 1/3\,(-4 \times p'_x) = 1/3\,(-4 \times 1)$$

$$(1/3)U = -1.333.$$

The value of (1/3) U equal to –1.333, is entered as shown in the figure.

Distribution Factors:

The k values of the different members meeting at the joint A are $(k)_a = 2(EI/L)a = EI/2$; $(k)_c = 2(EI/L)_C = 2(1.875\ EI/5) = 0.75\ EI$. The distribution factors (DF) are required to be computed for each joint. Considering the Joint A, the distribution factors (DF), are;

For the member "a", DF $= -\left(\dfrac{1}{2}\right)\dfrac{(k)_a}{(k)_a + (k_c)} = -.2$

For the member "b", $= -\left(\dfrac{1}{2}\right)\dfrac{(k)_c}{(k)_a + (k_c)} = -.3$

Similarly considering the joint B, the K values of the members meeting at the joint B area:

$K_b = 1.5\ (EI/L)_b = .375\ EI$, the 1.5 value is due to the end 1 of b being hinged.

$K_c = 0.75\ EI$

And the distribution factors (DF) for the joint B are:

For the member "b", DF $= -\left(\dfrac{1}{2}\right)\dfrac{k_b}{k_b + k_c} = -1/6$

For the member "c", DF $= -\left(\dfrac{1}{2}\right)\dfrac{k_c}{k_b + k_c} = -1/3$

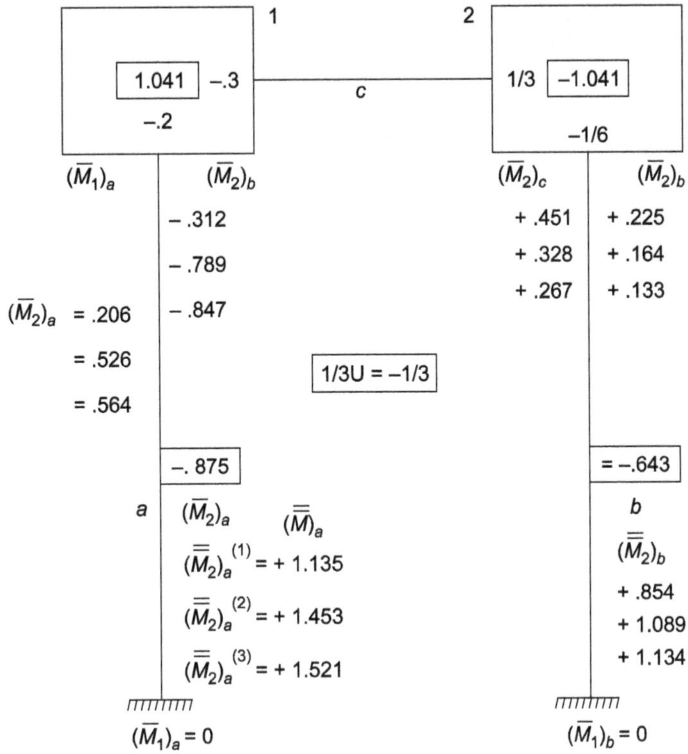

Evaluation of modified deformations by iteration (three-cycles)

$$(\overline{M}_1)_a = 2(EI/L)_a(\theta_1)_a$$

Typical Modified Deformations

$$(\overline{M})_a = 6(EI/L^2)_a(\delta_2)_a$$

where δ_2 is the relative transverse displacement of the end 2 of the member.

The distribution factors for the evaluation of $\overline{\overline{M}}$ values of the columns "*a*" & "*b*" are: for the column, *a*,

$$(DF)_a = -\frac{3}{2}\times\frac{k_a}{k_a+k_b} = -.875$$

And for the column, *b*, $(DF)_b = -\frac{3}{2}\times\frac{k_b}{k_a+k_b} = -.643$

These values are entered in the figure against the columns.

At this stage, we are in a position to carry out the iterations, similar to the "Gauss-Seidel" procedure.

First Cycle: Consider evaluation of the modified deformations in the sequence $(\overline{M}_2)_a$ and $(\overline{M}_1)_c$ for the joint *A*, and $(\overline{M}_2)_b$ and $(\overline{M}_2)_c$ for the joint *B*, from Eq. (10) derived for a general case.

$$(M_2)a = (-.2)\left[-m_A + (\overline{M}_1)_a + (\overline{M}_2)_c + (\overline{\overline{M}})_a\right]$$

$$\left(\overline{M}_2\right)_c = \left(\overline{\overline{M}}\right)_a = \left(\overline{M}_1\right)_a = 0$$

Substitute, $-m_A = 1.041$

$$\therefore \quad \left(\overline{M}_2\right)_a^{(1)} = (-.2)\ (1.041) = -.208$$

$$\therefore \quad \left(\overline{M}_2\right)_c^{(1)} = -.312$$

Considering the joint B

$$\left(\overline{M}_2\right)_b^{(1)} = \frac{1}{6}\left[\left(-m_B + (\overline{M}_1)\right)_c + (\overline{M}_1)_b + \left(\overline{\overline{M}}\right)_b\right] \qquad \text{Where } (\overline{M}_1)_b = 0$$

And we assume $\left(\overline{\overline{M}}\right)_b = 0$

$$\therefore \quad \left(\overline{M}_2\right)_b^{(1)} = -(1/6)(-1.041 - .312) = +.225$$

$$\left(\overline{M}_2\right)_c = -1/3(-1.353) = +.451$$

We have so far computed modified rotations, $\overline{M}$ for the first cycle, the modified linear translations, $\overline{\overline{M}}$ for the cycle one evaluated below.

$$\left(\overline{\overline{M}}\right)_a^{(1)} = -.875\left\{1/3U + (\overline{M}_1)_a + (\overline{M}_2)_a^{(1)} + (\overline{M}_1)_b + (\overline{M}_2)_b^{(1)}\right\}$$

Where $\left(\overline{M}_1\right)_a = \left(\overline{M}_2\right)_b = 0$, and the $\left(\overline{M}_2\right)_a$ and $\left(\overline{M}_2\right)_b$ are taken from the first cycle.

$$\therefore \quad \left(\overline{\overline{M}}\right)_a^{(1)} = +1.135$$

and similarly

$$\left(\overline{\overline{M}}\right)_b^{(1)} = -.643\ (-1.316) = 0.845$$

Taking the "latest" values of the modified deformations, new values of the modified deformations are computed. The Final moments are given in the following Table 8.6.

Table 8.6 *Evaluation of final moments*

Moments at	*Fixed end moments*	*Distributions moments, $\overline{M}$*	*Carry over moments, $\overline{M}$*	*Final moments after 3 cycles*	*Computer results after 10 cycles*
$(\overline{M}_1)_a$	0	–	-.564	.957	+.9559
$(\overline{M}_2)_a$	0	2 × -.564	0	.393	+.3946
$(\overline{M}_1)_b$	0	–	+.133	1.267	+ 1.262
$(\overline{M}_2)_b$	0	2 × .133	0	1.400	+1.386
$(\overline{M}_1)_c$	1.041	2 × (-.847)	.267	-.386	-.3946
$(\overline{M}_2)_c$	-1.041	2 × .267	-.347	-1.354	-1.386

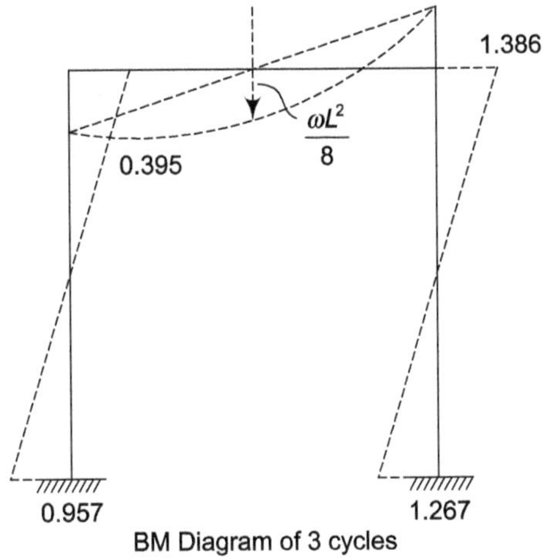

Fig. 8.19 *Illustration 1–Bending Moment Diagram due to Nodal Loads*

Illustration 2 Analyse simple portal frame Fig. 8.20 by the Kani's Method. The solution requires the 'Inter mediate Results.

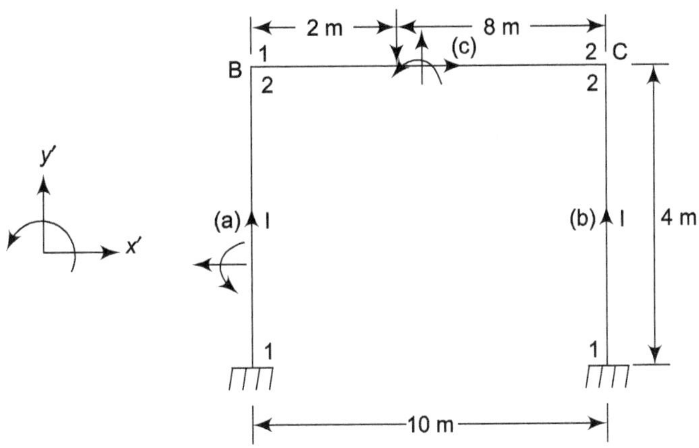

Fig. 8.20 *(a) Illustration 2–Geometry and Loading*, $P = 2\ T$

Results: From $\overline{M}$ and $\overline{\overline{M}}$ computed by Kani's algorithm.

$$(M_1)_a = 0 + -1.230 + .845 = -.385$$

$$(M_2)_a = 0 + 2(-1.23) + .845 = -1.615$$

$$(M_1)_c = 2.56 + 2(-.492) + .0430 = +1.615$$

$$(M_2)_c = -.61 + 2(.043) - .492 = -1.046$$

$$(M_2)_{b2} = 0 + 2(.103) + .845 = +1.051$$

$$(M_2)_{b2} = 0 + 0.103 + .845 = +.948$$

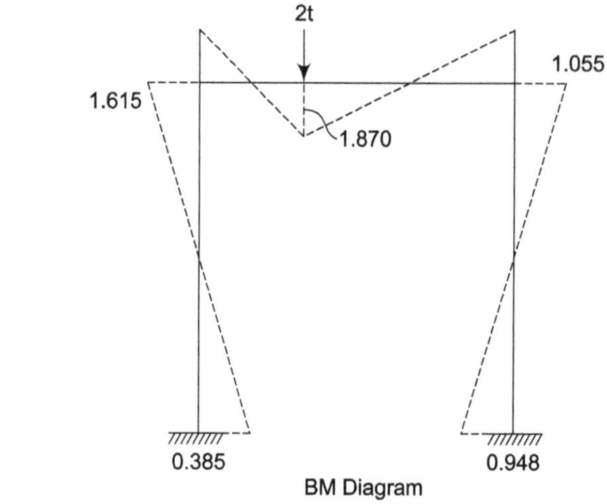

Fig. 8.20 *(b) Diagram, Illustration 2*

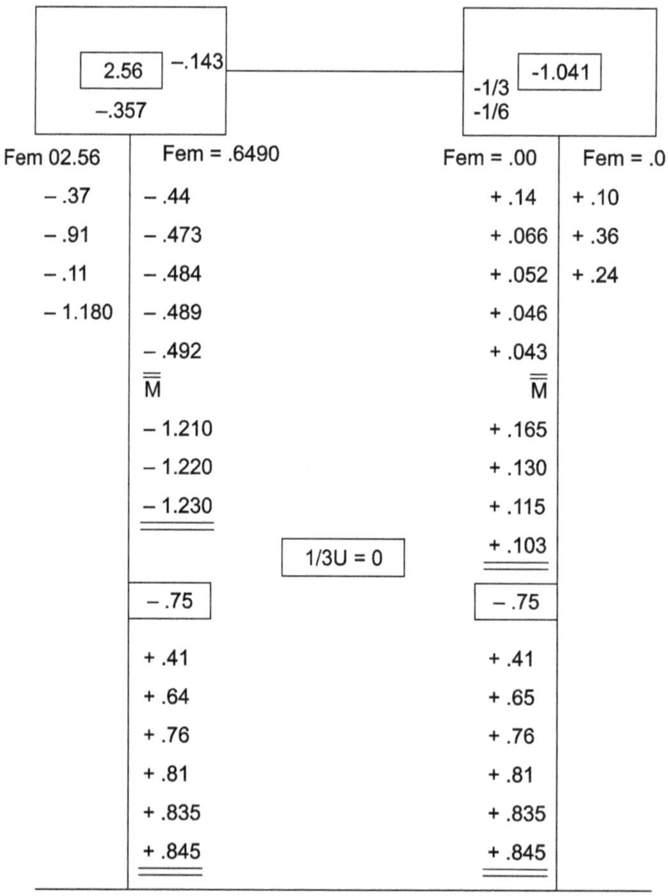

8.12 ANALYSIS OF THE FRAME TWO STOREY—SINGLE BAY BY KANI'S METHOD

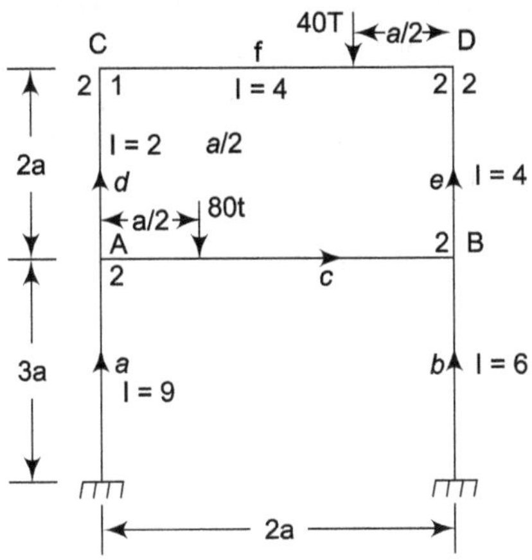

Fig. 8.21 *Single Bay–Two Storey*

1. Compute θ_A to θ_D from the $\overline{M}$ values, and from $\overline{\overline{M}}$ values δ_{AB} and δ_{CD}.
2. Draw elastic line.
3. Give the intermediate results by filling the blanks of the following table. Draw BM Diagram.

Member	*FEM*	$2\overline{M}$	$\overline{M}$	$\overline{M}$	*Final moment*
$(M_1)_a$	0				−1.36 a
$(M_1)_b$	0				−2.72 a
$(M_1)_c$	+5.62 a				+4.61 a
$(M_1)_d$	0				−2.00 a
$(M_2)_d$	0				−2.64 a
$(M_1)_f$	+3.75 a				−2.69 a
$(M_2)_f$	−11.25 a				+6.73 a
$(M_2)_e$	0				+6.54 a
$(M_1)_e$	0				+2.97 a
$(M_2)_c$	−1.87a				−2.52 a
$(M_2)_b$	0				−.40 a
$(M_1)_b$	0				−.20 a

If the $\overline{\overline{M}}$ values are not fully converged, the moment values may exhibit unbalancing of moments of the joints. The number of cycles of iteration depends on the convergence of $\overline{M}$ (on rotation) and $\overline{\overline{M}}$ (or sway) values. From the consideration of design accuracy, the un balances of the order of five to ten percent are acceptable.

8.13 TALL FRAMES WITH SWAY

A frame Fig. 8.22, is shown under "equivalent loading" system. If the columns bases are non-sinking and rigid, "vertical equivalent joint loads" are not to be considered in the case of the approximate analysis

(All members assumed in extensible) as they get directly transferred by columns to the foundations. The "equivalent moment loads and horizontal loads" are required to be considered. The above consideration is not valid if the column members are not vertical and they are sloping. In the present analysis, the effect of axial deformations on moments and shear of members is neglected. It is necessary to observe that for tall frames, we can continue to neglect axial deformations for their effect on shears and moments in the members of the frame. Such an approximation may be normally acceptable up to thirteen storeys. The axial deformations of girders do not depend on the height of the tiered building. It is local to floor, and in plane rigidity of a 'floor – slab' justifies the diaphragm action. Neglecting the axial deformations of the girders, we have the following equality for each girder line, say, G floor.

$$(\delta'_x)_G = (\delta'_x)_H = \cdots (\delta'_x)_K \qquad \text{Fig. 8.23}$$

Considering any typical floor G of the multistory frame, Fig. 4 the following stiffness equation for the moment loading m_I is obtained. The deformations affecting the moment load m_I are.

$$d = \{\theta_D\, \theta_H\, \theta_I\, \theta_J\, \theta_N\, \delta_{DI}\, \delta_{JN}\}, \qquad \text{Fig. 8.22}$$

(The parenthesis { } is used to represent a column matrix, even though the matrix is written as a row for the convenience of writing.)

$$M_I = [\overset{\theta_D}{(2EI/L)_{g3}}, \overset{\theta_H}{(2EI/L)_i}, \overset{\theta_I}{\{4EI/L)_{g3} + (4EI/L)_i + (4EI/L)_j + [+ (4EI/L)_{l3}\}},$$

$$+ \overset{\theta_J}{(2EI/L)_j}, \overset{\theta_N}{(2EI/L)_{l3}}, \overset{\delta_{DI}}{(6EI/L^2)_{g3}}, \overset{\delta_{IN}}{(6EI/L^2)_3}]d$$

Where d has already been defined. ...(7)

We have one equation per joint, based on the moment equilibrium. Therefore, there are as many equations as there are number of joints. For the unknown horizontal floor-deflections (sway displacements), which are in excess of the joint rotations, the additional equations are obtained from the "force equilibrium" conditions, Fig. 8.23

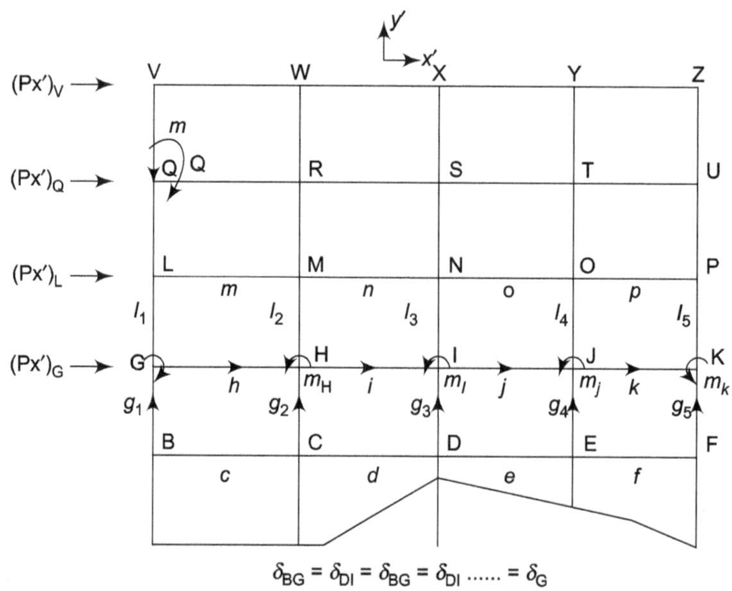

Fig. 8.22 *Equivalent Loads Co-ordinates System*

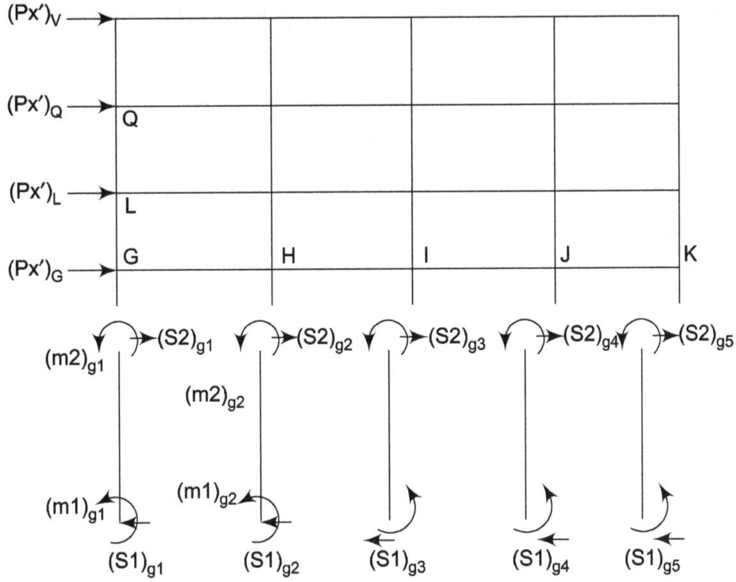

Fig. 8.23 *Free Body for Horizontal Force Equilibrium*

Free body shown is considered for force equilibrium.
From Equilibrium condition of the free body Fig. 8.23

$$\Sigma P'_x = \Sigma s$$

where $(\Sigma P'_x) = (p'_x)_G + (p'_x)_L + (p'_x)_Q + (p'_x)_V$, and

$$\Sigma s = (s_2)_{g1} + (s_2)_{g2} + (s_2)_{g3} + (s_2)_{g4} + (s_2)_{g5}$$

$$= \left(\frac{M_1 + M_2}{L}\right)_{g1} + \left(\frac{M_1 + M_2}{L}\right)_{g2} \cdots \left(\frac{M_1 + M_2}{L}\right)_{g5} = \sum_{i=1}^{5}\left(\frac{M_1 + M_2}{L}\right)_{gi}$$

Let *N* be any typical node and the joints to the bottom *(B)*, left *(L)*, right *(R)*, top *(T)*, be designated by *B, L, R, T* respectively and the corresponding members connected to the joint N be *b, l, r* and *t*, Fig. 8.22.

The arrow directions are in the ascending order of designation-letters" following bottom to top and left to right sequence.

The Eq. (7) may be rewritten, following the joint and member designation, Fig. 8.22, as

$$\overline{M}_I = (\overline{M}_1)_b + (\overline{M}_1)_1 + 2\left[k_b + k_1 + k_r + k_t\right] - (\overline{M}_2)_r + (\overline{M}_2)_t + \overline{\overline{M}}_b + \overline{\overline{M}}_t$$

Which gives?

$$k_1\theta_N = (\overline{M}_2)_l = \frac{-K_l}{2(k_b + k_l + k_r + k_t)}\left[-m_I + (\overline{M})_b + (\overline{M}_1)_l + (\overline{M}_2)_r + (\overline{M}_2)_t + \overline{\overline{M}}_b + \overline{\overline{M}}_t\right] \qquad \text{...(10)}$$

Where $K_l = \left(\frac{2\,EI}{l}\right)_l$, $\overline{\overline{M}}_b = \frac{6\,EI\delta_{BN}}{L^2_{\ b}}$ and $\overline{\overline{M}}_t = \frac{6\,EI\delta_{NT}}{l^2_{\ t}}$

Similar expressions can be written for $(\overline{M}_1)_r$, $(\overline{M}_2)_b$ and $(\overline{M}_1)_t$

Where the bracketed quantity will remain the same and only the numerator K_l of the multiplier $\frac{-K_l}{2\sum K_i}$, gets changed to k_r, k_b and k_t respectively.

In the case of continuous frame, moment equilibrium, Eq. (10), is not sufficient and we have to write force equilibrium equation, Eq. (8) corresponding to displacement δ_G or $\overline{\overline{M}}_b$ and $\overline{\overline{M}}_t$.

The Eq. (8) may be re-written, following the joint and member designation, Fig. 8.22, as

$$\sum p'_x = \sum_{i=1}^{5}\left(\frac{M_1+M_2}{L}\right)_{bi} \quad \text{... (11)}$$

Summation index "i" is, equal to the number of columns of the frame,

Where M_{12} and M_{21} are due to end deformation only and are given by

$$(M_1)_{bi} = \left(\frac{4EI\theta_1}{L}\right)_{bi} + \left(\frac{2EI\theta_2}{L}\right)_{bi} + \left(\frac{6EI}{L^2}\right)_{bi}\delta_{BN}$$

$$(M_2)_{bi} = \left(\frac{2EI\theta_1}{L} + \frac{4EI\theta_2}{L}\right)_{bi} + \left(\frac{6EI}{L^2}\right)_{bi}\delta_{BN}$$

$$\left(\frac{M_1+M_2}{L}\right)_{bi} = \left[\frac{6EI\theta_1}{L^2} + \frac{6EI\theta_2}{L^2}\right]_{bi} + \left(\frac{12EI}{L^2}\right)_{bi}\delta_{BN} \quad \text{...(12)}$$

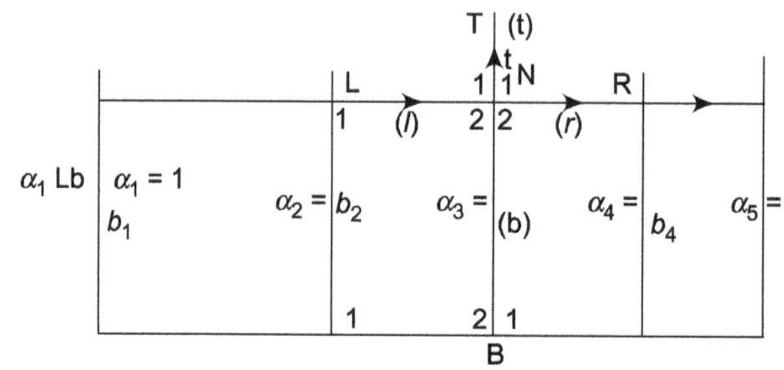

Fig. 5.5 *Typical Node Designation*

$$\sum P'_x = \sum_{i=1}^{5}\kappa\frac{3}{L_b}\left(\overline{M}_{12}+\overline{M}_{21}\right)_{bi} + \frac{6_{BN}}{L^2_b} - \sum_{i=1}^{5}\frac{k_{bi}}{\alpha_i^2}$$

Summation of all the floor loads at floor G and floors above.

The above result can now be used to express $\overline{\overline{M}}_b$, Eq. (13).

$$\overline{\overline{M}}_b = \left(\frac{6EI\delta_4}{L^2}\right)_{bi} = -\frac{2(EI)_b\alpha_i^2}{2\sum_{i=1}^{5}\frac{k_{bi}}{\alpha^2_l}}\left[-\sum p'_x + \sum_{i=1}^{5}\alpha^3_{L_b}\left(\overline{M}_{12}+\overline{M}_{21}\right)_{bi}\right]$$

$$\overline{\overline{M}}_b = \frac{-3k_b}{2\sum_{i=1}^{5} k_{bi}}\left[+\frac{1}{3}U+\sum_{i=1}^{5}(\overline{M}_{12}+\overline{M}_{21})_{bi}\right] \quad \text{... (13a)}$$

$$\text{where } U = \left[\sum L_b P_x'\right] \text{ and } K_b = \left(\frac{2EI}{L}\right)_b \quad \text{... (13b)}$$

The Eqs (10) and (13), are to be treated in the similar way as we did for the Gauss-Seidel procedure.

The Eqs (10) and (13) are the pivotal equations for the iteration procedure for frames with sway. The significant advantage of this procedure is that separate sway correction is not required. In the case of wind analysis, where lateral loads are dominant compared to vertical (gravity) loads, the convergence of iteration is faster by computing $\overline{\overline{M}}$ first, based on Eq. (13) (for the first step all $\overline{M}$ will be zero) followed by evaluation of $\overline{M}$ values based on Eq. (10). The numerical procedures are best explained by the examples.

8.12 Illustration 3: Analysis of a three storey and three bay frame.

With the pivotal equations developed, analyze the frame. The problem is general for illustrating the pivotal equations.

Find member end moments for the frame and loading. The equivalent joint loads are, $(p'_x)_A = (p'_x)_E = 2(P'_x)_I = 4\ t$, Fig. 8.25

and $m_A = m_E = 0$, $m_I = 1.33$ tm

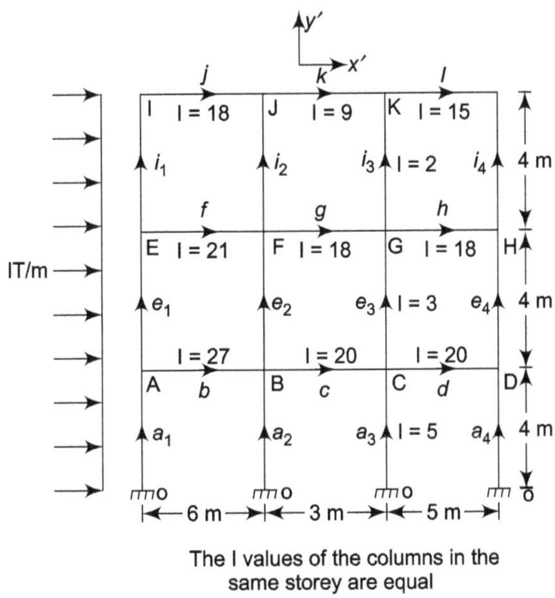

Fig. 8.25 *Illustration 3, Three Storey and Three Bay Frame*

The joint loads m_A and m_E and m_I are entered within the inner squares f the corresponding joints. (p'_x) values are not used directly (Eq. 13), and in its place $(-1/3U)$ values for different storey are

First Storey i.e., [1/3 U] = 1/3 (−4(2 + 4 + 4)) = −13.33

Below A line

Second Storey, [1/3U] = 1/3 (−4(2 + 4)) = − 8

Below E line

Third Storey, [1/3 U] = 1/3 (−4 × 2) = −2.67

Below I line

The above values of "1/3U" are entered in the three boxes corresponding to the three storeys, Fig. C.

The distribution factors, $\dfrac{-k_i}{2(k_b + k_i + k_r + k_t)}$, Eq. (10)

are required to be computed for each joint. Specimen calculations for a typical interior node G, are

$$\text{Distribution factor for the bottom member} = (1/2)\,\frac{2\left(\frac{3}{4}\right)}{2\left(\frac{3}{4}\right)+2\left(\frac{18}{3}\right)+2\left(\frac{18}{5}\right)+2\left(\frac{2}{4}\right)}$$

$$= -.75/21.7 = -.035$$

The value − .375 is written at the "bottom" and within the outer square of the joint *G*.

Similarly the distribution factors for

Left member = −6/21.7 = −.276

Right member = (−18/5)/21.7 = .166

Top member = (−2/4)/21.7 = −.023

are written at the left, right, and top of the joint *G*.

The distribution factors, $\dfrac{-3kb}{\sum_{i=1}^{4} k_{bi}}$ in Eq. (13) are required to be computed for each column of storey considered. Specimen calculations for columns of 2nd are:

$$\text{For column, } e_i,\ -(3/2)\times\frac{(2EI/L)_{ei}}{(2EI/L)_{ei}+(2EI/L)_{e2}+(2EI/L)_{e3}+(2EI/L)_{e4}} = -3/2x\frac{1}{1+1+1+1} = .375$$

As all the columns considered in the example are of equal height and same *EI* Value, the distribution factors, Eq. (13) for other columns,

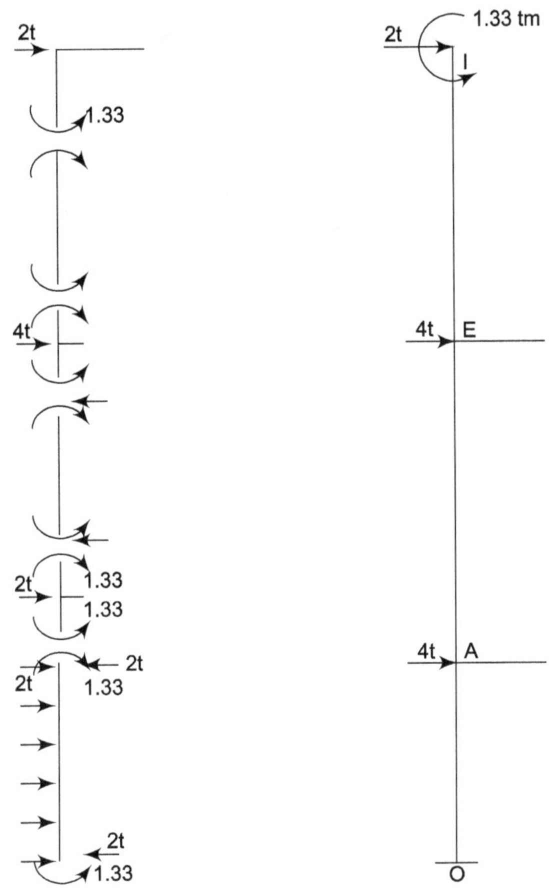

Fig. 8.25 *Equivalent Joint Loads*

e_2, e_3, e_4 are the same as the value computed for the column e_1. These values are written against each column in Fig. (c).

We are, at this stage in a position to carry out the iteration. For reasons of convergence, in the case of wind load, the $\overline{\overline{M}}$ values are computed first from Eq. (13). In the case of first cycle for $\overline{\overline{M}}$ values, assume M′ values to be zero.

The $\overline{\overline{M}}$ values are computed in the order of sequence, "bottom to top". Typical values or for the column e_1, e_2, e_3 and e_4 of second storey are the same (length and EI value same) and they are equal to

$$-.375 \times (-8.00) = +3.00$$

And the value +3.00 is entered on the "right hand" side of corresponding columns.

With the first approximated values of $\overline{\overline{M}}$, we proceed to find the $\overline{M}$ values of the first cycle. The $\overline{M}$ values are computed in order of sequence, "left to right", for a floor line, and floor lines are considered in the order of sequence, "bottom to top". The computation of $\overline{M}$ for a typical joint G (First Cycle) are

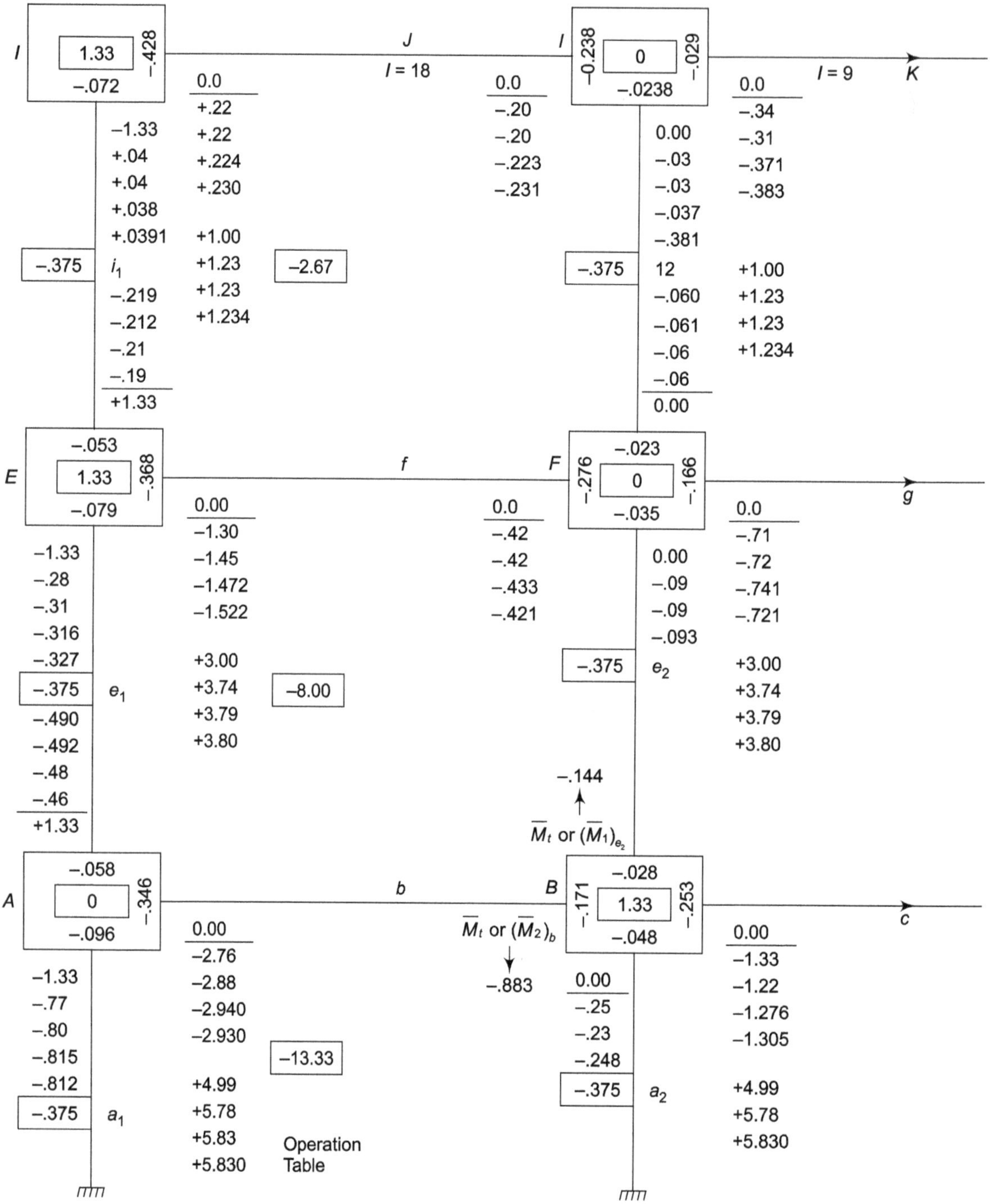

Operation Table

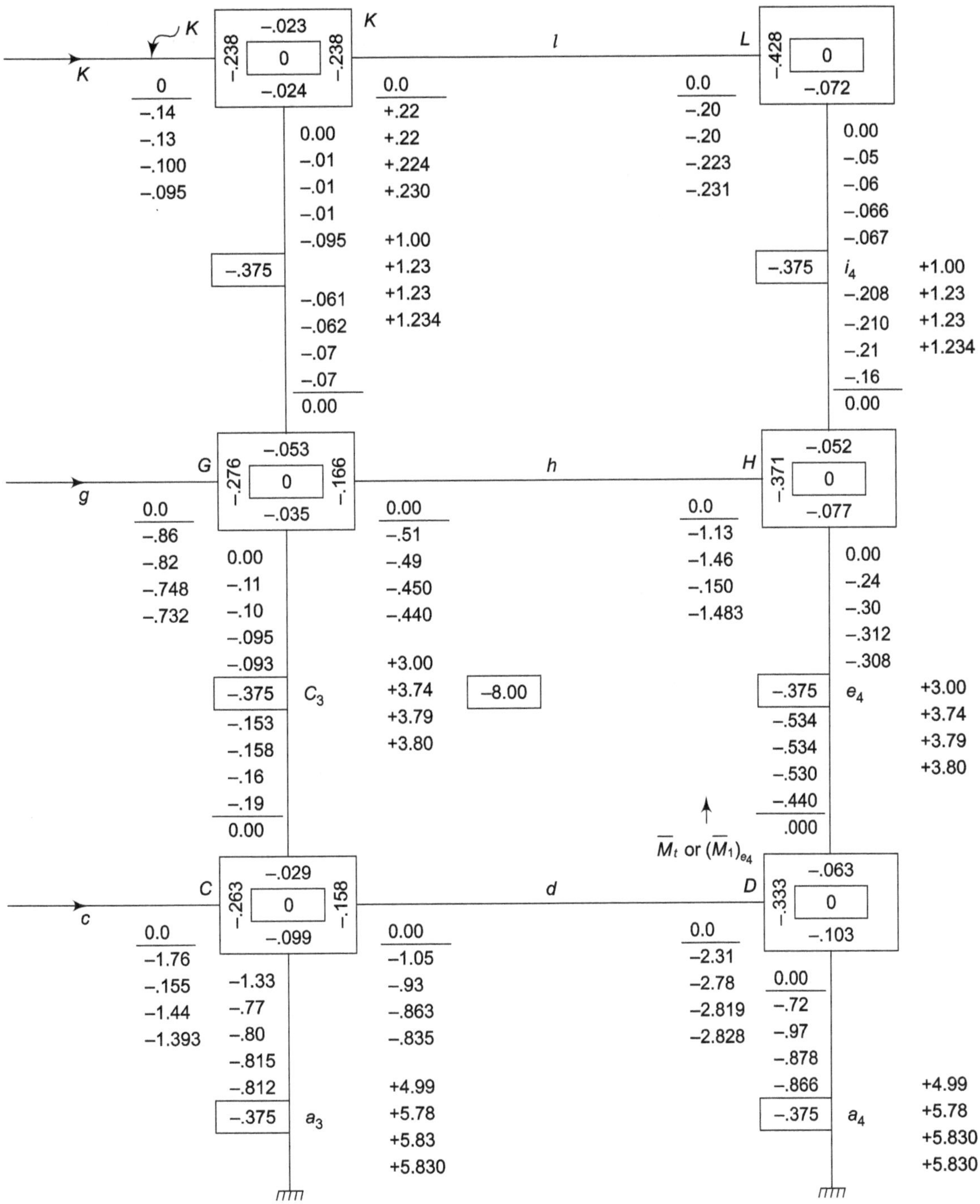
K
K
K
−.023
−.238
0
−.238
−.024
0
−.14
−.13
−.100
−.095
0.00
−.01
−.01
−.01
−.095
−.375
−.061
−.062
−.07
−.07
0.00
0.0
+.22
+.22
+.224
+.230
+1.00
+1.23
+1.23
+1.234
l
L
0.0
−.20
−.20
−.223
−.231
−.428
0
−.072
0.00
−.05
−.06
−.066
−.067
−.375
i_4
−.208
−.210
−.21
−.16
0.00
+1.00
+1.23
+1.23
+1.234
G
g
−.053
−.276
0
−.166
−.035
0.0
−.86
−.82
−.748
−.732
0.00
−.11
−.10
−.095
−.093
−.375
C_3
−.153
−.158
−.16
−.19
0.00
0.00
−.51
−.49
−.450
−.440
+3.00
+3.74
+3.79
+3.80
−8.00
h
H
0.0
−1.13
−1.46
−.150
−1.483
−.052
−.371
0
−.077
0.00
−.24
−.30
−.312
−.308
−.375
e_4
−.534
−.534
−.530
−.440
.000
+3.00
+3.74
+3.79
+3.80
$\overline{M}_t$ or $(\overline{M}_1)_{e_4}$
C
c
−.029
−.263
0
−.158
−.099
0.0
−1.76
−.155
−1.44
−1.393
−1.33
−.77
−.80
−.815
−.812
−.375
a_3
0.00
−1.05
−.93
−.863
−.835
+4.99
+5.78
+5.83
+5.830
d
D
0.0
−2.31
−2.78
−2.819
−2.828
−.063
−.333
0
−.103
0.00
−.72
−.97
−.878
−.866
−.375
a_4
+4.99
+5.78
+5.830
+5.830

Joint *G*

$$\overline{M}_b = -.035\left[-m_G + (\overline{M}_1)_{i3} + (\overline{M}_1)_g + (\overline{M}_2)_h + (\overline{M}_2)_{i3} + (\overline{\overline{M}})_{i3} + (\overline{\overline{M}})_{i3}\right]$$

For the first cycle $(\overline{M}_{21})_h$ $2(\overline{M}_{21})_{i3}$ being not known, are considered to be zero and the other values are

$$M_G = 0 \text{ (from data)}$$

$$(\overline{M}_1)_{e3} = -.19 \text{ (known because the joint C has been solved already)}$$

$$(\overline{M}_1)_g = -.71 \text{ (joint F has been solved already)}$$

$$\left.\begin{matrix}(\overline{\overline{M}}_{e3}) = +3.00 \\ (\overline{\overline{M}}_{13}) = +1.00\end{matrix}\right\} = +3.00$$ For better convergence, $\overline{\overline{M}}$ values are computed first.

$$\overline{M}_b = -.035\ [0 - .19 - .71 + 3 + 1] = -.035\ (3.1) = -.11$$

Similarly,

Left, L: $\overline{M}_L = -.276\,(3.1) = -.86$

Right, r: $\overline{M}_r = -.166\,(3.1) = -.51$

Top, t: $\overline{M}_t = -.023\,(3.1) = -.061$

Typical calculations of Cycle IV.

$$\left(\overline{\overline{M}}_{a1}\right) = -.375\left[-13.33 + \sum_{i=1}^{4}\left(\overline{M}_{12} + \overline{M}_{21}\right)_{a1}\right], \text{ and}$$

$$(\overline{M}_1)_{a1} = -.815, \text{ from cycle III}$$

$$(\overline{M}_2)_{a2} = -.242$$

$$(\overline{M}_2)_{a3} = -.267$$

$$(\overline{M}_2)_{a4} = -.878$$

Substituting the values of $\overline{M}$

$$(\overline{\overline{M}}_{a1}) = -375\ (-15.532) = 5.83$$

And for the example, $\overline{\overline{M}}_{a1} = \overline{\overline{M}}_{a2} = \overline{\overline{M}}_{a3} = \overline{\overline{M}}_{a4}$

Similarly,

$$\overline{\overline{M}}_{e1} = \overline{\overline{M}}_{e2} = \overline{\overline{M}}_{e3} = \overline{\overline{M}}_{e4} = -\ .375\left[-8.00 + \Sigma(\overline{M}_{12} + \overline{M}_{21})_{e1}\right]$$

where $(\overline{M}_1 + \overline{M}_2)_{e1} = -.492 - .316 = -.808$ (from Cycle III)

$$(\overline{M}_1 + \overline{M}_2)_{e2} = -.141 - .093 = -.234$$

$$(\overline{M}_1 + \overline{M}_2)_{e3} = -.158 - .095 = -.253$$

$$(\overline{M}_1 + \overline{M}_2)_{e4} = -.534 - .312 = -.846$$

Substituting the $\overline{M}$ values,

$$\overline{\overline{M}}_{e1} = \overline{\overline{M}}_{e2} = \overline{\overline{M}}_{e3} = \overline{\overline{M}}_{e4} = -.375\,(-10.141) = +3.80$$

And it can be shown similarly that $\overline{\overline{M}}_1 = +1.234$, $\overline{M}$ values for a typical joint, say joint I,

Joint I $\quad \overline{M}_b = -.072\,[-1.33 \underbrace{-.219}_{\text{IV Cycl.}} \underbrace{-.223}_{\text{III Cycl.}} \underbrace{-1.234}_{\text{IV Cycl.}}]$

$$\overline{M}_b = -.072 \times (-.538) = +.039$$

and

$$\overline{M}_r = (\overline{M})_j = -.428\,(-.538) = +.23$$

the final values are given in the Table 8.7 final moments.

Table 8.7 *Final moments*

Member	Fem	$\overline{M}$	$2\overline{M}$	M	Final Moment
$(M_1)_{a1}$	+1.33	–.812	0	+5.83	+6.348
$(M_2)_{a1}$	-1.33	0	2x(-312)	+5.83	+2.876
$(M_1)_{a2}$	0	–.248	0	+5.83	+5.582
$(M_2)_{a2}$	0	0	2x(–.248)	+5.83	5.334
$(M_1)_{a3}$	0	–.259	0	+5.83	5.571
$(M_2)_{a3}$	0	0	2x(–.259)	+5.83	5.312
$(M_1)_{a4}$	0	–.866	–	+5.83	4.964
$(M_2)_{a4}$	0	0	2x(–.866)	+5.83	4.098
$(M_1)_c$	0	–1.393	2x(–1.305)	0	–4.003
$(M_2)_c$	0	–1.305	2x(–1.393)	0	–4.091
$(M_1)_d$	0	–2.828	2x(–.835)	0	–4.498
$(M_2)_d$	0	–.835	2x(–2.828)	0	–6.491

(Contd.)

(Contd.)

$(M_1)_{e1}$	+1.33	–.327	2x(–.490)	+3.80	+3.823
$(M_2)_{e1}$	-1.33	–.490	2x(–.327)	+3.80	1.326
$(M_1)_{e2}$	0	–.090	2x(–.144)	+3.80	+0.020
$(M_2)_{e2}$	0	–.144	2x(–.090)	+3.80	+3.476
$(M_1)_{e3}$	0	–.093	2x(–.153)	+3.80	+3.401
$(M_2)_{e3}$	0	–.153	2x(–.093)	+3.80	+3.461
$(M_1)_{e4}$	0	–.308	2x(–.534)	+3.80	+2.424
$(M_2)_{e4}$	0	–.534	2x(–.308)	+3.80	+2.650
$(M_1)_f$	0	–.421	2x(–1.522)	0	-3.465
$(M_2)_f$	0	–1.522	2x(–.421)	0	-2.364
$(M_1)_g$	0	–.732	2x(–.721)	0	-2.174
$(M_2)_g$	0	–.721	2x(–.732)	0	-2.185
$(M_1)_h$	0	–1.483	2x(–.440)	0	-2.363
$(M_2)_h$	0	+.440	2x(–1.483)	0	-3.406
$(M_1)_{i1}$	+1.33	+.039	2x(–.219)	+1.23	+2.161
$(M_2)_{i1}$	-1.33	–.219	2x(+.039)	1.23	-.241
$(M_1)_{i2}$	0	–.038	2x(–.060)	1.23	+1.072
$(M_2)_{i2}$	0	–.060	2x(–.038)	1.23	+1.094
$(M_1)_{i3}$	0	–.095	2x(–.061)	1.23	1.013
$(M_3)_{i3}$	0	–.061	2x(–.095)	1.23	0.979
$(M_1)_{i4}$	0	–.067	2x(–.208)	1.23	0.747
$(M_2)_{i4}$	0	–.208	2x(–.067)	11.23	0.888
$(M_1)_j$	0	–.231	2x(–.230)	0	+0.229
$(M_2)_j$	0	+.230	2x(–.231)	0	-0.232
$(M_1)_k$	0	–.095	2x(–.383)	0	-0.861
$(M_2)_k$	0	–.383	2x(–.095)	0	-0.573
$(M_1)_l$	0	–.399	2x(–.095)	0	-0.589
$(M_2)_l$	0	–.095	2x(–.399)	0	-0.893
$(M_2)_b$	0	-2.93	2x(–.883)	0	-4.694

8.14 CONCLUDING REMARKS

If the girder members of a frame are prismatic but inclined, then elements of the matrix Eq. (7), are (from K', matrices) derived already, chapter on stiffness method. In the case of column members with inclinations, and axial deformations neglected for shears and moments, we take free body corresponding to Fig. 4, but this time for moments of joint forces (equivalent moment loads included) about "centre of rotation", and obtain and equation similar to the Eq. (13).

As in the case of continuous beam, if a base of a column member is hinged, the K_i values require to be modified to ¾ ($2E/L$) = ($1.5\ EI/L$).

The Kani's method is preferred from computational considerations. Kani's method combines joint rotations and sway displacement. Both the method are solutions by iteration of the system of slope deflection equations, but Kani's method, essentially solves for deformation and the moment distribution or cross method solves for moments (by applying correcting moments at each cycle of distribution and satisfy joint moment equilibrium).

The fast convergence of the "Kani's method" is essentially due to satisfying equilibrium at a joint (iterative values at a joint based on, say, Eq. 10) the far ends of members meeting at the joint are not considered fixed but the, values of joint rotation of the same cycle or the cycle previous to it, are considered. Finally, the method is self correcting, i.e. any computational mistake is eliminated after a few cycles. The moment distribution (MD) method has one distinct advantage over the "Kani's method" because it can be interpreted in physical terms i.e. balancing moment values at a joint and carrying over the effect of moments applied to far ends of members assumed fixed for the particular iteration. The Moment Distribution method solves for the moments – the values used for the design process.

In the case of moment distribution method, iteration is used for "moment equilibrium" only and for force equilibrium simultaneous equations are formulated to solve arbitrary sway. The iteration procedure of the "Kani's method" is based on satisfying simultaneously the moment equilibrium (Eq. 10) as well as force equilibrium, (Eq. 13), and thereby eliminates "sway analysis" and solution of simultaneous equations for the unknown sways.

In the case of frames with base columns hinged, typical equations based on moment and force equilibrium are required to be developed (similar to the equations (ii) and (iii) developed for beams on spring supports, p. 236). The moment equilibrium remains unaltered with the provision that for the hinged column, $K = 1.5\ EI/L$. The force equilibrium algorithm gets slightly altered. Similar modification is required in the case of unequal heights of the base column of frames.

Evaluation of deformations: As the "Kani's method" solves for deformation in the form of $\overline{M}$ and $\overline{\overline{M}}$, the actual end deformations are obtained by the relations:

$$\delta = \overline{\overline{M}}\left(\frac{l^2}{6EI}\right) = \overline{\overline{M}}\left(\frac{1}{3K}\right)$$

$$\theta = \overline{M}\left(\frac{l}{2EI}\right) = \frac{\overline{M}}{K}$$

8.15 EXTENSION OF KANI'S METHOD: AXIAL DEFORMATIONS OF COLUMN MEMBERS

It is possible to work out an algorithm, accounting for the axial deformation of the columns. But the suitability for the hand computations becomes too tedious to recommend its use. For such tall buildings where the columns cannot be assumed inextensible, the reliance may have to be on the computer software.

EXAMPLES FOR PRACTICE

Example 1

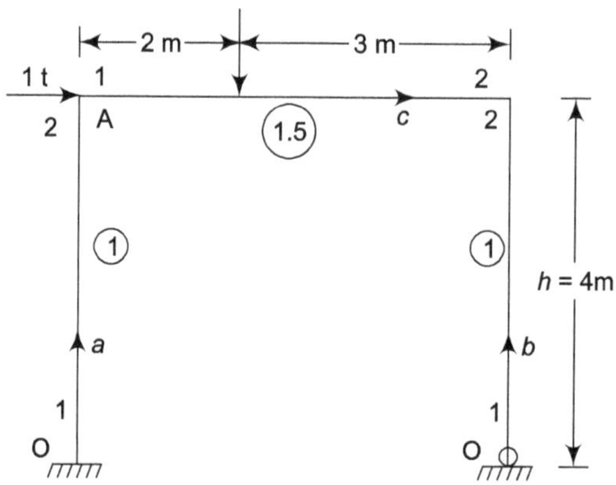

Ans. $(M_1)_a = -.4946$ tm

$(M_2)_a = -2.649$ tm

$(M_1)_b = 2.934$ tm

$(M_2)_b = 4.208$ tm

Example 2

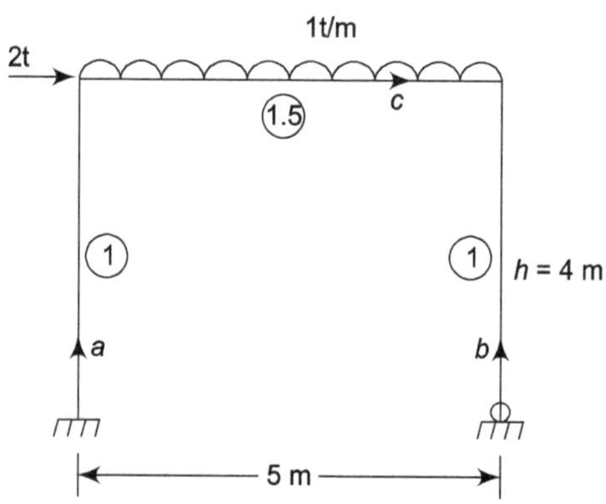

Ans. $(M_1)_a = +1.9118$ tm

$(M_2)_a = +.7892$ tm

$(M_2)_b = +2.77$ tm

Example 3

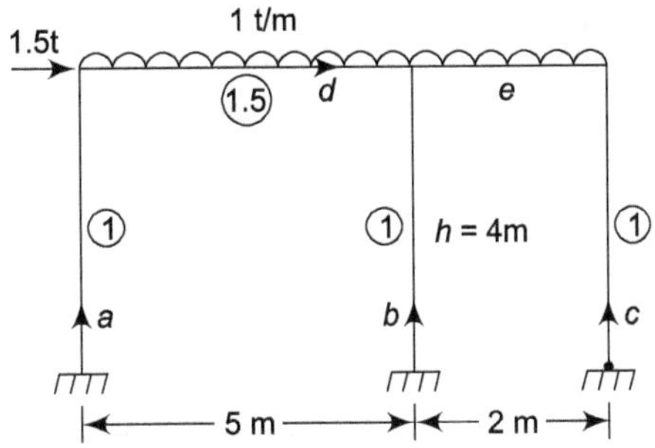

Ans. $(M_1)_a = 1.044$ tm; $(M_2)_a = .4607$ tm

$(M_1)_b = 1.493$ tm; $(M_2)_b = .589$ tm

$(M_2)_c = .6168$ tm

Example 4

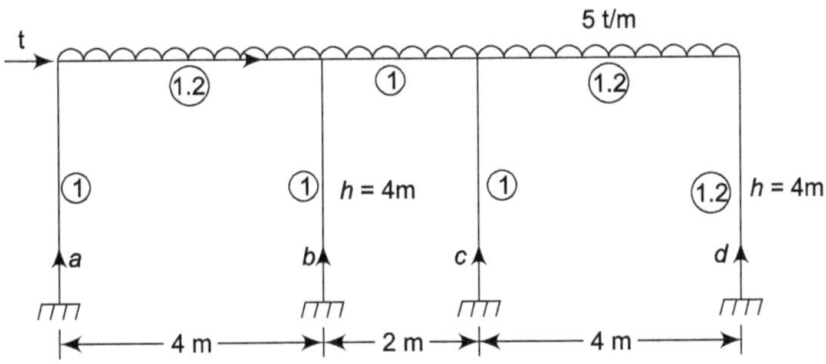

Ans. $(M_1)_a = .9233$ tm ; $(M_2)_a = .3888$ tm

$(M_1)_b = 1.266$ tm; $(M_2)_b = 1.237$ tm

$(M_1)_c = .9644$ tm; $(M_2)_c = .7138$ tm

$(M_1)_d = 1.335$ tm; $(M_2)_d = 1.213$ tm

Example 5

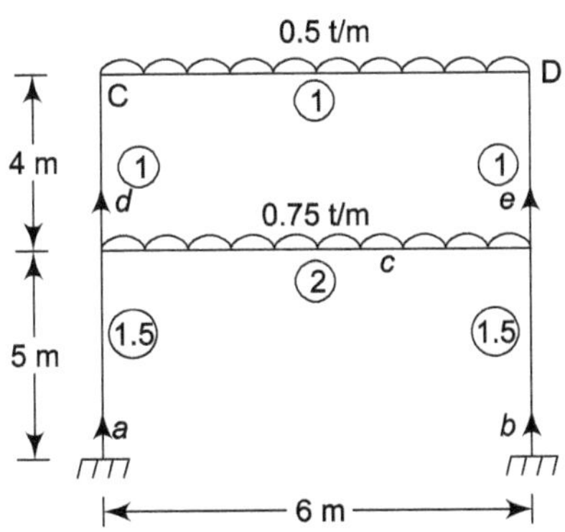

Ans: $(M_1)_a = .3938$ tm; $(M_2)_a = -.7875$ tm; $(M_1)_b = .3937$ tm; $(M_2)_b = +.7875$ tm

$(M_1)_d = -.9375$ tm; $(M_2)_d = -1.088$ tm;

$(M_1)_e = .9375$ tm; $(M_2)_e = 1.088$ tm

Example 6

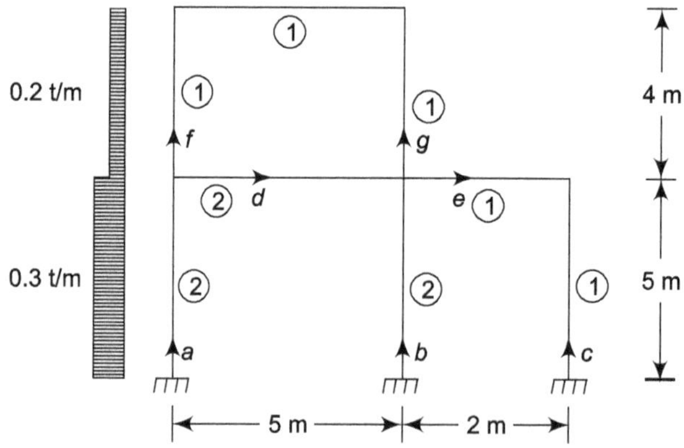

Ans: $(M_1)_a = 1.0368$ tm; $(M_2)_a = -.6294$ tm;

$(M_1)_b = .4540$ tm; $(M_2)_b = .0799$ tm

$(M_1)_f = 1.8357$ tm; $(M_2)_f = 1.1573$ tm;

$(M_1)_g = 1.515$ tm; $(M_2)_g = 1.253$ tm

Example 7

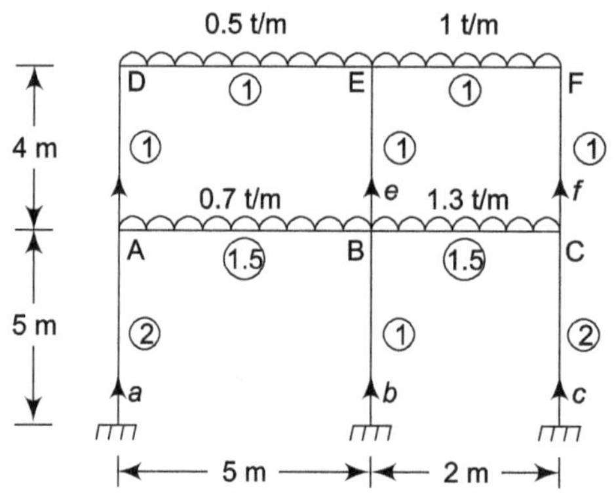

Ans: $(M_2)_a = -2.02$ tm; $(M_2)_a = -.5543$ tm;

$(M_2)_b = .1750$ tm; $(M_2)_b = .2749$ tm

$(M_1)_c = .1522$ tm; $(M_2)_c = .1542$ tm; $(M_1)_d = .527$ tm; $(M_2)_d = .6225$ tm

$(M_1)_e = .4224$ tm; $(M_2)_e = .4483$ tm;

$(M_2)_f = .1262$ tm; $(M_2)_f = .1526$ tm

Example 8

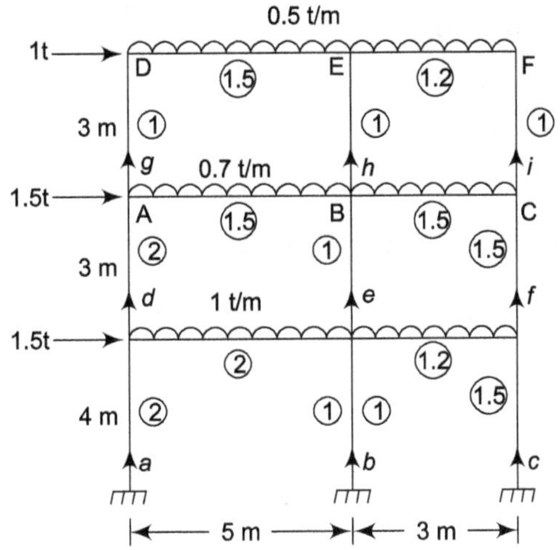

Ans: $(M_1)_a = .625$ tm; $(M_2)_a = -.5489$ tm;

$(M_1)_b = .9012$ tm; $(M_2)_b = .9030$ tm

$(M_1)_c = .8229$ tm; $(M_2)_c = .2966$ tm;

$(M_1)_d = .8101$ tm; $(M_2)_d = .3025$ tm

$(M_1)_e = 2.118$ tm; $(M_2)_e = 1.810$ tm;

$(M_1)_f = 1.503$ tm; $(M_2)_f = .9558$ tm

$(M_1)_g = 1.218$ tm; $(M_2)_g = 1.330$ tm;

$(M_1)_h = 2.895$ tm; $(M_2)_h = 3.077$ tm

$(M_1)_i = 1.654$ tm; $(M_2)_i = 1.826$ tm;

Example 9

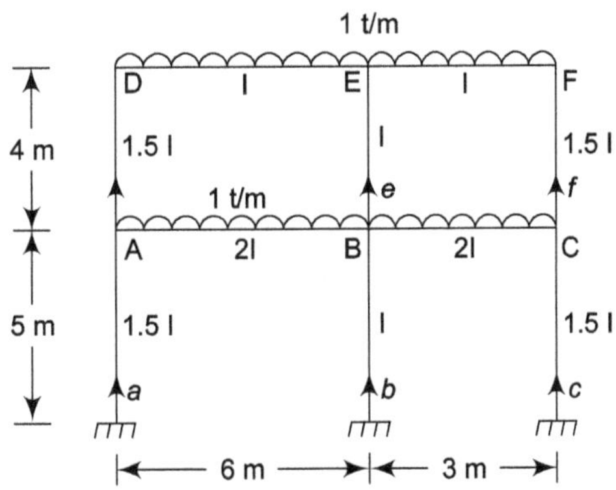

Ans: $(M_1)_a = .5867\ tm$; $(M_2)_a = -.5748$ tm;

$(M_1)_b = 1.159$ tm; $(M_2)_b = 1.153$ tm

$(M_2)_c = .5338$ tm; $(M_1)_d = .1665$ tm;

$(M_2)_d = .8658$ tm;

$(M_1)_e = 3.794$ tm; $(M_2)_e = 3.909$ tm;

$(M_1)_f = 2.928$ tm; $(M_2)_f = 2.578$ tm

REVISION PROBLEMS FOR THE DISPLACEMENT METHOD

1. Analyse by a method of your choice. Give justification for the choice.

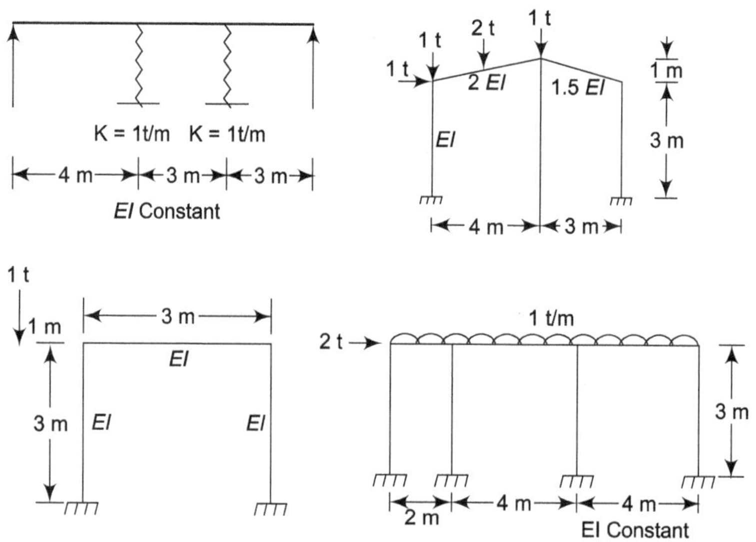

Analysis should include:
(i) B M Diagram.
(ii) Elastic Line, and Computation of deflections and rotations.

2. Analyse the following structures by slope deflection method. Draw B.M. and S.F. diagrams and sketch the elastic line 1.

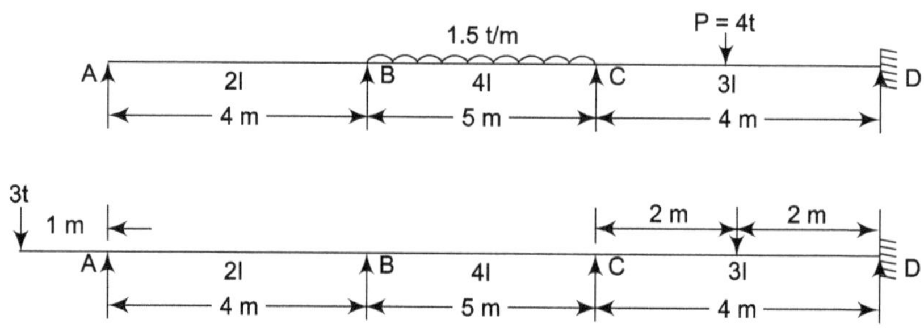

3. Solve 1 for a settlement of $\Delta = Pl^3/50\ EI$ at support B.

4.

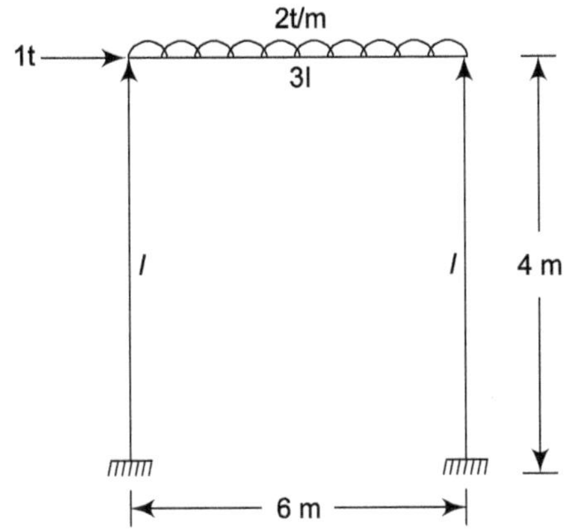

5.

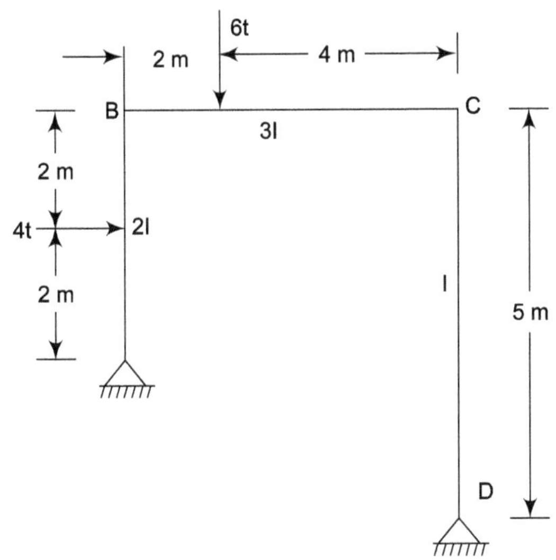

6.

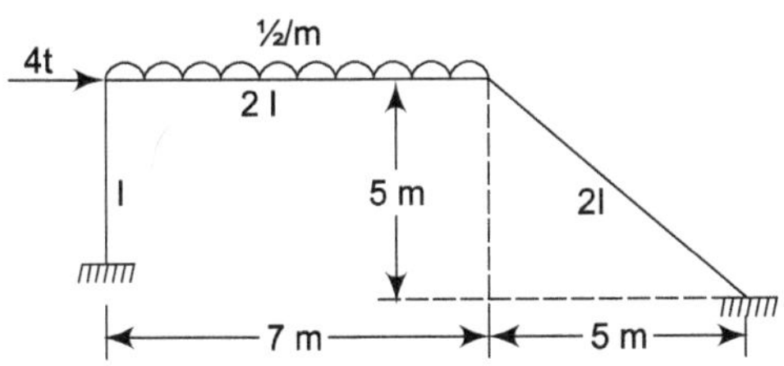

Write stiffness and load matrices. Give complete solution for problems 1, 2 by approximate analysis only.

7.(a)

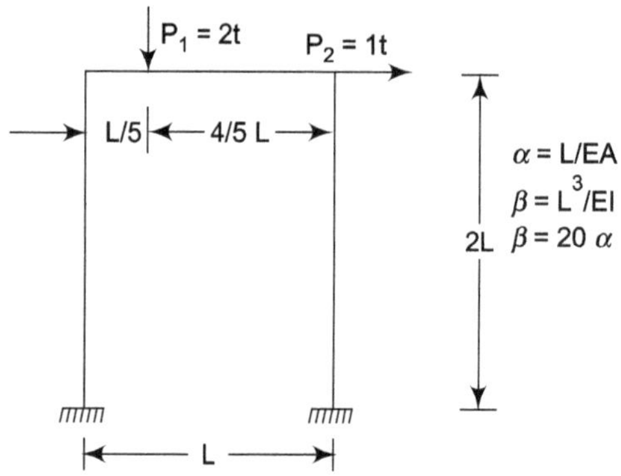

7.(b)

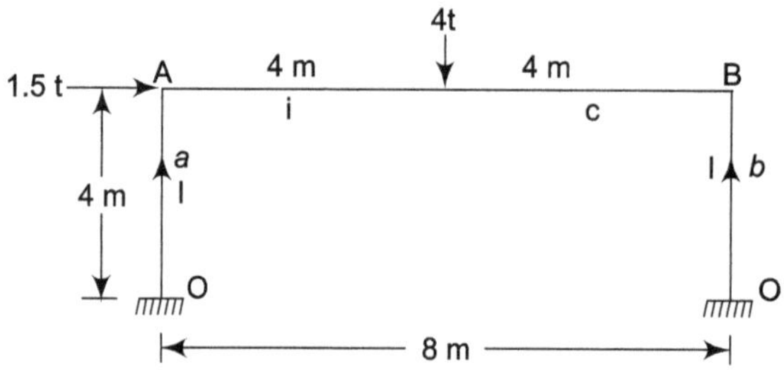

Analyse by moment distribution method.

Draw shear and moment diagrams and sketch the elastic line.

(i) Solve Question 1 to 6 of the previous sheet.

7.(c)

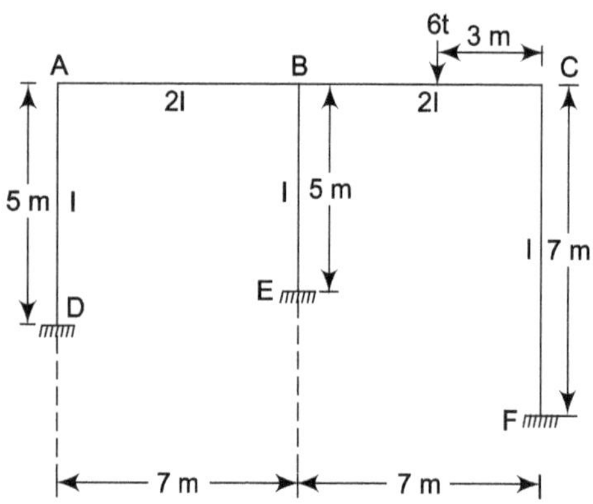

8.

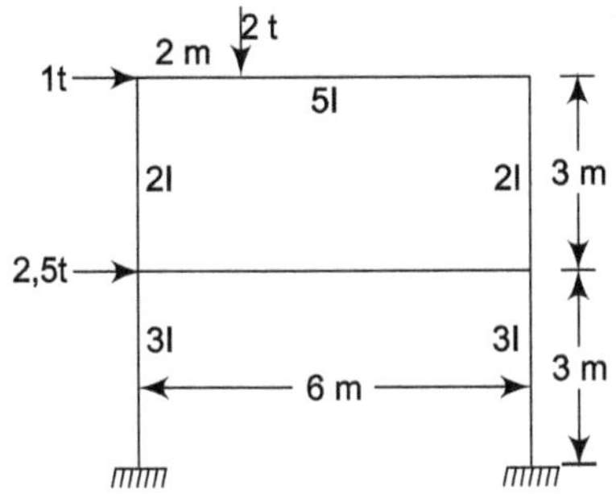
2 m
2 t
1t
5I
2I
2I
3 m
2,5t
3I
3I
3 m
6 m

9.

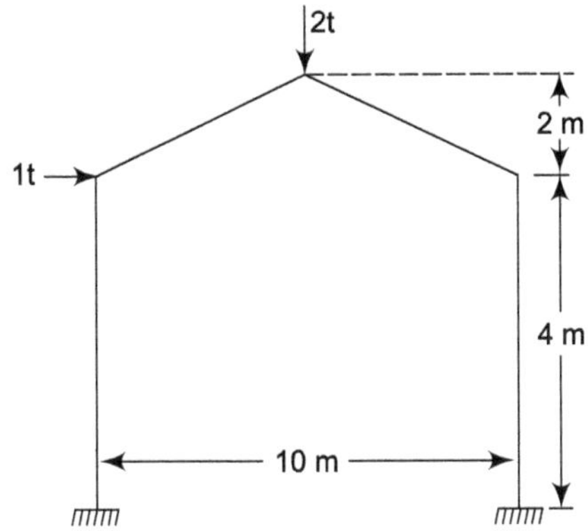
2t
2 m
1t
4 m
10 m

10.

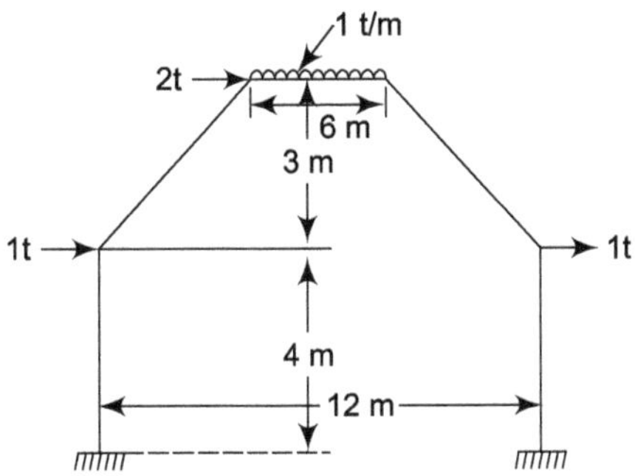
1 t/m
2t
6 m
3 m
1t
1t
4 m
12 m

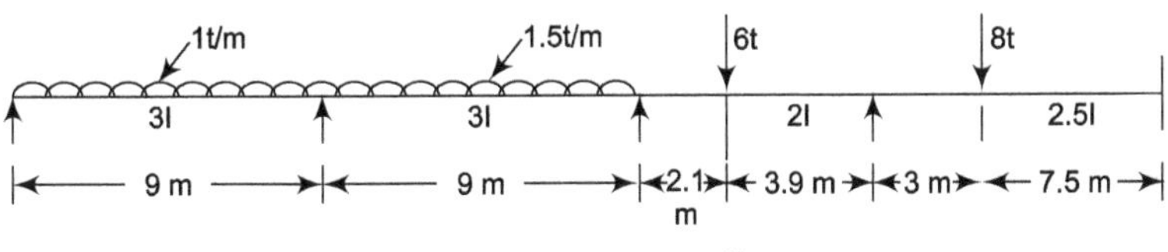
1t/m
1.5t/m
6t
8t
3I
3I
2I
2.5I
9 m
9 m
2.1 m
3.9 m
3 m
7.5 m

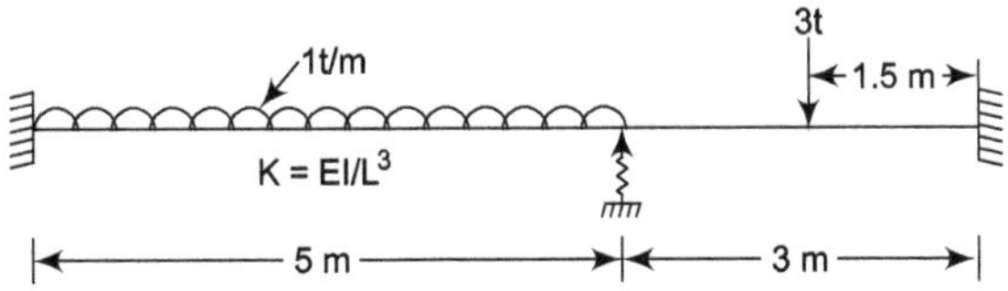
3t
1t/m
1.5 m
K = EI/L3
5 m
3 m

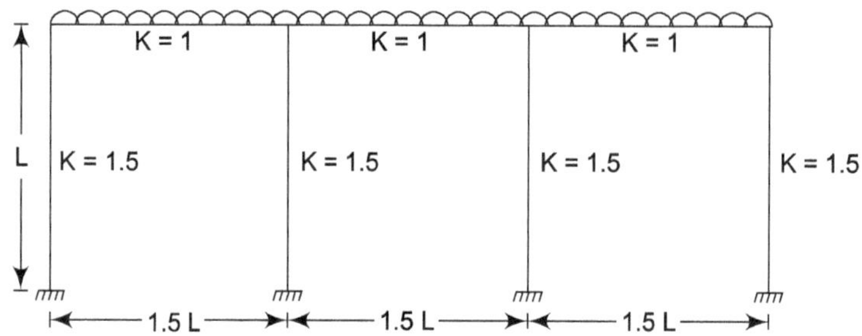

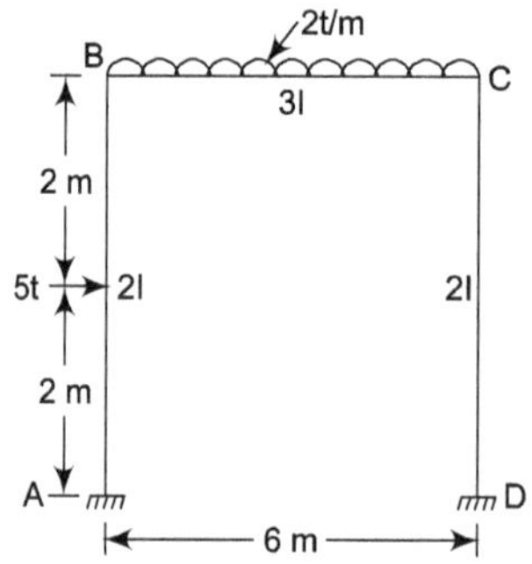

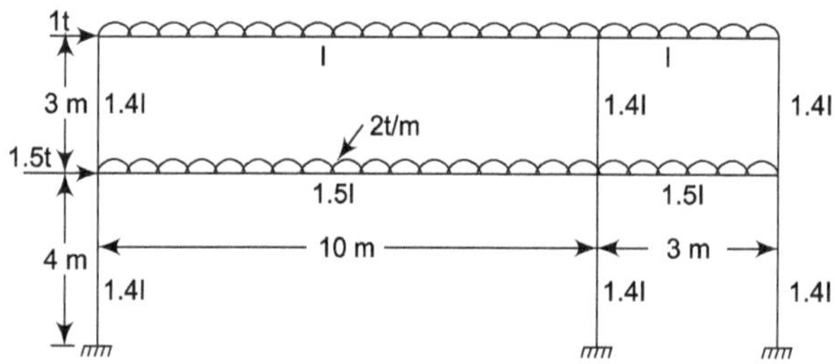

Rigid Diaphragm provided by the Floor Slabs because all frames undergo the same displacement d,

$$d_{xi} = d_i$$

d_x is the lateral displacement matrix of the entire building, an assemblage of frames. For the building considered, the number of frames is four. Adding the stiffness equations of all the frames.

$$P_x = K_x\, d_x, \text{ where } K_x \text{ is obtained by summing up the frame stiffness}$$

REFERENCES FOR CHAPTER 6, 7 AND 8

1. Wang P.C., *Numerical and Matrix Methods in Structural Mechanics*, John Wiley and Sons.
2. Livesley R.K., *Matrix Methods of Structural Analysis*, Pergamon Press.
3. Pestel, C.E. and Leckie, F.A., *Matrix Methods in Elastomechanics*, McGraw-Hill Book Co. Inc.
4. Kinney, J.S., *Indeterminate Structural Analsyis*, Addison–Wealey (Available in Indian Cheap edition).
5. Clough, R.W., Wilson, E.L., and King, I.P., Large capacity multi-storey frame analysis programs, *Journal of the Structural Division*, ASCE.
6. Horne M.R., *Progress in Solid Mechanics*, vol. 2, Ch. 7, North Holland, Amsterdom, 1961.
7. Metheson, J.A.L. Hypes Static Structures, Vol. 1, Academic Press Inc. New York.
8. Kani G., *Analysis of Multistory Frames*, London, Crosby Lock Wood, 1957.

Chapter 9: Introduction to Applied Dynamics and Design of Tall Buildings

Chapter

9

Introduction to Applied Dynamics and Design of Tall Buildings

9.1 GENERAL INTRODUCTION TO TALL BUILDINGS–TYPES

It is a typical example of a dedicated structure. The degrees of freedom being large, the numerical solutions are restricted in nature. Where ever appropriate, the theoretical formulations are explained with the help of flow charts as aids or bases to the Computer programming. The lift shafts, shear walls, and tubes are being used for moderately tall buildings, say, up to twenty storeys. The shear walls are the choice of the designers. The analysis of skeletal structures has already been presented in the previous chapters. In this Chapter, we limit the analysis particularly suited to the multi-storey frames. We continue to lay emphasis on the deflected shapes. The general assumptions are: (i) In the formulation of the analysis of tall building frames using shear walls for resisting lateral loads, shear walls are treated as wide –column skeletal members, and (ii) The floor slabs are justifiably assumed rigid in their own plane undergoing corresponding rigid body displacements, two in-plane translations, and one in plane rotation. The approximations in typical three dimensional buildings, referring to the two flexural rotations and vertical displacement at each node are used as the models for the analysis. Various approximations are appropriately grouped. The numerical modeling for the analysis of tall buildings is mainly based on the approximations [(1, 2, 3)] used by (i) Clough, Wilson, and King, and (ii) Weaver, and Nelson.

---

[1] Clough, Ray W., Wilson, Edward L., and King, Ian P, "*Large Capacity Multistory Frame Analysis Programs*", Journal of the Structural Division, ASCE Vol. 89, No. ST4, Proc. Paper 3592, August, 1963, pp179-204.

[2] Clough, Ray W., King, Ian P., and Wilson, Edward L., "*Structural Analysis of Multistory Building*", Journal of the Structural Division, ASCE Vol. 90, No. ST3, Proc. Paper 3925, June, 1964, pp19-34.

[3] William Weaver, Jr. and Mark F. Nelson, A. M., "*Three-dimensional Analysis of Tier Buildings*", Journal of the Structural Division, ASCE Proc. Paper 5019, December, 1966.

The topic of Tall Buildings is extensively researched and practiced. However, the three land mark papers for the analysis of Tall Buildings are the essential readings for any newly introduced researcher and design practitioner. The growth in construction activities has been phenomenal world over. The approximations are considered for developing the dynamic equations of equilibrium. The first few mode shapes corresponding to the horizontal degrees of freedom are the dominant displacements for earthquake design. The modern day tall building is a big structural system. Its maintenance is an important aspect. The maintenance and general health monitoring of a tall building is briefly discussed by applying the concept of the 'signature' of the buildings.

Types of Tall Buildings

The following are the types of buildings based on the Structural considerations:

- Framed building,
- Shear wall building,
- Framed cum shear wall building,
- Staggered panel building,

Lift shaft cum shear walls in a framed system,
Tube Buildings,
Bundled tube building, and
Super Tall Building (STB).

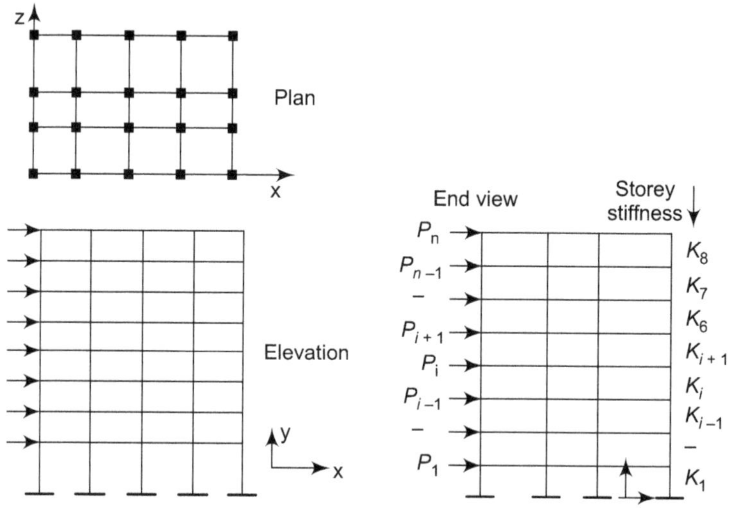

Fig. 9.1 *Typical Frame Building with the Two Views of Frames*

A common feature in the types is that the girders lie in the plane of the slabs. The slabs together with the girders are considered to behave like a rigid diaphragm in its own plane.

9.1.1 Shear Building

Consider a typical framed multi-storey building, Fig. 9.1. If the system of floor slabs and girders are assumed rigid diaphragm compared to the flexural stiffness of columns. The degrees of freedom at each floor consists of lateral lateral displacement and rotations at the intersection of columns with girders.

Because of its deformation characteristics such a framed building is also known as 'shear building'. Such a building for the study of its dynamic response is modeled as shown below. The material damping is not shown.

The floor masses are shown as point masses. The stiffness provided by the swaying of columns of a storey is shown as spring between the floor masses. This is the simplest model or idealization with floor loads shown. We study more such idealizations for static and dynamic analyses.

Floor level	(1)	–	$(i-1)$	$(i=4)$	$(i+1)$	–	$(n-1)$	$(n=8)$

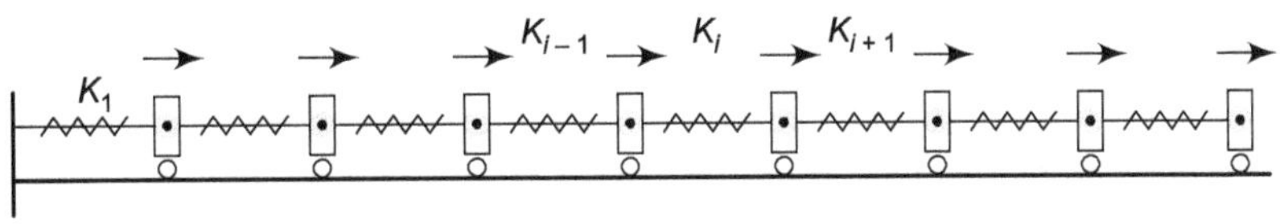

Fig. 9.2 *Shear Building and its Modeling for Dynamic Analysis*

9.1.2 Framed Buildings

Even though the slab together with the girders is considered to behave like a rigid diaphragm in its own plane, transversely the slab as well as beams deform out of their plane. In low rise buildings, the columns, when considered flexible in comparison to the deep beams together with slab deforms laterally with practically negligible rotations of the column ends at the joints. Such a building we referred as "Shear Buildings", Fig. 9.4. If there is a single load at the roof level, the deflected shape is linear. The deflected shapes enable us conceptualize and help us understand their behavior under lateral loads. The design philosophy is: Weak girder and strong columns, Fig. 9.3.

If the plan of a building is not symmetrical, under lateral loading caused by earthquake or wind, the tall building, the slabs undergo lateral displacements not only in the direction of the loads applied, but also twist (in plane rotation) and displacements perpendicular to the plane of the applied loading. If it is an earthquake ground motion, the accelerations take place in the plane of the slabs but also rotational accelerations in the direction of the twist in each plane of the slab. The inertial loadings, corresponding to the rotational accelerations, are obtained by the product of Moment of Inertia and rotational acceleration at each floor level. The expressions for the mass moment of inertia are developed later, refer Articles 9.8 and 9.9.

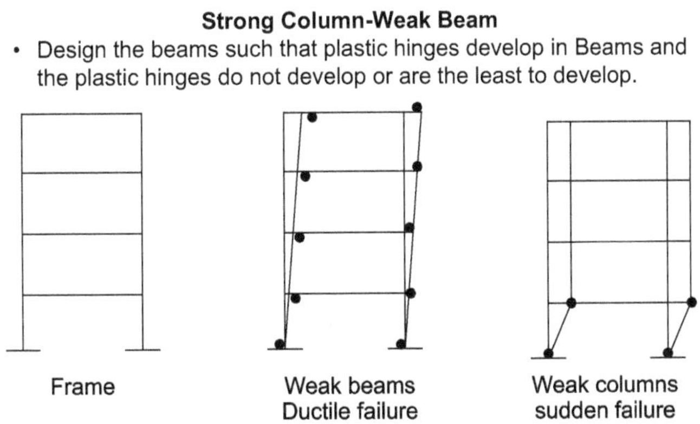

Fig. 9.3 *Building Frame with Strong Columns and Weak Beams*

The low rise buildings are more sensitive to the lateral loads caused by earthquake loads than the wind loads. As the number of storeys increases, say more than 20 storeys, the wind loading is more critical. Both the lateral loads are random dynamic loading. The dynamic equation of equilibrium, neglecting damping natural frequencies are evaluated. The mode shapes corresponding to the frequencies are important. The modal analysis, being shape based, the amplitudes are treated as single degree of freedom. It is bering mentioned here for an exposure. We will see more of this while analyzing these structures. An overall view is being attempted which is valid for other types also.

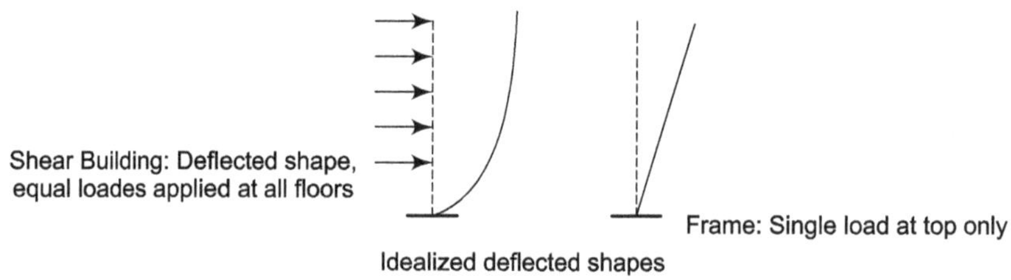

Fig. 9.4 *Comparison of Deflected Shapes: Loading Cases*

9.1.3 Shear Wall Building

With Shear walls only the deflected shape, as may be seen, is like that of a simple cantilever, Fig. 9.5(a).

9.1.4 Frame cum Shear Wall Buildings

Generally the frame is combined with the shear walls, Fig. 9.5. As a result there is an interaction between shear walls and frame. Such an interaction is shown in the Fig. 9.5(a). Lift shafts and the tube buildings, also have a deflected shape similar to a cantilever shape.

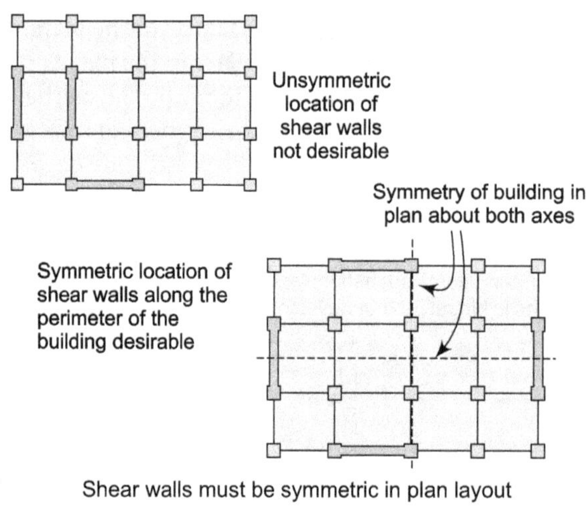

Fig. 9.5 *Symmetrical case Lacks rotational Symmetry even though Symmetrical about the Axes Individually*

The shear walls should be placed symmetrically to the extent possible. The symmetrical layout, Fig. 9.5 even though symmetrical about the individual axes, lacks rotational symmetry.

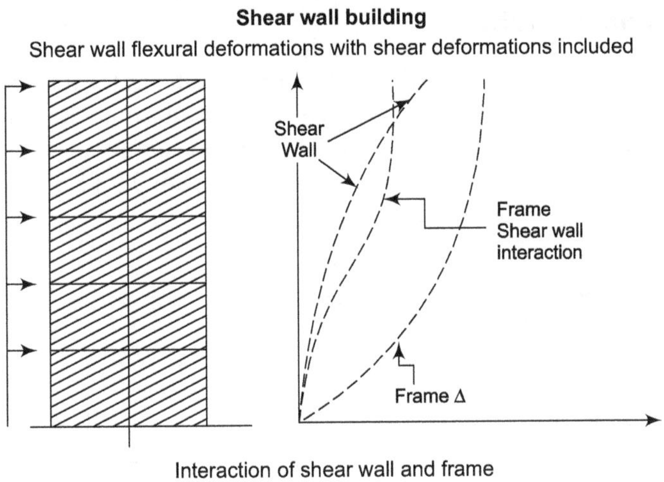

Fig. 9.5(a) *Symmetrical Case; Interaction between Frame and Shear Wall*

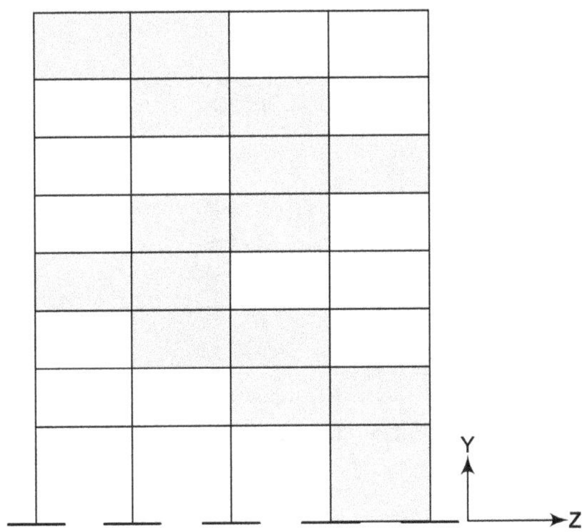

Fig. 9.6 *Frames with Staggered Panels*

9.1.5 Staggered Panel System

A disadvantage felt in regard to the continuous shear wall is the constraint in the functionality and flow of movement. The staggered paneling not only provides flexibility, it has been found to be more efficient than the conventional continuous shear walls. For the details, refer the two papers [4,5].

9.1.6 Lift Shafts cum Shear Walls in a Framed System

Their functioning is similar to that of shear walls. Even though, the shafts are provided for vertical 'traffic', their contribution to lateral stiffness is significant.

[4] Prasada Rao, K.N.V., and Seetharamulu, K. "Staggered Shear Panels in Tall Buildings," Journal of the Structural Division, ASCE Vol. 10, No. 5, May 1983, pp 1174–1193.

[5] Prasada Rao, K.N.V., Seethararmulu, K., and Krishnamoorthy, S., "Frames with Staggered Panels: Experimental Study'" ASCE Vol. 110, No. 5, May 1984, pp 1134–1148.

9.1.7 Tubes and Bundled Tubes

In the case of the tall buildings, say of the order of 100 story and above, the tubes are bundled. The Shear buildings, structurally speaking, is Bundled Tubes. The interior tubes parallel to wind or earthquake act as shear webs and ensure minimum shear lag, (Plane section remaining plane), and as a result reduce the quantity of unit structural steel, (120 kg/m^2) and fundamental period (of 7.8 seconds). Finally We may extend the concept to super tall buildings, Fig. 9.7.

Tube Buildings

John Hancock and World Trade Centre are typical examples

Plan of lower floors nine tubes, progressively reduced to two-an example of bundled tube-110 floors

World's Tall Structures

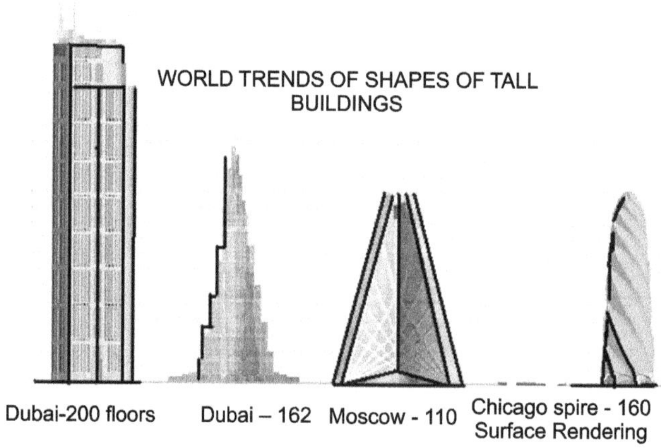

9.1.8 Super Tall Buildings

Such a system is recommended for tall buildings of the order of 150 to 250 storeys, Fig. 9.7, the corner vertical tubes are connected at an interval of about 10 storeys by horizontal boxes, about a storey deep at least. Such a system of the future is expected to be a self contained township.

A combination of four tube buildings

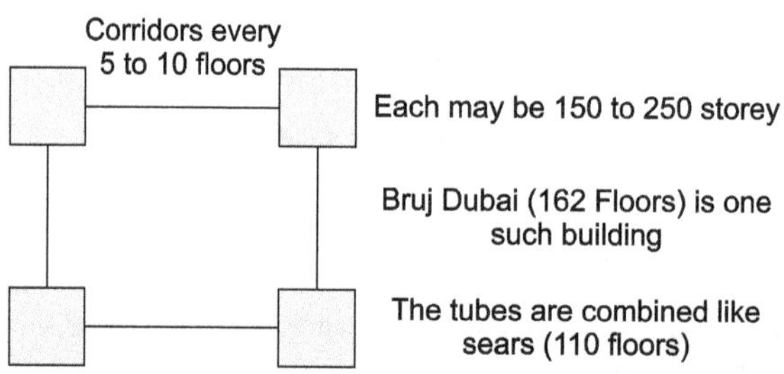

Fig. 9.7 *Super Tall Buildings*

9.2 BASICS OF DYNAMIC ANALYSIS

Before we move even to the analysis of tall Buildings under lateral loads, we familiarize with the basics of Dynamics. The vibrations and dynamic analysis of structures are extremely interesting. The advanced kinetics and sound vibrations may not be of interest to the structural engineers charged with the design of civil engineering structures. It may interest to those who are pursuing research in the topics of Structural engineering. There is yet another aspect related to the civil engineering structures and that is associated kinetics applied to human body as a structure. From the human body, we learn an interesting aspect of 'self stressing'. The self stressing can become an important wing of structural engineering added to the conventional loads applied externally, like, live loads, wind loading, mass loading in the form of earthquake loading, etc. The self stressing is a kind of *energization* of the structures for carrying the externally applied loads. In a limited sense, the application is referred as Tensegrity in the published literature. The simpler applications have been in the form of the tension structures and cable domes. The researchers in bio mechanics need to understand self stressing as energization enabling to carry the impending loads. It is being mentioned here as an appraisal with no intentions of covering these interesting aspects which are beyond the intended scope of the book. Let us move over to the basics.

In nature, it is rarely that the loading is totally static excepting the self weight.[6] Our strategy in adding dynamic analysis is with a limited objective of introducing the essentials of applied dynamics. And with the help of the relevant codes, an engineer should be able to design a good structure. A scattering exposure, however, is given to random vibrations.

We start with the Single degree freedom (SDF) systems, and their response to the time varying applied loads in time domain. Generalization to multi-degree freedom systems, leads us the Stiffness matrix and mass matrix. Natural frequencies and the corresponding mode shapes as eigen values lead us to the harmonic analysis. Housner's spectra are used for the computation of the idealized Earthquake loads. Thereafter, it is a static analysis. It is, therefore, a pseudo dynamic analysis. Similar provisions for the design against wind loads are used for the pseudo dynamic analysis based on Davenport's wind spectra and gust factor. In the case of wind loads acting on low rise buildings, we use simple provisions of the code. We need to pay attention to the lateral loads on tall buildings. The two analytical models are discussed for the analysis under the lateral loads. The effect of gravity loads for their effect on flexure is localized. The familiarity with the approximate methods is a good preparation for the solution by the matrix approach based on the displacement method.

After discussing the solution techniques, we go back to the dynamic analysis of SDF systems, but this time, frequency domain is preferred over the time domain. The approach of the frequency domain has become popular with the usage of Digital Fourier Transform (DFT) and Fast Fourier Transform (FFT). The scope of the coverage is limited to the use of FFT for the maintenance and repair diagnostics. The FFT is used as a signature of the Tall building at the time record is taken. It is similar to the ECG of the human being. The signature analysis is becoming increasingly popular with the maintenance engineers of Bridges and Tall Buildings.

9.2.1 Dynamics of Single Degree Freedom System

Dynamic analysis of single degree freedom; introduce bars, plates and similar SDF systems.

The following are some typical single degree Freedom Systems.

Example (i) It is a portal frame of an infinitely stiff girder of mass M, and the columns are of negligible mass with flexural rigidities equal to EI. The independent displacement is δ. The same assumption was used while modeling an eight storey shear building, (Fig. 9.2).

[6] Mass loading under the action of earth quakes is no more a static loading. The mass under the action of gravity is 'self weight'.

Example (ii) It is a linkage system consisting of two rigid rods of negligible mass are pin connected at B, and hinge supported at *A*. A point mass of mass *M* is roller supported at *C*. The centers of rotations (*CR*) for the bars '*a*' and '*b*' are shown in the Fig. (ii). The independent displacement is δ, Fig. 9.8.

Example (iii) It is a rectangular disk (plate) of finite dimensions '*a*' and '*b*'. It is supported on a hinge at *A* and Spring at *B*. The independent displacement is θ. The polar mass moment of inertia, *I* of the disc is considered along with the rotational acceleration for the inertial force.

The polar mass moment of inertia at its centroid is given by, $I_{zz} = \lambda(I_x + I_y)$. The I_{zz} value is transferred to the center of rotation at the hinge *A*.

All the above cases of single degree freedom are represented by the same equation of motion.

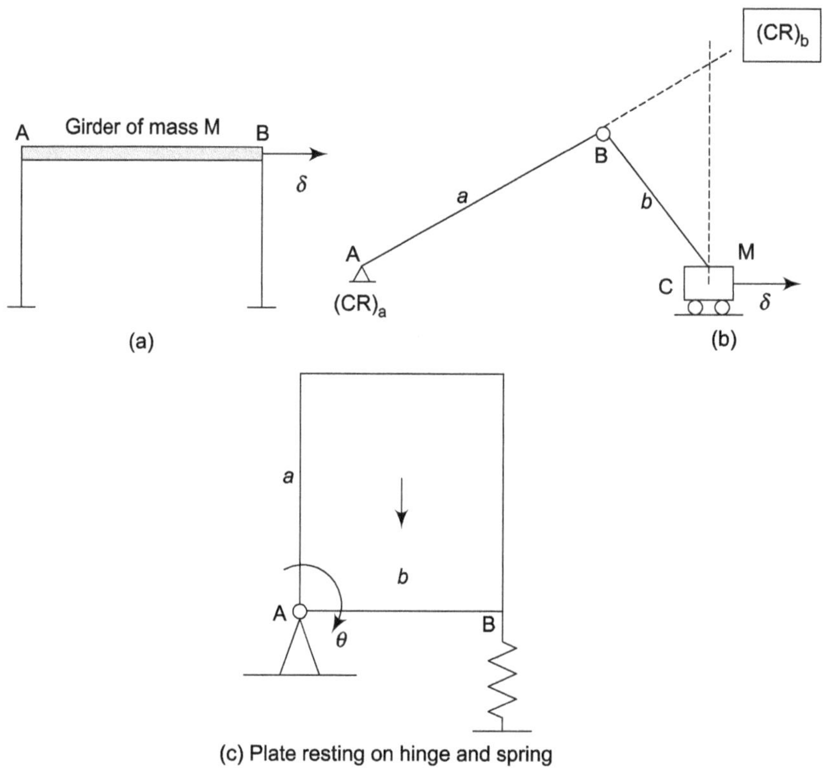

Fig. 9.8 *Some Examples of Single Degree Freedom Systems*

f_D Damping Force, assumed proportional to velocity

f_I Inertial Force, The direction as per the D'alemberts principle. The loading depends on acceleration and mass of the discrete particle. It is also known as'mass loading'.

f_S Spring Force. It is linearly proportional to the displacement and the spring constant.

Equation of motion based on equilibrium of the particle of mass, and D'alemberts principle for the inertial force.

$$f_I + f_D + f_S = p(t)$$

Using spring and damping constants,

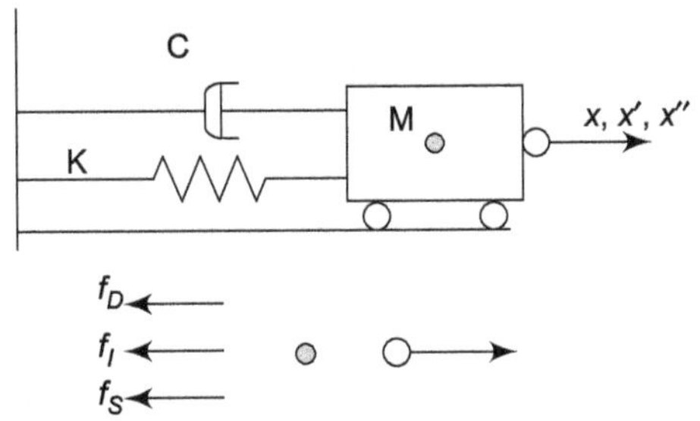

Fig. 9.9 *Single Degree Freedom System*

$$M\ddot{x} + C\ddot{x} + Kx = p(t)$$

Rotational accelerations and mass moment of Inertia: The two cases, Fig. 9.8 cases (b) and (c) involve translation as well as rotation. The rotational accelerations are multiplied by the respective mass moments of inertia for arriving at the inertial forces.

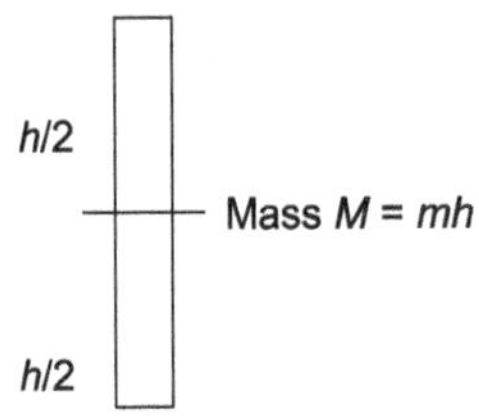

(i) Rod of uniform mass, *m*

Mass moment of inertia: The Fig. (ii) is a case of rods of uniform mass.

Uniform rod of mass, *M* is shown.

Total mass, $M = mh$

Moment of inertia, $I = \int_{-h/2}^{h/2} y^2 m\, dy$

$I = mh^3/12$

Or $I = Mh^2/12$

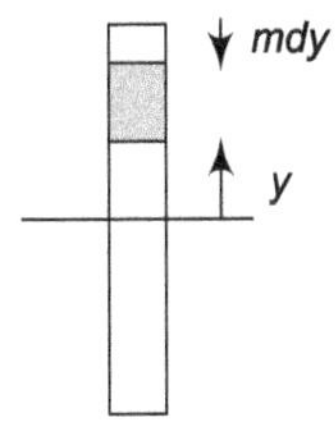

(ii)

Mass moment of Inertia of a uniform rectangular plate:

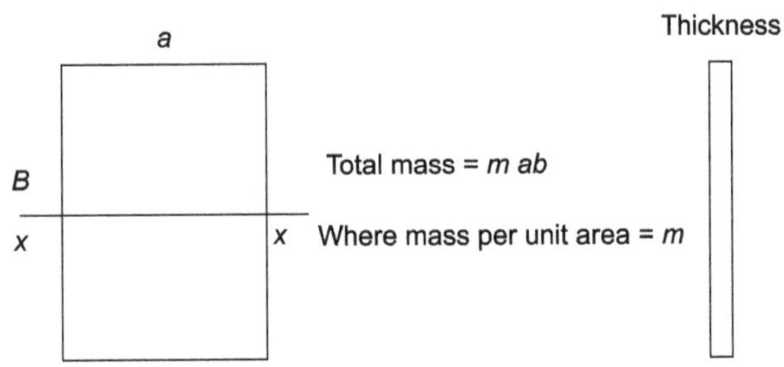

(iii) Rectangular plate of uniform thickness

Mass moment of Inertia, about the axis *xx*, $= m\, ab^3/12$

Polar moment of inertia $= I_{xx} + I_{yy} = m\, ab\, (a^2 + b^2)/12$

Mass moment of inertia of a circular slab:

It may be shown that the mass moment of inertia about the diameter of a circular plate = ¼ $m\, r^2$

Polar moment of inertia = ½ $m\, r^2$

Example 1: Write equation of motion of a rectangular plate, Fig. example 1

***Z*(*t*) is the independent variable.**

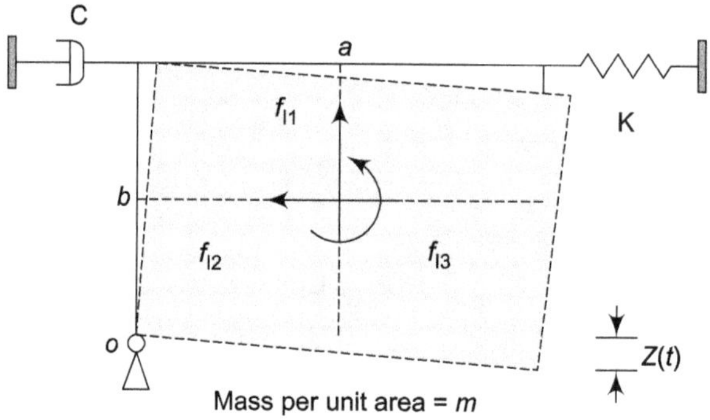

Fig. *Example 1*

Inertial forces:

$$f_{I1} = m\,ab\,\ddot{z}\,/2$$
$$f_{I2} = m\,ab\,\ddot{z}\,b/2a$$
$$f_{I3} = m\,ab\,(a^2 + b^2)\,\ddot{z}\,/12\,a$$

Spring Force:

$$f_s = zb\,K/a$$

Damping force proportional to velocity:

$$F_D = Cb\,\dot{z}\,b/a$$

The directions of these forces are taken from the D Alembert's principle.
Taking moments of all the forces about the center of rotation we get the equation of motion,

$M^*\,\ddot{z} + C^*\,\dot{z} + K^*\,z = p$, **which is the standard form with the new coefficient**

$$\mathbf{M^* = m\,a\,b\,(1/3 + b^2/3a^2),}$$
$$\mathbf{C^* = C\,b/a,\ and}$$
$$\mathbf{K^* = K\,b/a}$$

Example 2: Rocking vibration of a 3D block $a \times b \times d$.

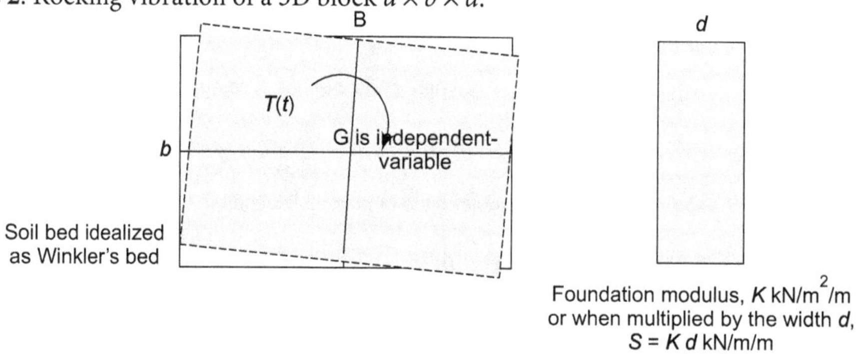

Mass per unit area, $m = \rho\, d$ where ρ is the density

$$I = m\,ab(a^2 + b^2)/12$$

1. Inertial force, $f_I = I\ddot{\theta}$
2. Spring force:

Due to soil (Winkler's spring idealization), the spring forces are:

$$f_{s1} = (ka^3\, d/12)\theta$$

Due to Mg,

$$f_{s2} = -\,(M\,g\,b/2)\theta$$

$$f_{s1} + f_{s2} = (Ka^3\, d/12 - M\,g\,b/2)\theta$$

The equation of motion:

$$I\ddot{\theta} + \underbrace{(Ka^3\, d/12 - Mg\,b/2)}_{K^*}\theta = T(t), \text{ (Damping not Considered)}$$

From the above it is obvious that any single degree freedom system can be reduced to the standard form:

$$M\ddot{\mathbf{x}} + C\,\dot{\mathbf{x}} + K\mathbf{x} = p(t)$$

Before we move over to the Multi Degree Freedom systems, it is important to know the effect of ground motion caused by the earthquakes, applied to a SDF system.

9.2.2 Earthquake Loading: Equation of Motion and Solution

It is caused by Ground Motion: Consider Fig. 9.4

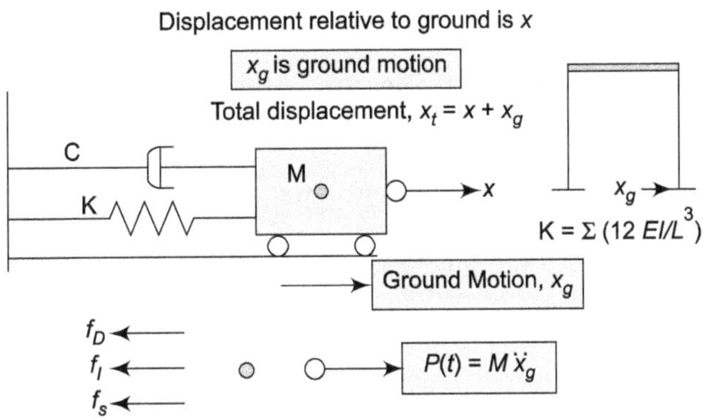

Fig. 9.10 *Response of Single Degree Freedom System due to Ground Motion*

A portal frame, with infinitely stiff girder supported on two columns of flexural rigidity EI and length L is replaced by our standard model. But this time in place of applied force $P(t)$, we are studying the effect of ground motion. The equation of motion after substituting for f_D, f_I, and f_S, we get

$$\mathrm{M}\,\ddot{\mathbf{x}}_t + \mathrm{C}\,\dot{\mathbf{x}} + \mathrm{K}\,\mathbf{x} = 0$$

Because we are interested in the relative displacement, **x**, substituting $\mathbf{x}_t = \mathbf{x} + \mathbf{x}_g$

$$\mathrm{M}\,\ddot{\mathbf{x}} + \mathrm{C}\,\dot{\mathbf{x}} + \mathrm{K}\,\mathbf{x} = -\,\mathrm{M}\,\ddot{\mathbf{x}}_g$$

The above is the standard form of a SDF system, where $P(t) = \mathrm{M}\,\ddot{\mathbf{x}}_g$

The earthquake loading being random, change of sign is acceptable.

Solution of the equation of motion and Damping

It is appropriate to introduce some basic properties of damping while attempting solution of the equation of motion,

$$M\ \ddot{x} + C\ \dot{x} + K x = P(t)$$

Let us take up the solution of the homogeneous equation:

$$M\ \ddot{x} + C\ \dot{x} + K x = 0$$

If the solution is assumed as $A\,e^{\lambda t}$ and recognizing that being a second order, differential equation the constant 'A' in fact represent two constants of integration. We obtain

$$M\lambda^2 + C\lambda + K = 0 \qquad \text{...(a)}$$

Dividing by M and substituting $C/M = 2\eta$ and $k/M = \omega^2$ for reasons that will be obvious to you soon,

$\lambda^2 + 2\,\eta\lambda + \omega^2 = 0$, and the roots are:

$$\lambda = \frac{-2\eta \pm \sqrt{(4\eta^2 - 4\omega^2)}}{2}$$

$$\lambda/\omega = -\xi \pm \sqrt{(\xi^2 - 1)} \text{ Dividing by } \omega \text{ and substituting } \eta/\omega = \xi$$

or

$$\lambda = -\xi\omega \pm \omega\sqrt{(\xi^2 - 1)}$$

Let us examine, the second term, $\omega\sqrt{(\xi^2 - 1)}$

We notice three situations,

Situation 1. $\xi < 1$, the quantity within the $\sqrt{}$ is negative. And the roots are:

$$\lambda = -\xi \pm i\,\omega\sqrt{(1 - \xi^2)}$$

The solution consists of sine - cosine terms, and therefore, oscillatory. The system is known as an **under damped** system,

From the λ values, we write the solution:

$$\mathbf{X}(t) = e^{-\xi\omega t}\,(\mathbf{A}\sin\omega_D t + \mathbf{B}\cos\omega_D t), \text{ where } \omega_D = \omega\sqrt{(1 - \xi^2)}$$

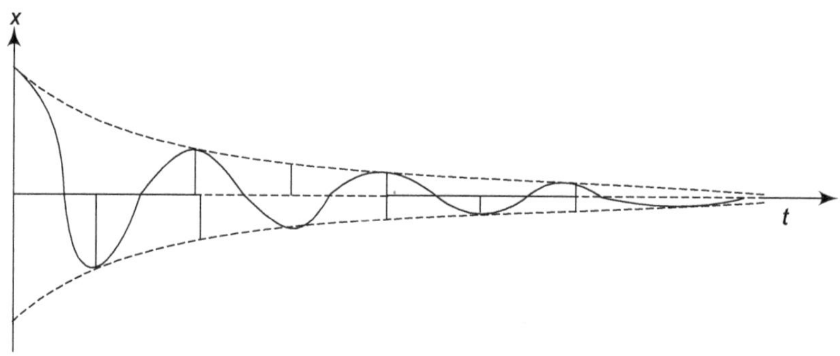

Fig. 9.11 *Under Damped Response of Single Degree Freedom System*

If we make the above substitutions, $\xi = \eta/\omega$ and $k/M = \omega^2$, the Eq. (a) may be rewritten as

$$\ddot{x} + 2\xi\,\omega\,\dot{x} + \omega^2 x = \frac{1}{m}P(t) \qquad \text{...(b)}$$

From the initial conditions, the constants A and B are found from the initial conditions at $t = 0$. With the constantans found, the solution is

$$\mathbf{X(t)} = e^{-\xi \omega t}\left(\frac{x_0 + x_0\ \xi\omega}{\omega_D}\sin \omega_D\ t + x_0 \cos w_D t\right) \qquad \ldots\text{(i)}$$

Situation 2. $\xi > 1$, the quantity within $\sqrt{}$ is positive and the response dies exponentially without any oscillations. It is an **over damped system**, and finally

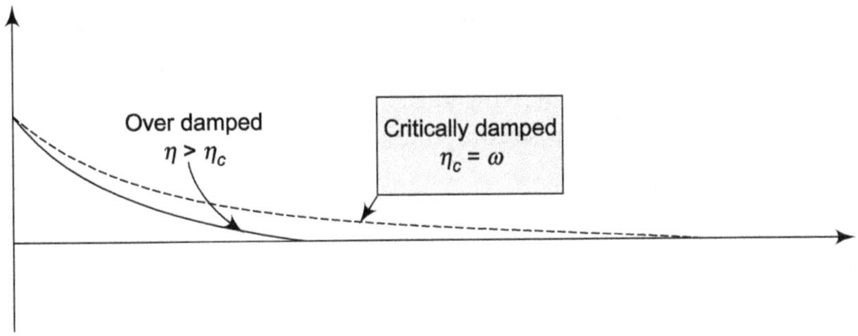

Fig. 9.11(a) *Response: Critically Damped and Over Damped System*

Situation 3. $\xi = 1$ is the dividing value. And such a damping is known as **critical damping.**

9.2.3 Solution of Equation of Motion – General Loading $p(\tau)$

In the following salient results are summarized.

(a) Duhamal Integral

Consider the loading $P(t)$. Let $P(\tau)$ be the load after Time τ. Treat the force acting over $d\tau$ equal to $P(\tau)$ $d\tau$ as an impulse loading.

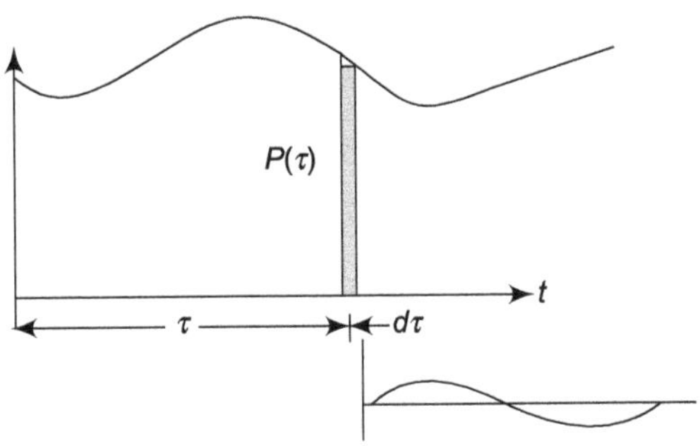

Fig. 9.12 *General Loading and Impulse Excitation*

Let $\Delta \dot{x}$ be the change in velocity caused by the impulse, $P(\tau)\ d\tau$ at time $t - \tau$.

$$\Delta \dot{x} = 1/m\ P(\tau)\ d\tau$$

and

$$\Delta x = \frac{1}{m\omega}\ P(\tau) \text{ sine } \omega(t - \tau)d\tau$$

For the case without damping,

$$x = \frac{1}{m\omega}\int_0^t P(\tau)\ \sin\ \omega(t-\tau)d\tau$$

For the case with damping,

$$x = \frac{1}{m\omega}\int_0^t P(\tau)\,e^{-\xi\omega(t-\tau)}\ \sin\ \omega_D(t-\tau)d\tau \qquad \text{...(ii)}$$

The above integral is known as **Duhamel Integral.**

For the numerical evaluation of the Eq. (ii), the details of the Simpson's quadrature, When based on modal analysis, the Multi-degree Freedom systems using the appropriate modal participation factor and the corresponding modal shape for arriving at the floor loads, is no different from a Single Degree Freedom system for the numerical response computations.*

(b) Runge Kutta Numerical Method

Runge Kutta method of solution: It is a numerical method for solving the differential equation, our equation of motion. The algorithm is given in the following. For details, read Scarborough for the numerical solution. There is not much of structural concept.

The Eq. (b) giving the equation of motion is reproduced in the following:

$$\ddot{x} + 2\xi\omega\,\dot{x} + \omega^2 x = \frac{1}{m}P(t) \qquad \text{...(b)}$$

$$\ddot{x} = \frac{1}{m}P(t) - \omega^2 x - 2\,\xi\omega\,\dot{x}$$

Consider the following figure for the loading.

Assume the initial conditions:

$$X_0 = 0.0,\ \dot{x}_0 = 0.0.$$

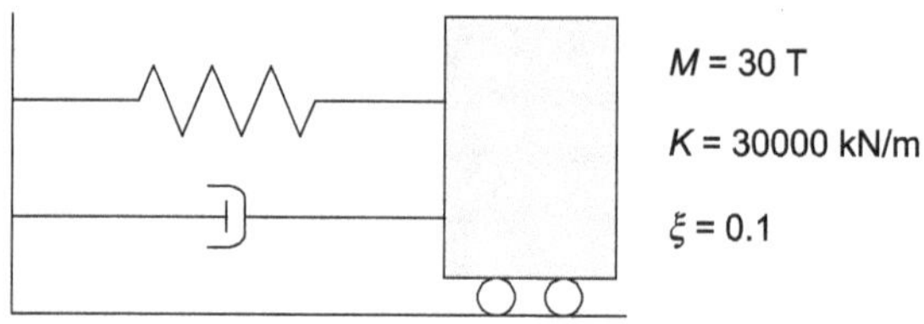

$\omega = \sqrt{K/M}$ = 31.62 rad/sc, f = 5 cps, Period, T = 0.2 sec

$2\xi\omega$ = 6.324. Take time interval, h = 0.05 sec << T ok

Fig. 9.13 *Single Degree System: Initial Condition Given*

Take time interval, h equal to 0.05 seconds and compute the response at the end of 0.4 sec. Extrapolate the response curve beyond 0.4 second.

$$\omega = \sqrt{\frac{K}{M}} = 31.62 \text{ rad/sec}, f = 5 \text{ CPS, Period } T = 0.2 \text{ sec.}$$

$$2\xi\omega = 6.324, \text{ take time interval, } h = 0.05 \text{ sec} << T \text{ ok.}$$

We may write, $\ddot{x} = f(t, x, \dot{x})$

Discretizing the values, consider a station, n. The Runge Kutta constants are:

$$K_1 = h\,f(t_n, x_n, \dot{x}_n)$$

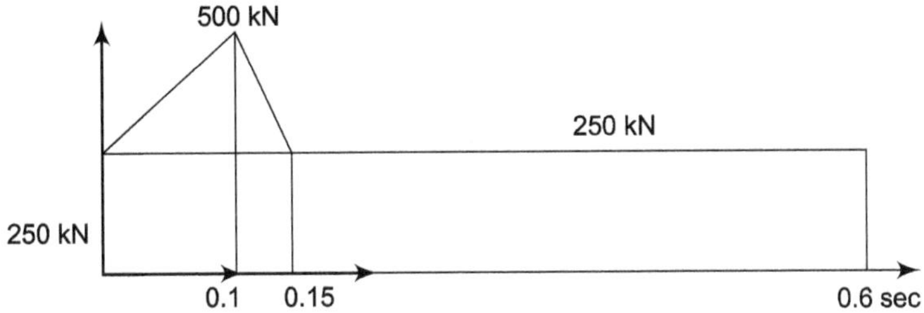

Fig. 9.14 *A case of Impulse Load Curve*

$$K_1 = h(p_i/m - \omega^2 x_i - 2\xi\omega\, \dot{x}_i)$$

$$K_2 = h\left[\frac{p_i + p_{i+1}}{2m} - \omega^2\left(x_i + \frac{h\dot{x}_i}{2} + \frac{hk_1}{8}\right) - 2\xi\omega\frac{\dot{x}_i + k_1}{2}\right]$$

$$K_3 = h\left[\frac{p_{i+1}}{m} - \omega^2\left(x_i + \frac{h\dot{x}_i}{2} + \frac{hk_1}{8}\right) - 2\xi\omega\frac{\dot{x}_i + k_1}{2}\right]$$

$$K_4 = h\,[\,p_{i+1}/m - \omega^2\,(x_i + h\,\dot{x}_i + hk_3/2\,) - 2\xi\omega\,(\,\dot{x}_i + k_3)]$$

$$\Delta x_i = h\,[\,\dot{x}_i + (k_1 + k_2 + k_3\,)/6\,],\ \dot{x}_{i+1} = \dot{x}_i + \Delta\dot{x}_i$$

Time, t sec.	p_i in N	K_1	K_2	K_3	K_4	Δx_i	$\Delta \dot{x}_i$	$\dot{x}_i$	$\dot{x}_i$
0.0	250							0.0	0.0
		0.4167	–0.3247	0.4434	0.0695	0.0099	0.3139		
0.05	375							0.0099	0.3139
		0.0307	–0.2719	–.1199	–.3579	0.0127	–.1851		
0.10	500							0.0226	0.1288
		–.3374	–.5480	–.7230	0.0563	–0.0070	–.4705		
0.15	250							0.0156	–.3417
		–.2553	0.2920	0.2055	0.2771	–.0151	0.1695		
0.20	250							0.0005	–.1722
		0.4461	0.4514	0.4506	0.1709	0.0026	0.4035		
0.25	250							0.0031	0.2313
		0.1885	–.1893	–.1296	–.1867	.0105	–.1060		
0.30	250							0.0136	0.2449
		–.3408	–.4867	–.4635	–.1540	0.0015	–.3991		
0.35	250							0.0151	–.1542
		–.2901	–.0138	–.0040	0.0021	–.0103	–.0539		
0.40	250							0.0048	–.2081

Observations from the response curve:

- The period as seen confirms equal to 0.2 second
- From t = 0.25 to 0.6 seconds, it is practically '**steady state**'
- The free sketching shows
- The maximum response occurs at t > 0.1 sec. It is approximately 0.024
- There is scope to improve the free sketching. It is good to anticipate and try sketching. Just like guessing Inflexion points and sketching deflected shape.

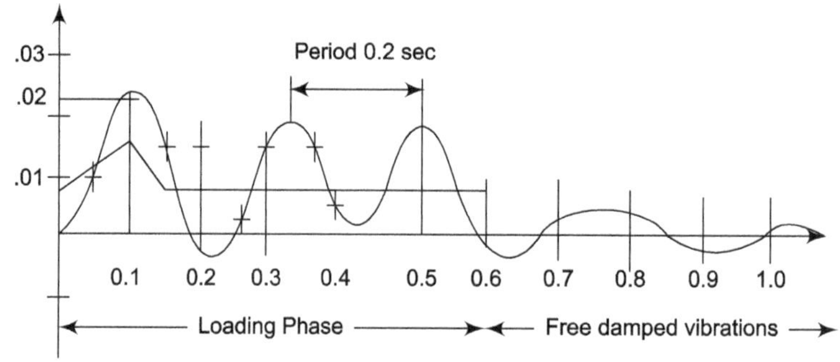

Fig. 9.15 *Load Curve and Response in Loading Phase and Free Vibrations there After*

Note: The Modal Analysis, which uses the mode shapes together with the associated Participation factors, we can analyze MDF systems. It, however, presupposes modal superposition and therefore elastic analysis. We also assume modal damping which may not be realized. The response using the 'step method' does not require superposition and damping 'C' is used unaltered. The method is valid for the inelastic response analysis.

(c) Step Method of Solution: Applicable to Inelastic Materials

Concept and back ground to the flow chart: Being a structural concept, the attempt should be to base the response analysis on the equation of motion.

- The initial conditions x_0, $\dot{x}_0$ are given. Acceleration is dependent on the force being applied. Henceforth we use the starting station as *i*. Therefore, use the equation of motion for finding the starting acceleration.
- We need increments of displacement, Δx, $\Delta\dot{x}$, and $\Delta\ddot{x}$.
- We assume the acceleration variation over a small time segment to be linear. The assumption fixes the variation of $\dot{x}$ to be parabolic and similarly x to be cubic so as to achieve linear variation of acceleration using this property.
- Use incremental equation of motion combined with the above property obtain Δx_i
- Calculate $\Delta\dot{x}_i$, and $\Delta\ddot{x}_i$
- Thus obtain new values of x, $\dot{x}$, and $\ddot{x}$.

Compute $\ddot{x}_i = [(-C\dot{x}_i - Kx_i) + p_i]$ From the initial conditions, the values on the RHS are known. Over a time interval $h = \Delta t$, the acceleration is assumed to be varying linearly, Fig. 9.15A(i)

$$\ddot{x}_\tau = \ddot{x}_i + \frac{\Delta\ddot{x}_i}{\Delta t}\tau \qquad \text{...(i)}$$

It follows from integration

$$\dot{x}_\tau = \dot{x}_I + \ddot{x}_i\tau + \frac{\Delta\ddot{x}_i}{\Delta t}\frac{\tau^2}{2} \qquad \text{...(ii)}$$

Or for $\tau = \Delta t$

$$\dot{x}_{i+1} = \dot{x} + \ddot{x}_i\,\Delta t + \Delta\ddot{x}_I\,\frac{(\Delta t)^2}{2} \qquad \text{...(iii)}$$

Difference:

From Eq (iii) $$\Delta\dot{x}_I = \ddot{x}_i\,\Delta t + \Delta\ddot{x}_I\,\frac{\Delta t}{2} \qquad \text{...(I)}$$

Similarly

$$x_\tau = x_i + \dot{x}_i \tau + \ddot{x}_i \frac{\tau_2}{i2} + \frac{\Delta \ddot{x}}{\Delta t} \frac{\tau^3}{6}$$

Or for $\tau = \Delta t$

$$X_{i+1} = X_i + \dot{x}_I \Delta t + \ddot{x}_I \frac{\Delta t_2}{!2} + \frac{\Delta \ddot{x}_i}{1} \frac{(\Delta t)^3}{6}$$

Or
$$\Delta x_{\text{i}} = \dot{x}_I \Delta t + \ddot{x}_I \frac{(\Delta t)^2}{!2} + \frac{\Delta \ddot{x}_i}{1} \frac{(\Delta t)^3}{6} \quad \ldots\text{(II)}$$

But use the above equation for evaluating $\Delta \ddot{x}_i$

Use of incremental equation of motion for evaluating Δx_i.

$$M(\Delta \ddot{x}_i) + C(\Delta \dot{x}_i) + K\,\Delta x_i = \Delta p_i$$

Substituting $\Delta \ddot{x}_I$ and $\Delta \dot{x}_I$ from equations (ii) and (i),

$$\Delta \overline{K}_i\, \Delta x_i = \Delta \overline{p}_i$$

where
$$\Delta \overline{K}_i = \left(\frac{6M}{\Delta t^2} + \frac{3c_i}{\Delta t} + K_i \right)$$

and
$$\Delta \overline{p}_I = \left\{ \Delta p_i + \left(\frac{6}{\Delta t} \dot{x}_i + 3\ddot{x}_i \right) M + C_i \left(\frac{\Delta t}{2} \ddot{x}_i + 3\dot{x}_i \right) \right\}$$

Compute:
$$\Delta x_i = \frac{\Delta \overline{p}_i}{\Delta \overline{\overline{K}}_i};$$
$$x_{i+1} = x_i + \Delta x_{\text{i}};$$

From Eqs (I) and (II),

$$\dot{x}_{i+1} = \dot{x}_i + \Delta \dot{x}_i;$$
$$\ddot{x}_{i+1} = \ddot{x}_i + \Delta \ddot{x}_i \quad \ldots\text{(III)}$$

Repeat till $i = n$ → Stop

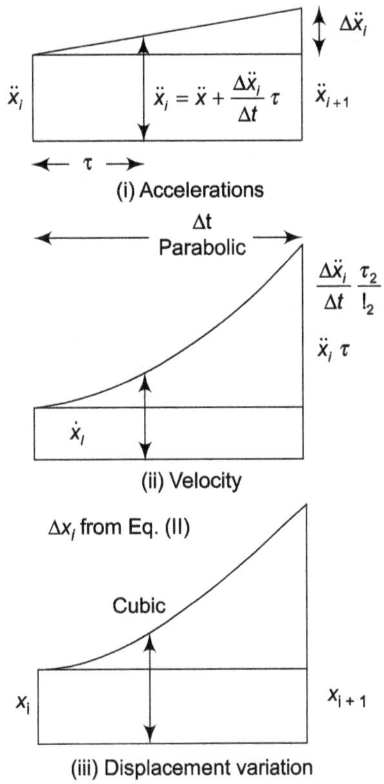

Fig. 9.15(a) *Acceleration Idealization*

We repeat the above by writing a flow chart in the following:

The **flow chart** for the method is:

1. Given the initial conditions, say $x_i = 0.0$, $\dot{x}_i = 0.0$ or any other given conditions, compute $\ddot{x}_1$ from the equation of motion.
2. For $i = 1$ to n, Compute Δx_i, $\Delta \ddot{x}_i$, and $\Delta \dot{x}$ from Eqs (iii), (ii), and (i) respectively
3. Compute x_{i+1}, $\dot{x}_{i+1}$, and $\ddot{x}_{i+1}$. For the case of inelastic response, consider the secant values from the inelastic K curve and inelastic C curve, Fig. 9.15(B)(at the back of this page)

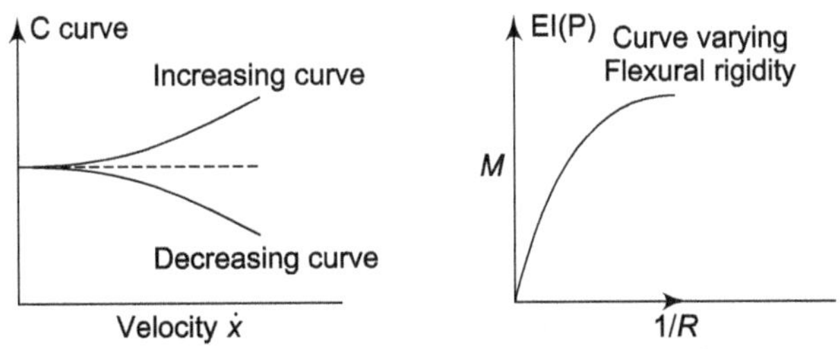

Fig. 9.15(b) *Damping and Inelasticity*

4. Compute NEW $\Delta\ddot{x}_i$, $\Delta\dot{x}$, and Δx_i. Iterate tillt convergence is achieved.

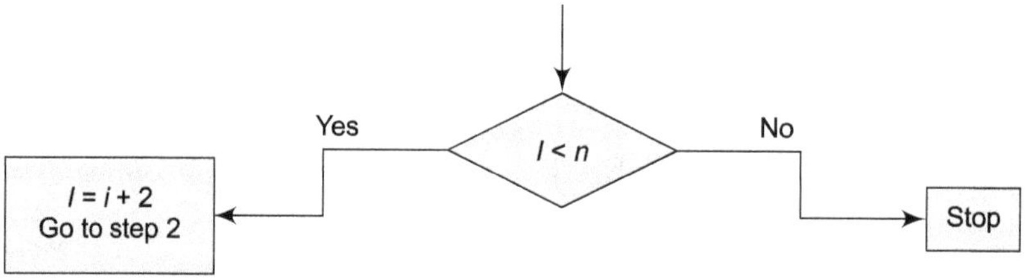

Refer Art. 9.5.2 for Dynamic Response of Multi Degree Systems.

Lateral loads on tall buildings: The lateral loads on tall buildings are quite often more crucial than gravity loads at the floors. The lateral loads are mainly: Wind Loads, and Earthquake Loads. The earthquake loads are caused by the ground motion. Unlike the earthquake loads, critical wind loads occur practically every year. The earthquake loads are based on the digitized data recorded for typical earthquakes occurred over decades. The return period of critical earthquake at a place may be any period between 30 to 100 years. But the earthquake loads deserve more attention because of the suddenness and resulting in catastrophic damages.

9.3 EARTHQUAKE LOADS AND RESPONSE

These are more critical than the wind loads for the tall buildings, which are low rise to medium rise buildings. Whereas the wind loads are more critical for the high rise structures including Towers.

What we should know while using any code for the earthquakes: The earthquake responses are 'calculations intensive'. If we are basing our design on

9.3.1 Housner's Spectra

Housner's spectral value which is a maximum acceleration of some typical earthquake records.

9.3.2 Mode Participation Factor

'Mode participation factor, **IS** 1893 (Part 1): 2002[7, 8] as a simplified equivalent to the mode participation factor. We thus arrive at the equivalent static loads at the floor levels for static analysis. The code also recommends the participation factor based on the mode shapes, and sometimes it is referred as dynamic analysis. With the floor loads so computed, the distribution based on mode shape, we proceed on as a static analysis. It is more appropriate to refer to the mode based approach as 'quassi dynamic' analysis. But if we use ground accelerations $\ddot{x}_g$ obtained from the EQ digitized instrumented data, we proceed on for the response calculations using the various numerical techniques. It is then a Dynamic Analysis.

9.4 DYNAMIC CHARACTERISTICS

Determination of natural frequencies and corresponding normal modes is the first step in the dynamic analysis of the structure, based on the superposition of the modes:

The total displacement is obtained as the sum of the modal components.

[7] Clough, R. W., and Penzien, J., "Dynamics of Structures", McGraw-Hill, New York, 1993.
[8] IS 1893 (Part 1); 2002, Bureau of Indian Standards, June 2002.

$$\delta_{(t)} = \sum_{i=1} \varphi_i \, \delta_i(t) \text{ Let } \delta_{(t)} = \phi\delta \qquad \ldots(1)$$

In which $\varphi = [\varphi_1, \varphi_2 \ldots \varphi_N]$ and $\Delta = [\delta_1, \delta_2 \ldots \delta_N]^t$

The representation is similar to the expansion using fourier series. The function ϕ_i replace the trigonometric terms sine and cosine series. The number of the modes, N is equal to the degrees of freedom or the number of independent displacements. As stated earlier, the modal superposition assumes the material is linearly elastic and the structure undergoes small deformations.

The equation of motion for a freely vibrating system is

$$[M]\ddot{\delta}(t) + C\dot{\delta}(t) + K\,\delta(t) = 0 \qquad \ldots(2)$$

or $M\,\phi\,\ddot{\delta} + c\,\phi\,\dot{\delta} + K\,\phi\,\delta = 0$

First mode: Consider the first model (δ_1, ϕ_1)

Premultiplying the above equation by φ_1^t and using the orthogonal property

$(\phi_i^t\,[M]\,\phi_1 = m_1 \text{ and } \phi_1^t\,[M]\,\phi_j = 0),\ j \neq 1$

$$m_1\ddot{\delta}_1 + c_1\dot{\delta}_1 + k_1\delta_1 = 0$$

Orthogonality is assumed for *C* also ...(3)

Which is the equation of motion for an equivalent single degree of freedom system? In the case of an undamped system the equation is $m_1\ddot{\delta}_1 + k_1\delta_1 = 0$ or for $\delta(t) = \delta \sin \omega t$

$$-\,m_1\omega^2\delta_1 + k_1\delta_1 = 0 \qquad \ldots(4)$$

From which $\omega = \sqrt{\dfrac{K_1}{m_i}}$ and is the circular frequency in the mode 1. The equation of motion for the undamped system is

$$-\,\omega^2[M]d + [K]\delta = 0$$

From the equation, the dynamic characteristics are determined by using either the stiffness matrix or the flexibility matrix of the structure. In either method the equation is first reduced to the standard form

$$[A]\,\{X\} = \lambda\{X\} \qquad \ldots(5)$$

Which is solved by using any of the available numerical schemes that are direct or iterative? When the stiffness matrix is used, it may be seen from the equation of motion obtained by pre-multiplying with K^{-1}

$$-\omega^2 K^{-1}[M]\delta + I\delta = 0$$

Or

$$[A - \lambda I]\,\delta = 0$$

where $\qquad A = K^{-1}[M] \text{ and } \lambda = \dfrac{1}{\omega^2}$

In the above, the determinant of $[A - \lambda I]$ gives the Eigen values and Eigen vectors. It is acceptable but A may not remain symmetrical even though *K* is symmetrical and *M* is not only symmetrical but also diagonal in the case of our study of Tall Buildings. The mass is mainly due to the floor slabs and for the deflections in the plane of the floor slabs, *M* is a diagonal matrix.

In order to enable a faster numerical algorithm, it is desirable that *A* is symmetrical in the equation

$$[A - \lambda I] = 0.$$

For achieving a symmetrical *A*,

Factorize = $L\,L^t$, where L is a lower triangular matrix. Substituting in the equation of motion.

$$-\omega^2[M]\,\delta + L\,L^t\delta = 0$$

Let $L^t\delta = \dot{x}$ leading to $\delta = (L^t)^{-1}x$

Making the above substitution

$$-\omega^2[M][L^t]^{-1}x + L\,x = 0$$

Premultiplying by L^{-1}, we get $[A - \lambda I] = 0$ or $Ax = \lambda x$.

where $A = L^{-1}\,[M]\,[L^t]^{-1} = L^{-1}\,[M]\,[L^{t-1}]^t$ is a symmetrical matrix and

$$\lambda = 1/\omega^2$$

9.4.1 Flexibility Matrix Approach for Dynamic Characteristics

It has been shown elsewhere that we directly obtain the flexibility matrix F. For achieving the eigen values and eigen vectors, pre-multiply the equation of motion by F.

$-\omega^2 F[M]\delta + \delta = 0$, we used the property $FK = 1$. It is much easier to factorize the diagonal matrix, $M = M^{1/2}\,M^{1/2}$ where $M^{1/2}$ consists of square roots of the diagonals m_{ii}. Letting $M^{1/2}\,\delta = x$ and $\delta = M^{1/2}$ x and premultiplying the equation of motion by $M^{1/2}$,

we obtain

$$[A - \lambda]x = 0$$

where, $[A] = M^{1/2}\,FM^{1/2}$ is symmetrical ...(6a)

and $\{x\} = M^{1/2}\,\delta$ ...(6b)

In both the procedures, presented above, the natural frequencies can be obtained commencing from the smallest value. As the first few natural frequencies and the corresponding eigen vectors only are important. We may not find all the higher frequencies and corresponding mode shapes. Each method is as effective as the other. In the first one, inversion is involved whereas in the second flexibility, coefficients are computed for as many loads cases as the number of storeys applying unit load at each floor level in turn. While solving for different load cases, only the element in load vector are altered and the other elements are not disturbed in the Gaussain elimination procedure and hence the computer time taken for formulating the lateral flexibility matrix is not much. The use of one or the other of the approaches is a matter of choice. In the present analysis, the flexibility matrix is used. Similar procedure is adopted for the conventional system analyzed by the wide column frame analogy. The eigen value problem is solved by 'Givens Householder method' which is a direct approach.

The equation of motion for a mode i subjected to a varying load, $p(t)$ is given by

$$m_i\ddot{\delta}_i + c_i\,\dot{\delta}_i + k_i\,\delta_i = \phi_i^t\,p(t) \quad \text{...(7)}$$

where $m_i = \phi_i^t\,[M]\,\phi_i$ and $k_i = \phi_i^t\,[K]\,\phi_i$

Dividing the Eq. (3.7) by m_1

$$\ddot{\delta}_i + \frac{c_i}{m_1}\dot{\delta}_i + \frac{K_i}{m_i}\delta_i = \frac{\phi_i^t\,p(t)}{\phi_i^t[M]\,\phi_i} \quad \text{...(8)}$$

The factor, $\dfrac{\phi_i^t}{\phi_i^t[M]\,\phi_i}$ on the right hand side of the above equation is known as the *"participation factor"* and when written in a summation form,

$$\frac{\sum_{r=1,N} \phi_{1,r}}{\sum_{r=1,N} m_r\,(Q_{i,r})^2} \qquad \ldots(9)$$

For ground motion x, the equation for a single degree of freedom system is

$$m\ddot{x} + c\,\delta + k\,\delta = 0 \qquad \ldots(10)$$

In which $x = x_g + \delta$ where x_g is the ground displacement.

Rewriting Eq. (9.10)

$$m\ddot{\delta} + c\,\delta + k\,\delta = -m\ddot{x}_g \qquad \ldots(11)$$

For a multi-degree of freedom system the equation is

$$\ddot{\delta} + \frac{c_i}{m_i}\dot{\delta}_i + \frac{k_i}{m_i}\delta_i + \frac{\phi_i^t[M]\{I\}}{\phi_i^t[M]\varphi_i}\ddot{x}_g \qquad \ldots(12)$$

From which $\ddot{\delta} + \frac{c_1}{m_1}\dot{\delta}_i + \frac{k_i}{m_i}\delta_i = \frac{\phi_i^t[M]\{I\}}{\phi_i^t[M]\phi_i}\ddot{x}_g$ ...(13)

The participation factor (FACT) for mode i when written in a summation form,

$$(\text{FACT})_i = \frac{\sum_{r=1,N} \phi_1, m_r}{\sum_{r=1,N} m_r\,(Q_{ir})^2} \qquad \ldots(14)$$

9.5 RESPONSE TO DYNAMIC LOADING

The response to dynamic loading is generally defined either by mode superposition or direct integration. The choice between the two is merely one of numerical effectiveness. The solutions obtained using the different procedures should be identical within numerical errors of the time integration scheme and rounding off errors in the computations. In the present response analysis, one method from each procedure is illustrated – (1) Simpson's rule for evaluating the Duhamel integral falling under mode superposition and (2) step by step method with $\beta = 1/6$ (based on linear variation of acceleration) in Newmark-β method falling under direct integration.

9.5.1 Simpson's Rule for Duhamel Integral

The Duhamel integral for a damped system is given by

$$\delta(t) = \frac{1}{M\omega_D}\int_0^t P(\tau)\,e^{-\xi\,\omega(t-\tau)}\sin\omega_D\,(t-\tau)\,d\tau \qquad \ldots(15)$$

Where damping ratio $\xi = c/(2\,m\omega)$ and c is damping factor. The equation is written in the following modified form for enabling recursive operation:

$$\delta(t) = A\sin g\,\omega_D t - B\cos\omega_D t \qquad \ldots(16)$$

In which

$$A = \frac{1}{M\omega_D}\int_0^t P(\tau)\frac{e^{\xi\omega T}}{e^{\xi\omega T}}\cos\omega_D\tau\,d\tau \qquad \ldots(17a)$$

and $$B = \frac{1}{M\omega_D}\int_0^t P(\tau)\frac{e^{\xi\omega T}}{e^{\xi\omega T}}\sin\omega_D\tau\,d\tau \quad ...(17b)$$

To evaluate the Duhamel integral by Simpson's rule, the recursive relation between A_i and A_{i+1} is written as

$$A_{i+1} = e^{-\xi\omega(2\Delta\tau)}A_i + \frac{\Delta\tau}{3}p(t_i)\cos\omega_D t_i$$

$$+\frac{\Delta\tau}{3}4p(t_i+\Delta\tau)e^{-\xi\omega\Delta\tau}\cos(t_i+\Delta\tau)+\frac{\Delta\tau}{3}p(t_{i+1})\cos\omega_D(t_{i+1}) \quad ...(18)$$

Similar recursive relation between B_i and B_{i+1} can be written.

The superposition presupposes material 'linear elastic' and small displacements.

The equation of motion is 'linear elastic' and small displacements.

9.5.2 Step Method—Multi Degree Freedom Systems

The matrix approach enables the analysis for multi degree freedom systems. Scalar approach was presented in Art. 9.2.3

$$[M]\ddot{\underline{\delta}}+[C]\dot{\underline{\delta}}+[K]\underline{\delta}=\underline{P}(t) \quad ...(19)$$

where

$$\underline{\delta}=[\delta_1 \;\; \delta_2 \ldots \delta_N]^t \quad ...(20a)$$

and $$\underline{P}=[P_1 \;\; P_2 \ldots P_N]^t \quad ...(20b)$$

The computations for defining the elements of [C] matrix corresponding to the ξ_1 values used for modal superposition are computed based on the damping assumed.

The following are the steps for obtaining the response at time t_{i+1} from the values given for the time station *i*.

Step 1 Compute $\ddot{\underline{\delta}}_i$ from the equation of motion

$$\ddot{\underline{\delta}}_1 = -[M]^{-1}[C]\dot{\underline{\delta}}_i+[K]\underline{\delta}_i+\underline{p}_i$$

Step 2 Compute $\Delta\underline{\delta}_i$ from $\Delta\underline{\tilde{p}}_i=[\tilde{k}_i]\Delta\underline{\delta}_1$

Where $\Delta\underline{\tilde{p}}_i=\Delta\underline{p}_i+[M]\left[\frac{6}{(\Delta t)}\dot{\underline{\delta}}_1+3\ddot{\underline{\delta}}_1\right]+C_i\left[3\dot{\underline{\delta}}_1+\frac{\Delta t}{2}\ddot{\underline{\delta}}_1\right]$

and $$[\tilde{k}_i] = [k_i]+\frac{6}{(\Delta t)^2}[M]+\frac{3}{\Delta t}C_2$$

C_i may vary with time.

Step 3 Compute $\dot{\underline{\delta}}_1$ from the equation

$$\Delta\dot{\underline{\delta}}_1 = \frac{-\Delta t}{2}\ddot{\underline{\delta}}_1-3\dot{\underline{\delta}}_1+\frac{3}{\Delta t}(\Delta\underline{\delta}_i)$$

At the end of the time interval Δt,

$$\underline{\delta_{i+1}}=\underline{\delta}_i+\Delta\underline{\delta}_i$$

And $$\underline{\dot{\delta}_{i+1}}=\dot{\underline{\delta}}_i+\Delta\dot{\underline{\delta}}_i$$

The operations are repeated from step 1 to 3 to enabling us to obtain the responses $\ddot{\underline{\delta}}$, $\dot{\delta}$, and $\underline{\delta}$.

The operations are repeated from step 1 to step 3 till we obtain the response $\ddot{\underline{\delta}}$, $\dot{\underline{\delta}}$, and $\underline{\delta}$ to a prescribed time.

For good results, the time interval should be about 1/6 to 1/5 of the shortest natural period, among the periods considered.

9.6 FRAME BUILDING AS A MULTI-DEGREE FREEDOM SYSTEM

Before a generalized analysis of the skeletal frames, we recall the simple case of "Shear Building'.

9.6.1 Shear Building

All masses including the slab are discretized at slab levels. Flexibility matrix for horizontal displacements only is developed. Let us go back to the simple case, a shear building, Fig. 9.2 which represents Fig. 9.1

The Fig. 9.2 is reproduced in the following.

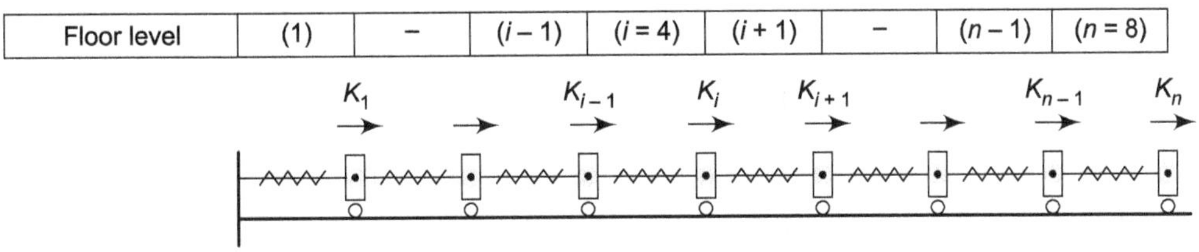

Fig. 9.16 *Generalization of a Shear Building*

The stiffness matrix in Force-deformations is similar to the tridiagonal form noticed in the case of continuous beams on non sinking supports. The same tridiagonal form will be noticed in the case of tall buildings. The storey of a tall building is treated as member and the size of the storey-stiffness is governed by the degrees of freedom in a floor. Let us begin with the shear building, Fig. 9.5. It is assumed that all frames in a tall building are identically loaded. Therefore, consider a single frame, Fig. 9.17.

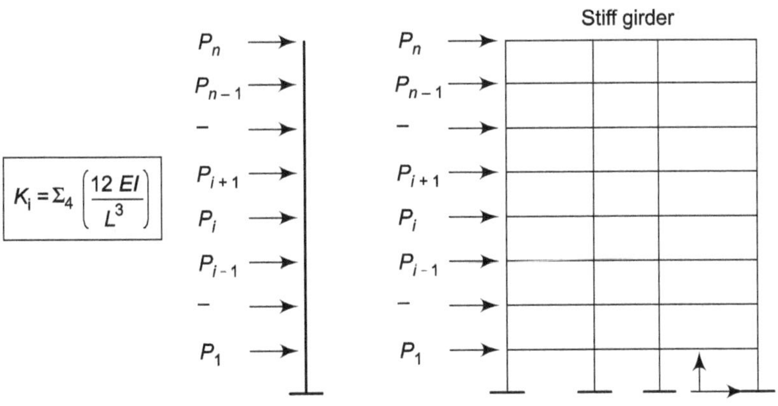

Fig. 9.17 *Single Frame under Lateral Floor Loading*

The stiffness equation, $\{P\} = [K]\{d\}$, when written in the expanded form,
The stiffness equation for a shear building establishes the pattern of a stiffness matrix.

$$
\begin{Bmatrix} P_1 \\ P_2 \\ P_3 \\ - \\ P_{i-1} \\ P_i \\ P_{i+1} \\ - \\ P_{n-2} \\ P_{n-1} \\ P_n \end{Bmatrix}
=
\begin{bmatrix}
K_1+K_2, & K_2 & 0 & 0 & 0 & 0 & 0 & 0 & 0 \\
K_2, & K_2+K_3, & K_3 & 0 & 0 & 0 & 0 & 0 & 0 \\
0 & K_3, & K_3+K_4, & 0 & 0 & 0 & 0 & 0 & 0 \\
- & - & - & - & - & - & - & - & - \\
 & & K_{i-1}+K_i, & K_i & 0 & & & & \\
 & & & K_i, & K_i+K_{i+1}, & K_{i+1} & & & \\
 & & & 0 & K_{i+1}, & K_{i+1}+K_{i+2} & & & \\
- & - & - & - & - & - & - & - & - \\
 & & & & & & K_{n-2}+K_{n-1}, & K_{n-1} & 0 \\
 & & & & & & K_{n-1}, & K_{n-1}+K_n, & K_n \\
 & & & & & & 0 & K_n, & K_n
\end{bmatrix}
\begin{Bmatrix} d_1 \\ d_2 \\ d_3 \\ - \\ d_{i-1} \\ d_i \\ d_{i+1} \\ - \\ d_{n-2} \\ d_{n-1} \\ d_n \end{Bmatrix}
$$

9.6.2 General Case of Tall Buildings

Analytically, it is similar to the stiffness matrix of a shear building. The scalar elements get replaced by the submatrices, and the scalar load gets replaced by a vector consisting of join moments and vertical loads in addition to a simple floor load. It is so simple as stated. It will be more obvious while developing the theory in the following pertaining to Frame-building.

9.7 FRAME BUILDING

There are two analytical models as stated already. In the following we take the CWK model for plane frames.

9.7.1 Plane Frame CWK Model

Instead of directly going from the members (and associated member axes) to the complete structure, we introduce 'storey', a substructure, as an intermediate member.

Skeletal member assemble stiffness matrix of a storey treating storey stiffness as a member-stiffness assemble the stiffness of the entire structure.

Our attempt will be to minimize axes transformations. While choosing the axes of skeletal members, we have been using the right hand rule, and as far as possible system axes were chosen conforming to the majority member axes. In the case of continuous beams, the member axes were conforming to the system axes, and therefore no axes transformations. Because of the nature of tier buildings, the member axes of a storey (substructure) are the same as the axes of the total multi-storey frame. In the CWK model, storey stiffness is a plane substructure; whereas **WN model,** the entire 3-D storey is treated as a member. In the following, y-axes of the substructure is vertical downwards.

The assumption of the inextensibility of girders is justified because of the rigid body displacements of the floor slabs. The axial deformation of the columns, however, is considered for their flexural effects on the girders.

9.7.2 Some Concepts about Distribution of Total Lateral Load

In the preceding section we had obtained, the force deformation relation pertaining to horizontal displacements in the plane of the slabs.

$\overline{Px} = \overline{Kx}\ \overline{\delta x}$, where $\overline{Kx}$ was obtained, when we sifted $\overline{\delta x}$ and θ separately. The stiffness $\overline{Kx}$ may be fully populated. [9]

Suppose, if it were diagonal (It is hypothetical, it is not so even in a shear buiding), one can isolate floor slab, represent it by a rigid bar, and the lateral stiffness of the building by a set of springs. Assume the building to be symmetrical and we know earthquake loading on the entire building. Refer the figure in the following. The spring reactions are the loads $(p_{zi})_j$ acting on the 'rigid bar' substituting the slab i. If δ_{zi} is the deflection of our hypothetical rigid bar, the spring reactions are: $(p_{zi})_j = (k_{zi})_j\ \delta_{zi}$ 'the springs are j =1 to 5 in the figure, Fig. 9.18.

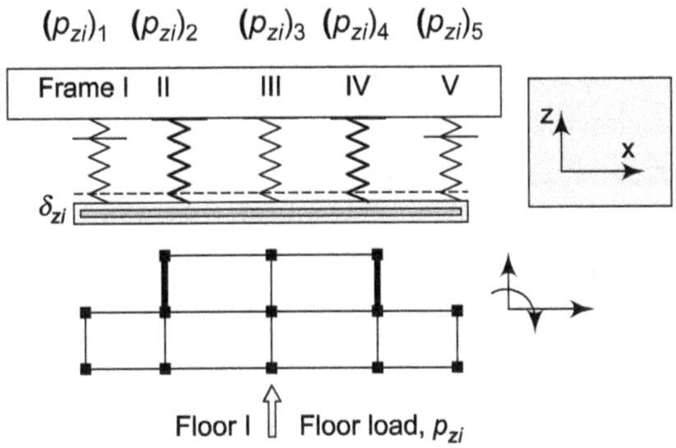

Fig. 9.18 *Distribution of Total Lateral Loading, Symmetrical System*

The load , $P_{zi} = \sum_{1}^{s}(p_{zi})j$. The summation is over j, and $(p_{zi})_j = (k_{zi})_j\ \delta_{zi}$

The slabs in the building cannot be isolated. Each slab is connected to the slabs at top and at bottom. Therefore, we compute

- $\overline{K}_{zI}$ for each frame I, and
- $\overline{K}_{zt}$ of the entire building system is given by

$$\overline{K}_{zt} = \sum_{1}^{V}\overline{K}_{zI},$$, and the stiffness equation is given by,

$$\overline{\overline{P}}_{zt} = \overline{K}_{zt}\,\overline{\delta}_{z}$$, where $\overline{\delta}_{z}$ is the same for each frame (Our example frame I to V)

[9] Note: While developing theory for each frame x was used for z.

$$\overline{K}_{zt} = \sum_{1}^{V} \overline{K}_{zI}, \quad \text{and}$$

The code gives the earthquake load $\overline{\overline{P}}_{zt}$. The only unknown is given by

$$\overline{\delta z} = (\overline{K}_{zt})^{-1}\ \overline{\overline{P}}_{zt}$$

In the case of an unsymmetrical building, the rigid body displacements of each slab consist of $\overline{\delta}$, $\overline{\delta}_z$, and θ.

9.8 ANALYSIS OF SINGLE PLANE FRAME—CWK MODEL

The CWK model was presented by Clough, Wilson and King for analysis of Tall Buildings. In the process, tridiagonal formulation of the stiffness matrix of a complete structure is not obvious with the exception of continuous beams. If we introduce an intermediate stage of stiffness matrix such that we first find out a stiffness matrix for a 'storey unit', consisting of girder or floor line and the set of columns below. At this stage, based upon the "Storey Stiffness" matrix as a sub matrix when we assemble the stiffness matrix for the complete multi-storey structure, it results into a tridiagonal matrix. In this chapter, the development of the tridiagonal matrix will be discussed, starting from single-straight-member stiffness. The orientation or system coordinates is slightly changed by taking Y' coordinate in the downward direction, and a sense of X' to Y' (i.e., clockwise, following right hand rule) is maintained to fix the Y direction or member coordinates. The member arrow directions are fixed following, "left to right" and "top to bottom".

Typical member stiffness matrix for girders. A typical stiffness matrix for a girder member is given by Eq. (9.1). The Girders are assumed inextensible.

$$\begin{bmatrix} S_2 \\ M_1 \\ M_2 \end{bmatrix}_g = \begin{bmatrix} +\dfrac{12EI}{L^3} & -\dfrac{6EI}{L^2} & -\dfrac{6EI}{L^2} \\ -\dfrac{6EI}{L^2} & \dfrac{4EI}{L} & \dfrac{2EI}{L} \\ -\dfrac{6EI}{L^2} & \dfrac{2EI}{L} & \dfrac{4EI}{L} \end{bmatrix}_g = \begin{bmatrix} \delta_{y2} - \delta_{y1} \\ \theta_1 \\ \theta_2 \end{bmatrix}_{(\delta_M)g} \qquad \text{...(9.1)}$$

And a typical matrix for a column member is given by Eq. (9.2) and Fig. 9.4.1

$$\underset{(p_m)_c}{\begin{bmatrix} P_{x2} \\ P_{y2} \\ m_1 \\ m_2 \end{bmatrix}_c} = \underset{(K_m)_c \quad (4x\,4)}{\begin{bmatrix} \dfrac{EA}{L} & 0 & 0 & 0 \\ 0 & \dfrac{12EI}{L^3} & -\dfrac{6EI}{L^2} & -\dfrac{6EI}{L^2} \\ 0 & -\dfrac{6EI}{L^2} & \dfrac{4EI}{L} & \dfrac{2EI}{L} \\ 0 & -\dfrac{6EI}{L^2} & \dfrac{2EI}{L} & \dfrac{4EI}{L} \end{bmatrix}_c} = \underset{(\delta_m)_c}{\begin{bmatrix} \delta_{x2} - \delta_{x1} \\ \delta_{y2} - \delta_{y1} \\ \theta_1 \\ \theta_2 \end{bmatrix}_c} \qquad \text{...(9.2)}$$

9.8.1 Connection Matrix and Stiffness

Based on the compatibility conditions at joints and floor lines, the compatibility equation for the two storeys or Fig. 9.19.1 & 2 is given by the Eq. (9.3).

$$
\begin{bmatrix} [\delta]_{Top\ 18\times 1} \\ \begin{bmatrix} \delta_{y2}-\delta_{y1} \\ \theta_1 \\ \theta_2 \end{bmatrix}_{g1} \\ \begin{bmatrix} \delta_{y2}-\delta_{y1} \\ \theta_1 \\ \theta_2 \end{bmatrix}_{g2} \\ \begin{pmatrix} \delta_{x2}-\delta_{x1} \\ \delta_{y2}-\delta_{y1} \\ \theta_1 \\ \theta_2 \end{pmatrix}_{b1} \\ \begin{pmatrix} \delta_{x2}-\delta_{x1} \\ \delta_{y2}-\delta_{y1} \\ \theta_1 \\ \theta_2 \end{pmatrix}_{b2} \\ \begin{pmatrix} \delta_{x2}-\delta_{x1} \\ \delta_{y2}-\delta_{y1} \\ \theta_1 \\ \theta_2 \end{pmatrix}_{b3} \end{bmatrix}
\begin{bmatrix}
8 & 9 & 10 & 11 & 12 & 13 & 14 & 15 & 16 & 17 & 18 & 19 & 20 & 21 \\
0 & -1 & 1 & 0 & 0 & 0 & 0 & 0 & 0 & 0 & 0 & 0 & 0 & 0 \\
0 & 0 & 0 & 0 & 1 & 0 & 0 & 0 & 0 & 0 & 0 & 0 & 0 & 0 \\
0 & 0 & 0 & 0 & 0 & 1 & 0 & 0 & 0 & 0 & 0 & 0 & 0 & 0 \\
0 & 0 & -1 & 1 & 0 & 0 & 0 & 0 & 0 & 0 & 0 & 0 & 0 & 0 \\
0 & 0 & 0 & 0 & 0 & 1 & 0 & 0 & 0 & 0 & 0 & 0 & 0 & 0 \\
0 & 0 & 0 & 0 & 0 & 0 & 1 & 0 & 0 & 0 & 0 & 0 & 0 & 0 \\
0 & -1 & 0 & 0 & 0 & 0 & 0 & 0 & 1 & 0 & 0 & 0 & 0 & 0 \\
1 & 0 & 0 & 0 & 0 & 0 & 0 & -1 & 0 & 0 & 0 & 0 & 0 & 0 \\
0 & 0 & 0 & 0 & 1 & 0 & 0 & 0 & 0 & 0 & 0 & 0 & 0 & 0 \\
0 & 0 & 0 & 0 & 0 & 0 & 0 & 0 & 0 & 0 & 0 & 1 & 0 & 0 \\
1 & 0 & 0 & 0 & 0 & 0 & 0 & -1 & 0 & 0 & 0 & 0 & 0 & 0 \\
1 & 0 & 0 & 0 & 0 & 0 & 0 & 0 & 0 & -1 & 0 & 0 & 0 & 0 \\
0 & 0 & 0 & 0 & 0 & 1 & 0 & 0 & 0 & 0 & 0 & 0 & 0 & 0 \\
0 & 0 & 0 & 0 & 0 & 0 & 0 & 0 & 0 & 0 & 0 & 0 & 1 & 0 \\
0 & 0 & 0 & -1 & 0 & 0 & 0 & 0 & 0 & 0 & 1 & 0 & 0 & 0 \\
1 & 0 & 0 & 0 & 0 & 0 & 0 & -1 & 0 & 0 & 0 & 0 & 0 & 0 \\
0 & 0 & 0 & 0 & 0 & 0 & 1 & 0 & 0 & 0 & 0 & 0 & 0 & 0 \\
0 & 0 & 0 & 0 & 0 & 0 & 0 & 0 & 0 & 0 & 0 & 0 & 0 & 1
\end{bmatrix}
\begin{bmatrix} [\delta]_{Top\ 18\times 1} \\ \begin{bmatrix} (\delta_x')_F \\ (\delta_Y')_{J1} \\ (\delta_Y')_{J2} \\ (\delta_Y')_{J3} \\ \theta_{J1} \\ \theta_{J2} \\ \theta_{J3} \end{bmatrix} \\ FLR \\ \begin{bmatrix} (\delta_x')_B \\ (\delta_Y')_{J1} \\ (\delta_Y')_{J2} \\ (\delta_Y')_{J3} \\ \theta_{J1} \\ \theta_{J2} \\ \theta_{J3} \end{bmatrix} \\ BOT \\ - \\ - \end{bmatrix}
$$

FLR 18 X 1 $\qquad (A_b)_{18\times 14}$

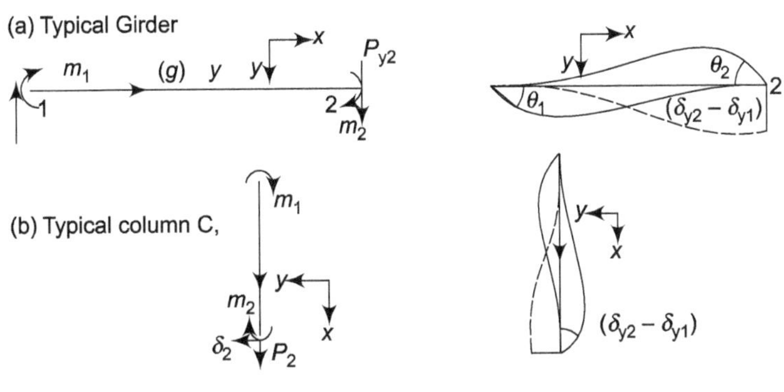

Fig. 9.19(a) *Members of a Typical Storey*

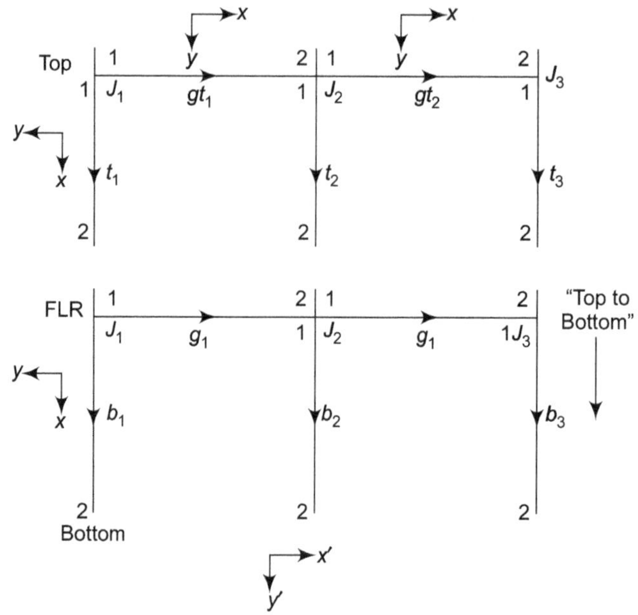

Fig. 9.19(b) *Storey Treated as a Member*

Analysis based on Clough, Wilson and King: It is a land mark paper on Tall Building (1964).

Obtain stiffness matrix of a typical storey member, Fig. 9.20. Neglect axial deformation of girder for their effect on moments of shears. In the hierarchy wise, we are moving from the line as a member to the storey as a member.

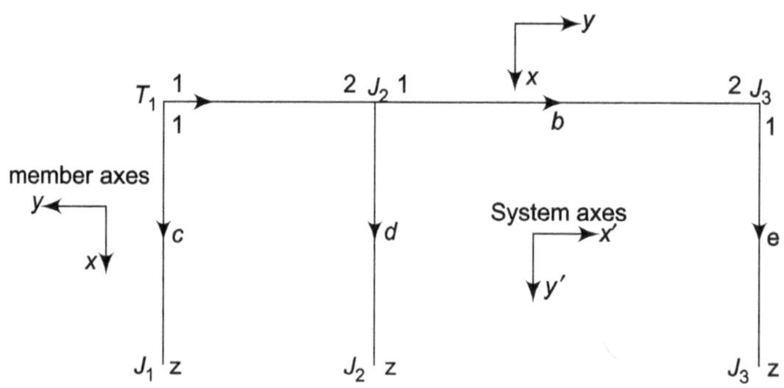

Fig. 9.20 *Typical Storey Member, m*

The displacement components at ends f and b are

$$\left\{\delta'_x, (\delta'_y, \theta)_{J1}, (\delta'_y, \theta)_{J2}, (\delta'_y, \theta)_{J3}\right\}_M = d_f$$

$$\left\{\delta'_x, (\delta'_y, \theta)_{J1}, (\delta'_y, \theta)_{j2}, (\delta'_y, \theta)_{J3}\right\}_{M+1} = d_b$$

The total number of end deformation is 14. For the size of connection matrix ($n \times 14$), the value of n depends on reference line of member coordinate axis.

The following three cases consider different reference lines.

(a) Line joining end points in un-deformed position as member reference line.

[A] matrix in relation $d_m{}' = Adj$

$$
\begin{bmatrix}
\begin{bmatrix}\delta_{y2}-\delta_{y1}\\ \theta_1\\ \theta_2\end{bmatrix}_a\\
\begin{bmatrix}\delta_{y2}-\delta_{y1}\\ \theta_1\\ \theta_2\end{bmatrix}_b\\
\begin{bmatrix}\delta_{x2}-\delta_{x1}\\ \delta_{y2}-\delta_{y1}\\ \theta_1\\ \theta_2\end{bmatrix}_c\\
\begin{bmatrix}\delta_{x2}-\delta_{x1}\\ \delta_{y2}-\delta_{y1}\\ \theta_1\\ \theta_2\end{bmatrix}_d\\
\begin{bmatrix}\delta_{x2}-\delta_{x1}\\ \delta_{y2}-\delta_{y1}\\ \theta_1\\ \theta_2\end{bmatrix}_e
\end{bmatrix}
=
\underbrace{\begin{bmatrix}
\begin{bmatrix}0&-1&0&1&0&0&0\\0&0&1&0&0&0&0\\0&0&0&0&1&0&0\end{bmatrix}\\
\begin{bmatrix}0&0&0&-1&0&1&0\\0&0&0&0&1&0&0\\0&0&0&0&0&0&1\end{bmatrix}\\
\begin{bmatrix}0&-1&0&0&0&0&0\\1&0&0&0&0&0&0\\0&0&1&0&0&0&0\\0&0&0&0&0&0&0\end{bmatrix}\\
\begin{bmatrix}0&0&0&-1&0&0&0\\1&0&0&0&0&0&0\\0&0&1&0&1&0&0\\0&0&0&0&0&0&0\end{bmatrix}\\
\begin{bmatrix}0&0&0&0&0&-1&0\\1&0&0&0&0&0&0\\0&0&0&0&0&0&1\\0&0&0&0&0&0&0\end{bmatrix}
\end{bmatrix}}_{A_f}
\underbrace{\begin{bmatrix}
\begin{bmatrix}[0]\end{bmatrix}\\
\begin{bmatrix}[0]\end{bmatrix}\\
\begin{bmatrix}0&+1&0&0&0&0&0\\-1&0&0&0&0&0&0\\0&0&0&0&0&0&0\\0&0&1&0&0&0&0\end{bmatrix}\\
\begin{bmatrix}0&0&0&1&0&0&0\\-1&0&0&0&0&0&0\\0&0&0&0&0&0&0\\0&0&0&0&1&0&0\end{bmatrix}\\
\begin{bmatrix}0&0&0&0&0&1&0\\-1&0&0&0&0&0&0\\0&0&0&0&0&0&0\\0&0&0&0&0&0&1\end{bmatrix}
\end{bmatrix}}_{A_b}
\begin{bmatrix}
\begin{bmatrix}\delta_x\\ \begin{bmatrix}\delta'_y\\ \theta\end{bmatrix}_{J_1}\\ \begin{bmatrix}\delta'_y\\ \theta\end{bmatrix}_{J_2}\\ \begin{bmatrix}\delta'_y\\ \theta\end{bmatrix}_{J_3}\end{bmatrix}_f\\
\begin{bmatrix}\delta'_x\\ \begin{bmatrix}\delta'_y\\ \theta\end{bmatrix}_{J_1}\\ \begin{bmatrix}\delta'_y\\ \theta\end{bmatrix}_{J_2}\\ \begin{bmatrix}\delta'_y\\ \theta\end{bmatrix}_{J_3}\end{bmatrix}_b
\end{bmatrix}
$$

Alternatively in the partitioned form

$$
\begin{bmatrix}(d_m)_a\\(d_m)_b\\(d_m)_c\\(d_m)_d\\(d_m)_e\end{bmatrix}
=
\begin{bmatrix}A_{1f}&A_{1b}\\A_{2f}&A_{2b}\\A_{3f}&A_{3b}\\A_{4f}&A_{4b}\\A_{5f}&A_{5b}\end{bmatrix}
\begin{bmatrix}(d^t)_f\\ \\ -\\ \\ (d^t)_b\end{bmatrix}
$$

The corresponding in matrix is given by

$$
K_m=\begin{bmatrix}
(k_m)_{3\times3}&&&&\\
&(k_m)_{3\times3}&&&\\
&&{}^{*}(k_m)_{4\times4}&&\\
&&&(k_m)_{4\times4}&\\
&&&&(km)_{4\times4}
\end{bmatrix}_{18\times18}
$$

The typical girder and column member stiffness matrices are:

$$(k_m)_a = \begin{bmatrix} 12EI/L^3 & -6EI/L^2 & -6EI/L^2 \\ * & 4EI/L & 2EI/L \\ * & * & 4EI/L \end{bmatrix}_a$$

and

$$(k_m)_d = \begin{bmatrix} AE/L & 0 & 0 & 0 \\ * & 12EI/L^3 & -6EI/L^2 & -6EI/L^2 \\ * & * & 4EI/L & 2EI/L \\ * & * & * & 4EI/L \end{bmatrix}_d$$

respectively.

The stiffness matrix, $K_J = A^t\, k_m\, A$

The matrix k_j, when partitioned referring to the ends f and b, is expressed as

$k_j = \begin{bmatrix} K_{ff} & K_{fb} \\ K_{bf} & K_{bb} \end{bmatrix}$; where $K_{bf} = (K_{fb})^t$, and K_{ff}, K_{fb} and K_{bb} are matrices.

For the three column lines, the size of these matrices is 7×7.

The assembly of K_{ff}, K_{fb}, K_{bf} and K_{bb} constituting K_j matrix is given below.

For large number of members constituting a storey it is not convenient to write the entire K_m matrix, where in the partitioned form only the diagonal submatrices are not nulls. It is, therefore, desirable to partition the matrices. In this case

$$[K]_{\text{storey}} = \begin{bmatrix} K_{ff} & K_{fb} \\ K_{bf} & K_{bb} \end{bmatrix} = \begin{bmatrix} A_{1f} & A_{1b} \\ A_{2f} & A_{2b} \\ A_{3f} & A_{3b} \\ A_{4f} & A_{4b} \\ A_{5f} & A_{5b} \end{bmatrix} = \begin{bmatrix} (k_m)_{3\times 3} & & & & \\ & (k_m)_{3\times 3} & & & \\ & & (k_m)_{4\times 4} & & \\ & & & (k_m)_{4\times 4} & \\ & & & & (k_m)_{4\times 4} \end{bmatrix} [A]$$

And it can be shown that

$K_{ff} = (A_{if})^t\, (K_m)_{ii}\, A_{if}$, where i is a dummy index to be summed following summation convention.

in the expanded form.

$K_{ff} = (A_{f1})^t\, (km)_{11}\, A_{if} + (A_{f2})^t\, (km)_{22}\, A_{2f} + (A_{f3})^t\, (km)_{33}\, A_{3f} + (A_{f4})^t\, (km)_{44}\, A_{4f} + (A_{f5})^t\, (km)_{55}\, A_{5f}$

And similarly

$$K_{fb} = (A_{if})^t\, (km)_{ii}\, A_{ib}$$
$$K_{bb} = (A_{ib})^t\, (km)_{ii}\, A_{ib}$$

(b) Tangent to the end of a member as reference line. In this case, we are taking relative deformation of end 2 with respect to end 1. Typical girder and column member stiffness are

$$(K_m)_{\text{girder}} = \begin{bmatrix} 12EI/L^3 & -6EI/L^2 \\ -6EI/L^2 & 4EI/L \end{bmatrix} = \text{and } (K_m)_{\text{column}} = \begin{bmatrix} AE/L & 0 & 0 \\ 0 & 12EI/L^3 & -6EI/L^2 \\ 0 & -6EI/L^2 & 4EI/L \end{bmatrix}$$

The reference lines for members are as shown in Fig. 9.21 and the connection matrix giving compatibility of d_m and d_j is given below.

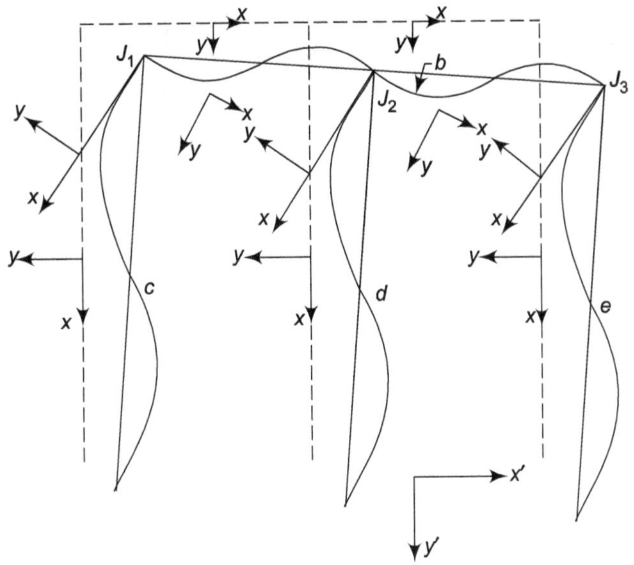

Fig. 9.21 *Deflected Shape of a Storey with Member Axes Shown*

'A' Matrix

$$[1 \quad 2 \quad 3 \quad 4 \quad 5 \quad 6 \quad 7 \quad 8 \quad 9 \quad 10 \quad 11 \quad 12 \quad 13 \quad 14]$$

$$\begin{bmatrix} \begin{bmatrix} \delta_{y2}-\delta_{y1} \\ \theta_2-\theta_1 \end{bmatrix}_a \\ \begin{bmatrix} \delta_{y2}-\delta_{y1} \\ \theta_2-\theta_1 \end{bmatrix}_b \\ \begin{bmatrix} \delta_{x2}-\delta_{x1} \\ \delta_{y2}-\delta_{y1} \\ \theta_2-\theta_1 \end{bmatrix}_c \\ \begin{bmatrix} \delta_{x2}-\delta_{x1} \\ \delta_{y2}-\delta_{y1} \\ \theta_2-\theta_1 \end{bmatrix}_d \\ \begin{bmatrix} \delta_{x2}-\delta_{x1} \\ \delta_{y2}-\theta_{y1} \\ \theta_2-\theta_1 \end{bmatrix}_e \end{bmatrix} = \begin{bmatrix} 0 & -1 & -l_a & 1 & 0 & 0 & 0 & 0 & 0 & 0 & 0 & 0 & 0 & 0 \\ 0 & 0 & -1 & 0 & 1 & 0 & 0 & 0 & 0 & 0 & 0 & 0 & 0 & 0 \\ 0 & 0 & 0 & -1 & -l_b & 1 & 0 & 0 & 0 & 0 & 0 & 0 & 0 & 0 \\ 0 & 0 & 0 & 0 & -1 & 0 & 1 & 0 & 0 & 0 & 0 & 0 & 0 & 0 \\ 0 & -1 & 0 & 0 & 0 & 0 & 0 & 0 & 1 & 0 & 0 & 0 & 0 & 0 \\ 1 & 0 & -l_c & 0 & 0 & 0 & 0 & -1 & 0 & 0 & 0 & 0 & 0 & 0 \\ 0 & 0 & -1 & 0 & 0 & 0 & 0 & 0 & 0 & 1 & 0 & 0 & 0 & 0 \\ 0 & 0 & -1 & 0 & 0 & 0 & 0 & 0 & 0 & 0 & 1 & 0 & 0 & 0 \\ 1 & 0 & 0 & 0 & -l_d & 0 & 0 & -1 & 0 & 0 & 0 & 0 & 0 & 0 \\ 0 & 0 & 0 & 0 & -1 & 0 & 0 & 0 & 0 & 0 & 0 & 1 & 0 & 0 \\ 0 & 0 & 0 & 0 & 0 & -1 & 0 & 0 & 0 & 0 & 0 & 0 & 1 & 0 \\ 0 & 0 & 0 & 0 & 0 & 0 & -l_e & -1 & 0 & 0 & 0 & 0 & 0 & 0 \\ 0 & 0 & 0 & 0 & 0 & 0 & -1 & 0 & 0 & 0 & 0 & 0 & 0 & 1 \end{bmatrix} \begin{bmatrix} \begin{bmatrix} \delta'_x \\ \begin{bmatrix} \delta'_y \\ \theta \end{bmatrix}_{J1} \\ \begin{bmatrix} \delta'_y \\ \theta \end{bmatrix}_{J2} \\ \begin{bmatrix} \delta'_y \\ \theta \end{bmatrix}_{J3} \end{bmatrix}_f \\ \begin{bmatrix} \delta \\ \begin{bmatrix} \delta'_y \\ \theta \end{bmatrix}_{J1} \\ \begin{bmatrix} \delta'_y \\ \theta \end{bmatrix}_{J2} \\ \begin{bmatrix} \delta'_y \\ \theta \end{bmatrix}_{J3} \end{bmatrix}_b \end{bmatrix}$$

(c) Connection matrix by taking reference line – as the line joining the end points in deformed state.

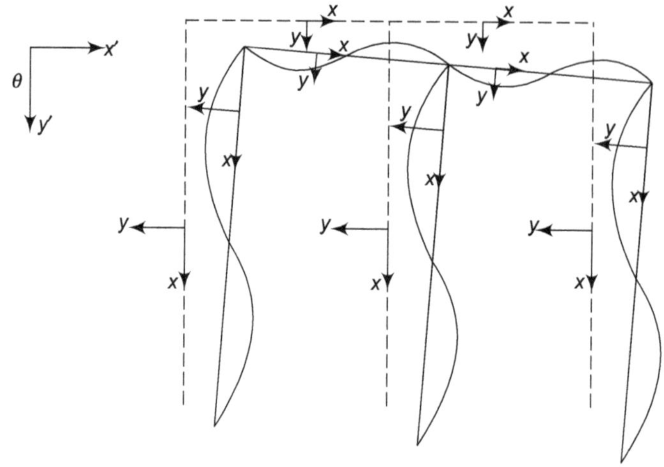

Fig. 9.22 *Line Joining the End Points in Deformed Position as Member Axes*

'A' Matrix

$$
\begin{bmatrix}
\begin{pmatrix} \theta_1 \\ \theta_2 \end{pmatrix}_a \\
\begin{pmatrix} \theta_1 \\ \theta_2 \end{pmatrix}_b \\
\begin{pmatrix} \delta_{x2}-\delta_{x1} \\ \theta_1 \\ \theta_2 \end{pmatrix}_c \\
\begin{pmatrix} \delta_{x2}-\delta_{x1} \\ \theta_1 \\ \theta_2 \end{pmatrix}_d \\
\begin{pmatrix} \delta_{x2}-\delta_{x1} \\ \theta_1 \\ \theta_2 \end{pmatrix}_e
\end{bmatrix}
=
\begin{bmatrix}
0 & \frac{-1}{L_a} & 1 & \frac{1}{L_a} & 0 & 0 & 0 & 0 & 0 & 0 & 0 & 0 & 0 & 0 \\
0 & \frac{-1}{L_a} & 0 & \frac{1}{L_a} & 1 & 0 & 0 & 0 & 0 & 0 & 0 & 0 & 0 & 0 \\
0 & 0 & 0 & \frac{-1}{L_b} & 1 & \frac{1}{L_b} & 0 & 0 & 0 & 0 & 0 & 0 & 0 & 0 \\
0 & 0 & 0 & \frac{-1}{L_b} & & \frac{1}{L_b} & 1 & 0 & 0 & 0 & 0 & 0 & 0 & 0 \\
0 & -1 & 0 & 0 & 0 & 0 & 0 & 0 & 1 & 0 & 0 & 0 & 0 & 0 \\
\frac{-1}{L_c} & 0 & 1 & 0 & 0 & 0 & 0 & \frac{1}{L_c} & 0 & 0 & 0 & 0 & 0 & 0 \\
\frac{-1}{L_c} & 0 & 0 & 0 & 0 & 0 & 0 & \frac{1}{L_c} & 1 & 0 & 0 & 0 & 0 & 0 \\
0 & 0 & 0 & 0 & -1 & 0 & 0 & & 0 & 1 & 0 & 0 & 0 & 0 \\
\frac{-1}{L_c} & 0 & 0 & 0 & 1 & 0 & 0 & \frac{1}{L_c} & 0 & 0 & 1 & 0 & 0 & 0 \\
\frac{-1}{L_c} & 0 & 0 & 0 & 0 & 0 & 0 & \frac{1}{L_c} & 0 & 0 & 0 & 1 & 0 & 0 \\
0 & 0 & 0 & 0 & -1 & 0 & 0 & & 0 & 0 & 0 & 0 & 1 & 0 \\
\frac{-1}{L_c} & 0 & 0 & 0 & 0 & 0 & 1 & \frac{1}{L_c} & 0 & 0 & 0 & 0 & 0 & 0 \\
\frac{-1}{L_c} & 0 & 0 & 0 & 0 & 0 & 0 & \frac{1}{L_c} & 0 & 0 & 0 & 0 & 0 & 1
\end{bmatrix}_{13\times 14}
\begin{bmatrix}
\begin{bmatrix} \delta'_x \\ \begin{pmatrix} \delta'_y \\ \theta \end{pmatrix}_{J1} \\ \begin{pmatrix} \delta'_y \\ \theta \end{pmatrix}_{J2} \\ \begin{pmatrix} \delta'_y \\ \theta \end{pmatrix}_{J3} \end{bmatrix}_f \\
\begin{bmatrix} \delta'_x \\ \begin{pmatrix} \delta'_y \\ \theta \end{pmatrix}_{J1} \\ \begin{pmatrix} \delta'_y \\ \theta \end{pmatrix}_{J2} \\ \begin{pmatrix} \delta'_y \\ \theta \end{pmatrix}_{J3} \end{bmatrix}_b
\end{bmatrix}
$$

$$K_J = \begin{bmatrix} K_{ff} & K_{fb} \\ K_{bf} & K_{bb} \end{bmatrix}$$

$$A_{13\times 14}$$

K_{ff} K_{fb}

$$K_j = \left[\begin{array}{ccccccc|ccccccc}
\left(\frac{12}{l_c^3}+\frac{12}{l_d^3}+\frac{12}{l_c^3}\right)EI & 0 & -\frac{6EI}{l_c^2} & 0 & -\frac{6EI}{l_d^2} & 0 & -\frac{6EI}{l_e^2} & -\frac{12EI}{l_c^3}-\frac{12EI}{l_d^3}-\frac{12EI}{l_e^3} & 0 & -\frac{6EI}{l_c^2} & 0 & -\frac{6EI}{l_d^2} & 0 & -\frac{6EI}{l_e^2} \\
0 & \frac{12EI}{l_a^3}+\frac{EA}{l_c} & \frac{6EI}{l_a^3} & -\frac{12EI}{l_a^3} & \frac{6EI}{l_a^2} & 0 & 0 & 0 & -\frac{EA}{l_c} & 0 & 0 & 0 & 0 & 0 \\
-\frac{6EI}{l_c^2} & \frac{6EI}{l_a^2} & \frac{4EI}{l_a}+\frac{EI}{l_c} & -\frac{6EI}{l_a^2} & \frac{2EI}{l_a} & 0 & 0 & -\frac{6EI}{l_c^2} & 0 & \frac{2EI}{l_c} & 0 & 0 & 0 & 0 \\
0 & -\frac{12EI}{l_a^3} & -\frac{6EI}{l_a^2} & \frac{12EI}{l_a^3}+\frac{12EI}{l_b^3}+\frac{EA}{L} & -\frac{6EI}{l_a^2}+\frac{6EI}{l_b^2} & -\frac{12EI}{l_b^3} & -\frac{6EI}{l_b^2} & 0 & 0 & 0 & -\frac{EA}{L} & 0 & 0 & 0 \\
-\frac{6EI}{l_d^2} & \frac{6EI}{l_a^2} & \frac{2EI}{l_a} & -\frac{6EI}{l_a^2}+\frac{6EI}{l_b^2} & \frac{4EI}{l_a}+\frac{4EI}{l_b}+\frac{4EI}{l_d} & -\frac{12EI}{l_b^2} & \frac{6EI}{l_b} & -\frac{6EI}{l_d^2} & 0 & 0 & 0 & \frac{2EI}{l_d} & 0 & 0 \\
0 & 0 & 0 & -\frac{12EI}{l_b^3} & -\frac{6EI}{l_b^3} & +\frac{12EI}{l_b^3}+\frac{EA}{L} & -\frac{6EI}{l_b^2} & 0 & 0 & 0 & 0 & 0 & -\frac{AE}{L} & 0 \\
-\frac{6EI}{l_c^2} & 0 & 0 & \frac{6EI}{l_b^2} & \frac{2EI}{l_b} & -\frac{6EI}{l_b^2} & \frac{4EI}{l_b}+\frac{4EI}{l_c} & -\frac{6EI}{l_c^2} & 0 & 0 & 0 & 0 & 0 & \frac{2EI}{l_e} \\
\boxed{\frac{1}{l_c^3}+\frac{1}{l_d^3}+\frac{1}{l_e^3}}(-12EI) & 0 & \frac{6EI}{l_c^2} & 0 & \frac{6EI}{l_d^2} & 0 & \frac{6EI}{l_e^2} & \frac{12EI}{l_c^3}+\frac{12EI}{l_d^3}+\frac{12EI}{l_e^3} & 0 & \frac{6EI}{l_c^2} & 0 & \frac{6EI}{l_d^2} & 0 & \frac{6EI}{l_e^2} \\
0 & -\frac{EA}{l_c} & 0 & 0 & 0 & 0 & 0 & 0 & \frac{EA}{L} & 0 & 0 & 0 & 0 & 0 \\
-\frac{6EI}{l_c^2} & 0 & \frac{2EI}{l_c} & 0 & 0 & 0 & 0 & \frac{6EI}{l_c^2} & 0 & \frac{4EI}{l_c} & 0 & 0 & 0 & 0 \\
0 & 0 & 0 & -\frac{AE}{L} & 0 & 0 & 0 & 0 & 0 & 0 & \frac{AE}{L} & 0 & 0 & 0 \\
-\frac{6EI}{l_d^2} & 0 & 0 & 0 & \frac{2EI}{l_d} & 0 & 0 & 0 & 0 & 0 & 0 & \frac{4EI}{l_d} & 0 & 0 \\
0 & 0 & 0 & 0 & 0 & -\frac{AE}{L} & 0 & 0 & 0 & 0 & 0 & 0 & \frac{AE}{L} & 0 \\
-\frac{6EI}{l_e^2} & 0 & 0 & 0 & 0 & 0 & \frac{2EI}{l_e} & 0 & 0 & 0 & 0 & 0 & 0 & \frac{4EI}{l_e}
\end{array}\right]$$

K_{bf} K_{bb}

Stiffness Matrix, $K_j = A^T k_m A$

Typical girder and column stiffness are: $\begin{bmatrix} 4EI/L & 2EI/L \\ 2EI/L & 4EI/L \end{bmatrix}$; $= \begin{bmatrix} AE/L & 0 & 0 \\ 0 & 4EI/L & 2EI/L \\ 0 & 2EI/L & 4EI/L \end{bmatrix}$ Eq. (9.4a)

$$\underset{\text{member coordinates, } m}{\begin{bmatrix} \delta_T \\ \delta_B \end{bmatrix}} = \underset{\text{system coordinates, } S}{A \begin{bmatrix} \delta_T \\ \delta_F \\ \delta_B \end{bmatrix}} \quad \ldots(9.4a)$$

Or

$$\underset{18 \times 1}{\left[\delta_B\right]_m} = A_b \underset{14 \times 1}{\begin{bmatrix} \delta_B \\ \delta_B \end{bmatrix}_S} \quad \ldots(9.4b)$$

And the "system stiffness matrix" for the storey, Fig. 6.2, is

$$\begin{pmatrix} P_F \\ F_B \end{pmatrix}_3 = \underbrace{\begin{pmatrix} k_{ff} & k_{fb} \\ k_{bf} & k_{bb} \end{pmatrix} \begin{pmatrix} \delta_F \\ \delta_B \end{pmatrix}_S} \quad \ldots(9.5)$$

Where K = $A'_b\, K_m\, A$, and K_m is given by the relation

$$K_m = \underset{P_m}{\begin{bmatrix} (P_m)_{g1} \\ (P_m)_{g2} \\ (P_m)_{b1} \\ (P_m)_{b2} \\ (P_m)_{b3} \end{bmatrix}} \underset{K_m}{\begin{bmatrix} (k_m)_{g1} & & & & \\ & (k_m)_{g2} & & & \\ & & (k_m)_{b1} & & \\ & & & (k_m)_{b2} & \\ & & & & (k_m)_{b3} \end{bmatrix}} \underset{\delta_m}{\begin{bmatrix} (\delta_m)_{g1} \\ (\delta_m)_{g2} \\ (\delta_m)_{b1} \\ (\delta_m)_{b2} \\ (\delta_m)_{b3} \end{bmatrix}} \quad \ldots(9.6)$$

The expanded form of matrices $(P_m)_g$, $(P_m)_b$, $(\delta_m)_g$, $(\delta_m)_b$, $(K_m)_g$ and $(K_m)_b$ are given by the Eqs (9.1) and (9.2).

9.8.2 Stiffness Matrix and Recursion Algorithm

The Eq. (9.5) can be generalized to give a stiffness matrix for the entire structure. Let the storey member be designated from the top to bottom be 1, 2, 3, … and corresponding girder lines be G_1, G_2, G_3 … referring to any typical girder line G_1, deformations for the top floor, δ_τ; floor under consideration, δ_{Fi}; and bottom floor, δ_B are shown in the Fig. 9.22.

If the floor nodes are designated by J_1, J_2, J_3, we express δ_{Qi} and P_{Qi} by the matrix expressions, Eq. (9.7).

$$\delta_{Gi} = \begin{bmatrix} (\delta'_x)_{g1} \\ (\delta'_y)_{J1} \\ (\delta'_y)_{J2} \\ (\delta'_y)_{J3} \\ (\theta)_{J1} \\ (\theta)_{J2} \\ (\theta)_{J3} \end{bmatrix}; P_{Gi} = \begin{bmatrix} (P'_x)_{g1} \\ (P'_y)_{J1} \\ (P'_y)_{J2} \\ (P'_y)_{J3} \\ (m)_{J1} \\ (m)_{J2} \\ (m)_{J3} \end{bmatrix} \text{ Summation of all the horizontal forces } P'_x \quad \ldots(9.7)$$

At this stage we are in a position to write the stiffness matrix K for the entire structure and the form of which is "TRIDIAGONAL", Eqs (9.8) and (9.9)

The load matrix P_{Gi} can be obtained from the section ship $P_{Gi} = A^t P_m$

As an illustration the first element of P_{Gi} from the above Eq. (9.7) to given by

$$-\left[(P_{y2})_{t1} + (P_{y2})_{t2} + (P_{y2})_{t3}\right] + (P_{y1})_{b1} + (P_{y2})_{b2} + (P_{y3})_{b3}$$

It can be verified from equilibrium, that $(P'_y)_{Gi}$ is equal to the above statement.

Similarly the second element $(P'_y)_{J1} = -(P_{y2})_{g1} - (P_x)_{b1}$

$$\begin{array}{c} \\ \begin{bmatrix} P_{G1} \\ P_{G2} \\ P_{G3} \\ P_{G(i-1)} \\ P_{Gi} = P_F \\ P_{G(i+1)} = P_s \\ - \\ - \end{bmatrix} \end{array} = \begin{array}{c} (\delta)_{G1} \quad (\delta)_{G2} \quad TG_{(i-1)} \qquad F_{Gi} \qquad BG_{(i+1)} \\ \begin{bmatrix} (K_{ff})_1 & (K_{fb})_1 & & & & \\ (K_{ft})_2 & (K_{ff})_2 & (K_{fb})_2 & & & \\ & & (K_{ft})_{G(i-1)} & (K_{ff})_{G(i-1)} & (K_{fb})_{G(i-1)} & \\ & & & (K_{ft})_{Gi} & (K_{ff})_{Gi} & (K_{fb})_{Gi} \\ & & & (K_{ft})_{G(i+1)} & (K_{ff})_{G(i+1)} & (K_{fb})_{G(i+1)} \end{bmatrix} \end{array} \begin{bmatrix} \delta_{G1} \\ \delta_{G2} \\ \delta_{G3} \\ - \\ - \\ \delta_{G(i-1)} \\ \delta_{G1} = \delta_F \\ \delta_{G(i+1)} \\ - \\ - \end{bmatrix} \quad \ldots(9.8)$$

Submatrices other than shown are null

where $(K_{fb})_1 = (K_{ft})_2$,

Or in general $(K_{fb})_i = (K_{ft})_{i+1}$ and $(K_{ff})_i = (K_b)_{i-1} + (K_f)_i$ ...(9.9)

A typical equation for the P_{Gi} is given by Eq. (9.10)

$$P_{Gi} = P_F = (K_{ft})_{Gi}\,\delta_T + (K_{ff})_{Gi}\,\delta_F + (K_{fb})_{Gi}\,\delta_B$$

$$P_{Gi} = (K_{fb})_{G(i-1)} + (K_{ff})_{Gi} + (K_{fb})_{Gi} \quad \ldots(9.10)$$

The Eq. (9.10), is a pivotal equation for recursion to be used to obtain the unknown deformation, δ_{Gi} in the "Tridigonal Matrix" equation, Eq. (9.8). Starting from the first two equations or P_{G1} and P_{G2}

$$P_{G1} = (K_{ff})_1\, \delta_{G1} + (K_{fb})_1\, \delta_{G2} \qquad \text{...(i)}$$

$$P_{G2} = (K_{fb})_1\, \delta_{G1} + (K_{ff})_2\, \delta_{G2} + (K_{fb})_2\, \delta_{G3} \qquad \text{...(ii)}$$

From Eq. (i) $\delta_{G1} = (K_{ff})_1\, \mathrm{P}_{G1} - (K_{ff})_1^{\ -1}\, (K_{fb})_1\, \delta_{G2}$

Or

$$\underbrace{P_{G2} - (K_{fb})_1\, (K_{ff})_1^{-1}\, P_{G1}}_{\bar{P}_{G2}} = \underbrace{\left[(K_{ff})_2 - (K_{ff})_1\, (K_{fb})_2\right]}_{(\bar{P}_{ff})_2}\delta_{G2} + (K_{fb})_2\, \delta_{G3} \qquad \text{...(9.11)}$$

Letting

$$\left.\begin{aligned} \bar{P}_{G2} &= P_{g2} - (K'_{fb})_1\, (K_{ff})_1^{-1}\, P_{g1} \\ (\bar{K}_{ff})_2 &= [(K_{ff})_2 - (K'_{fb})_1\, (K_{ff})^{-1}\, (K_{fb})_1] \end{aligned}\right\} \qquad \text{...(9.12)}$$

$$\bar{P}_{G2} = (\bar{K}_{ff})_2\, \delta_{G2} + (K_{fb})_2\, \delta_{G3}$$

In general,

$$\bar{P}_{G1} = P_{G1} - (K'_{fb})_{i-1}\, (K_{ff})_{i-1}^{-1}\, P_{G(i-1)}$$

$$(\bar{K}_{ff})_i = (K_{ff})_i - (K_{fb'})_{i-1}\, (K_{ff})_{i-1}^{-1}\, (K_{fb})_{i-1} \qquad \text{...(9.13)}$$

$$(\bar{P}_G)_i = (\bar{K}_{ff})_i\, \delta_{Gi} + (K_{fb})_i\, \delta_{G(i+1)}$$

9.8.3 Displacements by Back Substitution

The last row of the matrix equation, Eq. (9.8), for the floor, n, Fig. 9.4 gives

$$\bar{P}_{Gn} = (\bar{K}_{ff})_n\, \delta_{Gn} + (K_{fb})_n\, \delta_{G(n+1)}$$

But $\delta_{G(n+1)} = 0$, Fig. 9.4

$\therefore$

$$\bar{P}_{Gn} = (\bar{K}_{ff})_n\, \delta_{Gn} \qquad \text{...(9.14a)}$$

And

$$\delta_{Gn} = (K_{ff})_n^{-1}\, \bar{P}_{Gn} \qquad \text{...(9.14b)}$$

For the case of multistory building or 3 columns lines the inversion required is for a matrix $(K_{ff})^{-1}$, which is only 7×7 size matrix.

The values of δ_{Gn} obtained from the Eq. (9.14), when back substituted, $\delta_{G(n-1)}$ can be found from the following relations.

$$\bar{P}_{G(n-1)} = (K_{ff})_{n-1}\, \delta_{G(n-1)} + (K_{fb})_{n-1}\, \delta_G \qquad \text{...(9.15)}$$

Back substituting the value of δ_{Gn}

$$\bar{P}_{G(n-1)} = (\bar{K}_{ff})_{n-1}\, \delta_{G(n-1)} + (K_{fb})_{n-1} \delta_G (K_{ff})_n^{\ -1}\, \bar{P}_{Gn}$$

$$\delta_{G(n-1)} = (\bar{K}_{ff})_{n-1}\left[\bar{P}_{G(n-1)} + (K_{fb})_{n-1} (K_{ff})_n^{\ -1}\, P_{Gn}\right] \qquad \text{...(9.16)}$$

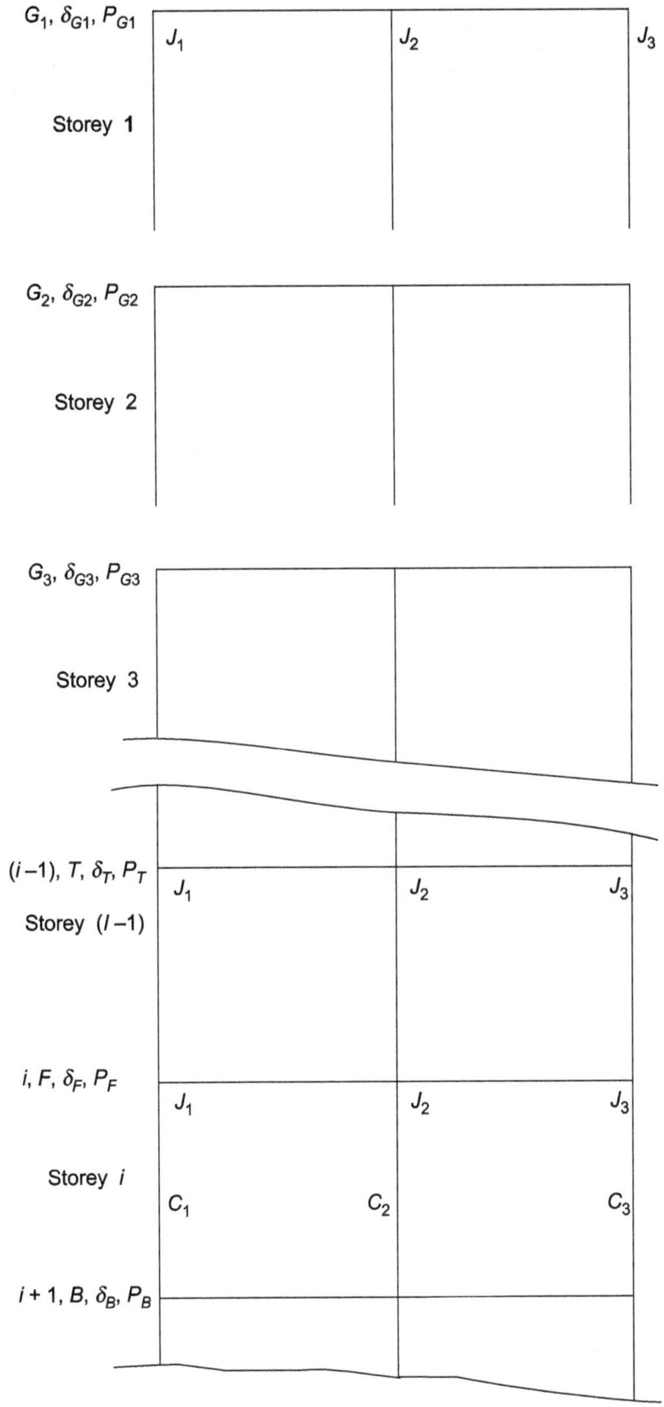

Fig. 9.23 *Girder Line or Floor Designation G_i, Joint Designation, J_i, and Column Designation C_i*

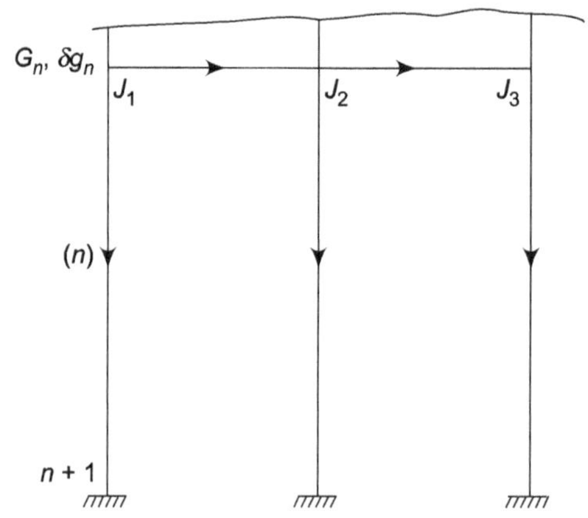

Fig. 9.24 *Ground Floor Storey, n*

During the process of back substitution, any typical deformation δ_{Gi} is evaluated from the following relation

$$\delta_{Gi} = (K_{ff})_i^{-1}\left[\bar{P}_{Gi} - (K_{fb})_i\, \delta_{G(i+1)}\right] \quad \ldots(9.17)$$

$\delta_{G(i+1)}$ in the above relation is the back substitution. Following the procedure of the back substitution, all the deformations vectors are computed in the sequence.

$\delta_{Gn}, \delta_{G(n-1)}, --, \delta_{Gi} --, \delta_{G2}, \delta_{G1}$. Based upon compatibility condition, Eq. (9.3), $[\delta_m = A_\delta]$ system, member deformations δ_m are evaluated. Finally the member end forces are obtained from the relation Eq. (9.6).

9.8.4 General Remarks about Tall Building Characteristics

The nature of recursion in the above formulation is suitable for coding a computer program. The normal procedure of inverting a complete matrix will severely limit the analysis even for a moderately tall building, say 13 storeys.

For 13 storeys and 6 column lines, No. of joints = 78, and stiffness matrix will be of size (224 × 224) and storing of all the elements 224 × 224 is for a building which is moderately and without lift shafts, etc., where finite elements are extensively used. The data input, using the recursion method when programmed may consist of

1. Number of storeys.
2. Column lines.
3. Bay width listed in the sequence "left to right".
4. Column heights of different storeys, listed "top to bottom".
5. Moments of inertia in the sequence "left to right" and "top to bottom" for girders and columns.
6. Cross-sectional areas for columns.
7. Equivalent or generalized joint and floor loading (These may be obtained by separate sub routine to convert given loading to generalized loading). Loading cases.
8. Modulus of elasticity.

The output data may consist of

1. System deformations.
2. Girder moments and shears.
3. Column-axial forces moments and shears.

Using the recursion procedure the computer time varies only as a linear function of the number of storeys of a structure and whereas with the normal procedure of a complete stiffness matrix the computer time is proportional to the cube of the number of unknowns.

If the loading cases to be considered are large, the recursion procedure is superior to the iteration method discussed in Chapter 3.

The sequence of programming is suggested in the following:

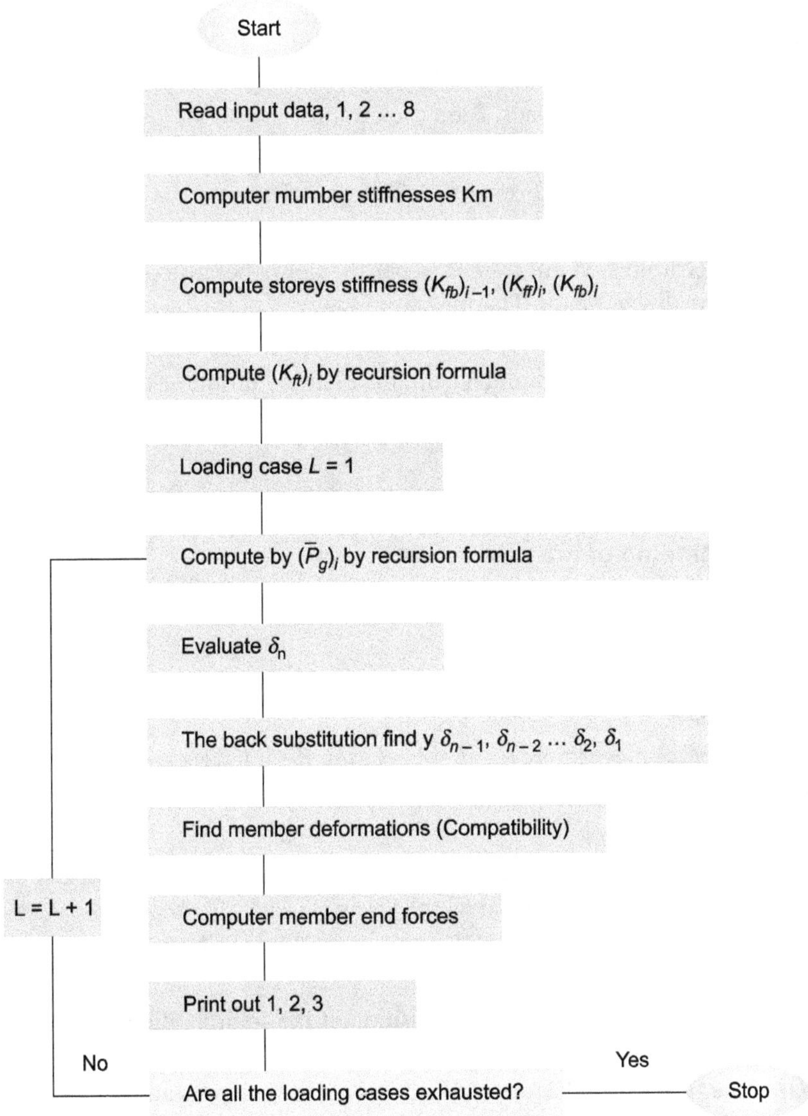

The above flow chart for static analysis of a framed building. The important omissions are:

(a) Combination of more than one frame. The aspect is conceptually explained in the following and sharing of lateral loads between them.
(b) Consideration of in-plane rigidity of the floor slabs for sharing of the total earthquake-lateral load between the frames.
(c) Evaluation of not only the lateral displacement of each floor but also the rotation of slabs.
(d) Effect of torsion stiffness of the column is neglected.

9.8.5 Distribution of Lateral Earthquake Load and Its Distribution between Frames

Symmetric positioning of frames. From the codal provisions, we know $P_T = \{P_{T1}, P_{T2}, \ldots, P_{Tn}\}$. The frame analysis, presently assumed, frame load for frame (i) as given

Whereas from the known, P_T,

$$P_T = P_{(1)}, P_{(2)} + \ldots P_{(s)} = \{P_i\}$$

Our objective is to find the lateral loads, P_i on each frame.

From the preceding analysis in the relation,

$P_{(i)} = K_{(i)}\ \delta_h$, Horizontal deflection δ_h ignores axial deformations of the girders.

We know $K_{(i)}$. It is required to find $P_{(i)}$ from $P_{(T)}$

In the case of symmetrically laid out frames as per the figure because of the implane rigidity of the floor slabs, δ_h is same for all the frames. Therefore,

$$P_T = (\Sigma K_i)\delta_h = K_T\ \delta_h, \text{ where } K_T = \Sigma K_{(i)}$$

P_T being known and $K = \Sigma K_{(i)}$ is computed from the member properties, we find

$$\delta_h = (K_T)^{-1}\ P_T$$

And the individual frame loading, $P_{(i)} = K_{(i)}\ \delta_h$

The member forces are obtained from the member stiffnesses.

Unsymmetrical Positioning of Frames

Referring to Fig. 9.25, δ_h is known as drift. In the case of the unsymmetrical laid out frames, we find out,

CG of floor slab, and

CF, center of stiffness.

From the preceding section, we know $K_{(i)}$ in the relation $P_{(i)} = K_{(i)}\ (\delta_h)_i$

By inverting $K_{(i)}$ we get the flexibility matrix

$$F_{(i)} = [K_{(i)}]^{-1}$$

P_{TX}, P_{TY}, and $P_{T(m)}$ are known from CG and CF. A reference point may be CG or any other convenient point, whose x and z coordinates are same at all the levels.

The Problem is thus made amenable to the force method. The compatibility conditions are provided by δ_{hx}, δ_{hy} and θ of the individual slab.

It is always advisable to have symmetrical disposition of the column and shear walls and thus avoid twist.

We recall that in the "CWK" model torsional stiffness of the columns and shear walls is neglected.

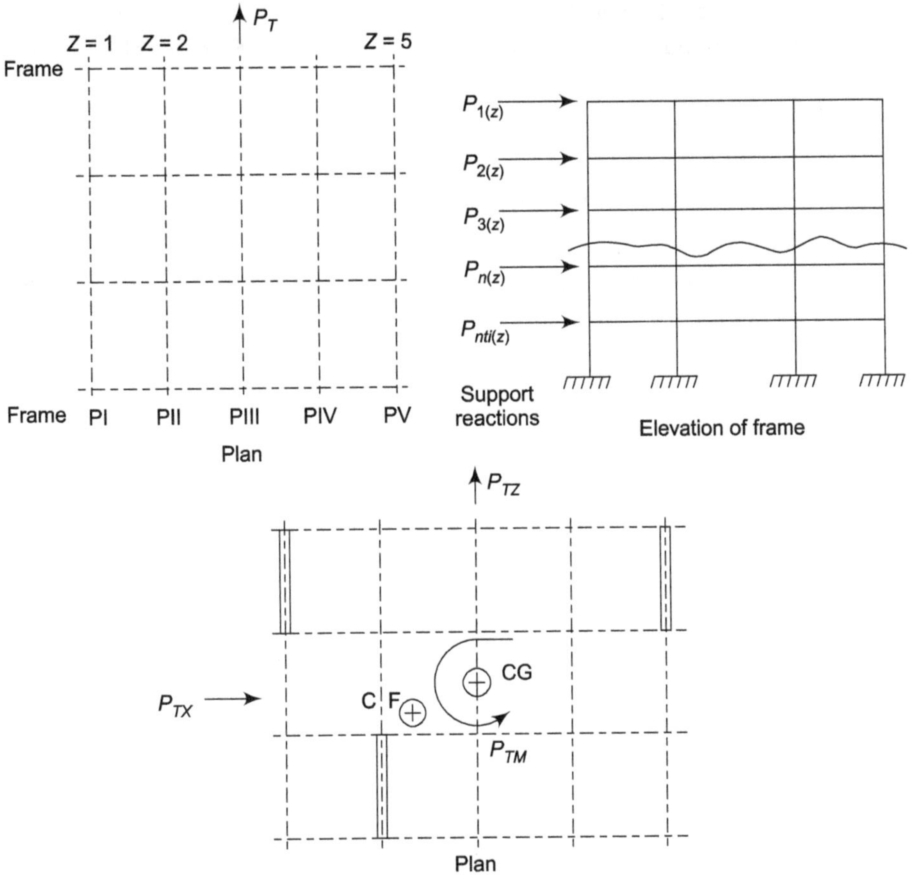

Fig. 9.25 *Un-symmetrical Layout of Frames*

9.9 THREE DIMENSIONAL FRAME BUILDING—WN MODEL

Refer Fig. 9.26.

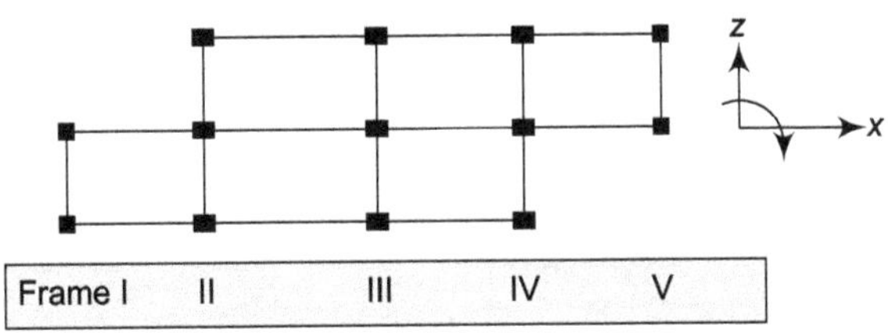

Fig. 9.26 *Plan of a 3-D Building*

If the column lines, NC remain the same for its entire height, number of storeys, *ns*. The building is unsymmetrical in both the directions. For any lateral load applied, the degrees of freedom are three per floor.

Slab in-plane rigid body displacements = 3,

and excluding the the column end deformations δx, δz, and twist θ, out of plane deformations are δy, θx, and θz.

The storey stiffness,

$$\begin{Bmatrix} p_f \\ p_b \end{Bmatrix} = \begin{Bmatrix} K_{ff} \; k_{fb} \\ K_{bf} \; k_{bb} \end{Bmatrix} \begin{Bmatrix} d_f \\ d_b \end{Bmatrix} \begin{Bmatrix} 1 \\ 2 \end{Bmatrix} \begin{matrix} f \text{ floor} \\ b \text{ bottom} \end{matrix}$$

The size of each sub matrix = 3 times NC + 3

Being a tier building, it results in a tri-diagonal matrix of total size of the global matrix is given by:

$$ns\,(3 \times NC + 3)$$

Numerical algorithm given by the authors, inversions are limited to the size of the unknowns per floor line and it sifts in-plane rigid body displacements from the flexural deformations (two rotations and one vertical displacement per column line). The reader is referred to the paper by the authors, Weaver, Nelson and Manning for the details.

Development of Stiffness Matrix of Tall Buildings WN Model[10]

In the case of **WN model**, the number of unknowns is equal to:

Three rigid body in plane displacements per slab + Three times number of column lines = (3 + 3 * NC) *NS. The **WN model** considers columns to be torsionally stiff. A comprehensive flow chart for computer analysis is given[1].

9.10 SHEAR WALL AND STOREY STIFFNESS OF TALL BUILDING SYSTEM

The typical alternatives are sketched, Fig. 9.27

Shear wall is treated as a wide Column. The wide column has the large depth in the plane of bending, Fig. 9.7

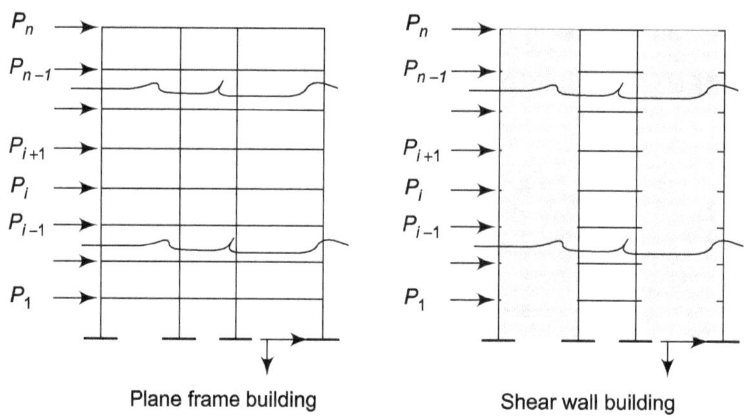

Fig. 9.27 *Alternatives, Plane Frame and Shear*

[10] Weaver, W. Jr, and Nelson, M. F., "*Dynamics of Tier Buildings*," Journal of Structural Division, ASCE, Vol. 94, No. EM6, December 1968, pp 1455–1474.

Member stiffness of a shear wall as a wide column, and storey stiffness

The member axes of a typical shear wall or wide column member *i*, are shown in the Fig. 9.7. It may be verified that the member stiffness matrix, excluding the rigid body displacements, and accounting for the shear deformations, is

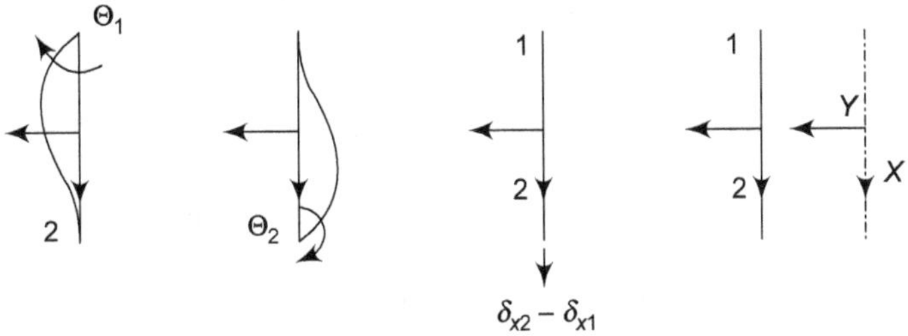

Fig. 9.28 *Independent Member Deformations, Excluding Rigid Body Displacements*

$$k_m = \begin{pmatrix} (2EI/L)\dfrac{2+\beta}{1+2\beta} & (2EI/L)\dfrac{1-\beta}{1+2\beta} & 0 \\ 0 & (2EI/L)\dfrac{2+\beta}{1+2\beta} & 0 \\ 0 & 0 & AE/L \end{pmatrix}$$

where $$\beta = \frac{6EI}{L^2 \bar{A}G}$$

$\bar{A}$ is the shear area of the cross section

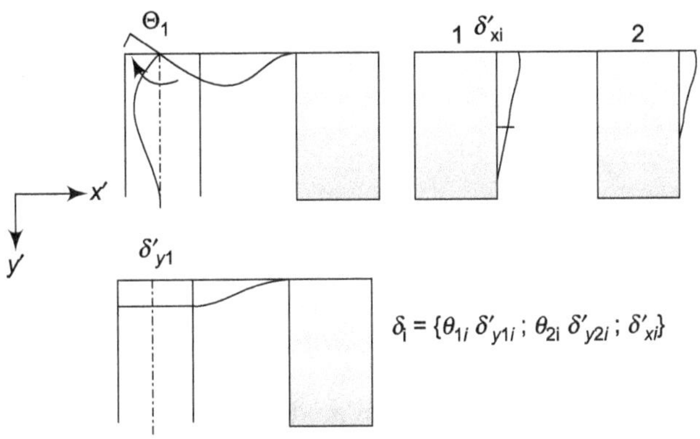

Fig. 9.29 *Deformations of the Nodes in a Floor Line*

Floor deformations δ'_{y2i} and $\theta_{2i,}$ may be drawn similarly.

Degrees of freedom per floor line: Ref Fig. 9.6. The skeletal frame has number of column lines, $NC = 4$. Each node undergoes rotation and vertical displacement. The girder line being inextensible; all the joints in a floor undergo one lateral displacement.

The degrees of freedom per floor line = 2 * NC + 1 = 9.

For the corresponding shear wall frame, the number of column lines is reduced to two only. The shear wall building has the degrees of freedom = 5. Flexural deformations of the girder linking, the two shear walls takes place within the free length. The girders are rigidly attached to the shear walls. And the hypothesis of plane section remaining plane after deformations is assumed to be valid for the wide columns also. The end of the girder undergoes the same rotation as that of the column at the node, as shown in the Fig. 9.29.

These properties are utilized for writing the connection matrix, **A**.

9.11 GENERAL CASE OF STOREY STIFFNESS: SHEAR WALL CUM FRAME

We considered the cases of either a single frame building or a single shear wall cum frame building. Let us now examine shear wall cum framed plane building, Fig. 9.9. We continue with the plane shear wall cum frame. The earthquake codes give the lateral loading for the entire building. If the entire shear wall cum frame system has identical properties, the applied lateral load on the total system is shared equally. Let us consider one such shear wall cum frame building. Henceforth a single shear wall cum frame, will be referred as SF frame. The analysis as presented in Article 9.3.2, even though the system or global stiffness matrix is tridiagonal it has a mix of rotations equal to the number of column lines (NC), vertical displacements are also equal to NC, and one lateral displacement per floor. All the unknowns, and their number is equal to (2* NC +1)* NS. The theory is covered separately, Article 9.4. The paper gives an algorithm to sift the lateral displacement from the other displacement.

The earthquake loading is a mass loading. And the floor slabs being mass intensive, the loads are at the slab levels. The 'horizontal ground motion' is our main concern, and therefore, the horizontal loading'.

If the building system is symmetrical, but not identical frames, one of the recommendations by the authors is to arrange them side by side connected by rigid links. The approach increases the number of unknowns significantly. Their alternative algorithm is good but the CWK model does not consider torsion in columns. From these considerations, it is felt that if 3-D analysis is required, the Weaver-Nelson (WN) model may be a better option. Let us continue with the CWK model for a single SF frame, neglecting torsional as well as out of plane stiffness.

Examine, Fig. 9.30. It has three column lines including one 'wide–column' line of a twenty storey building.

There being four such frames, the earthquake loads computed based on the code provisions, are uniformly reduced by a factor ¼.

9.12 SIFTING OF $\bar{\delta}_x$ FROM θ AND DYNAMIC ANALYSIS

Often it is required to find natural frequencies and modes of vibration of the structural systems. In the case of grids loaded transversely vertical vibrations are important. In the case of multistorey buildings, because of negligible vertical deformations as compared to horizontal sway, horizontal vibrations are crucial. The slabs at floor levels of multistorey frames justify lumped masses for horizontal vibrations. This assumption of lumped masses is not quite valid in the case of transversely loaded grid systems where the slab-mass is distributed over the entire area of the grid system vibrating in vertical direction. The mass and moments of inertia are required for finding inertial forces and inertial torques respectively. For writing equilibrium equations, the inertial forces are considered using D' Alembert's principle. In the following discussion for tall buildings, inertial forces are considered for the mode shapes corresponding to the natural frequencies, and these are used for determining earthquake loads at different floor levels. The dynamic response is presented separately.

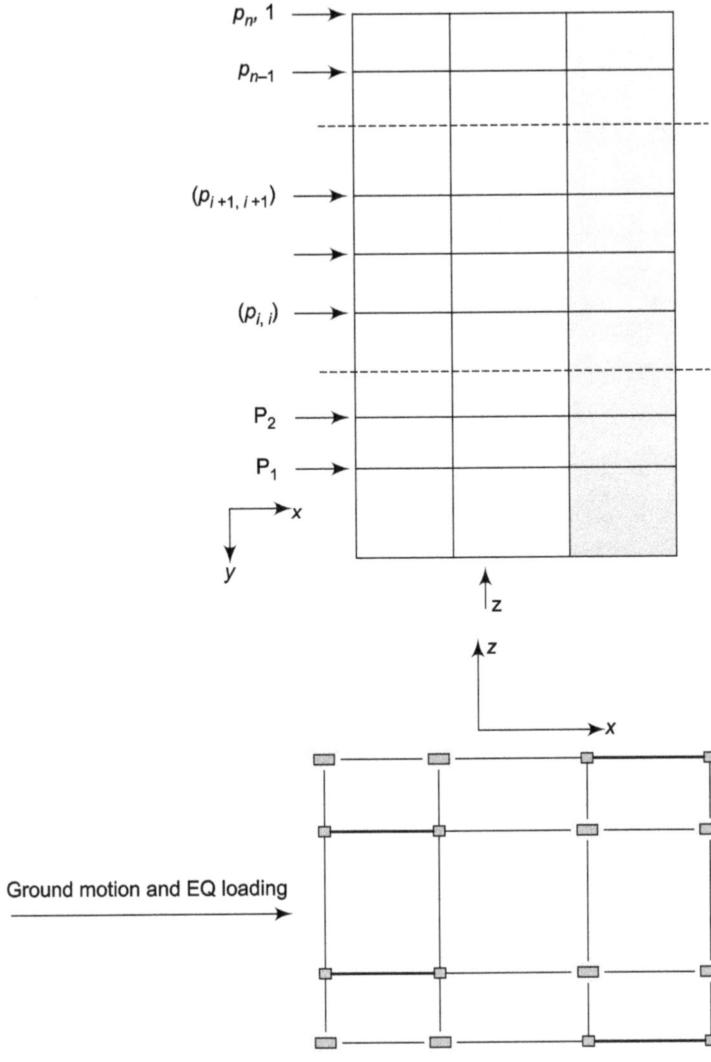

Fig. 9.30 *Shear Wall cum Frame*

Consider the earthquake load in x-direction. The SF frames are structurally identical. The Eq. (9.8) is an assembled stiffness matrix of a multistorey frame. The load and deformation sub-matrices (P_G and δ_G) of individual floor lines include moments and rotations together with forces and displacements. As we are considering horizontal lateral vibrations only, a stiffness matrix connecting lateral forces and corresponding displacements is to be obtained first. To achieve this, premultiply either side of the matrix equation, Eq. (9.8), by a suitable matrix D such that the vertical force and moment elements are separated from the lateral force elements acting at nodes. A typical D matrix for a three column line (Fig. 9.2) multistory frame is explained in the following.

$$\bar{P} = \{\bar{m}, \bar{P}_x\},$$

where $$\bar{m} = \left\{\left((p'_y)_{j1}, (p'_y)_{j2}, (p'_y)_{j3}, m_{j1}, m_{j2}, m_{j3}\right)_G, m_{G2} \ldots\right\}$$

$$\bar{p}_x = \left\{\left((p'_x)_{G1}, (p'_x)_{G2}, (p'_x)_{G3}\right), \ldots\right\}$$

Similarly it can be shown that $d = D^t \bar{d}$

Where $$\bar{d} = \{(\theta)_{G1,} (\theta)_{G2} \ldots \vdots (\delta x)_{G1,} (\delta x')_{G2} \ldots\} = \{\bar{\theta}, \bar{\delta}_x\}$$

$$\bar{\theta} = \{(\theta)_{G1}, (\theta)_{G2} \ldots\},\ \bar{\delta}_x = \{(\delta'_x)_{G1}, (\delta'_x)_{G2} \ldots\}$$

$$\theta_G = \{(\delta'_y)_{J1}, (\delta'_y)_{J2}, (\delta'_y)_{J3}, \theta_{J1}, \theta_{J2}, \theta_{J3}\}$$

Vertical displacements are combined with the rotations.

The matrix equation, Eqs (9.8), (9.18) may be re-written as

$$DP = (D\,K\,D^t)\,\bar{d}$$

Or

$$\bar{P} = K'\bar{d},$$

where $$K' = D\,K\,D^t \qquad \ldots(9.19)$$

When the matrix equation, Eq. (9.19) is suitably partitioned,

We get

$$\begin{bmatrix} \bar{m} \\ \cdots \\ \bar{p}_x \end{bmatrix} = \begin{bmatrix} K'_{\theta\theta} & \vdots & K'_{\theta x} \\ \cdots & & \cdots \\ (K'_{\theta x})^t & & K'_{xx} \end{bmatrix} \begin{bmatrix} \bar{\theta} \\ \cdots \\ \bar{\delta}_x \end{bmatrix} \qquad \ldots(9.20)$$

For a frame of, say, 10 storeys and 3 nodes or column lines, the size of the sub matrices of Eq. (9.20) are given in Fig. 9.31

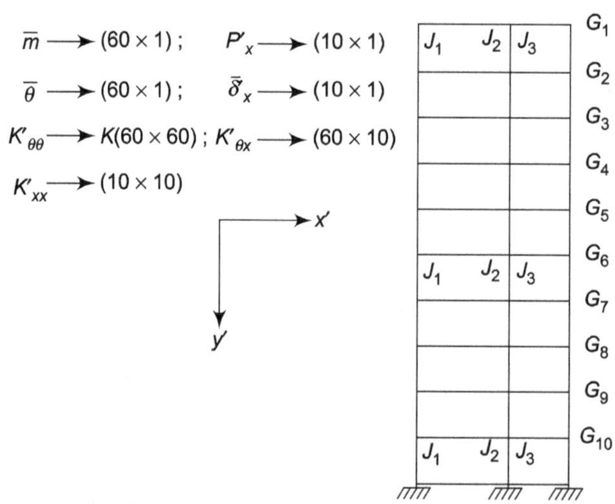

Fig. 9.31 *A typical Ten Storey Frame*

Expanding the Eq. 9.20

$$\bar{m} = K'_{\theta\theta}\,\bar{\theta} + K'_{\theta x}\,\bar{\delta}_x\,,\; \bar{p}_x = (K'_{\theta x})^t\,\bar{\theta} + K'_{xx}\,\bar{\delta}_x\,,$$

From the first equation, $\bar{\theta} = (K'_{\theta\theta})^{-1}\,(\bar{m} - K'_{\theta x}\,\bar{\delta}_x)$ when substituted

In the second equation

$$\bar{P}_x = \bar{K}\,\bar{\delta}_x \qquad \ldots(9.21)$$

Where

$$\bar{P}_x = \bar{P}_x - (K'_{\theta x})^t\,(K'_{\theta\theta})^{-1}\,\bar{m} \text{ and}$$

$$\bar{K} = k_{xx} - (K'_{\theta x})^t\,(K'_{\theta\theta})^{-1}\,K'_{\theta x}$$

For natural frequencies of the structural system, we write

$$M\ddot{\bar{\delta}}_x + \bar{K}\,\bar{\delta}_x = 0$$

Where M is a mass matrix. If ω is the circular frequency

$$(-M\omega^2 + \bar{K})\,\bar{\delta}_x = 0 \qquad \ldots(9.22)$$

Putting $\lambda = \dfrac{1}{\omega^2}$

$$(M - \lambda\,\bar{K})\,\delta_x = 0 \qquad \ldots(9.23)$$

In the Eq. (9.23), $(M - \lambda\,\bar{K})$ may be transformed into a form $(A - \lambda I)$ and its determinant when equated to zero, Eq. (9.24), gives the eigen values, λ

$$|(A - \lambda I)| = 0 \qquad \ldots(9.24)$$

From the values of λ, the circular frequencies, ω and the modes of vibration $(\bar{\delta}_x)$ are evaluated.

9.13 ALTERNATIVE APPROACH BY FLEXIBILITY MATRIX, F

We do not invert $\bar{K}$; instead we analyze for the "unit loading" and obtain the flexibility matrix, F. The use of the approach has been demonstrated through illustrations elsewhere.

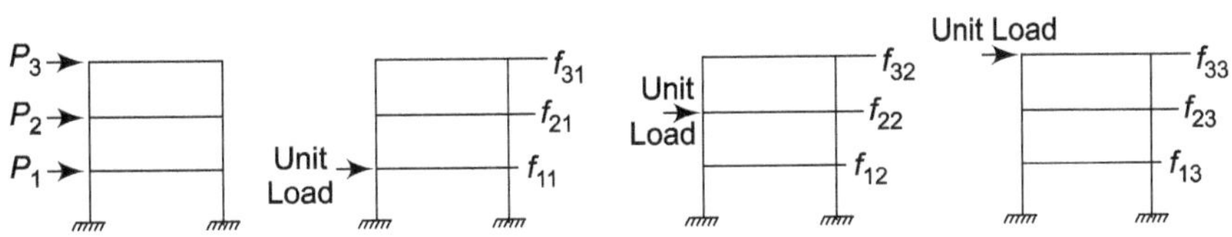

Fig. 9.32 *Unit Load Approach for obtaining Flexibility Matrix for for Flexibility Matrix*

For natural modes of vibration, flexibility matrix is preferred. For the above frame,

$$F = \begin{bmatrix} f_{11} & f_{12} & f_{13} \\ f_{21} & f_{22} & f_{23} \\ f_{31} & f_{32} & f_{33} \end{bmatrix}$$

The development given in the following is applicable to 3D buildings also.

The SF frame is analyzed for one load at a time of unit value. The deflections, say, for P_1 equal to unity give the first column of the flexibility matrix **F**. The analysis is performed for each individual loading case separately, their number equal to the degrees of freedom of the lateral displacements. The assemblage of the n number of column-vectors fully defines the flexibility matrix, **F** matrix. Except the floor-rigid–body displacements all other degrees of freedom are eliminated.

Thus, **d** = **F** P. The inversion of **F** gives stiffness, $\mathbf{K_h}$ of the SF frame involving lateral forces and corresponding lateral displacements.

$$\mathbf{P_h} = \mathbf{K_h}\,\boldsymbol{\delta}_\mathbf{h}$$

The **F** matrix will be used for the frequencies and the mode shapes.

Flexibility Matrix-Illustration

The software, MATLAB has been used for the standard operations of Eigen values and Eigen vectors, which are our Frequencies and corresponding Model shapes respectively.

Example of a five storey SF frame is taken from the old manual of Staadpro:
Slabs and shear walls 6 in thick,
Beams and columns: 18 in × 12 in,
The columns include the bulbs at the edges of shear wall, and
The girders include the girders at the junction of slab with wall at each floor.
Material is concrete with poison's ratio = 0.16
Storey height = 15 ft.
Column and shear wall bases fixed.
All girders 20 ft long.

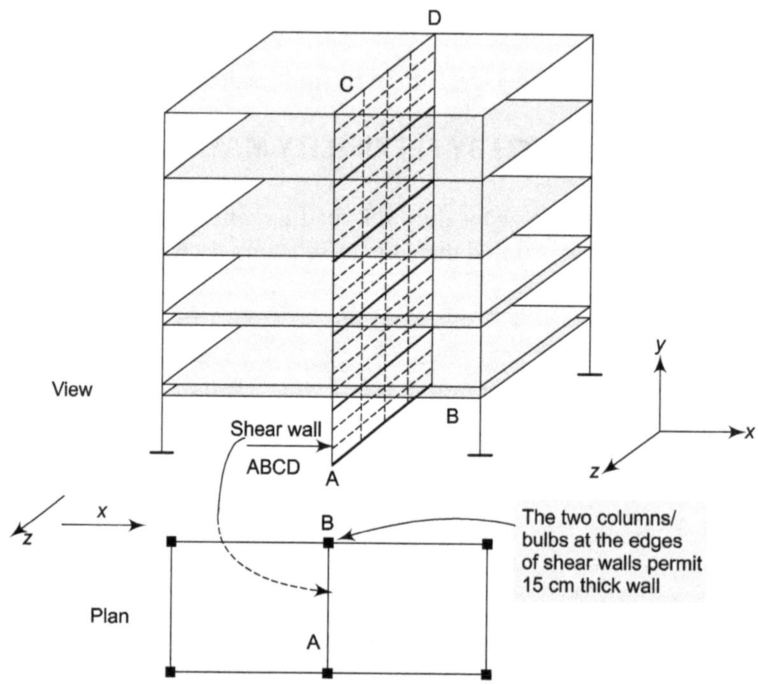

Fig. 9.33 *Statement of Problem*

Mass calculations should be as representative as possible. Earthquakes are caused by the ground motion over which we do not have any control but the more important contributing factor, the mass, should be minimized. If we can design light structures, it must be a pleasure watching an earthquake taking place. No wonder earthquake loading is labeled as mass loading.

For the example chosen, Fig. 9.32, the masses are contributed by:

Slab:	26.77 t
Walls (12 cm), excluding shear wall:	26.34
Beams:	14.28
Columns:	6.23
Finishing:	4.46
Shear wall	6.59
Live load:	3.72
	88.18 t

Roof: It does not include Live Load and Brick wall; Shear wall and columns contribute for half storey height; less LL: Net roof load = 64.98 2 t

Discretizing the masses at floor level,

$$\text{Mass matrix (diagonal matrix), } \mathbf{m} = \frac{1}{g}\lceil 88.18, 88.18, 88.18, 88.18, 64.98 \rfloor$$

Vibrations in x-direction:

The building being more flexible in **x-direction**, it has larger period of vibration.

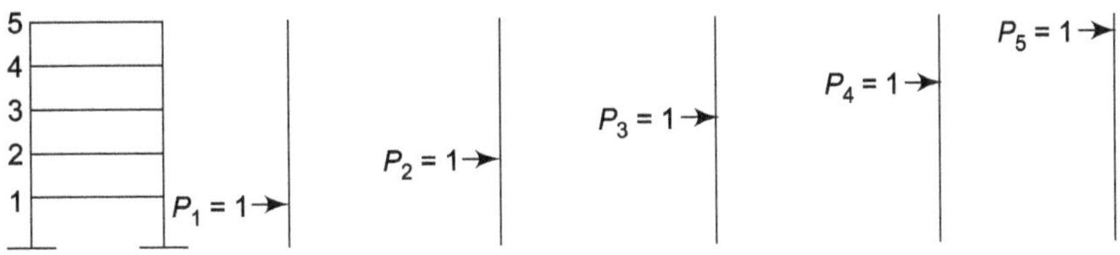

The deflection due to the above loading, constitute the flexibility matrix, **F**

Deflection in inches:

$$\mathbf{F} = \begin{array}{ccccc} \overrightarrow{p_1} & \overrightarrow{p_2} & \overrightarrow{p_3} & \overrightarrow{p_4} & \overrightarrow{p_5} \\ p_{1=1} & p_{2=1} & p_{3=1} & p_{4=1} & p_{5=1} \end{array}$$

$$F = \begin{pmatrix} .00601 & .00762 & 0.00782 & .00786 & .00789 \\ .00762 & .1678 & 0.01878 & .01909 & .01919 \\ .00782 & .01878 & 0.02840 & .03054 & .03095 \\ .00786 & .01909 & 0.03054 & .04034 & .04265 \\ .00789 & .01919 & 0.03095 & .04265 & .05288 \end{pmatrix}$$

Following the above procedure, except the rigid body floor displacements all other degrees of freedom are eliminated without compromising accuracy.

9.14 FLEXIBILITY MATRIX, F-EIGEN VALUES AND MODE SHAPES

For a single degree of freedom system, we had developed the equation of motion:

$$M\ddot{\mathbf{x}} + C\dot{\mathbf{x}} + K\mathbf{x} = p(t) \quad \text{(i)}$$

It shown that $\omega_D = \omega\sqrt{(1-\xi^2)}$

For the range of ξ values for the engineering materials (a maximum of 0.20), it is << 1,

$$\boxed{\omega_D \approx \omega}$$

Damped frequency ≃ frequency without damping

We are interested in frequencies and mode shapes for the case of horizontal ground motion. Natural period and mode shapes are found for free vibrations without damping. The eq. (i) is simplified to:

$$M\ddot{x}_h + K x_h = 0 \quad \text{(ii)}$$

Where K is obtained by inverting F, $K = F^{-1}$

Premultiplying the Eq. (iii) by $K^{-1} = F$

$$F M \ddot{x}_h + \text{I} = 0, \text{ where I is unit matrix} \quad \text{(iii)}$$

The vector x_h consists of horizontal displacements only.

The example above was used to obtain F matrix based on computer analysis, and it was performed for all the degrees of freedom. For the example above, the horizontal deflections (in inches), defines the flexibility matrix, F.

Dropping subscript for convenience of writing, equation is rewritten in the following.

$$M\ddot{x} + Kx = 0$$

Writing the solution

$$x = \bar{x} \sin \omega t, \text{ we get}$$

$$\frac{1}{12g}[FM] - 1/\omega^2 I = 0,$$

$$A - \frac{12\,g}{\omega_2} I = 0$$

or

$$A - \lambda I = 0, \quad \text{(iv)}$$

where $\lambda = \dfrac{12\,g}{\omega_2}$ and $A = [\text{F M}]$

It is a standard form for evaluating the Eigen values and mode shapes. Using MATLAB[11],

(1) Eig(a) = {9.8600 1.0269 0.3400 0.1683 0.1120},

(2) And corresponding mode shapes are:

Φ_1	Φ_2	Φ_3	Φ_4	Φ_5
0.1304	−0.3734	0.5457	−0.5747	0.3961
0.3030	−0.5888	0.2546	0.3594	−0.5530
0.4522	−0.3369	−0.5031	0.2211	0.5641
0.5579	0.1923	−0.2980	−0.5703	−0.4187
0.6127	0.6028	0.5436	0.4079	0.2091

[11] MATLAB is used for frequencies and mode shapes. STAADPRO is used for the analysis of the frame.

9.14.1 Modal Participation Factor (a Scalar)

The modal participation factor, $\Gamma_i = \dfrac{\Sigma W_i \Phi_i}{\Sigma W_i \Phi_i^2}$. Γ suggests participation from each mode. The relevance and derivation will be given separately.

Modal participation Factors are computed in the following table.

Floors from bottom	*Weight W_i*	*Mode coefficient, Φ_i*	$W_i \Phi_i$	$W_i \Phi_i^2$	$\Gamma_i \Phi_i$
1	88.18	.1304	11.4987	1.4994	.274
2	88.18	.303	26.7185	8.0957	.637
3	88.18	.4522	39.8750	18.0315	.950
4	88.18	.5579	49.1956	27.4462	1.173
5	64.98	.6128	39.8132	24.3936	1.288
			Σ = 167.1010	Σ = 79.4664	

$$\Gamma_1 = 167.1010/79.4664 = 2.1028$$

Similarly, other participation factors are computed.

$$\Gamma = 2.1028, -.7327, .4335, -.2803, .1414$$

9.14.2 Modal Contributions

The modal contributions are given by $\Gamma_i \{\Phi_i\}$.

It will be shown that the Participation factor $\Gamma_{i\,\text{giv}}$ is only an indicator of the modal participation of a given loading, whereas modal contribution is given by $\Gamma_i \{\Phi_i\}$.

The $\Gamma_i \Phi_i$ are tabulated in the following.

W_i	$\Gamma_1 \Phi_1$	$\Gamma_2 \Phi_2$	$\Gamma_3 \Phi_3$	$\Gamma_4 \Phi_4$	$\Gamma_5 \Phi_5$	
64.98	1.2884	-.4417	.2357	-.11434	.0298	Σ = *
88.18	1.1732	-.1409	-.1292	.15987	-.0596	1.003
88.18	0.9509	.2468	-.2181	-.06198	.0803	.998
88.18	0.6371	.4314	.1104	-.10075	-.0787	.998
88.18	0.2742	.2736	.2366	.1611	.0564	1.002

Σ = * The summations should add to 1.000. The variations from 1.000 are acceptable.

Some national codes recommend using the first three modes to represent any given loading. In the following we notice the contributions from the first three modes may not be adequate. Let us consider truncation after the first three modes. $\Sigma \Gamma_i \Phi_i$, summation i =1 to 3

[1.0824, .9031 .9796 1.178 .7844]. There is a large deviation from unity in the last term. Let us see the improvement, if any, if the fourth mode is added. The contributions from: the first four modes are $\Sigma \Gamma_i \Phi_i$

[.968, 1.063, .918, 1.077, .946]. The values suggest use of the first four modal loading.

The computations for the evaluation of the time period, Table below.

$$\lambda = \frac{12g}{\omega_2}$$

Λ	ω^2	Ω	*Frequency, f*	*Period*
9.86	38.94	6.24	0.993	T_1 =1.01**
1.0269	373.94	19.34	3.08	0.32
0.3400	1129.41	33.61	5.34	0.186
0.1683	2281.64	47.77	7.60	0.131
0.1120	3428.57	58.55	9.32	0.107

Plot the first three mode shapes[12] in x direction: Fig. 9.34

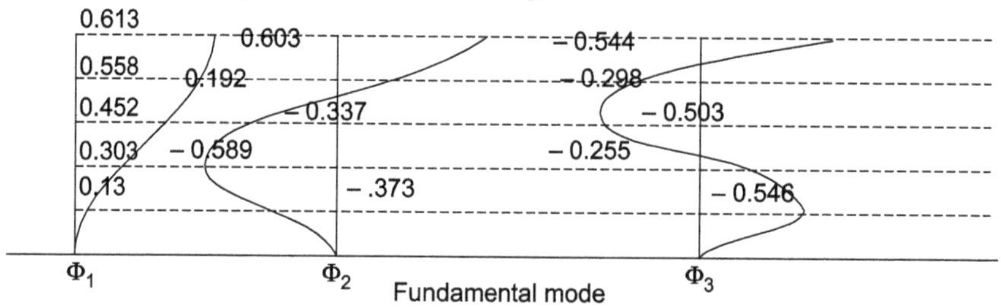

Fig. 9.34 *First Three Mode Shapes*

** Sears tower, 110 storeys, $T = 7.8$ sec.

Repeat the above analysis for the vibrations in the z-direction

Vibrations in z- direction:

$$\text{Flexibility matrix, } F = \begin{pmatrix} 0.00019 & .00026 & .00033 & .00039 & .00046 \\ * & .00064 & .00093 & .00122 & .00151 \\ * & * & .00168 & .00231 & .00295 \\ * & * & * & .00354 & .00466 \\ * & * & * & * & .00650 \end{pmatrix}$$

1. Eig(a) = {0.8951 0.0424 0.0101 0.0037 0.0046 }

   And the corresponding mode shapes are:

2. Mode shapes are:
$$\begin{pmatrix} -0.0631 & -0.3851 & 0.6044 & 0.4739 & 0.4604 \\ -0.1922 & -0.5719 & 0.2902 & -0.6614 & -0.2647 \\ -0.3595 & -0.4626 & -0.4356 & 0.5114 & -0.3832 \\ -0.5443 & 0.0008 & -0.3602 & -0.2580 & 0.6590 \\ -0.7305 & 0.5574 & 0.4807 & 0.0999 & -0.3699 \end{pmatrix}$$

3. Modal participation factor, $\Gamma_i = \dfrac{\Sigma W_i \phi_i}{\Sigma W_i \phi_i^2}$.

   The reader may like to repeat the calculations for vibrations in z-direction.

4. Modal contributions: The modal contributions are given by $\Gamma_i \{\Phi_i\}$

   The reader may like to check the adequacy of the first three modes based on the modal contributions.

[12] No change of sign is the 'Fundamental mode' only one 'segment'. Every change of sign gives rise to additional 'segment'. The two segments' of ϕ_2 is the second mode, and so on for the higher modes.

5. Compute the time period.
   Plot of mode shapes in z direction:
   In the following the numerical values are plotted floor wise, Fig. 9.35.

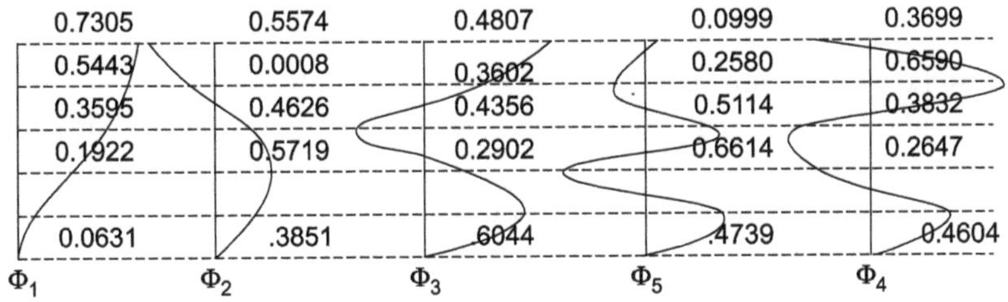

Fig. 9.35 *First Five Mode Shapes*

The **'MATLAB'** computed the fifth mode before the fourth mode. The observation is based on the last mode shape consists of four 'segments'; therefore it is labeled as fourth mode. The mode shape labeled as 5 consists of five 'segments'.

Being a small size problem, no attempt was made to achieve a symmetrical matrix for computing the Eigen values and mode shapes. Instead we worked with the unsymmetrical 'A' matrix.

Some observations: Coupling of frequencies

- It is not desirable to design a building with two frequencies almost equal. The closeness of frequencies is known as '*coupling of frequencies*'. One of the reasons for the failure of Tacoma Narrow suspension bridge is the coupling of frequencies. Our structure does not have any coupling effect.
- Mode shapes conform to the expectations.

9.15 MISCELLANEOUS ASPECTS OF DYNAMICS

9.15.1 Symmetrical 'A' Matrix for Eigen Values

A matrix was used earlier, in the relation, $|A - \lambda I| = 0$

We rewrite the Eq. (iii) in the following

$\mathbf{M}\,\ddot{\mathbf{x}} + \mathbf{K}\,\mathbf{x} = 0$, we had substituted $\mathbf{x}_h$ by x.

Using F matrix:

$$\mathrm{F\,M}\,\ddot{\mathbf{x}} + \mathbf{x} = 0$$

Substituting $\mathbf{x} = \bar{x}\,\sin \omega t$

$$-c\,\mathrm{F}\,M\bar{x} + \bar{x} = 0 \text{ or}$$

Substituting $M = M^{1/2}\,M^{1/2}$

If M is a diagonal matrix, $M^{1/2}$ is obtained simply by taking the square roots of the diagonal scalar elements. The idealization of discredited masses at the floor levels gives rise to a diagonal matrix.

$$-\omega^2\,F\,M^{1/2}\,M^{1/2}\,\bar{x} + \bar{x} = 0$$

Let $M^{1/2}\,\bar{x} = \mathbf{y}$

$-\omega^2\,\mathrm{F}\,M^{1/2}\mathbf{y} + M^{-1/2}\,\mathbf{y} = 0$, pre multiplying by $M^{1/2}$, we get

$$-\omega^2 A + \mathbf{y} = 0$$

Where $A = M^{1/2}\,\mathbf{F}\,M^{-1/2}$

For a non-trivial solution, the characteristic equation is:

$$|A - \lambda I| = 0, \text{ where } \lambda = 1/\,\omega^2 \quad \text{(v)}$$

Comparing with Eq. (v) , the above Eq. (v') has achieved two objectives,

1. From computational economy, the algorithm of evaluating the eigen values for symmetrical A matrices are less time consuming, and
2. Because the algorithm starts with the smallest λ value, we evaluate the values of periods commencing with the largest.
3. We had seen in the example above one could truncate without having to consider all the mode shapes.

Using K matrix , we get the characteristic equation:

$$\text{I}\,|A - \omega_2 I| = 0,$$

Where $\text{A} = M^{-1/2}\,\mathbf{K}\,M^{-1/2}$

The disadvantage of the approach is:

- An additional step of inverting F,
- Computations started with the smallest period, which is not our objective.

9.15.2 Orthogonality of the Mode Shapes

We continue to ignore damping because the periods and corresponding mode shapes are not affected by damping. For the case of free vibrations and say, harmonic 'n', substituting $\Phi_n\, x_n$ for $\mathbf{x}$ and ω_n for ω, we get

$$-\,\omega_2 M\,\Phi_n\, x_n + K\,\Phi_n\, x_n = 0, \text{ or } K\,\Phi_n = \omega^2{}_n M\,\Phi_n$$

Pre-multiplying with $\Phi^t{}_r$

$$\Phi^t{}_r K\,\Phi_n = \omega^2{}_n \Phi^t{}_r M\,\Phi_n \quad \text{(a)}$$

Similarly it can be shown by switching dummy r and dummy n

$$\Phi^t{}_n K\,\Phi_r = \omega^2{}_r \Phi^t{}_n M\,\Phi_r \quad \text{(b)}$$

Subtracting (b) from (a)

$$(\omega_n^2 - \omega_r^2)\,\Phi_n^t\ M\,\Phi r = 0 \quad \text{(c)}$$

Here we have used the property that for the symmetrical matrices,

$$\Phi^t{}_r M\,\Phi_n = \Phi^t{}_n M\,\Phi_r$$

From Eq. (c), when $n = r$, the non-zero value of $\Phi_r^t\ M\,\Phi_r = m_r$ (i)

And when $n \neq r$, $\Phi_n^t\ M\,\Phi_r = 0$ (ia)

From Eq. (a) $\Phi_r^t\ K\,\Phi_r = \omega_n^2\ m_r$, or

$$\Phi_n^t K\,\Phi_r = k_r = \omega_n^2\ m_r \quad \text{(ii)}$$

The 'm_r' and 'k_r' are the modal mass and the modal stiffness respectively.

The Eqs (i) and (ii) define orthogonality of two shape functions, Φ_r and Φ_n

We use this property of orthogonality extensively in structural engineering. It is discussed with reference to the trigonometric functions.

9.16 HOUSNER'S AVERAGE SPECTRA AND EARTHQUAKE CODAL LOADS

We had already seen while discussing the effect of ground motion, that the equation of motion is:

$$M\ddot{x} + C\dot{x} + Kx = -M\ddot{x}g \tag{I}$$

The negative sign for the random ground motion my be dropped.

Assume **x** as superposition of mode shape with respective amplitudes. Treat each mode shape as independent

$$\mathbf{x} = \mathbf{\Phi}\,\mathbf{z}\sin\omega t, \text{ where } \mathbf{z} = \left\{\bar{x}_1\ \bar{x}_2\ \bar{x}_3\ \ldots\ \bar{x}_n\right\},$$

z is a column matrix, and $\bar{x}_1$ is a scalar.

$$\mathbf{\Phi} = [\mathbf{\Phi}_1\ \mathbf{\Phi}_2\ \mathbf{\Phi}_3 \ldots \mathbf{\Phi}_n].$$

We are familiar with $\mathbf{\Phi}_i$ corresponding to frequency ω_i

Housner suggested use of maximum values of $[\ddot{x}g]_{\max}$. He considered average of the four typical earthquakes, and it is known as acceleration spectra. The codes give reduced spectra, A_h based on the ductility of the structural members and uncertain return period. The reduced spectra is based on the partial damage but no collapse of the tall buildings. We may assume the masses to be discretized at the slab level, and for the horizontal vibrations, $\mathbf{x_h}$

The Eq. (I) is rewritten in the following:

$$\mathbf{\Phi}\ddot{z} + C\mathbf{\Phi}\dot{z} + \mathbf{K}\mathbf{\Phi}\mathbf{z} = \lceil M \rfloor A_h \begin{pmatrix}1\\1\\.\\1\end{pmatrix}$$

The common term sin ωt is cancelled from either side.

We pre-multiply by $\{\mathbf{\Phi}_r\}^t$ and taking advantage of the orthogonality of the shape functions and ignoring the damping term,

$$-\omega_r^2 m_r z_r + k_r z_r = \{\mathbf{\Phi}_r\}^t \begin{pmatrix}m_1\\m_2\\.\\m_3\end{pmatrix} A_h$$

Dividing by m_r, and substituting the value of $m_r = \mathbf{\Phi}_r^t\ M\,\mathbf{\Phi}_r$ in the above equation,

We get,

$$\ddot{z}_r + \omega_n^2 z_r = \frac{\sum_i m_i \mathbf{\Phi}_{ir}}{\sum_i m_i \mathbf{\Phi}_{ir}^2} A_h$$

The term $\Gamma_r = \dfrac{\sum_i m_i \mathbf{\Phi}_{ir}}{\sum_i m_i \mathbf{\Phi}_{ir}^2}$ is known as Modal Participation factor, P_r

'r' designates the rth mode.

And contribution by the mode r, is given by $\Gamma_r \mathbf{\Phi}_r$

9.17 APPROXIMATE METHODS—EIGEN VALUES AND MODE SHAPES

There are methods which are amenable to hand computations. These are iterative techniques. The methods assume the structure to be shear buildings (infinitely stiff rigid slab not only in its own plane but out of plane also). The methods are:

- Stodola method. An initial assumption is made of the vibration mode shape. It is adjusted iteratively. The frequency vibration is determined from the equation of motion.
- Holzer's method. An initial assumption of the vibration frequency. It is adjusted iteratively by satisfying the boundary condition. The Holzer method is synonymous with the 'transfer matrix method'[13] for shear buildings.

It is felt that the approximation may be used for four to six storeys. For the taller buildings, it becomes tedious and the accuracy is doubtful. In the present era of computers, these approximate methods are definitely out dated. Use of modal analysis is more appropriate for the tall buildings where superposition is admissible, that is, the assumption of linear elasticity is valid. The methods (or techniques), however are conceptual.

9.18 MONITORING OF TALL BULDINGS—FFT ANALYZER

Before we discuss maintenance aspect of tall buildings, it is required to learn about the FFT analyzer which is based on the dynamics in frequency domain. A prerequisite to the frequency domain is the response study in time domain using Fourier series.

9.18.1 Fourier Series in 'Spacial' and Time Coordinates

An approximated shape of the sum of first four terms is:

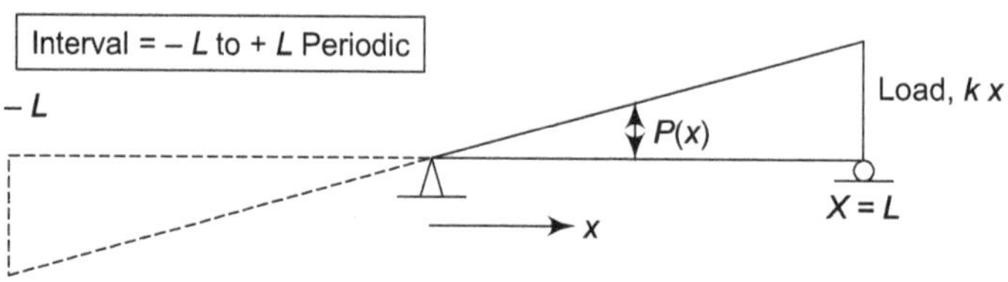

Represent $p(x)$ loading as trigonometric series:

$$p(x) = \frac{1}{2}a_0 + \sum_0^{\alpha}\left(a_n \cos\frac{n\pi x}{L} + b_n \sin\frac{n\pi x}{L}\right)$$

$$a_n = 0, \text{ and } b_n = (-1)^{n+1}$$

$$P(x) = 2k\left(\sin\frac{n\pi x}{L} - \frac{1}{2}\frac{2n\pi x}{L} + \frac{1}{3}\sin\frac{3n\pi x}{L} - \frac{1}{4}\sin\frac{4n\pi x}{L}\right)$$

An approximated shape of the sum of first four terms is:

[13] Pestel, Edward C., and Leckie, Frederick A., *Matrix methods in elastomechanics*, McGraw-Hill, New York.

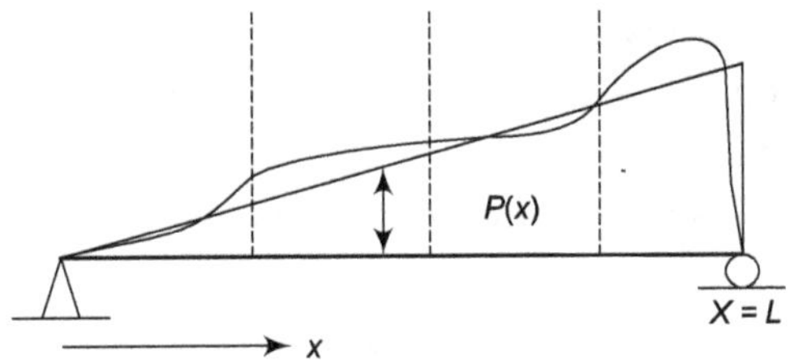

9.18.2 Fourier Series in Spacial Coordinates

Represent $p(x)$ loading as trigonometric series:

$$p(x) = \frac{1}{2}a_0 + \sum_0^{\alpha}(a_n \cos nx + b_n \sin nx)$$

$$a_n = \frac{1}{\pi}\int_{-\pi}^{\pi} f(x)\cos nx\, dx$$

$$b_n = \frac{1}{\pi}\int_{-\pi}^{\pi} f(x)\sin nx\, dx$$

While obtaining the above results, we used the property,

$$\left.\begin{aligned}
&\int_{-\pi}^{\pi} f(x)\sin mx \cos nx\, dx = 0, \text{ in general} && \text{(a)}\\
&\int_{-\pi}^{\pi} f(x)\cos mx \cos nx\, dx = 0, \text{ when } m \neq n, \text{ and} && \text{(b)}\\
&\int_{-\pi}^{\pi} \cos^2 nx\, dx = \pi && \text{(c)}
\end{aligned}\right\}$$

The Eqs (a), (b), and (c) define the orthogonality property of the trigonometric function.

9.18.3 Fourier Series and Periodic Loading

Let us consider a general loading. We assume the same to be periodic.

A case of general loading

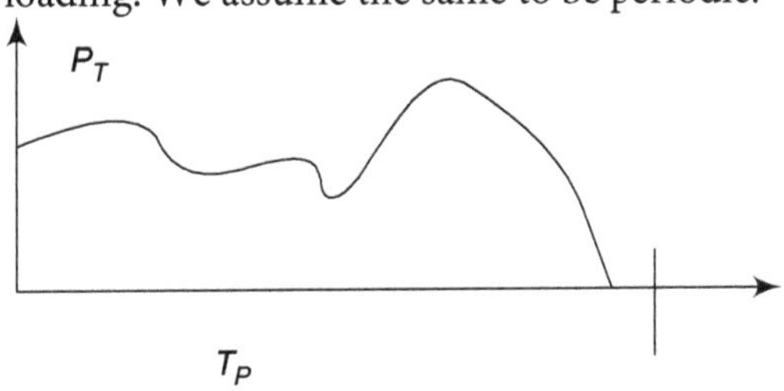

Introduce periodicity by assuming a time longer than the duration of the general loading.

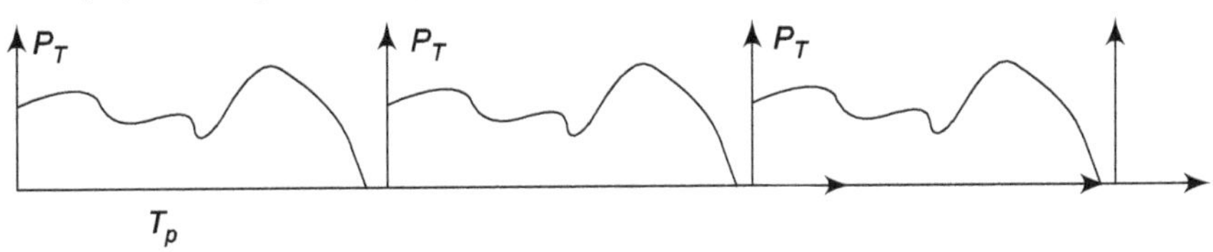

The subsequent steps are outlined in the following:

- Fourier series solution for the idealized periodic loading.
- Solution in exponent form.
- Dynamic steady state response in frequency domain in place of time domain.

Response:

Loading as a Fourier transform pair: $p(t) = \frac{1}{2\pi}\int_{-\infty}^{\infty} c(\bar{\omega}) \exp(i\bar{\omega}t)\, d\bar{\omega}$, where

$$c(\bar{\omega}) = \int_{-\infty}^{\infty} p(t) \exp(-i\bar{\omega}t)\, dt$$

Response is also obtained in the integral form. For details refer Clough and Penzien.

$$X(t) = \frac{1}{2\pi}\int_{-\infty}^{\infty} H(\bar{\omega})\, c(\bar{\omega}) \exp(-i\bar{\omega}t)\, d\bar{\omega}$$

There are two limitations for obtaining the dynamic response.

- Fourier transforms of general applied loadings are not available, and
- Evaluation of the resulting integral for obtaining the response $x(t)$ may be tedious, if not feasible.

  However numerical procedures have been evolved by the development of Discrete Fourier Transform (DFT) and Fast Fourier Transform (FFT) analysis.

  FFT Analyzers with the built in software are commercially available. Suffice it to say that with FFT algorithm, the Frequency Domain Approach has revolutionized the field of Applied Structural Dynamics.

9.19 MAINTENANCE OF TALL BUILDINGS

Maintenance Monitoring

1. Visual

Inspecting for Visual Defects-not possible to inspect the entire structure like bridges and tall buildings.

Absolutely essential but supplementary monitoring is necessary

2. Localized Recording

Discrete measurements at identified places for δ, ε cracks, static and fatigue.

Non-Destructive Testing (NDT).

Knowledge is necessary regarding precise locations where failure occurs.

3. Global Signatures

Integrity related to the entire structure assessed through signatures, vibrational monitoring.

When monitored periodically and systematically no catastrophic collapses can ever occur if this global technique is adhered to.

It is like ECG of a heart patient.

9.20 INSTRUMENTATION FOR MONITORING OF TALL BUILDINGS

A qualitative sketching of spectra from a transducer is sketched, Fig. 9.36. Any deterioration in the strength of the bridge gets indicated by the changes in the frequencies (change of stiffness) and energy ordinates *s(f)*. The corresponding interpretation is higher peaks and lowering of the peaks respectively. New peaks may appear as a result of redistribution of internal forces.

The frequencies corresponding to the peaks are the natural frequencies. Any change of frequency, generally lowering of the frequencies, need to be watched. Any tendency of coalescing of the frequencies may lead to coupling – it is a hazard and measures should be taken to separate them by altering structural properties.

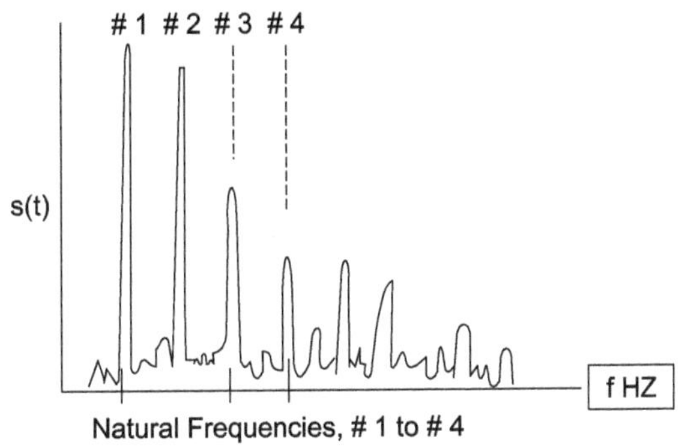

Fig. 9.36 *Signature from a Transducer*

Prototype Monitoring for Overall Performance

FFT analysis for signatures.
A bridge is qualitatively shown as a case study.
The span lengths and their number Vibrational Sensors are permanently installed. About Sixteen to Twenty four Accelerometers are not unusual for sensing the vibrations.
The accelerometers are clustered into modules, one cluster for each span.
For the bridge decks the Analytical models are based on skeletal idealization as well as based on Finite Elements.
Similar monitoring is done for Tall Buildings, where the sensors are installed in typical floors.

Results:
Monitoring of any likely changes in signature will establish general health of the Bridge-Any changes in signatures will lead to the points of local distress, fatigue cracks, plastic hinges, etc.
About Four lowest natural Frequencies are considered to be adequate.

The signatures establish, if any:

- Dominance of Flexure mode in x-direction
- Dominance of Flexure mode in y-direction
- Dominance of torsion mode
- The spectra are collected during occupancy and high wind loadings,
- Any reduction in the frequencies is a reflection on reduction in the stiffness of the structure due to cracking or major distress. The nature of frequency, flexural, torsion, etc. suggests the nature of damage and perhaps helps in location of distress in a big building system, and
- Continuous collection of data to verify the building's integrity. It is like ECG of human beings.

9.21 SOME FEATURES OF TALL BUILDINGS

Low rise vs. High Rise Buildings, for equal floor areas and (floor index) land cost is not in purview.

The low unit structural steel 117.36 kg/m^2 is indicative only pertaining to basic structural frame steel.

Tall Buildings 1950-2007

Tallest Office Structures

- First Wisconsin Center, Milwauki, Wisconsin: 42 Storey 13#/Sqft. 63.57 Kg/Sqm
- Sears Tower, Chicago,
- Illinois: 109-Storey 24#/Sqft.
  117.36 Kg/Sqm
- World Trade Center (Late) 110 Storey
- Taiwan, Super Tall Building
- Future STB Communication through Helicopters
- There is no significant additional material requirement in low rise buildings(less than six storey) when lateral loads are also considered. The marginal increase in maximum moments at the joints can be accounted for without any significant increase in quantities.
- As the buildings become taller (12 storey and above), the consideration of lateral loads result in additional quantities.
- Elaborate fire protection and the associate cost increases with height.
- For buildings taller than 20 storeys, wind loading tends to be more critical. The concept of smart buildings may save some material. But the net effect is larger material for the same floor indices.
- The higher values of moments and deflections due to increasing heights require provision of shear walls and larger lift shafts to enable arresting larger lateral deflections.
- The taller buildings for the same floor index tend to be tower like and thereby the use of tubes and bundled tubes are not uncommon. For such heights wind is more governing than earthquake.
- Occupancy up to about six storeys in city environment, the occupancy is mainly by offices and commercial establishments which fetch higher financial returns.
- Management of lift traffic becomes an important aspect of design.
- Cost of services including plumbing and supply connections is much higher and maintenance more demanding.
- Tall Buildings as "Smart" buildings: The lateral loads, particularly the wind loads become severe when the tower like buildings is taller than 50 storeys. Provision of masses with dashpot and springs are not unusual.

IMPORTANT REFERENCES FOR CHAPTER 6 TO 9

Recent papers are more computer oriented than any introduction of newer concept.

1. Wang, P.C., *Numerical and Matrix Methods in Structural Mechanics*, John Wiley and Sons.
2. Livesley, R.K., *Matrix Methods of Structural Analysis*, Pergamon Press.
3. Pestel, C.E. and Leckie, F.A., *Matrix Methods in Elastomechanics*, McGraw-Hill Book Co. Inc.

4. Kinney, J.S., *Indeterminate Structural Analsyis*, Addison – Wealey (Indian edition available).
5. Clough, R.W., Wilson, E.L., and King, I.P., Large Capacity Multi-storey Frame Analysis Programs, *Journal of the Structural Division*, ASCE.
6. Merchant W., *The Structural Engineer*, 32, 185, 1954.
7. Merchant W., *The Structural Engineer*, 33, 84, 1955.
8. Horne M.R., *Progress in Solid Mechanics*, vol. 2, Ch. 7, North Holland, Amsterdam, 1961.
9. Metheson, J.A.L. *Hyper Static Structures*, Vol. 1, Academic Press Inc. New York.
10. Kani G., *Analysis of Multistory Frames London*, Crosby Lock Wood, 1957.

APPENDIX: MATHEMATICAL PRELIMINARIES

Appendix

Mathematical Preliminaries

A.1 MATRICES

Matrix formation of simultaneous equations

In the subject of structural analysis based on "displacement method" or "force method". We get linear simultaneous equations, Eq. (A.1), for the unknowns x_1, x_2, …, where the x_i are unknown "displacements" or unknown "forces".

$$a_{11}x_1 + a_{12}x_2 + a_{13}x_3 \dots + a_{1j}x_j + \dots + a_{1n}x_n = k_1$$
$$a_{21}x_1 + a_{22}x_2 + a_{23}x_3 \dots + a_{2j}x_j + \dots + a_{2n}x_n = k_2$$
$$\cdots\cdots\cdots\cdots\cdots\cdots\cdots\cdots\cdots\cdots\cdots\cdots$$
$$\cdots\cdots\cdots\cdots\cdots\cdots\cdots\cdots\cdots\cdots\cdots\cdots$$
$$a_{11}x_1 + a_{12}x_2 + a_{13}x_3 \dots + a_{1j}x_j + \dots + a_{1n}x_n = k_i$$
$$\cdots\cdots\cdots\cdots\cdots\cdots\cdots\cdots\cdots\cdots\cdots\cdots$$
$$a_{n1}x_1 + a_{n2}x_2 + a_{n3}x_3 \dots + a_{nj}x_j + \dots + a_{nn}x_n = k_n \tag{A.1}$$

The Eq. A.1 may be written in the following form,

$$[A]\,[X] = [K] \tag{A.2}$$

Where $[A]$, $[X]$ and $[K]$ are matrices, which are simple arrays of elements defined by relations, Eq. (A.3).

$$[X] = \begin{bmatrix} x_1 \\ x_2 \\ \vdots \\ x_i \\ \vdots \\ x_n \end{bmatrix} \quad \text{and} \quad [K] = \begin{bmatrix} k_1 \\ k_2 \\ \vdots \\ k_i \\ \vdots \\ k_n \end{bmatrix}$$

The size of a matrix is fixed by the number of rows and columns of the array of elements. The matrix $[A]$ is generally a "square matrix" of size $n \times n$, where the first n is the number of the rows and the second n is the number of columns. The sizes of matrices $[X]$ and $[K]$ are $n \times 1$ each.

The Eq. (A.2) represents Eq. (A.1), only after we define the matrix multiplication $[A]$ $[X]$ and the equality of matrix relations.

Equality relations

Any two matrices A and B are equal if

(i) The sizes are same

(ii) Every elements of A is equal to the corresponding element of B, i.e.

$$a_{ij} = b_{ij} \tag{A.4}$$

Before we proceed to define matrix multiplication suggested above it is necessary to define rules of addition and subtraction.

Addition and Subtraction

The sum C of two matrices of "the same size", $m \times n$, Eq. (A.5)

$$A + B = C \tag{A.5}$$

is obtained by the addition of elements a_{ik} and b_{ik} to obtain corresponding element c_{ik} of C, i.e.,

$$c_{ik} = a_{ik} + b_{ik} \tag{A.5a}$$

In the case of subtraction, Eq. (1.6)

$$[A - B = C] \tag{A.6}$$

The elements of C matrix are obtained by the relation

$$c_{ik} = a_{ik} + (-\, b_{ik}) = a_{ik} - b_{ik} \tag{A.6a}$$

Matrix Multiplication

Referring to the Eqs (A.1) and (A.2) any typical equation $(a_{i1}x_1 + a_{i2}x_2 + a_{i3}x_3 \ldots a_{ij}x_j + \ldots a_{in}x_n + \ldots = k_i)$ is obtained by multiplying the elements of ith row of $[A]$ matrix with the elements of column matrix $[X]$ and then establish the equality condition of two matrices, Eq. (A.4).

Matrix multiplication rule can be generalized for any two matrices A and B to get the resulting matrix C, i.e.,

$$[A]\,[B] = [C] \tag{A.7}$$

The number of columns of A are the same as the number of rows of B, i.e., matrices A and B of size $(m \times n)$ and $(n \times p)$ respectively when multiplied will give a resulting matrix C of size $(m \times p)$. The elements of C matrix are given by

$$c_{ij} = \sum_{k=1}^{n} a_{ik}\, b_{kj} \tag{A.7a}$$

If an index (k) is repeated, the summation symbol "Σ" need not be used, and the mere repetition of the index "k" is treated as the summation for all possible values of k. Following the summation conventions:

$$c_{ij} = a_{ik}\, b_{kj} \quad k = 1, 2 \ldots n$$

The operation is explained by the following scheme. Let matrix A of (4×3) size and matrix B of (3×2) be multiplied. $m = 4$, $n = 3$, $p = 2$. The resulting matrix C will be of size (4×2).

$$\begin{bmatrix} a_{11} & a_{12} & a_{13} \\ a_{21} & a_{22} & a_{23} \\ a_{31} & a_{32} & a_{33} \\ a_{41} & a_{42} & a_{42} \end{bmatrix} \begin{bmatrix} b_{11} & b_{12} \\ b_{21} & b_{22} \\ b_{31} & b_{32} \end{bmatrix} = \begin{bmatrix} (a_{1k}\, b_{k1}) & (a_{1k}\, b_{k2}) \\ (a_{2k}\, b_{k1}) & (a_{2k}\, b_{k2}) \\ (a_{3k}\, b_{k1}) & (a_{3k}\, b_{k2}) \\ (a_{4k}\, b_{k1}) & (a_{4k}\, b_{k2}) \end{bmatrix} \tag{A.8}$$

$$A \qquad B \qquad = \qquad C$$

The following is the expanded form of the matrix C, Eq. (1.8a)

$$C_{ik} = a_{ij} bjk \tag{A.8a}$$

It may be observed from the rule (A.5), (A.5a) for addition and subtraction that, $A + B = B + A$, but it does not follow from (A.7) and (A.8) that the order of factors in the product $A\,B$ can be interchanged even when the matrices are square. When the matrices are not square it is just not possible to multiply when the order is interchanged. It is left to the reader to verify the statement by a numerical example.

Null Matrix

A null matrix consists only of zero elements.

Diagonal Matrix and Unit Matrix

It is a square matrix the elements of which are zero except for those on the main diagonal. A "unit matrix" or "identity matrix" is a particular case of a diagonal matrix, in which the elements on the main diagonal are unity, and all other elements are zero.

$$\text{Identity Matrix} = I = \begin{vmatrix} 1 & 0 & 0 & 0 & \dots & 0 \\ 0 & 1 & 0 & 0 & \dots & 0 \\ 0 & 0 & 1 & 0 & \dots & 0 \\ 0 & 0 & 0 & 1 & \dots & 0 \\ \dots & \dots & \dots & \dots & \dots & \dots \\ 0 & 0 & 0 & \dots & \dots & 1 \end{vmatrix} \dots \tag{A.9a}$$

The non-zero elements of diagonal matrices are written within symbol $\lceil \dots \rfloor$.

Thus $$I = \lceil 1 \quad 1 \quad 1 \quad 1 \quad \dots \quad 1 \rfloor \tag{A.9b}$$

Transposed Matrix

The transpose $[A]^t$ of a matrix A is formed by changing rows to columns. For example, if

$$A = \begin{bmatrix} 1 & 2 & 3 & 4 \\ 5 & 6 & 7 & 8 \\ 9 & 10 & 11 & 12 \end{bmatrix}$$

Then

$$A^t = \begin{bmatrix} 1 & 5 & 9 \\ 2 & 6 & 10 \\ 3 & 7 & 11 \\ 4 & 8 & 12 \end{bmatrix} \tag{A.10}$$

It may be verified that the transpose of $[A]^t$ is again A.

Symmetric Matrix

It is square matrix having the property, $A^t = A$.

Orthogonal Matrix

If $[A]$ is an orthogonal matrix then $A\,A^t = I$. Direction cosine matrices of rotational transformation are examples of orthogonal matrices.

A.2 SINGULAR MATRIX AND RANK OF A MATRIX

If the determinant of a square matrix is zero, then the matrix is called a singular matrix. If the highest size of a non-zero determinant of a singular matrix A is r (i.e., the sub matrix of size $r \times r$) then the matrix A is said to be of rank r.

Augmented Matrix

If $[A]$ and $[C]$ are the matrices representing simultaneous equations, $A\,X = C$, where X gives the unknown elements, the augmented matrix A_g is given by introducing an additional column C along with the arrays of the coefficient matrix $[A]$.

i.e.,
$$A_g = \begin{bmatrix} a_{11} & a_{21} & \cdots & \cdots & a_{n1} & c_1 \\ a_{21} & a_{21} & \cdots & \cdots & a_{n2} & c_2 \\ \cdots & \cdots & \cdots & \cdots & \cdots & \cdots \\ \cdots & \cdots & \cdots & \cdots & \cdots & \cdots \\ a_{n1} & a_{n2} & \cdots & \cdots & a_{nn} & c_n \end{bmatrix} \tag{A.11}$$

A.3 FUNDAMENTAL THEOREM: RANK OF MATRICES

Fundamental Theorem

Let r and r' be the ranks of coefficient matrix $[A]$ and augmented matrix $[A, C]$ of a set of m linear equations for n unknowns, that is,

$$\underset{(m \times n)}{\mathrm{A}} \quad \underset{(n \times 1)}{\mathrm{X}} = \underset{(n \times 1)}{C}$$

The theorem states that (i) if $m = n = r = r'$, we have a unique solution for X, (ii) if $r = r' < (m = n)$, we assign arbitrary values to $n - r$ unknowns and then solve for the remaining r unknowns (iii) if $r' > r$, the equations are inconsistent and no solution is possible. In the case (ii), we have infinite number of solutions due to the infinite choices of the arbitrary values assigned to the $(n - r)$ unknowns.

Example A.1 Obtain the elements of unknown's matrix X

Solution:

(a)
$$\begin{bmatrix} 3 & -1 & 3 \\ 1 & 1 & -3 \\ 5 & -1 & 3 \end{bmatrix} \begin{bmatrix} x_1 \\ x_2 \\ x_3 \end{bmatrix} = \begin{bmatrix} 4 \\ -1 \\ 7 \end{bmatrix}$$

$\therefore$ Determinant, $[A] = 0$. The rank of coefficient matrix is not equal to 3.

It may be verified to be equal to 2.

Rank of augmented matrix:

$$\text{Augmented matrix} = \begin{bmatrix} 3 & -1 & 3 & 4 \\ 1 & 1 & -3 & -1 \\ 5 & -1 & 3 & 7 \end{bmatrix}$$

It may be verified that its rank, $r' = 3$.
The equations are inconsistent and no solution is possible.

(b) $$\begin{bmatrix} 2 & -1 & 3 \\ 1 & 1 & -3 \\ 5 & -1 & 3 \end{bmatrix} \begin{bmatrix} x_1 \\ x_2 \\ x_3 \end{bmatrix} = \begin{bmatrix} 4 \\ -1 \\ 7 \end{bmatrix}$$

In this case $r = r' = 2$ and assuming $x_3 = k$ (which suggests infinite number of solutions), $x_1 = 1$ and $x_2 = 3k - 2$.

Rank of matrix is an important consideration in the Force Method or Flexibility Method of Structural Analysis. With the help of the "fundamental theorem" it will be shown elsewhere that the evaluation of rank gives all the information regarding degree of indeterminacy, consistency of equations and whether the structural system is stable or not. An example is given in Chapter 2.

A.4 MINOR, COFACTOR AND ADJOINT OF MATRICES

MINOR:

Minor M_{ij} of a matrix A is the determinant of the remaining matrix obtained after deleting ith row and jth Column:

If $$A = \begin{bmatrix} 1 & 2 & -1 \\ 2 & -1 & 1 \\ 3 & 1 & 1 \end{bmatrix},$$

$$M_{11} = \det \begin{bmatrix} -1 & 1 \\ 1 & 1 \end{bmatrix} = -2$$

Similarly $M_{12}, M_{13}, M_{21} \ldots M_{33}$ may be found.

COFACTOR:

Cofactor $$c_{ij} = (-1)^{i+j} M_{ij}$$

ADJOINT OF MATRICES:

Adjoint, A_d, of a matrix $[A]$ is obtained by replacing the elements a_{ij}, i.e., it is the transposed matrix of the cofactors of elements of $[A]$.

Thus $$A_d = \begin{bmatrix} A_{11} & A_{21} & A_{31} \\ A_{12} & A_{22} & A_{32} \\ A_{13} & A_{23} & A_{33} \end{bmatrix}$$

Where A_{ij} are the cofactors of elements a_{ij} of matrix $[A]$.

A.5 EIGEN VALUES OF A SQUARE MATRIX

If a set simultaneous equations are of the following form,

$$a_{11}x_1 + a_{12}x_2 + a_{13}x_3 \ldots + a_{1n}x_n = \lambda x_1$$
$$a_{21}x_1 + a_{22}x_2 + a_{23}x_3 \ldots + a_{2n}x_n = \lambda x_2$$
$$\cdots\cdots\cdots\cdots\cdots\cdots\cdots\cdots$$
$$a_{n1}x_1 + a_{n2}x_2 + a_{n3}x_3 \ldots + a_{nn}x_n = \lambda x_n \tag{A.13}$$

Which may be written in the matrix form

$$A\,X = \lambda X$$

Or $$[A - \lambda I]X = 0, \tag{A.13a}$$

Then a non-trivial solution (i.e., $X \neq 0$) is possible,

Only if, $|A - \lambda I| = 0$, i.e.,

$$\begin{vmatrix} a_{11} - \lambda & a_{11} & \ldots & a_{1n} \\ a_{21} & a_{22} - \lambda & \ldots & a_{2n} \\ \ldots & \ldots & \ldots & \ldots \\ a_{n1} & a_{n2} & \ldots & a_{nn} - \lambda \end{vmatrix} = 0$$

When expanded, we get the characteristic equation,

$$\lambda^n + c_1\lambda^{n-1} + c_2\lambda^{n-2} + \ldots + c_n = 0 \tag{A.14}$$

The roots $\lambda_1, \lambda_2 \ldots, \lambda_n$ of the characteristic Eq. (A.14) are known as the "eigen values" or "characterstic values" of matrix A. A systematic evaluation of all the Eigen values and the corresponding unknowns $\{x_1, x_2, \ldots, x_n)$ called as Eigen Vectors, is given by P.C. wang[1].

A.6 REAL SYMMETRIC MATRIX

The coefficients of "real symmetric matrix" are all real. An important property of the real symmetric matrix is that its eigen values λ_i are always real.

Another important property of a symmetric matrix is derived from its quadratic form and it is defined by Eq. A.15.

Quadratic form of matrix $[A] = X^t A\, X$, where X is a column matrix. (A.15)

If the quadratic form of a matrix $[A]$ is always positive for any non-zero column vector X, then the matrix A is known as "positive definite". In structural theory it can be shown that energy = $\frac{1}{2} d^t\, k\, d$, where k is a stiffness matrix and d is the deformation matrix. Therefore $d^t\, k\, d$ is, the quadratic form of the stiffness matrix k. If k is positive definite, then the structures are stable and the simultaneous equations are well conditioned. Alternatively, a matrix is positive definite if its determinant and all the determinants of their principal minors are positive. The stiffness matrix and the flexibility matrix are positive definite, and this property is important in the solution of problems by the stiffness or flexibility method. The 'flexibility method' is described in Chapter 2, and 'Displacement Method' in Chapters 6 & 7.

---

[1] Wang, P.C. "Numerical and Matrix Methods in Structural Mechanics with Application to computers", John Wiley & Wiley & Sons, Inc. N.Y., 1960.

A.7 INVERSION OF MATRICES AND SOLUTION OF SIMULTANEOUS EQUATIONS

In structural analysis the matrices are generally banded or sparse. The method of solution will be discussed elsewhere for the particular case of multistory buildings.[2] Any book on matrix methods describes the matrix inversion, and standard routines for inversion or solution of simultaneous equations. Brief explanations of some of the methods are given while solving the numerical examples in the following.

Cramers rule and inversion of a matrix by Adjoint method:

Using summation convention, the Cramers rule can be expressed by the following relation,

$$x_j = A_{ij} b_i / a$$

Where "$A_1 ij$ "&" a" are the cofactors and determinant respectively of a given matrix (a_{ij})

and $$a_{ij}\, x_j = b_i$$

or $$A\, X = B$$

in the expanded form,

$$x_j = \frac{\begin{vmatrix} a_{11} a_{12} & \cdots & a_{1,\, j-1}\, b_1 & \cdots & a_{1n} \\ a_{21} a_{22} & \cdots & a_{2,\, j-1}\, b_2 & \cdots & a_{2n} \\ \cdots & \cdots & \cdots & \cdots & \cdots \\ a_{n1} a_{n2} & \cdots & a_{n,\, j-1}\, b_n & \cdots & a_{nn} \end{vmatrix}}{a}$$

Where "| |" is the symbol for determinant.

Thus

$$x_1 = \frac{1}{a}[A_{11}\; A_{21} \cdots A_{n1}]\begin{bmatrix} b_1 \\ b_2 \\ b_n \end{bmatrix}$$

$$x_2 = \frac{1}{a}[A_{12}\; A_{22} \cdots A_{n2}]\, B$$

....

$$x_j = \frac{1}{a}[A_{1j} A_{2j} \cdots A_{nj}]\, B$$

or
$$\begin{bmatrix} x_1 \\ x_2 \\ \vdots \\ x_n \end{bmatrix} = \frac{1}{a}\begin{bmatrix} A_{11} & A_{21} & \cdots & A_{n1} \\ A_{12} & A_{22} & \cdots & A_{n2} \\ A_{13} & A_{23} & \cdots & A_{n3} \\ \cdots & \cdots & \cdots & \cdots \\ A_{ij} & A_{2j} & \cdots & A_{nj} \\ \cdots & \cdots & \cdots & \cdots \\ A_{1n} & A_{2n} & \cdots & A_{nn} \end{bmatrix}\begin{bmatrix} b_1 \\ b_2 \\ \cdot \\ \cdot \\ \cdot \\ \cdot \\ b_n \end{bmatrix} \quad \text{(A.16)}$$

Thus $[A]^{-1} = \dfrac{1}{a}$ (Adjoint of A)

[2] Chapters 8 and 9.

Example

Find inverse of a matrix $[A]$ by Adjoint method.

$$A = \begin{bmatrix} 2 & 1 & 1 \\ 3 & 4 & 2 \\ 1 & 2 & 5 \end{bmatrix}$$

$$|A| = a = 2(20 - 4) - 1(15 - 2) + 1(6 - 4) = 21$$

Adjoint of A, $$A_g = \begin{bmatrix} A_{11} & A_{12} & A_{13} \\ A_{21} & A_{22} & A_{23} \\ A_{31} & A_{32} & A_{33} \end{bmatrix}^t$$

$$= \begin{bmatrix} 16 & -13 & 2 \\ -3 & 9 & -3 \\ -2 & -1 & 5 \end{bmatrix}^t = \begin{bmatrix} 16 & -3 & -2 \\ -13 & 9 & -1 \\ 2 & -3 & 5 \end{bmatrix}$$

And $$A^{-1} = \frac{1}{a}[A_g]$$

Inverse of a matrix by linear transformations

The procedure is better explained by solving the following example

Find the inverse of $A = \begin{bmatrix} 1 & 2 & 3 \\ 2 & -1 & 1 \\ 3 & 1 & -1 \end{bmatrix}$

We write $A = I\,A$

Choose T_1, T_2, T_3 as given below,

$$T_1 = \begin{bmatrix} 1/a_{11} & 0 & 0 \\ -a_{12}/a_{11} & 1 & 0 \\ -a_{13}/a_{11} & 0 & 1 \end{bmatrix}; \quad T_2 = \begin{bmatrix} 1 & -b_{12}/b_{22} & 0 \\ 0 & 1/b_{22} & 0 \\ 0 & -b_{32}/b_{22} & 1 \end{bmatrix};$$

Elements a_{11}, a_{12}, a_{13} belong to matrix $[A]$ Elements b_{12}, b_{22}, b_{32} belong to matrix $[B] = T_1 A$

$$T_3 = \begin{bmatrix} 1 & 0 & -\dfrac{c_{13}}{c_{33}} \\ 0 & 1 & -\dfrac{c_{23}}{c_{33}} \\ 0 & 0 & \dfrac{1}{c_{33}} \end{bmatrix}$$

Elements c_{13}, c_{23}, c_{33}

belong to Matrix $C = T_2 B$

Continuing the linear transformation using T_1, T_2 and T_3 we get

$$T_3\, T_2\, T_1\, A = I$$

or $$[A]^{-1} = T_3\, T_2\, T_1$$

In this case

$$T_1 = \begin{bmatrix} 1 & 0 & 0 \\ -2 & 1 & 0 \\ 3 & 0 & 1 \end{bmatrix};\quad T_2 = \begin{bmatrix} 1 & \frac{2}{5} & 0 \\ 0 & -\frac{1}{5} & 0 \\ 0 & -1 & 1 \end{bmatrix};\quad T_3 = \begin{bmatrix} 1 & 0 & \frac{1}{5} \\ 0 & 1 & \frac{1}{5} \\ 0 & 0 & -\frac{1}{5} \end{bmatrix}$$

Thus

$$A^{-1} = T_3\, T_2\, T_1 = \frac{1}{5}\begin{bmatrix} 0 & 1 & 1 \\ 1 & -2 & 1 \\ 1 & 1 & -1 \end{bmatrix}$$

Inverse of Symmetric matrices by factorization

The procedure outlined herein is valid, in general, for "hermitian"[3] positive definite matrices. Symmetric matrices usually encountered in structural problems is a particular case. This procedure of factorizing a given matrix, $[A]$, into triangular matrices is also known as "Square Root" method.

The examples (a) and (b) explain the procedure

(a) $$A = \begin{bmatrix} 4 & 3 & 2 & 1 \\ 3 & 4 & 1 & 3 \\ 2 & 1 & 5 & 2 \\ 1 & 3 & 2 & 6 \end{bmatrix}$$

(b) same as (a) with the only change that a_{44} is to be taken equal to 4.

Example (a)

Step 1. Taking advantage of symmetry there are only ten independent elements for a symmetric matrix of size (4×4).

Let

$A = L\,L'$, Where L is a lower triangular matrix and L' is transpose of L

$$L = \begin{bmatrix} l_{11} & 0 & 0 & 0 \\ l_{21} & l_{22} & 0 & 0 \\ l_{31} & l_{32} & l_{33} & 0 \\ l_{41} & l_{42} & l_{42} & l_{44} \end{bmatrix}$$

[3] For the definitions, refer any book on Matrix Algebra

We have

$$\begin{bmatrix} 4 & 3 & 2 & 1 \\ 3 & 4 & 1 & 3 \\ 2 & 1 & 5 & 2 \\ 1 & 3 & 2 & 6 \end{bmatrix} = \begin{bmatrix} l_{11} & 0 & 0 & 0 \\ l_{21} & l_{22} & 0 & 0 \\ l_{31} & l_{32} & l_{33} & 0 \\ l_{41} & l_{42} & l_{42} & l_{44} \end{bmatrix} \begin{bmatrix} l_{11} & l_{21} & l_{31} & l_{41} \\ 0 & l_{22} & l_{32} & l_{42} \\ 0 & 0 & l_{33} & l_{43} \\ 0 & 0 & 0 & l_{44} \end{bmatrix}$$

Evaluate l_{ij} from the equality relation in the following sequence.

$a_{11} = (l_{11})^2$ Choose positive or negative value of l_{11}. From $a_{13} = l_{11}\, l_{21}$ evaluate l_{21}. Follow it by the equality condition for the elements in the sequence a_{13}, a_{14}; a_{22}, a_{23}, a_{24} ; a_{33}, a_{34} ; a_{44}

Solution

$$(l_{11})^2 = a_{11} = 4 \quad \text{Say,} \quad l_{11} = +2$$

$$l_{21}\, l_{11} = a_{12} = 3 \rightarrow l_{21} = \frac{3}{2};\ l_{11}\, l_{31} = 2 \rightarrow l_{31} = 1;\ l_{11}\, l_{41}$$

$$= a_{14} \rightarrow l_{41} = \frac{1}{2},\ (l_{21})^2 + (l_{22})^2 = a_{22} = 4 \rightarrow (l_{22}) = \frac{\sqrt{7}}{2}$$

$$l_{21}\, l_{31} + l_{22}\, l_{32} = a_{23} = 1 \rightarrow l_{32} = -\frac{1}{\sqrt{7}}$$

$$l_{21}\, l_{41} + l_{22}\, l_{42} = a_{24} = 3 \rightarrow l_{42} = \frac{9}{2\sqrt{7}}$$

$$(l_{31})^2 + (l_{32})^2 + (l_{33})^2 = a_{33} = 5 \rightarrow l_{33} = \frac{3\sqrt{3}}{\sqrt{7}}$$

$$l_{31}\, l_{41} + l_{32}\, l_{42} + l_{32}\, l_{43} = a_{34} = 2 \rightarrow l_{43} = \frac{5}{\sqrt{21}}$$

$$(l_{41})^2 + (l_{42})^2 + (l_{43})^2 + (l_{44})^2 = a_{44} = 6 \rightarrow l_{44} = \frac{\sqrt{5}}{\sqrt{3}}$$

Step 2. From the relation

$$A = L\,L',\ A^{-1} = (L')^{-1}\, L^{-1}$$

Let $\qquad A^{-1} = R$

Post multiply by L to get $RL = (L')^{-1}$

If we know $(L')^{-1}$, once again from equality considerations the elements of R can be evaluated. The next step of the problem is to evaluate $(L')^{-1}$."Inverse of a triangular matrix is also a triangular matrix."

Fortunately it can be seen that in the third step for the evaluation, we will be requiring to know only the leading diagonal nonzero elements of $(L')^{-1}$. And in the case of triangular matrix the leading diagonal elements of the inverse are simply the reciprocals of diagonal elements of the given triangular matrix.

Further, from the consideration that inverse of a triangular matrix is also a triangular matrix,

$$(L')^{-1} = \begin{bmatrix} 2 & \frac{3}{2} & 1 & \frac{1}{2} \\ 0 & \frac{\sqrt{7}}{2} & -\frac{1}{\sqrt{7}} & \frac{3}{2\sqrt{7}} \\ 0 & 0 & \frac{3\sqrt{3}}{\sqrt{7}} & \frac{5}{\sqrt{21}} \\ 0 & 0 & 0 & \frac{\sqrt{5}}{\sqrt{3}} \end{bmatrix}^{-1}$$

$$(L')^{-1} = \begin{bmatrix} \frac{1}{2} & ? & ? & ? \\ 0 & \frac{2}{\sqrt{7}} & ? & ? \\ 0 & 0 & \frac{\sqrt{7}}{3\sqrt{3}} & ? \\ 0 & 0 & 0 & \frac{\sqrt{3}}{\sqrt{5}} \end{bmatrix}$$

Step 3. Evaluation of the elements or $R = A^{-1}$ from the relation

$$R\,L = (L')^{-1}$$

$$= \begin{bmatrix} r_{11} & r_{12} & r_{13} & r_{14} \\ (r_{12}) & r_{22} & r_{23} & r_{24} \\ (r_{13}) & (r_{23}) & r_{33} & r_{34} \\ (r_{14}) & (r_{24}) & (r_{34}) & r_{44} \end{bmatrix} = \begin{bmatrix} 2 & 0 & 0 & 0 \\ \frac{3}{2} & \frac{\sqrt{7}}{2} & 0 & 0 \\ 1 & -\frac{1}{\sqrt{7}} & \frac{3\sqrt{3}}{\sqrt{7}} & 0 \\ \frac{1}{2} & \frac{3}{2\sqrt{7}} & \frac{5}{\sqrt{21}} & -\frac{\sqrt{5}}{\sqrt{3}} \end{bmatrix}$$

$$= \begin{bmatrix} \frac{1}{2} & ? & ? & ? \\ 0 & \frac{2}{\sqrt{7}} & ? & ? \\ 0 & 0 & \frac{\sqrt{7}}{3\sqrt{3}} & ? \\ 0 & 0 & 0 & \frac{\sqrt{3}}{\sqrt{5}} \end{bmatrix}$$

The elements r_{ij} within ( ) follow from symmetry. From equality considerations evaluate r_{ij} in the following sequence.

$r_{44} = 0.600;$ $r_{43} = r_{34} = 0.333;$

$r_{24} = r_{42} = -0.867;$ $r_{41} = r_{14} = 0.667;$

$r_{33} = 0.444;$ $r_{23} = r_{32} = 0.555;$

$r_{13} = r_{31} = -0.555;$ $r_{22} = 1.844$

$r_{12} = r_{21} = -1.444;$ $r_{11} = 1.444$

$$R = (A)^{-1} = \begin{bmatrix} 1.444 & -1.444 & -0.555 & 0.667 \\ -1.444 & 1.844 & 0.555 & -0.867 \\ -0.555 & 0.555 & 0.444 & -0.333 \\ 0.667 & -0.867 & -0.333 & 0.600 \end{bmatrix}$$

(b) In this example a_{44} is changed to 4, to show that factorized triangular matrices may consist of imaginary elements.

$$A = \begin{bmatrix} 4 & 3 & 2 & 1 \\ 3 & 4 & 1 & 3 \\ 2 & 1 & 5 & 2 \\ 1 & 3 & 2 & 4 \end{bmatrix}$$

Step 1. $A = L\,L'$

By matching the elements and solving with a proper sequence

$$L = \begin{bmatrix} 2.0 & 0.000 & 0.000 & 0.000 \\ 1.5 & 1.323 & 0.000 & 0.000 \\ 1.0 & -0.378 & 1.963 & 0.000 \\ 0.5 & 1.702 & 1.093 & +0.577i \end{bmatrix}$$

Step 2. Compute the diagonal elements only, which are reciprocals of diagonal elements of (L').

$$(L')^{-1} \begin{bmatrix} 0.50 & ? & ? & ? \\ 0 & 0.755 & ? & ? \\ 0 & 0 & 0.50 & ? \\ 0 & 0 & 0 & -1.732i \end{bmatrix}$$

Step 3. $RL = [L']^{-1}$

$$R = A^{-1} = \begin{bmatrix} -3.060 & * & * & * \\ 4.325 & -5.640 & * & * \\ 1.700 & -2.320 & -0.664 & * \\ -3.390 & +4.320 & +1.660 & -3.0 \end{bmatrix}$$

The elements (*) follow from symmetry.

The determinant of $[A]$ is the square of the product of the diagonal elements of $[L]$

Example for the solution of unknowns of a system of simultaneous equations.

$$A\,X = b$$

$[A]$ & b are given:

$$A = \begin{bmatrix} 4 & 3 & 2 & 1 \\ 3 & 4 & 1 & 3 \\ 2 & 1 & 5 & 2 \\ 1 & 3 & 2 & 4 \end{bmatrix} ; b = \begin{bmatrix} 2 \\ 4 \\ 3 \\ 1 \end{bmatrix}$$

Step 1. It is same as the example (a),

Step 2. Let $L'X = Y$ in the relation $L\,L'\,X = b$

We then have $LY = b$; $[L]$ and b being known,

Y can be evaluated.

$$\begin{bmatrix} 2.0 & 0.000 & 0 & 0 \\ 1.5 & 1.323 & 0 & 0 \\ 1.0 & -0.378 & 1.963 & 0 \\ 0.5 & 1.702 & 1.093 & 0.577i \end{bmatrix} \begin{bmatrix} y_1 \\ y_2 \\ y_3 \\ y_4 \end{bmatrix} = \begin{bmatrix} 2 \\ 4 \\ 3 \\ 1 \end{bmatrix}$$

By matching the elements, we get the values in the sequence given below:

$$y_1 = 1; y_2 = 1.890; y_3 = 1.383; y_4 = 7.240\,i;$$

Step 3. We find the elements of $[X]$, by considering the previous substitution

$$L'X = y$$

Or
$$\begin{bmatrix} 2.0 & 1.50 & 1.000 & 0.500 \\ 0 & 1.32 & -0.378 & 1.702 \\ 0 & 0 & 1.963 & 1.093 \\ 0 & 0 & 0 & 0.577i \end{bmatrix} \begin{bmatrix} x_1 \\ x_2 \\ x_3 \\ x_4 \end{bmatrix} = \begin{bmatrix} 1.000 \\ 1.890 \\ 1.383 \\ 7.240 \end{bmatrix}$$

The elements of X are evaluated in the sequence of the results given below

$$x_4 = 12.40; x_3 = -6.20; x_2 = -16.30; x_1 = 12.72;$$

The result $\quad X = \{12.72 \quad -16.30 \quad -6.20 \quad 12.40\}$

Inverse of un-symmetric matrices

Find inverse of the given un-symmetric matrix A

$$A = \begin{pmatrix} 2 & 1 & 1 \\ 3 & 4 & 2 \\ 1 & 2 & 5 \end{pmatrix}$$

For a matrix of $n \times n$, all the n^2 elements suggest the total number of unknown elements (considering lower and upper matrices together) to be n^2.

$$L = \begin{pmatrix} 1 & 0 & 0 \\ l_{21} & 1 & 0 \\ l_{31} & l_{32} & 1 \end{pmatrix} ; \qquad U = \begin{pmatrix} u_{11} & u_{12} & u_{12} \\ 0 & u_{22} & u_{22} \\ 0 & 0 & u_{33} \end{pmatrix}$$

$$\begin{bmatrix}\text{Unknown}\\ \text{elements}\end{bmatrix} = \frac{n^2 - n}{2} = 3 \qquad \begin{bmatrix}\text{Unknown}\\ \text{elements}\end{bmatrix} = \frac{n^2 + n}{2} = 6$$

Total number of unknown elements = $n^2 = 9$

Step 1. $A = LU$

$$\begin{pmatrix} 2 & 1 & 1 \\ 3 & 4 & 2 \\ 1 & 2 & 5 \end{pmatrix} = \begin{pmatrix} 1 & 0 & 0 \\ l_{21} & 1 & 0 \\ l_{31} & l_{32} & 1 \end{pmatrix}\begin{pmatrix} u_{11} & u_{12} & u_{12} \\ 0 & u_{23} & u_{22} \\ 0 & 0 & u_{33} \end{pmatrix}$$

$$A \qquad = \qquad L \qquad\qquad U$$

By comparing the elements on either side of the equality

$$a_{11} = u_{11}; \quad a_{12} = u_{12}; \quad a_{13} = u_{13};$$

$$a_{21} = (\text{Second row of } L)\ (\text{First column of } U)$$

$$= l_{21}\, u_{11} \text{ or } l_{21} = \frac{a_{21}}{u_{11}}, l_{21} = \frac{3}{2}$$

The other elements are found in the following sequence

$$u_{22} = \frac{5}{2}; u_{23} = \frac{1}{2}; l_{31} = \frac{1}{2}; l_{32} = \frac{3}{5}; u_{33} = \frac{21}{5};$$

i.e.,

$$L = \begin{pmatrix} 1 & 0 & 0 \\ \frac{3}{2} & 1 & 0 \\ \frac{1}{2} & \frac{3}{5} & 1 \end{pmatrix}; \quad U = \begin{pmatrix} 2 & 1 & 1 \\ 0 & \frac{5}{2} & \frac{1}{2} \\ 0 & 0 & \frac{21}{5} \end{pmatrix}$$

Step 2. $A^{-1} = U^{-1}\, L^{-1}$

Post-multiply by L

$$A^{-1}\, L = U^{-1}$$

Find U^{-1} before evaluating A^{-1}

For finding U^{-1} compare

$$U\, U^{-1} = \text{I}$$

U and I being known, evaluate the elements of U^{-1}

$$U^{-1} = \begin{pmatrix} \frac{1}{2} & -\frac{1}{5} & -\frac{2}{21} \\ 0 & \frac{2}{5} & -\frac{1}{21} \\ 0 & 0 & \frac{5}{21} \end{pmatrix}$$

Step 3.

If $R = A^{-1}$

Then $RL = U^{-1}$ as explained in step 2.

L and U^{-1} being known, evaluate R in the sequence

$$r_{33} = \frac{5}{21};\quad r_{23} = -\frac{1}{21};\quad r_{13} = -\frac{2}{21};\quad r_{32} = -\frac{1}{7};$$

$$r_{22} = \frac{3}{7};\quad r_{12} = -\frac{1}{7};\quad r_{31} = \frac{2}{21};\quad r_{21} = -\frac{13}{21};\quad r_{11} = \frac{16}{21}$$

And thus

$$A^{-1} = R = \begin{pmatrix} \frac{16}{21} & -\frac{1}{7} & -\frac{2}{21} \\ -\frac{13}{21} & \frac{3}{7} & -\frac{1}{21} \\ \frac{2}{21} & -\frac{1}{7} & \frac{5}{21} \end{pmatrix}$$

Solution for unknown X, when a coefficient matrix A is not symmetric

Obtain X for given C

$$\begin{pmatrix} 2 & 1 & 1 \\ 3 & 4 & 2 \\ 1 & 2 & 5 \end{pmatrix} \begin{pmatrix} x_1 \\ x_2 \\ x_3 \end{pmatrix} = \begin{pmatrix} 2 \\ 3 \\ 4 \end{pmatrix}$$

$$A \quad X = C$$

Theory:

$$A = LU \quad \text{Step 1}$$

$$AX = C$$

$$LUX = C$$

Let $UX = Y$ Step 3

Then $LY = C$ Step 2

Step 1. Same as step 1 of example covered under inverse of un-symmetric.

Step 2. $LY = C$

$$Y = \{2 \quad 0 \quad 3\}$$

Step 3. $UX = Y$

$$X = \left\{\frac{5}{7}, -\frac{1}{7}, \frac{5}{7}\right\}$$

The following techniques for evaluation of unknown and inverse are excluded though connected with this chapter.

1. Solution of linear equations $A\,X = C$ by
   (a) Iteration and
   (b) Relaxation or incremental iteration, are linked with an important consideration of convergence. Any book on numerical analysis covers these aspects.

   The techniques (a) and (b) are discussed in detail, refer Part A and B.
2. Inversion of matrix by partitioning is a tedious procedure. However, this procedure is recommended for some problems to minimize storage in computers. Techniques more powerful than simple partitioning technique are covered elsewhere in connection with tridiagonal matrices and solution of unknowns in structural problems.[4]

Linear transformations with permutation matrix

The factorization technique for inversion is one of the most suitable methods. The linear transformations discussed herein have the biggest disadvantage of (i) storing the product of linearly transforming matrices $T_1, T_2, T_3 \ldots$ and (ii) rearrangement of rows.

In the case of large matrices re-arrangement of rows is very essential, so that the leading diagonal elements are dominant. In the following, "permutation matrix" (defined while solving an example below) is used along with the linear transformations to avoid excessive use of computer storage. Most of the inversion "subroutines" use the "permutation matrix" technique. This procedure eliminates the disadvantages, (i) and (ii). Let us consider the matrix

$$A = \begin{pmatrix} 1 & 2 & 3 \\ 2 & -1 & 1 \\ 3 & 1 & -1 \end{pmatrix}$$

Step 1. Scan the first column and find the numerically largest element. In this example it is: $a_{31} = 3$. Store the position of a_{31}, i.e., 1st column and 3rd row, in the permutation matrix, P, of size $(2 \times n)$ where n is the size of matrix, A.

$$P = \begin{pmatrix} 1 & \wedge \\ 3 & \wedge \end{pmatrix} \begin{matrix} \leftarrow \text{Write column position (1)} \\ \leftarrow \text{Write row position (3)} \end{matrix}$$

Augment $[A]$ matrix by an additional $(n + 1)^{\text{th}}$ column. The additional column has unity in the row containing the largest element.

$$\begin{pmatrix} 1 & 2 & 3 & \vdots & 0 \\ 2 & -1 & 1 & \vdots & 0 \\ 3 & 1 & -1 & \vdots & 1 \end{pmatrix}$$

Step 2. Do the linear transformation such that the pivotal element is unity and the other elements of the same column are reduced to zero. We get

$$\begin{pmatrix} 0 & \frac{5}{3} & \frac{10}{3} & -\frac{1}{3} \\ 0 & -\frac{5}{3} & \frac{5}{3} & -\frac{2}{3} \\ 1 & \frac{1}{3} & -\frac{1}{3} & \frac{1}{3} \end{pmatrix}$$

[4] Chapter 9

Cancel the first column, i.e., we shift the column 2, 3, 4 to the column position 1, 2 and 3

$$b = \begin{pmatrix} \frac{5}{3} & \frac{8}{3} & -\frac{1}{3} \\ \boxed{-\frac{5}{3}} & \frac{5}{3} & -\frac{2}{3} \\ \frac{1}{3} & -\frac{1}{3} & -\frac{1}{3} \end{pmatrix}$$

Step 3. Similar to the step 1, scan the first column and find the position of the largest element. If the largest element happens to be in the same position as in the previous steps, then locate the nest largest. In this example, the numerically largest pivotal element can be either 5/3 or – 5/3 which occupy the positions b_{11} and b_{21} and it has excluded the 3rd row element considered already. Let us take b_{21} as the pivotal element. The pivotal element b_{21} is obtained from a_{22}. The P matrix will get its second column defined as

$$P = \begin{pmatrix} 1 & 2 & \wedge \\ 3 & 2 & \wedge \end{pmatrix} \begin{matrix} \rightarrow \text{Column Position (1)} \\ \rightarrow \text{Row Position (3)} \end{matrix}$$

Augment the matrix b by a column matrix, $\{0 \quad 1 \quad 0\}$. The second position is unity because $\left(-\frac{5}{3}\right)$ belongs to the 2nd row, and thus we have,

$$\begin{pmatrix} \frac{5}{3} & \frac{8}{3} & -\frac{1}{3} & 0 \\ -\frac{5}{3} & \frac{5}{3} & -\frac{2}{3} & 1 \\ \frac{1}{3} & -\frac{1}{3} & \frac{1}{3} & 0 \end{pmatrix}$$

Repeat step 2, with the only change that the linear transformations reduce all the elements of the first column to zero except the pivotal element which is reduced to unity.

$$\begin{pmatrix} 0 & 5 & -1 & 1 \\ 1 & -1 & \frac{2}{5} & -\frac{3}{5} \\ 0 & 0 & \frac{3}{15} & +\frac{1}{5} \end{pmatrix}$$

Shifting the columns one position to the left, we get

$$\begin{pmatrix} 5 & -1 & 1 \\ -1 & \frac{2}{5} & -\frac{3}{5} \\ 0 & \frac{1}{5} & \frac{1}{5} \end{pmatrix}$$

Step 4. The last pivotal element of the first column would belong to the row not considered so far. It so happens that it is (the value is 5) the largest element also. Its actual position in the $[A]$ matrix is in the first row of 3rd column. Therefore, the completely defined P matrix is

$$\begin{pmatrix} 1 & 2 & 3 \\ 3 & 2 & 1 \end{pmatrix} \begin{matrix} \rightarrow \text{Column Position} \\ \rightarrow \text{Row Position} \end{matrix}$$

The matrix of step 3 is now augmented by an additional column with unity in the same row as that of the pivotal element,

$$\begin{pmatrix} 5 & -1 & 1 & 1 \\ -1 & \frac{2}{5} & -\frac{3}{5} & 0 \\ 0 & \frac{1}{5} & \frac{1}{5} & 0 \end{pmatrix}$$

Step 5. Doing the linear transformations to reduce the first column to a value $\{1 \quad 0 \quad 0\}$ we get.

$$\begin{pmatrix} 1 & -\frac{1}{5} & \frac{1}{5} & \frac{1}{5} \\ 0 & \frac{1}{5} & -\frac{2}{5} & \frac{1}{5} \\ 0 & \frac{1}{5} & \frac{1}{5} & 0 \end{pmatrix}$$

Shift the column to the left to get $\frac{1}{5}\begin{pmatrix} -1 & 1 & 1 \\ 1 & -2 & 1 \\ 1 & 1 & 0 \end{pmatrix}$

The above is the permuted inverse and requires, and we be "unscrambled".
The interpretation of the permutation matrix is as follow

$$\text{Column} \downarrow \begin{pmatrix} 1 & 2 & 3 \\ 3 & 2 & 1 \end{pmatrix} \uparrow \text{row}$$

The permutation matrix is flanked by the downward arrow for columns and upward arrow for rows. It suggest that column wise:

1st Column	2nd Column	3rd Column
is ↓ changed to	is ↓ changed to	is ↓ changed to
3rd column	2nd column	1st column

$$\frac{1}{5}\begin{pmatrix} 1 & 1 & -1 \\ 1 & -2 & 1 \\ 0 & 1 & 1 \end{pmatrix}$$

Rowwise : Considering the right flanking arrow it interprets as follows

3rd row	2nd row	1st row
is ↓ changed to	is ↓ changed to	is ↓ changed to
1st row	2nd row	3rd row

And we get

$$\frac{1}{5}\begin{pmatrix} 0 & 1 & 1 \\ 1 & -2 & 1 \\ 1 & 1 & -1 \end{pmatrix}$$

Which is the required inverse.

The above operations can be programmed on computers by appropriate post-multiplying matrices for changing the column positions, and the rows are changed by pre-multiplying. For example, if the first column is required to be changed to contain elements of the 3rd column a null matrix of the same size as the given matrix is modified by replacing zero by unity in the position of 1st column and 3rd row.

$$\begin{pmatrix} a_{11} & a_{12} & a_{13} \\ a_{21} & a_{22} & a_{23} \\ a_{31} & a_{32} & a_{33} \end{pmatrix}\begin{pmatrix} 0 & 0 & 0 \\ 0 & 0 & 0 \\ 1 & 0 & 0 \end{pmatrix} = \begin{pmatrix} a_{13} & 0 & 0 \\ a_{23} & 0 & 0 \\ a_{33} & 0 & 0 \end{pmatrix}$$

Similarly if the second column is not going to get changed, unity is entered in (2, 2) position of the post-multiplying matrix. For the last case of 3rd column elements to get replaced by the elements of the first column, enter unity in the position given by the 3rd column and first row. Thus for the column operations the post-multiplying matrix is given by

$$T = \begin{pmatrix} 0 & 0 & 1 \\ 0 & 1 & 0 \\ 1 & 0 & 0 \end{pmatrix}$$

The above matrix gives an orthogonal transformation. The un-scrambling of permuted inverse is achieved by post-multiplying "permuted A^{-1}" by T^t. Similarly for undoing the row permutations pre-multiply by T^t. In the case of example chosen $T = T^t$.

In general the following gives the rule for interpreting the permutation matrix

$$P = \begin{bmatrix} p_1 & p_2 & p_3 & p_4 & \cdots\cdots & p_n \\ r_1 & r_2 & r_3 & r_4 & \cdots\cdots & r_n \end{bmatrix}$$

If $\quad p_3 = 2$ and $r_3 = 3$

$t_{23} = 1$ and all the other elements in the p_3 column are equated to zero.

It can be verified that for the example considered

$$\frac{1}{5}T^t\begin{pmatrix} 1 & 1 & -1 \\ 1 & -2 & 1 \\ 1 & 1 & 1 \end{pmatrix}T^t$$; gives the required result.

Example (*a*) Find $(A)^{-1}$ by using permutation matrix approach.

$$A = \begin{bmatrix} 3.7 & 5.2 & 9.1 & -2.8 \\ 2.5 & -3.8 & 7.4 & 8.7 \\ 9.5 & 4.4 & 6.8 & 3.0 \\ 1.6 & -1.9 & -2.4 & -7.5 \end{bmatrix}$$

Example (b) If C is given $\{5.1 \quad 8.2 \quad 13.4 \quad -9.8\}$

In the relation $\quad A\,X = C$

Find X by considering coefficient matrix $[A]$ of Ex. (a)

Ans. $\{.501, \quad .994, \quad .133, \quad 1.112\}$

Index

F

G

H

I

J

K

L

M

N

O

P

R

S

T

U

V

www.ingramcontent.com/pod-product-compliance
Lightning Source LLC
LaVergne TN
LVHW062159200926
848484LV00006B/1285
9780071835855